INFRARED SOLAR PHYSICS

INTERNATIONAL ASTRONOMICAL UNION
UNION ASTRONOMIQUE INTERNATIONALE

INFRARED SOLAR PHYSICS

PROCEEDINGS OF THE 154TH SYMPOSIUM OF THE
INTERNATIONAL ASTRONOMICAL UNION,
HELD IN TUCSON, ARIZONA, U.S.A., MARCH 2–6, 1992

EDITED BY

D.M. RABIN
J.T. JEFFERIES

National Solar Observatory,
National Optical Astronomy Observatories,
Tucson, Arizona, U.S.A.

and

C. LINDSEY

Solar Physics Research Corporation,
Tucson, Arizona, U.S.A.

KLUWER ACADEMIC PUBLISHERS
DORDRECHT / BOSTON / LONDON

Library of Congress Cataloging-in-Publication Data

```
International Astronomical Union. Symposium (154th : 1992 : Tucson,
    Ariz.)
    Infared solar physics : proceedings of the 154th Symposium of the
International Astronomical Union, held in Tucson, Arizona, U.S.A.,
March 2-6, 1992 / edited by D.M. Rabin and J.T. Jefferies and C.
Lindsey.
        p.    cm.
    Includes index.
    ISBN 0-7923-2522-2 (alk. paper)
    1. Sun--Congresses.   2. Infrared astronomy--Congresses.
3. Astrophysics--Congresses.    I. Rabin, D. M.    II. Jefferies, John
T.   III. Lindsey, C.   IV. Title.
QB520.I554   1992
523.7--dc20                                                    93-33129
```

ISBN 0-7923-2522-2

Cover picture: Terra cotta Sun face.

Published on behalf of
the International Astronomical Union
by
Kluwer Academic Publishers, P.O. Box 17, 3300 AA Dordrecht, The Netherlands.

Kluwer Academic Publishers incorporates
the publishing programmes of
D. Reidel, Martinus Nijhoff, Dr W. Junk and MTP Press.

Sold and distributed in the U.S.A. and Canada
by Kluwer Academic Publishers,
101 Philip Drive, Norwell, MA 02061, U.S.A.

In all other countries, sold and distributed
by Kluwer Academic Publishers Group,
P.O. Box 322, 3300 AH Dordrecht, The Netherlands.

Printed on acid-free paper

All Rights Reserved
© 1994 International Astronomical Union

No part of the material protected by this copyright notice may be reproduced or utilized in any form or by any means, electronic or mechanical including photocopying, recording or by any information storage and retrieval system, without written permission from the publisher.

Printed in the Netherlands

TABLE OF CONTENTS

Preface by *R. W. Noyes*	xi
Foreword	xii
Committees and Supporting Organizations	xiii
Acknowledgements	xiv
Participants	xv
J. T. Jefferies Overview of Infrared Solar Physics	1

1. INFRARED DIAGNOSTICS OF THE SOLAR ATMOSPHERE AND SOLAR ACTIVITY

T. R. Ayres The Cold Heart of the Solar Chromosphere	11
P. Foukal and T. Moran Properties of Faculae from Observations Near the Opacity Minimum	23
A. Kučera and E. A. Baranovsky A Solar Plage Model	29
E. H. Avrett, J. M. Fontenla, and R. Loeser Formation of the Solar 10830 Å Line	35
H. P. Jones Interpreting Recent Observations of He I 10830 Å	49
J. W. Harvey and W. C. Livingston Variability of the Solar He I 10830 Å Triplet	59
B. Fleck, F.-L. Deubner, D. Maier, and W. Schmidt Observations of Solar Oscillations in He I 10830 Å	65
K. L. Harvey Observations of Dynamic Events in He I λ 10830	71
V. Levkovsky, P. Papushev, and R. Salakhutdinov An Investigation of IR Triplet He I 10830 Å Profiles in Active Regions and the Quiet Chromosphere	77
C. K. Kumar and J. Davila Potential IR Observations of the Solar Corona	81

C. Lindsey 85
  The Sun in Submillimeter Radiation

P. Kaufmann, E. Correia, J. E. R. Costa, and A. M. Zodi 93
  Far Infrared and Submillimeter Continuum Observations of Solar Flares: Justifications and Prospects for Ground-Based Experiments

V. Petrosian 103
  Submillimeter and Far Infrared Emission from Solar Flares

A. Falchi, R. Falciani, and P. Mauas 113
  Infrared and Submillimeter Diagnostics of Activity and Flares

E. Correia, P. Kaufmann, and A. Magun 125
  The Observed Spectrum of Solar Burst Continuum Emission in the Submillimeter Spectral Range

M. R. Kundu, S. M. White, N. Gopalswamy, and J. Lim 131
  Interferometry of Solar Flares at 3-mm Wavelength

2. Infrared Observations of the 1991 Total Solar Eclipse

T. A. Clark 139
  Eclipse Observations of the Extreme Solar Limb at Submillimeter Wavelengths

D. E. Jennings, D. Deming, G. McCabe, R. Noyes, G. Wiedemann, and F. Espenak 151
  12-μm Observations at the 1991 Eclipse

M. W. Ewell, Jr., H. Zirin, J. B. Jensen, and T. S. Bastian 161
  850 μm Observations of the 11 July 1991 Total Solar Eclipse

S. M. White and M. R. Kundu 167
  Observations of the 1991 Eclipse at 3.5 mm Wavelength

T. A. Clark, D. A. Naylor, G. J. Tompkins, and C. Lindsey 173
  Near IR Observations of the 11 July 1991 Total Solar Eclipse from Mauna Kea, Hawaii

E. V. Tollestrup, G. G. Fazio, J. Woolaway, J. Blackwell, and K. Brecher 179
  Infrared Images of the Sun During the July 11, 1991 Solar Eclipse

J. R. Kuhn, H. Lin, P. Lamy, S. Koutchmy, and R. N. Smartt 185
  IR Observations of the K and F Corona During the 1991 Eclipse

R. M. MacQueen, K.-W. Hodapp, and D. N. B. Hall 199
  Infrared Coronal Observations at the 1991 Solar Eclipse

Y. Suematsu, H. Fukushima, and Y. Nishino 205
  On the Coronal and Prominence Structures Observed at the Total Solar Eclipse of 11 July 1991

V. Rušin, M. Rybanský, M. Minarovjech, and T. Pintér   211
    The White-Light, Far Red (600–700 nm) and Emission Coronae at the July 11, 1991 Eclipse

A. Sanchez-Ibarra, M. Cisneros-Molina, G. Hinojosa-Palafox,   217
F. Cisneros-Peña, J. Guerrero de la Torre, M. Norzagaray-Cosío, and C. Tapia-Fonllem
    The Structure of the White-Light Corona at the 1991 Eclipse

3. INFRARED PERSPECTIVES ON ATMOSPHERIC DYNAMICS

R. F. Stein and Å. Nordlund   225
    Subphotospheric Convection

S. Koutchmy   239
    The Infrared Granulation — Observations

S. Keil, J. Kuhn, H. Lin, and K. Reardon   251
    Simultaneous IR and Visible Light Measurements of the Solar Granulation

T. A. Darvann   259
    Measurements of Horizontal Flows in 1.6 μm Granulation

V. A. Kotov and S. Koutchmy   265
    On Sunspot and Facular Contrast Variations Near 2 μm and 4 μm

T. Leifsen   271
    Solar 5 minute Oscillations at 2.23 μm

L. V. Didkovsky and V. A. Kotov   277
    Ground-Based Near-Infrared Observations of Global Solar Oscillations

Li Rufeng, Ye Binxum, Chen Hailin, Liu Shaohua, Deng Bailian,   283
Ma Jagu, H. A. Hill, and P. H. Oglesby
    Solar Oscillations Instrument at an Infrared Wavelength of 1.6 μm at Yunnan Observatory

J. W. Cook   287
    Magnetic Fields, Oscillations, and Heating in the Quiet Sun Temperature Minimum Region from Ultraviolet Observations at 1600 Å

4. Infrared Atomic Physics and Line Formation

E. S. Chang 297
Atomic Physics of the 12 μm and Related Lines

R. J. Rutten and M. Carlsson 309
The Formation of Infrared Rydberg Lines

E. H. Avrett, E. S. Chang, and R. Loeser 323
Modeling the Infrared Magnesium and Hydrogen Lines from Quiet and Active Solar Regions

M. Carlsson and R. J. Rutten 341
Computation of Infrared Hydrogen Lines

D. Hoang-Binh 347
New Atomic Data for Mg I Lines

D. Hoang-Binh and H. van Regemorter 353
On the Ion Broadening of the 12 μm lines of Atomic Magnesium

W. G. Schoenfeld, E. S. Chang, and M. Geller 359
High-l Rydberg Lines of Fe I in the ATMOS Spectra: $4f$–$5g$, $5g$–$6h$ ...

R. T. Boreiko, T. A. Clark, D. A. Naylor, and J. R. Busler 365
High-n Hydrogen Lines in Solar Infrared Spectra from Balloon-borne, Mauna Kea, and ATMOS Observations

D. A. Naylor, G. J. Tompkins, T. A. Clark, G. R. Davis, and W. D. Duncan 371
Solar Submillimeter and Millimeter Spectroscopy between 7 and 30 cm^{-1} from the James Clerk Maxwell Telescope

5. Magnetic Fields and Infrared Magnetometry

D. Deming, T. Hewagama, D. E. Jennings, G. McCabe, and G. Wiedemann 379
Vector Magnetometry Using the 12-μm Emission Lines

S. K. Solanki 393
Properties of Magnetic Features from the Analysis of Near-Infrared Spectral Lines

O. Steiner 407
Theoretical Models of Magnetic Flux Tubes: Structure and Dynamics

P. Maltby 423
The Thermal and Magnetic Structure of Sunspots

S. H. Saar   437
    Infrared Measurements of Stellar Magnetic Fields

D. Rabin   449
    Near Infrared Imaging Magnetometry

M. Bünte, O. Steiner, S. K. Solanki, and V. J. Pizzo   459
    Flux Tube Shredding and Its Infrared Signature

D. Degenhardt and B. W. Lites   465
    The Structure of Umbral Fluxtubes

S. K. Solanki, I. Rüedi, W. Livingston, and H. U. Schmidt   471
    1.5 μm Observations and the Depth of Sunspot Penumbrae

G. Kopp and D. Rabin   477
    A Magnetic Field Strength vs. Temperature Relation in Sunspots

T. A. Darvann and S. Koutchmy   483
    The IR Contrast of Magnetic Elements Obtained from High Spatial Resolution Observations at 1.6 μm

K. Sinha   489
    Diagnostic Tools for Sunspots: the Molecules C_2, MgH and TiO

S. H. Saar   493
    New Infrared Measurements of Magnetic Fields on Cool Stars

6. THE INFRARED SPECTRUM

E. Biémont   501
    Atomic Spectroscopy in the Infrared

C. B. Farmer   511
    The ATMOS Solar Atlas

R. L. Kurucz   523
    Synthetic Infrared Spectra

R. Blomme, A. J. Sauval, and N. Grevesse   533
    Line Shifts and Asymmetries in the IR Solar Spectrum

N. Grevesse, A. J. Sauval, and R. Blomme   539
    Solar Abundances of C, N, and O

S. Johansson, G. Nave, M. Geller, A. J. Sauval, and N. Grevesse   543
    Analysis of Very High Excitation Fe I Lines $(4f - 5g)$ in the Solar Infrared Spectrum

A. J. Sauval and N. Grevesse   549
    The Sun as a Laboratory Source for IR Molecular Spectroscopy

7. INFRARED TECHNOLOGY AND THE FUTURE

F. Roddier and J. E. Graves 557
    Prospects in Adaptive Optics for Solar Applications

S. T. Ridgway 567
    Solar Optical Interferometry

O. Engvold 579
    The Near-Infrared Capabilities of LEST

W. Livingston 589
    A 4-meter McMath Telescope for the Infrared

D. Y. Gezari 595
    The Applicability of a 5–18 μm Array Camera to Solar Imaging

R. N. Smartt, S. Koutchmy, and J.-C. Noëns 603
    Near-IR Solar Coronal Observations with New-Technology Reflecting Coronagraphs

PREFACE

It is a truism in astronomy that new spectral windows and new techniques open new eras of discovery; the solar infrared is no exception. The infrared window to the Sun is scarcely "new", of course – its first scientific exploration dates from 1800, when William Herschel discovered invisible solar radiation redward of what our eyes can see. But the complete opening of this window is just now going on. The opacity of the Earth's atmosphere has been surmounted by going to space, and on the ground new detectors and spectrographic techniques have made observations possible that were only a dream a decade ago. This volume, and the Symposium it records, show what has been accomplished in the past few years, but equally importantly, they point the way to what can be done in the next.

A fair question at the outset is, "Why do we care?" After all, we already have many windows through which to examine the Sun, that most-studied of all astronomical objects. In some respects our view is more precise through the infrared window; for example, we can observe spectral lines with Zeeman sensitivity several times greater than any line in the visible. We can also see slightly different solar regions than through other windows; for example, our line of sight penetrates deepest into the solar photosphere in the near infrared. And the physical conditions where the solar infrared radiation is formed make for a different perspective; thus, the near-linear weighting of the Planck function with temperature, or the fact that the continuum and many spectral lines are formed essentially in local thermodynamic equilibrium, give a contrasting and complementary view to that at shorter wavelengths. There are drawbacks of course – notably the loss of angular resolution as we observe at longer wavelength. But the net advantages of the new window have amply rewarded the effort of opening it to our view.

Solar researchers often describe their field as "mature" – a shorthand for explaining why it is such hard work to make major advances in solar research, compared to the abundant flow of discoveries in less thoroughly studied areas. The opening of the infrared window makes solar research, for a while at least, a little less mature, and pleasantly so, since the joy of discovery is what motivates all science. For solar research there is added pleasure in relating the new discoveries to an already detailed picture of the Sun that has been built up over the years, so that very specific hypotheses can be formulated and critically tested by the new data.

The Symposium came at a good time. First, there was the fortuitous circumstance of the 1991 solar eclipse, which occurred over one of the world's best infrared sites. Second, there is the remarkable progress in infrared instrumentation and detectors, which has been gathering momentum over the past few years and shows no signs of slowing down. And third, recent theoretical advances – in radiative transfer, in convection and magnetohydrodynamics, in atomic physics – have led to exciting interactions between infrared observations and theory. The excitement that pervades the newly young field of infrared solar physics is evident in this record of its first IAU Symposium.

Robert W. Noyes

FOREWORD

To our knowledge, IAU Symposium No. 154 was the first international meeting devoted to Infrared Solar Physics. That the first meeting was an IAU symposium, rather than a colloquium or regional meeting, testifies to the explosive growth of this discipline during the last five years.

The meeting was held in Tucson, Arizona, some 80 km from the McMath-Pierce Telescope on Kitt Peak. Designed with infrared observations in mind, the McMath-Pierce Telescope has been the venue of many discoveries in infrared solar physics, such as the discovery of the 12-μm emission lines, the first direct evidence for kilogauss magnetic knots as shown by the Zeeman-sensitive iron line at 1.56 μm, and the discovery of five-minute oscillations and thermal inhomogenity in the temperature minimum region as revealed by the 4.7-μm vibration-rotation bands of carbon monoxide. Today, the National Solar Observatory and its visiting scientists pursue a wide range of infrared studies, both at the McMath-Pierce facility and at the NSO facilities on Sacramento Peak. For all these reasons, NSO was proud to host IAU Symposium No. 154. However, these proceedings leave no doubt that infrared solar physics is now a global enterprise, with vital contributions from many countries – from the ground, from balloons, and from space. More than 95 scientists from 16 countries participated in the symposium.

We thank the members of the Scientific Organizing Committee for their successful efforts in designing the scientific program and inviting the speakers. The symposium would not have been possible without the financial support of the IAU and the organizations listed below. We have also listed at least some – those whom faulty memories can recall – of the many individuals who made important contributions to the success of the meeting and the production of these proceedings. Finally, it gives us particular pleasure to recognize the indefatigable efforts of the other members of the Local Organizing Committee, among whom Ann Barringer must be singled out as the linchpin that held us together during the frenetic days leading up to the meeting.

Douglas Rabin   John Jefferies   Charles Lindsey

SPONSORING IAU COMMISSIONS

12 Solar Radiation and Structure (principal sponsor)
10 Solar Activity
36 Theory of Stellar Atmospheres

SCIENTIFIC ORGANIZING COMMITTEE

E. Biémont
T. A. Clark
D. Deming (Chair)
R. Falciani
P. Kaufmann
I. Kim
S. Koutchmy
C. Lindsey
D. Rabin
S. Solanki

LOCAL ORGANIZING COMMITTEE

A. Barringer
M. Giampapa
J. Jefferies
G. Kopp
C. Lindsey
D. Rabin (Chair)
J. Wagner

SUPPORTING ORGANIZATIONS

International Astronomical Union
National Solar Observatory
National Optical Astronomy Observatories
National Aeronautics and Space Administration
National Science Foundation
Infrared Laboratories, Inc.

ACKNOWLEDGEMENTS

Imo Appenzeller	Landessternwarte Heidelberg-Königstuhl
Jeannette Barnes	National Optical Astronomy Observatories
Emmerson Bartlomé	National Solar Observatory
Jacqueline Bergeron	Institut d'Astrophysique
Trisha Boyka	National Optical Astronomy Observatories
William Ditsler	Kitt Peak National Observatory
Joyce DuHamel	National Optical Astronomy Observatories
Steven Grandi	National Optical Astronomy Observatories
Mark Hanna	National Optical Astronomy Observatories
Paul Hartmann	National Solar Observatory
Tom Kinman	Kitt Peak National Observatory
Greg Ladd	National Solar Observatory
John Leibacher	National Solar Observatory
Eric Low	Infrared Laboratories, Inc.
Elly McFadden	Holiday Inn Palo Verde
Wyant Morton	University of Arizona
Terry Nash	Pima Community College
Mike Peralta	National Optical Astronomy Observatories
Claude Plymate	National Solar Observatory
Maurice Skones	University of Arizona
Monique Orine	IAU, Paris
Seth Tuttle	National Science Foundation
William Wagner	National Aeronautics and Space Administration
Sidney Wolff	National Optical Astronomy Observatories

PARTICIPANTS

Lawrence S. Anderson	University of Toledo, USA
Grant Athay	High Altitude Observatory, USA
Eugene H. Avrett	Harvard-Smithsonian Center for Astrophysics, USA
Thomas R. Ayres	University of Colorado, USA
K.S. Balasubramaniam	National Solar Observatory, USA
Ann Barringer	National Solar Observatory, USA
Emile Biémont	Université de Liège, Belgium
Ronny Blomme	Koninklyke Sterrenwacht van Belgie, Belgium
James W. Brault	National Solar Observatory, USA
Edward S. Chang	University of Massachusetts, USA
Thomas A. Clark	University of Calgary, Canada
Howard Cohl	National Solar Observatory, USA
John Cook	Naval Research Laboratory, USA
Emilia Correia	Escola Politecnica Saõ Paulo, Brazil
Tron A. Darvann	University of Olso, Norway
Detlev Degenhardt	Universitäts-Sternwarte Göttingen, Germany
Drake Deming	NASA Goddard Space Flight Center, USA
Leonid Didkovsky	Crimean Astrophysical Observatory, Ukraine
George A. Dulk	University of Colorado, USA
Oddbjørn Engvold	University of Oslo, Norway
Malcolm Ewell, Jr.	California Institute of Technology, USA
Roberto Falciani	University of Florence, Italy
Crofton Farmer	Jet Propulsion Laboratory, USA
Giovanni G. Fazio	Harvard Smithsonian Center for Astrophysics, USA
Bernhard Fleck	Universität Würzburg, Germany
Peter V. Foukal	Cambridge Research and Instrumentation, USA
Daniel Gezari	NASA Goddard Space Flight Center, USA
Mark Giampapa	National Solar Observatory, USA
J. Elon Graves	University of Hawaii, USA
Nicolas Grevesse	Université de Liège, Belgium
Craig Gullixson	National Solar Observatory, USA
Paul Hartmann	National Solar Observatory, USA
Jack Harvey	National Solar Observatory, USA
Karen L. Harvey	Solar Physics Research Corporation, USA
Samuel L. Hensel, Jr.	Electronic Space Systems Corporation, USA
D. Hoang-Binh	Observatoire de Paris, France
Klaus W. Hodapp	University of Hawaii, USA
Hugh S. Hudson	University of California San Diego, USA
David Jaksha	National Solar Observatory, USA
John T. Jefferies	National Solar Observatory, USA
Donald Jennings	NASA Goddard Space Flight Center, USA
Sveneric Johansson	University of Lund, Sweden
Harrison Jones	NASA Southwest Solar Station, USA
Richard Joyce	Kitt Peak National Observatory, USA

Pierre Kaufmann	Escola Politecnica Saõ Paulo, Brazil
Stephen L. Keil	Air Force Phillips Laboratory, USA
Iraida S. Kim	Sternberg State Astronomical Institute, Russia
Greg Kopp	National Solar Observatory, USA
Valeri Kotov	Stanford University, USA
Serge Koutchmy	Institut d'Astrophysique, France
Jeffrey Kuhn	Michigan State University, USA
Krishna Kumar	Howard University, USA
Mukul R. Kundu	University of Maryland, USA
Robert L. Kurucz	Harvard-Smithsonian Center for Astrophysics, USA
Yolande Leblanc	University of Colorado, USA
John Leibacher	National Solar Observatory, USA
Haosheng Lin	Michigan State University, USA
Charles Lindsey	Solar Physics Research Corporation, USA
William Livingston	National Solar Observatory, USA
Eric Low	Infrared Laboratories, Inc., USA
Günter Lustig	Institut für Astronomie Graz, Austria
Robert M. MacQueen	Rhodes College, USA
Per Maltby	University of Oslo, Norway
Ciro Marmolino	Universita di Napoli, Italy
George McCabe	NASA Goddard Space Flight Center, USA
M. R. McPherson	Michigan State University, USA
Laurence November	National Solar Observatory, USA
Pavel G. Papushev	Institute of Solar-Terrestrial Physics, Russia
Vahé Petrosian	Stanford University, USA
Claude Plymate	National Solar Observatory, USA
Douglas Rabin	National Solar Observatory, USA
Kevin Reardon	Williams College, USA
Steven Ridgway	Kitt Peak National Observatory, USA
François Roddier	University of Hawaii, USA
Li Rufeng	Yunnan Observatory, Kunming, China
V. Rušin	Astronomical Institute, Czechoslovakia
Robert Rutten	Sterrekundig Instituut, The Netherlands
Steven Saar	Harvard-Smithsonian Center for Astrophysics, USA
Antonio Sanchez-Ibarra	Centro de Investigacion en Fisica, Mexico
A. Jacques Sauval	Observatoire Royal de Belgique, Belgium
Kiyoto Shibasaki	Nobeyama Radio Observatory, Japan
Ian Short	University of Toronto, Canada
K. Sinha	Uttar Pradesh State Observatory, India
Raymond Smartt	National Solar Observatory, USA
Sami Solanki	Institut für Astronomie Zürich, Switzerland
Robert Stein	Michigan State University, USA
Oskar Steiner	Kiepenheuer-Institut für Sonnenphysik, Germany
Yoshinori Suematsu	National Astronomical Observatory, Tokyo, Japan
Eric V. Tollestrup	Harvard-Smithsonian Center for Astrophysics, USA
Jeremy Wagner	National Solar Observatory, USA

Lloyd Wallace	Kitt Peak National Observatory, USA
Stephen White	University of Maryland, USA
Guenter Wiedemann	European Southern Observatory, Germany
Hubertus Wöhl	Kiepenheuer Institut für Sonnenphysik, Germany
Igor Zayer	Lockheed Palo Alto Research Laboratory, USA
Harold Zirin	California Institute of Technology, USA

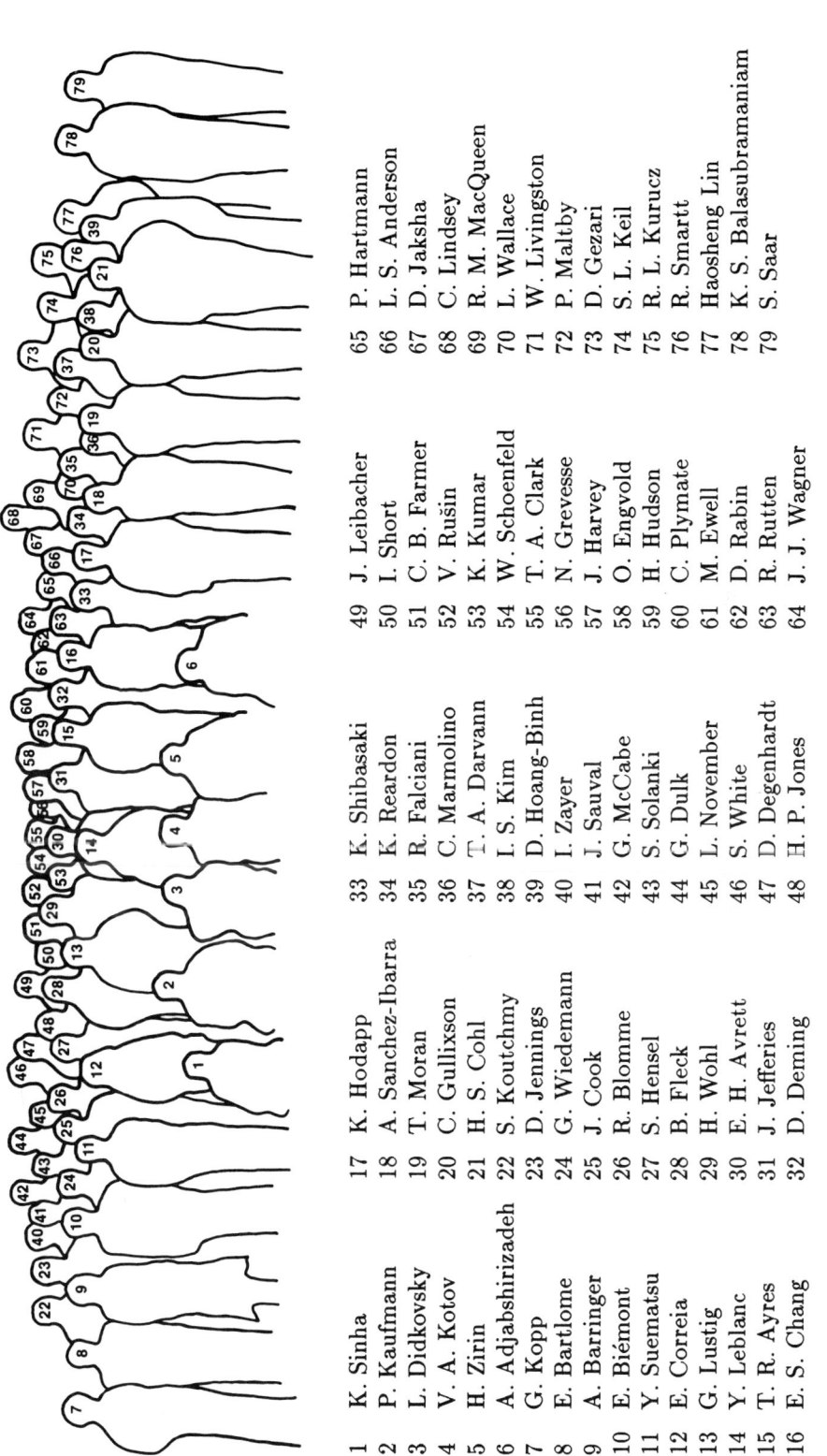

1	K. Sinha	17	K. Hodapp	33	K. Shibasaki	49	J. Leibacher	65	P. Hartmann
2	P. Kaufmann	18	A. Sanchez-Ibarra	34	K. Reardon	50	I. Short	66	L. S. Anderson
3	L. Didkovsky	19	T. Moran	35	R. Falciani	51	C. B. Farmer	67	D. Jaksha
4	V. A. Kotov	20	C. Gullixson	36	C. Marmolino	52	V. Rušin	68	C. Lindsey
5	H. Zirin	21	H. S. Cohl	37	T. A. Darvann	53	K. Kumar	69	R. M. MacQueen
6	A. Adjabshirizadeh	22	S. Koutchmy	38	I. S. Kim	54	W. Schoenfeld	70	L. Wallace
7	G. Kopp	23	D. Jennings	39	D. Hoang-Binh	55	T. A. Clark	71	W. Livingston
8	E. Bartlome	24	G. Wiedemann	40	I. Zayer	56	N. Grevesse	72	P. Maltby
9	A. Barringer	25	J. Cook	41	J. Sauval	57	J. Harvey	73	D. Gezari
10	E. Biémont	26	R. Blomme	42	G. McCabe	58	O. Engvold	74	S. L. Keil
11	Y. Suematsu	27	S. Hensel	43	S. Solanki	59	H. Hudson	75	R. L. Kurucz
12	E. Correia	28	B. Fleck	44	G. Dulk	60	C. Plymate	76	R. Smartt
13	G. Lustig	29	H. Wohl	45	L. November	61	M. Ewell	77	Haosheng Lin
14	Y. Leblanc	30	E. H. Avrett	46	S. White	62	D. Rabin	78	K. S. Balasubramaniam
15	T. R. Ayres	31	J. Jefferies	47	D. Degenhardt	63	R. Rutten	79	S. Saar
16	E. S. Chang	32	D. Deming	48	H. P. Jones	64	J. J. Wagner		

OVERVIEW OF INFRARED SOLAR PHYSICS

JOHN T. JEFFERIES

National Solar Observatory, National Optical Astronomy Observatories,*
P. O. Box 26732, Tucson, Arizona 85726, U.S.A.

Abstract. A brief overview of the field of infrared solar physics is given with particular reference to the topics to be considered during the Symposium.

Key words: infrared: stars – Sun: atmosphere – Sun: general

1. Introduction

My task is to set a stage for what is to follow during the rest of the week, to convey some sense of solar infrared astronomy – what it is, what it has achieved, where it is going, and why it is important. I shall try to do that – but selectively and briefly, as is appropriate for an overview.

I think that this is the first major meeting on solar IR astronomy and we are, of course, gratified that the IAU thought it sufficiently important to grant it the status of a Symposium. Clearly I agree with their assessment; it *is* important. It is also an exciting area to be associated with, as it gives us a new perspective on nature, and fresh insights into old problems – often overturning our preconceptions and prejudices in the process. We will hear a lot about a wide range of science which IR studies have illuminated – of new light on models of the quiet Sun, new insights on atomic excitation, new horizons in solar magnetometry, new ways to look at solar oscillations, new potential for studying solar inhomogeneities, new perspectives on the outer atmosphere derived from eclipse observations, new levels of precision in abundance determinations. Some old friends also reappear in new garments – such as the 10830 line of helium which derives new vitality from improved observational techniques – though it seems that we still struggle, as we have for 40 years at least, to find a satisfying accounting for it. We shall hear plenty about all these things, and more, as the week unfolds.

So the solar IR is vigorous and wide-ranging in its importance. A number of specific characteristics combine to make it so, and I may perhaps take a minute or two to review these, dwelling first on the continuum and second on the lines.

2. The Infrared Continuum

The first thing to say about the IR continuum is that there is a lot of it. Our Symposium defines it as covering the range from 1 μm to 1 mm or so. That is a factor of 1000 in wavelength, compared with the factor of less than 2 which we have to work with in the visible. The same physical processes control the continuum opacity over nearly all this range – namely free-free interactions involving an electron and a neutral H atom or a positive ion. We understand these processes, we know quite precisely their dependence on wavelength and particle density. The wide range of opacity which the IR opens up allows us to span a correspondingly wide range of

---

* Operated by the Association of Universities for Research in Astronomy, Inc., under cooperative agreement with the National Science Foundation.

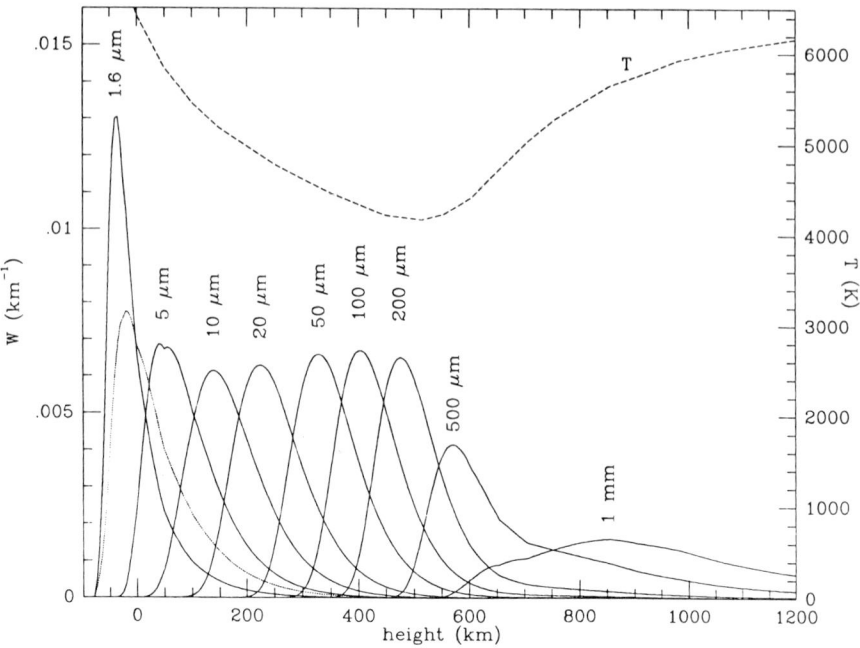

Fig. 1. Weighting functions for several wavelengths in the infrared for the VAL IIIC model atmosphere. The dotted curve is for 500 nm continuum radiation and the dashed curve shows the temperature distribution for the model.

atmospheric conditions as we scan through the IR. This is illustrated in Figure 1, which shows the contribution functions for a selection of wavelengths from 1.6 to 800 μm, for a standard solar model, VAL III-C, (Vernazza et al., 1981), whose temperature distribution is shown at the top of the Figure. Over this range we sample from the lowest accessible level in the photosphere up through the temperature minimum – more than 1000 km as compared to the 40 km or so which is accessible to us in the visible continuum. We can, of course, go further into millimeter wavelengths using the new facilities on Mauna Kea (the Caltech sub-mm Observatory and the JCMT) to which solar astronomers have been granted access – Ewell et al. (1992), Lindsey et al. (1990), Lindsey and Jefferies (1991), and Lindsey et al. (1992). Thus there are very good reasons, as Figure 1 shows, why modelling of the solar atmosphere has drawn so heavily on observations at longer wavelengths, and why one would continue to place emphasis on improving them in accuracy, in spatial resolution, and in the wavelength range which is covered. However, to achieve this diagnostic power has not been straightforward. Much of the range is only accessible from above the upper atmosphere or from space, as Figure 2 illustrates, so that balloons, aircraft, or space vehicles are needed to cover the full wavelength range of the infrared. To achieve the level of spatial resolution needed for critical studies of the Sun, we shall have to go to much larger apertures (or baselines) than are

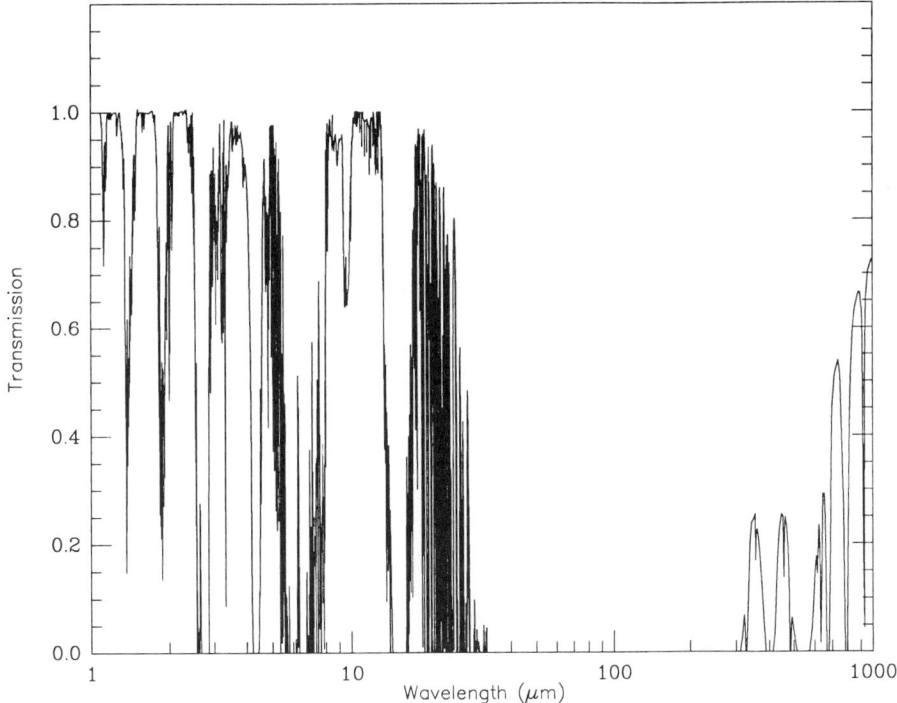

Fig. 2. Atmospheric transmission for Mauna Kea (1.2 mm water-vapor) – from Traub and Stier (1976).

presently available. Even so, much has been achieved in experiments such as those of Eddy et al. (1969), Rast et al. (1978), at the shorter wavelengths, and by such observers as Noyes, Beckers and Low (1968), Clark et al. (1971), Kundu (1971), Gezari et al. (1973), Lindsey and Hudson (1976), in the sub-millimeter to millimeter wavelength region. A first idea of the potential which increased resolution will hold for clarifying the structure of the Sun over the wide range of heights spanned by the 'infrared' is given in the images shown in Figure 3 taken from the JCMT at 350 μm and 1.3 mm – Lindsey and Jefferies (1991). Clearly the continuum intensity over the solar disk from the short IR through to the millimeter wavelengths is a fundamental and powerful diagnostic tool for inferring conditions in the Sun, and we shall hear a lot about this during the Symposium.

Another incisive application of IR continuum observations of the Sun is found in eclipse studies – both of the extreme limb and of the outer reaches of the atmosphere which can be especially well observed on those occasions. These aspects, too, will be well represented in our discussions here. As an illustration, Figure 4, compiled from Lindsey et al. (1986), Roellig et al. (1991), and Lindsey et al. (1992), shows how the Sun's intensity varies at the limb for several wavelengths in the IR, together with the predictions of a standard model. Clearly there is a discrepancy between

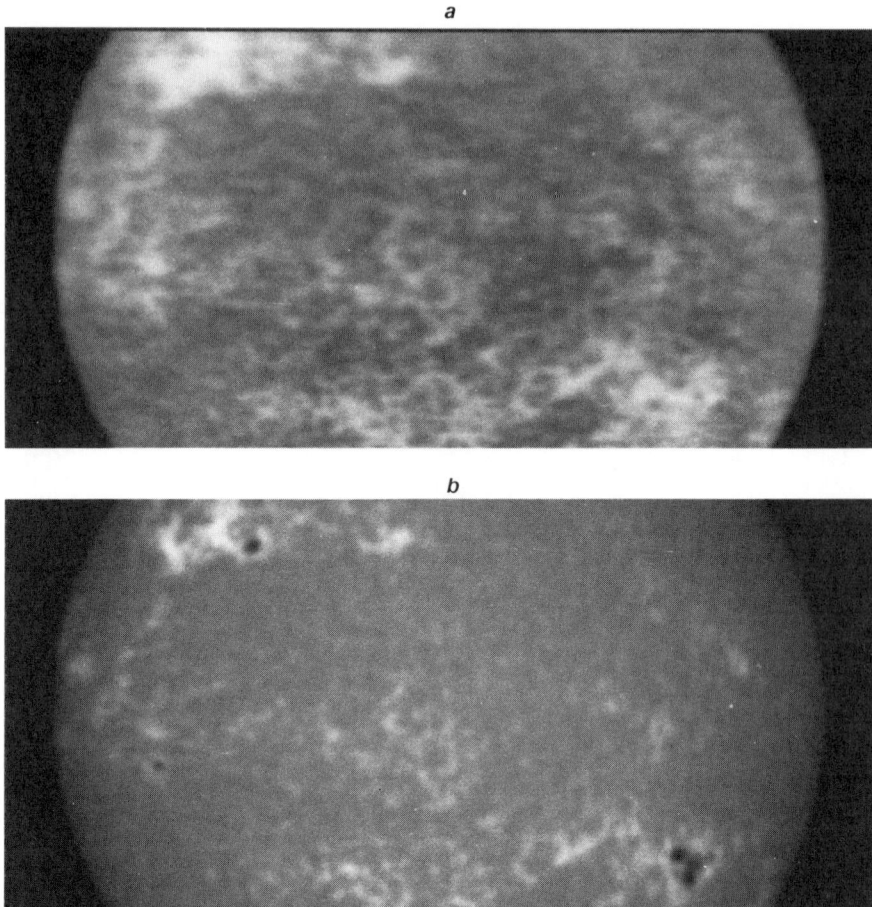

Fig. 3. Solar images at 850 μm (panel a), and in the Ca II K-line (panel b).

the observational data and the theory which, in fact, only grows worse as the wavelength increases up through 1.3 mm. This is probably a result of our neglect of inhomogeneities, as millimeter observers have been pointing out since the 1950's – see, for example, Hagen (1954), Coates (1958) Simon and Zirin (1969), Lantos and Kundu (1972), Kalaghan (1974).

There is a more subtle, and as yet barely touched, diagnostic possibility which the IR offers, namely its potential for inferring the nature of the inhomogeneities found in the solar atmosphere. Because, as I mentioned earlier, the IR flux from the Sun reflects the temperature linearly, the radiation from a composite region will contain, coded in its intensity, information about the temperature of the different components of the source – in the nature of a linear average over the fluctuations. This is a peculiar advantage of the IR – by contrast, at shorter wavelengths, the hotter elements tend to dominate the measured intensity giving us little to work

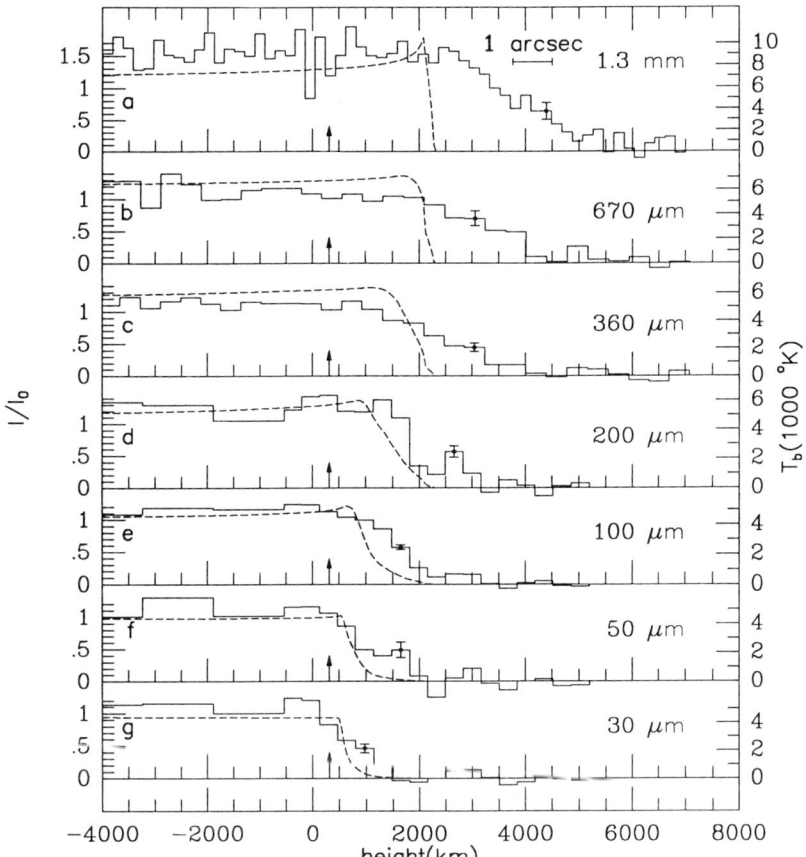

Fig. 4. Limb brightness profiles determined by occultation at total solar eclipses. Panel a is for 1991 July 11 (Lindsey et al. 1992,); b and c are for 1987 March 18 (Roellig et al. 1991); and d, e, f, and g were for the 1981 July 31 eclipse (Lindsey et al. 1986.)

with in determining the temperature characteristics of the composite structure. There are encouraging prospects, indeed, that with infrared data at several wavelengths we will be able to infer some essential characteristics of this sub-telescopic structure.

So the IR continuum is a flexible and powerful diagnostic tool. Its formation is simple and well understood, so we can make predictions with a good deal of confidence. We have an extended wavelength range to work with, so we can span a wide range of conditions. And perhaps we shall be able to make progress on inhomogeneous diagnostics.

3. The Infrared Line Spectrum

What about the lines? What can they tell us? One clear and decided advantage to the IR is in vector magnetometry where the higher splitting opens a range of new possibilities. An early application of the IR advantage was made by Harvey and Hall (1975), using the 1.56 μm, line of Fe I; this approach was subsequently used by Stenflo et al. (1987) and has been developed, at the hands of Rabin and his collaborators (see, e.g., Rabin and Graves, 1989), into a very powerful instrument at the McMath telescope. The discovery of the 12-μm emission lines by Brault and Noyes (1984) opened a further avenue for high-accuracy magnetometry of the Sun which has been developed by Deming and his collaborators (Deming et al. 1988), and by Hewagama et al. (1989), also using the McMath telescope. In particular the large splitting often allows us to measure the magnetic *field* directly in situations where only the magnetic *flux* is measured using visible line magnetometers. That this difference is critical is due to the fact that most (if not all) of the solar flux is found to be concentrated in flux-tubes and the area they occupy in different solar features is at best poorly known – thus we cannot with any confidence estimate the actual field strength without going into the IR. In fact we may be able to go further and obtain the distribution of field strengths in the sampled area, especially using lines in the 12-μm range where the splitting is very large. This whole area of research is very active and full of promise; we will hear much more about this in the future.

To turn to another aspect – high resolution IR spectroscopy from the ground gave us the 12-μm lines – Brault and Noyes (1984) – with their strong emission cores and a host of intriguing problems to go along with them. In Figure 5, which is actually taken from the *ATMOS* spectra (Farmer and Norton, 1989), we show some other lines in the same sequence. Many of us had thought that all these lines, arising in transitions between high-lying levels of neutral Mg, would be formed in LTE. Adjacent levels must be closely coupled, by electron collisions, with each other and with the ionized state and so we would seem to have all the conditions needed for an LTE population. Also the core emission seemed just to be reflecting the temperature increase into the chromosphere. It was really a pretty and consistent picture and early computations – Lemke and Holweger (1987) – seemed to bear this out. But this conclusion was wrong; as studies by Chang et al. (1991), and Carlsson et al. (1992) have shown, these lines are not formed in LTE in the Sun and the core emission has nothing to do with the chromospheric temperature increase. However, I cannot say that I understand the theoretical results as deeply as I would like and I look forward to hearing, at this meeting, the latest on this intriguing question. In the meantime we have the data on several other lines of Mg from the beautiful *ATMOS* spectra– and it will be a further challenge to the theorists to explain all of it at the same time. It was, of course, the *emission* character of the 12-μmlines that excited interest in their formation. No such interest would have attached to them had they been in absorption. Yet the super-excitation, which underlies the emission phenomenon, is driven by the local conditions where the line is formed, and for a differnt line, formed at a different location, we shall as likely find an *under*-excitation, giving rise to an absorption line—equally far from LTE, equally

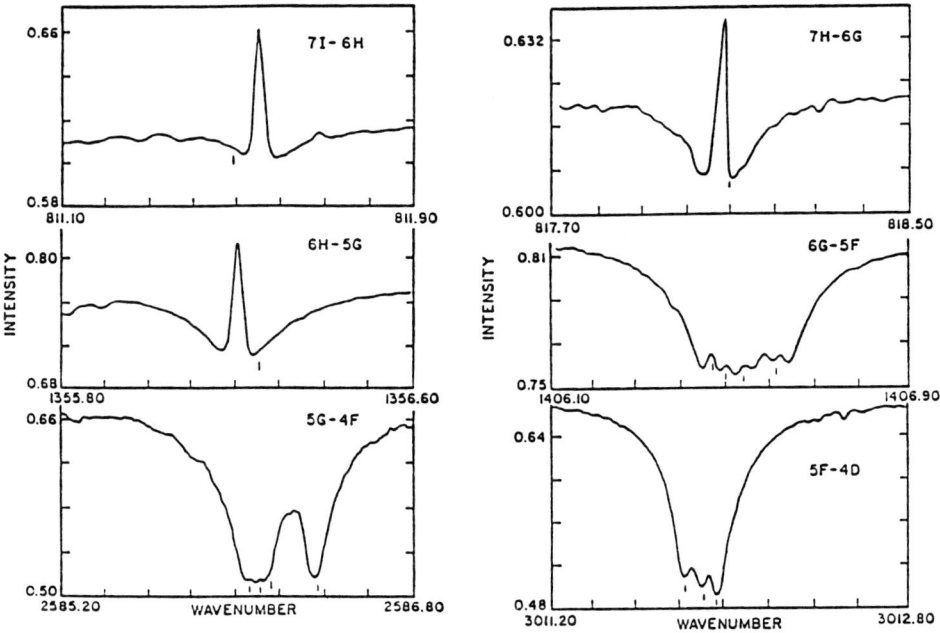

Fig. 5. *ATMOS* profiles of several lines of Mg I.

anomalous, and leading equally to error if interpreted as in LTE. The Mg I studies have shown us that, when dealing with atomic or molecular lines in the mid- to far-IR, whether in emission or absorption, we must approach their interpretation with great care, and not trust inferences (*e.g.* of temperature or abundance) based on LTE without a most careful analysis to ensure the validity of this assumption.

4. Concluding Remarks

There are many other aspects of the solar IR spectrum which we shall be hearing about during the week – determination of isotopic abundances, solar oscillations, high-lying lines of H, the 10830 line – all exciting new work opened up by the IR.

One final note. The solar IR is a burgeoning field with a bright and promising future – and as such it brings its own demands for the facilities, instruments, and support needed to develop the potential of the field. Some ideas will come out at this meeting – *e.g.*, the plans for a bigger McMath, new arrays, adaptive optics to combat the diffraction disadvantage of the IR. There can be little doubt that the pressure for such specialized capabilities will only increase, and the arguments for satisfying them become more compelling, as the successes of solar IR astronomy mount. I look forward to a stimulating few days learning more about those successes.

References

Brault, J.W. and Noyes, R.W.: 1984, *Astrophys. J. (Letters)* **269**, 61.
Carlsson, M., Rutten, R.J., and Shchukina, N.G.: 1992, *Astron. Astrophys.* **253**, 567.
Chang, E.S., Avrett, E.H., Mauas, P.J., Noyes, R.W., and Loeser, R.: 1991 *Astrophys. J. (Letters)* **379**, 79.
Clark, T.A., Courts, G.R., and Jennings, R.E.: 1971, *Phil. Trans. A.* **270**, 55.
Coates, R.J.: 1958, *Astrophys. J.* **128**,83.
Deming, D., Boyle, R.J., Jennings, D.E., and Wiedemann, G.: 1988, *Astrophys. J.* **333**, 978.
Eddy, J.A., Lena, P.J., and MacQueen, R.M.: 1969, *Solar Phys.* **10**, 330.
Ewell, M.W., Zirin, H., Jensen, J., and Bastian, T.: 1993, these proceedings.
Farmer, C.B., and Norton, R. (eds.): 1989, *A High-Resolution Atlas of the Infrared Spectrum of the Sun and Earth from Space*, vol. 1, The Sun, NASA RP-1224.
Gezari, D.Y., Joyce, R.R., and Simon, M.: 1973, *Astron. Astrophys.* **26**, 409.
Hagen, J.P.: 1954, *J. Geophys. Res.* **59**, 158.
Harvey, J.W., and Hall, D.N.B.: 1975, *Bull. Amer. Astron. Soc.* **7**, 459.
Hermans L., and Lindsey, C.: 1986, *Astrophys. J.* **310**, 448.
Hewagama, T., Jennings, D. Deming, D., Boyle, R., and Zipoy, D.M.: 1989, *Bull. Amer. Astron. Soc.* **21**, 839.
Kalaghan, P.M.: 1974, *Solar Phys.* **39**, 315.
Kundu M.K.: 1971, *Solar Phys.* **21**, 130.
Lantos, P., and Kundu, M. 1972, *Astron. Astrophys.* **21**, 119.
Lemke, M., and Holweger, H.: 1987, *Astron. Astrophys.* **173** 375.
Lindsey, C.A., and Hudson, H.: 1976, *Astrophys. J.* **203**, 753.
Lindsey, C.A., Becklin, E.E., Orrall, F.Q., Werner, M.W., Jefferies, J.T., and Gatley, I.I.: 1986, *Astrophys. J.* **308**, 448.
Lindsey, C.A., Jefferies, J.T., Clark, T.A., Harrison, R.A., Carter, M.K., Watt, G., Becklin, E.E., Roellig, T.L., Braun, D.C., Naylor, D.A., and Tomkins, G.J.: 1992, *Nature* **392**, 739.
Lindsey C.A., and Jefferies, J.T.: 1991, *Astrophys. J.* **383**, 443.
Noyes, R. W., Beckers, J. M. and Low, F. J.: 1968, *Solar Phys.* **3**, 36.
Rabin, D.M. and Graves J.E.: 1989, *Bull. Amer. Astron. Soc.* **21**, 854.
Rast, J., Kneubühl, F.K., and Müller, E.A.: 1978 *Astron. Astrophys.* **68**, 229.
Simon, M., and Zirin, H.: 1969, *Solar Phys.* **9**, 317.
Stenflo, J.O., Solanki, S.M., and Harvey, J.W.: 1987 *Astron. Astrophys.* **173**, 167.
Traub W., and Stier M.T.: 1987, *Appl. Optics* **15**, 364.
Vernazza, J.E., Avrett, E.H., and Loeser, R.: 1981 *Astrophys. J. Suppl.* **45**, 635.

PART 1

INFRARED DIAGNOSTICS OF THE SOLAR ATMOSPHERE AND SOLAR ACTIVITY

THE COLD HEART OF THE SOLAR CHROMOSPHERE

T. R. AYRES

*Center for Astrophysics and Space Astronomy, University of Colorado,
Campus Box 389, Boulder, CO 80309, U.S.A.*

Abstract. The early 1990's heralded the deployment of vastly improved space instruments in the ultraviolet (HST) and X-ray (ROSAT) bands, where thermal inhomogeneities in high-excitation chromospheres and coronae are seen in their most favorable light. The infrared spectrum provides a key complementary view of inhomogeneities, but only recently has begun to be seriously exploited for studies of the solar chromosphere and the outer atmospheres of other late-type stars.

Key words: infrared: stars – molecular processes – stars: chromospheres – stars: coronae – Sun: atmosphere

1. Introduction

Thermal inhomogeneities are a fact of life in the high-excitation outer atmospheres of late-type stars. The existence of severe contrasts in plasma conditions from point to point in the chromosphere and higher-altitude layers provides a strong motivation — and an endless source of frustration — for observers and theorists alike. On the one hand, the dazzling array of fine structure in the outer atmosphere of the Sun challenges the creativity of solar observers, faced with the daunting prospect of trying to resolve sub-arcsecond features from the ground; on the other hand it cruelly tantalizes stellar physicists who can record only the disk-average spectra of their subjects. Furthermore, theorists would prefer to work with spherically-symmetric, laterally-homogeneous models, rather than the messy reality of the solar chromosphere-corona.

Nevertheless, the inhomogeneous character of stellar outer atmospheres is not simply a trivial annoyance that Nature foists upon us; rather it is the fundamental signature of the heating and cooling processes that give rise to the decidedly non-classical high-altitude layers. A clear understanding of the heretofore elusive energization mechanisms requires a concerted effort to dissect the physical properties of the inhomogeneities. That effort necessarily must be rooted in the ultraviolet and higher-energy emissions that form preferentially in the heated gas. At the same time, infrared studies — particularly of low-excitation species like molecules — can shed light on the ambient atmosphere in which the chromospheric and coronal structures are embedded. The contrasting — and sometimes outwardly inconsistent — pictures must be reconciled in order to obtain a complete description of the inhomogeneous solar outer atmosphere; and to provide potential spectral diagnostics for the far less tractable, but equally important, stellar case.

Those of you expecting a detailed discussion of solar inhomogeneities will be disappointed because I will focus here mostly on stars. Those of you who are expecting an extensive treatise on carbon monoxide — my favorite toxic molecule — also will be disappointed because I will focus primarily on wavelengths other than the infrared and species other than molecules. And, those of you expecting a comprehensive review of the literature will be disappointed as well, because I will focus on some very new results that have not yet been submitted for publication. The general theme of my presentation is that we find ourselves in an exciting time, where the advent of new instruments – particularly the space-borne variety like HST and ROSAT – are providing an unprecedented view of stellar chromospheres and coronae. The infrared promises equally dramatic advances when the present technical revolution in panoramic detectors and cryogenic grating spectrometers moves from the instrument shops into the solar (and night-time) observatories.

2. The RIASS Coronathon

No solar physicist would challenge the statement: "The solar outer atmosphere is thermally inhomogeneous." Ground-based Ca II filtergrams and FUV/X-ray images taken from satellites and sounding rockets provide incontrovertible proof that the solar chromosphere-corona is dominated by strong contrasts in temperature and density on very fine spatial scales. Nevertheless, other stars cannot be resolved in the same direct way as the Sun, so the evidence in favor of solar-like structures is largely circumstantial.

In conjunction with the all-sky X-ray survey performed by the Röntgensatellit (ROSAT) between August 1990 and January 1991, there was a major effort to explore the empirical properties of stellar coronae in a large, minimally-biased sample of late-type stars. A cooperative observing program ("RIASS") between the International Ultraviolet Explorer (IUE) and the ROSAT all-sky survey team resulted in contemporaneous far-UV spectroscopy (primarily of the crucial C IV $\lambda 1550$ emission) and 0.1–2.4 keV soft X-ray detections of about fifty carefully-chosen "coronal" stars (Ayres *et al.*, 1992). Figure 1 illustrates a correlation diagram depicting the dependence of the high-excitation coronal flux on the lower-excitation transition-zone (TZ) emission. X-ray upper limits are represented by the smaller symbols.

The solar-type (F9 and later) main-sequence and subgiant stars generally follow a 3/2 power law between their coronal and TZ emissions (*cf.* Ayres, Marstad, and Linsky, 1981), although the mid-F stars appear to deviate from the main trend somewhat (cf., Simon and Drake, 1989) and are deficient in low-activity objects. Our Sun would appear near the bottom of the MS distribution. The ordinary giants (luminosity class III) display a more complicated behavior. The cooler giants (G3 and later) tend to follow a 3/2 power law like that of the MS stars; but many of the warmer giants – virtually all $\approx 2 - 3 \mathcal{M}_\odot$ Hertzsprung-gap stars – fall on a secondary relation displaced about an order of magnitude in $\mathcal{R}_X$ below the MS line: this is the "X-ray deficiency" described by Simon and Drake (1989). It is also worth noting that despite the relatively normal (*i.e.*, MS-like) behavior of the G3–K0 giants, the archetype red giant Arcturus (α Boo: K1 III) falls at least two orders of magnitude below the MS trend (based on a nondetection in an 18.6 ks ROSAT pointing reported by Ayres, Fleming, and Schmitt [1991]; and a marginal detection of C IV in a coadded IUE spectrum totalling about 18.9 ks of exposure.) Arcturus is significant because it represents a surrogate for the distant future of the low-mass stars ($\mathcal{M} \lesssim 1 \mathcal{M}_\odot$) in general, and our Sun in particular. Most of the other class-III giants in the diagram are more massive ($\mathcal{M} \gtrsim 2 \mathcal{M}_\odot$), and have evolved from progenitors that were A- and B-type stars on the MS.

The late-type supergiants (luminosity classes I and II) seem to follow the same "X-ray deficient" trend as the Hertzsprung-gap giants; correction for reddening

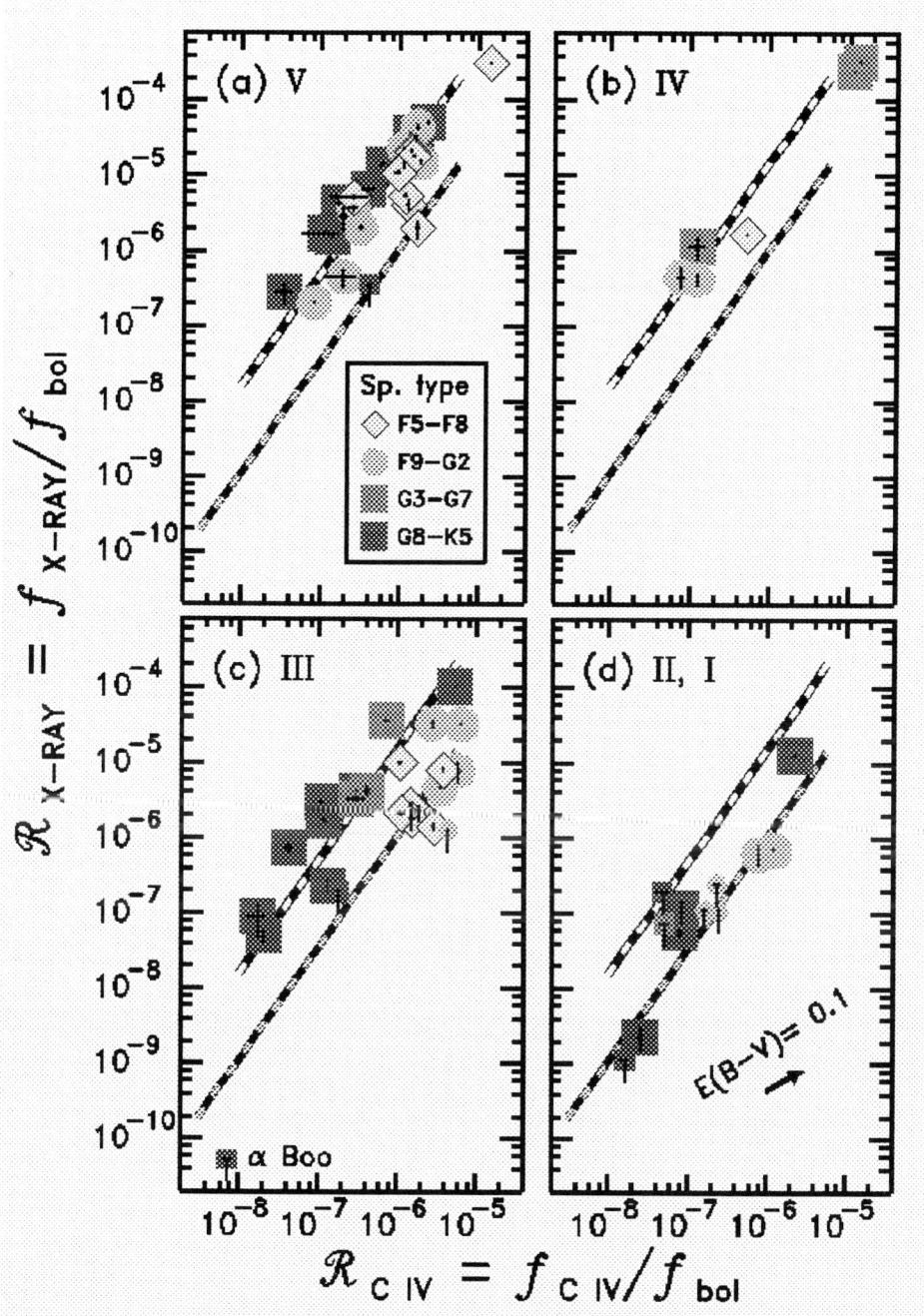

Fig. 1. Correlation diagram from RIASS "Coronathon".

would move the luminous stars even further away from the MS line. The supergiants all are massive objects ($\mathcal{M} \gtrsim 5\mathcal{M}_\odot$) which only recently have moved into the cool half of the H–R diagram from the upper main sequence (O- and B-types), although most probably are crisscrossing on "blue loops" of their convoluted evolutionary trajectories. The RIASS diagram is significant, because for the first time it directly compares two classes of supergiants that previously were considered unrelated. On the one hand are the "overactive" yellow supergiants like β Cam (G0 Ib) and β Dra (G2 Ib) which are strong coronal X-ray and FUV emission sources; on the other hand are the so-called "hybrid" supergiants (*e.g.*, Hartmann, Dupree, and Raymond, 1980) that show much weaker C IV emission and evidence for significant mass-loss as deduced from their red-asymmetric chromospheric Mg II features. Prior to ROSAT, there was only a marginal detection of coronal X-rays from the optically-brightest hybrid — α Triangulum Australae (K2 IIb-IIIa) — by EXOSAT (Brown *et al.*, 1991). The hybrids were considered unusual at the time of their discovery because signatures of coronal emission (namely the proxy C IV) and mass-loss are essentially mutually exclusive among the red giants (Linsky and Haisch, 1979). Arcturus is a case in point. Now, however, the sensitive ROSAT survey has revealed that the two types of supergiants fall on the same X-ray/C IV trend, suggesting a deeper connection between their ostensibly disparate coronal properties.

I don't want to make too much of the existence of two distinct X-ray/C IV relations: I've already gotten into trouble with other kinds of bifurcations! Possibly one is witnessing the operation of two different modes of the hydromagnetic "dynamo" (*e.g.*, Parker, 1970). One mode likely applies to MS stars such as the Sun that are moderate-to-fast rotators with relatively deep convective envelopes; the other mode likely applies to stars that either are fast rotators with shallow convective layers (like the Hertzsprung-gap giants), or slow-to-moderate rotators with deep convection zones (like the supergiants in general).

I present the RIASS correlation diagrams for their didactic value: to impress upon you the remarkable range of the normalized X-ray and C IV fluxes found among MS stars. The former runs over about three decades from the quiet G/K dwarfs to hyperactive RS CVn-type binaries, while the latter runs over more than two decades. The ranges increase to seven and three decades, respectively, if one includes the depths of the "coronal graveyard" (*i.e.*, Arcturus). The X-ray and C IV emissions on the Sun arise almost exclusively in magnetic structures, from the network bright points to large-scale active regions. The surface coverage by the major source of quiet-Sun emission — the network elements — is only a few percent. Thus, it seems natural to explain the enormous increase in $\mathcal{R}_X$ and $\mathcal{R}_{C\ IV}$ from the quiet to active stars simply in terms of an increased surface coverage of some fundamental "quantum" of activity, say magnetic flux tubes. Compelling as the notion is, the fact that the power law connection between X-rays and C IV is *steeper* than unity suggests the quantum itself changes character in the more active stars, perhaps owing to an increasing dominance of active regions at the expense of the network.

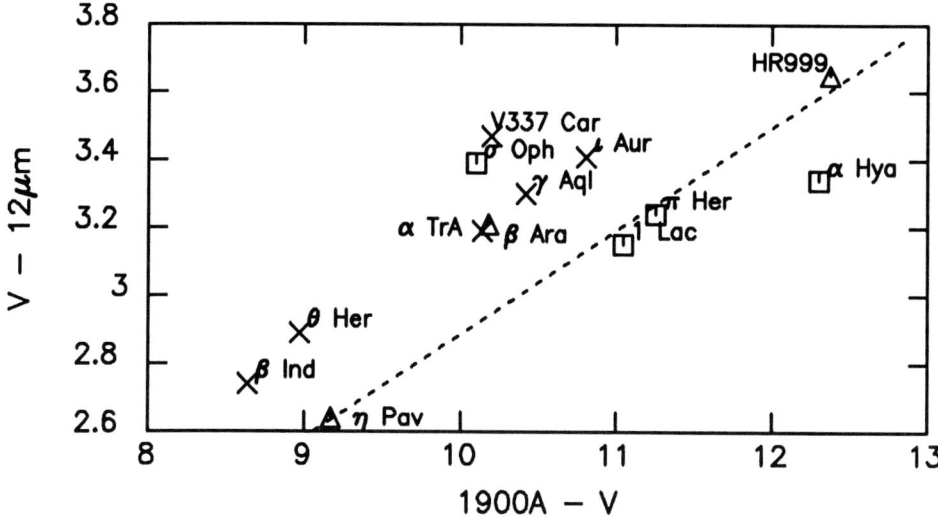

Fig. 2. IRAS/IUE two-color diagram for the "hybrid" (crosses) and normal (symbols) K bright giants. Bluer UV colors are to left.

3. The IRAS/IUE Connection

One of my students, R. B. Burton, has determined narrow-band colors at 12 μm and 1900 Å for more than 400 late-type stars observed by both the Infrared Astronomical Satellite and the International Ultraviolet Explorer. Our original motivation was to ferret out previously unknown warm companions of luminous cool stars, exploiting the extreme thermal leverage provided by the widely separated wavelength bands, and the high precision and consistency of the spacecraft-measured photometric indices.

One intriguing group with regard to the IRAS/IUE connection are the hybrid chromosphere bright giants, which I mentioned previously. Several years ago I noticed that the archetype hybrid — α TrA — has a surprisingly bright 1900 Å continuum in IUE spectra compared to the normal K-giant Arcturus, which has similar optical colors. I speculated that the UV excess was the result of a previously unseen F-type MS companion, and that some of the curious "hybrid" aspects of α TrA might be due to its putative binary nature (Ayres, 1985).

Burton's (1992) new study has raised serious doubts concerning my previous interpretation, by demonstrating that – among homogeneous groups of stars ranging from the K bright giants to the RS Canum Venaticorum binaries – the 1900 Å continuum excess shows a positive correlation with activity (measured by $\mathcal{R}_{C\ IV}$ for example). Figure 2 illustrates that most of the hybrids have a significant UV excess with respect to "normal" K bright giants (dotted curve); Figure 3 demonstrates that most of the hybrids are overactive as well. Burton interprets the continuum excess to be due to the photospheric emission of active regions, which very likely are

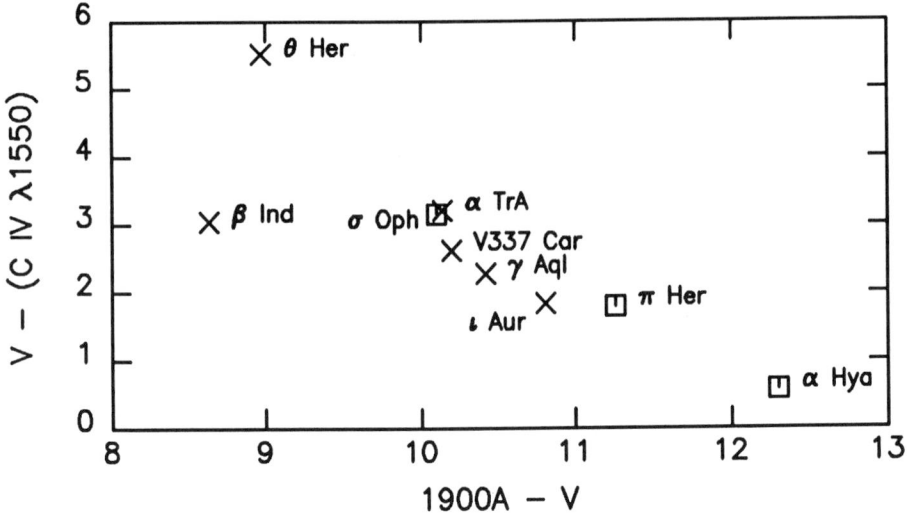

Fig. 3. Correlation of enhanced C IV emission with UV color excess.

warmer at mid-photospheric altitudes (where the 1900 Å continuum arises) than the undisturbed atmosphere. If true, the 1900 Å emission (easily recorded by the IUE) potentially can serve as a proxy measure of the surface coverage by facular areas, thereby providing a new indirect diagnostic for chromospheric inhomogeneities on other stars. Here, again, the shortwavelength emission has proved to be the superior tracer of chromospheric inhomogeneities, while the infrared has served primarily as a pivot for the UV thermal lever.

4. CO in the Sun

I already have discussed solar carbon monoxide *ad nauseam* in recent reviews (*e.g.*, Ayres, 1990, 1991) and do not wish to belabor the issue. In short, the excessively dark cores of the strong mid-IR (4.8 μm) $\Delta v = 1$ lines of CO close to the solar limb indicate the presence of remarkably cool material at high altitudes, probably well within the chromosphere itself. The fundamental question for those of us studying chromospheric inhomogeneities is how pervasive the cool component might be, and what role it plays in the energy balance of the outer atmosphere.

If cool gas occupies a significant volume of the chromosphere above the height of the traditional $T_{\min}$, then simulations of the energy balance of those layers based on conventional chromospheric thermal profiles can be very misleading. In particular, if much of the "action" that ultimately results in chromospheric emission occurs on small spatial scales, while the bulk of the remaining atmosphere is essentially passive, then homogeneous treatments of the radiative energy balance derived from spatially-averaged spectra will be triply wrong: initially in the empirical evaluation of surface fluxes; secondly in the inference of the temperature-pressure stratification

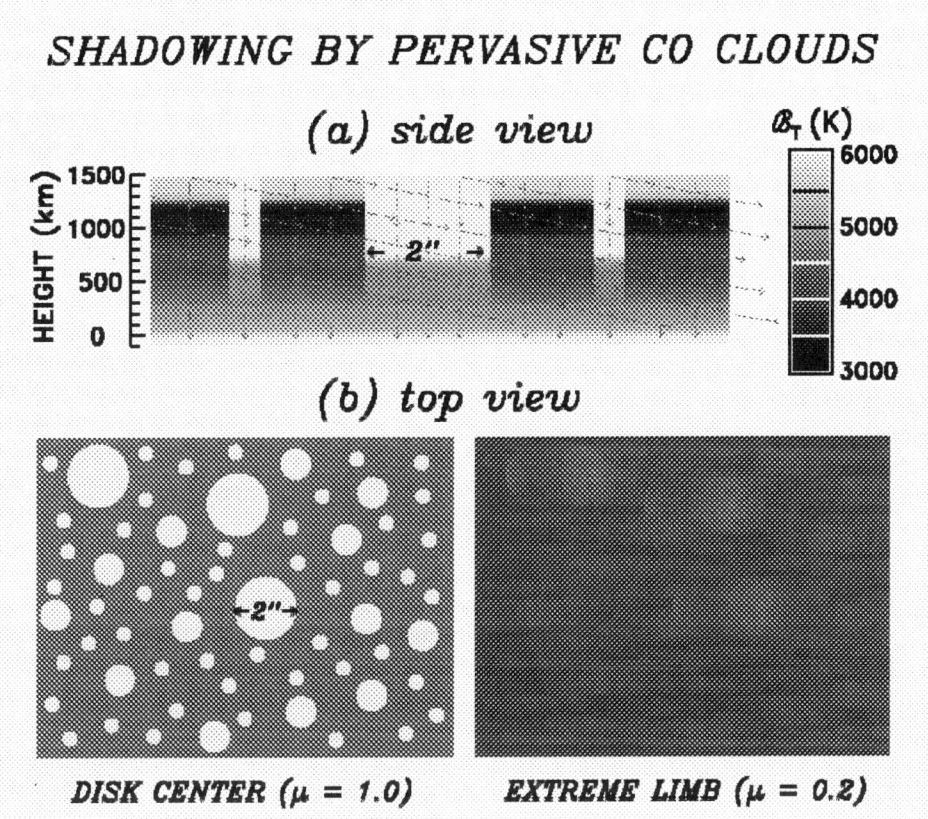

Fig. 4. Thermal models (panel [a]) and 2-D RT simulations (panel [b]) for the classical thermal bifurcation scenario: sparsely-distributed hot columns embedded in a cool near-RE matrix. The pervasive (75% coverage) cool component has a high-altitude *cloud* of very cold gas, above which is a steep rise of temperature into the chromospheric "canopy". In panel (b) the simulated spectroheliogram is for the strong 3−2 R14 line. The limb view was corrected for foreshortening to emphasize the *shadowing* phenomenon.

of the chromosphere; and finally in the numerical evaluation of the net radiative losses, from which the mechanical heating function is derived.

The existing studies of the solar 4.8 μm bands have had little or no spatial resolution and thus cannot resolve the issue directly. Ca II and CN filtergrams, as well as sounding-rocket images of the 1600 Å continuum, reveal a general *bright-point* character to the chromosphere, but are not sensitive to the cool component. I have conducted a series of 2-D radiative transfer experiments to assess the ability of the strong CO $\Delta v = 1$ lines to probe the conditions in the "cool clouds" (see Ayres, 1991). Figure 4 depicts the scenario I believe to be true, namely a chromosphere composed of small-scale bright points embedded in an otherwise cool "cloud deck" itself sandwiched between the warm photosphere (up to 500–600 km) and the hot chromospheric "canopy" ($\approx$ 1300 km). However, the simulations suggest that *shadowing* by even a small amount of CO in the chromosphere can overwhelm any hot material for limb lines of sight. Thus, we can reliably deduce the *thermal profile* of the cool component from the limb spectrum of the CO 4.8 μm bands, but not its *filling factor*.

The issue will remain controversial until the surface of the Sun is imaged directly in the CO fundamental lines. The NSO McMath main spectrograph currently is being reconfigured to carry a large IR grating that will be capable — in conjunction with the NSO infrared camera — of obtaining definitive stigmatic spectra of the 4.8 μm bands with diffraction-limited spatial resolution. The engineering project is scheduled for completion in 1992.

5. CO in the Stars

Studies of the Sun always will transcend those of other stars in technical sophistication, because the brightness of our stellar neighbor makes practical many kinds of instrumentation that would utterly fail in the low light conditions of night-time astronomy. Nevertheless, the nature of chromospheric inhomogeneities is not likely to be divined solely from the one example presented us by the Sun, just as a single slice of a biological specimen cannot qualify as a thorough dissection.

The history of stellar observations of the 4.8 μm CO bands is relatively brief, owing to the lack of sensitivity of current spectrometers. The instrument of choice has been the 1.4-m FTS at the NOAO Mayall 4-m telescope on Kitt Peak. One of the most important studies with the 4-m FTS also was one of the first: the observation by Heasley *et al.* (1978) of the archetype red giant Arcturus. The lack of chromospheric emission reversals in the deep CO absorptions — contrary to the predictions of the best-available models — prompted the authors to propose a multi-component scenario in which only 25% of the outer atmosphere was truly chromospheric; the remaining 75% was allotted to a cool (extended) photosphere in radiative equilibrium. The Heasley *et al.* observation was one of the original motivations for the "thermal bifurcation" hypothesis I introduced a few years later (Ayres, 1981). Unfortunately, Arcturus was about the only normal late-type star bright enough in the thermal IR to be recorded successfully by the Mayall FTS.

In the past few years, cryogenically-cooled spectral isolators — pioneered by the balloon instrumentation group at the NASA Goddard Space Flight Center (D. Jennings, D. Deming, G. Wiedemann, and colleagues) — have extended the effective sensitivity of the Mayall FTS dramatically. For his thesis work, Wiedemann (1989) was able to obtain high-resolution, high-S/N 4.8 μm spectra of a number of late-type giants and a few dwarfs and subgiants. Figure 5 illustrates some of the better examples. A theoretical basis for the analysis of such spectra has been given by Wiedemann and Ayres (1991). Notice, for example, the lack of measurable CO absorption in the F-type subgiant Procyon; the increased complexity of the CO spectrum in the four giants compared with the Sun; and the somewhat shallower and narrower absorptions of β Gem compared with α Boo and α Tau (the spectrum of α Aur is similar to that of β Gem, when the nearly featureless spectrum of the fast-rotating secondary star [F9/G0] is subtracted).

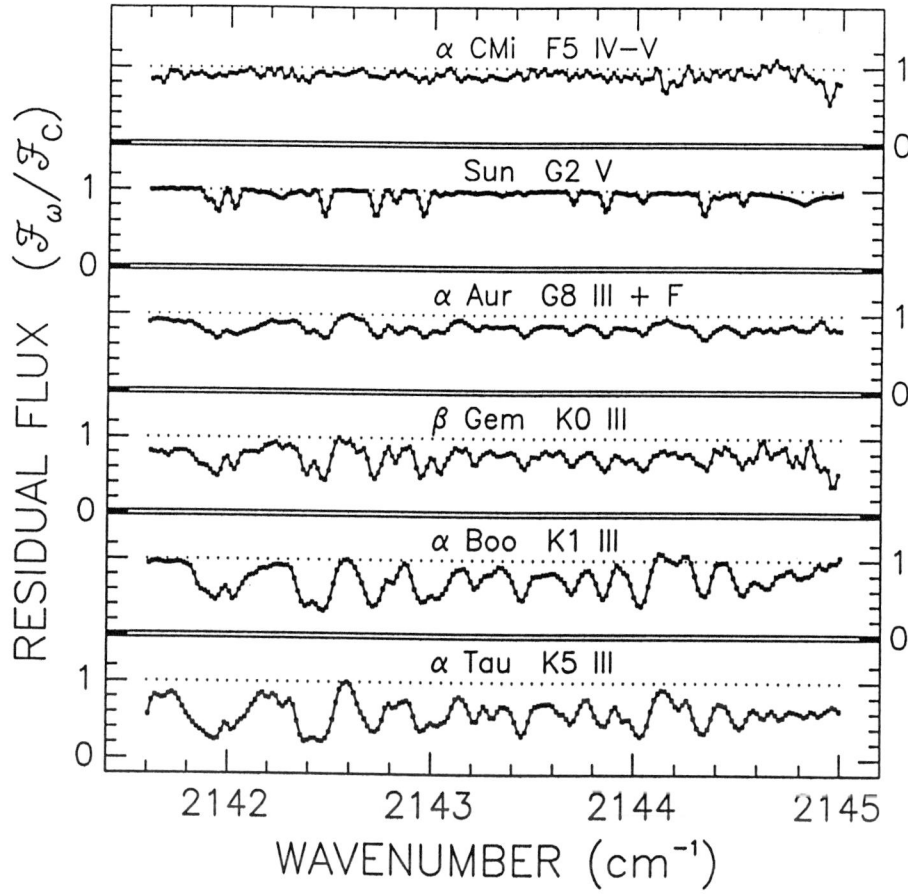

Fig. 5. Residual flux spectra of CO $\Delta v = 1$ bands in bright late-type stars.

The lack of CO absorptions in Procyon is consistent with the numerical simulations of Muchmore and Ulmschneider (1985) which indicate that the CO "cooling catastrophe" I proposed in the 1981 paper will not develop in stars with effective temperatures only a few hundred degrees warmer than the Sun. The additional complexity of the red-giant spectra compared with that of the Sun is due to the strengthening of the secondary isotopic bands, $^{13}C^{16}O$ and $^{12}C^{18}O$, in the chemically-evolved atmospheres. In fact, the isotopic richness of the CO spectrum of α Aur suggests that the slow-rotating primary is a post-helium-flash "clump" giant, rather than a "first-crosser" like the secondary star. Finally, the small but significant differences between the CO spectra of the coronal giant β Gem and the two noncoronal red giants probably are a symptom of a more pervasive cool "neophotospheric" component in the less active stars.

The results of Wiedemann's work are tantalizing, but unfortunately the number of objects accessible even with the Goddard postdisperser on the Mayall FTS (or other FTS's like that at the Canada-France-Hawaii Telescope) is quite limited: the prospects for broad surveys like those performed in the FUV and X-ray bands are dim. However, new stellar instruments promise to eclipse the FTS-era studies. The vanguard of the new generation are cryogenically-cooled grating spectrographs coupled with IR cameras. A good example is the "CSHELL" recently introduced at the NASA Infrared Telescope Facility. Others are under development at NOAO, ESO, GSFC, and elsewhere. While the resolving power of the grating instruments still is a factor of two below the $\omega/\Delta\omega \approx 80,000 - 100,000$ considered desirable for dwarf-star work, a vast array of problems can be attacked in the giant and supergiant regions. Furthermore, the impetus towards higher resolving powers, driven by studies of surface magnetism (*e.g.*, Saar, these proceedings) and circumstellar velocity fields (*e.g.*, Tsuji, 1988), will inevitably bring the dwarf stars within reach as well.

6. Conclusions

An old Chinese curse succinctly describes the present state of astronomy: "May you live in interesting times." Today's revolution in instrumentation across the entire electromagnetic spectrum far surpasses that of a decade ago when the likes of IUE, *Einstein*, IRAS, the VLA, and the M&M FTS's (McMath and Mayall) provoked a period of "paradigm perestroika" in our conceptions of the nature of chromospheric and coronal activity and their underlying physical inspiration. Some of the discoveries from the new generation of instruments are just beginning to appear, particularly on the more mature fronts of UV and X-ray astronomy. Nevertheless, the rapid pace of technical progress in the thermal infrared promises an equally glowing future in studies of the complementary aspects — like the CO clouds — of solar and stellar inhomogeneities.

I close by reiterating the conclusion of the review I gave in Heidelberg two years ago: A major future innovation would be the construction of a general-purpose large-aperture infrared solar telescope to permit high spatial resolution studies of the dynamics and evolution of surface structure and embedded magnetic activity. A group at the National Solar Observatories, based on a suggestion by R. W. Noyes, currently is exploring the feasibility of *doubling* the aperture of the McMath telescope by substituting a 4-m primary and a 6-m heliostat for the existing optics (see W. Livingston, these proceedings): the dry Kitt Peak site is excellent for infrared work; the all-reflecting, unobscured design of the McMath telescope minimizes thermal background; and the cost savings would be substantial compared with an entirely new facility. The unprecedented, unique view of the Sun afforded by a large-aperture IR solar telescope would fuel a major revolution in our understanding of the physical processes that shape the dynamic, chaotically-structured layers above the disarmingly placid visible surface of our nearby star.

Acknowledgements

This work was supported by grants from the National Science Foundation and the National Aeronautics and Space Administration.The observations described in

Fig. 1 were obtained in conjunction with the RIASS program, under the overall direction of W. Wamsteker, and with the considerable help of "designated observers" at the NASA IUE Observatory (GSFC) and its ESA counterpart at Vilspa. The observations of Fig. 5 were obtained at the Kitt Peak site of NOAO (and NSO), operated by AURA under a cooperative arrangement with the NSF.

References

Ayres, T. R.: 1981, *Astrophys. J.* **244**, 1064.
Ayres, T. R.: 1985, *Astrophys. J. (Letters)* **291**, L7.
Ayres, T. R.: 1990, in J. O. Stenflo (ed.), 'Solar Photosphere: Structure, Convection, and Magnetic Fields', *Proc. IAU Symposium* **138**, 23.
Ayres, T. R.: 1991, in P. Ulmschneider, E. R. Priest, and R. Rosner (eds.), 'Mechanisms of Chromospheric and Coronal Heating', Springer-Verlag, Heidelberg, p. 228.
Ayres, T. R., Fleming, T. A., and Schmitt, J. H. M. M.: 1991, *Astrophys. J. (Letters)* **376**, L45.
Ayres, T. R., Marstad, N. C., and Linsky, J. L.: 1981, *Astrophys. J.* **247**, 545.
Ayres, T. R., et al.: 1992, *The RIASS Coronathon: Contemporaneous X-ray and Far-Ultraviolet Measurements of a Review-Panel-Limited Sample of Normal Late-Type Stars*, in preparation.
Burton, R. B.: 1992, *Hidden Binary Systems: the IRAS/IUE Connection*, M. A. Thesis, University of Colorado.
Brown, A., Drake, S. A., Van Steenberg, M., and Linsky, J. L.: 1991, *Astrophys. J.* **373**, 614.
Hartmann, L., Dupree, A. K., and Raymond, J. C.: 1980, *Astrophys. J. (Letters)* **236**, L143.
Heasley, J. N., Ridgway, S. T., Carbon, D. F., Milkey, R. W., and Hall, D. N. B.: 1978, *Astrophys. J.* **219**, 790.
Linsky, J. L., and Haisch, B. M.: 1979, *Astrophys. J.* **229**, L27.
Muchmore, D. O., and Ulmschneider, P.: 1985, *Astron. Astrophys.* **142**, 393.
Parker, E. N.: 1970, *Ann. Rev. Astron. Astrophys.* **8**, 1.
Simon, T., and Drake, S. A.: 1989, *Astrophys. J.* **346**, 303.
Tsuji, T.: 1988, *Astron. Astrophys.* **197**, 185.
Wiedemann, G. R.: 1989, *Development of a Postdispersion System for Astronomical Infrared Observations with Fourier Spectrometers and its Application in the Study of the CO Fundamental Bands in F, G and K Stars*, Ph. D. thesis, University of Munich (MPE Report 214).
Wiedemann, G., and Ayres, T.: 1991, *Astrophys. J.* **366**, 277.

PROPERTIES OF FACULAE FROM OBSERVATIONS NEAR THE OPACITY MINIMUM

P. FOUKAL

CRI, Inc., 21 Erie Street, Cambridge, MA 02139, U.S.A.

and

T. MORAN

NASA/Goddard Space Flight Center, Greenbelt, MD 20771, U.S.A.

Abstract. Imaging of active regions in continuum around 1.6 μm shows that many facular regions are less bright than the photosphere when observed nearer to disk center than $\mu = \cos\theta \sim 0.75$. The contrast of these dark faculae increases with magnetic flux above a threshold of approximately 2×10^{18} Mx. This explains why not all faculae are dark at 1.6 μm, since the magnetic flux density in many regions of bright Ca K plage emission falls below this threshold. After correction for blurring, the typical contrast value is about 4-5%, so the brightness temperature deficit is about 130 K. Faculae are brighter than the photosphere at 1.63 μm nearer to the limb than $\mu \sim 0.5$. The negative contrast of dark faculae may arise from cooling of the surrounding photosphere, or from increased visibility of cool layers of the facular flux tube itself. Quantitative comparison of these IR data with MHD models awaits calculation of flux tube contrasts at realistic angular resolution.

Key words: infrared: stars – MHD – Sun: faculae, plages – Sun: photosphere

1. Introduction

Low photospheric continuum opacity near 1.6 μm enables observations of the deepest photospheric layers to be made in that wavelength region. Observations with comparable depth penetration may in principle be possible in the blue, due to higher temperature sensitivity of the Planck function at shorter wavelengths (Ayres, 1989), but the density of Fraunhofer lines in that spectral range makes it difficult to isolate the continuum, even with a spectrograph. The difference in geometrical depth between optical depth unity at 0.5 μm and 1.6 μm is only about 35 km (Vernazza, Avrett, and Loeser, 1976), but the energy balance of the sun changes rapidly between convective and radiative transport in this short interval, so the increased penetration is of great astrophysical interest.

Photometric measurements of facular contrast at this opacity minimum in the IR can yield important new information on the energy balance in these small-diameter magnetic flux tubes, and in their immediate surroundings. Semi-empirical models of faculae are largely based on the behavior of their photometric contrast at various positions in the profiles of lines such as Ca K (*e.g.*, Chapman, 1984; Walton, 1987). Wide-band photometry in the visible, especially near the limb, has been used in attempts to distinguish between different models of facular excess brightness away from disc center (Muller, 1975; Lawrence *et al.*, 1988; Wang and Zirin, 1987). Two-color photometry has been used to measure the temperature gradient in the facular atmosphere near $\tau_{0.5} = 1$ (Foukal and Duvall, 1985; Elste, 1985). Observations near 1.6 μm are of particular interest because the semi-empirical models are most weakly constrained at the deepest layers, where most of the facular radiation arises.

Early measurements of facular contrasts near the opacity minimum at 1.64 μm were made by Worden (1975) using a spectrometer and a single InSb detector in

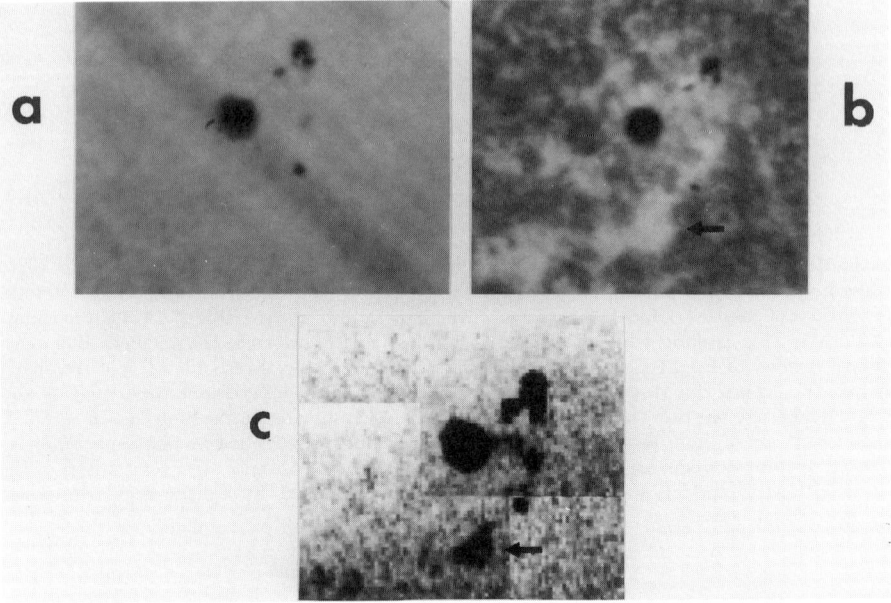

Fig. 1. Images of an active region in a) red continuum at 6558 Å; b) Ca K; and c) 1.63 μm. From Foukal et al. (1990).

a raster technique. His 2-D images of the quiet sun indicated a negative contrast of network faculae at 1.64 μm, while the same regions had positive contrast in a chromospheric line at 1.71 μm. This result was confirmed in the first images of active regions at 1.63 μm, made using a platinum silicide array and a narrow band filter (Foukal, Little, and Mooney, 1989). Better observations of facular contrast at 1.63 μm were obtained using the NOAO InSb infrared array (Foukal et al., 1990). Examination of these images showed that while many faculae (indicated by Ca K plage emission) were dark at 1.63 μm, others showed little or no contrast. The most recent analysis, discussed below (Moran, Foukal, and Rabin, 1992), has since shown that the facular contrast at 1.63 μm is correlated with magnetic flux.

In this overview, the results of facular contrast measurements at 1.63 μm are summarized in Section 2. An analysis of the most recent observations is given in Section 3, and a discussion of the results and their relevance for flux tube models is reviewed in Section 4.

2. Results of Facular Photometry at 1.63 μm

Figure 1 shows a particularly clear example of dark facular structures. The arrows indicate a structure that is bright in Ca K, dark at 1.63 μm, but does not correspond to any spots that are visible in the red continuum. The contrast of such structures measured near disc center is tyically 1-2% (Foukal et al., 1990). We stress that the invisibility of these structures in the red continuum is *not* simply a display artifact, since microdensitometry of such λ6558 spectroheliogram negatives has shown that such 1-2% intensity depressions over the spatial scale of the dark faculae seen in Figure 1 are easily detectable with a S/N ratio of better than 5:1 (Foukal, Little, and Mooney, 1989). After correction for seeing and scattering, the true contrast value is found to be at least 4-5% (Moran, Foukal, and Rabin, 1992).

Observations at various disc positions show that the 1.63 μm facular contrast passes through zero around $\mu = \cos\theta \sim 0.75$, and becomes positive again near the limb (Foukal et al., 1990). Observations near 1.2 μm, close to disc center, indicate that the contrast of dark faculae is significantly reduced at that wavelength (Moran, Foukal, and Rabin, 1992).

3. Analysis

To investigate the finding that many, but not all, faculae observed near disc center were dark at 1.63 μm, facular images such as those in Figure 1 were coaligned with Kitt Peak magnetograms (Moran, Foukal, and Rabin, 1992). The structures were found to fall into three categories. The first consists of structures that are seen in the magnetograms, but have negligible contrast in the visible and at 1.63 μm. The second contains structures which have negligible contrast in the visible and negative contrast at 1.63 μm. These are the dark faculae. The third category contains structures which have negative contrast in the visible and at 1.63 μm (spots and pores). The structures of the first and second categories are bright in Ca K. The contrast and magnetic flux of each structure is plotted in Figure 2. We see that, at magnetic flux values corresponding to NSO magnetograph readings below about 375 gauss, the contrast is below 0.2%, the limit of detection. Points below this cut-off belong to category 1. They represent faculae that are bright in Ca K, but whose magnetic flux is below 2×10^{18} Mx. For magnetic fluxes above this cutoff, the contrast increases monotonically with the magnetic field. This holds for structures of categories 2 and 3.

4. Discussion

To summarize, faculae observed at 1.63 μm tend to appear dark near disc center, are not visible near $\mu \sim 0.75$, but become bright toward the limb. The corrected contrast values of 4-5% near disc center indicate a temperature deficit of about 130 K. Faculae at 1.2 μm have contrast below detection limits (0.2%) near disk center, consistent with measurements at similar opacity and spatial resolution in the visible region of the spectrum. Contrast of dark faculae at 1.63 μm is found to be correlated with magnetic flux; below a flux level of 2×10^{18} Mx contrast is below 0.2% and above this cutoff contrast increases nearly linearly with flux.

Measurements of the contrast of small (sub-arcsecond) network pores observed at high spatial resolution in the green show a similar cut-off and linear dependence on magnetic flux (Zirin and Wang, 1992). This suggests that these "invisible sunspot" structures might be similar to those which are seen as dark faculae at

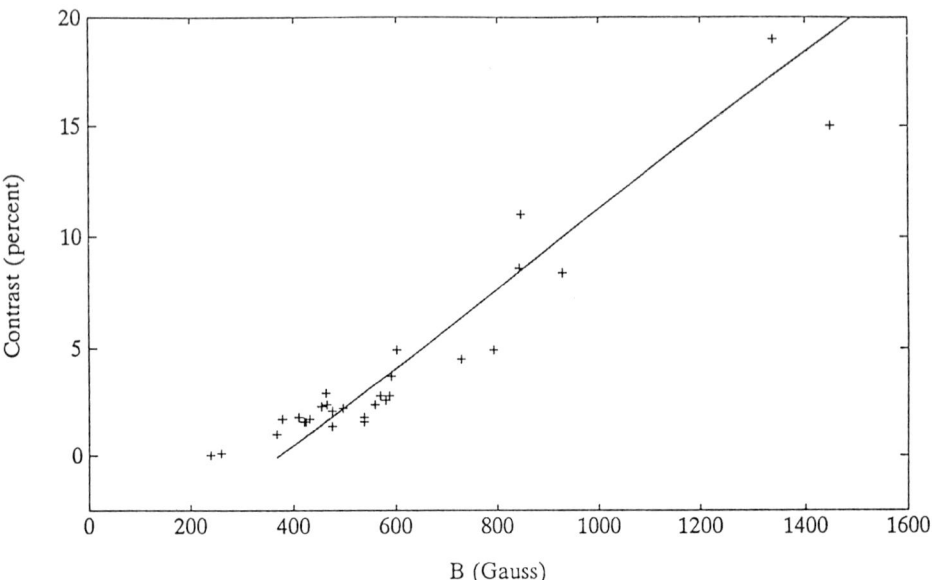

Fig. 2. A plot of 1.63 μm contrast versus NSO magnetic field strength for 29 selected structures (from Moran, Foukal, and Rabin, 1992).

1.63 μm. It appears that when the surrounding region is averaged into the measurement (*i.e.*, at lower spatial resolution), the contrast at 1.63 μm remains below unity while the visible contrast averages to zero.

MHD models have been constructed to explain facular contrast as a function of contributions from a hot, bright tube core, and from the relatively darker surrounding region. The surroundings are cooled below photospheric values by radiation into the (less opaque) flux tube (Deinzer *et al.*, 1984; Knölker and Schüssler, 1988; Spruit, 1976). Model parameters include a convective suppression factor, density reduction factor, field strength and tube diameter. For a particular choice of parameters, the model predicts facular contrast as a function of heliocentric angle and wavelength. The flux tube itself seems to be brighter than the photosphere in these models, even at 1.63 μm, so the observation of a dark facula may imply that the cooling of the flux tube surroundings becomes increasingly evident deeper in the photosphere (or the model may be in error).

Comparison between such models and observations of the kind described here could, in principle, provide important new insight into the temperature structure of plasma inside the flux tube and the region just outside it. But care must be exercised in such comparisons, since the calculations refer to flux tubes of diameter far below the angular resolution limit of the observations. Quantities that might be relatively insensitive to spatial resolution, like the ratio of contrasts at two wavelengths widely separated in H⁻ opacity, are more likely to prove useful than absolute contrasts at any one wavelength. To make progress, it will be important for modellers to generate observable quantities related to flux tubes *including* their surroundings, so that the new results of IR measurements with 2-3 arcsec resolution, such as those reported here, can be usefully interpreted.

References

Ayres, T.: 1989, *Solar Phys.* **124**, 15.
Chapman, G.: 1984, *Astrophys. J.* **232**, 923.
Deinzer, W., Hensler, G., Schüssler, M., and Weisshaar, E.: 1984, *Astron. Astrophys.* **139**, 435.
Elste, G.: 1985, in *Theoretical Problems in High Resolution Solar Physics*, ed. H.Schmidt, MPA Report 212, p. 185.
Foukal, P., and Duvall, T.: 1985, *Astrophys. J.* **296**, 739.
Foukal, P., Little, R., Graves, J., Rabin, D., and Lynch, D.: 1990, *Astrophys. J.* **353**, 712.
Foukal, P., Little, R., and Mooney, J.: 1989, *Astrophys. J. (Letters)* **336**, 33.
Knölker, M., and Schüssler, M.: 1988, *Astron. Astrophys.* **202**, 275.
Lawrence, J., Chapman, G., and Herzog, A.: 1988, *Astrophys. J.* **324**, 1184.
Moran, T., Foukal, P., and Rabin, D.: 1992, *Solar Phys.* **142**, 35.
Müller, R.: 1975, *Solar Phys.* **45**, 105.
Spruit, H.: 1976, *Solar Phys.* **50**, 269.
Vernazza, J., Avrett, E., and Loeser, R.: 1976, *Astrophys. J. Suppl.* **30**, 1.
Walton, S.: 1987, *Astrophys. J.* **312**, 909.
Wang, H., and Zirin, H.: 1987, *Solar Phys.* **110**, 281.
Worden, P.: 1975, *Solar Phys.* **45**, 521.
Zirin, M., and Wang, M.: 1992, *Astrophys. J. (Letters)* **385**, 27.

A SOLAR PLAGE MODEL

A. KUČERA

Astronomical Institute of the Slovak Academy of Sciences,
CS-059 60 Tatranská Lomnica, Czechoslovakia

and

E. A. BARANOVSKY

Crimean Astrophysical Observatory, 334413 Crimea, Ukraine

Abstract. An investigation of a solar plage based on observations of UV, visible, and IR spectral lines is presented. Using model calculations, we have established distributions of the main physical parameters (temperature, density, turbulent velocity) in both plage and quiet-Sun atmospheres. Details of the models are presented and the role of magnetic pressure is discussed.

Key words: infrared: stars – Sun: chromosphere – Sun: faculae, plages

1. Introduction

The problems of the heating of the solar atmosphere and of solar activity still remain of central interest in solar physics. Previous investigations of the physical properties of solar plages were carried out by many of authors (*e.g.*, Shine and Linsky (1972, 1974), Ayres *et al.* (1986) and LaBonte (1986a,b). Observations in the far UV spectral region and their interpretations [for example, Kelch and Linsky (1978), Baranovsky and Severny (1979), Basri *et al.* (1979), Lemaire *et al.* (1981)] have shown that physical conditions are complicated inside the solar plages and many physical influences – like magnetic fields, velocity fields, a mechanical energy fluxes and radiative losses – play important roles in both the energy and pressure balance of the plage atmosphere. The parameters of the 'average' plage have been obtained through the investigation of many plages. These studies have shown that a credible approach requires complex calculations of line formation. This work describes a further attempt to observe and model solar plages using continua and lines (including the Ca II IR triplet) that are formed over a range of height in the model atmosphere.

2. Observations and data reduction

Low-spatial resolution spectral observations have been made using the Horizontal Solar Telescope with Spectrograph (HSTS) at Stara Lesna Observatory. The main parameters of the HSTS were presented in Kučera *et al.* (1990). A set of nine well-calibrated photographic line profiles of Hα, Hβ, Hγ, Ca II H and K and the IR triplet (at 849.8 nm, 854.2 nm and 866.2 nm), and HeD$_3$ were obtained in both active and quiet regions. The observation lasted 14 minutes (without the calibration measurements); the slit height was the equivalent of 237″.5 and its width 0″.3. All line profiles have been normalized to their nearby continua. We followed the method of Jebsen and Mitchel (1978) to estimate the continua in the UV region for Ca II H and K lines.

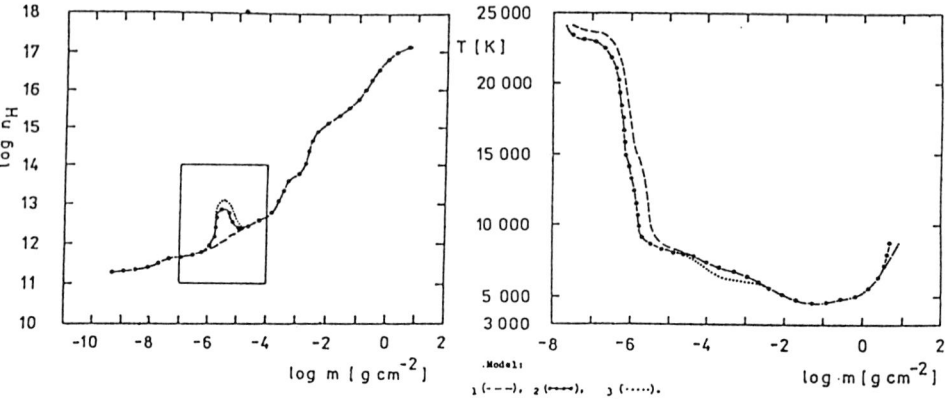

Fig. 1. Hydrogen number density and temperature as a function of mass column densities.

3. Modelling

We used a non-LTE computational program based on the Equivalent Two Level Atom (ETLA) approach. The following assumptions were adopted throughout our calculations:

1. A plane-parallel atmosphere without illumination by coronal radiation.
2. Complete redistribution (CRD) in the line transfer calculations.
3. The atmospheric parameters depend on height only.
4. Atomic models consist of 9 levels and continuum for hydrogen, of 6 levels and continuum for Ca II ion, and of 13 levels and continuum for Helium.

The computational method used is outlined by Baranovsky and Stepanian (1976, 1979), Baranovsky and Severny (1979), and Kučera et al. (1990) – it follows the approach of Avrett and Loeser (1969).

4. Results and Discussion

Three models of plage atmosphere were finally selected based on the agreement obtained between the computed profiles and the observations, which fit the observational profiles well. The models differ from each other in the hydrogen number density in the higher layers of the chromosphere as well as in the depth dependence of the temperature. The models are illustrated in Figure 1. Computed and observed profiles of Ca II K, Ca II IR 866.2 nm and He D_3 are shown in Figure 2 with corresponding values of the macroturbulent velocity. (For HeD_3 the differences between plage and quiet Sun are shown.)

TABLE I
Plage Model No. 3

N	h (km)	m (g cm^{-2})	$\tau_{L\alpha}$	T (K)	v (km s^{-1})	n_H (cm^{-3})	n_e (cm^{-3})
1	1211.3	1.670-24	2.496-3	40850	7.0	1.900+11	1.872+11
2	1211.3	5.327-10	4.996-3	39850	7.0	1.900+11	1.865+11
3	1211.3	1.456-9	1.000-2	38950	7.0	2.010+11	1.972+11
4	1211.2	2.973-9	2.001-2	37450	7.0	2.250+11	2.212+11
5	1211.2	5.210-9	4.006-2	35450	7.0	2.380+11	2.320+11
6	1211.1	8.534-9	8.018-2	34040	7.0	2.650+11	2.604+11
7	1211.0	1.319-8	1.605-1	31930	7.0	2.720+11	2.648+11
8	1210.9	1.837-8	3.212-1	29120	7.0	3.050+11	2.996+11
9	1210.8	2.388-8	6.428-1	26610	7.0	3.440+11	3.375+11
10	1210.7	2.973-8	1.287	24400	7.0	3.770+11	3.682+11
11	1210.6	3.707-8	2.575	23300	7.0	4.240+11	4.148+11
12	1210.4	5.010-8	5.154	23180	7.0	4.370+11	4.282+11
13	1210.0	7.665-8	1.032+1	23050	7.0	4.500+11	4.406+11
14	1209.3	1.331-7	2.065+1	22830	7.0	4.760+11	4.672+11
15	1208.0	2.455-7	4.133+1	22370	7.0	5.420+11	5.323+11
16	1205.9	4.409-7	8.271+1	21140	7.0	5.950+11	5.836+11
17	1204.1	6.346-7	1.655+2	17480	7.0	6.610+11	6.463+11
18	1202.6	8.083-7	3.313+2	14580	7.0	7.250+11	7.051+11
19	1200.9	1.025-6	6.632+2	13470	7.0	8.250+11	7.889+11
20	1198.9	1.316-6	1.327+3	12050	7.0	9.400+11	8.626+11
21	1197.5	1.598-6	2.657+3	10600	7.0	1.400+12	1.129+12
22	1197.0	1.804-6	5.317+3	9390	7.0	3.500+12	1.773+12
23	1196.7	2.071-6	1.064+4	8940	7.0	9.000+12	2.468+12
24	1196.4	2.505-6	2.130+4	8780	6.5	1.050+13	2.301+12
25	1196.0	3.307-6	4.263+4	8640	6.5	1.300+13	2.305+12
26	1195.2	4.826-6	8.533+4	8430	6.0	1.000+13	1.565+12
27	1193.1	7.715-6	1.708+5	8230	6.0	6.000+12	8.895+11
28	1184.9	1.346-5	3.418+5	7970	6.0	2.460+12	3.587+11
29	1160.4	2.455-5	6.841+5	7830	6.0	2.960+12	3.443+11
30	1122.3	4.559-5	1.369+6	7510	6.0	3.660+12	2.703+11
31	1065.7	8.484-5	2.741+6	7100	6.0	4.620+12	2.131+11
32	981.7	1.587-4	5.485+6	6600	5.0	5.910+12	2.178+11
33	891.9	2.989-4	1.098+7	6300	5.0	1.270+13	2.876+11
34	835.6	5.695-4	2.197+7	6150	5.0	4.500+13	4.042+11
35	771.2	1.096-3	4.398+7	6000	4.5	5.280+13	4.072+11
36	705.1	2.121-3	8.802+7	5800	4.4	1.330+14	5.083+11
37	673.0	4.058-3	1.762+8	5450	4.3	5.900+14	8.094+11
38	644.1	7.899-3	3.526+8	5150	4.2	1.000+15	1.019+12
39	608.6	1.533-2	7.057+8	4820	4.1	1.510+15	1.306+12
40	561.9	3.073-2	1.413+9	4560	4.0	2.200+15	1.780+12
41	500.0	5.778-2	2.827+9	4470	3.9	3.200+15	2.285+12
42	427.4	1.136-1	5.659+9	4540	3.8	6.000+15	2.484+12
43	353.2	2.255-1	1.333+10	4670	3.6	1.200+16	2.723+12
44	280.5	4.492-1	2.267+10	4900	3.5	2.500+16	3.475+12
45	206.3	9.028-1	4.537+10	5120	3.3	4.800+16	5.218+12
46	125.3	1.854	9.080+10	5900	3.1	9.200+16	1.846+13
47	13 3	3.808	1.817+11	7300	3.0	1.180+17	2.238+14
48	-152.7	7.298	3.638+11	8600	3.0	1.355+17	1.265+15

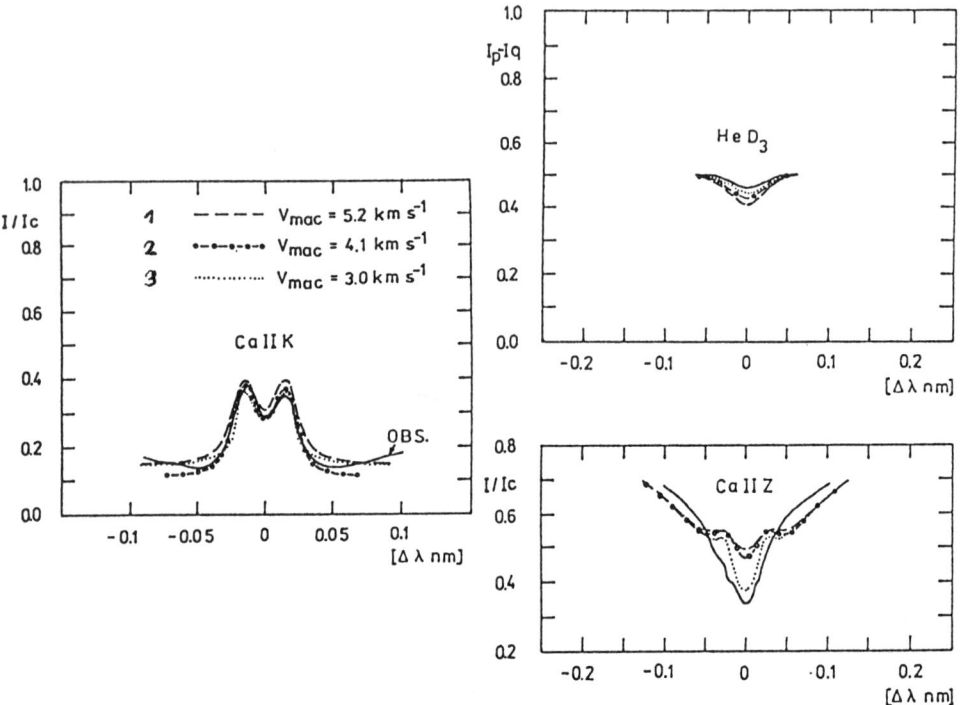

Fig. 2. Computed and observed intensity profiles of the plage for Ca II K, He D$_3$ and infrared Ca II Z lines.

We can see that model 3 best fits the observational profiles. (The double reversal in the internal parts of the observational profile of the IR Ca II line is weak due to seeing over a long exposure time.) The principal parameters of the model 3 are listed in Table I. As shown in Figure 1, the model assumes a rapid enhancement of the hydrogen number density over a small range in height. From Table I we determine that at these heights the gas pressure ($P_1 = NkT + \rho v^2/2$) exceeds the hydrostatic pressure, ($P_2 = mg$, where g is gravitational acceleration).

To balance the plage atmosphere we assume the additional magnetic pressure. In the higher layers of the atmosphere (where the preasure is relative low) the magnetic field may play an important role as a source of the pressure and at the same time its geometrical configuration can limit the expansion of the solar plasma. It may enhance the number density at the crucial height. A magnetic field in the range 5–100 G is enough to supply the pressure needed to balance the atmosphere.

The possibility of significantly affecting the hydrogen number density by magnetic fields must be further examined. Certainly it would be useful to take into account all possible pressure components in the calculations of pressure balance in the solar atmosphere. To verify the above stated hypothesis, more precise observations are needed, which are currently under way.

Acknowledgements

This work has been supported under Grant GA 494/1991 by the Slovak Academy of Sciences.

References

Avret, E.H., Loeser, R.: 1969, *Smithsonian Astrophys.Obs Spec.Rept.* **303**.
Ayres, T.R., Testerman, L. and Brault, J.W.: 1986, *Astrophys.J.* **304**, 542.
Baranovsky, E.A., Severny, A.B.: 1979, *Izv.Krimsk.Astrofiz.Obs.* **60**, 99.
Baranovsky, E.A., Stepanian, N.N.: 1976, *Izv.Krimsk.Astrofiz.Obs.* **55**, 14.
Baranovsky, E.A., Stepanian, N.N.: 1979, *Izv.Krimsk.Astrofiz.Obs.* **60**, 135.
Basri, G.S., Linsky, J.L., Bartoe, J.D.F., Brueckner, G. and Van Hooser, M.E.: 1979, *Astrophys.J.* **230**, 924.
Jebsen, D.E., Mitchel, Jr., E.: 1978, *Solar Phys.* **57**, 309.
Kelch, W.L., Linsky, J.L.: 1978, *Solar Phys.* **58**, 37.
Kučera, A., Rybák, J., Minarovjech, M., Novocký, D., Saniga, M.: 1990, *Astrophys. Space Sci.* **17**, 281.
Kučera, A., Scherbakova, Z., and Baranovsky, E.: 1990, in P. Ulmschneider, E.R. Priest, and R. Rosner (eds.), *Mechanisms of Chromospheric and Coronal Heating*, p. 109.
LaBonte, B.J.: 1986a, *Astrophys J. Suppl.* **62**, 229.
LaBonte, B.J.: 1986b, *Astrophys J. Suppl.* **62**, 241.
Lemaire, P., Goutterbroze, P., Vial, J.C. and Artzner, G.E.: 1981, *Astron. Astrophys.* **103**, 160.
Shine, R.A., Linsky, J.L.: 1972, *Solar Phys.* **25**, 357.
Shine, R.A., Linsky, J.L.: 1974, *Solar Phys.* **39**, 49.

FORMATION OF THE SOLAR 10830 Å LINE

E. H. AVRETT

*Harvard-Smithsonian Center for Astrophysics, 60 Garden Street,
Cambridge, MA 02138, U.S.A.*

J. M. FONTENLA

The University of Alabama in Huntsville, Huntsville, AL 35899, U.S.A.

and

R. LOESER

*Harvard-Smithsonian Center for Astrophysics, 60 Garden Street,
Cambridge, MA 02138, U.S.A.*

Abstract. One-dimensional hydrostatic-equilibrium models are shown here for faint, average, and bright components of the quiet Sun, and for a plage region, describing in each case how the atmosphere is stratified through the photosphere, chromosphere, and transition region up to a temperature of 10^5 K. The observed coronal line radiation is assumed to be the inward incident radiation at the 10^5 K boundary. This coronal radiation penetrates into the upper chromosphere causing sufficient helium ionization to populate the lower level of the He I 10830 Å line, producing optically-thin absorption of the photospheric continuum at 10830 Å. The amount of absorption, which is proportional to the optical thickness of the upper chromosphere in the 10830 line, depends on 1) the strength of the coronal lines at wavelengths in the He I 504 Å ionizing continuum, and 2) the density and geometrical thickness of the upper chromosphere. The computed 10830 Å line is shown for the four atmospheric models and for three values of the coronal illumination. The calculated off-limb 10830 intensity distribution shows a minimum in the low chromosphere and a maximum at roughly 2000 km above the photosphere, in general agreement with observations, indicating that this is the predominant height of the transition region over most of the solar surface.

Key words: He I 10830 Å – infrared: stars – line: formation – Sun: atmosphere

1. Introduction

In this paper we show how the solar helium 10830 Å line is formed in various model calculations. We use one-dimensional models to represent faint, average, and bright regions of the quiet Sun and to represent a plage region. We treat these components separately, ignoring horizontal interactions, since the horizontal size of the smallest observed features is large compared to the vertical extent of the region in which the spectrum is formed.

Each component model is constructed as follows: in the photosphere and chromosphere we start with an assumed distribution of temperature *vs.* height and use the hydrostatic equilibrium equation to determine the distribution of pressure *vs.* height. A turbulent pressure contribution is included based on the microturbulent velocity inferred from the Doppler widths of lines formed at various heights in excess of thermal Doppler widths. Our models do not include magnetic forces.

We solve the equations of radiative transfer and statistical equilibrium to determine the atomic and molecular number densities and internal radiation intensities, and we use energy-balance equations to determine the temperature distribution in the transition region, as explained below. Then we calculate the emergent spectrum and compare it with solar observations. The differences between the observed and computed spectra are analyzed, and then the assumed temperature distribution is modified to yield a new computed spectrum that is in closer agreement with the observations. The process is continued until good agreement with available observations is obtained. Thus the observed spectrum "determines" the corresponding temperature distribution.

The observed increase in the brightness temperature of the solar spectrum at far infrared wavelengths, $\lambda > 150$ μm and at ultraviolet wavelengths, $\lambda < 160$ nm, corresponds to a chromospheric rise in temperature starting about 500 km above the height where the disk-center visible continuum is formed. Models based on observations in the extreme ultraviolet (Vernazza, Avrett, and Loeser 1981) indicate that the temperature continues to increase gradually, reaching a value of about 8000 K at a height near 2000 km, where hydrogen begins to be partially ionized and to radiate substantial amounts of energy in the Lyman lines and continuum. Lyman α radiative cooling prevents any further *gradual* temperature rise, but there is enough mechanical heating in these and higher layers to ionize hydrogen completely. These effects cause the atmosphere to have a *transition region* only a few kilometers thick separating the layers where $T = 10^4$ and 10^5 K.

The transition region differs from the chromosphere not only because of its very large temperature and ionization gradients, but also because the radiated energy comes from mechanical energy deposited at higher temperatures and transported downward, rather than from any substantial amount of mechanical energy deposited locally. The transition-region models we discuss here are those of Fontenla, Avrett, and Loeser (1990, 1991, 1992). The temperature distribution in the transition region is not adjusted to match observations, as in the photosphere and chromosphere, but is computed theoretically assuming energy balance, using the boundary conditions near 10^4 K from the chromospheric model, and including the effects of particle diffusion. The downward flow of energy at the 10^5 K boundary is adjusted to balance the radiative energy losses at temperatures between 10^5 and 10^4 K.

In the transition region the upward diffusion of hydrogen atoms causes Lyman α radiation to be emitted at higher temperatures than would occur under conditions of local statistical equilibrium without diffusion. An important transport mechanism in the lower transition region is the recombination energy associated with pairs of protons and electrons that diffuse downward. Differential mass flows also may play a significant role in transporting energy, but we find that assuming hydrostatic equilibrium and accounting for particle diffusion gives computed hydrogen and helium spectra that are consistent with the observations.

The excitation and ionization of He I and He II in the transition region and upper chromosphere are considered in detail by Fontenla *et al.* 1992, hereafter FAL. We find that particle diffusion in the transition region is as important for helium as for hydrogen. In the upper chromosphere the ionization of helium is particularly sensitive to coronal line radiation in the $\lambda < 504$ Å range, while such coronal lines have a much smaller effect on hydrogen.

The He I 10830 Å line is formed almost entirely in the upper chromosphere. The absorption of infrared continuum radiation at 10830 Å by He I $2s\ ^3S$–$2p\ ^3P^0$ transitions at this wavelength depends on the integrated He I $2s\ ^3S$ number density along the line of sight (*i.e.*, on the integrated optical depth of the transition). We show in Section 3 that this integrated optical depth, τ, depends on the strength of the illuminating coronal radiation, the density at the top of the chromosphere, and the geometrical extent of the upper chromosphere.

2. Model Atmospheres

Before discussing the helium calculations we describe our atmospheric models and comment on their applicability to the inhomogeneous solar atmosphere. As noted earlier the abrupt temperature increase in the transition region occurs where hydrogen first becomes significantly ionized. For a more gradual chromospheric temperature rise than the average one, corresponding to a smaller amount of mechanical heating, the transition region would occur at a greater height. For higher chromospheric temperatures, corresponding to greater mechanical heating, the transition region would be located lower, closer to the photosphere. The calculations indicate, however, only moderate differences in the transition region heights for cell-center, average, and bright-network components of the quiet Sun: 2240, 2200, and 2015 km, respectively. Our plage model has the transition region at a height of 1740 km. Figure 1 shows the temperature distributions for our quiet-Sun component models A (cell-center), C (average), and F (bright network) and for our plage model P. (These atmospheric models are tabulated in FAL.)

Various off-limb observations show evidence of chromospheric and transition-region material at considerably greater heights than indicated in Figure 1. Our hydrostatic models are based on balancing the effects of gravity and hydrostatic pressure, and give the atmospheric stratification in the absence of any dynamical forces that could push the transition region to greater heights in certain areas. An analogous model of the Earth's ocean would be a plane surface, with water below and air above. This one-dimensional model might seem unrealistic to a surfer or to a sailor in a storm, but it would still be a useful model for many purposes.

Our view of how the calculated model stratification may apply to the Sun is indicated in Figure 2 (from FAL). The dark-shaded region is a cross-section of the transition region between the chromosphere and the corona. The height of the transition region may vary irregularly due to dynamical forces not included in our hydrostatic calculation, but over much of the solar surface this height may be close to the hydrostatic value.

Disk spectra obtained near $\mu = 1$ and the temperature distributions determined from them may be insensitive to these irregularities, especially if they occupy a small fraction of the solar surface. However, limb spectra obtained near $\mu = 0$ (*i.e.*, along the horizontal direction in Figure 2) would tend to intersect these high-lying features. Such a horizontal line of sight may be optically thick or optically thin depending on wavelength.

Consider first a wavelength for which a horizontal ray passing above the average height of the transition region becomes optically thick by intersecting only a small number of high-lying features. The ray becomes optically thin only at a still greater height, and the transition region will appear to be located at that height.

Now consider a wavelength for which the optical thickness is of order unity along the ray tangent to the average height of the transition region. The high-lying features are transparent at this wavelength, and the transition region will appear

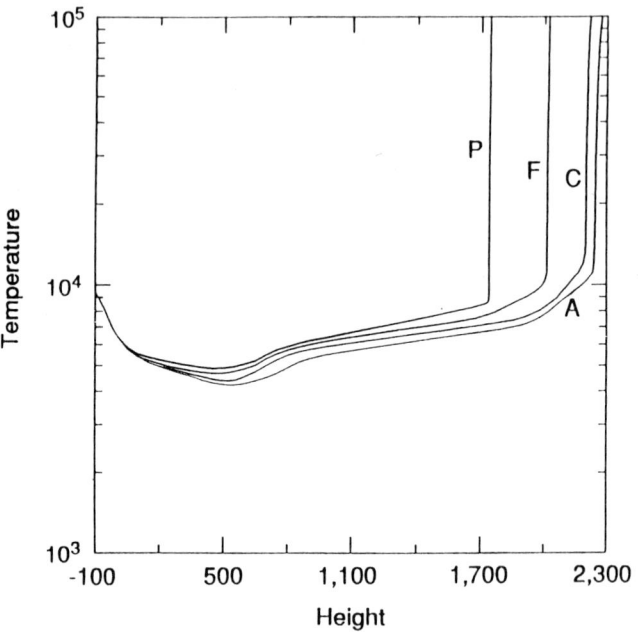

Fig. 1. Temperature (K) vs. height for models A, C, F, and P. Height is in km above the photospheric level where the vertical optical depth is unity in the continuum at 5000 Å.

in this case to be located at its actual average height.

At the center of the 10830 Å line we find that the optical thickness is of order unity along the ray tangent to $h = 2100$ km (where $T \approx 10^4$ K) for model C. Limb observations of this line should therefore be useful in establishing the average height of the transition region. Our calculated off-limb results are shown in Section 5.

3. Model Calculations and Disk Profiles

The properties of the 10830 Å line depend mainly on the density and extent of the upper chromosphere and on the coronal radiation that illuminates this region in the $\lambda < 504$ Å ionizing continuum. Recombination following He I ionization causes the $2s\ ^3S$ level to be populated, and causes $2s\ ^3S$–$2p\ ^3P^0$ absorption of the 10830 Å continuum on the disk.

For the incident radiation we use the observed solar irradiances in the extreme ultraviolet compiled by Tobiska (1991). He lists separate estimates of the irradiance contributions from coronal source regions and from the deeper layers (transition region and chromosphere) differentiated statistically rather than by spatial observations, since they are based on full-disk measurements. Tobiska gives data for times

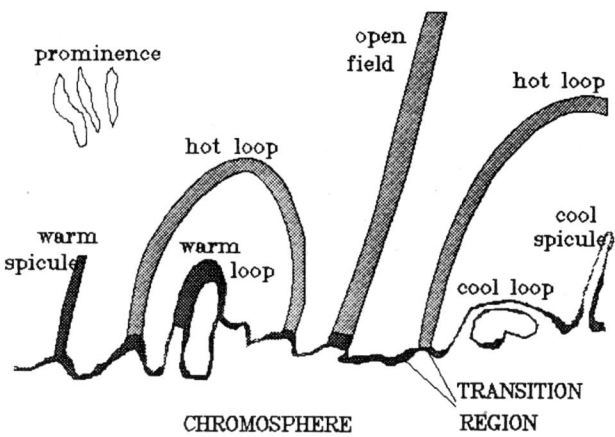

Fig. 2. Some of the magnetic structures observed in the low corona. The sketch indicates cool material (T about 10^4 K) in the chromosphere and in spicules, loops, and prominences; warm material (about 10^5 K) in the transition region and in loops and spicules; and hot material (about 10^6 K) in the loops and open fields.

of low and moderate solar activity, and we have used these results to determine the corresponding mean intensity at the solar surface for coronal source regions only. The results are listed in Table 6 of FAL.

We use the low- and moderate-activity values for our models A and C respectively. For models F and P we use 3 times the moderate activity values. In addition to model C we show the results for models CL and CH which are the same as C except that CL uses the low-activity incident intensities and CH uses the higher values, as in models F and P.

Figure 3 shows the calculated number density of the $2s\ ^3S$ level (in cm^{-3}) as a function of height for models CL, C, and CH. The 10830 line-center optical thicknesses between 1400 and 2200 km in the three cases are 0.042, 0.17, and 0.37, respectively.

The corresponding disk-center absorption lines are shown in Figure 4. The line has three components with relative strengths 0.5556, 0.3333, and 0.1111 at wavelength displacements 0, −0.0898, and −1.249 Å, respectively. The first two are blended together producing a slightly asymmetric result while the third component is well separated. These profiles clearly show the effect of different values of the coronal radiation.

We now consider the relative behavior for models A, C, F, and P. Figure 5 shows the calculated $2s\ ^3S$ number density vs. height in each case. The transition region is at a different height for each model. Models CH and F have the same incident

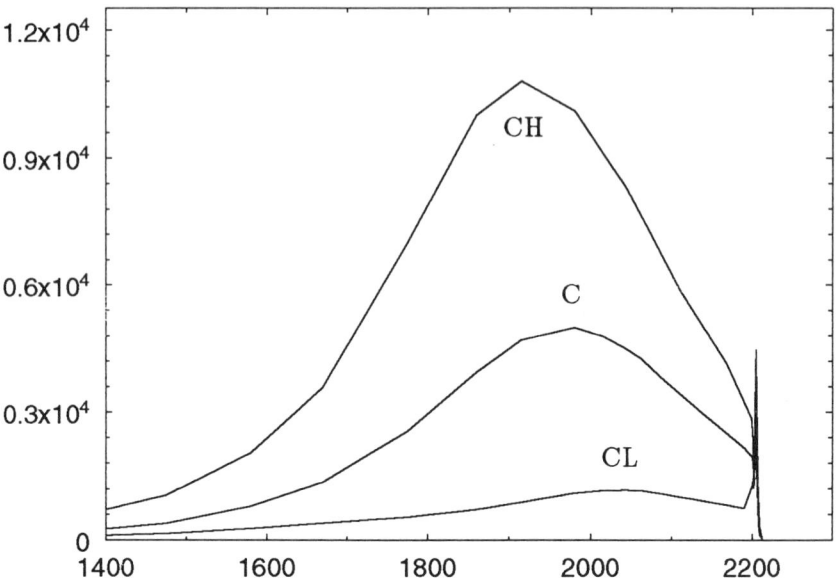

Fig. 3. Number density (cm^{-3}) of the $2s\ ^3S$ level of He I vs. height (km) calculated for models CL, C, and CH.

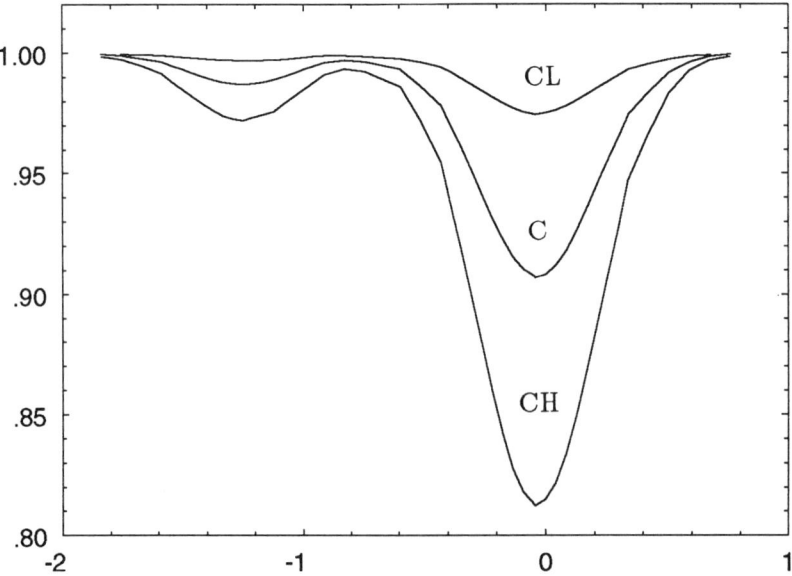

Fig. 4. Disk-center residual intensity of the 10830 Å line vs. $\Delta\lambda$ in Å for models CL, C, and CH.

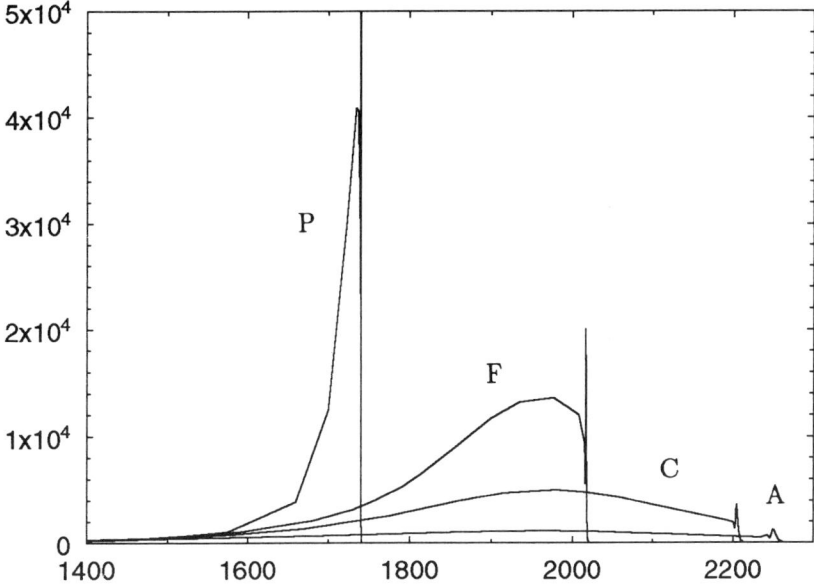

Fig. 5. Number density (cm^{-3}) of the $2s\ ^3S$ level of He I vs. height (km) calculated for models A, C, F, and P.

radiation, and the maximum $2s\ ^3S$ number density is larger for model F than for model CH (see Figure 3), but the integrated model-F number density is smaller because the transition region is located deeper, and because the attenuation of the ionizing $\lambda < 504$ Å radiation in the lower chromosphere is roughly the same in the two cases. The integrated model P number density is less than that of model F for the same reasons. The total line center optical depths for models A, C, F, and P are 0.054, 0.17, 0.26, and 0.15, respectively. The corresponding disk-center profiles are shown in Figure 6. Note that the profile for model P lies between the profiles for models A and C.

These results can be described by the simple equation

$$I_0 = I_c e^{-\tau} + S(1 - e^{-\tau}) \qquad (1)$$

where τ is the line center optical thickness of the chromosphere, S is the line source function (which we find to be essentially constant, and equal to about $0.4 \times I_c$), I_c is the continuum intensity at 10830 Å, and I_0 is the line center intensity. In Table I we give the values of these quantities for the various models. The values of I_0/I_c correspond to the residual intensities shown in Figures 4 and 6.

Models A and CL have the same coronal illumination, and the model A chromospheric densities are lower than in model CL, but the chromosphere extends to greater heights in model A. As a result there is slightly more line absorption in the case of model A.

As noted earlier, models CH, F, and P also have the same coronal illumination, and these three models have progressively larger densities at a common location in

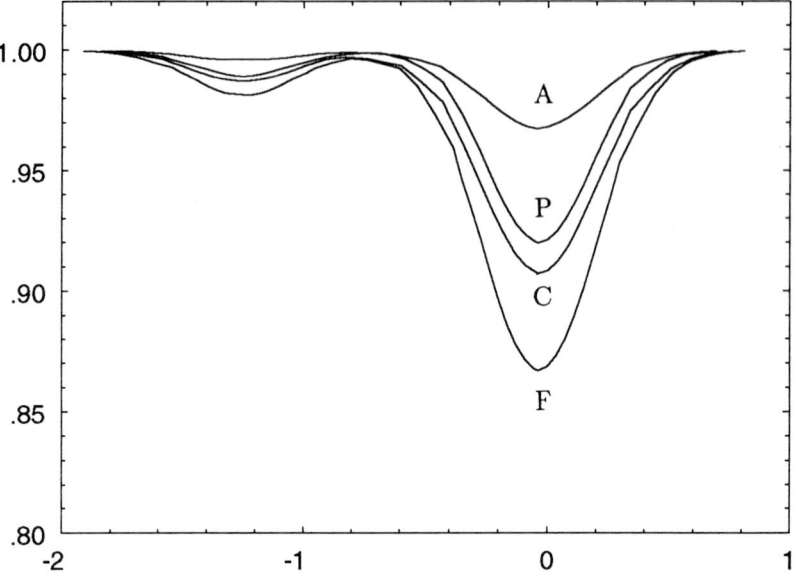

Fig. 6. Disk-center residual intensity of the 10830 Å line vs. $\Delta\lambda$ in Å for models A, C, F, and P.

the chromosphere (at 1700 km for example), but they have progressively smaller amounts of 10830 Å absorption because the geometrical thickness of the chromosphere is progressively smaller, as shown in Figure 5.

The reduction in the height of the transition region (*i.e.*, in the geometrical thickness of the chromosphere) for the brighter component models thus has a substantial effect on the calculated line absorption.

4. Center-to Limb Behavior

From Table I the line center vertical optical thickness of the chromosphere is calculated to be 0.17 for model C. The chromospheric optical thickness increases for disk positions toward the limb, and the line center intensity I_0 decreases relative to the continuum value I_c. The calculated intensity values from center to limb for the six models are listed in Table II.

The values for the fractional radius 1.0 in this table correspond to a line of sight at the limb which is tangent to height $h = 0$, where the vertical optical depth is unity in the continuum at 5000 Å. (We use plane geometry to determine the atmospheric stratification, and then spherical geometry in these intensity calculations.)

5. Calculated Limb Emission

We give the following calculated results at and above the limb: 1) the integrated intensity within a 1 Å band centered on the strongest 10830 Å component, and

TABLE I
Parameters in Equation 1[a]

model	τ	S	I_c	I_0	I_0/I_c
A	0.054	1.6	4.00	3.87	0.968
CL	0.042	1.6	4.01	3.91	0.975
C	0.17	1.6	4.01	3.63	0.905
CH	0.37	1.6	4.01	3.26	0.813
F	0.26	1.7	4.03	3.50	0.868
P	0.15	1.8	4.13	3.80	0.920

[a] S, I_c, and I_0 have the units 10^{-5} ergs cm^{-2} s^{-1} sr^{-1} Hz^{-1}

TABLE II
Line Center (I_0) and Continuum (I_c) Intensities[a] vs. Radial Disk Position

model		\multicolumn{6}{c}{fractional radius}					
		0	0.5	0.8	0.9	0.95	1.0
A	I_0	3.87	3.70	3.32	3.02	2.76	1.78
	I_c	4.00	3.84	3.49	3.21	2.98	2.04
	I_0/I_c	0.97	0.96	0.95	0.94	0.93	0.87
CL	I_0	3.91	3.75	3.38	3.09	2.82	1.88
	I_c	4.01	3.85	3.50	3.23	3.00	2.12
	I_0/I_c	0.98	0.97	0.97	0.96	0.94	0.89
C	I_0	3.63	3.45	3.02	2.72	2.43	1.63
	I_c	4.01	3.85	3.50	3.23	3.00	2.12
	I_0/I_c	0.91	0.90	0.87	0.84	0.81	0.77
CH	I_0	3.26	3.06	2.63	2.30	2.04	1.49
	I_c	4.01	3.85	3.50	3.23	3.00	2.12
	I_0/I_c	0.81	0.79	0.75	0.71	0.68	0.70
F	I_0	3.50	3.32	2.89	2.57	2.30	1.60
	I_c	4.03	3.87	3.53	3.27	3.06	2.26
	I_0/I_c	0.87	0.86	0.82	0.79	0.75	0.71
P	I_0	3.80	3.63	3.25	2.95	2.69	1.80
	I_c	4.13	3.98	3.66	3.42	3.23	2.50
	I_0/I_c	0.92	0.91	0.89	0.86	0.83	0.72

[a] in 10^{-5} ergs cm^{-2} s^{-1} sr^{-1} Hz^{-1}

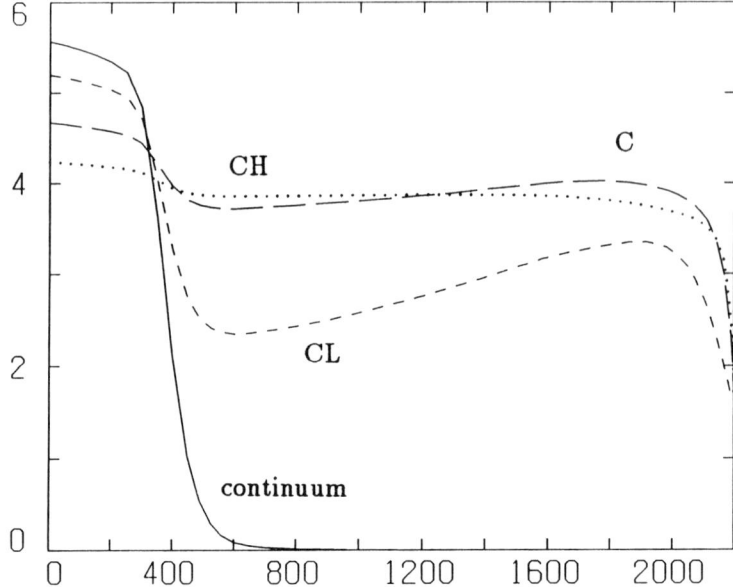

Fig. 7. Off-limb intensity in 10^5 ergs cm^{-2} s^{-1} sr^{-1} within a 1 Å pass-band centered on the strongest component of the 10830 Å line vs. tangential height above the photosphere in km, for models CL (short dashes), C (long dashes), and CH (dots). The solid curve shows the corresponding intensity in the nearby continuum within a 1 Å pass-band.

2) the integrated intensity within a nearby 1 Å continuum band. For simplicity these will be called line and continuum intensities, respectively. Figure 7 shows the off-limb intensities for models CL, C, and CM plotted *vs.* tangential height h (defined above). The continuum intensity, which is the same for the three models, exceeds the line intensity at $h = 0$ but then decreases rapidly with height as the continuum becomes transparent. The chromosphere has a line optical thickness of order unity along the limb tangent ray and becomes optically thick as the tangent height increases. The chromosphere becomes optically thin again only for rays tangent to the top of the chromosphere and above. The curves corresponding to models C and CH in Figure 7 are similar because in these two cases the chromosphere is optically thick along rays tangent to heights in the range 500–2000 km, while the model CL line intensity shows a lower minimum value because the low chromosphere is more transparent.

6. Off-Limb Observations

We are not aware of any published observations that can be compared directly with the results in Figure 7. However, the 10830 Å line is very similar to, but stronger than, the He I D_3 line at 5875.68 Å. (The 10830 and D_3 transitions take place between the levels $2s\ ^3S$–$2p\ ^3P^0$ and $2p\ ^3P^0$–$3d\ ^3D$, respectively.) The observed brightness distribution above the limb is reported to be very similar for the two lines

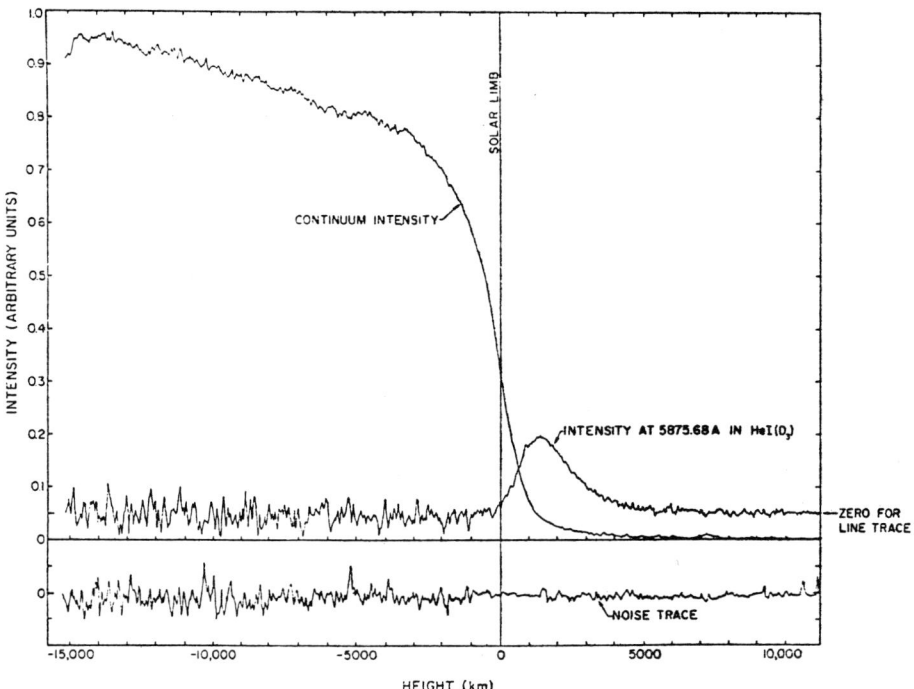

Fig. 8. Residual intensity *vs.* height above the limb in the He I D_3 line at 5875.68 Å obtained by subtracting the continuum variation at the limb from that in the line. The continuum variation and a residual noise trace are shown for reference. No seeing corrections were made in the data shown. From White (1963).

(S. Koutchmy, private communication). Both show emission above the limb that extends to $h \approx 2000$ km, the height of the upper chromosphere in the hydrostatic models.

Figure 8 shows the D_3 emission above the limb measured by White (1963). The residual line emission shown in this figure is obtained by subtracting the corresponding continuum intensity in the same narrow bandwidth used for the line. The continuum distribution is plotted for reference. There is essentially no D_3 absorption on the disk, so that subtracting the continuum gives a good measure of the line emission corrected for the effects of scattered light. Since the line-center 10830 intensity is less than the continuum value on the disk, subtracting the continuum would cause negative residual values for $h < 300$ km.

Eclipse observations of the D_3 line have been reported by Athay and Menzel (1956), Gulyaev (1971,1972), and Lifshitz *et al.* (1976). Zirin (1975) published filtergrams showing D_3 emission above the limb at a level of about 10% of the central disk intensity. He found the emission to be reduced by a factor of 2 to 3 in coronal holes. Koutchmy and Avrett (1989) reported similar results based on observations by Koutchmy at Sacramento Peak Observatory: the D_3 emission has a maximum at a height of about $h = 1800$ km except in coronal holes where there is little extended emission above the limb.

In Sections 3 and 4 of this paper we have shown the calculated behavior of the 10830 Å line on the disk, but we have not attempted to compare these results with observations. He I 10830 Å observations are discussed in the papers of Jones (1993), Fleck et al. (1993), Harvey and Livingston (1993), and Harvey (1993) in these proceedings.

7. Conclusions

Our model calculations indicate that the line absorption clearly depends on the coronal illumination. However, the computed profiles corresponding to different spatial components of the atmosphere also depend on the geometrical thickness of the upper chromosphere and on the density of this region.

We give results based on one-dimensional hydrostatic-equilibrium models of faint, average, and bright components of the quiet Sun and of a plage region. In these four cases the transition region is located at successively lower heights, i.e., the chromosphere is successively reduced in extent. From a theoretical viewpoint the only way to avoid this effect is for the brighter component models to be subject to successively larger dynamic or Lorentz force contributions, in order to increase their chromospheric scale heights.

A consequence of the decrease in chromospheric thickness, despite the increase in density at the top of the chromosphere, is a decrease in 10830 absorption. It should be possible to use observations at other wavelengths to distinguish between 1) this decrease in 10830 absorption in regions that are brighter at other wavelengths due to chromospheric heating, and 2) the decrease due to diminished EUV coronal illumination.

In our model calculations we arbitrarily chose the coronal illumination for model F and for model P to be 3 times that of model C. We found the 10830 line for model P to be weaker than that of model C because of the reduced geometrical thickness of the chromosphere in the model P calculation. Since observations of plage regions show more absorption at 10830 Å than do observations of quiet regions, the model P calculation discussed here needs to be modified either by increasing the chromospheric scale heights in some way, or by increasing the coronal illumination. We believe that an increase in illumination is the change that is primarily needed.

It would be useful to have quantitative off-limb measurements of the 10830 Å line to compare with the results in Figure 7. We expect such observations to resemble those of the similar but weaker D_3 line: to show emission that extends out to roughly 2000 km, as in Figure 7, and that weakens in coronal hole regions. The observed intensity distribution is not expected to have the sharp drop-off at 2200 km shown in Figure 7, but to have a more gradual decrease (as in Figure 8) corresponding to the fraction of chromospheric and transition-region material located at greater heights.

Acknowledgements

This research was supported by NASA grants NSG-7054 and NAGW-2096.

References

Athay, R. G., and Menzel, D. H.: 1956, *Astrophys. J.* **123**, 285.
Fleck, B., Deubner, F. -L., Maier, D., and Schmidt, W.: 1993, these proceedings.
Fontenla, J. M., Avrett, E. H., and Loeser, R.: 1990, *Astrophys. J.* **355**, 700.
Fontenla, J. M., Avrett, E. H., and Loeser, R.: 1991, *Astrophys. J.* **377**, 712.
Fontenla, J. M., Avrett, E. H., and Loeser, R.: 1992, *Astrophys. J.* in press.
Gulyaev, R. A.: 1971, *Solar Phys.* **18**, 410.
Gulyaev, R. A.: 1972, *Solar Phys.* **24**, 72,
Harvey, J., and Livingston, W.: 1993, these proceedings.
Harvey, K.: 1993, these proceedings.
Jones, H. P.: 1993, these proceedings.
Koutchmy, S., and Avrett, E. H.: 1989, unpublished conference manuscript available from the authors.
Lifshits, M. A., Akimov, L. A., Belkina, I. L., and Dyatel, N. P.: 1976, *Solar Phys.* **49**, 315.
Tobiska, W. K.: 1991, *J. Atmos. Terr. Phys.*, **53**, 1005.
Vernazza, J. E., Avrett, E. H., and Loeser, R.: 1991, *Astrophys. J. Suppl.* **45**, 635.
White, O. R.: 1963, *Astrophys. J.* **138**, 1316.
Zirin, H.: 1975, *Astrophys. J.* **199**, L63.

INTERPRETING RECENT OBSERVATIONS OF He I 10830 Å

HARRISON P. JONES

*NASA/Goddard Space Flight Center, Laboratory for Astronomy and Solar Physics,
Greenbelt, MD 20771, U.S.A.*

Abstract. Spectra-spectroheliograms obtained in the 10830 Å line with the NASA/NSO Spectromagnetograph are analyzed to produce images in equivalent width, line depth, velocity, and continuum intensity. These and other images imply that if large-scale deficits in coronal irradiation induce weakening of the 10830 line, the ionizing radiation must be produced in close enough spatial proximity to the region of formation for 10830 to allow sharply defined spatial boundaries. The complex morphology of the images shows the importance of underlying chromospheric structure in providing a highly variable "substrate" which is modulated by varying illumination. Clear evidence is seen of He I 10830 Å formation in magnetic loops where the velocity field, particularly at footpoints, directly affects the excitation of the line. Steady flows as seen by Lites *et al.* (1985) are also observed, particularly near the limb. The observations and radiative transfer calculations are consistent with a simple, optically thin "cloud" model of line formation, and observed line profiles in weak network elements compare favorably with predictions from mean models.

Key words: He I 10830 Å – infrared: stars – line: formation – Sun: atmosphere – techniques: spectroscopic

1. Introduction

The 10830 Å line of He I is seen in emission just above the solar limb and in weak absorption on the disk with strong spatial and temporal variations. Since the atomic transition occurs between the lowest metastable levels of orthohelium which are both energetically far above ($\sim$ 20 eV) and radiatively isolated from the true ground state (parahelium), excitation of the line occurs primarily as recombination from He II. The precise mechanisms responsible for the formation of the He I 10830 Å line are still a matter for considerable speculation and controversy. However, the rich phenomenology evident in 10830 Å images and the well-documented correlations of many 10830 Å features with EUV and soft X-ray phenomena are ample evidence of the importance of the line in solar physics.

Radiative transfer calculations with mean models (Avrett, 1991; Avrett *et al.*, 1993) place the formation of the line near the interface between the upper chromosphere and the lower transition region and show the sensitivity of the lower-level population to ionizing radiation produced in the surrounding transition region and corona and to chromospheric density. These calculations are consistent with the observed limb emission, the well established visibility of coronal holes in 10830 Å spectroheliograms, the high correspondence between 10830 Å spectroheliograms and soft X-ray images of the Sun, and the good temporal correlations between integrated 10830 Å "fluxes" and other proxies for EUV irradiance (Harvey, 1984, 1993).

In this paper, some early 10830 Å observations from a new instrument, the NASA/NSO Spectromagnetograph (SPMG: Jones *et al.*, 1992), are described which may help to quantify the mechanisms for He line formation and their observational consequences. The SPMG is now in routine operation at the NSO/Kitt Peak Vacuum Telescope (KPVT) and will soon replace the aging 512-channel Diode Array Magnetograph (Livingston *et al.*, 1976 a,b). Instead of the two-slit Babcock configuration of the old instrument, the SPMG images resolved, two-dimensional, long-slit spectra over a several-Ångstrom bandpass on a CCD detector. Special video processing electronics accumulate and record or reduce the resolved line profiles synchronously with spatial scanning of the solar image across the entrance slit of a spectrograph. Thus the many monochromatic images and quantities derived from them are strictly simultaneous and precisely registered. Fortunately, the detector (Texas Instruments 241 chip) responds well enough at 1.083 μm without special cooling to produce far fewer "streaks" from slow "fixed" pattern variations than the Diode Arrays of the old instrument.

2. Observations

The observations described here were taken in conjunction with a rocket flight of the NASA/Goddard Solar Extreme Ultraviolet Rocket Telescope and Spectrograph (SERTS)-4 (Neupert et al., 1992) on May 7, 1991. Figure 1 shows a full-disk magnetogram and 10830 Å spectroheliogram taken with the SPMG at the KPVT prior to the rocket flight. Comparison of the SPMG 10830 Å and magnetic data with EUV spectra and images obtained by the SERTS instrument is in progress and will be reported separately as appropriate. This paper treats the SPMG data alone.

Various solar features seen in 10830 Å are labeled for reference in Figure 1. When seen in negative contrast, coronal holes appear as slightly brighter (lower absorption) regions in 10830 Å equivalent width images which overly unipolar magnetic fields; three large coronal holes, labeled CH_{1-3} are marked in Figure 1. A similar region of slightly enhanced brightness which is labeled FC overlies a region of weak magnetic field between two unipolar regions of opposite polarity and is probably a "filament channel" (McCabe and Mickey, 1981). Outside of coronal holes and filament channels, the magnetic field patterns can be seen as network-like, weak absorption features in the 10830 Å image. Concentrated dark points which tend to be associated with X-Ray bright points (Harvey, 1985) can be seen at various points as can dark filament- and loop-like structures. As discussed below, the small bright features in the large southwest active region are artifacts of high line-of-sight motions.

Even though coronal holes as seen in Figure 1 and other He 10830 Å data are low contrast phenomena, they are consistently outlined by sharp, if irregular, boundaries. Lites et al. (1985) point out that if the He 10830 Å contrast in coronal holes is a result of reduced excitation by EUV and soft-Xray radiation incident from above, this radiation must originate close to the He I line-forming region. Otherwise the spatial signature would be more diffuse. However, the boundaries often appear as well-formed network and could indicate a density enhancement surrounding the base of the coronal hole. Either explanation argues for thermodynamic manifestations of coronal holes to be at least as deeply rooted as the chromosphere/transition-region boundary.

The full-disk images were followed by ten He 10830 Å equivalent width area scans of the target region for the SERTS-4 instrument at one image every 15 minutes. Although no dramatic event was recorded in these area scans and they are not shown here, rapid viewing of the images in sequence leaves an impression of a very "busy" field of view with many rapid but small-scale changes occurring. This

INTERPRETING RECENT OBSERVATIONS OF He I 10830 Å

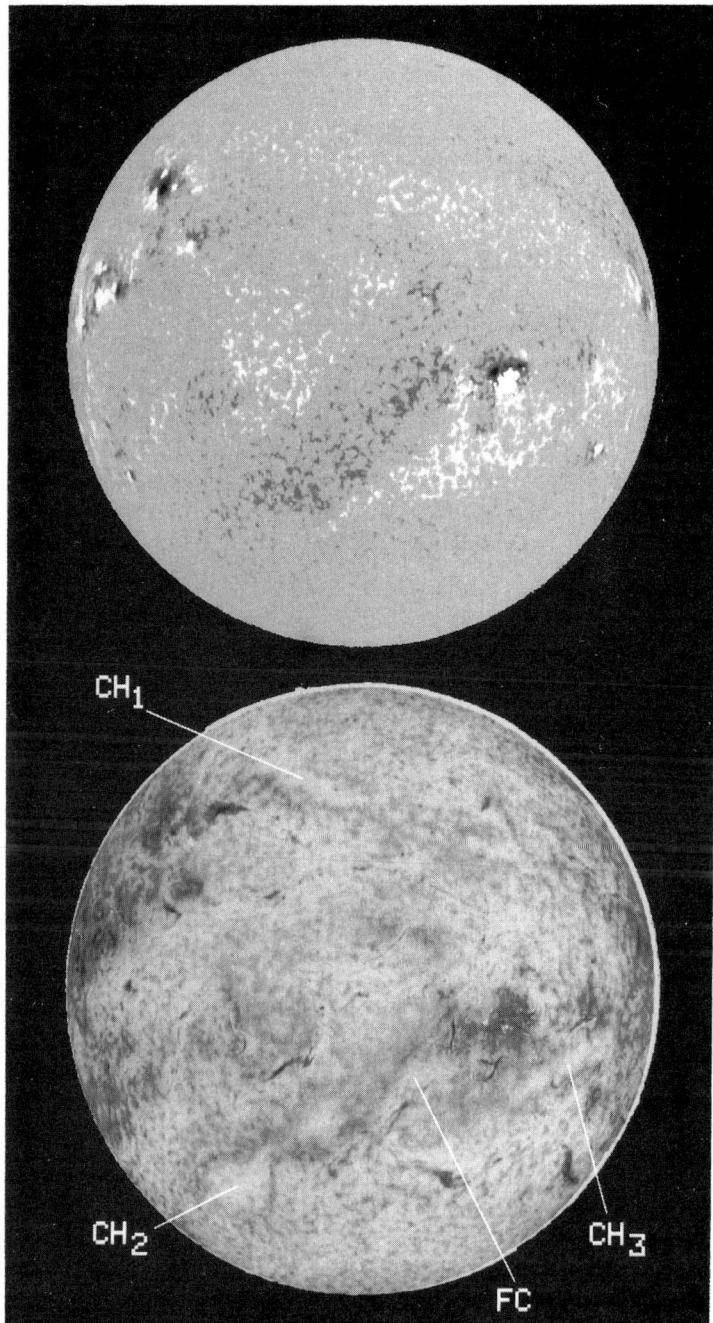

Fig. 1. Upper: full-disk magnetogram (Fe I, 8688 Å), 7 May 1991, 15:36-16:26 UT. Lower: full-disk 10830 Å equivalent width spectroheliogram, 16:40-17:22 UT, shown with negative contrast; labeled features are described in text

is particularly evident in some of the dark filamentary structures where absorption features appear to move back and forth along lines of force.

Figure 2 shows continuum intensity (top), line-of-sight velocity, a sample spectral section along a row marked by horizontal fiducials on each image, equivalent width, and line-center depth (bottom) derived from the He 10830 Å spectra-spectroheliogram. The labeled vertical lines point to the positions of selected features along the spectral section whose line profiles are shown in Figure 3.

The continuum image shows contamination from the weakest member of the 10830 Å triplet (10829.1 Å). The velocities are quite large compared to photospheric values with the grey scale covering the range from -8 to +7 km s^{-1}. Both line depth and equivalent width are shown here in negative contrast (small values appear brighter) as in the conventional display of NSO/KP 10830 Å spectroheliograms. Notice the nearly identical appearance of the line depth and equivalent width– a correlation which is easily explained if the lines are optically thin as discussed below. The morphology in and near the active region is complicated and clearly suggestive of line formation in magnetic loops and filaments. The regions of highest velocity tend to coincide with the apparent footpoints of these magnetic features at locations which are often marked by increased absorption or apparent emission (see below).

The contrast of the spectrum is enhanced by a non-linear grey-scale mapping; the line is quite weak (typically with depths of just a few percent of the continuum) at most places in the image. All of the images (not the spectra) have had center-to-limb trends and residual streaks from imperfect fixed pattern correction removed by simple curve fitting techniques. The labeled fiducials point to: A) a background or "quiet sun" feature; B) a network feature; C) a short active-region filament or loop; D) an artificial "emission" feature (see below); and E) a long-lived, network-like velocity structure as seen, for example by Lites *et al.* (1985) near the limb.

3. Interpretation

Line profiles for the five labeled features of Figure 2 are shown in Figure 3. The horizontal bars above the graph delimit two different wavelength bands which were used during the course of this analysis to determine the continuum intensity. The profiles of Figure 3 and the images of Figure 2 were derived using the two lower continuum bands. An immediate difficulty can be seen to arise in high-velocity features such as "D" where the line has been shifted partially into one of the continuum bands; the artificially induced reduction in continuum intensity forces part of the line into apparent emission and reduces both the line depth and equivalent width. The upper single band was used for an initial analysis of the data and is similar to the "two-slit" algorithm used for the synoptic full-disk spectroheliogram. In this case, the effect is enough to produce a negative (emission line) equivalent width. The difference in equivalent width images with the different determinations of continuum is shown in Figure 4. The single-band determination shows several artificial emission features and a number of small, low contrast "speckles" which are wholly absent in the image with the two-band determination of continuum. For future data taken with the SPMG, the continuum will be selected from spectral

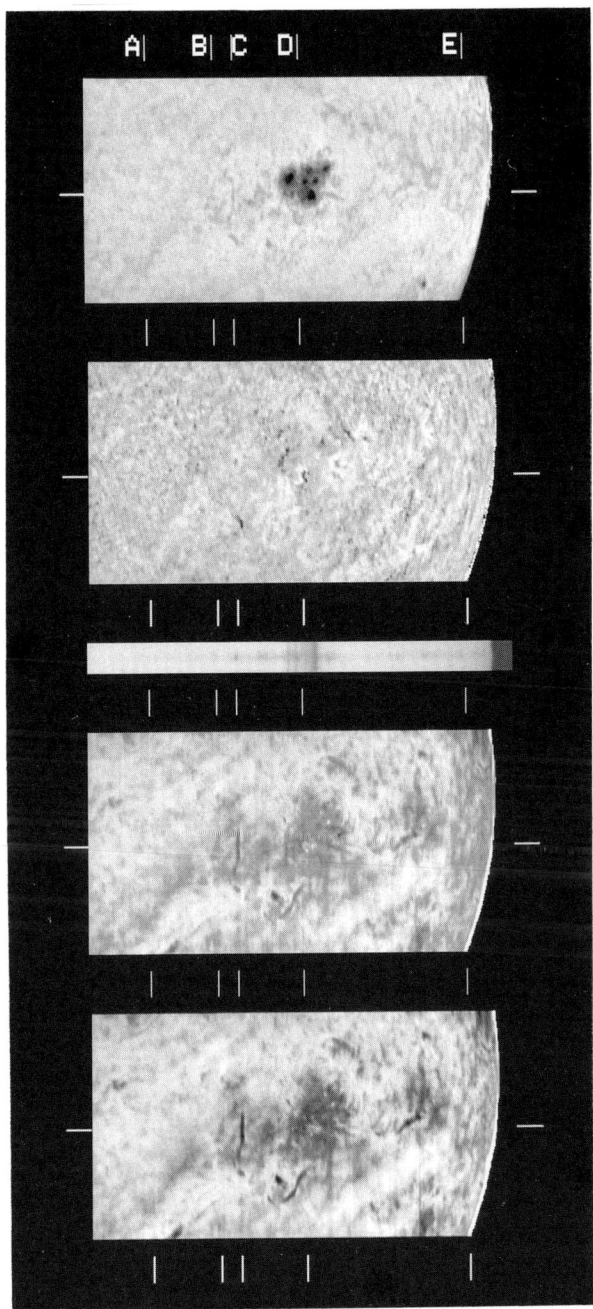

Fig. 2. Continuum intensity (*top*), velocity, spectral section (λ increasing from bottom to top), equivalent width, and line depth (*bottom*) images derived from a He 10830 Å spectra-spectroheliogram. The image row for the spectrum and specific labeled features are marked at the borders (see text).

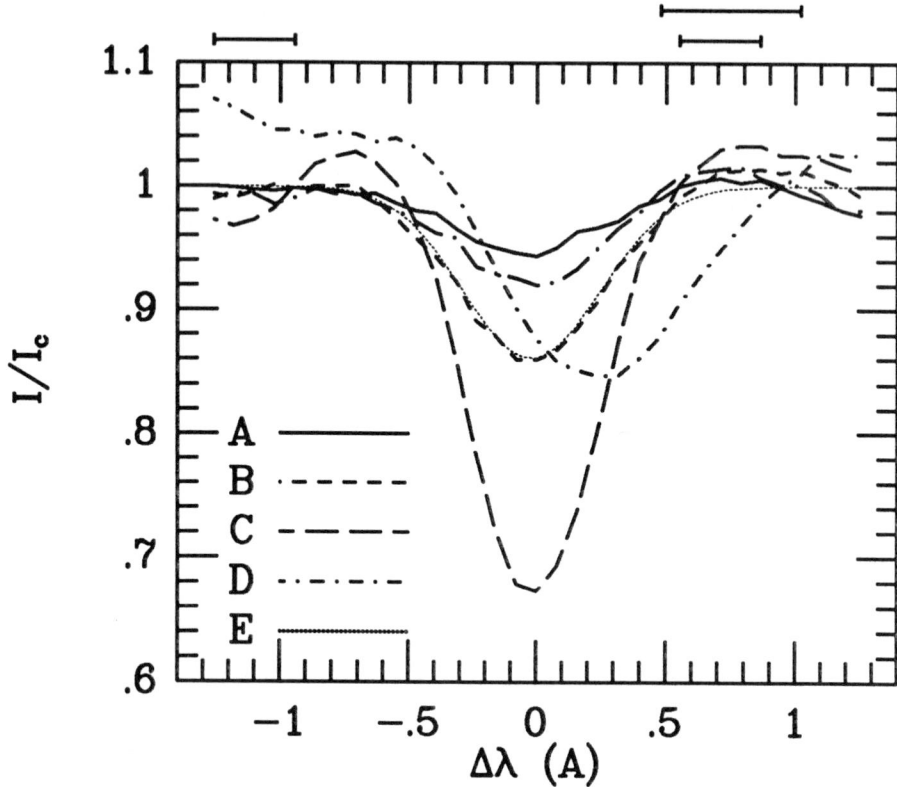

Fig. 3. Line profiles for features labeled in Figure 2. The light dotted line near the profile for feature "B" is a fit as described in the text. Horizontal bars above the graph delimit two different continuum bands used for analysis

region which is far enough removed to be unaffected by Doppler shifts.

There is a weak water vapor line almost exactly coincident with the 10830 Å line (Breckenridge and Hall, 1973; see also Harvey, 1993) whose effective Doppler width is typically 60 mÅ (much smaller than that for the He line) and which can increase the central depth by the order of 1% of the continuum (a noticeable effect where the line is weak). No account of water vapor has been taken here, but the qualitative results are not seriously affected.

While no attempt is made in this paper to produce detailed self-consistent non-LTE models based on the data, some approximate quantitative constraints can be determined by noting that both the observations and average models are consistent with a "cloud" model of line formation (Grossman-Doerth and von Uexküll 1971, 1973); that is, along any line of sight the region of appreciable He I 10830 Å absorption and emission is confined to a narrow spatial domain well above and separated

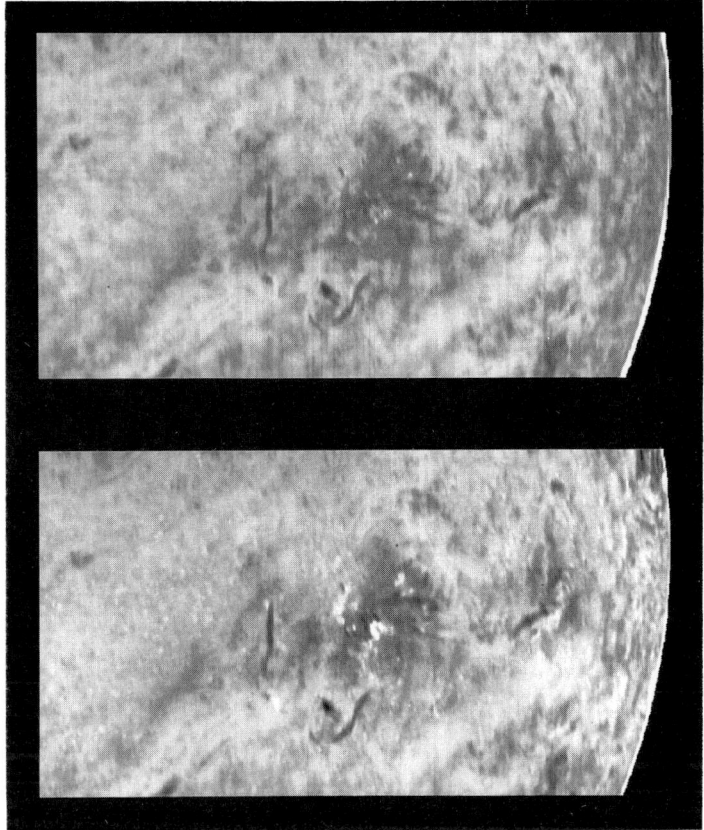

Fig. 4. Equivalent width images calculated using two continuum bands (top) and a single continuum band (bottom) as shown in Figure 3.

from the photosphere which produces the background continuum. If the line source function is approximately constant and independent of wave length in this cloud of line-forming material, then the equation of transfer has the particularly simple solution

$$I(\lambda) \approx I_c \exp(-\tau(\lambda)) + S(1 - \exp(-\tau(\lambda))). \qquad (1)$$

In terms of line-depth, $d(\lambda) \equiv 1 - I(\lambda)/I_c$, equation (1) leads simply to

$$d(\lambda) = (1 - S/I_c)(1 - \exp(-\tau(\lambda))) \rightarrow (1 - S/I_c)\tau(\lambda), \qquad (2)$$

where the limit $(\tau(\lambda) \ll 1)$ is valid for optically thin clouds.

The solar absorption feature at 10830 Å is actually the blend of two members of the $2\ ^3S$–$2\ ^3P$ triplet separated by $\delta = 0.091$ Å with the statistical weights of the longer to the shorter wave-length lines having the ratio $r = 5/3$. Thus, if τ_0 denotes the sum of all the line-center optical depths of each of the Zeeman sub-states for the two lines, we may write

$$\tau(\lambda) = [(1-\alpha)\phi(\lambda - \lambda_1) + \alpha\phi(\lambda - \lambda_1 + \delta)]\tau_0 \tag{3}$$

where λ_1 is the line center wavelength of the longer-wavelength line, $\alpha = r/(1+r)$, and the profile coefficient ϕ is expressed in terms of Doppler width λ_D as approximately

$$\phi(\lambda) = \exp(-(\lambda/\lambda_D)^2). \tag{4}$$

Anticipating the observed fact that $\delta/\lambda_D < 1$, one finds that τ and, hence, d have a maximum at wavelength

$$\lambda^* = \lambda_1 - \alpha\delta \tag{5}$$

from which it follows that the equivalent width, $w = \int_0^\infty d(\lambda)d\lambda$, is related to maximal line depth $d^* = d(\lambda^*)$ by

$$w/d^* = \sqrt{\pi}\lambda_D/[1 - \alpha(1-\alpha)(\delta/\lambda_D)^2]. \tag{6}$$

Finally, one may use equation (6) to find the Doppler width from the observed equivalent width and maximum line depth. One Newton-Raphson iteration with an initial guess of $\lambda_D^o = w/(\sqrt{\pi}d^*)$ of the cubic equation for λ_D which results from equation (6) yields

$$\lambda_D \approx \frac{w}{\sqrt{\pi}d^*}[1 - \pi\alpha(1-\alpha)\delta^2/(w/d^*)^2]. \tag{7}$$

The light dotted line which nearly coincides with the profile for feature "B" in Figure 3 is the calculated "fit" of the sum of two Gaussians, appropriately weighted and shifted according the multiplet structure for the 10830 Å line, with a Doppler width as determined by equation (7); clearly the approximations made in the preceding development lead to consistent representations of the data.

Table I shows the equivalent widths, line depths, and Doppler widths calculated from equation (7) for each of the labeled features of Figures 2 and 3 together with the same parameters derived graphically from the model calculations of Avrett (1991). The Doppler width from equation (7) is shown in Å, as a velocity (km s^{-1}), and as the temperature (K) required to produce the observed width from purely thermal motions. An obvious feature of the observed profiles is that they are much too broad to be explained by purely thermal motions if the models are even approximately correct. The wavelength profile coefficient for optical depth as computed from the height variations of temperature and lower level population shown by Avrett (1991) without macroturbulence or microturbulence, for example, shows an effective Doppler width corresponding to a temperature of about 8300 K. Thus substantial spatially unresolved velocities, with amplitudes approximating the range of resolved velocities, are required to explain the width of the line.

One should not take comparisons between the model and observed results too seriously, since the models were not constructed to explain these data. Although the model profiles look qualitatively very much like observed ones, the observed

TABLE I
Line Profile Parameters

Feature	w(Å)	d^*	λ_D Å	km s^{-1}	K
Observations					
A: Quiet Sun	0.045	0.055	0.456	12.6	38700
B: Network	0.095	0.141	0.373	10.3	25900
C: AR Filament	0.174	0.328	0.292	8.1	15800
D: High-Velocity	0.069	0.152	0.250	6.9	11600
E: Limb Network	0.060	0.078	0.430	11.9	34300
Models (Avrett, 1991)					
Cell Interior:	0.008	0.022	0.210	5.8	8200
Average Sun:	0.020	0.033	0.345	9.6	22100
Bright Network:	0.035	0.060	0.328	9.1	20000
Plage:	0.048	0.090	0.300	8.3	16700

profiles tend to be both deeper and broader than the calculated ones, suggesting that the observed features have higher density and more extreme "turbulence" than the models. The recent calculations of Avrett et al. (1993), however, show larger central line depths and correspond better with the observations. Also, the models clearly do not (nor were they intended to) represent the darkest features of the observations such as the numerous filaments and loops. These features are perhaps more aptly approached by radiative transfer methods which have been applied to prominences (Heasley et al., 1974; Heasley and Milkey, 1976, 1978).

The shapes of the profiles are all consistent with optically thin, Gaussian lines, and it is not possible to infer both the line source function and the optical depth from the observed profile which, from equation (2), depends only on the product. However, some limits can be established because He I 10830 Å features appear darker than the underlying continuum so that $0 < S < I_c$. Thus, line-center optical depths do not exceed a few tenths [$S = 0$ in Eq. (2)]. Moreover, if the temperature regime of the models is even roughly correct, the line is underexcited since the source function must be less than the Planck function of about 5000 K typical of the photosphere ($S = I_c$ in equation (2)), but the line-forming plasma has typical high-chromospheric electron temperatures in excess of 8000 K.

4. Conclusions

The observations presented here lend support to a dual mechanism for formation of He I where coronal radiation enchances recombination of He II in a highly structured chromosphere. Large velocities of He-forming material can drastically alter the appearance of equivalent-width or line-depth images if the continuum is taken too close to the nominal line center. Thus, in spite of apparent emission in NSO/KP synoptic data, the He I 10830 Å line is always seen in absorption on the disk except possibly in flares. The line appears to be optically thin everywhere on the disk, and unresolved velocity fields approaching 10 km s^{-1} are needed to explain the observed line widths. The observed profiles appear typically to be both broader and deeper than those calculated by Avrett (1991). More detailed modeling and comparison of this and other data with nearly simultaneous and co-spatial EUV and soft X-ray data is planned or in progress and should lead to a more quantitative understanding of some of the phenomena described here.

Acknowledgements

The author is pleased to acknowledge many useful conversations with J. Harvey, K. Harvey, and E. Avrett in preparing this presentation, and T. Duvall for his assistance with the observations. NSO/Kitt Peak data used here are produced cooperatively by NSF/NOAO, NASA/GSFC, and NOAA/SEL, and the research is partially funded by Supporting Research and Technology grants from the Space Physics Division of the NASA Office of Space Science and Applications.

References

Avrett, E. H.: 1991, in *Workshop on the Solar Electromagnetic Radiation Study for Solar Cycle 22*, R. F. Donnelly, ed., NOAA ERL, Boulder.
Avrett, E. H., Fontenla, J. M., and Loeser, R.: 1993, these proceedings.
Breckenridge, J. B. and Hall, D. N. B.: 1973, *Solar Phys.* **28**, 15.
Grossman-Doerth, U. and von Uexküll, M.: 1971, *Solar Phys.* **20**, 31.
Grossman-Doerth, U. and von Uexküll, M.: 1971, *Solar Phys.* **28**, 333.
Harvey, J.W.: 1984, in *Solar Irradiance Variations on Active Region Time Scales*, B.J. LaBonte et al. eds., NASA CP-**2310**, 197.
Harvey, J. W.: 1993, these proceedings.
Harvey, K. L.: 1985, *Aust. J. Phys.* **38**, 875.
Heasley, J. N., Mihalas, D., and Poland, A. I.: 1974, *Astrophys. J.* **192**, 181.
Heasley, J. N. and Milkey, R. W.: 1976, *Astrophys. J.* **210**, 827.
Heasley, J. N. and Milkey, R. W.: 1978, *Astrophys. J.* **221**, 677.
Jones, H. P., Duvall, T. L. Jr., Harvey, J. W., Mahaffey, C. T., Schwitters, J. E., and Simmons, J. E.: 1992, *Solar Phys.* **139**, 211.
Lites, B. W., Keil, S. L., Scharmer, G. B., and Wyller, A. A.: 1985, *Solar Phys.* **97**, 35.
Livingston, W. C., Harvey, J., Pierce, A. K., Schrage, D., Gillespie, B., Simmons, J., and Slaughter, C.: 1976a, *Appl. Optics* **15**, 33.
Livingston, W. C., Harvey, J., Slaughter, C., and Trumbo, D.: 1976b, *Appl. Optics* **15**, 40.
McCabe, M. K. and Mickey, D. L.: 1981, *Solar Phys.* **73**, 59.
Neupert, W. N., Epstein, G. L. and Thomas, R. J.: 1992, *Solar Phys.* **137**, 87.

VARIABILITY OF THE SOLAR He I 10830 Å TRIPLET

J. W. HARVEY and W. C. LIVINGSTON

National Solar Observatory, National Optical Astronomy Observatories,*
P. O. Box 26732, Tucson, AZ 85726, U.S.A.

Abstract. The He I 10830 Å triplet gives a unique view of the solar chromosphere. Digital spectroheliograms have been made regularly since early 1974 using this line and the NSO Vacuum Telescope on Kitt Peak. For many purposes (detection of coronal holes, giant two-ribbon flares, and dark point events) these images are sufficient. A Sun-as-a-star signal is also produced by averaging all the pixels in each daily image. To calibrate this 'irradiance' signal in terms of line equivalent width, a comparison is made with integrated sunlight spectrophotometric measurements obtained less frequently. After correction for the effects of water vapor blends, we find a linear relation between the two measurements. The daily averages have been assembled into a time series covering nearly two solar cycles. This time series shows cycle modulation of about ±30% and rotational modulation of about ±10%. The general variation is similar to that of other activity indices but with some interesting small differences. Since images are available, it has been possible to decompose the full disk index into components due to plages, filaments, coronal holes and background. At all times during the cycle, most of the signal comes from the background but most of the variability from plages.

Key words: He I 10830 Å – infrared: stars – Sun: activity – Sun: chromosphere

1. Introduction

Neutral helium lines are particularly interesting because they arise exclusively from chromospheric material while at the same time ultraviolet radiation from the overlying corona contributes significantly to their formation (*e.g.*, Avrett, these proceedings). Several papers at this symposium deal with the He I 10830 Å triplet and a recent review of observations was presented by Shcherbakov and Shcherbakova (1990). Uniquely among ground-accessible spectrum lines, observations of the Sun using helium lines show the signatures of coronal features such as holes, bright points (Harvey *et al.*, 1975) and the heated foot points of large magnetic structures associated with mass ejections (Harvey, Sheeley, and Harvey, 1986; Harvey, these proceedings). However, these coronal signatures are rather subtle; helium images are primarily excellent pictures of the chromosphere. The strongest helium line observable from the ground is the 10830 Å line (actually a triplet). A potentially interesting line at 2.058 μm is blocked by strong telluric absorption. A program of daily, digital 10830 Å spectroheliograms was started in 1974 using a 512-channel magnetograph (Livingston *et al.*, 1975a) and a 70-cm vacuum telescope on Kitt Peak (Livingston *et al.*, 1975b). This paper is an overview of some of the results and a preliminary attempt to decompose a full-disk averaged signal into its various sources.

2. Observations

A full-disk observation consists of four 512-arcsec-wide scans of a solar image across the entrance slit of a 10.4-m spectrograph. It takes approximately 40 minutes to

---

* Operated by the Association of Universities for Research in Astronomy, Inc., under cooperative agreement with the National Science Foundation.

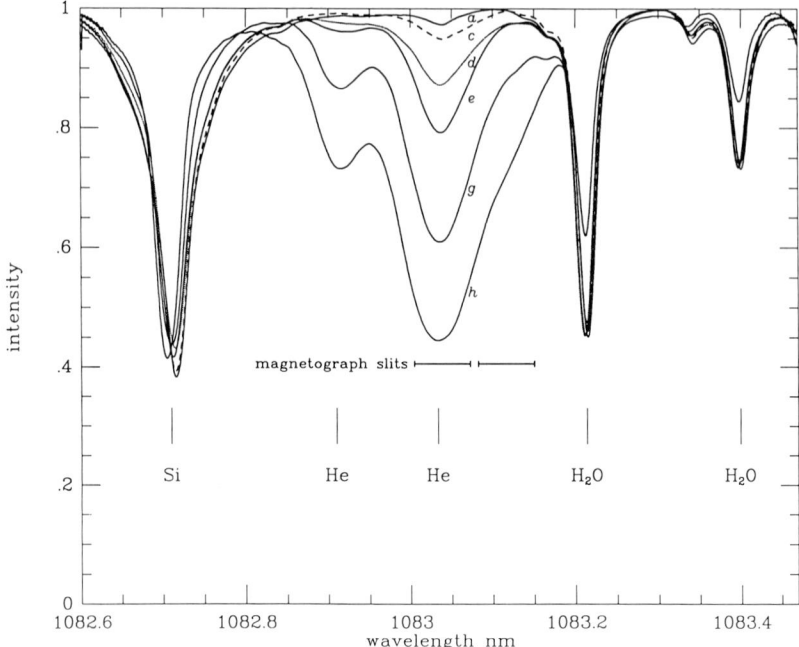

Fig. 1. Disk spectra of various solar features. The spectral windows used for the daily observations are indicated. Features a and c are light parts of the background, d is a typical network element, and $e - h$ are filaments and plages.

make these scans. During the scans, at each arcsec on the disk, two spectral windows are sampled. A signal is calculated that is the ratio of the difference to the sum of these samples. Figure 1 shows these windows along with the spectra of various solar features. Each window has a width of 0.68Å and there is a 0.16Å gap between them. The (non-ideal) choices of windows and processing algorithm were fixed in 1974 by mechanical, optical and computer constraints associated with the magnetograph observations.

By a fortunate accident, both spectral windows include water vapor contributions that change nearly identically, and the calculated signal is almost free of water vapor contamination (*cf.* Breckinridge and Hall, 1973). Because the helium line is usually weak, the calculated signal is also largely free of photospheric contributions. In particular, sunspots and granulation do not appear in the images.

The calculated signals are recorded on magnetic tape and processed later in various ways. The first step is to remove streaks and limb darkening from each scan line by subtracting a model with parameters determined by weighted least-squares fitting. Next, the images are filtered by a spatial frequency notch filter. At this point, the original data are discarded and the processed images are useful by themselves for many purposes. Further processing includes remapping each image to a latitude-longitude format at reduced resolution and producing a Sun-as-star

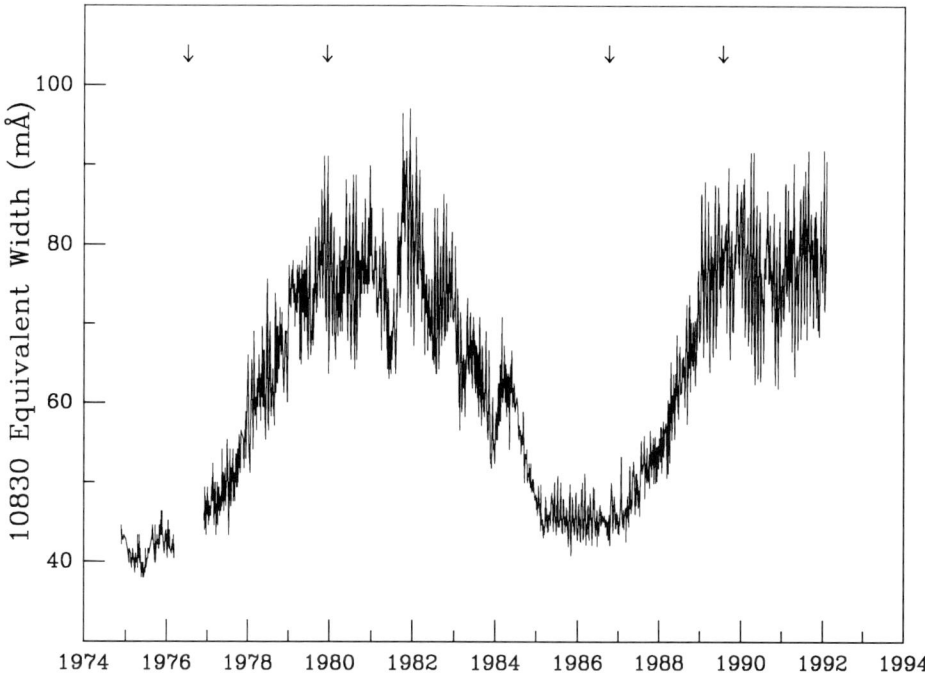

Fig. 2. Daily measurements of the disk-averaged strength of the 10830Å line 1974-1992. Arrows indicate the times of sunspot number maxima and minima.

signal by averaging all the disk measurements. These two reductions are the basis of the rest of this paper. All of the helium data are archived and available to the research community.

The disk-averaged signal (except for minor effects of the destreaking) indicates the strength of the helium line if the Sun were observed as a star, without any spatial resolution. This signal is instrumentally stable in time but is uncalibrated. We transform the instrumental scale to equivalent width by comparing it with high accuracy, monthly measurements of the integrated Sun equivalent width corrected for blends of water vapor lines (Livingston, Wallace, and White, 1988; Livingston et al., 1991). Comparison of the two sets of measurements shows a linear relation between them. The best-fit linear relation is used to calibrate the daily measurements with a formal systematic uncertainty of a few mÅ. The same calibration can be applied to the resolved maps but this requires a rather large extrapolation for strong 10830 features. This calibration assumes that the line equivalent width is proportional to line strength and that line shape changes are not important. This is unlikely to be true in detail and is probably the main source of scatter between the two sets of measurements.

Figure 2 shows the daily, disk-averaged measurements reduced to equivalent width units. Studies of earlier portions of these data have been published (Harvey, 1981, 1984). The main results of Figure 2 are that the helium signal averages 60 mÅ over the solar cycle with a ±30% cycle modulation and a roughly ±10% rotational modulation. To first order, the signal can be represented rather well by a single, biased sine wave. Curiously, although helium images are unlike solar images in any other spectral region of which we are aware, the disk-averaged helium signal is better correlated with variability in Lyα, 208 nm and total solar irradiance corrected for sunspots than other available solar activity indices (Lean, 1990; Donnelly et al., 1985; Foukal and Lean, 1988; Willson and Hudson, 1991). This probably indicates observational defects in other chromospheric indices rather than a superiority of the helium indicator. Nevertheless, the result prompted us to investigate what solar features produce the helium full-disk signal. We are able to do this because resolved images are used to construct the signal.

3. Decomposition of the 10830Å Disk-averaged Signal

For this first attempt to decompose the 10830 full-disk signal we used the daily images mapped into latitude-longitude format. These maps are routinely averaged with weights into single maps of each Carrington rotation. In digital form, a 7-year time series of helium line strengths vs. latitude along the central longitude meridian (6×10^6 values) was convolved with a weighting function that synthesized a daily full-disk average. This synthesis is plotted in the upper panel of Figure 3. It agrees well with the actual daily measurements included in Figure 2. Small differences can be attributed to low-pass temporal filtering that is part of the construction of the synthetic signal.

We decomposed the signal by assigning various ranges of the helium signal in spatially resolved maps (and also in corresponding maps of magnetic field strength) to various features and then synthesizing the daily signal from these various ranges. There is an unsatisfactory subjectivity and uncertainty about this procedure that we plan to reduce in future work. With that caveat, we first note that the variations of the individual components are principally due to changes in the fractional areas rather than strength variations. A notable exception is the background component which rises during high activity despite a decrease of the area of the disk classified as background. On a time scale of years, all the components except coronal holes follow the general rise of solar activity. It is clear that the background component always produces most of the 10830 disk-averaged signal. Its 20% increase from minimum (1986) to maximum (1989) may simply be due to a poor job of eliminating faint plage remnants, or the rise may be due to general enhancement of the background. The 27-day periodicity present during high activity suggests that the first explanation is part of the story but more detailed work is needed to determine how much. Notice that most of the variability on scales of years and 27 days comes from plages. Filaments contribute a surprisingly large fraction of the variability, presumably because they are the most intense features seen in neutral helium lines.

Following Foukal and Lean (1988), it is interesting to speculate about the level of solar irradiance that would be reached under prolonged conditions of no activity.

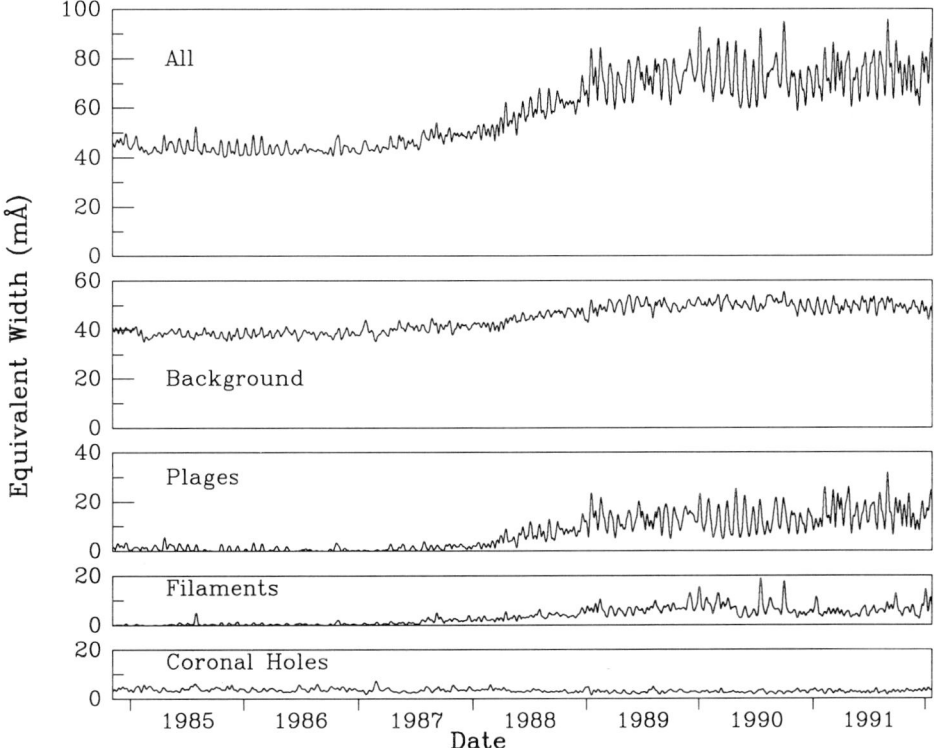

Fig. 3. Full disk signal synthesized from synoptic maps and decomposed into 4 sources.

Since the 10830 disk-averaged signal follows measurements of solar irradiance (corrected for sunspots) very well, and a decrease of 40 mÅ between 1982 and 1986 was associated with a decrease of sunspot-corrected solar irradiance of about 0.17%, we estimate that another 0.2% drop of irradiance might be possible if the activity that produces the 10830 line vanished entirely. In other words, the level of solar irradiance during the activity minimum of 1986 might be 0.2% higher than the level accompanying a prolonged cessation of activity. The spatial decomposition of the helium signal indicates that the background is the only possible source of such a drastic change. It is not at all clear that the relation between helium line strength and solar irradiance established for plages also holds for the small network elements in the background. Thus our irradiance change estimate should be considered as purely speculative.

Acknowledgements

The NSO/Kitt Peak data used here are produced cooperatively by NSF/NOAO, NASA/GSFC, and NOAA/SEL. We are grateful to numerous observers who have made the nice set of data during the past 18 years.

References

Breckinridge, J. B. and Hall, D. N. B.: 1973, *Solar Phys.* **28**, 15.
Donnelly, R. F., Harvey, J. W., Heath, D. F., and Repoff, T. P.: 1985, *J. Geophys. Res.* **90**, 6267.
Foukal, P. and Lean, J.: 1988, *Astrophys. J.* **328**, 347.
Harvey, J., Krieger, A. S., Timothy, A. F., and Vaiana, G. S.: 1975, *Osserv. Mem. Oss. Astrofis. Arcetri* **104**, 50.
Harvey, J. W.: 1981, in S. Sofia (ed.), *Variations of the Solar Constant*, NASA CP 2191, p. 265.
Harvey, J. W.: 1984, in G. Chapman, H. Hudson, and B. La Bonte (eds.), *Workshop on Solar Variability on Active Region Time Scales*, NASA CP 2310, p. 197.
Harvey, K. L., Sheeley, N. R., Jr., and Harvey, J. W.: 1986, in P. A. Simon, G. Heckman, and M. A. Shea (eds.), *Solar Terrestrial Workshop Proceedings Meudon 1984*, NOAA, Boulder, p. 198.
Lean, J.: 1990, *J. Geophys. Res.* **95**, 11933.
Livingston, W., Donnelly, R. F., Grigoryev, V., Demidov, M. L., Lean, J., Steffen, M., White, O. R., and Willson, R. C.: 1991, in A. N. Cox, W. C. Livingston, and M. S. Matthews (eds.), *Solar Interior and Atmosphere*, Univ. Arizona Press, Tucson, p. 1109.
Livingston, W. C., Harvey, J., Slaughter, C., and Trumbo, D.: 1975a, *Appl. Optics* **15**, 40.
Livingston, W. C., Harvey, J., Pierce, A. K., Schrage, D., Gillespie, B., Simmons, J., and Slaughter, C.,: 1975b, *Appl. Optics* **15**, 33.
Livingston, W. C., Wallace, L., and White, O. R.: 1988, *Science* **240**, 1765.
Shcherbakov, A. G. and Shcherbakova, Z. A.: 1990, in I. Tuominen, D. Moss, and G. Rüdiger (eds.), *The Sun and Cool Stars: activity, magnetism, dynamos*, Springer, Berlin, p. 252.
Willson, R. C. and Hudson, H. S.: 1991, *Nature* **351**, 42.

OBSERVATIONS OF SOLAR OSCILLATIONS IN He I 10830 Å

B. FLECK, F.-L. DEUBNER and D. MAIER
*Institut für Astronomie und Astrophysik der Universität Würzburg
Am Hubland, 8700 Würzburg, Germany*

and

W. SCHMIDT
Kiepenheuer-Institut für Sonnenphysik, Schöneckstr. 6, 7800 Freiburg, Germany

Abstract. In continuation of a series of studies devoted to the dynamics of the solar photosphere and chromosphere we have attempted to further extend the range of heights in the atmosphere towards the transition region by including observations of the He I 10830 line. We have recorded simultaneous time series of He I 10830 and Mg I 8807 spectra in the quiet solar atmosphere using the echelle spectrograph at the German Vacuum Tower Telescope in Izaña, Tenerife. The velocity signal derived from the Doppler shifts of He 10830 clearly reveals oscillatory motions. The intensity of He 10830, on the other hand, is hardly affected by the oscillations. In the cell interior the 3-min oscillations prevail. Longer periods are found ain the cell boundaries of the chromospheric network where He absorption is enhanced. The V–V phase difference spectrum between the oscillations of He 10830 and those of Mg 8807 confirms previous observations of a non-propagating component that dominates the acoustic wave spectrum in the chromosphere.

Key words: infrared: stars – Sun: atmosphere – Sun: oscillations

1. Introduction

Despite the accumulation of a quite extensive observational data base, and despite considerable theoretical efforts made in the last 25 years several fundamental issues concerning the dynamics of the solar chromosphere are still unsettled, as for instance the heating mechanism (see Ulmschneider *et al.*, 1991), the driving mechanism of spicules (see, *e.g.*, Athay, 1986), the Ca K "bright point" phenomenon (see Rutten and Uitenbroek, 1991), and wave propagation behaviour (running or standing waves?). The two major obstacles on the way to a better understanding of these phenomena are: *a)* the intricate small-scale nature of the chromospheric fine structure which can only be resolved under very good seeing conditions, and *b)* the lack of optically thin lines (in the visible) that could be used as a reliable diagnostic without the uncertainties imposed by radiative transfer effects, and the broad complex line contribution functions of the commonly used lines such as H α, Ca II H and K and the lines of the Ca II infrared triplet. Both obstacles could be easily by-passed by observing UV lines from space (as was done, *e.g.*, by the OSO 8 mission). At present, however, there is no space-borne solar observatory in orbit, and – the OSL project being postponed to the end of this decade – there will be no space-borne solar telescope ready in the near future capable of tackling these problems.

The present paper deals with the last of the questions mentioned above, that of the wave propagation behaviour in the solar chromosphere. In previous studies we have observed the Ca II infrared lines at 8542 and 8498 Å and the Ca II K line (Fleck and Deubner, 1989; Deubner and Fleck, 1990; Deubner *et al.*, 1992). There we arrived at some inconsistent results, conceivably due to the observed resonance lines being optically thick. To overcome these difficulties and to gain further information from higher layers in the chromsophere we decided to extend our observational database by including time series of He I 10830 to serve as a reliable velocity diagnostic of the upper chromosphere. Similar motives led Lites (1986) to use He 10830 for his studies on chromospheric sunspot oscillations. Venkatakrishnan et al. (1992) studied the formation of spicules by analyzing spatio-temporal fluctuations of He 10830, and Zhang et al. (1991) investigated oscillations of quiescent filaments from observations in this line. He 10830 spectroheliograms, obtained at Kitt Peak National Observatory on a routine basis, were frequently used for comparisons with coronal structures such as coronal holes or X-ray bright points (*e.g.* Harvey and Sheeley, 1977; Kahler *et al.*, 1983; Golub *et al.*, 1989). Here we are using this line for the first time to study oscillatory motions in the quiet upper chromosphere.

2. Observations and Data Reduction

A first set of observations was collected on November 11, 1990 with the echelle spectrograph of the Vacuum Tower Telescope in Izaña, Tenerife. The OSL brassboard CCD camera with a 1024×1024 array of $(18.3\,\mu m)^2$ pixels was used in the 2×2 binning mode to obtain time series of He 10830 spectra alternatively in two positions, 30" apart in a quiet region of the disk center. The exposure time was 3 s, and the cycle time for each pair of spectra 24 s. The total duration of the time series was approximately 1.5 hours. The echelle spectrograph was used in the 21st order with an entrance slit of 85" length and $300\,\mu m$ width ($\approx 1.4"$) With a second 1024×1024 CCD camera, spectra of Mg I 8806.8 Å were recorded synchronously with the He spectra. Seeing was fair.

The light level in the He spectra yielded typical continuum readings of 400 counts, *i.e.*, about 10% of the saturation level of the OSL camera. Therefore, special care had to be taken in performing the dark current and flat field corrections.

From the corrected spectra we deduced the continuum intensity $I_c(x,t)$, and Doppler shift $V(x,t)$ as well as minimum intensity $I_l(x,t)$ of the line profiles by means of a 4th order polynomial fits applied to the line cores. The line intensity was normalized with respect to the continuum intensity ($I = I_l/I_c$). Random excursions of the telescope and guiding errors in the direction of the slit were eliminated by using crosscorrelation techniques. Finally, the $x - t$ wave patterns were analyzed by applying standard Fourier methods (power, crosspower, phase and coherence spectra).

3. Results and Discussion

In Figure 1 we have displayed the spatio-temporal velocity and intensity fluctuations of He 10830 for one of the two time series. The velocity signal exhibits a pronounced oscillatory behaviour, similar to that observed in typical chromospheric lines (like Ca K, or the Ca IR lines). Longer periods are found in the cell boundaries of the chromospheric network (enhanced He absorption), compared to the cell interior where the 3-min oscillations prevail (*cf,. e.g.,* Damé *et al.*, 1984; Deubner and Fleck, 1990). The intensity fluctuations of the core of the He line exhibit remarkably

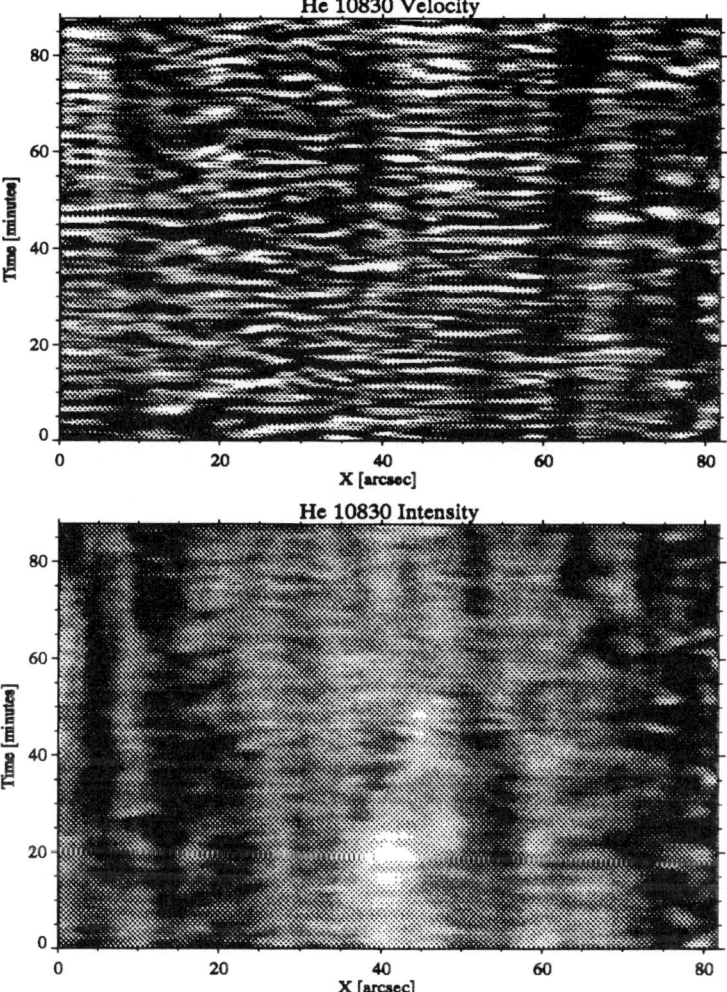

Fig. 1. Spatio-temporal fluctuations of velocity and brightness in the He I 10830 line. Maximum grey scale contrast corresponds to $\pm 2\,\text{km s}^{-1}$, and to $0.93 < I_l/I_c < 0.98$, respectively.

less periodicity than the corresponding velocity fluctuations, and in particular also considerably less periodicity than the brightness fluctuations of the chromospheric lines mentioned before. This striking impression is confirmed qualitatively by Figure 2 where we have displayed the temporal power, phase and coherence spectra of the velocity and brightness fluctuations depicted in Fig. 1. The velocity power spectrum reveals two broad peaks centered at about 3.5 and 5.5 mHz, whereas the intensity power spectrum decreases monotonically.

The coherence spectrum in Fig. 2 indicates reliable V–I phases up to about 6 mHz. In this frequency range we measure a V–I phase of approximately $-120°$,

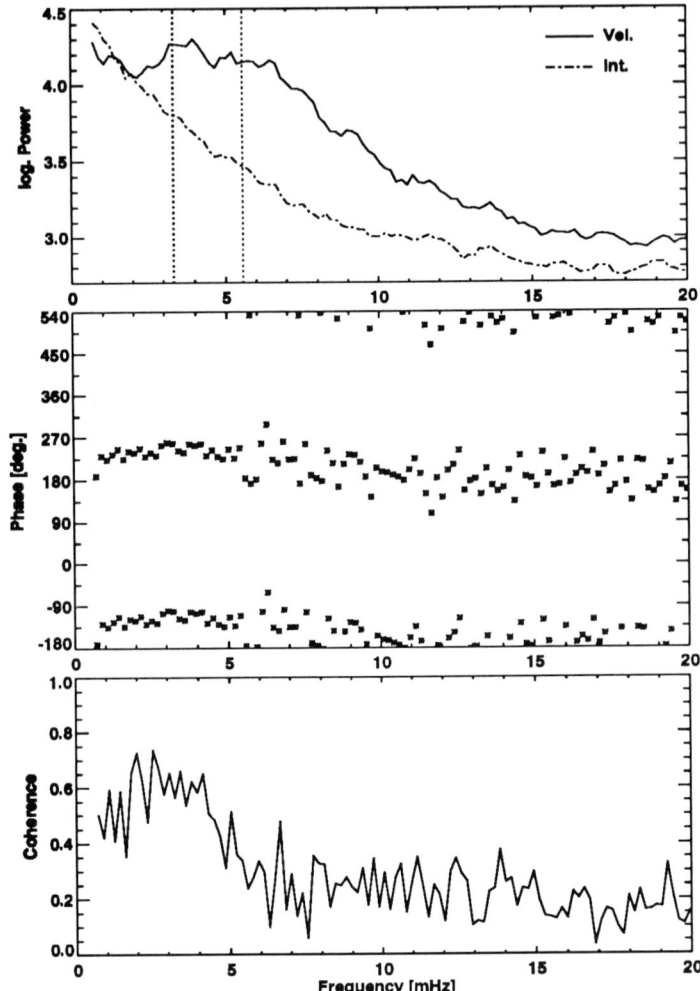

Fig. 2. Power, phase and coherence spectra from the velocity and brightness fluctuations displayed in Fig. 1

brightness preceding upward motion. This result contrasts strongly with the value of −90° expected for adiabatic evanescent oscillations in the chromosphere. We suggest that this behaviour is related to the particular conditions, in the He I absorbing layer, for excitation of the lower level of the 10830 transition, and to the back-radiation from coronal UV lines in particular.

Fig. 3 depicts the power spectra of the velocity fluctuations of He 10830 and Mg 8807 together with the corresponding V–V phase difference and coherence spectra. The Mg line is formed near the temperature minimum at about 500 km above $\tau_{5000} = 1$. As expected, its power spectrum peaks around 3.3 mHz with a slight shoulder at 5.5 mHz. A difference of almost 2 orders of magnitude between the high

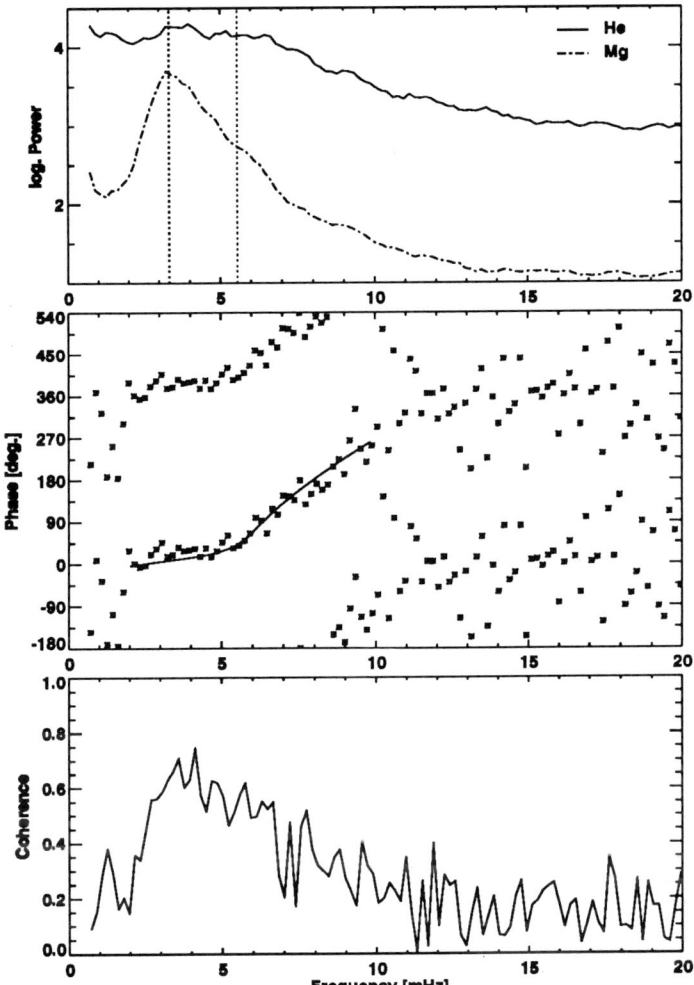

Fig. 3. Power, phase and coherence spectra from the velocity fluctuations of He 10830 and Mg 8807.

frequency levels of the He and the Mg power spectra is partially due to the higher noise amplitude in the He spectra (photon noise plus instrumental effects causing additional uncertainties in determining the Doppler shift of such a broad line).

The coherence lies above noise level in the frequency interval from about 2 to 10 mHz, indicating reliable phases in this range in which we have superimposed a theoretical V–V phase difference spectrum according to the theory of acoustic-gravity waves of Souffrin (1966). The assumed parameters are sound speed $c_s = 6 \text{ km s}^{-1}$, mean horizontal wave number $k_x = 0.8 \text{ Mm}^{-1}$, relaxation time $\tau_R = 50 \text{ s}$, and, most importantly, a height difference Δz between the two line forming layers of only 550 km. The latter parameter is essential for the slope of the spectrum in the acoustic regime above 6 mHz. Assuming that He 10830 is formed in the upper chromosphere at about 1800 km above $\tau_{5000}=1$ and Mg 8807 near the temperature minimum, one would expect a slope according to $\Delta z \approx 1300$ km, *i.e.* about 2.5 times steeper than observed. One conceivable explanation for this striking discrepancy could be that the assumed line formation heights are incorrect, i.e. that the lines are formed only about 550 km apart. This explanation, however, appears highly unlikely. Another explanation might be that the phase speed of acoustic waves is about 3 times higher in the chromosphere than in the photosphere, conceivably due to the influence of magnetic fields. However, this explanation does not appear very satisfactory either (for details see Mein, 1977; Fleck and Deubner, 1989; or Fleck, 1991). As a third alternative we suggest a non-propagating pattern of motions in the chromosphere, composed of either standing waves generated by reflection at the transition region (see ref. above), or of harmonics of the chromospheric resonance frequency (Deubner *et al.*, 1992 in prep.). Accordingly the phase difference builds up only in the interval in between the Mg formation height, *i.e.* the temperature minimum, and the "magic height" at about 1000 km (see Fleck and Deubner, 1989; or Fleck, 1991). Above this altitude the whole atmosphere oscillates in phase (or – depending on height – in antiphase).

4. Conclusions

Our preliminary results demonstrate the feasibility of studies of the upper chromospheric dynamics using the He 10830 line as a new diagnostic, even in quiet regions with only weak absorption. A more detailed analysis (including detailed studies of the wave forms measured in He 10830) based on new observations obtained under better seeing conditions is underway.

References

Athay, R.G.: 1986, in P. A. Sturrock (ed.), *Physics of the Sun*, Vol. 2: The Solar Atmosphere, Reidel, Dordrecht, p. 51.
Damé, L., Gouttebroze, P., Malherbe, J.M.: 1984, *Astron. Astrophys.* **130**, 331.
Deubner, F.-L., Fleck, B.: 1990, *Astron. Astrophys.* **228**, 506.
Deubner, F.-L., Fleck, B., Kossack, E.: 1992, in preparation.
Fleck, B.: 1991, *Thesis*, Univ. Würzburg.
Fleck, B., Deubner, F.-L.: 1989, *Astron. Astrophys.* **224**, 245.
Golub, L., Harvey, K.L., Herant, M., Webb, D.F.: 1989, *Solar Phys.* **124**, 211.
Harvey, J.W., Sheeley, N.R.: 1977, *Solar Phys.* **54**, 343.
Kahler, S.W., Davis, J.M., Harvey, J.W.: 1983, *Solar Phys.* **87**, 47.
Lites, B.W.: 1986, *Astrophys. J.* **301**, 1005.
Mein, N.: 1977, *Solar Phys.* **52**, 283.
Rutten, R.J., Uitenbroek, H.: 1991, *Solar Phys.* **134**, 15.
Souffrin, P.: *Ann. Astrophys.* **29**, 55.
Ulmschneider, P., Priest, E.R., Rosner, R. (eds.): 1991, *Mechanisms of Chromospheric and Coronal Heating*, Springer, Berlin.
Venkatakrishnan, P., Jain, S.K., Singh, J., Recely, F., Livingston, W.C.: 1992, *Solar Phys.* **138**, 107.
Zhang, Y., Engvold, O., Keil, S.L.: 1991, *Solar Phys.* **132**, 63.

OBSERVATIONS OF DYNAMIC EVENTS IN He I $\lambda 10830$

KAREN L. HARVEY

Solar Physics Research Corporation, 4720 Calle Desecada, Tucson, AZ 85718, U.S.A.

Abstract. The characteristics of large-scale two-ribbon flares and dark points observed in He I $\lambda 10830$ are summarized and compared with observations of these phenomena at other wavelengths and with the underlying magnetic field.

Key words: He I 10830 Å – infrared: stars – Sun: flares – Sun: magnetic fields

1. Introduction

Full-disk He I $\lambda 10830$ spectroheliograms have been taken by the National Solar Observatory (NSO) since early 1974 on a daily basis using a 512-channel magnetograph at the Vacuum Telescope/Kitt Peak (Livingston *et al.*, 1976. Though the primary use the He I $\lambda 10830$ spectroheliograms is for the identification of coronal holes, a variety of phenomena are being studied both in these data and in high time-resolution He I observations of a portion of the solar disk. This paper summarizes the characteristics of two types of events observed at this wavelength: (1) on a large scale, *Two-Ribbon Flare Events* and (2) on a small scale, *Dark Points*. Further details of these phenomena are described in Harvey, Sheeley, and Harvey (1987), Harvey (1985, 1991), and Harvey and Sheeley (1992).

2. Two-Ribbon Flare Events Observed in He I $\lambda 10830$

Erupting quiescent filaments, termed *Disparition Brusque*, often are followed by two-ribbon flares or weak brightenings along the filament channel (see Švestka, 1976, and references therein). Such events also are seen, often more easily, in He I $\lambda 10830$ spectroheliograms. A total of 90 two-ribbon events were identified in the daily NSO He I $\lambda 10830$ full-disk spectroheliograms from early 1974 to June 1985. To determine the timing and history of these 90 events, additional data were examined. These include: (1) Hα filtergrams, from *Photographic Journal of the Sun, Osservatoire Astronomica di Roma, Solar Geophysical Data*, and flare patrol films from Sacramento Peak Observatory, Big Bear Solar Observatory (BBSO), and the SOON system, and (2) filament observations from *Pulkovo Solar Data Bulletin, Cartes Synoptiques de la Chromosphere Solaire et Catalogues des Filaments et des Centres d'Activité*.

2.1. Results

Most of the He I flare events identified occur outside active regions, such as the two-ribbon event of 27 October 1988 shown in Figure 1. In all events, the ribbons appear *dark* in He I $\lambda 10830$ and straddle the magnetic inversion line previously occupied by the erupted filament. The two-ribbon event begins tens of minutes to a few hours after the onset of a filament eruption; the two ribbons separate with time at speeds of 0.1 to 2 km s^{-1}. Their spatial scale is 10^5 to 10^6 km; the largest extend over $\sim$9% of the Sun's surface. The dark He I ribbons generally correspond to weak

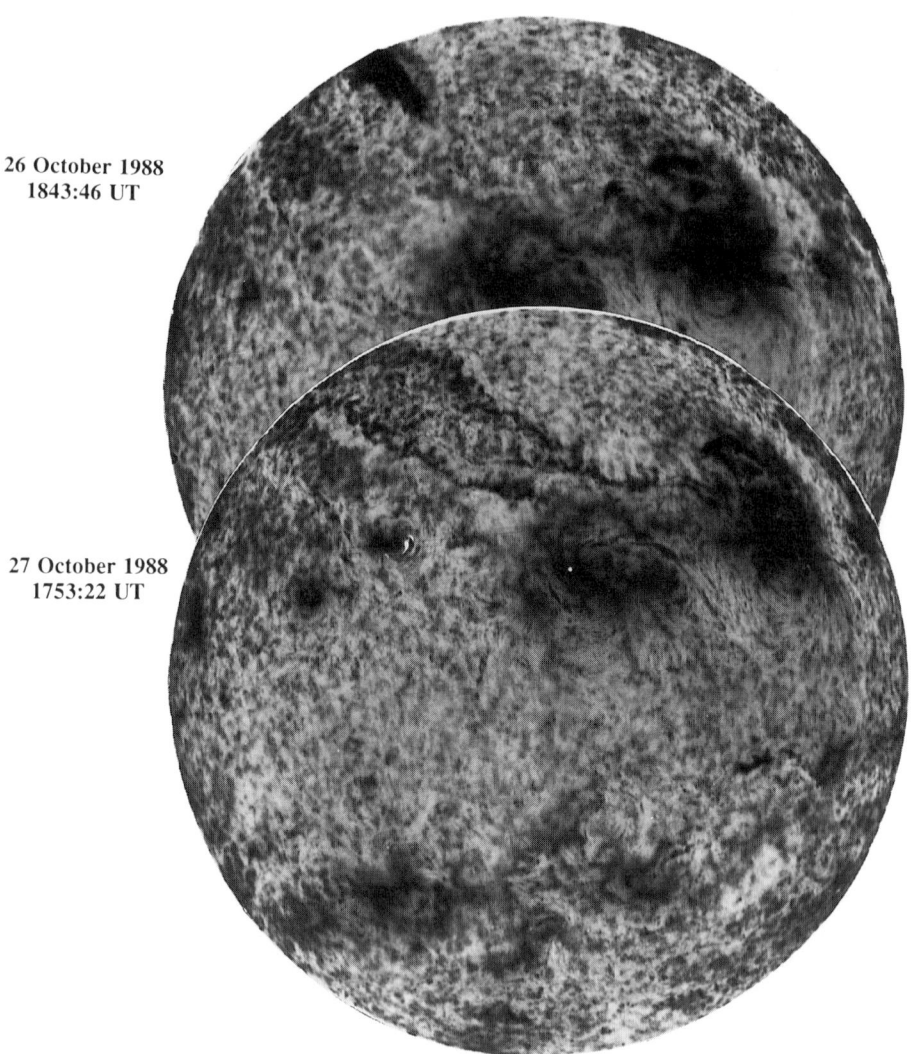

Fig. 1. NSO He I λ10830 spectroheliograms showing a filament near the north-east limb on 26 October 1988. This filament erupted on 27 October sometime between 1043 and 1346 UT and was followed by an extensive two-ribbon flare event and the expansion of the northern polar coronal hole down to the edge of the northmost ribbon. The *dark* He I ribbons (*arrows*) extended almost 10^6 km along the $H_\parallel = 0$ line. North is at the top and East is to the left.

TABLE I
Association of Two-Ribbon Events with Other Solar Activity

Two-Ribbon Events Associated with:	%	Timing w/r Ribbon Onset	
Coronal Mass Ejection	96%	1.9 hours	before
Type I/Noise Storm	73%	0.8 hours	before
Long-Duration 1-8Å X-ray Event	98%	0.4 hours	before
GRF 10 cm-λ Radio Burst	87%	0.0 hours	after
Type II Radio Burst	11%	0.6 hours	after
Geomagnetic Storm	43%	2.5 days	after

Hα brightenings; only 10 of the 90 events were reported as flares. The ribbons are first detected in He I λ10830 and then within minutes in Hα. The rise time of these events is on average 2.1 hours, ranging from 0.3 to 6.3 hours; their durations range from 3 to >64 hours, with 70% exceeding 20 hours. Their duration is longer in He I λ10830 than in Hα. With 64% of the two-ribbon events, coronal holes enlarge or form, these changes are transient, lasting from 1 to 6 days.

The association and timing of other solar activity with these two-ribbon events is summarized in Table 1.

The characteristics of these two-ribbon events are identical to those of major, more energetic, two-ribbon flares in active regions, though they are weaker, longer-lived, and larger. They are not a separate class of flares, but are at one end of a wide size spectrum of two-ribbon flares. The comparatively low level of associated chromospheric activity indicates that these events may be primarily coronal.

It is concluded that the two-ribbon events are associated with a large-scale restructuring of the coronal magnetic fields that leads initially to the eruption of a filament and a coronal mass ejection. As the coronal mass ejection and filament move outward, the coronal magnetic fields stretch and open. Below this, the magnetic fields reconnect resulting in heating in the reforming coronal loops and enhancements at the loop footpoints. This picture is consistent with the formation of arcades of X-ray loops at the site of recently erupted filaments (Webb *et al.*, 1976; Kahler, 1977; Rust and Webb, 1977), and the enhancements at their footpoints seen in He I λ304 (Svestka, 1976; Harvey and Sheeley 1992), a line showing the same structures as in He I λ10830 (Harvey and Sheeley, 1977). The release of energy during a two-ribbon event, though at low levels, persists for extended periods (tens of hours) and over extensive areas of the Sun ($\leq$9% of its surface area).

3. Dark Points Observed in He I λ10830

Observations of the quiet sun in He I λ10830 reveal the presence of many small-scale dark structures of less than 30''. These structures are called dark points (Harvey, 1985) and, in many cases, are found to correspond to X-ray bright points (Harvey *et al.*, 1975).

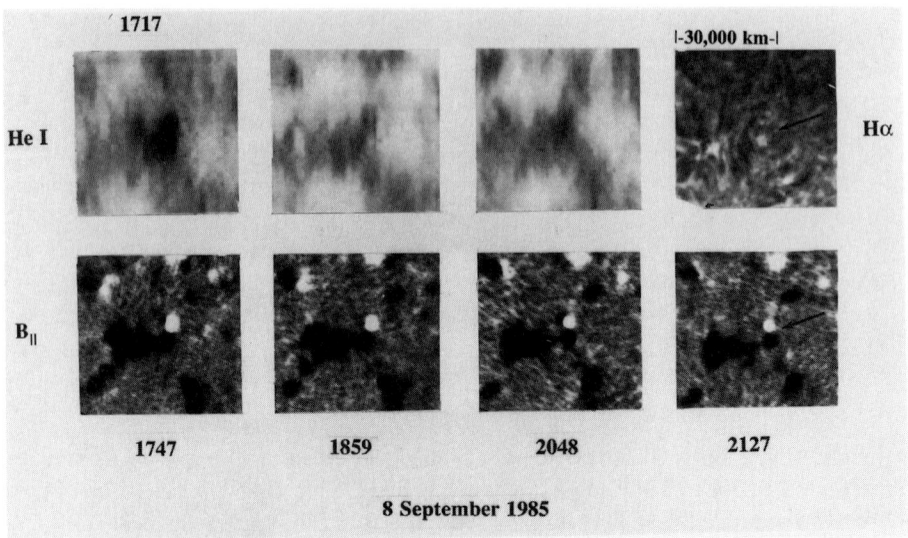

Fig. 2. Sequence of He I λ 10830 spectroheliograms and BBSO magnetograms on 8 September 1985 showing a dark point and the associated cancelling magnetic bipole. Fibrils connecting the opposite polarity magnetic elements of this magnetic bipole can be seen in the Hα filtergram at 2127 UT.

He I λ10830 dark points were studied in relation to the evolution of the photospheric longitudinal component of the magnetic field and chromospheric structures using simultaneous time-sequence images in He I λ10830 (NSO) and Hα lines (BBSO), 20 cm radio emission (Very Large Array), and of the magnetic field (BBSO) in selected areas of the quiet Sun. The cadence of these data are at least 3 min with a spatial resolution of 2–3″.

3.1. RESULTS

The spatial scale of dark points observed in He I λ10830 are 8″ to 40″; their lifetimes range from a few minutes to hours. Rapid and strong intensity variations often occur in these structures, lasting from 9 to 40 minutes. These events are accompanied by brightenings in Hα (Harvey, 1991), C IV (Porter *et al.*, 1987), and in 20-cm radio emission (Habbal and Harvey, 1988). The characteristic flare-like intensity variations and their association with surges and eruptions of small filaments (Habbal and Harvey, 1988; Harvey, 1991) suggest that such events are flares, be it on a small scale. Almost all He I dark points are located at sites of magnetic bipoles. Two-thirds of these are cancelling magnetic bipoles, *i.e.*, where magnetic flux in the opposite polarity magnetic elements is decreasing (Fig. 2). One-third of the dark points are associated with the emergence of an ephemeral region. An investigation of simultaneous time-sequence He I, magnetic field, and Hα observations indicates that

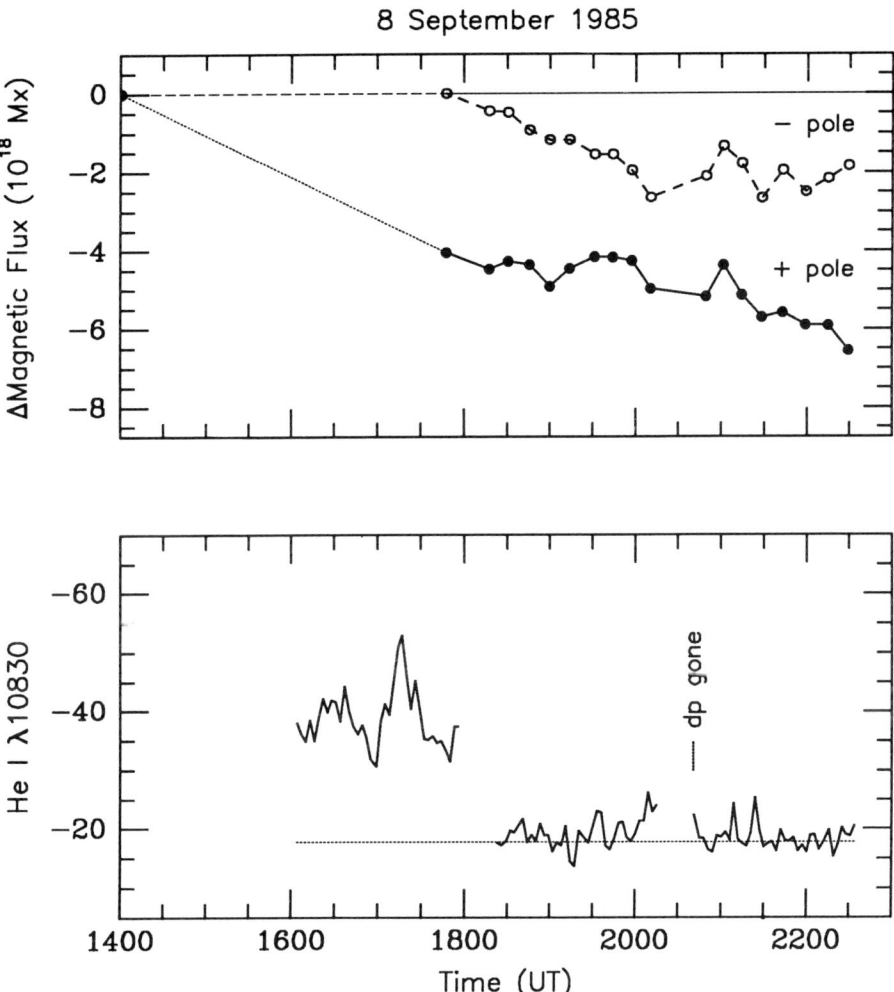

Fig. 3. Time variation of the He I λ10830 intensity (scale inverted) of the dark point in Figure 2 and the change in magnetic flux in the associated the cancelling bipole. The magnetic flux loss in this cancelling bipole is 8×10^{18} Mx over an 8.5 hr period, continuing hours after the associated He I dark point disappears.

the cancellation or emergence of magnetic flux is not a sufficient condition for the occurrence of He I dark points. Many dark points are observed only during a fraction of the time of the magnetic field interaction, such as the event shown in Figure 2. Here, the opposite polarity network elements of the associated magnetic bipole continue to cancel several hours after the "end" of an identifiable dark point (Fig. 3). Such events suggest that their formation and evolution result from local magnetic field reconnection, rather than from the disappearance or emergence of magnetic bipoles. Further evidence of this comes from the observation (1) of the almost immediate reconnection with the nearby opposite polarity network of the magnetic fields of an ephemeral region as soon as it began emerging (Harvey, 1991), and (2) of Hα fibrils or filaments that connect the opposite polarity network elements in a cancelling bipole (Fig. 2). Once the fields have become completely reconnected, no further heating occurs and the dark point and flares cease.

Though there is a good correspondence between dark points observed He I $\lambda 10830$ and X-ray bright points (Harvey et al., 1975), the association found is not one-to-one (Golub et al., 1989). However, this result must be viewed with caution as both He I dark points and X-ray bright points show rapid intensity variations on time scales of minutes, and there are very few simultaneous observations of these structures to reliably determine their detailed relation.

Acknowledgements

The analyses of the He I $\lambda 10830$ observations was done while a Visitor at the National Solar Observatory, National Optical Astronomy Observatories (operated by the Association of Universities for Research in Astronomy, Inc., under cooperative agreement with the National Science Foundation). This work is supported in part through NSF Grant ATM-8319589 and under a contract with Lockheed Missiles and Space Company, Inc., through NASA Contract NAS8-37334. The NSO/Kitt Peak data used here are produced cooperatively by NSF/NOAO, NASA/GSFC, and NOAA/SEL.

References

Golub, L, Harvey, K. L., Herant, M., and Webb, D. F.: 1989, *Solar Phys.* **124**, 211.
Habbal, S. R. and Harvey, K. L.: 1988, *Astrophys. J.* **326**, 988.
Harvey, J. W., Kriger, A. S., Timothy, A. F., and Vaiana, G. S.: 1975, *Osservazioni e Memorie Osservatorio de Arcetri* **104**, 50.
Harvey, J. W. and Sheeley, N. R., Jr.: 1977, *Solar Phys.* **54**, 343.
Harvey, K. L.: 1985, *Australian Journal of Physics* **38**, 875.
Harvey, , K. L.: 1991, in Y. Uchida, R.C. Canfield, T. Watanabe, E. Hiei (eds.), *Flare Physics in Solar Activity Maximum 22*, Lecture Notes in Physics No. 387, Springer-Verlag, Berlin, p. 62.
Harvey, K. L. and Sheeley, N. R., Jr.: 1992, in preparation.
Harvey, K. L., Sheeley, N. R., Jr., and Harvey, J. W.: 1987, in P. A. Simon, G. Heckman, M. A. Shea (eds.), *Proceedings of the Solar Terrestrial Predictions Workshop* held in Meudon, France, June 1984, p. 198.
Livingston, W. C., Harvey, J., Slaughter, C., and Trumbo, D.: 1976, *Appl. Optics* **15**, 40.
Porter, J. G., Moore, R. L., Reichmann, E. J., and Harvey, K. L.: 1987, *Astrophys. J.* **380**, 1987.
Rust, D. M. and Webb, D. F.: 1977, *Solar Phys.* **54**, 403.
Švestka, Z.: 1976, in *Solar Flares*, D. Reidel Publishing Company, Dordrecht-Holland, p. 229.
Webb, D. F., Krieger, A. S., and Rust, D. M.: 1976, *Solar Phys.* **48**, 159.

AN INVESTIGATION OF IR TRIPLET He I 10830 Å PROFILES IN ACTIVE REGIONS AND THE QUIET CHROMOSPHERE

V. LEVKOVSKY, P. PAPUSHEV and R. SALAKHUTDINOV

Institute of Solar-Terrestrial Physics, Irkutsk 33, P.O. Box 4026, 664033, Russia

Abstract. This paper gives a description of the technique and equipment used to make solar observations in the near IR region with a coronagraph. Examination of active-region spectra in the neighborhood of the He I 10830 Å line confirm that the the line is enhanced in sunspot umbrae. The line profiles of umbrae and faculae differ qualitatively. Flares show a decrease in the line depth in comparison to faculae, and in flare kernels the line undergoes an emission reversal. Simultaneous observations of spectra of chromospheric spicules at two heights above the solar surface reveal that the half-width of the 10830 line increases with height.

Key words: He I 10830 Å – infrared: stars – Sun: activity – Sun: flares – telescopes

1. Introduction

Since the mid-70s, at the Sayan Solar Observatory Coronagraph of the Institute of Solar-Terrestrial Physics (formerly SibIZMIR), observations of solar spectra have been carried out in the near IR region using different types of image tubes. The best results were achieved by use of image tubes with magnetic focusing, installed within the chamber a diffraction spectrograph which enables spectra to be obtained in the He I 10830 line with 1.0 Å/mm dispersion and 0.03 Å spectral resolution.

The advantage of image tubes with magnetic focusing is that, owing to the high resolving power of the photocathode and the luminescent screen (up to 60 lp/mm), the use of a strong, homogeneous magnetic field allows geometrical aberrations to be eliminated. We used Helmholtz coils to produce a focusing magnetic field, of 3.5 kG, with 1% homogeneity. Because of the large magnetic field strength, the electron Larmor radius inside the image tube turned out to be less than image resolution, and the device was insensitive to variations in external conditions. Spectra from the flourescent screen were recorded using fast optics ($f/1.1$) imaging onto photographic film. The resulting resolution on film was about 20 μm. The use of contact photography by means of an optical-fiber disk results in substantially degraded image quality due to inhomogeneous transmission through the optical fiber section. The spectra obtained were digitized using an automated microdensitometer, with a density-measurement error of 0.02 D and a scanning step of 5 μm.

The observing program was aimed at a study of the behavior of the He I 10830 triplet in different features on both the solar limb and disk. In this case we tried to take advantage of the fact that rather stringent requirements on light scattering are satisfied for the optical elements in the coronagraph; the instrumental scattered light is less than 0.5% (Nikolsky and Sazanov, 1967). With this instrument we obtained spectra of the He I 10830 line in sunspot umbrae and in chromospheric spicules and flares.

TABLE I

Results of Observations and Analysis of Two Sunspots

Sunspot Area	r_λ^s	$r_\lambda^{s\ a}$	r_λ^{pl}	$\Delta\lambda_{0.5}^s$ (Å)	$\Delta\lambda_{0.5}^{pl}$ (Å)	$(\frac{W_v}{W_n})^s$	$(\frac{W_v}{W_n})^{pl}$
35×10^{-6}	0.93 ± 0.02	0.95	0.78	1.0	0.68	1.3	2.1
50×10^{-6}	0.94 ± 0.02	0.94	0.83	1.4	0.62	1.2	1.5

^{a} reduced value.

2. Observational Results

2.1. Sunspot Umbrae

Fay et al. (1972) were among the first to detect the enhancement of the absorption line He I 10830 in a sunspot umbra. As they report, the spectrum was recorded photoelectrically; in this case the spectral region around 10830 Å was recorded in a time of about 1 minute (spectral resolution 0.1 Å, time constant 0.8 second). A question arises as to whether the sunspot line profile may be contaminated by plage due to blurring of the image over long exposure time.

In order to address this question using the device described above, we obtained spectrograms with 0.5–1.0 second exposure. Because of the low level of instrumental stray light in the coronagraph, we considered it possible to address the effect of image blurring by itself upon the profiles, using the method suggested by Stepanov (1957). Results of observations and analysis of two sunspots are presented in Table I (Borodina and Papushev, 1979). Indices s and pl stand for the sunspot umbra and plage respectively, the other symbols being standard. As is evident from Table I, the He I 10830 line is indeed present in the sunspot umbra, and its depth is increased in comparison to the quiet photosphere, for which $r = 98\%$. Fay et al. (1972) obtained residuals, r, of 0.83 and 0.72 for sunspots with fractional hemispheric areas, S, of 70×10^{-6} and 100×10^{-6} respectively. Comparing our values of 0.95 and 0.94 for $S = 35 \times 10^{-6}$ and $S = 50 \times 10^{-6}$, one is justified in believing that the dependence of r on the sunspot umbral area found by Fay et al. (1972) is real. The asymmetry of the He I 10830 profile, caused by its triplet character, is less in a sunspot umbra than in plage.

2.2. Chromospheric Flares

Available data on the behavior of the helium IR triplet line in solar active regions suggest that this line is useful not only for studying the physical conditions in flare chromospheres but also for predicting the occurrence of a flare, as well as for concluding that a flare has recently occurred in a particular active region. For this purpose, we have observed and examined spectrograms of active regions in the He I 10830 line during the period of the Solar Maximum Year, 1980–1982.

Typical flare behavior of the IR triplet He I 10830 is exemplified by the flare of importance 1B in active region 474 (see *Solar Geophysical Data*) of 1982 July 17 at 02:11:30 UT. The flare lasted about 30 minute During this time interval we recorded the He I 10830 line profiles in different spatial features in the flare. Simultaneously with the 10830 spectrograms, Hα filtergrams were taken at line center from the spectrograph slit monitor using a Hallé filter. Our 10830 line-profile observations had better time resolution (0.1 second exposures) than previous observations of the line in flares.

In flares the depth of the He I 10830 line becomes smaller, compared to the plage line depth, by 10–20%. In flare kernels the line goes into emission. The line spends about one-third of the total flare lifetime in emission. The profiles show the 10830 emission core to be significantly weaker in the flare kernel than is the core of Hα.

Following the disappearance of the emission kernel of the flare, the depth of the 10830 line where the kernel was located decreases by 10–20%. The half-width of the emission core of the line in the flare kernel varies from $\Delta\lambda_{0.5} = 0.8$ Å to $\Delta\lambda_{0.5} = 0.8$Å.

2.3. CHROMOSPHERIC SPICULES

Investigations by Papushev (1977) and Papushev and Salakhutdinov (1989) indicate a possible transition of spicular material evolving from a "cold" state to a "hot" state and, thus, the formation of a transition region between the chromosphere and the corona. To address this problem it seems appropriate to study spicule evolution simultaneously in several spectral lines, which is a difficult task when observations are made at the coronagraph with a single objective lens, except in the case of the spectral blend consisting of the lines H_8 and teh neighboring He line.

Extensive data from spicule observations in different spectral lines have been published in the literature to date. In particular, according to Zirker (1962) the halfwidths of the hydrogen lines Hα and Hβ decrease with height in spicules, while the halfwidth of the He I D_3 line increases. We have attempted to measure the He I 10830 line at two height levels in chromospheric spicules simultaneously, since 10830 is the most intense helium emission line. Again, we benefit from the substantially lower level of stray light due to atmospheric turbulence in the near IR region than in the visible.

Results of analysis of line profiles from 12 spicules at two height levels are listed in Table II. We find an increase in halfwidth with height, but our halfwidths are greater than those obtained by Nikolsky (1965) and Livshitz and Demkina (1965). The larger halfwidths are the subject of further investigation. We are also investigating the evolution of spicules in the He I 10830 line over time.

Our results suggest a rise in the temperature of the spicule with height, and a thermal emission mechanism of the middle and upper chromosphere in the He I 10830 line.

TABLE II

Halfwidths of 10830 Profiles of Spicules at Two Heights

Halfwidth (Å)	
Height (±1000 km)	
4000 km	7000 km
2.89	2.96
2.41	2.96
2.02	2.89
2.34	2.88
2.27	2.62
2.27	2.55
2.20	2.62
2.13	2.95
2.14	2.82
2.15	2.97
2.28	3.23
2.40	2.48

3. Conclusions

In this report we have presented some preliminary results of an investigation of the He I 10830 line profile in the Sun. We plan to publish the results after a full analysis of the observational data.

References

Borodina, O. A. and Papushev, P. G. 1979, *Pisma v A. Zh.* **5**, 620.
Fay, T. D., Wyller, A. A. and Yun, H. S. 1972, *Solar Phys.* **23**, 58.
Livshitz, M. A. and Demkina, L. B. 1965, *Soln. Dannye, No. 10*, 55.
Nikolsky, G. M. 1965, *Astron. Zh.* **42**, 86.
Nikolsky, G. M. and Sazanov, A. A. 1967, *Astron. Zh.* **44**, 426.
Papushev, P. G. and Salakhutdinov, R. T. 1989, *Proceedings of the XIII Consultation Meeting on Solar Phys.* **1**, 263.
Stepanov, V. E. 1957, *Soobshch. Gos. Astron. In-Ta Im. Shternberga, no. 100*, 3.
Zirker, J. B. 1962, *Ap. J.* **136**, 250.

POTENTIAL IR OBSERVATIONS OF THE SOLAR CORONA

C. KRISHNA KUMAR

Department of Astronomy, Howard University, Washington, DC 20059, U.S.A.

and

JOSEPH DAVILA

Code 682, NASA/Goddard Space Flight Center, Greenbelt, MD 20771, U.S.A.

Abstract. We examine the potential for observations of the solar corona using forbidden lines in the near infrared.

Key words: infrared: stars – Sun: corona

1. Introduction

Observations of IR emission lines beyond 1 μm wavelength in the spectrum of the corona were first made during eclipses in the 1960's by Mangus and Stockhausen (1966) and by Münch et al. (1967). There have been no reports of subsequent investigations of the IR spectrum of the corona. In view of the developments in IR technology a re-examination of the potential IR studies of the corona was felt worthwhile. The sources of IR lines in the coronal spectrum and the issues involved in studying them are discussed below, followed by a section on the possible contributions to coronal physics and other fields.

There are two sources of IR lines. They are: (i) the excited fine structure levels of the ions in the coronal material, and (ii) fine structure levels of atoms and ions of the material shed by sun-grazing comets. The comet source is available infrequently (perhaps 1 or 2 days a year) but it offers possibly the only opportunity for measuring coronal magnetic fields (see discussion in last section).

2. Sources of IR Lines

2.1. Coronal Material

In the corona proper, above the transition region, highly charged ions with ionization energies near 500 eV dominate the ionization equilibrium. Most of these ions do not have fine structure levels in the ground state that can emit in the IR. One exception is the 1.43 μm line of Si X. Cosmically underabundant elements, $\log(X/H) < -7.0$, may have appropriate levels but they are not considered here. In the transition region the lower ionization stages can emit many IR lines.

2.2. Sungrazing Comets

Sungrazing comets pass through the corona and shed neutral gases and dust particles. The dust particles evaporate to produce atoms which along with the gas phase atoms are ionized to higher charge states until they reach ionization equilibrium with the coronal matter. This process can take hours to days before equilibrium is reached. Therefore the comet material will exhibit all the ionization stages seen in the transition region except that this will happen in the corona. These lower-ionization stages can result in a rich IR line spectrum. Table I lists the ions of the cosmically more abundant species expected to be in comets which will emit IR lines from their ground state fine structure levels.

There are two type of sungrazing comets. They are: (i) the bright Ikeya-Seki (1965) which appear very seldom (every 50 years or so) and, (ii) the dimmer pygmies which averaged one a year over the last decade. These pygmies were discovered (Sheely et al.1982, MacQueen and St. Cyr 1991) using the coronagraphs on SOLWIND and SMM satellites. They have perihelion distances $q < 5\ R_\odot$ with many falling into the sun. None of the sungrazers discovered on SMM images were found on groundbased coronograph pictures (MacQueen and St. Cyr 1991). Their peak surface brightness is $2 \times 10^{-8}\ B_\odot$ at 530 nm. This is two orders of magnitude lower than the detection limits of existing groundbased coronographs operating in the visual. However in the case of SOLWND-1, the brightest known pygmy sungrazer, a post-perihelion enhancement of the corona was detected by the SOLWIND coronograph. This appears to have been detected in a ground based spectrum also, taken using a coronagraph (Chocol et al. 1983). The observers claim to have detected lines of Si II and Ni II. Therefore it may be possible in some bright cases to observe the spectra of sungrazers from the ground if one knows where to look for them. In the IR at least an order of magnitude decrease in the level of the atmospherically scattered light is possible (in units of $B_\odot$). If, in addition, the coronagraph is located on an aircraft the sky brightness drops to $10^{-7}\ B_\odot$ at least (Newkirk and Eddy 1964). This makes the spectroscopy of sungrazing comets, using an aircraft based coronagraph, especially attractive. Whether the cometary lines can be detected depends on the unknown mass loss rate of the comet. The reported detection by Chocol et al. (1983) of the cometary lines is encouraging. Attempts should be made to observe the spectra of the sungrazing comets. To do this one needs to find the comet first.

3. Observational Constraints

If space-based coronagraphs continue to be the sole means for discovering sungrazing pygmy comets one has at most a few hours or a day before they fall into the sun or go outside 10 $R_\odot$. This is because the coronagraph's field of view is limited to say 10 $R_\odot$ centered on the sun. Real time information is needed to execute spectroscopic observations from the ground or from aircraft equipped with coronagraphs. Such real time detection of the comets was not possible with the SOLWIND and SMM instruments, but it may be with the proposed LASCO coronagraph. This instrument has an outer limit of 30 $R_\odot$ and can give 2 or 3 day's notice so that even balloon-based observations can be made.

Münch et al. (1967) measured radiance of the purely coronal line at 1.43 μm to be about $10^{-4}\ B_\odot$ along a path 0.2 $R_\odot$ above the limb. It should be possible to measure the Si X line, out to 3 $R_\odot$, from the ground and further out from an aircraft. In this connection it is worth noting that the measured sky brightness at 2.2 μm from a balloon coronagraph was $10^{-12}\ B_\odot$ (MacQueen, 1968).

In summary, the Si X line at 1.43 μm is probably bright enough to use for mapping the corona out to 3 $R_\odot$ using a ground-based coronagraph and IR imager or more likely with an aircraft mounted system. A balloon-based system will definitely extend the coverage to greater distances from the sun's limb. Spectroscopic observations of sungrazing comets may also be possible. To do this real time detection of the comet from space based coronagraphs is necessary until other methods of discovering them are developed.

4. Scientific Expectations

Finally, what are the scientific reasons for the IR observations of comets?

- The possibility that IR coronagraphs may have lower scattered light background than the visual band instruments, makes imaging of the corona using the 1.43 μm light of Si X attractive. It may extend the observable distance from the limb beyond that possible with the green line. It will, at least, supplement the observations in the visual.

- The ions expected to appear in the spectra of sungrazing comets cover a large range in ionization energies, from a few eV to 500 eV. The relative concentrations of the ions can yield information about ionization rates in different parts of the corona.

- The spectra of the sungrazers can be used to derive the chemical abundances of cometary material, including the non-volatile elements. Until the Halley fly-by, such information was not available.

- One of the ions expected in the cometary debris is Ca II. The upper levels emitting the triplet near 850 nm will be excited by sunlight. The lifetime of this level is appropriate for measuring the magnetic fields in the 0.1 G to 1 G by means of the Hanle effect. This may be the only spectroscopic method for measuring coronal magnetic fields.

TABLE I
IR line-emitting ions in the spectra of sungrazing comets

Emitting in 1-10 μm range	Emitting in 10-100 μm range
Mg IV,V,VII,VIII	N III
Si VI,VII,IX,X	O III,IV
Ca IV,V,VII,VIII	Ti IV
Ti VI,VII	Cr III,IV,V,VI
Cr V	Fe V,VI
Fe VII,VIII	Ni VII
Ni VII,VIII	

References

Chocol, D., Rušin, V., Kulcar, L., and Vanysek, V.: 1983, *Astrophys. and Sp. Sci.* **91**, 71.
Macqueen, R.M.: 1968, *Ap.J.* **154**, 1062.
Macqueen, R.M., and St. Cyr, O.C.: 1991, *Icarus* **90**, 96.
Mangus, J., and Stockhausen, R.: 1966, NASA Rept. X-614-66-29.
Münch, G., Neugebauer, G., and McCammon, D.: 1967 *Astrophys. J.* **149**, 681.
Sheeley, N.R., Howard, R.A., Koomen, M.J., and Michels, D.J.: 1982, *Nature*, **300**, 239.
Newkirk, G. and Eddy, J.: 1964 *J. Atmos. Phys.*, **21**, 34.

THE SUN IN SUBMILLIMETER RADIATION

CHARLES LINDSEY

Solar Physics Research Corporation, 4720 Calle Desecada, Tucson, AZ 85718, U.S.A.

Abstract. Continuum observations in the far IR have given us a broad spectrum of new and powerful diagnostic utilities for the solar atmosphere. The infrared continuum is formed in LTE with thermal free electrons by free-free interactions. This gives us a flexible and accurate atmospheric thermometer that has made infrared measurements fundamental to modeling of the quiet solar medium for more than two decades. The submillimeter and millimeter continua are particularly useful with respect to thermal diagnostics of the low chromospheric temperature minimum, where non-radiative heating of the solar medium becomes clearly manifest. Modern submillimeter telescopes and instrumentation on Mauna Kea, in Hawaii, are now revolutionizing solar observations in the submillimeter spectrum, giving us the first observations of detail on the scale of the chromospheric supergranular network, sunspots and prominences. These observations are showing us a remarkable and unexpected view of thermal structure that emerges as one probes to successively higher levels above the chromospheric temperature minimum.

Key words: heating – infrared: stars – Sun: chromosphere – Sun: filaments – sunspots

1. Introduction

The submillimeter spectrum is clearly defined as excluding wavelengths longward of 1 mm, but its short-wavelength limit is somewhat arbitrary. The terrestrial atmosphere blocks almost all radiation between 30 and 300 μm, (see Figure 2 of Jefferies, 1993, this volume). This occlusion seems to provide a rough discriminator for astronomers whereby 300 μm is definitely considered submillimeter while 30 μm falls in the far-infrared domain and is rarely thought of as submillimeter. However, the 30–300 μm band is observable from the stratosphere or above, and is of great interest in solar diagnostics. We will consider wavelengths greater than 30 μm to be submillimeter, for the purpose of this discussion, but a clear boundary is actually not established.

The submillimeter solar spectrum, at first sight, appears to exude a certain blandness. As wavelength increases, spectral lines weaken leaving only a few diffuse emission lines longward of 100 μm (*e.g.*, Boreiko and Clark 1986; Naylor *et al.* 1993 find no lines longward of 300 μm). The continuum offers spatial information, but at a high cost, for diffraction reduces spatial resolution. Nevertheless, the submillimeter spectrum does present us with an exceptionally powerful diagnostic perspective. It has served a unique role as a thermal probe of the chromospheric medium for the quiet Sun. And now, modern large submillimeter telescopes are opening this diagnostic to a broad range of other interesting chromospheric structure. A brief review of the mechanisms important in submillimeter radiative transfer in the solar atmosphere offers some insight into this role.

2. Continuum Emission and Absorption

The primary mechanism of continuum emission, and thus also opacity, throughout of the infrared spectrum longward of 2 μm is collisions of free electrons with atoms and ions, *i.e.*, bremsstrahlung. For collisions with neutral hydrogen atoms, the

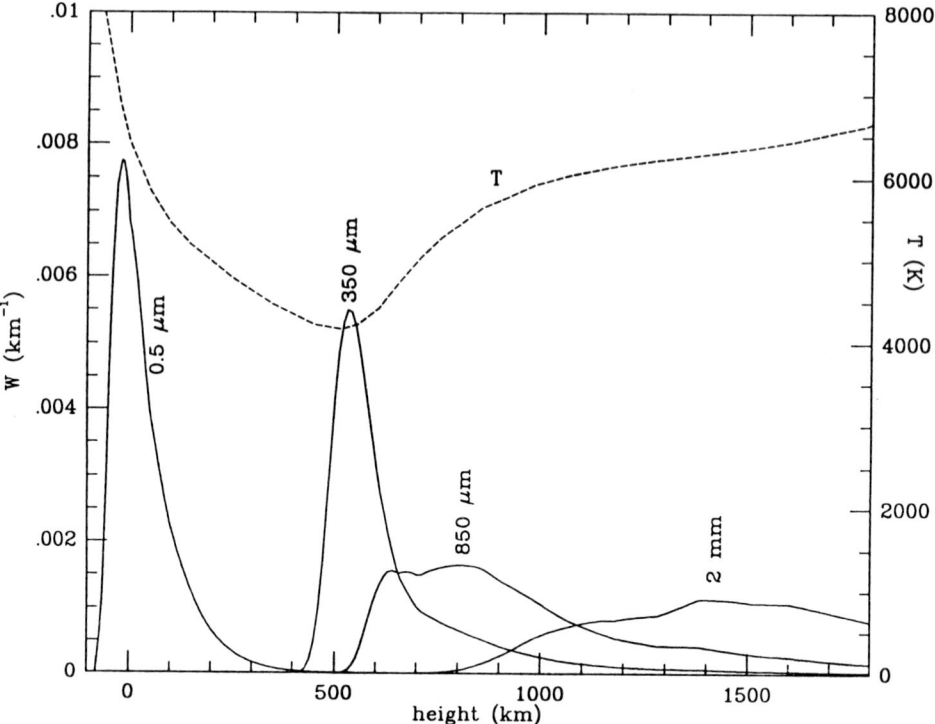

Fig. 1. Weighting functions for 350, 850 and 2,000-μm continuum radiation are plotted against the temperature profile (dashed curve) of model C of VAL III.

process is termed H⁻ free-free emission, after a formalism that considers the free electron together with the neutral atom as an unbound H⁻ ion both before and after the interaction. H⁰ free-free emission, $i.e.$, electrons colliding with protons (or any other singly charged ion), begins to dominate H⁻ free-free emission when the atmospheric medium is more than about a half percent ionized. Free-free opacity increases strongly with wavelength, λ, in the infrared. For H⁻ free-free opacity, it is proportional to λ^2, and this rule also serves as a rough approximation for H⁰ free-free opacity. Figure 1 shows weighting functions for 350, 850 and 2,000 μm radiation for atmospheric model C of Vernazza, Avrett and Loeser 1981 (VAL III C). The shorter wavelengths, centered in atmospheric windows accessible from Mauna Kea, are formed by chromospheric levels at which the temperature begins to increase with height.

Because the emission arises from thermal collisions, the infrared continuum is formed essentially in thermal equilibrium with the local thermal free electrons that give rise to it. The source function is the Planck function, which at solar temperatures follows the Rayleigh-Jeans law and is therefore simply proportional to temperature. It is useful to keep in mind that in a non-uniform medium the visible and ultraviolet Planck functions at solar temperatures are weighted strongly toward hotter elements, so much so that images at the shorter wavelengths may represent mostly a rather small fraction of the very hottest gas. In the infrared, one can usually suppose that the continuum intensity roughly represents the temperature of gas at about unit optical depth, rather than optically thin gas that has a disproportionately large source function. All of these qualities make the infrared and submillimeter continuum surely the best known chromospheric thermometer.

3. Recent Submillimeter Observations

I will bypass a large amount of interesting past work at submillimeter wavelengths to look over a sample of recent exciting results from the world's largest submillimeter telescope, the 15-m James Clerk Maxwell Telescope (JCMT) on Mauna Kea. Figure 3 of Jefferies (1993) shows an image covering most of the solar disk in 850 μm radiation (Panel a) next to a simultaneous image of the Sun in the K-line of singly ionized calcium (Panel b). There are striking similarities. Both images clearly show the chromospheric supergranular network. In this Figure, the submillimeter contrast (Rayleigh-Jeans) has been enhanced to match the greater K-line contrast. Lindsey and Jefferies (1991a) find the distribution of infrared intensities from the supergranular network to be consistent with that expected from the VAL III (1981) models.

Notwithstanding clear similarities, there are some clear differences between the 850 μm image and the K-line image: In particular, the K-line image is strongly limb darkened, where the 850 μm image is slightly limb brightened, a quality first inferred observationally by Noyes, Beckers and Low (1968) and mainly attributable to a general increase in chromospheric temperature with height. The extreme limb brightnesses measured by occultation during eclipse are reviewed by Clark (1993) in this volume. (For a summary glance see Figure 4 of Jefferies 1993, also in this volume). The strong limb-darkening seen in the K-line reflects an increasing departure of the K-line source function from local thermal equilibrium (LTE) with increasing height, as decreasing opacity allows the calcium ions to radiate more efficiently into space. This is one of the stronger of the non-LTE manifestations that are shared by all chromospheric lines in the visible, and we now know that at least some photospheric lines far into infrared depart from LTE (see Chang 1993 and Rutten and Carlson 1993, in this volume).

For a smooth atmosphere, there is a simple formalism relating the infrared continuum limb brightness profile to the vertical temperature gradient and similarly to the disk-center temperature as a function of wavelength. Where the λ^2 opacity rule is accurate, for instance, the brightness temperature, T_b, of radiation emerging at incidence θ is simply a function of the product λ^2/μ:

$$T_b(\lambda, \mu) = f(\lambda^2/\mu), \tag{1}$$

where $\mu = \cos\theta$ (Simon and Zirin, 1969). These relationships had already lead to predictions of strong limb brightening in radio measurements in the 1950s, an effect that never clearly materialized for wavelengths of order 1 cm or greater. Equation (1) is not satisfied because the chromosphere is not smooth. This realization lead to models incorporating rough structure such as spicules as early as Coates (1958).

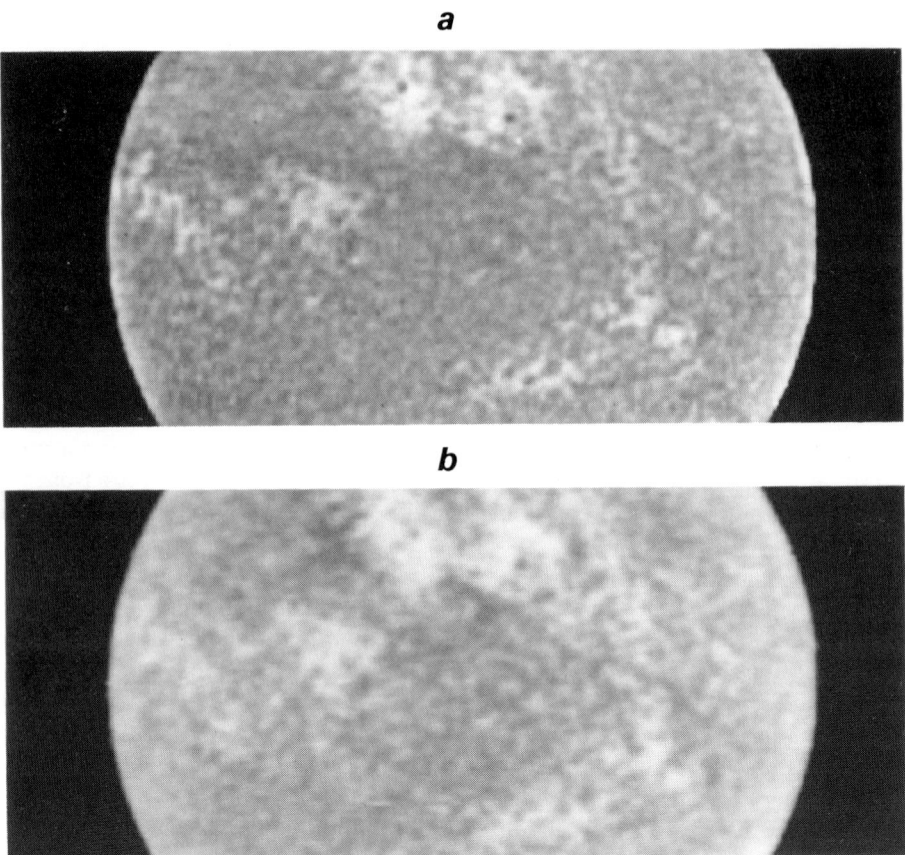

Fig. 2. Images of the Sun in 350-μm (Panel a) and 1,200 μm radiation (Panel b) made on 1991 Feb 9 at ∼2200 UT and ∼2500 UT respectively. Vertical and horizontal fiducials indicate the positions of large sunspots.

The best available review of the limb-brightness problem remains that of Simon and Zirin (1969). The submillimeter continuum seems to be unique in exhibiting clear limb brightening.

An indication of how heated chromospheric structure emerges above the temperature minimum level is found in Figure 2, which shows images of the Sun in 350- and 1200-μm radiation taken a few hours apart. Network boundaries that appear relatively faint and thin at 350 μm brighten and appear to expand to a greater filling in 1200-μm radiation. However, at least some part of this expansion is attributable to diffraction.

3.1. SUNSPOTS

Vertical and horizontal fiducials in Figure 2 show the locations of two large sunspots. They appear nearly invisible at these wavelengths, distinguishable from the quiet Sun only by contrast with surrounding plage. Even large sunspots appear mostly just by contrast against surrounding brighter plage. Occasionally they are as bright as the plage themselves, as seen in Figure 3 of Jefferies (1993) in this volume, where a group of spots registers as very dark in the K-line (lower right of Panel b) but resembles bright plage in the 850 μm continuum (lower right of Panel a). This comparison further illustrates the drastic departure from LTE of visible chromospheric lines, which generally register large sunspots as quite dark. The best known sunspot models have low chromospheres considerably cooler than the quiet Sun temperature minimum, resulting in a dark submillimeter sunspot. The models that come closest to satisfying the submillimeter observations seem to be those of Maltby et al. (1993). It now seems that even these models will need a considerably hotter lower chromospheres to fit the new submillimeter and millimeter observations, requiring a strong heating mechanism in those regions with the most intense magnetic fields.

3.2. LARGE-SCALE VARIATIONS IN THE QUIET SUN

While the 850 μm images clearly show the quiet-Sun supergranular network (see Figure 3 of Jefferies, 1993, in this volume), they also show large-scale quiet-Sun intensity variations that the K-line does not. There are extended regions up to 10% darker than the more typical quiet Sun, and these seem to have a weak tendency to border active regions. They are not apparent at 350 μm (Fig. 2a), but are pronounced at 1,200 μm (Fig. 2b). One is tempted to think of such a region as a sort of chromospheric hole. However, preliminary estimates indicate that even these darkened regions have a greater brightness temperature in 1,200 μm radiation than the quiet Sun in 350 μm radiation, thus, they evidently are not "chromospheric holes," but perhaps simply regions of reduced heating. While not seen in the calcium K-line, similar features are clearly seen in the calcium infrared triplet and weakly in the sodium D-lines, where large active regions tend to be surrounded by darkened halos.

3.3. FILAMENTS AND PROMINENCES

Filaments seem to be nearly invisible against the solar disk at submillimeter wavelengths. This strongly suggests material roughly equal in temperature to the low chromosphere, ~5,500 K, which is considerably cooler than most prominence models but not all (see, e.g., Hirayama 1985 and Hirayama, Nakagomi and Okmoto 1979). We know that filaments have a substantial optical depth in the submillimeter continuum, because they are clearly seen as prominences in this radiation. Kosugi Ishiguro and Kiyoto (1986) clearly see filaments as dark against the solar disk at 3 mm, where the disk brightness is considerably higher than 5,500 K. Thus, submillimeter and radio observations are accumulating a strong case in favor of prominences being of mostly relatively cool material.

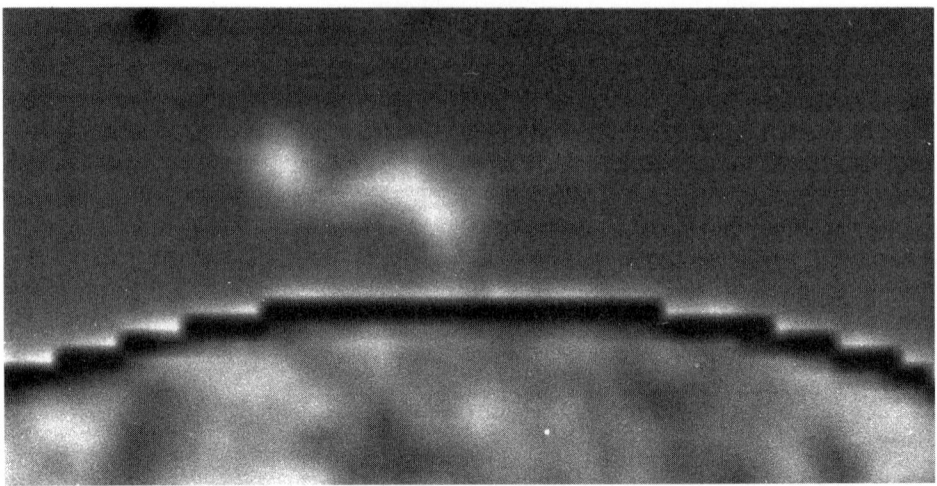

Fig. 3. A large active prominence in 1,300 μm radiation on 1991 July 10 at 18 00 UT. The prominence was raster-scanned by the 15 m JCMT 23 hr before a total solar eclipse over the facility on Mauna Kea.

Figure 3 shows a large prominence in 1.3-mm radiation. This prominence shows a peak brightness temperature of ∼2,500 K. Rough estimates, assuming an optical path length through the prominence comparable to its horizontal dimensions and a temperature of 5,500 K suggest a free electron density of order 10^{10} cm^{-3} (Harrison et al. 1993).

3.4. CHROMOSPHERIC DYNAMICS

Infrared thermal diagnostics are as applicable to chromospheric dynamics as to quasi-static modeling. Kaufmann et al. (1993) and Correia et al. (1993) explain the need for submillimeter observations in solar flare diagnostics. Observations even down to ms time scales are desired. The practical problem reduces to dedicating a substantial submillimeter facility to a program to monitor active regions for long enough to get a good sampling of large-flare observations.

Far infrared diagnostics have already given us very interesting thermal diagnostics of the response of the chromospheric medium to compressional perturbations due to the five-minute oscillations in the quiet Sun. Figure 4, taken from Kopp et al. (1992), shows time series of local brightness oscillations measured at 50, 100 and 200 μm from the Kuiper Airborne Observatory, made simultaneously with Doppler observations of the same region in the sodium D_1 line from the Stokes Polarimeter at the Mees Solar Observatory (see Mickey 1985) on 1987 May 14. There is a high correlation between velocity and submillimeter intensity, with intensities leading velocity by varying phase shifts. Observations in visible lines, particularly those of Lites, Chipman and White (1982) have shown similar phase shifts between velocity and core brightness of certain members of the calcium infrared triplet, particularly Ca II 8498 Å. Kopp et al. 1992 use the measured infrared phase-shifts to

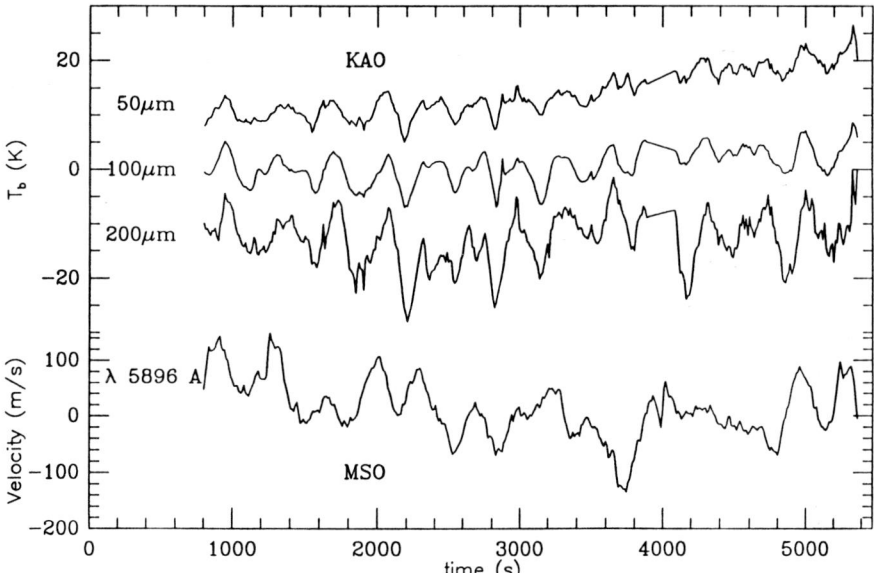

Fig. 4. Time-series showing brightness variations in 50, 100 and 200 μm continua (top three plots) in response to five-minute oscillations monitored by Doppler variations in the Sodium D₁ line λ5896 Å (bottom plot).

characterize the low chromosphere by thermal relaxation at various rates over a range of heights spanning the temperature minimum. They find relaxation rates that decrease rapidly with height, suggesting a simple adiabatic response of the chromospheric medium above about 1,000 km. In the low chromosphere, these rates appeared unexpectedly large at the time the observations were made, but now seem reasonably consistent with recent theoretical work by Anderson (1992). Again, the interpretation of the infrared continuum is greatly simplified by the emission being always in LTE with the thermal free electrons.

4. Summary

The infrared continuum remains central to thermal diagnostics of the solar photosphere and chromosphere. Modern large submillimeter telescopes and instrumentation are opening a broad new range of solar phenomena to submillimeter observations, and are opening us to a remarkable, new perspective on chromospheric structure. The LTE source function of the infrared continuum promises to facilitate our interpretation of the infrared in terms of atmospheric models, especially where rough, inhomogeneous atmospheric structure is important, as it is in the chromosphere (see Lindsey and Jefferies, 1991b).

With more advanced instrumentation now being developed, we can expect submillimeter observations to give us considerably more insight into dynamical qualities of the chromosphere in its response to acoustic and other hydro-mechanical perturbations, such as the five-minute oscillations, and to transient events, such as flares. Even the observation of a clear lack of submillimeter emission from a moderately large flare would be interesting (see Kaufman 1993). Good flare observations in this region are problematical for the moment, but modern array detectors now being developed will soon make them practical. It seems clear that we have only obtained a first glimpse of the potential of submillimeter diagnostics at this juncture.

Acknowledgements

This review presents original results from the James Clerk Maxwell Telescope, run by the Joint Astronomy Center, based at the University of Edinburgh. This work was also supported by NSF Grant ATM-9122073. The author made extensive use of facilities at the National Solar Observatory, where he holds a visiting appointment.

References

Anderson L.: 1992, private communication.
Boreiko, R.T. and Clark, T.A.: 1986, *Astron. Astrophys.*, **157**, 353.
Chang, E. S.: 1993, these proceedings.
Clark, T. A.: 1993, these proceedings.
Coates, R. J.: 1958, *Astrophys. J.* **128**, 83.
Correia, E., Kaufmann, P. and Magnum, A.: 1993, these proceedings.
Harrison, R. A., Carter, M. K., Clark, T. A., Lindsey, C., Jefferies, J. T., Sime, D. G., Watt, G., Roellig, T. L., Becklin, E. E., Naylor, D. A. , Tomkins, G. J. and Braun, D. C.: 1993, *Astron. Astrophys.*, in press.
Jefferies, J. T.: 1993, these proceedings.
Hirayama, T.: 1985, *Solar Phys.*, **100**, 413.
Hirayama, T., Nakagomi, Y. and Okmoto, T.: 1979 in E. Jensen, P. Maltby, and F. Q. Orrall (eds.), 'Physics of Solar Prominences', *Proc. IAU Colloq.* **44**, 48.
Kaufman, P., Correia, E., Costa J. E. R. and Zodi, A. M.: 1993, these proceedings.
Kopp, G., Lindsey, C., Roellig, T. L., Werner, M. W., Becklin, E. E., Orrall, F. Q., Jefferies, J. T.: 1992, *Astrophys. J.* **388**, 203.
Kosugi, T., Ishiguro, M. and Kiyoto, S. K.: 1986, *Pub. Astr. Soc. Japan*, **38**, 1.
Lindsey, C. and Jefferies, J. T.: 1991(*a*), *Astrophys. J.*, **349**, 286. (radiative transfer).
Lindsey C. and Jefferies, J. T.: 1991(*b*), *Astrophys. J.*, **383**, 443. (chromospheric supergranular network).
Lites, B. W., Chipman, E. G. and White, O. R.: 1982, *Astrophys. J.*, **253**, 367.
Maltby, P.: 1993, these proceedings.
Mickey, D. L.: 1985, *Solar Phys.*, **97**, 223.
Naylor, D. A.: 1993, these proceedings.
Noyes, R. W., Beckers, J. M. and Low, F. J.: 1968, *Solar Phys.* **3**, 36.
Rutten R. and Carlsson, M.: 1993, these proceedings.
Simon, M. and Zirin, H.: 1969, *Solar Phys.*, 9, 317.
Vernazza, J. E., Avrett, E. H., and Loeser, R.: 1981, *Astrophys. J. Supp.*, **45**, 635.

FAR INFRARED AND SUBMILLIMETER CONTINUUM OBSERVATIONS OF SOLAR FLARES: JUSTIFICATIONS AND PROSPECTS FOR GROUND-BASED EXPERIMENTS

P. KAUFMANN, E. CORREIA, J. E. R. COSTA and A. M. ZODI

Centro de Radio-Astronomia e Aplicações Espaciais, CRAAE/EPUSP,
C.P. 8174, 05508 - São Paulo, SP, Brazil

Abstract. Solar flare observations in the sub-mm spectral bands are essentially non-existent. There is evidence that some solar bursts exhibit a spectral component rising in intensity towards wavelengths shorter than 3 mm, displaying fast sub-second pulses at different repetition rates. On the other hand, the spectral features of white light flares are also unknown in the infra-red range of frequencies. In both wavelength ranges the physics of the emission processes may involve particles accelerated to high energies. The diagnostics of solar flare continuum emission in the IR and sub-mm spectral regions will provide crucial tests on various flare models and bring some clues on the initial primary energy release mechanisms. We propose the construction and operation of a ground-based telescope, operating at two sub-millimeter wavelengths (at about 210 GHz and 405 GHz), with high time resolution (one millisecond), capable of determining the spatial position of burst emission centroids with high definition (a few arcseconds) using the multiple beam technique. Final installation and operation at a high-altitude site in the Argentinian Andes mountains are planned in a joint cooperation with Argentina's Instituto de Astronomia y Fisica del Espacio, IAFE (M. Rovira and associates) and Complejo Astronomico El Leoncito, CASLEO, San Juan (H. Levato and associates); and Switzerland's University of Bern, Institute of Applied Physics, IAP, Bern (A. Magun and associates).

Key words: infrared: stars – Sun: flares – Sun: radio radiation

1. Introduction

Recent discoveries indicate that it is essential to complete the diagnostic of flare emission in the sub-millimetric and infrared range of wavelengths. However such measurements have never been made with instruments sensitive enough and capable of detecting rapid variations. Previous suggestions are reviewed showing that this knowledge has a crucial importance for the understanding of the primary processes of particle acceleration in plasma instabilities. We propose a project to diagnose solar flares at two sub-mm wavelengths located in atmospheric windows, with a time resolution of 1 millisecond and spatial definition, for burst emission sites, of a few arcseconds. Two frequencies are being considered: 210 GHz (using three receivers for a multibeam imaging capability), and a single receiver at 405 GHz to extend the spectral information. The system configuration at 210 GHz is similar to that used successfully at 48 GHz by the Itapetinga/Bern group (Georges et al., 1989; Herrmann, 1990; Costa et al., 1991), but at a much higher frequency and with an antenna scaled down to a smaller physical size. One single reflector, about 1.5 m in diameter, will concentrate solar radiation into the 210 GHz three feed-horn arrangement and the 405 GHz receiver. The proposed instrument can be built with a modest budget. Initially we plan two years of continuous operations at a site of exceptional quality in Argentinas's Andes Cordillera, known as El Leoncito, part of CASLEO (Complejo Astronomico El Leoncito), near San Juan, Argentina.

2. The Flare Electromagnetic Spectrum

Figure 1 shows a simplified description of what is known about the electromagnetic continuum emission of a solar flare (after Kaufmann, 1988). Spectral curves I, II, III, IV and V are quite well known from observations and are explained by a number of existing models.

The band including sub-millimetric and infrared wavelengths is poorly known and essentially unexplored for flare emission. Spectral curves A, B and C are model computations. Spectral curve III shows the quiet Sun blackbody emission. Curve IV shows the optical-continuum of the white-light flares (WLF), which appears in some bursts. It is difficult to measure this, since its intensity is four orders of magnitude smaller than the quiet Sun emission. The nature of the WLF emission is still controversial; it can be found in both large and small flares (Cliver et al., 1983; Henoux and Aboudarham, 1991). Irrespective of the physical mechanism which produces WLF, it seems to imply the acceleration of very energetic particles (Najita and Orrall, 1970).

In the UV and soft X-ray bands there are many measurements, and the emission consists of a superposition of several mechanisms, producing the continuum and discrete emission lines. The current models attribute the hard X-ray emission (curve V) to the bremsstrahlung of mildly relativistic electrons initially accelerated. For photon energies larger than about 400 keV, the gamma radiation appears in the continuum as well as in discrete lines that result from nuclear reactions; this has been fully confirmed by the gamma ray spectrometer on the SMM satellite in the 21st solar cycle. The production of these high energy photons requires extremely energetic particles: protons with tens to hundreds of MeV or some GeV, or electrons from MeV to hundreds of MeV. The only known mechanisms to reach such high energies are based on second-step accelerations (like Fermi acceleration). It has been shown, however, that gamma ray continuum and line emission can appear in the first few seconds of the initial phase of a flare, for which there is no proposed explanation so far (Forrest and Chupp, 1983; Rieger, 1991; Henoux and Aboudarham, 1991). Recent results from the BATSE experiment on board the GRO satellite, which has a sensitivity more than one order of magnitude better than the SMM HXRBS detector, have indicated the presence of well-defined hard X-ray pulses of 100 ms in duration (Machado et al., 1992. Such fast time scales for hard X-rays cannot be easily explained by electron bremsstrahlung and may require a review of the concepts of the particle beam production in the very first phase of solar flares.

It is quite surprising that no solar flare measurements have yet been made in the sub-mm and IR bands. Until a few years ago, the acceleration of electrons up to ultrarelativistic energies was not conceivable in the first phase of a burst. They could appear as a result of acceleration in a second step, such as the first solar gamma-ray space observations suggested. According to Najita and Orrall (1970), Ohki and Hudson (1975), and others, the early appearance of these particles in great number would produce IR and sub-mm emission by bremsstrahlung and thermal interactions (curves A and B in Fig. 1, for the optically thin and thick cases, respectively). Therefore flare measurements in the sub-mm and IR bands never had a justified priority.

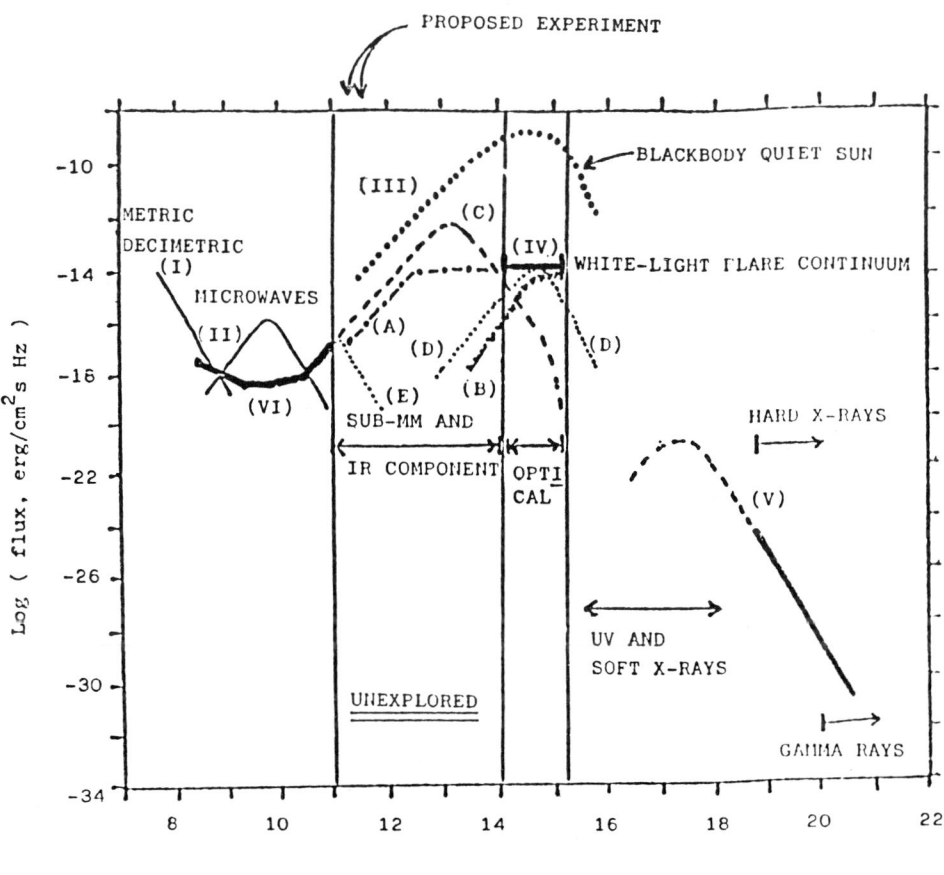

Fig. 1. The electromagnetic spectrum of solar-burst continuum emissions (discussed in the text). The flare spectrum in the frequency band covering the sub-mm and IR is essentially unknown. Curves I - VI refer to observations. All others refer to models: A, B Ohki and Hudson, 1975; C Kaufmann *et al.*, 1986; D Stein and Ney, 1963; Shklovsky, 1964; and E de Jager *et al.*, 1987.

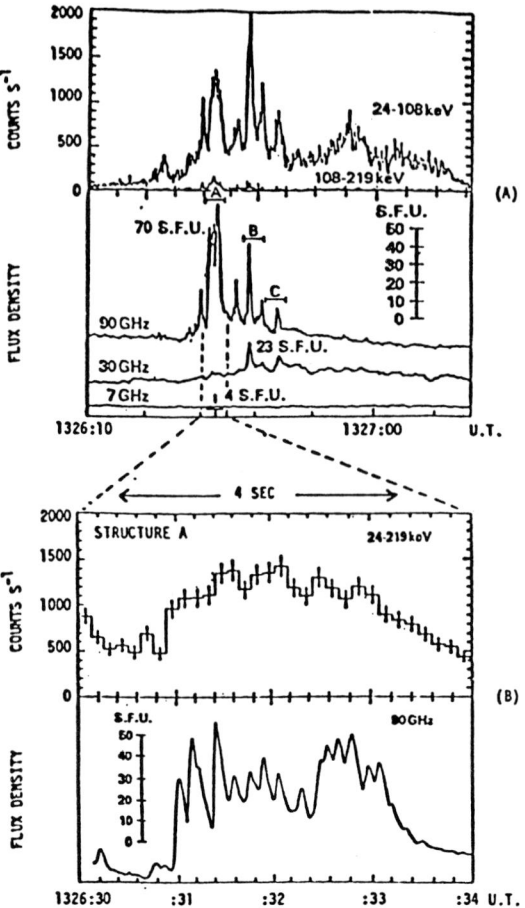

Fig. 2. The extraordinary solar burst of 21 May 1984 which exhibited various time structures, with 90 GHz emission, coincident with hard X-ray emission, much more intense than at 30 GHz and lower microwave frequencies. Each structure consisted of multiple fast pulses, with rise times smaller than 30 milliseconds (see expansion at bottom (B)). This result is hard to explain with current models (Kaufmann et al., 1986).

Recent observations, however, have entirely changed our perspective. Curve VI in Figure 1 shows the spectrum of an extraordinary solar flare that occurred on May 21, 1984. This event shows the flux intensity increasing with frequency at least up to about 100 GHz (Kaufmann et al., 1985; Correia and Kaufmann, 1987) (Fig. 2). It consists of very rapid pulses (30 ms) with start times coincident with the hard X-ray emission. This spectral component needs to be better defined in the sub-mm and IR bands (frequencies larger than 100 GHz). If this emission component is part of a thermal spectrum (curve A or B in Fig. 1), the time scales are too short by many orders of magnitude, so that this mechanism is not acceptable.

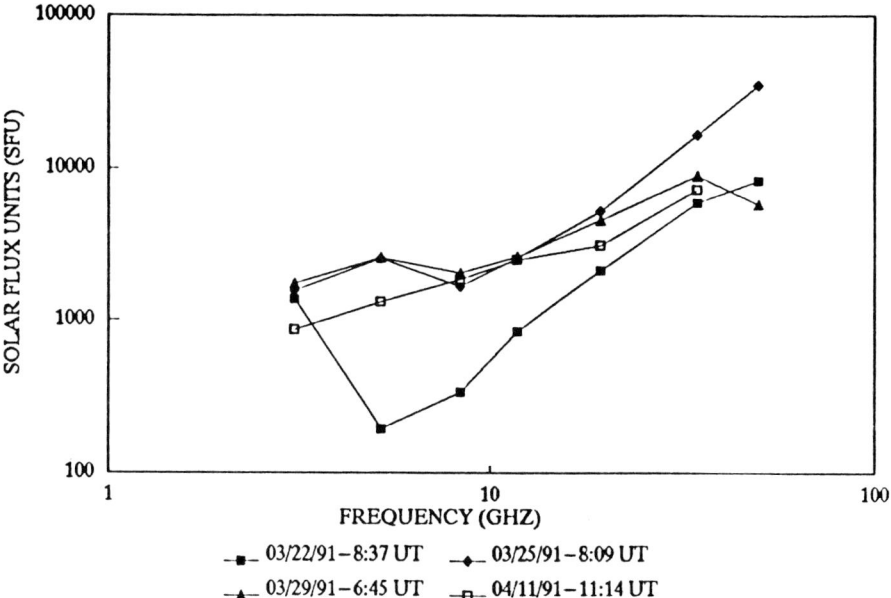

Fig. 3. More recent examples of solar bursts showing spectra with intensity increasing for frequencies larger than 19 GHz (Correia et al., 1992).

In fact, spectra similar to curve VI (Fig. 1) or with fluxes rising with frequency towards 100 GHz have been found for other solar bursts in the past (Hachenberg and Wallis, 1961; Croom, 1970; Cogdell, 1970 ; Shimabukuro, 1970, 1972; Akabane et al., 1973; Zirin and Tanaka, 1973). The data, however, were obtained with poor sensitivity and time resolution. Recently Correia et al. (1992) have shown that such events are not so rare. Analyzing spectra of 115 events with fluxes larger than 100 sfu at 35 and 50 GHz, obtained by the Bern Observatory (Magun et al., 1991), they found that more than 25 percent of the bursts had rising intensities for frequencies above 35 GHz (Fig. 3), and that 50 percent displayed nearly flat spectra in that range of frequencies. The later may belong to a class of events found by Hachenberg and Wallis (1961), interpreted by Ramaty and Petrosian (1972) as due to free-free absorption of gyrosynchrotron emission from non-thermal electrons.

It has generally been believed that the time scales for the energy conversion processes in flares are of the order of tens to hundreds of seconds, which is now recognized as entirely incorrect. Early radio observations at decimetric wavelengths had, in fact, already suggested the presence of very fast time structures in solar bursts (Butz et al., 1976; Droge, 1977; Slottje, 1978). In the last (21st) solar cycle, millisecond time structures, multiple and discrete, were observed in solar bursts in microwaves and mm-waves (Kaufmann et al., 1975; 1980; 1984). They were also suggested as being present in hard X-rays (Kiplinger et al., 1983), and concurrently in hard X-rays and microwaves (Takakura et al., 1983). Models intending to explain the energy conversion in flares as a result of multiple, rapid and discrete micro-bursts, eventually quasi-quantized in energy, have been suggested (Kaufmann et al., 1978; Loran et al., 1985; Sturrock et al., 1985; de Jager et al., 1987). These ideas are now being confirmed by more recent observations obtained at decimetric and microwave wavelengths, being associated with the "fragmentation" of energy release in solar flares (Alaart et al., 1991; Qijin et al., 1991; Benz, 1991). A striking confirmation of the concept is being obtained by the initial results from high-sensitivity hard X-ray solar flare observations from the GRO satellite (Machado et al., 1992). It is likely, however, that solar flare continuum emission at hard X-ray, mm, sub-mm, and IR wavelengths does not have a direct physical connection to the emission observed at decimetric and centimetric wavelengths.

Almost three decades ago, white-light continuum emission in flares was suggested to be part of a synchrotron radiation spectrum (Stein and Ney, 1963) produced by ultrarelativistic electrons, with a spectral maximum in the IR and visible. Shklovsky (1964) suggested that, associated with the ultrarelativistic electrons, a large number of IR photons would necessarily be produced. These early models have not survived, however the reasons for their rejection must be revisited. The time scales, then believed to be hundreds of seconds, did not require the rapid acceleration of particles to such high energies. The early evidence of flare-associated hard X-rays, known from a few rocket measurements, raised the suggestion of inverse-Compton processes operating in those events (Shklovsky, 1964; Zheleznyakov, 1970). This concept was further explored by Korchak (1971) and Brown (1976), but it did not prosper, mostly because of the long time scales (10 to 1000 seconds) then believed to apply and for which the bremsstrahlung mechanism was sufficient to explain the observed X-rays (Acton, 1964). Another reason was the complete absence of flare diagnostics at sub-mm and IR wavelengths, a situation that still remains today.

The recent observations mentioned above impose severe boundary conditions on the interpretation of the explosive processes occurring in the first phase of bursts in solar plasmas. The presence of ultrarelativistic electrons in this phase is a possibility that again becomes likely. Other mechanisms for energy losses producing X-rays, like inverse-Compton scattering in very dense and compact sources of ultrarelativistic electrons, may again become a plausible possiblity (Kaufmann et al., 1986; McClements and Brown, 1986). According to this concept, at least for certain flares, the primary process would involve multiple and discrete events with ultrarelativistic particles accelerated and slowed down in few tens of milliseconds (not measurable by most experiments). This would result immediately in beams of mildly relativistic electrons, whose behaviour is better described and interpreted. This area of research can be clarified with measurements in the sub-mm and IR bands.

3. High Altitude Ground-Based Submillimeter Solar Telescope (SST)

For a practical low cost investigation, we consider ground based solar observations from a high-altitude site, at frequencies in the sub-mm range, through the atmo-

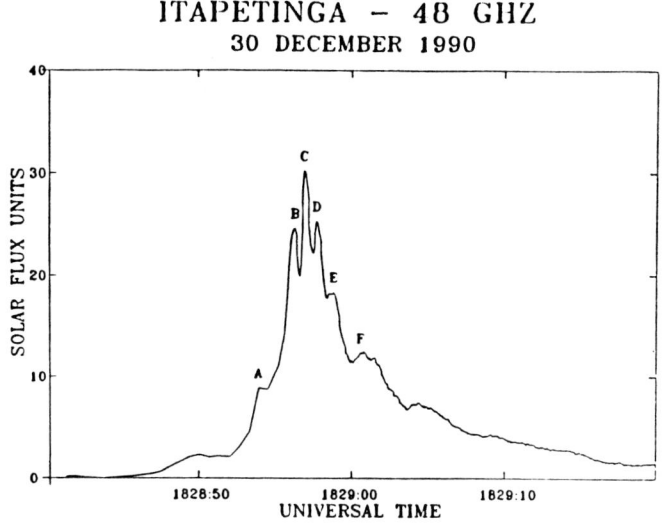
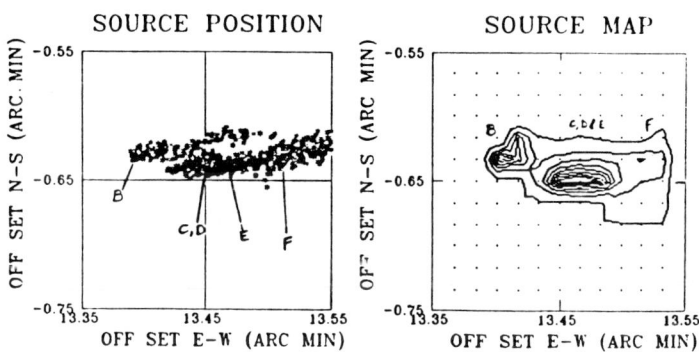

Fig. 4. December 30, 1990 solar burst observed at Itapetinga Radio Observatory at 48 GHz. In the boxes we show the emission center positions at different times marked in the event time profile. This is an example of a well localized event. The position variation stays inside an area smaller than 12 × 12 arcsec (Costa et al., 1991).

spheric transmission windows centered at 210 GHz and 405 GHz. Inspecting the spectra shown in Figure 1, we see that data obtained at those two frequencies will allow unambiguous inferences relative to different models of production of the observed radiation.

We are considering a mountain site, located in the Argentinian Andes at 2500-m altitude, known as El Leoncito near the city of San Juan. It is part of the Complejo Astronomico El Leoncito (CASLEO), and already enjoys an excellent local infrastructure.

Theoretical estimates of atmospheric transmission have been made by the atmospheric physics group of the University of Bern, Switzerland (Kampfer, 1991), using typical meteorological, atmospheric, and precipitable-water-vapor data known for El Leoncito (Filloy and Arnal, 1991; Levato, 1991). The water-vapor content is generally smaller than 4 mm from March to December, being on average less than 2 mm, and very often there are minima of about 0.5 mm. The average increases to 5 mm from January to March. Furthermore, El Leoncito exhibits clear days 80% of the year. Kampfer (1991) estimated a central frequency at 405 GHz for the upper window, as well atmospheric transmissions for various conditions. For the extremes of atmospheric conditions the predicted zenith transmissions at 405 GHz range from 32% to 78%, which indicates that the El Leoncito site presents fairly good atmospheric conditions for extended periods during the year.

For the sub-mm telescope design there are compromises to be followed in order to obtain enough sensitivity (about 1 sfu), with high time resolution (milliseconds), with beamwidths large enough to cover a solar active center (arcminutes) and then allowing dynamic imaging, at least in one frequency. These specifications can be met using a 1.5-m Cassegrain reflector. The antenna effective aperture and telescope performance can be determined using major planets, with sufficient integration time to detect them.

Finally, based on the successful concept of dynamic imaging of bursts observed at 48 GHz, using the Itapetinga 13.7-m antenna (Herrmann, 1990; Costa *et al.*, 1991), we are considering the use of three beams at 210 GHz, overlapping by about (HPBW/2). As a solar burst evolves it is possible to describe the centroid of emission, and its displacements, with a few arcseconds accuracy, and time resolution of milliseconds with such a system.

Acknowledgements

CRAAE is jointly operated by the Universities of Sao Paulo, Campinas and Mackenzie, and the National Institute for Space Research. This program is being partially supported by Sao Paulo State Foundation for Research (FAPESP).

References

Acton, L.W.: 1964, *Nature*, **204**, 64.
Akabane, K., Nakajima, H., Ohki, K., Moriyama, F., Miyaji, T.: 1973, *Solar Phys.* **33**, 431.
Alaart, M.A.F., van Nieuwkoop, J., Slottje, C., and Sondaar, L.H.: 1990, *Solar Phys.*, in press.
Benz, A.: 1991, *IAU Colloquium 133*.
Brown, J.C.: 1976, *Phil. Trans. R. Soc. Lond.*, **281**, 473.
Butz, M., Hirth, W., Furst, E., and Harth, W.: 1976, *Sond. Kleinheubacher Berichte*, **19**, 345.
Cliver, E.W., Kahler, S.W., McIntosh, P.: 1983, *Astrophys. J.* **264**, 699.
Cogdell, J.R.: 1972, *Solar Phys.* **22**, 147.
Correia, E. and Kaufmann, P.: 1987, *Solar Phys.* **111**, 143.
Correia, E., Kaufmann, P., and Magun, A.: 1993, these proceedings.
Costa, J.E.R., Correia, E., Kaufmann, P., Herrmann, R., Magun, A.: 1991, *IAU Colloq. 133*.
Croom, D.L.: 1970, *Solar Phys.* **15**, 414.
Droge, F.: 1977, *Astron. Astrophys.* **57**, 285.
de Jager, C., Kuijpers, J., Correia, E., Kaufmann, P.: 1987, *Solar Phys.* **110**, 317.
Filloy, E.M., Arnal, E.M.: 1991, display at 21st IAU General Assembly, Buenos Aires.
Forrest, D.J., Chupp, E.L.: 1983, *Nature*, **395**, 291.

Georges, C.B., Schaal, R.E., Costa, J.E.R., Kaufmann, P., Magun, A.: 1989, *IEEE* Cat. no. 89TH0260-0, **II**, 447.
Hachenberg, O., and Wallis, G.: 1961, *Zs.f. Astrophys.*, **52**, 42.
Henoux, J.C., Aboudarham, J.: 1991, *IAU Colloq. 133*.
Herrmann, R.: 1990, *Licenciate Thesis, Institute of Applied Physics, University of Bern, Switzerland*.
Kampfer, N.: 1991, private communication. Kaufmann, P., Iacomo Jr., P., Koppe, E.H., Santos, P.M., Schaal, R.E., and Blakey, J.R.: 1975, *Solar Phys.***45**, 189.
Kaufmann, P., Piazza, L.R.; Schaal, R.E., Iacomo, P.: 1978, *Ann. Geophys.*, **34**, 105.
Kaufmann, P., Strauss, F.M., Opher, R., Laporte, C.: 1980, *Astron. Astrophys.***87**, 58.
Kaufmann, P., Correia, E., Costa, J.E.R., Dennis, B.R., Hurford, G.J., Brown, J.C.: 1984, *Solar Phys.***91**, 539.
Kaufmann, P., Correia, E., Costa, J.E.R., Zodi Vaz, A.M., and Dennis, B.R.: 1985, *Nature*, **313**, 380.
Kaufmann, P., Correia, E., Costa, J.E.R., and Zodi Vaz, A.M.: 1986, *Astron. Astrophys.***157**, 11.
Kaufmann, P.: 1988, *Adv.Space Res.*, **48**, (11)39.
Kiplinger, A.L., Dennis, B.R., Emslie, A.G., Frost, K.J., and Orwig, L.E.: 1983, *Astrophys. J.***265**, 199.
Korchak, A.A.: 1971, *Solar Phys.***18**, 284.
Levato, H.: 1991, private communication.
Loran, J.M., Brown, J.C., Correia, E., and Kaufmann, P.: 1985, *Solar Phys.***97**, 363.
Machado, M., Schwartz, R., Fishman, G., Meegan, C., Paciesas, W. and Wilson, R.: 1992, *Solar Phys.*(Submitted).
McClements, K.G., and Brown, J.C.: 1986, *Astron. Astrophys.***165**, 235.
Magun, A., Fuhrer, M., Grater, M., Herrmann, R., Rolli, E., and Marti, A.: *Report No 58, University of Bern, IAP*.
Najita, K., and Orrall, F.Q.: 1970, *Solar Phys.***15**, 176.
Ohki, K., and Hudson, H.S.: 1975, *Solar Phys.***43**, 405.
Qijin, F., Qin, Z., Gao, Z., Li, Z., Huang, G., Jin, S., Cheng, Y., Huo, C., and Wei, S.: 1991, *Proc. Chantilly Flares Workshop*, 77.
Ramaty, R. and Petrosian, V.: 1972, *Astrophys. J.***178**, 241.
Rieger, E.: 1991, *IAU Colloquium 133*.
Shimabukuro, F.I.: 1970, *Solar Phys.*, **15**, 424.
Shimabukuro, F.I.: 1972, *Solar Phys.*, **23**, 19.
Shklovsky, J.: 1964, *Nature*, **202**, 275.
Slottje, C.: 1978, *Nature*, **275**, 520.
Stein, W.A., and Ney, E.P.: 1963, *J. Geophys. Res.*, **68**, 65.
Sturrock, P.A., Kaufmann, P., Moore, R.L., and Smith, D.: 1985, *Solar Phys.*, 94, 341.
Takakura, T., Kaufmann, P., Costa, J.E.R., Dagaonkar, K., Ohki, K., and Nitta, N.: 1983, *Nature*, **302**, 317.
Zirin, H. and Tanaka, K.: 1973, *Solar Phys.*, **32**, 173.
Zheleznyakov, V.V.: 1970, *Radio Emission of the Sun and Planets*, Pergamon Press.

SUBMILLIMETER AND FAR INFRARED EMISSION FROM SOLAR FLARES

VAHÉ PETROSIAN

CSSA, Stanford University, Stanford, CA 94305, U.S.A.

Abstract. The mechanisms for emission in the submillimeter and far-infrared (10^{11} and 10^{13} Hz) regions by solar flares and expected fluxes at these frequencies are described and evaluated. These inferences are based on observations of flare emission at other frequencies and on models for these emissions. In the impulsive phase, non-thermal synchrotron emission by electrons responsible for > 10 MeV gamma-ray emission can give rise to significant radiation in the 10^{11} to 10^{13} Hz region from large flares. Free-free or thermal gyrosynchrotron from the hot plasma responsible for the gradual soft X-ray emission can produce significant radiation in the 10^{11} to 10^{13} Hz range. However, only radiation in the lower end of this range would have a brightness temperature exceeding the quiet sun brightness.

Key words: infrared: stars – Sun: flares –Sun: particle emission – Sun: radio radiation – Sun: X-rays, gamma rays

1. Introduction

Electromagnetic radiation from solar flares has been observed from tens of MHz frequencies to GeV energies with a notable gap in the 10^{11} to 10^{14} Hz range. I will attempt to estimate the expected fluxes in this range concentrating on the lower end of this range, *i.e.* in the submillimeter and far infrared region (submm and FIR, respectively). Although I will not discuss such emissions from active regions, some of the extrapolations used here, especially those in connection with thermal emission, will be applicable to active regions as well.

In Section 2 I will present an overall description of the relevant observations. In Section 3 I will describe the general features of models which can successfully explain these features and indicate what the predictions of such models are for emission in the Submm and FIR range. In section 4 I will describe some recent observations and indicate how they can be extrapolated to predict the fluxes in Submm and FIR range. Finally, I will present a brief summary in section 5.

2. Overview of Flare Emission

Figure 1 shows a schematic spectrum of flare emission from 10^8 to 10^{24} Hz where I plot, as a function of the logarithm of frequency ν, the logarithm of the product of frequency and flux density, $F(\nu)$. The flare emission has been divided into two parts, which are usually referred to as "Impulsive" and "Gradual" phase emissions, typically lasting 1 to 100 seconds, and minutes to hours respectively. Such a division, of course, is not clearly defined and does not apply to all flares. A more clear distinction between these two phases, the one which I shall adopt here, is division by the type of radiation mechanism. By the Impulsive phase I will refer to non-thermal emission by accelerated electrons (accelerated ions will not be important in my discussion) while the Gradual phase will refer to thermal emission by plasma heated or evaporated as a result of energy deposition by accelerated particles or by other processes (such as plasma turbulence or shocks).

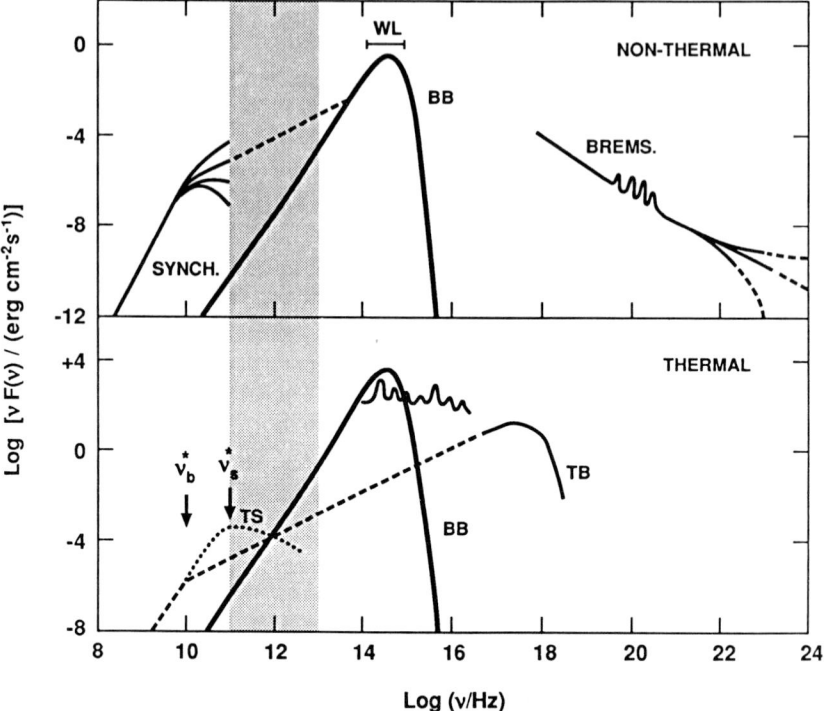

Fig. 1. A schematic presentation of typical strong solar flare spectra over a wide range of frequencies. The upper panel represents the impulsive phase, *i.e.* non-thermal emission, and the lower panel gradual thermal emission. Solid lines are observed, dashed and dotted lines are based on model extrapolations. The heavy solid line represent black body (BB) emission at 5800 K from 1 and 10^4 arcsec² area in the upper and lower panels, respectively. The wiggly portions indicate the presence of nuclear (upper panel) and atomic (lower panel) lines. WL stands for white-light emission.

It should also be noted that the relative and absolute values of fluxes shown in Figure 1 are very rough measures of the actual values. This is because, first, there is no such thing as a typical value of flux at any frequency. Distributions of fluxes obey power law forms extending over many decades whenever such distributions are available. Second, even though there is a strong correlation between emission at different frequencies, this correlation is only statistical in the sense that the distribution of flux ratio at different frequencies generally has a large dispersion. The solid lines in Figure 1 show observations for a relatively large flare. The dashed lines are extrapolations based on models described next.

3. Basic Models

It is generally believed that the source of energy of flares is the current-carrying magnetic field and that this energy is released by some reconnection mechanism.

Observations require that a substantial fraction, if not all, of this energy goes into acceleration of electrons and ions either directly or via generation of plasma turbulence or shocks. If the plasma density is high in the acceleration region, some of this energy can be dissipated quickly and heat the plasma. The model which has become almost standard is that the bulk of accelerated particles are injected into a magnetic loop and that the interaction of these particles with the flare plasma in the loop produces the spectra shown in Figure 1. The interactions are many but the most important ones are interactions with large scale magnetic field geometry, Coulomb collision, synchrotron radiation at high energies (see *e.g.*, McTiernan and Petrosian 1990), and fairly uncertain but probably significant interactions with plasma turbulence (Miller and Ramaty 1989; Hamilton and Petrosian, 1992). For analyses of these processes one requires knowledge of the variation, along the length s of the loop, of the plasma density $n(s)$, the magnetic field $B(s)$, and the turbulence energy density $w(s)$. For analysis of spatially unresolved observations, *i.e.* those integrated over the whole loop, the significant parameters are the column depth, N_{tr}, from the acceleration region (presumably at the top of the loop) to the transition region, the field convergence $d\ln B/ds$ and the ratio of the turbulence energy to the magnetic field energy $8\pi w(s)/B^2$.

The primary radiation processes sometimes are not the same as the primary energy loss mechanisms. We now describe these for the two phases mentioned above.

3.1. Impulsive Phase Radiation

At the heart of the impulsive phase lies the *Hard X-ray Emission* ($\geq$ 20 KeV), which is believed to be due to thick-target bremsstrahlung emission by the accelerated electrons. The observed spectra can be fit to a power law $F(k) \propto k^{-\gamma}$ where $F(k)$ is the flux of photons (photons s^{-1} cm^{-2} keV^{-1}) at photon energy k. This implies an accelerated electron energy spectrum $f(E) \propto E^{-\gamma-1}$. This model for the hard X-ray emission seems to be consistent with most of the other observations, in addition to the spectrum. It can explain temporal variations (Kiplinger *et al.* 1984, Lu and Petrosian 1988), and the few available spatially resolved observations (Brown, Hayward and Spicer 1981; Leach and Petrosian 1983; see also Kundu and Woodgate 1986), and polarization observations (Leach, Emslie, and Petrosian 1985).

The *Gamma-ray emission* ($>$ 300 keV) is also partly due to electron bremsstrahlung and, in the one to seven MeV range, partly due to nuclear line excitation by protons and ions. Emission at $>$10 MeV has been observed for a few flares. This is also believed to be due to electron bremsstrahlung, although pion decay gamma-rays could also contribute at these energies. Observational results that support the bremsstrahlung model are the distribution of fluxes, spectral indices and particularly the variations of these distributions or their moments across the solar disk (see *e.g.*, Rieger *et al.* 1983; Vestrand *et al.* 1987). Such variations can be attributed to the anisotropy of the bremsstrahlung emission, which increases with increasing photon energies (Petrosian 1985). Consequently, the analysis of gamma-ray emission provides strong constraints on the model parameters (McTiernan and Petrosian 1991). An important feature of the gamma-ray observations which has significant bearing on the expected submm and FIR emission is the spectral flattening at higher energies which implies similar flattening of the spectrum of accelerated electrons. This flattening is evident in the aforementioned statistical analysis but also is present in direct observations of the so-called electron-dominated events (Rieger and Marschhäuser 1991), and in the spectrum of electrons seen in the interplanetary medium (Dröger et al. 1989).

The third emission process is that in the *microwave range*, which is believed to be synchrotron emission with emissivity, $j(\nu) \propto \nu^{-\gamma/2}$, from the coronal portion of the loop, in contrast to hard X-ray and gamma-ray emission which are radiated mainly from the footpoints. The very close correlation between the microwave and hard X-ray fluxes is the primary evidence that the accelerated electrons responsible for X-ray and gamma-ray emission produce the microwave radiation as well. There have been many attempts to fit the simultaneous hard X-ray and microwave radiation to the thick target model. The references to, and discussion of, those studies can be found in Lu and Petrosian (1990), who show that such a fit is possible for loops with length $L \simeq 10^9$ cm and with magnetic field $B \simeq 350$ to 600 G.

The energy E (in units of mc^2) of a relativistic electron emitting at frequency ν is

$$E \approx (300 \text{ G}/B)^{1/2} (\nu/10^9 \text{ Hz})^{1/2}. \tag{1}$$

At $B = 500$ G electrons with $E \geq 8$ produce the highest frequency observed microwave radiation ($\sim 10^{11}$ Hz). These electrons produce bremsstrahlung gamma-rays of ≤ 4 MeV. It is, therefore, important to know the photon spectrum at > 4 MeV for prediction of emission at higher frequencies (10^{11} to 10^{13} Hz).

It should be noted that interpretation of the microwave emission is not as straightforward as that of the hard X-rays because microwaves are subject to various absorption processes and the geometry of the magnetic field plays an important role in their emission and absorption. At high frequencies where synchrotron emission in optically thin the spectrum is not well known and can have the different forms shown in Figure 1. These factors clearly will add to the uncertainties in prediction of fluxes in the submm and FIR range.

The *White light* emission (indicated as WL in Figure 1) is another radiative signature of the impulsive phase which, unfortunately, is not well understood. We shall, therefore, not discuss the relevance of this observation to possible submm and FIR emission.

3.2. THERMAL EMISSION

As shown in Figure 1, primary thermal emission ranges from the optical (10^{14} Hz) to the soft X-ray region (≤ 10 keV). Most of this radiation is believed to be coming from a plasma heated and evaporated by the energy deposition of non-thermal electrons and ions. The yield of non-thermal electrons to bremsstrahlung radiation is small ($Y \simeq 10^{-5}$) so that most of the energy is deposited in the plasma via Coulomb collisions and is subsequently radiated during the thermal phase. Thus the total emission during this phase is considerably larger than that during the impulsive phase.

Given the observed thermal emission, we can extrapolate it to the FIR and submm region. The thermal plasma, responsible for the observed emission in the UV to X-ray range, can reach temperatures of 10^8 K, emitting thermal bremsstrahlung (free-free) and bound-bound and bound-free photons. The free-free emission at a few KeV can be extrapolated to the microwave range as shown by the dashed line in Figure 1. However, at these frequencies the competing process of thermal gyrosynchrotron emission could also be significant (dotted line). We shall return to this in the next section.

4. Some Recent Observations

As pointed out above, Submm and FIR emission during the impulsive non-thermal phase most likely will be due to synchrotron radiation by relativistic electrons. These particles also produce gamma rays. Therefore, the most direct information about the Submm–FIR range comes from observations of gamma-rays and from extrapolation of the millimeter observations at $\nu < 10^{11}$ Hz. Recently there have been some new instruments in operation which have extended the impulsive-phase observation into these regions.

4.1. Millimeter Observations

The first indications that there may be significant emission at millimeter (and Submm) wavelengths came from observations of flat microwave spectra in some flares (among which there were some strong ones). Attention was drawn to these kinds of bursts first by Ramaty and Petrosian (1972) and later by Kaufmann et al. (1986). Figure 2 shows some of these spectra. To my knowledge, the latter authors were the first to observe millimetric emission from flares. The flux in one flare observed by them (not shown in Figure 2) continues to rise and reaches its maximum at the highest frequency observed, 90 GHz.

More recently the Maryland group, using BIMA, has observed many flares at 86 GHz during the June 1991 campaign of the Max 91 program (Kundu et al. 1991). An excellent example of this work is presented in Figure 3 where it is shown that during the impulsive phase the time profile of 86 GHz radiation matches exactly that of gamma-rays detected by the BATSE instrument of Compton-GRO. In addition, there is significant 86 GHz emission after the impulsive phase with a slower time evolution similar to that of GOES soft X-ray emission. The post-impulsive radiation is from the heated plasma, the X-rays are from thermal bremsstrahlung emission, while the millimetric radiation comes from either thermal bremsstrahlung or thermal gyrosynchrotron emission. Some of the spectra observed by BIMA and other telescopes are also flat (White et al. 1992).

4.2. Gamma-ray Observations

Observations by the GRS instrument on SMM during cycle 21 were the first to provide information about gamma-ray emission at energies greater than 10 MeV. Flares with such emission show a strong concentration toward the limb of the sun (Rieger et al. 1983). Analysis of GRS observations of cycle 22 flares has revealed the

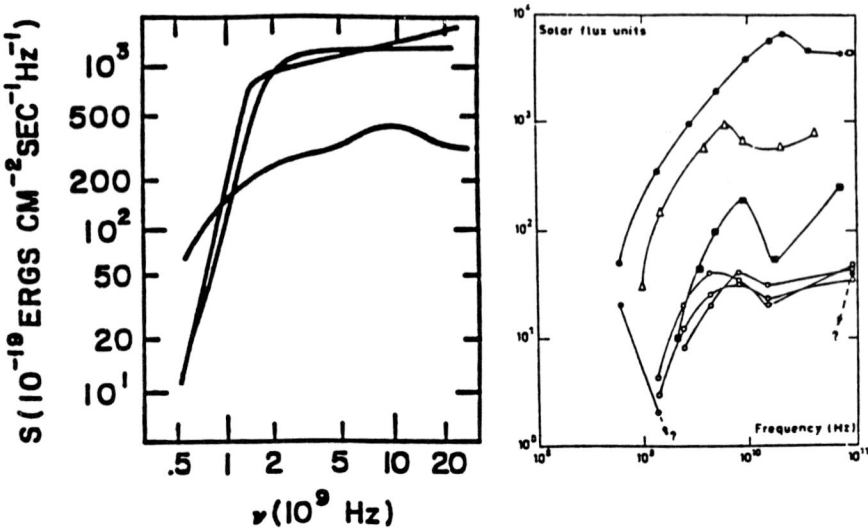

Fig. 2. Some examples of flat microwave spectra of flares. Left and right panels are from compilations by Ramaty and Petrosian (1972), and Kaufmann *et al.* (1986), respectively.

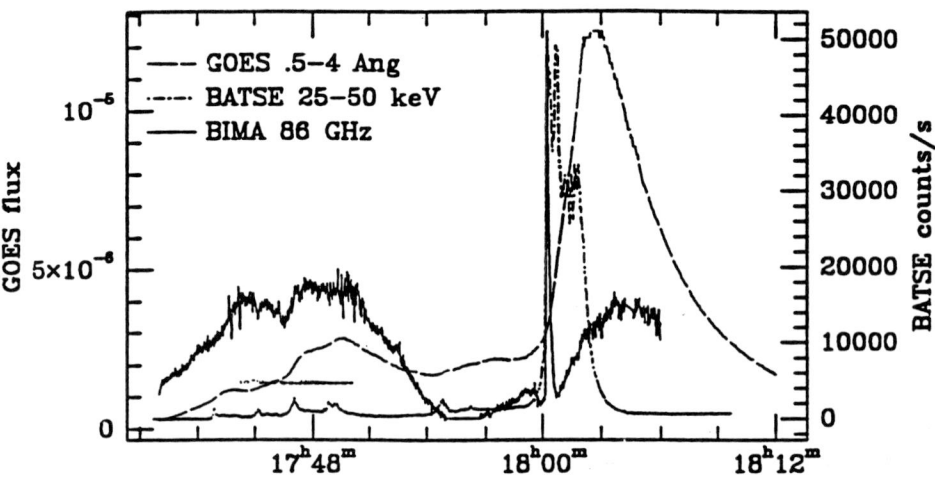

Fig. 3. The evolution of emission around 18 UT on 1991 June 13 seen in soft X-rays recorded by the GOES satellite (8 - 25 keV; long-dashed line), in 25 - 50 keV X-rays detected by BATSE (short-dashed line) and at 86 GHz recorded by BIMA (solid line). The gamma-ray (0.1 to 0.3 MeV) time profile by BATSE, not shown here, has also a single sharp spike coinciding with the first spike of hard X-ray (25 - 50 keV) and with the spike of the BIMA 86 GHz emission. The microwave peak flux is 6×10^{-19} erg/(cm^2 s Hz). [From Kundu *et al.*(1992).]

existence of so-called electron dominated flares, which show no sign of nuclear line emission but a continuous emission up to ~50 MeV with a sharp drop-off beyond it (Rieger and Marschhäuser 1991). These suggest electron bremsstrahlung as a source of the high energy gamma-rays. The sharp cut-off at ≥ 50 MeV cannot be attributed to electron transport (synchrotron losses) or radiative transfer ($e^{\pm}$ pair production by gamma-rays) effects which become more important at higher photon energies. This cutoff must, therefore, reflect a steepening in the distribution of the accelerated electrons with $E > 10^2$ (Petrosian and McTiernan 1991). This indicates that we should expect synchrotron emission up to $\nu \approx 10^{13}(B/300G)Hz$.

The latest gamma-ray observation from the Gamma-1 instrument and from EGRET on Compton-GRO have shown flare emissions up to 1 or 2 GeV. Kocharov et al. 1991) interpret the Gamma-1 observation of the March 20, 1991 flare as due to electron bremsstrahlung and the June 15, 1991 flare as due to a combination of electron bremsstrahlung at low energies and π° decay at higher energies. The EGRET spectrum of the June 11, 1991 flare has spectral index $\gamma = 2.2$ and extends to 2 GeV. If this is an indication of the existence of electrons with energies > 100 MeV, we would expect synchrotron emission up to 10^{14} Hz. We should note that these and some of the flares observed by GRS in Cycle 22, unlike the flares with >10 MeV emission of Cycle 21, are not localized near the limb. This may be an indication of emission by π° decay which is expected to be more isotropic than electron bremsstrahlung. Alternatively, a pure electron bremsstrahlung model is possible if the magnetic field guiding the accelerated has a large non-radial component. In either case, whether the accelerated-electron spectrum is truncated at $\geq$ 100Mev or extends to 1 Gev, we expect synchrotron emission well into the Submm and FIR range.

4.3. ESTIMATED FLUXES

As mentioned in the Introduction, the distribution of the observed fluxes of flares at all energies obeys a power law with no indication of a typical size. The Submm and FIR fluxes will most probably have similar distributions with most flares having fluxes at the instrumental threshold. In view of this, I now estimate the expected fluxes for some of the largest observed flares.

The *Impulsive Phase* fluxes can be estimated by either extrapolating the observed millimeter flares to higher frequencies or from consideration of the relative yield of synchrotron and bremsstrahlung radiation by relativistic electrons. These give similar results. In the first of these methods, the uncertainty lies in the spectral index one assumes in extending the fluxes to $\nu > 10^{11}$Hz. The flat spectrum flares shown in Figure 2 provide the best possibility of observing in the Submm and FIR range. The extrapolation to higher frequencies of these spectra depends on the model for such flat spectra. For a discussion of various possibilities see Ramaty and Petrosian (1973) and White et al. (1992).

Briefly, it seems unlikely that the flat spectra are produced by thermal bremsstrahlung. If the emission at higher frequencies is due to optically thin synchrotron emission, then the spectrum will depend on the spectral index of accelerated electrons. A flat spectrum would require a very hard electron spectrum ($\gamma \approx 0$ in the thick target model), while for most hard X-ray flares and for the gamma-ray flares describe above the power law index $\gamma > 2$. This would indicate a synchrotron spectral index < -1, *i.e.*, a declining spectrum. In the model proposed by Ramaty and Petrosian (1973), where the flat portion of the spectrum is due to free-free absorption of synchrotron emission, the spectrum could remain flat out to higher frequencies. The spectrum could even rise as in the Inverse Compton model proposed by Kaufmann *et al.* (1990). Taking the middle ground, the extrapolation of the spectrum of large flares shown in Figure 2 would indicate an expected Submm-FIR flux as large as 10^3 solar flux units or 10^{-16} erg/(cm^2 s Hz). Such events, of course, would be rare, occurring, on the average, approximately once per month.

The gamma-ray flares observed by Compton-GRO and Gamma-1 mentioned above indicate emission of about 10^{-8} to 10^{-9} erg/(cm^2 s) above 100 MeV. Extending the analysis of Lu and Petrosian (1989) to higher frequencies and electron energies we can estimate the expected flux at $\nu \approx 10^{13}$ Hz. Such photons will be emitted by 100 MeV particles in a magnetic field of $B = 500$G. The yield of synchrotron radiation from relativistic electrons of energy E spiralling with a 45° pitch angle along a length L of the coronal portion of the loop will be $Y_s \sim 10(B/500\text{G})^2(L/2 \times 10^9 \text{cm})E$, while the bremsstrahlung yield at gamma-ray energies when the electron reaches the photosphere will be $Y_b \approx 10^{-5} E \ln E$ (see for example, Petrosian 1985). This and the observed gamma-ray flux would imply an expected synchrotron emission at $\nu \sim 10^{13}$ Hz of about 10^{-16} erg/(cm^2 s Hz), which is similar to the above estimation based on the observed millimetric flux.

Note that the photospheric emission (black body at 6000 K) at 10^{13} Hz is 4.5×10^{-18} erg/(cm^2 s Hz) arcsec^2. Therefore, a source size and angular resolution less than 5 arcsec is needed for the flare emission to exceed the background (see Figure 1).

The *Gradual, Thermal Emission* at Submm wavelengths can be estimated based on the observed soft X-ray fluxes. The uncertainty here lies in the fact that thermal gyrosynchrotron and thermal bremsstrahlung are two possible mechanisms for emission in the radio and IR range. The soft X-ray observations yield values for the electron temperature, T, and the emission measure, EM. At very low frequencies the optical depth will be larger than one. At these frequencies the spectrum will be proportional to ν^2. For $\nu > \nu_b^*$, the frequency where the optical depth is one, the spectrum of thermal bremsstrahlung radiation will be nearly flat $[\nu F(\nu) \propto \nu]$, eventually joining the exponentially decreasing soft X-ray spectrum at 10^{18} Hz (see Figure 1). Gyrosynchrotron emission, when optically thin ($\nu > \nu_s^*$), will deviate from the Rayleigh-Jeans spectrum approximately as $\exp[-(\nu/\nu_{crit})^{1/3}]$ (see Petrosian 1983, and McTiernan and Petrosian 1984). It then follows that if $\nu_s^* \leq \nu_b^*$, gyrosynchrotron radiation will be less intense than bremsstrahlung at all frequencies, while if $\nu_s^* > \nu_b^*$, which is the case depicted in Figure 1, gyrosynchrotron emission will exceed bremsstrahlung for some range of frequencies above ν_b^*. For a more detailed discussion of this matter see Figure 2.22 and the related text in Harding, Petrosian, and Teegarden, 1984.) The exact range of frequencies where this occurs depends on the relative values of ν_b^*, ν_s^*, and ν_{crit}, which are functions of temperature, density, magnetic field, and size of the emitting plasma. Ignoring some slowly-varying logarithmic terms, we may write

$$\left(\frac{\nu_s^*}{\nu_b^*}\right) \approx 0.4 \left(\frac{B}{500 \text{ G}}\right) \left(\frac{T}{10^8 \text{ K}}\right)^{15/8} \left(\frac{10^{10} \text{ cm}^{-3}}{n}\right) \left(\frac{10^9 \text{ cm}}{L}\right)^{1/2}. \quad (2)$$

so that for large flares with emission measure $EM \approx n^2 L^3$ exceeding 10^{51} cm^{-3} this ratio will be less than one and gyrosynchrotron radiation will be unimportant. In that case the surface brightness temperature will be given by $T_B = T(\nu/\nu_b^*)^{-2}$ which will be greater than the background brightness temperature if $T \geq 3 \times 10^7$ and $\nu < 70 \nu_b^*$. For $\nu_b^* = 10^{10}$ Hz the gradual phase thermal emission would be detectable up to 10^{12} Hz, just barely into the submillimeter range.

Thus, we conclude that even though one can be optimistic about detecting impulsive phase emission from large flares into the Submm and FIR range, it appears difficult to detect thermal emission from the weaker and more common flares unless the plasma temperature and magnetic field are very high ($B > 10^3$ G, $T \geq 10^8$ K).

5. Discussion and Summary

I have described the relationship between the Submm and FIR emission and that at other photon energy ranges. The impulsive phase emission most likely will be dominated by synchrotron radiation of electrons accelerated to extreme relativistic energies. Extrapolation to higher frequencies of observed millimetric fluxes, a thick-target model calculation, and observed gamma-ray fluxes predict 10^{-16} erg/(cm^2 s Hz) of radiation in the 10^{13} Hz range from the largest flares. The flare brightness will then exceed the solar black-body brightness only in compact flares (angular simeq$\theta < 5''$ or linear size $L < 3500$ km). The thermal emission in the Submm-FIR range will most likely be dominated by free-free (*i.e.* thermal bremsstrahlung) emission which for an electron temperature of 3×10^7 K will have a brightness exceeding the 6000 K black body emission only below 10^{12} Hz.

One may wonder whether such observations of flares are needed when we already can predict the fluxes at these frequencies. In addition to the fact that we always must test theoretical predictions by observations, as is generally the case with new observations, some surprises may be in store for us. For example, the impulsive-phase emission at Submm-FIR may be comparable to the much larger white-light emission rather than millimetric radio emission. Or some other yet unknown mechanism may produce strong radiation in this range during the impulsive or gradual phases.

Even ignoring these possibilities, higher spatial resolution is possible in the Submm-FIR range as compared to very high energy gamma-rays which would be helpful in clarifying the acceleration and radiation site for relativistic electrons. Furthermore, higher-spectral-resolution observations and possible polarization measurements in the Submm-FIR range as compared to the X- and gamma-ray ranges can further clarify the emission mechanisms and set more rigid constraints on model parameters.

Acknowledgements

This work was partially supported by NASA NAGW 1976 and NSF ATM 90-11628.

References

Brown, J.C., Hayward, J. and Spicer, D.S.: 1981, *Astrophys. J. (Letters)* **245**, L91.
Dennis, B.R. and Zarro, D.M.: 1992, in press.
Dröge, W., Meyer, P., Evanson, P. and Moses, D.:1989, *Solar Phys.* **121**, 95.
Hamilton, R.J. and Petrosian, V.: 1992, *Astrophys. J.*, in press.
Harding, A.K., Petrosian, V. and Teegarden, B.J.: 1986, in E. Liang and V. Petrosian (eds.), *Gamma-ray Bursts*, AIP Conference Proceedings 141 (Stanford, CA), p. 115.
Kaufmann, P., Correia, E., Costa, J.E.R. and Zodi Vaz, A.M.: 1986, *Astron. Astrophys.* **157**, 11.
Kiplinger, A.L., Dennis, B.R., Frost, K.J. and Orwig, L.E.: 1984, *Astrophys. J. (Letters)* **287**, L105.
Kocharov, G.E., Kocharov, L.G., Kovaltsov, G.A. and Guelenko, V.G.: 1991, preprint. To be published in *Nuclear Astrophysics*.
Kundu, M.R. and Woodgate, B.: 1986, *Energetic Phenomena in the Sun*, NASA Conference Publ. 2439, chapter 3.
Kundu, M.R., White, S.M., Gopalswamy, N. and Lim, J.: 1992, preprint; see also *Bull. Amer. Astron. Soc.* **24**, 783.
Leach, J. and Petrosian, V.: 1983 *Astrophys. J.* **269**, 715.
Leach, J., Emslie, A.G. and Petrosian, V.: 1985, *Solar Phys.* **96**, 331.
Lee, T. and Petrosian, V.: 1992, *Bull. Amer. Astron. Soc.* **24**, 794.
Lu, E.T. and Petrosian, V.: 1988, *Astrophys. J.* **327**, 405.
Lu, E.T. and Petrosian, V.: 1989, *Astrophys. J.* **338**, 1122.
Miller, J.A. and Ramaty, R.: 1989, *Astrophys. J.* **344**, 973.
McTiernan, J.M. and Petrosian, V.: 1990, *Astrophys. J.* **359**, 524.
McTiernan, J.M. and Petrosian, V.: 1991, *Astrophys. J.* **379**, 381.
Neupert, W.M.: 1968, *Astrophys. J. (Letters)* **153**, L59.
Petrosian, V.: 1973, *Astrophys. J.* **186**, 291.
Petrosian, V.: 1981, *Astrophys. J.* **251**, 727.
Petrosian, V.: 1985, *Astrophys. J.* **299**, 987.
Petrosian, V. and McTiernan, J.M.: 1983, *Physics of Fluids*, **26**, 3023.
Petrosian, V. and McTiernan, J.M.: 1991, *Bull. Amer. Astron. Soc.* **23**, 1044.
Ramaty, R. and Petrosian, V. : 1972,*Astrophys. J.* **178**, 241.
Rieger E., Reppin, C., Kanbach, G., Forrest, D.J., Chupp, E.L., and Share, G.H.: 1983, Proc. 18th Internat. Cosmic Ray Conf., **4**, 79.
Rieger, E., Marschhäuser, H.: 1991, in R.M. Winglee and A.L. Kiplinger (eds.), *MAX 91/SMM Solar Flares, Observations and Theory*, Proc. MAX91 Workshop No. 3 (Boulder, CO), p. 68.
Vestrand, W.T., Forrest, D.J., Chupp, E.L., Rieger, E., and Share, G.H.: 1987, *Astrophys. J.* **322**, 1010.
White, S.M., Kundu, M.R., Bastian, T.S., Gary, D.E., Hurford, G.J., Kucera, T. and Bieging, J.H.: 1992, *Astrophys. J.* **384**, 656.

INFRARED AND SUBMILLIMETER DIAGNOSTICS OF ACTIVITY AND FLARES

A. FALCHI

Osservatorio Astrofisico di Arcetri, Largo Fermi 5, I-50125 Florence, Italy

R. FALCIANI

Dipartimento di Astronomia, Universitá di Firenze, Largo Fermi 5, I-50125 Florence, Italy

and

P. MAUAS

Osservatorio Astrofisico di Arcetri, Largo Fermi 5, I-50125 Florence, Italy

Abstract. We give a critical review of the observations of solar activity in the IR and sub-mm range, which are quite scarce, except for the Fe I triplet at 1.56 μm and the Mg I emission lines at 12.32 μm. These, however, are mainly intended for solar magnetic field studies rather than physical diagnostics on activity phenomena. We compute the emission in some continuuum windows and in some detectable Paschen and Brackett lines in two extreme flare models, viz. a "chromospheric" and a white-light flare model. The utility of the Paschen and Brackett lines as diagnostics of the atmospheric state is questionable since more information can be obtained more easily by observing the higher Balmer lines. On the other hand, observations in various continuum windows can be of high scientific value and efficiency. We also discuss possible coordination with simultaneous visible observations, in order to increase the diagnostic efficiency of a prospective observing run.

Key words: infrared: stars – Sun: activity – Sun: flares

1. Introduction

This paper is organized in 3 different sections: Section 2 contains a synoptic review of papers dealing with IR and sub-mm observations in the field of solar activity; Section 3 discusses the computed emission of flares in various continuum windows and in some detectable Paschen and Brackett lines, for different flare models, and Section 4 presents a possible set of IR and sub-mm observables which, as part of a wider observing program, are suitable for solar activity and flare studies.

2. Review of Existing Observations

No particular attention will be given in the present review to the results obtained for the He I 1.083 μm line, for the Fe I triplet at 1.565 μm and for the Mg I 12.32 μm emission lines. These spectral features have been extensively analyzed in particular sessions of this symposium and the details can be found in other papers of these proceedings. In the following list of works, by no means exhaustive, particular attention was given to the most recent results.

2.1. Sunspots

10 μm (continuum) Turon and Lena (1970) performed at the Kitt Peak National Observatory (KPNO) the first IR observations of sunspots using pinhole photometry.

- **1.215; 1.670; 1.54; 1.733; 2.086; 2.34 μm (continuum)** Albregtsen and Maltby (1981) obtained with refined pinhole-photometry techniques a very important set of data used to build up the umbral "Maltby model" (Maltby et al., 1986). Albregtsen et al. (1984) studied the umbral limb-darkening for 22 large sunspots and found a decrease in the umbral/photospheric intensity ratio towards the limb. They confirmed that sunspot intensity varies more or less linearly with the phase of the solar cycle.

- **1.711 - Mg I, 1.672 - Al I; 1.199, 1.609, 1.538, 2.182 μm - Si I.** Van Ballegooijen (1984) used the KPNO FTS to measure these lines in a sunspot umbra and he derived models of the umbra. The high-excitation lines in this set require a hotter model than the low-excitation lines and the continuum and $\Delta T \approx$ 460 K at $\tau_{5000} = 3$ between hot and cool components.

- **850 μm (continuum)** At the J.C. Maxwell Telescope (JCMT) in Hawaii, with $\lambda/\Delta\lambda \sim 10^3$, Lindsey et al. (1990) found that spots are slightly darker than the quiet Sun, and surrounded by a brighter plage area.

- **12.32 μm - Mg I.** These lines were not seen in umbrae but were detected as fully Zeeman split in penumbrae by Brault and Noyes (1983). They measured $|B| \sim$ 900 − 1200 G in the penumbra. No Evershed outflow was found by Deming et al. (1988), who only found Evershed inflow at the limb.

- **1.5648 μm - Fe I (g = 3.0) and 1.5652 μm - Fe I (g = 1.53).** Livingston (1991) found that molecular blends strongly limit the utility of these lines for sunspot studies.

2.2. Faculae

- **10.5; 18.5; 25.0 μm (continuum).** With pinhole scans (10-20″) at the Mauna Kea 2.24 m telescope Lindsey and Heasley (1981) found a contrast $\Delta I/I \leq$ 0.02 (much smaller than predicted!). The estimated filling factor was found to be $f \leq$ 0.1 (smaller than derived from the Mg-II lines data).

- **1.627 μm (continuum)** Foukal et al. (1989, 1990) found a negative contrast using the 58 × 62 InSb–KPNO array with a spectral resolution $\Delta\lambda = 40$ Å.

2.3. Supergranular structure

- **1.17 μm; 1.64 μm (continuum); 1.7108 μm - Mg I.** Worden (1975) measured a ΔT ranging from 50 K at 1.64 μm (formed at the deepest observable layer) to 500 K at 1.7108 μm (formed at the low chromosphere).

- **850 μm (continuum).** Lindsey et al. (1990) and Lindsey and Jefferies (1991) at the JCMT, with resolution $\lambda/\Delta\lambda \sim 10^3$, found that the network is approximately correlated with Ca II structures. Assuming a mean value for the brightness temperature T_b of 5400 K, they found for the supergranular network 5050 K $< T_b <$ 5750 K.

2.4. PROMINENCES

10 - 20 μm. For high hydrogen lines (observed with the KPNO FTS), Zirker (1985) found results consistent with known scenarios. No Stark wings were measured on these IR lines.

2.5. EMERGING FLUX REGIONS

12.32 μm - Mg I. Zirin and Popp (1990) measured a line-center contrast $I/I_{\rm quiet} \sim$ 1.13, implying $|\mathbf{B}| \sim 700$ G, and a total line contrast $[\int I\, d\lambda]/[\int I\, d\lambda]_{\rm quiet} \leq 3$.

2.6. PLAGES

0.4 - 0.8 - 1.2 mm. With a pinhole bolometer, 3' resolution, Beckman and Clark (1973) found evidence that T_b rises with height.

850 μm (continuum). At the JCMT Lindsey et al. (1990) determined $\Delta T_b \sim$ 1100 K.

12.32 μm - Mg I. Brault and Noyes (1983) for the first time resolved the Zeeman splitting in plages.
Deming et al. (1988) measured a mean value for $|\mathbf{B}|$ of 400 G.
Zirin and Popp (1990) found a line-center contrast $I/I_{\rm quiet} \sim$ 1.06–1.12. For an old plage the total line contrast $[\int I\, d\lambda]_{\rm plage}/[\int I\, d\lambda]_{\rm quiet} = 1$ and $|\mathbf{B}| \sim$ 200 to 400 G; for a young plage they found a contrast of 3.0 and $300 < |\mathbf{B}| < 800$ G.

1.5648 and 1.5652 μm - Fe I. Rabin et al. (1991) determined $|\mathbf{B}| \sim 800$–1600 G using the new two-dimensional near-IR magnetometer equipped with the KPNO 58 × 62 InSb array with 2″ resolution.
Livingston (1991) found $|\mathbf{B}| \sim 1000$–2000 G using the main McMath with 4″ resolution.

2.7. FLARES

4.1 mm (73 GHz). Akabane et al. (1973) determined that the time profile of the H_α and mm-bursts approximately coincide; this behaviour was not found for the cm bursts.

3.3 mm (90 GHz). Shimabukuro (1970, 1972) found that gradual and impulsive bursts were well correlated (in shape and time) with soft X-ray (SXR) bursts (2-12 Å). They could not explain this correlation with the values of emission measure and electron temperature derived from SXR bursts.
Kaufmann et al. (1986) and Costa and Kaufmann (1986) showed that very fast pulses (~ 60 ms) at mm wavelengths were well correlated with hard X-rays (HXR) bursts. A spectral flattening was found at mm wavelengths. Their interpretation was that the mm-wavelength emission was due to synchrotron radiation and the HXR emission to inverse Compton scattering. This interpretation was rejected by McClements and Brown (1986), who suggested a compact source (~ 350 km) with $|\mathbf{B}| \sim 1.4$ to 2.0 kG, and $N_e \sim 10^{11}$ cm^{-3}

and $T_e \sim 5 \times 10^8$ K. Their interpretation was that the mm-wavelength emission was due to gyrosynchrotron radiation and the HXR emission to bremsstrahlung. The need for higher spatial and temporal resolution at mm wavelengths was stressed.

3.5 mm (86 GHz). Kundu et al. (1992) reported interesting observations obtained with the BIMA interferometer (1-2" resolution) during the 1991 Solar Maximum campaign. They found two phases in mm-burst emission: a) a nonthermal impulsive phase correlated with HXR (25-100 kev) bursts, which is believed to prove the existence of a MeV electron beam (but sometimes the mm-spike has a 5-10 s delay with respect to the HXR spike); b) a thermal gradual phase correlated with the SXR behaviour. Other possible scenarios are also presented. Evidence for pre-flare heating at mm wavelengths is found.

20 and 350 μm (continuum). The first pioneering attempts to detect IR emission from flares were due to Hudson (1975) at the 152 cm IR telescope on Mt. Lemmon. Negative results were obtained, probably due to the lack of strong activity and to instrumental limitations including pinhole rastering.

12.32 μm - Mg I. At the KPNO–FTS, Deming et al. (1990) performed the first observations of a flare, with spatial and temporal resolutions of $\sim 5''$ and 2 minutes respectively. The 2 ribbon flare 3B/X5.7 of October 29, 1989 was observed 25 minutes after the HXR peak but simultaneously with the SXR maximum emission. The continuum intensity contrast $\Delta I/I_{quiet} = 0.07\pm 0.03$, was explained with a temperature enhancement $\Delta T \sim 300$ K at $h \sim 400$ km corresponding to $\tau_{5000} \sim 10^{-2}$. They found that $|\mathbf{B}|$(penumbra) $\sim$ 2.1 kG and $|\mathbf{B}|$(umbra) $\sim |\mathbf{B}|$(penumbra) $- 0.2$ kG. No spatial or temporal variations in $|\mathbf{B}|$ were detected. Variations of $\gamma = \arctan(\mathbf{B}_{\parallel}/\mathbf{B})$ probably reflect relative intensity variations of the σ/π Zeeman components, possibly related to magnetic reconnection processes, viz. to the formation of post-flare loops.

We would like to point out the importance, and the significance for the understanding of flares, of the results obtained by Deming et al. (1990) and by Kundu et al. (1992), which were made possible by new high-quality instrumentation. The latter paper, particularly, has been discussed during this meeting and we refer to that presentation for further details.

We would like to comment on these results:
- The diagnostic capabilities of IR and sub-mm observations for solar activity and flare studies have been almost totally neglected, mainly due to the lack of suitable, *dedicated* panoramic instruments.
- The observations of active phenomena in the 1.56 μm Fe I and 12.32 μm Mg I lines are only a by-product of observations intended to study the magnetic fields.
- IR observations can now be used effectively to study active phenomena, since they have attained reached the quality of observations in the visible (2-D arrays, etc.).

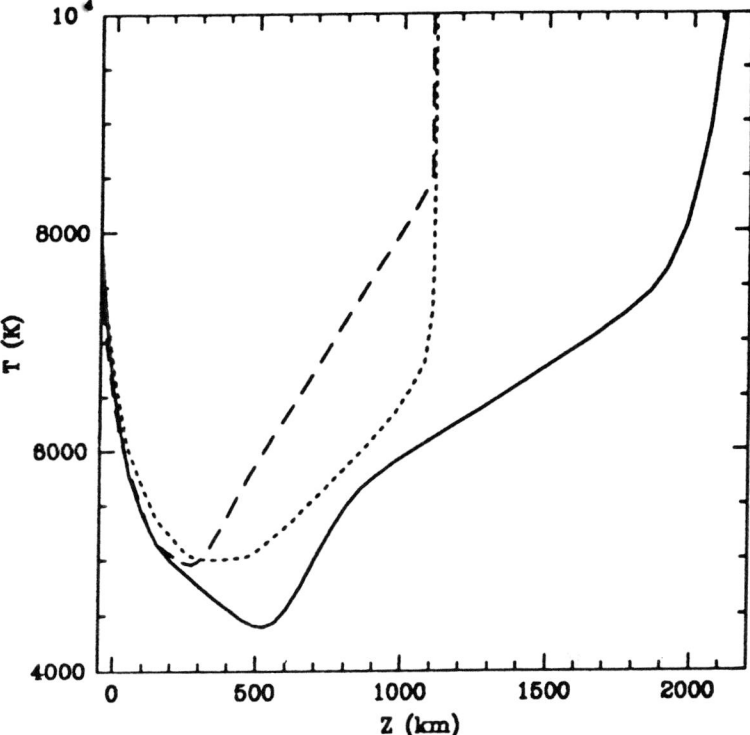

Fig. 1. Electron temperature as a function of height for several atmospheric models. *Full line:* Quiet Sun (Avrett, 1985); *dashed line:* Intense "chromospheric" flare model FL2 (Machado et al., 1980); *dotted line:* White-light flare model FLA (Mauas et al., 1990).

3. Diagnostic Capability of Some IR and Sub-mm Spectral Features

To see which IR and sub-mm observations are better suited to study activity phenomena, we evaluate the sensitivity of the different observables to changes in atmospheric models. We concentrate our attention on flare studies, since their contrast is larger (and hence more easily detectable) than for other active phenomena, even if their unpredictability may represent a real observing complication. Moreover, a wide variety of sufficiently reliable flare models (both semi-empirical and theoretical) is available.

We note the following continuum windows:

- 1.17 μm because optical depth unity at that wavelength corresponds to optical depth unity at 0.5 μm;

- 1.627 μm, the minimum of the continuum absorption coefficient;

- 2.2 μm. The "classical" continuum IR window. At this wavelength a "negative" flare was observed in the case of the flare star AD Leo (Byrne, 1989 and references therein). The absorption dip was measured concurrently with the flare emission in the visible and radio, and so far no reasonable explanation has been offered. We believe that this kind of observation should be repeated for other flare stars, in particular the Sun.

- The 3.8, 5.0 and 10.0 μm regions which offer clean continuum windows observable with ground-based optical techniques;
- The 30, 50, 100, and 200 μm regions, not reachable from the ground but important potentially for indicating the position of the flare transition region (TR) and corona;
- The 350, 850, 1000, and 3000 μm continuum windows, which can be reached from the ground and with a spatial resolution of several minutes of arc.

We considered 4 detectable hydrogen lines in the near IR (where panoramic detectors and narrow–band filters or spectrometers may be available): P_β at 1.2818 μm, P_γ at 1.0938 μm, Br_α at 4.0512 μm and Br_γ at 2.1655 μm. The computations have been performed with the NLTE code PANDORA (Vernazza, Avrett and Loeser 1973, 1981). All canonical mechanisms for the continuum opacity have been considered (including H^-, H_2^+, and Rayleigh and Thompson scattering). We also included the opacity due to the 1.7×10^7 weak atomic and molecular lines compiled by Kurucz (1985), as explained by Avrett, Machado, and Kurucz (1986).

For the present exploratory work we considered three atmospheric models:

- the modified VAL-C model (Avrett, 1985) for the quiet Sun;
- an intense flare model FL2 (Machado *et al.*, 1980) capable of reproducing many chromospheric and transition region features;
- a white-light flare model FLA (Mauas *et al.*, 1990) with smaller chromospheric temperatures but a hotter photosphere.

The height distribution of the electron temperature for the three models is given in Figure 1.

3.1. The Sensitivity of the IR Continuum Windows to Flare Models

The computed intensities in the different continuum windows for the three models are listed in Table I. The physical significance of the results of our numerical simulations, quoted in Table I, can be summarized as follows:

- The 1 to 5 μm range is formed approximately around optical depth unity at 0.5 μm,
- No negative-flare signature around 2.2 μm is noted (this fact adds weoght to arguments for repeating the IR continuum observations of stellar flares),
- Continuum observations in the 1 to 10 μm range can give valuable information on the physical processes occurring in a white-light flare (WLF). Since, the only opacity source is H_{ff}^- (the contribution of the Brackett continuum is negligible, $\sim$ 2 to 8%), the change in the observed emission during the WLF should originate in the photosphere.
- 350 and 850 μm continuum radiation originates at the low chromosphere in the quiet Sun, and at the transition region in a flare. Thus a very high contrast (easy to measure) would be observed at these wavelengths during a flare.

TABLE I

IR continuum window sensitivity to flare models. QS is the quiet sun model (Avrett, 1985), FL2 is the intense chromospheric flare model (Machado et al., 1980), FLA is the white-light flare model (Mauas et al., 1990); h and T are the height and the temperature of the layer where most of the emergent radiation originates. $C = I/I_{QS}$ is the computed emission contrast.

	QS		FL 2			FLA		
$\lambda(\mu m)$	h	T	h	T	C	h	T	C
1.17	0	6250	0	6420	1.1	0	6700	1.1
1.63	-20	6980	-25	6910	1.0	-25	7200	1.1
2.2	0	6520	0	6420	1.0	0	6700	1.1
3.8	50	5790	50	5840	1.0	50	6010	1.1
5.	50	5790	50	5840	1.1	50	6010	1.1
10.	150	5150	150	5140	1.2	150	5370	1.1
30.	300	4770	1025	8120	1.6	275	5030	1.1
50.	350	4660	1075	8360	1.9	1110	9840	1.1
100.	400	4560	1098	8487	2.1	1110	9840	1.3
200.	490	4410	1100	8500	2.8	1110	9840	1.7
350.	560	4430	1102	22000	3.9	1110	9840	2.1
850.	980	5900	1103	35000	5.9	1112	17300	3.1
1000.	980	5900	1103	35000	6.5	1112	17300	3.4
3000.	1580	6900	1105	150000	13.6	1114	43400	6.3

3.2. THE SENSITIVITY OF SOME HYDROGEN LINES TO FLARE MODELS

In Figure 2 we show the computed profiles of P_β, P_γ, Br_α and Br_γ for the atmospheric models plotted in Figure 1. In Table II we list the computed contrasts, with respect to the quiet Sun, of IR and Balmer lines for both flare models at different spectral resolutions.

We see that even with an infinite resolution, the contrast at the line center for the IR lines is smaller than for the Balmer lines. Note that the broad-band line contrast, corresponding to the flare model FLA, is mainly due to the continuum emission enhancement caused by the temperature increase in photospheric layers, rather than to real line emission. Furthermore, since all of these lines are not formed in local thermodynamic equilibrium, the Paschen and Brackett lines (corresponding to transitions between higher levels) are harder to interpret than the Balmer lines. If we also consider the experimental simplicity and the huge variety of observing tools available for the Balmer lines (in the visible range) in comparison with the IR lines, we conclude that, at present, the IR hydrogen lines do not help for a better understanding of solar activity.

4. Conclusions

The key ingredient for a better understanding of the basic mechanisms determining solar activity is the careful measurement of the velocity and the magnetic fields

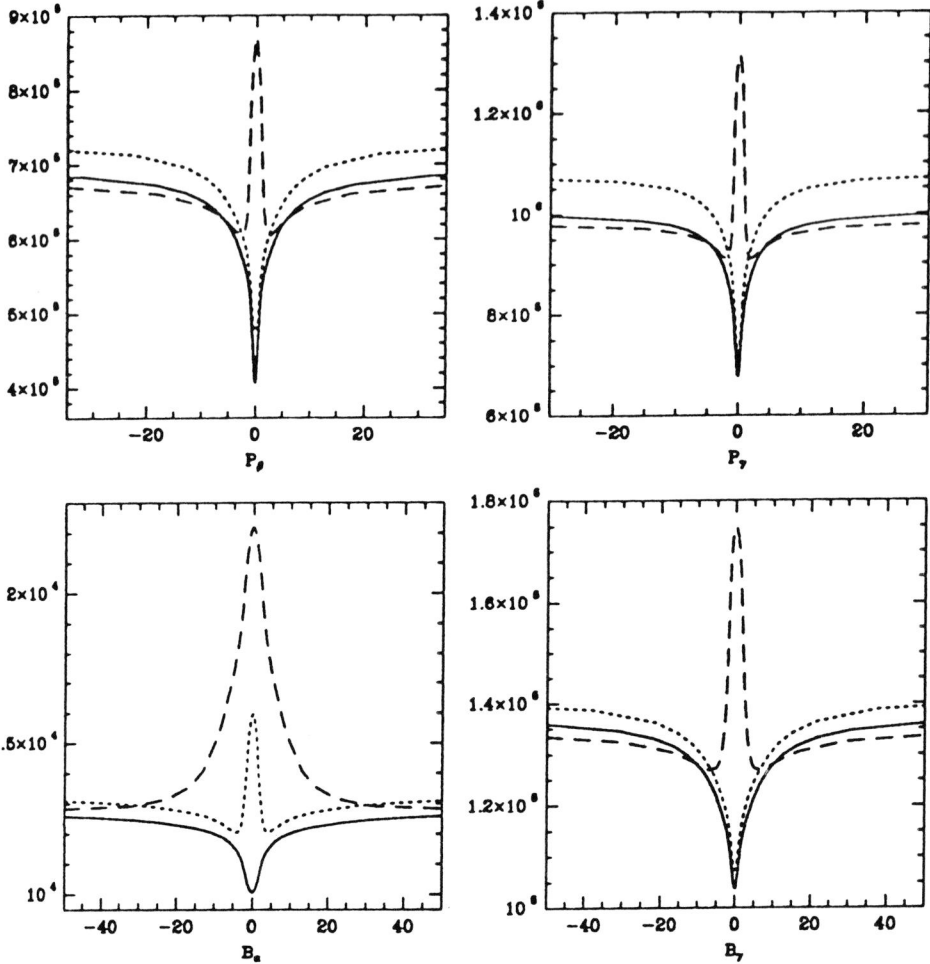

Fig. 2. Computed profiles of several hydrogen lines for different atmospheric models. *Full line:* Quiet Sun (Avrett, 1985); *dashed line:* Intense "chromospheric" flare model FL2 (Machado et al., 1980); *dotted line:* White-light flare model FLA (Mauas et al., 1990).

and their interactions at high spatial and temporal resolution, at different heights in the atmosphere (*e.g.* Falciani, 1988; Falchi et al., 1991). Moreover, an efficient and well-planned set of photospheric and chromospheric observations has the goal of determining the spectral signatures correponding to the different physical processes (particle acceleration and subsequent bombardment, EUV and X-ray irradiation, thermal conduction, *etc.*). For example, very recent results (Falchi et al., 1992; Cauzzi et al., 1992;) indicate the presence of a strong velocity field on some metallic lines one minute before the start of an HXR burst. In a very small area ($\approx 3''$) a blue shift of the order of 15 km s^{-1} lasts for one minute. The rapid evolution of this blue-shifted Doppler component in a small spatial patch offers a new, evolving,

TABLE II

Hydrogen line sensitivity to flare models. FL2 is the intense chromospheric flare model (Machado et al., 1980), FLA is the white-light flare model (Mauas et al., 1990); C_0 is the contrast I/I_{QS} computed in the line center and C_{20}, C_{40} are the contrasts integrated over ±10 and ±20 Å from line center.

	FL2			FLA		
	C_0	C_{20}	C_{40}	C_0	C_{20}	C_{40}
P_β	2.1	1.1	1.0	1.2	1.1	1.1
P_γ	2.4	1.1	1.0	1.0	1.1	1.1
Br_α	2.2	1.6	1.3	1.6	1.1	1.1
Br_γ	1.7	1.1	1.1	1.1	1.0	1.0
H_α	8.9	-	-	7.2	-	-
H_β	9.6	-	-	6.5	-	-
H_γ	7.0	-	-	3.8	-	-
H_δ	5.4	-	-	2.4	-	-

scenario for the development of the chromospheric flare.

We would like to conclude by proposing a coordinated observing program in which IR observations may represent an important diagnostic tool. We characterize this program by different spectral ranges to help to define possible future collaborations between different research groups.

VISIBLE RANGE

One- and two-dimensional spectroscopy in several spectral lines would allow us to estimate, not only the thermal structure of the atmosphere, but also the velocity and magnetic fields at different heights. Resolutions of $\Delta t \sim 2$ to 20 s and $\lambda/\Delta\lambda \sim 10^5$ would be required.

IR LINES

Observations of the 1.56 μm - Fe I and 12.32 μm - Mg I lines would allow us to measure magnetic fields with greater precision than can be obtained in the visible lines. The 1.083 μm line is a very important diagnostic of the excitation of He I.

IR CONTINUUM

Observations at 1.17, 1.627, 2.2, 3.8 and 10 μm may allow us to estimate the maximum penetration of the flare and to distinguish between photospheric and purely chromospheric white-light flares.

Observations in the first three continuum windows may represent optimum use of the new NSO/KPNO 128×128 IR array, which may reach resolutions of the order of $\Delta x, \Delta y \sim 1''$, $\Delta t \sim 2$ to 5 s, $\lambda/\Delta\lambda \sim 10^3$.

Observations in the 350 and 850 μm continuum windows can be used to monitor low intensity events, due to the high contrast expected at these wavelenghts. With the presently-available observational resources, resolutions of the order of Δx, $\Delta y \sim 6''$, $\lambda/\Delta\lambda \sim 10^3$, $\Delta t \sim 0.5$ s per pixel (in raster mode) can soon be reached.

MILLIMETER WAVELENGTH RANGE

The recent results obtained with instruments with high spatial resolution (*e.g.*, JCMT, BIMA, Caltech IR facility) are the best and most convincing proofs that observations at this spectral range can supply evidence of the propagation of very high energy electrons.

MICROWAVE RANGE

High-time-resolution (≤ 1 s) dynamic spectral observations can supply valuable information on the non-thermal processes acting in the observed region

It is quite obvious that, to complete this ideal data set, space-born information on SXR, HXR, γ-rays and in-situ particle detection should be available.

We believe that at this stage of our understanding of solar activity, very little can be learned without a coordinated set of observations in the different spectral ranges. With this objective in mind, it would be necessary to solve a large set of problems, for example, coordination of the pointing and orientation of the field of view of the different instruments involved.

Acknowledgements

We greatly acknowledge the possibility to use the computer program Pandora, kindly supplied by Dr. E. Avrett.

References

Albregtsen, F., and Maltby, P.: 1981, *Solar Phys.* **71**, 269.
Albregtsen, F., Joras, P.B., and Maltby, P.: 1984, *Solar Phys.* **90**, 17.
Akabane, K., Nakajima, H., Ohki, K., Moriyama, F., and Miyaji, T.: 1973, *Solar Phys.* **33**, 431.
Avrett, E.H.: 1985, in B.W. Lites (ed.), *Chromospheric diagnostics and modelling*, NSO Summer Workshop, p. 67.
Avrett, E.H., Machado, M.E., and Kurucz, R.L.: 1986, in D.F. Neidig (ed.), *The lower atmosphere of solar flares*, NSO Summer Workshop, p. 216.
Beckman, J.E., and Clark, C.D.: 1973, *Solar Phys.* **29**, 25.
Brault, J., and Noyes,R.: 1983, *Astrophys. J. (Letters)* **269**, L61.
Byrne, P.B.: 1989, in *Solar and Stellar Flares*, IAU Colloq. 104, *Solar Phys.* **121**, 61.
Cauzzi, G., Falchi, A., Falciani, R., and Smaldone, L.A.: 1992, in ESA Workshop *Solar Physics and Astrophysics at Interferometric Resolution*, ESA SP-344 (in press).
Costa, J.E.R., and Kaufmann, P.: 1986, *Solar Phys.* **104**, 253.
Deming, D., Boyle, R.J., Jennings, D.E., and Wiedemann, G.: 1988, *Astrophys. J.* **333**, 978.
Deming, D., Hewagama, T., Jennings, D.E., Osherovich, V., Wiedemann, G., and Zirin, H.: 1990, *Astrophys. J. (Letters)* **364**, L49.
Falciani, R.: 1988, *Adv. Space Res.* **8**, 11.
Falchi, A., Falciani, R., and Smaldone, L.A.: 1991, *Adv. Space Res.* **11**, 85
Falchi, A., Falciani, R., and Smaldone, L.A.: 1992, *Astron. Astrophys.* **256**, 255.
Foukal, P., Little, R., and Mooney, J.: 1989, *Astrophys. J. (Letters)* **336**, L33.
Foukal, P., Little, R., Graves, J., Rabin, D., and Lynch, D.: 1990, *Astrophys. J.* **353**, 712.
Hudson, H.S.: 1975, *Solar Phys.* **45**, 69.
Kaufmann, P., Correia, E., Costa, J.E.R., and Zodi Vaz, A.M.: 1986, *Astron. Astrophys.* **157**, 11.
Kundu, M.R., White, S.M., Gopalswamy, N., and Lim, J.: 1992, in *2nd GRO Science Workshop*, (in press).
Kurucz, R.L.: 1985, *Bull. AAS* **17**, 640.
Lindsey, C., and Heasley, J.N.: 1981, *Astrophys. J.* **247**, 348.
Lindsey, C.A., Yee, S., Roellig, T.L., Hills, R., Brock, D., Duncan, W., Watt, G., Webster, A., and Jefferies, J.T.: 1990, *Astrophys. J. (Letters)* **353**, L53.

Lindsey, C.A., and Jefferies, J.T.: 1991, *Astrophys. J.* **383**, 443.
Livingston, W.: 1991, in L.J. November (ed.), *Solar Polarimetry*, 11th NSO Summer Workshop, p. 356.
Machado, M.E., Avrett, E.H., Vernazza, J.E., and Noyes, R.W.: 1980, *Astrophys. J.* **242**, 336.
Mauas, P.J.D., Machado, M.E., and Avrett, E.H.: 1990, *Astrophys. J.* **360**, 715.
Maltby, P., Avrett, E.H., Carlsson, M., Kjeldseth-Moe, O., Kurucz, R.L., and Loeser,R.: 1986, *Astrophys. J.* **306**, 284.
McClements, K.G., and Brown, J.C.: 1986, *Astron. Astrophys.* **165**, 235.
Rabin, D., Jaksha, D., Plymate, C., and Wagner, J.: 1991, in L.J. November (ed.), *Solar Polarimetry*, 11th NSO Summer Worshop, p. 361.
Shimabukuro, F.I.: 1970, *Solar Phys.* **15**, 424.
Shimabukuro, F.I.: 1972, *Solar Phys.* **23**, 169.
Turon, P.J., and Lena, P.J.: 1970, *Solar Phys.* **14**, 112.
Van Ballegooijen, A.A.: 1984, *Solar Phys.* **91**, 195.
Vernazza, J.E., Avrett, E.H., and Loeser, R.: 1973, *Astrophys. J.* **184**, 605.
Vernazza, J.E., Avrett, E.H., and Loeser, R.: 1981, *Astrophys. J. Suppl.* **45**, 635.
Worden, S.P.: 1975, *Solar Phys.* **45**, 521.
Zirin, H., and Popp, B.: 1990, *Astrophys. J.* **340**, 571.
Zirker, J.B.: 1985, *Solar Phys.* **102**, 33.

THE OBSERVED SPECTRUM OF SOLAR BURST CONTINUUM EMISSION IN THE SUBMILLIMETER SPECTRAL RANGE

E. CORREIA and P. KAUFMANN

Centro de Radioastronomia e Aplicações Espaciais, EPUSP, CP 8174,
05508 São Paulo, Brazil

and

A. MAGUN

Institute of Applied Physics, Division of Solar Observations, University of Bern,
Sidlestrasse 5, CH-3012 Bern, Switzerland

Abstract. Observations of solar flares at high frequencies suggest that a considerable fraction of the events present flat or even increasing flux spectra at frequencies above 35 GHz. This imposes restrictions on the gyrosynchrotron emission mechanism and source parameters. We analysed a sample of 115 microwave events in order to investigate their spectra at peak flux. The analysis shows that about 50% of the sample exhibits a flat and 25% an increasing spectrum between 19 and 35 GHz. This class of events is significant and must be considered in the models. In order to better define the characteristics of this class of events it is necessary to carry out observations at frequencies well above 50 GHz with high time resolution and high sensitivity.

Key words: infrared: stars – Sun: radio radiation – Sun: flares

1. Introduction

Solar flare observations above 80 GHz are very rare and therefore only little is known about the spectrum above this frequency. It was believed that flares produce only negligible gyrosynchrotron emission at frequencies above 30 GHz. In the past two decades there were only a few solar patrol observations at frequencies up to 100 GHz. Occasional observations with high spatial resolution and high sensitivity were carried out only to measure the limb darkening and brightness temperature of quiet Sun. Most flare observations were made with low sensitivity and a time resolution not better than 100 milliseconds. During some of those observations strong events were detected suggesting an increasing flux towards higher frequencies (summarized in Kaufmann et al., 1986). Of a sample of 40 strong events (observed at Bern) which were correlated with gamma-ray emission above 300 keV (observed from SMM), 19 microwave spectra showed a constant or monotonically rising flux and very broadband spectral peaks up to 35 GHz (Crannell and Magun et al., 1984). It has also been noted that the hardest X-ray spectra are associated with a very low Microwave Richness Index (MRI) at 9 GHz (Bai and Dennis, 1985). In 1984, using the large antenna of Itapetinga Radio Observatory in Saõ Paulo (for the first time), an event with an increasing spectrum up to 100 GHz was observed (Kaufmann et al., 1985, Correia and Kaufmann, 1987). It suggests that the spectral maximum was well above 100 GHz. The time characteristics of this event impose severe restrictions to the gyrosynchrotron interpretation, especially when the turnover frequency of the spectra is above 120 GHz (Kaufmann et al., 1986; de Jager et al., 1987).

The question is: Are these events a rare phenomenon on the Sun?

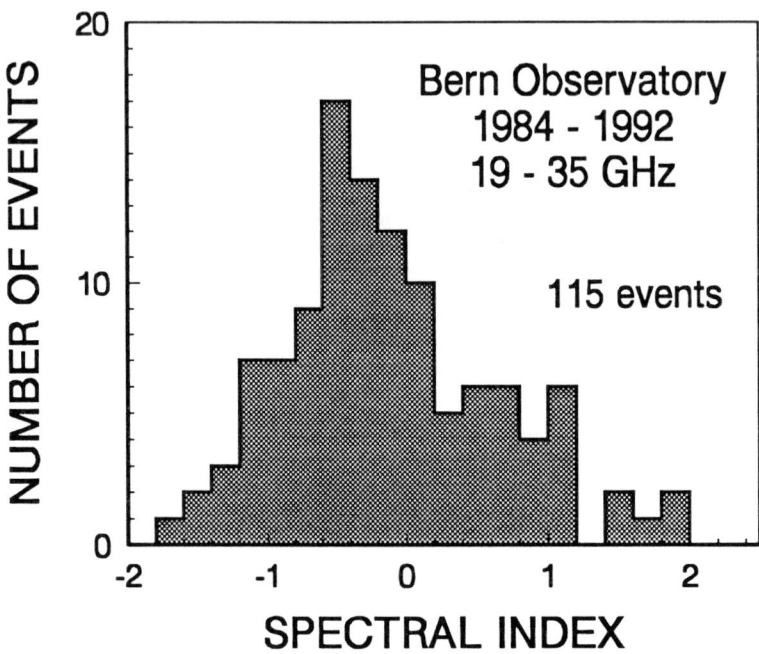

Fig. 1. Histogram showing the distribution of the spectral index, at frequencies between 19 and 35 GHz for the 115 events that had fluxes above 100 sfu.

2. Statistical Analysis

This analysis is based on a sample of bursts that were observed from 1984 to 1992 by patrol telescopes at the Solar Radio Observatory of the University of Bern, Switzerland. By selecting only those events with fluxes above 100 s.f.u. at 19 and 35 GHz a sample of 115 events resulted, whose spectral indices are shown in the Figure 1. The spectral indices were obtained from the linear slope between 19 and 35 GHz. Approximately 50% of the events exhibit a spectrum with a spectral index around zero and 25% above zero indicating a flat or increasing spectrum respectively.

The scatter diagram of the spectral index versus the heliographic longitude (Figure 2) indicates that no directivity is associated with the spectral characteristics of the events.

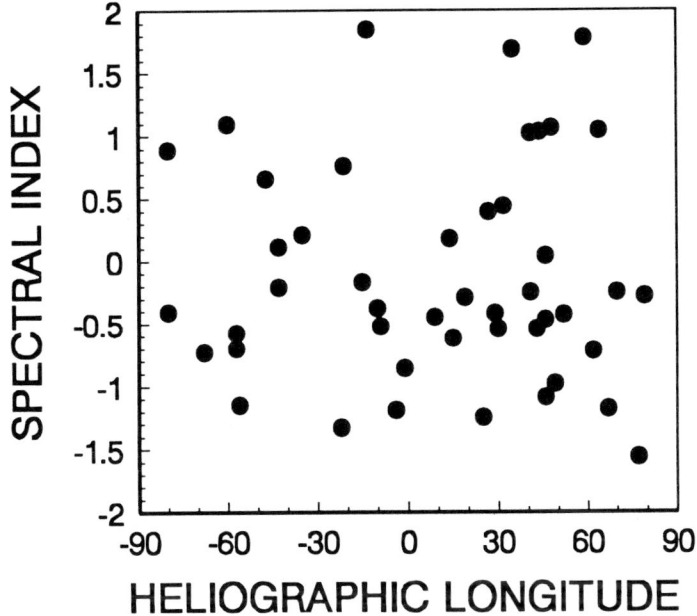

Fig. 2. Scatter diagram of Spectral Index and Heliographic longitude. Only events occurring during the years 1990 through 1992, the most significant part of the sample, have been included.

3. Discussion

Little information exists about the characteristics of events at frequencies above 30 GHz. This is especially true for weak events where the sensitivity of patrol instruments is insufficient for the observation of fluxes at higher frequencies. One of the earliest examples of flat spectra in the range 2-20 GHz (Hachenberg and Wallis, 1961), was interpreted as caused by free-free absorption of gyrosynchrotron emission of nonthermal electrons (Ramaty and Petrosian, 1972). The earlier data, however, were very poor in sensitivity and time resolution. The best event observed so far, with spectral coverage up to 100 GHz, occurred on May 21th, 1984. It was observed by the Itapetinga Radio Telescope with 1 ms time resolution and 0.01 sfu sensitivity and showed a spectrum increasing to frequencies above 100 GHz. The most notable characteristics of this event were flux variations with a relative amplitude of about 50% with unusually short durations of only 50 milliseconds. From the increasing spectrum a turnover frequency above 100 GHz was derived (Correia and Kaufmann,

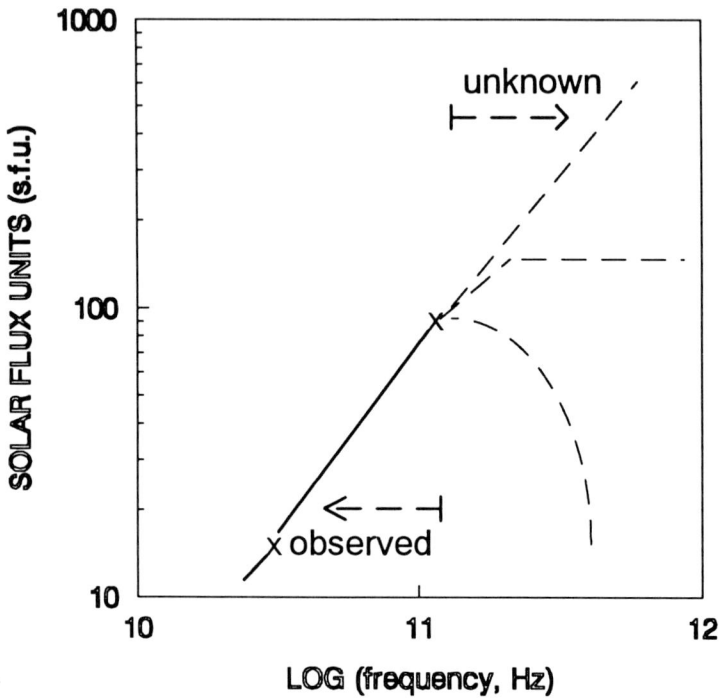

Fig. 3. Possible spectral behavior ar frequencies above 100 GHz.

1987). The combination of these two features is difficult to interpret as arising in gyrosynchrotron emission. Depending on the spectral features at frequencies above 100 GHz, *i.e.*, whether the spectrum is flat or increasing (Figure 3), it may be necessary to invoke another emissiom mechanism or to specify the plasma conditions better.

Thus we conclude from our analysis of more than 100 microwave spectra that there is a significant class of events with spectral peaks at frequencies well above 30 GHz. Furthermore it is worth noting that , according to Bai and Dennis (1985), the hardest X-ray spectra show a very low MRI that could mean that during the most energetic events the spectral maximum is shifted towards higher frequencies. In order to study this class of events and the relevant radiation mechanisms it is important to perform burst observations, especially at frequencies above 100 GHz, with high time resolution and high sensitivity.

Acknowledgements

CRAAE is jointly operated by the Universities of São Paulo, Campinas, and Mackenzie, and by the National Institute for Space Research. One of the authors (E.C.) received partial support from IAU and CNPQ to attend this meeting.

References

Bai, T., Dennis, B.R.: 1985, *Astrophys. J.* **292**, 699.
Correia, E., Kaufmann, P.: 1987, *Solar Phys.* **111**, 143.
Crannell, C.J., and Magun, A.: 1984, *Bull. Amer. Astron. Soc.* **16**, 537.
de Jager, C., Kuijpers, J., Correia, E., Kaufmann, P.: 1987, *Solar Phys.* **110**, 317.
Hachenberg, O., Wallis, G.: 1961, *Zs. f. Astrophys.* **52**, 42.
Kaufmann, P., Correia, E., Costa, J.E.R., Zodi Vaz, A.M.: 1986, *Astron. Astrophys.* **57**, 11.
Kaufmann, P., Correia, E., Costa, J.E.R., Zodi Vaz, A.M., Dennis, B.R.: 1985, *Nature* **331**, 380.
Ramaty, R., and Petrosian, V.: 1972, *Astrophys. J.* **178**, 241.

INTERFEROMETRY OF SOLAR FLARES AT 3-mm WAVELENGTH

M. R. KUNDU, S. M. WHITE, N. GOPALSWAMY and J. LIM

Department of Astronomy, University of Maryland, College Park, MD 20742, U.S.A.

Abstract. We describe a set of millimeter interferometric observations of solar flares carried out in conjunction with GRO experiments during the 1991 June Campaign of the Max'91 Program. We show evidence that millimeter emission probes the most energetic (MeV) electrons in solar flares; we also find that in the same flare there can be both impulsive nonthermal and gradual thermal millimeter emission. Millimeter emission usually occurs at the steep rise phase of the hard X-ray emitting electrons (25-100 KeV). There appears to exist some delay between BIMA mm-emission onset and GRO-BATSE 25-100 KeV X-ray emission. Both results have implications for the particle acceleration process.

Key words: Sun: flares – Sun: particle emission – Sun: radio radiation

1. Introduction

Most millimeter-wave observations of solar flares have been hampered both by poor spatial resolution (because of the lack of synthesis interferometers) and by poor sensitivity (since the flux from the Sun's thermal emission is so high at millimeter wavelengths). The number of reported observations of bursts at millimeter wavelengths is relatively small, and there are none for which true high-spatial-resolution imaging data have been reported (see, *e.g.*, White and Kundu 1992).

The usefulness of millimeter-wave observations lies in the fact that they are sensitive to both the highest energy electrons in flares as well as to cool material in the chromosphere. The emission of solar flares at millimeter wavelengths is of great interest both in its own right and because it is generated by the very energetic electrons which also emit gamma rays. Since high-resolution imaging at gamma ray energies is not presently possible, millimeter observations can act as a substitute. In addition, the millimetric emission is undoubtedly optically thin (except possibly for a small class of very large flares which have spectra rising beyond 100 GHz; Kaufmann *et al.* 1985), and thus is not subject to the same ambiguities of interpretation which are prevalent in the study of solar microwave bursts. It can be used as a powerful diagnostic of the energy distribution of electrons in solar flares and its evolution, and of the magnetic field.

We have carried out high-spatial-resolution millimeter observations of solar flares using the Berkeley-Illinois-Maryland Millimeter Array (BIMA). At the present time BIMA consists of only three elements, which is not adequate for mapping highly variable solar phenomena, but is excellent for studies of the temporal structure of flares at millimeter wavelengths at several different spatial scales. The first dedicated observations of solar flares with a high-spatial-resolution millimeter-wavelength interferometer were made by us in March 1989 (Kundu *et al.* 1990) when a solar active region with the greatest X-ray flare production rate yet recorded appeared on the disk of the Sun. Our observations represented an improvement of an order of magnitude in both sensitivity and spatial resolution compared with previous solar observations at these wavelengths. We found that most of the flares occurring within the field of view during the March 1989 observations were detected at millimeter wavelengths, including both very impulsive and longer-duration events.

Based on these early observations it appeared that millimeter burst sources were not much smaller than microwave sources; generally source sizes were in excess of 2″, but in some events sources may be ∼ 1″. Thermal gyrosynchrotron models for the radio emission in the impulsive phase could be ruled out because the flux at millimeter wavelengths was too high.

2. Observations During the June 1991 Campaign

During the period of June 8–13, 1991, BIMA was in its most compact configuration. As mentioned earlier, the array presently has three antennas, providing three baselines. In effect each baseline provides a one dimensional spatial Fourier transform of the brightness distribution on the sky. For a point source the phase gives us positional information, while the amplitude is the total flux of the source. For an extended source the response of the interferometer depends on the relationship of the source dimension to the fringe spacing of the Fourier transform: only if the source dimension is smaller than about one-third of the fringe spacing do we expect the amplitude variability to correspond to the total flux in the source. For the 1991 June observations the fringe spacing ranged from ∼ 30″ EW to ∼ 45″ NS; such large fringe spacings should guarantee that the correlated amplitude corresponds to the total flux of the source, at the expense of information about small spatial scales. These observations were made as part of the Max '91 Campaign involving the Compton Gamma Ray Observatory (GRO) along with other groundbased and space borne facilities.

3. Results

Figure 1 presents the GRO-BATSE (Burst and Transient Spectrometer Experiment) data along with 86 GHz data for a burst observed on June 13, 1991, ∼ 17:57–18:06 UT. The flare evolution is shown for the lowest BATSE energy channel, 25–50 keV. In Figure 2 we have plotted the time profile of the BATSE 0.1–0.3 MeV channel and the BIMA 86 GHz profile to the same scale. There were no significant counts above the background in the 0.3–1 MeV channel. The remarkable feature of this flare is that while the overall profile in the 25–50 keV range is relatively simple, with a sharp rise to a plateau level containing a number of sub–peaks followed by a rapid decay, in the higher-energy channels we see a sharp spike right at the beginning of the flare. We and others (Ramaty 1969; Ramaty and Petrosian 1972) have argued that millimeter emission in the impulsive phase should be due to gyrosynchrotron emission from the most energetic electrons in a flare, and on this basis one would predict that the millimeter emission would also show a sharp spike in the onset phase. This is exactly what is seen. The close similarity of the 0.1–0.3 MeV and 86 GHz profiles provides convincing evidence that the millimeter emission is indeed due to the most energetic electrons in a solar flare: the same energetic electrons that produce gamma ray continuum. The importance of this result is that we can obtain high–spatial–resolution information on source structures

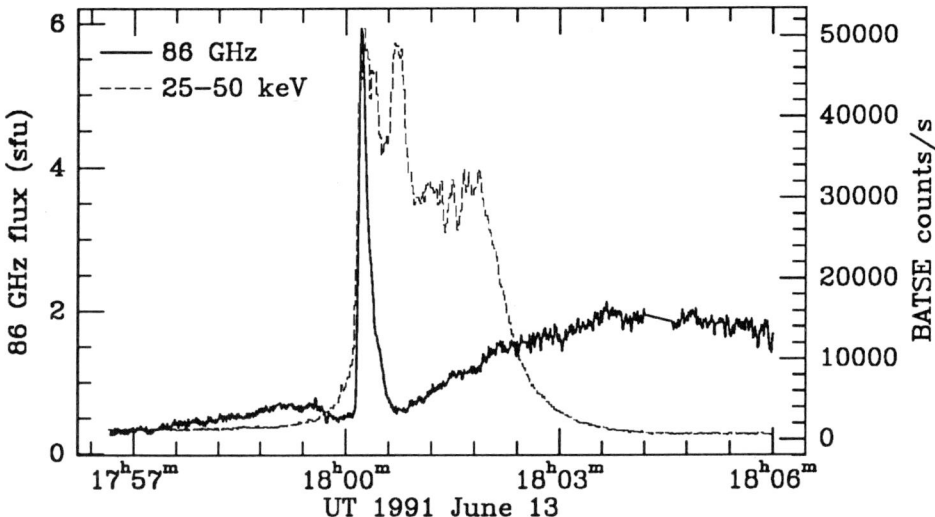

Fig. 1. Comparison of the time profiles of the 1991 June 13 18 UT flare emission in the 25–50 keV hard X-rays seen by BATSE (broken line) and the 86 GHz emission recorded by BIMA (solid line).

with millimeter-interferometer observations, which is not presently possible directly with gamma-ray observations.

Figure 1 shows further that the sharp rise in the 25–50 keV range coincides exactly with the 86 GHz rise, although there is some emission in the 25–50 keV range just prior to the impulsive rise which has no signature at 86 GHz. However, unlike the 25–50 keV burst which has a plateau lasting ~ 3 min and containing another major peak, the mm emission drops abruptly to almost its pre-burst level. This figure also emphasizes that, in addition to the impulsive phase coinciding with the gamma ray emission, the millimeter emission shows an extended component which rises slowly and continues beyond 1806 UT. Thus in this event we clearly see two easily-distinguishable components in the millimeter emission: an impulsive component coinciding in time with the impulsive gamma ray emission, and a gradual extended component with no gamma ray counterpart. We also note the small dip in the 86 GHz emission immediately prior to the impulsive rise, which seems to be a common feature of pre-flare activity in millimeter-wave emission.

Figure 3 shows an event in which impulsive 25–50 keV emission was seen, but there was very weak impulsive-phase emission at 86 GHz, even though the count rates at 25–50 keV were well above the levels of some small events which produced both 25–50 keV and 86 GHz impulsive-phase emission. The 86 GHz profile in this case shows gradual mm burst emission corresponding roughly with the GOES soft X-ray emission, although the BATSE data show several sharp peaks. In this case the hard X-ray rise corresponds to the rise phase of both mm and soft X-ray emission. A number of 1991 June flares, both with and without 86 GHz impulsive phases,

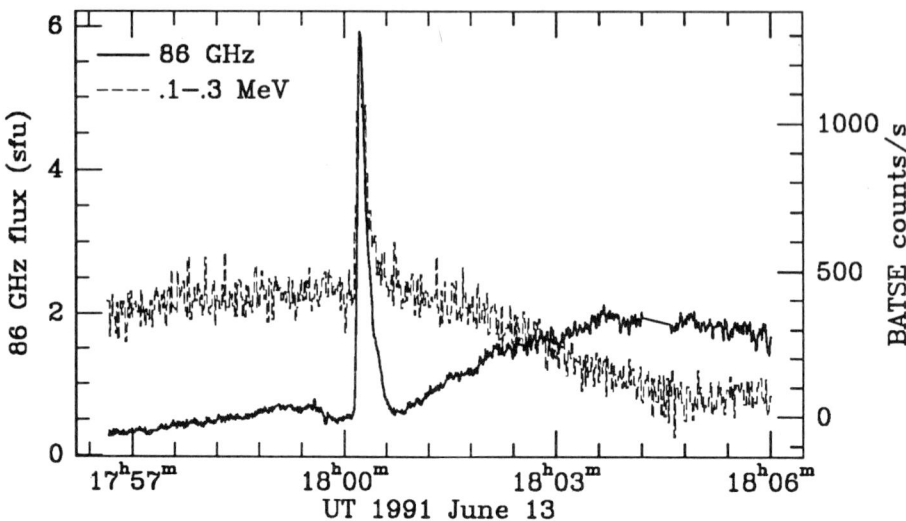

Fig. 2. Comparison of the time profiles of the 1991 June 13 18 UT flare emission in 0.1–0.3 MeV hard X-rays (broken line) and the 86 GHz emission (solid line).

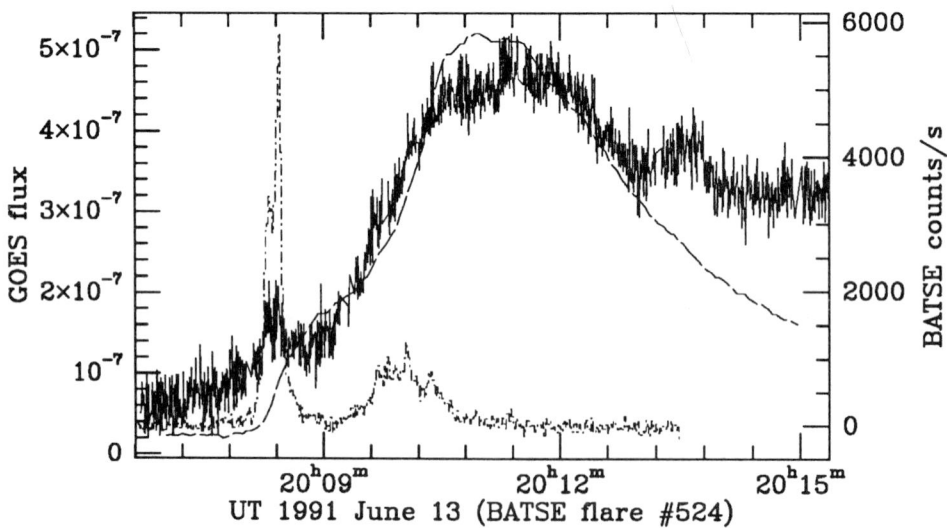

Fig. 3. The evolution of emission for a flare at around 20 UT on 1991 June 13 seen in soft X-rays recorded by the GOES satellite (8–25 keV; long-dashed line), in 25–50 keV hard X-rays (BATSE data, short-dot-dashed line) and at 86 GHz (BIMA data, solid line).

showed a thermal phase at 86 GHz which was similar in profile to the GOES data, including the June 13 18 UT event. We also found evidence that for some simple impulsive events with a very sharp rise there is a clear delay of several seconds between the 25–50 keV impulsive rise and the 86 GHz rise. Since we can associate the 86 GHz emission with high-energy electrons (> 300 keV), the implication is that there is a delay between the production of the 25–50 keV electrons and the > 300 keV electrons. Such delays can be attributed to the need to accelerate electrons from low to high energies, and can be used to constrain candidate emission mechanisms.

4. Conclusions

Our 1991 June observations permit us to make the following general conclusions.

1) There are two phases in millimeter burst emission: a nonthermal impulsive phase and a thermal gradual phase. Both phases are often observed in the same flare.

2) Impulsive-phase millimeter burst emission probes MeV electrons produced during flares. Some flares show no nonthermal impulsive phase at millimeter wavelengths, although they seem to show it at 25–50 keV. These gradual phase flares correspond well to the GOES soft X-ray emission.

3) Millimeter emission usually occurs at the steep rise phase of the hard X-ray emitting electrons (25–100 keV). There appears to exist some delay between BIMA mm-emission onset and BATSE 25–100 keV X-ray emission. Both results have implications in the particle acceleration process.

Acknowledgements

This work was made possible by GRO Phase I Guest Investigator grant NASA-NAG-W-1541. Solar research at Maryland is also supported by NSF grant ATM 90-19893 and NASA grants NAG-5-1540, and NAG-W-2172. Scientific research at U. Md. with BIMA is supported by NSF grant AST 91-00306.

References

Kaufmann, P., Correia, E., Costa, J.E.R., Vaz, A.M.Z., Dennis, B.R.: 1985, *Nature*, **313**, 380
Kundu, M.R., White, S.M., Gopalswamy, N., Bieging, J.H., Hurford, G.J.: 1990, *Astrophys. J. (Letters)*, **358**, L69.
Ramaty, R.: 1969, *Astrophys. J.*, **158**, 753.
Ramaty, R., Petrosian, V.: 1972, *Astrophys. J.*, **178**, 241.
White, S.M., Kundu, M.R.: 1992, *Solar Phys.*, in press.

PART 2

INFRARED OBSERVATIONS OF THE 1991 TOTAL SOLAR ECLIPSE

ECLIPSE OBSERVATIONS OF THE EXTREME SOLAR LIMB AT SUBMILLIMETER WAVELENGTHS

T. A. CLARK

Physics Department, University of Calgary, Calgary, Alberta T2N 1N4, Canada

Abstract. This paper will review the use of solar eclipses in the study of the extreme solar limb at sub-millimeter and millimeter wavelengths. This approach has been used to overcome the severe limitation imposed by diffraction upon the resolution attainable by direct solar limb scans at these wavelengths. Strong absorption by water vapor in the Earth's lower atmosphere has necessitated the use of telescopes at high altitude sites or in jet aircraft. Data from several of these experiments will be reviewed, including those from the recent James Clerk Maxwell Telescope observation at a wavelength of 1.3 mm of the eclipse of 11 July 1991. In view of the success of recent measurements in improving the spatial resolution with this technique, several of the ultimate limitations placed upon it by lunar surface roughness and by diffraction at the lunar limb are outlined.

These observations have demonstrated the inadequacy of present phenomenological solar atmospheric models at sub-millimetric source heights. Newer models have been developed to fit the observed extension, brightening and detailed structure of the solar limb by attempting to include the structure of the chromospheric network and its spicular field, and their relative success in doing so will be discussed.

Key words: eclipses – infrared: stars – Sun: atmosphere – Sun: chromosphere

1. Introduction

This review covers one specific technique for the study of the solar chromosphere, the detailed examination of the solar limb at millimeter and sub-millimeter wavelengths from mountain altitudes or aircraft during solar eclipses. The reasons for this specific approach can be summarized briefly.

Observations made at the extreme solar limb can provide a direct and precise attribution of height to measured parameters within the solar atmosphere. Disk-center measurements which are utilized in the construction of atmospheric models such as far IR and UV brightness temperatures are averages over relatively broad source functions. Furthermore, attribution of height to such disk-center measurements is model-dependent and most models (*e.g.*, VAL models, Vernazza *et al.*, 1976, 1981) assume gravitational-hydrostatic equilibrium and an unstructured atmosphere, conditions which appear not to hold in the chromosphere. Height above some reference level is measured directly in limb observations, even though the measurements are also averages, this time along a tangential path. In the case of eclipse experiments, this height measurement is provided by careful timimg of the occultation.

Sub-millimeter and millimeter radiation is measured since this radiation originates from heights which bracket the important temperature mimimum and the 6000 K plateau regions within the chromosphere. This predominantly continuum radiation is almost certainly formed in LTE, the main opacity sources, H^- and H free-free processes, are well understood, and the source functions are directly related to electron temperature. These factors are in distinct contrast to those for the UV radiation which also originates in this region. In this case, the radiation is primarily line emission, and source heights for these lines are more difficult to determine.

These observations must be made from high mountain or aircraft altitudes since moisture in the Earth's lower atmosphere almost totally absorbs radiation from about 20 to 300 μm. Even in windows beyond this, at 350, 450, 850 and 1100 μm, and at high and dry mountain sites, transmission is low, spectrally structured and highly variable. For eclipse observations, this situation leads to the requirement for upper-atmospheric observations where one has careful control over the location of the instrumentation. This effectively rules out balloon-borne or spacecraft experiments but this requirement can be met by jet aircraft and, for longer wavelengths at least, high mountain sites. An added bonus for fast aircraft platforms, whenever the geometry is favorable, is the possibility of extension of eclipse observing time by flying in the direction of motion of the Moon's shadow.

Observations are made in eclipses in order to overcome the severe resolution limit placed upon solar limb-scanning techniques by diffraction even at the world's largest millimeter telescopes. Spatial resolution for these telescopes is still only on the order of supergranular cell dimensions or of the depth of the chromosphere. Thus, direct single-dish limb scans cannot provide high resolution information. Interferometric techniques can increase this scanning resolution significantly, but these methods are difficult, particularly at shorter wavelengths, and resolution still falls short of that attained by eclipse methods. In contrast, even with modest time resolution, observation of a solar eclipse can enhance this resolution to the point where detailed structure can be inferred within the chromospheric layers. This advantage becomes particularly significant where one is forced by atmospheric transmission to consider aircraft observations, for which the telescope diameter restriction is much more severe.

2. Early Observations

Early work in the far IR and sub-millimeter by limb scanning produced mixed results. These experiments showed in general that the Sun showed far less limb brightening than was predicted by the single-stream phenomenological chromospheric models which had been assembled to account for the available data in a self-consistent way (*e.g.*, HSRA; Gingerich, 1971). This discrepancy between model and observation was attributed to "roughness" in the chromosphere on a scale of about 1 scale height (Simon and Zirin, 1969), and was generally related to the spicules within the chromospheric network, since spicules had already been identified and characterized from Hα photographs as existing above the far IR source height regions.

Much of the early millimeter work, both eclipse-related and scanning, notably by Shimabukuro and colleagues at Aerospace Corporation, (Shimabukuro, 1975), Zirin and his group (*e.g.*, Marsh *et al.*, 1981; Horne *et al.*, 1981; Wannier *et al.*, 1983), Kundu and his team at Maryland, (Ahmad and Kundu, 1981), Labrum and colleagues in Australia (Labrum *et al.*, 1978), the Queen Mary College, London group (*e.g.*, Newstead, 1969, Beckman *et al.*, 1973, Ade *et al.*, 1974), and Lindsey and others (Lindsey and Hudson, 1976, Lindsey *et al.*, 1981, 1984), led to discordant results, but some consensus emerged for weak limb brightening which increased with wavelength at sub-millimeter wavelengths.

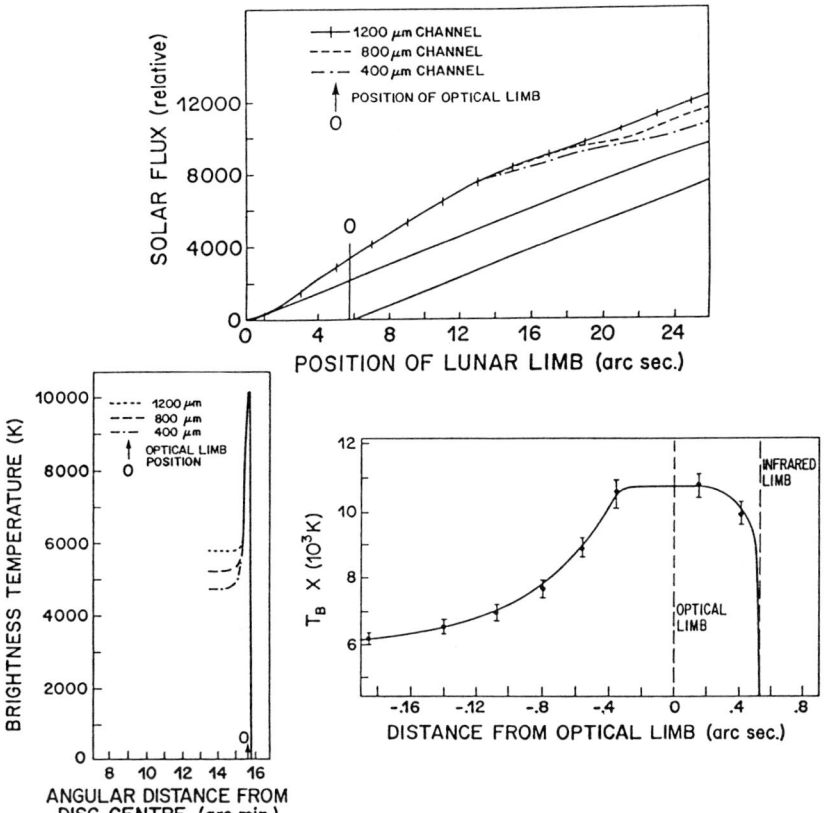

Fig. 1. (a) The eclipse curve for the 3 wavelengths as a function of time at 4th contact during the 30 June 1973 total solar eclipse, as measured from the Concorde aircraft by Beckman et al. (1975). (b) Derived limb profiles for the 3 wavelengths, showing the same brightness temperature at each wavelength, (c) Expanded limb brightness profile for 1.2mm. (From Beckman et al., 1975)

Meanwhile, a unique and innovative airborne eclipse experiment at millimeter and sub-millimeter wavelengths produced a somewhat perplexing result. Beckman and his colleagues from Queen Mary College, London, mounted a Michelson interferometer and optics on the prototype supersonic Concorde aircraft in 1973 and, along with several other groups, were able to enjoy the remarkable extension of eclipse totality to 70 minutes over Africa (Beckman, 1973). Their eclipse curve shown in Figure 1, apparently to good signal-to-noise level, was interpreted to indicate that an intense "spike" with peak intensity of more than 2 times that of the solar continuum extended from 4" inside to 6" outside the optical solar limb, with the same brightness temperature at each of the 3 wavelengths.

This somewhat surprising result from an admittedly very difficult experiment, prompted a more modest airborne experiment (Clark and Boreiko, 1982) to confirm

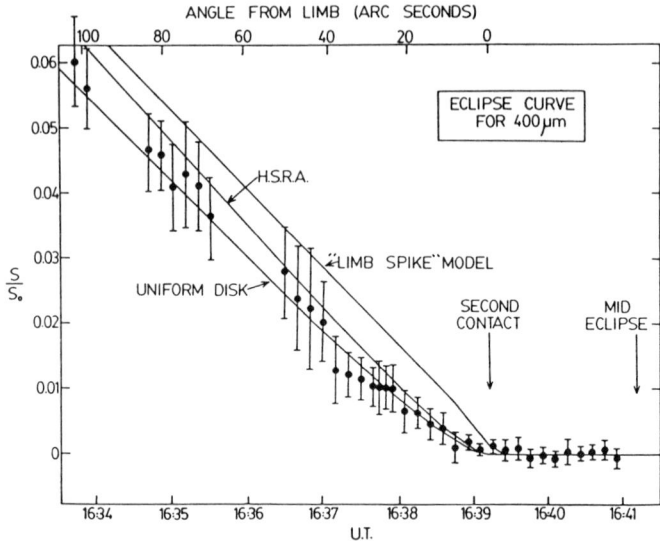

Fig. 2. Eclipse curve from the 400 μm photometer within 100″ of second contact in the 26 Feb 1979 eclipse, observed from the NASA Lear Jet Observatory. This curve is compared to model curves for a uniform disk (with no limb extension), an HSRA disk and a disk containing a limb "spike" (From Clark and Boreiko, 1982).

or test the result, using the NASA Lear Jet Observatory. A multichannel photometer was used in which individual optics focussed the Sun and adjacent Moon during the eclipse onto several separate detectors, to reduce the effect of steering jitter and beam pattern upon the observed eclipse curve. Though the 400 μm data shown in Figure 2 were somewhat noisy, no evidence was seen of an intense limb spike. This eclipse curve was best fitted to a solar profile with about 1/3 of the limb brightening predicted by the HSRA model (Gingerich et al., 1971). A subsequent annular eclipse experiment (Clark and Boreiko, 1980) was flown on the NASA Galileo II aircraft with a similar 400 μm photometer and confirmed this result, though the somewhat noisy data demonstrated again the severe restriction of this small-optics full-sun-imaging approach.

3. Kuiper Airborne Observatory Experiments

Then began a remarkable series of very careful airborne experiments by a team headed by Becklin and Lindsey, in which the 0.9 m telescope of the NASA Kuiper Airborne Observatory was pressed into service to observe both the 1981 and 1988 eclipses over the Pacific Ocean (Lindsey et al., 1983, 1986; Roellig et al., 1991). The telescope was screened with polyethylene and equipped with a 4-detector filter photometer (extended to 7 detectors in 1988), to measure individual bands in the far IR and sub-millimeter range. The major advantage of this technique over whole-sun imaging was the ability to observe the points of eclipse contact with the

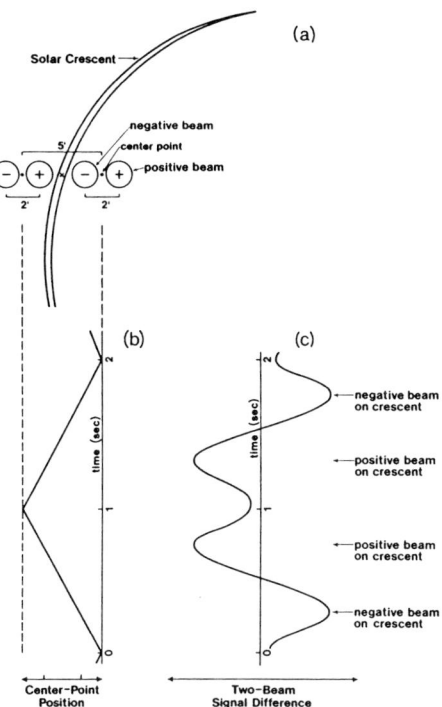

Fig. 3. (a) Schematic depiction of the 2-beam chopping-scanning geometry used by Lindsey et al. (1983, 1986) to observe the 1981 and 1988 eclipses from the Kuiper Airborne Observatory. (b) The 0.5 Hz triangular scan-pattern of the center-point of the 2-beam chop. (c) The resulting 2-beam difference signal profile as the positive and negative beams pass over the solar crescent. (From Lindsey et al., 1983)

100″ diffraction-limited beam of the telescope. However, a direct "staring" mode would carry with it a significant disadvantage. Telescope tracking errors across the diffraction-limited beam would result in significant intensity variations unrelated to the eclipse, even when two-beam chopping from one beam on the solar limb to an adjacent sky area 2′ away was used to remove atmospheric and telescope emission. A special observing technique was therefore developed and used both in these 2 flights and subsequently on the James Clerk Maxwell Telescope (JCMT) on Mauna Kea in the 11 July 1991 eclipse. The telescope was linearly scanned back and forth across the remnant solar crescent in a manner outlined in Figure 3, to produce the typical eclipse signal shown in Figure 4. The envelope of peaks represents the center-sample of the beam-pattern across the remaining solar crescent, or the equivalent integral eclipse curve.

Careful removal of the remnant Moon radiation and reconstruction of the eclipse curve produced the initial indication of significant limb extension at these wavelengths, of about the same magnitude as that measured by Beckman and Ross

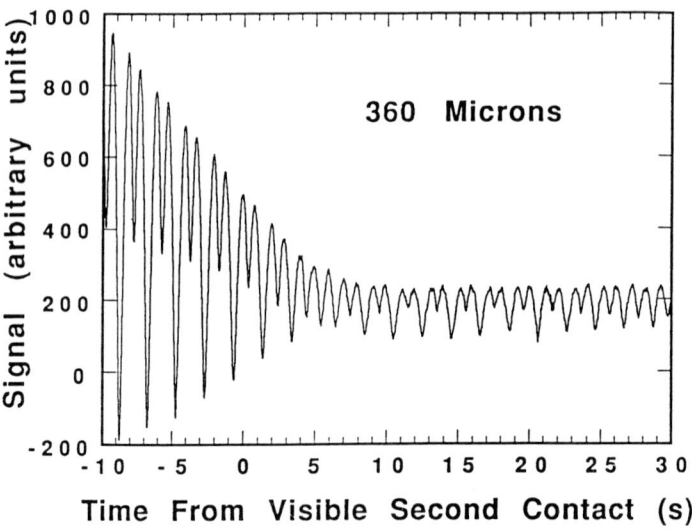

Fig. 4. Typical second-contact eclipse signal curve at 360 μm, uncorrected for moon emission, from the 1988 eclipse. (From Roellig et al., 1991)

(1975), but with no indication of the intense limb spike (Figure 5). There is evidence of some limb brightening in these records, of the order of 2% at 30 μm and 22% at 200 μm between −3000 and +350 km from the visible limb, normalized to disk center.

4. Modeling

Modeling of this extended and limb-brightened limb by Lindsey and his colleagues have explored several avenues. Initially, a "stretched" VAL (Vernazza et al., 1973) model was fitted to the profiles (Hermans and Lindsey, 1973) by stretching the entire chromosphere by a factor of 2.4 in height while maintaining the vertical temperature-optical depth relationship, thereby violating gravitational hydrostatic equilibrium in a "smooth" atmosphere. Electron densities were lowered to maintain the $T - \tau$ relation, while other densities were lowered in a commensurate manner to maintain ionization equilibrium. This density change would also affect Lyman-α and Lyman continuum emission however, and is probably unacceptable as a realistic model, even though the model fits the observed far IR data.

Lindsey (1987) and Braun and Lindsey (1987) used techniques developed for LTE calculations in a "rough" atmosphere to compute a model chromosphere consisting of a plane-parallel temperature minimum region out to 1000 km and randomly distributed cylindrical spicules interspersed within transparent coronal material above this height. This model was able to fit both the limb-brightened far-IR eclipse data and the relatively flat limb distribution at 2.6 mm, measured interferometrically by Wannier et al. (1983), by using a constant spicular temperature of

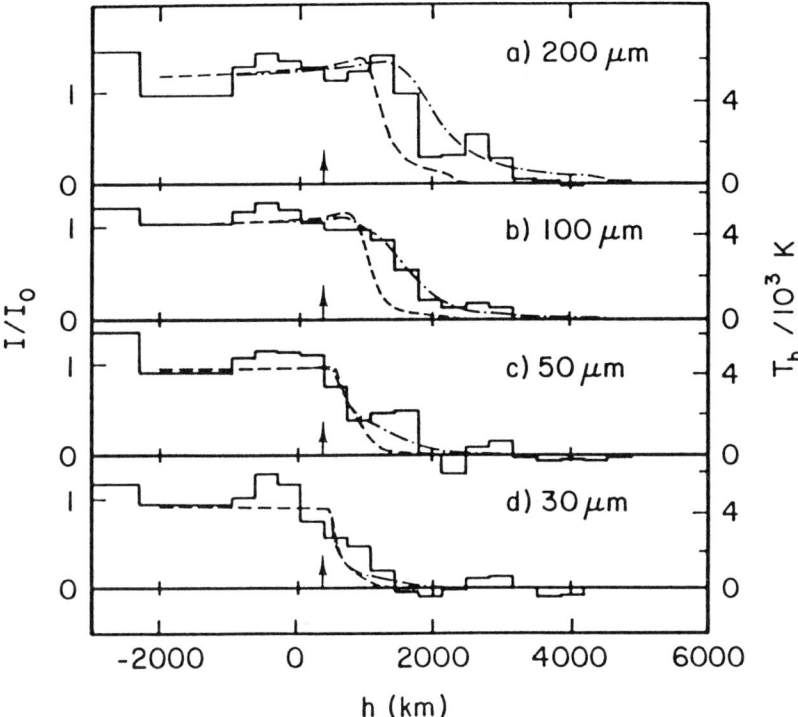

Fig. 5. Far IR solar limb distributions from the 1981 eclipse, compared to limb brightness profiles for an unperturbed VAL model (1973) (dashed curve), and a "stretched" VAL model, (From Hermans and Lindsey, 1986).

about 7000K and a selected spicular distribution with height. This lack of significant brightness enhancement at the limb at 2.6 mm [and also at 1.3 mm, (Horne et al., 1981)] is a significant argument in favor of the conclusion that the temperature of spicular material must be significantly lower than that derived by Beckers (1972).

Other analysis (Jefferies and Lindsey, 1988, and Lindsey and Jefferies (1990)) has provided statistical methods to account for spicule distributions within the chromosphere, and to generate an equivalent "smooth" atmosphere with appropriate opacity and source functions to fit both the submillimeter eclipse measurements and the 2.6 mm profiles of Wannier et al. (1983).

Roellig and his colleagues (Roellig et al., 1991) have recently published the analysis of the 1988 eclipse flight and have used a modified Braun and Lindsey (1987) spicule model with slightly lower electron density, to match their solar limb profiles, as shown in Figure 6. There is still disagreement between the modeled limb profiles and the observed results at the extreme limb at these wavelengths, the model predicting a small limb spike. This is probably a consequence of the abrupt transition in the model between a smooth atmosphere below 1000 km and the assumed spicular structure above this altitude, and the power law assumption for spicular number density.

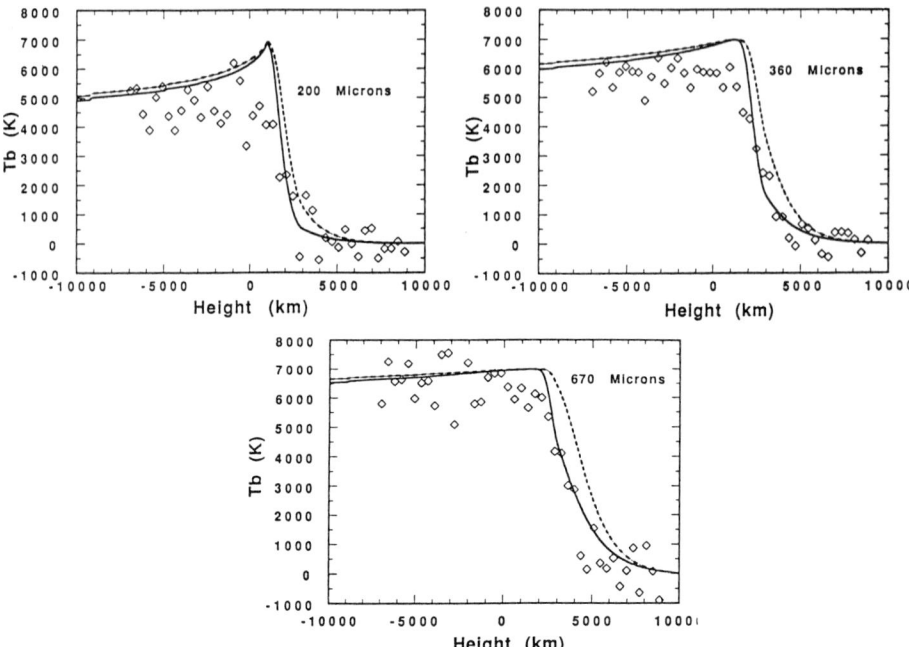

Fig. 6. The observed and model limb profiles at 200, 360 and 670 μm, with the Braun and Lindsey model shown by the dashed lines and the modified Roellig model shown by the solid lines. (From Roellig et al., 1991)

5. 11 July Eclipse Results

The 11 July 1991 total eclipse provided a unique opportunity to utilize the two large sub-millimeter and millimeter telescopes, the 10 m California Submillimeter Observatory (CSO) and the 15 m JCMT in Hawaii for extreme solar limb observations since the eclipse shadow passed almost directly over Mauna Kea. Ewell et al. (these proceedings) describe their observations at 850 μm on the 10-m CSO in which they find an extension of the limb and 10% limb brightening at this wavelength. They account for these results without invoking complex spicular structure by using a non-hydrostatic equilibrium model chromosphere with higher electron density than that predicted by the VAL model.

The 15 m JCMT was used by an international collaboration to monitor the solar limb occultation at 1.3 mm (Lindsey et al., 1992) during this eclipse. Here, the advantage of a large and stable telescope within the eclipse path was exploited, using a similar scanning technique to that developed for the airborne observations but without beam chopping to avoid beam-pattern effects. The secondary telescope mirror was scanned in a 40″, 1 Hz triangular wave across the remaining solar crescent to sample it at equivalent spatial intervals of 0.26″, providing unprecedented resolution at this wavelength.

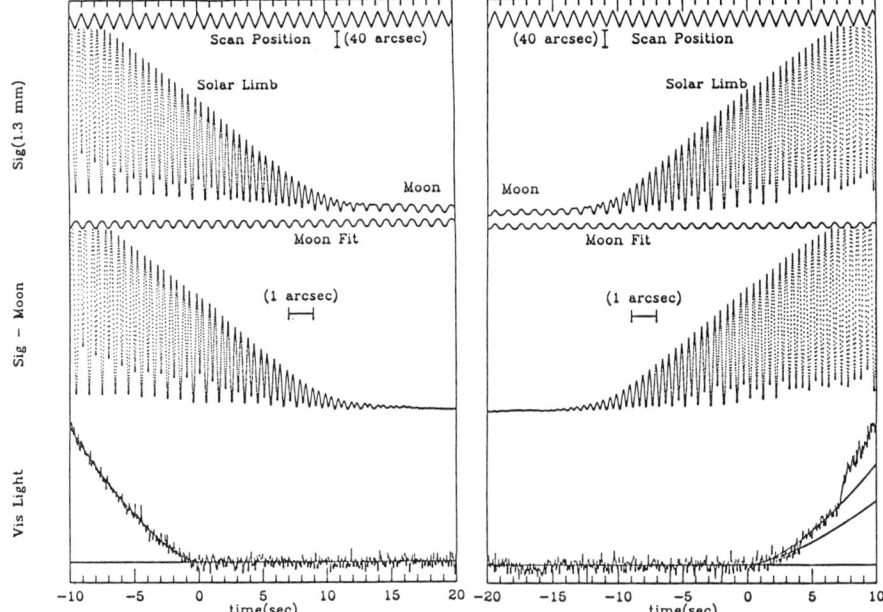

Fig. 7. JCMT scans of the disappearing solar crescent during 2nd and 3rd contacts of the 11 July 1991 solar eclipse at 1.3mm. The secondary mirror scan pattern is shown above these scans, while the visible light occultation, measured from the same site, is shown below each millimeter scan. The signal from the Moon was extracted from these scans before deriving the limb profiles.

Figure 7 shows the eclipse scans for 2nd and 3rd contacts, along with simultaneous visual scans from the same site. Again, these profiles indicate significant extension of the solar limb beyond the visible limb, in this case by about 5", with faint emission existing up to 8" from the visible limb. The second and third contact limb profiles of Figure 8 were derived from these occultation curves using the method outlined in Lindsey et al. (1986), with intensity normalized to disk center intensity. The limbs are seen to be extended, and the intensity at this limb is about 50% brighter than the quiet sun at disk center and falls off slowly beyond the millimeter limb, with an equivalent intensity scale height of about 1900 km.

6. Limits Imposed by Lunar Limb Shape and Diffraction.

The calculated limit placed upon the resolution of the occultation technique for this specific eclipse by the irregularity of the lunar limb at the position of each contact is shown in Figure 8 – Lindsey et al., 1992. This amounts to a broadened "scanning function" with a width of about 0.3". Also shown in Figure 8 is the calculated Fresnel diffraction pattern for solar radiation diffracted by the lunar limb. While representing fundamental limits to this technique, neither of these effects play any significant role in the interpretation of the present data.

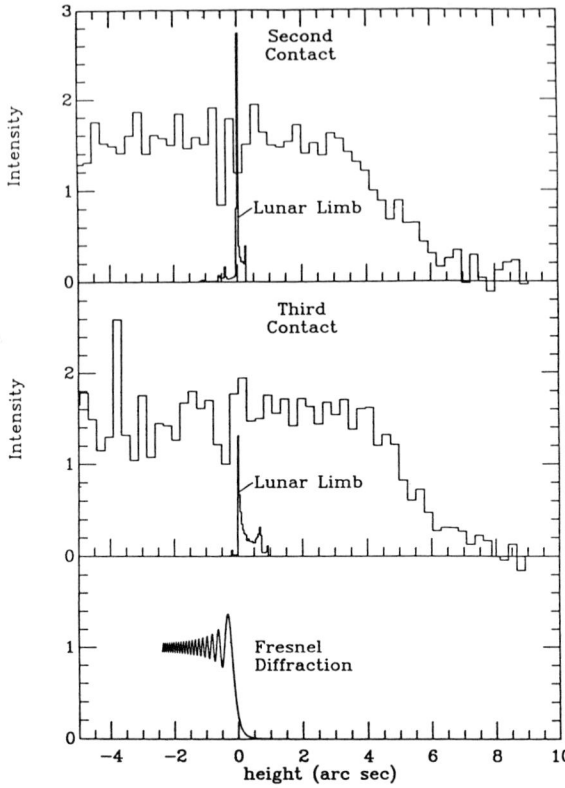

Fig. 8. Brightness profiles of the 1.3 mm solar limb at 3rd contact, computed from the occultation record of Figure 7, showing the extension of the limb beyond the visible limb and the slow decay of intensity at the millimeter limb. Also shown are the effects on the equivalent beam pattern of the rough lunar limb and Fresnel diffraction at that limb.

7. Conclusions

The use of the relatively sharp edge of the Moon's limb as a shutter to examine the distribution of brightness of the solar limb at submillimeter and millimeter wavelengths during solar eclipses has produced solid evidence of an extension of the solar limb, well beyond that predicted by homogeneous models at these wavelengths, which increases with wavelength as shown in the summary diagram of Figure 9. If spicules are the cause of this extension, then the increase is probably an indication of increasing opacity within the cooler (6000–7000K) spicular material, since the tangential line of sight intersects more than 1 spicule on average below a height of 3000 km above the visible limb (Michard, 1974). This tangent height occurs at a wavelength of about 1 mm. The extension beyond this wavelength and the slow fall-off of intensity at the limb is probably governed by the scale height of the number density of spicules and, at the longer wavelengths, by the increasing emission of coronal material.

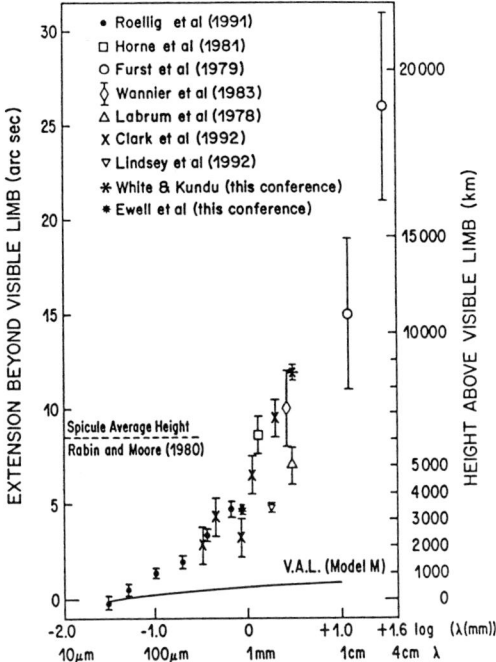

Fig. 9. Summary of measurements of limb extension as a function of the logarithm of wavelength, including recent limb-scan observations with the JCMT (Clark et al., 1992).

The variability of the size of the extension at a wavelength of about 1 mm in Figure 9 might be due to experimental and systematic error in the individual measurements but is much more likely to be due to the clumpiness of spicules on the solar surface and the resultant variability of the number of spicules in a particular line of sight, a conclusion also reached by Wannier et al. (1983) on the basis of interferometric scans of the solar limb at different solar position angles at 2.6 mm.

Puzzling features in these data which have recently been discussed by Foukal (1992) are that there appears to be no significant millimeter emission above the heights to which the Balmer-α-emitting spicules reach, and that the millimeter temperature of this spicular material appears to be so low. This indicates that the mechanism which propels the spicular material upwards does so with little dissipation of energy and heating, and that, once at this maximum height, the plasma ceases to emit both Balmer-α and millimeter radiation. Foukal concludes that this observation is important in judging the effectiveness of certain proposed models for chromospheric heating by the release of spicule potential energy within this layer (e.g., Athay and Holzer, 1982).

Thus, these latest measurements on the extended and weakly limb brightened millimeter solar limb place important constraints upon model development for this solar atmospheric layer. This is the important region from which the majority of the solar UV radiation arises and in which so-far unidentified non-radiative mechanisms are providing the source of heating, at least in the chromospheric network surrounding supergranular cells. These results will have to be included in a self-consistent manner along with the extensive disk-center and strong-line and UV eclipse measurements in more refined models of the chromosphere.

References

Ade, P.A.R., Rather, J.D.G. and Clegg, P.E.: 1974, *Astrophys. J.* **187**, 389.
Ahmad, I.A. and Kundu, M.R.: 1981, *Solar Phys.* **69**, 273.
Athay, R.G. and Holzer, T.: 1982, *Astrophys. J.* **255**, 743.
Beckers, J.: 1972, *Ann. Rev. Astron. Astrophys.* **10**, 73.
Beckman, J.E.: 1973, *Nature* **246**, 411.
Beckman, J.E., Lesurf, J.C.G. and Ross, J.: 1975, *Nature* **254**, 38.
Beckman, J.E., Clark, C.D. and Ross, J.: 1973, *Solar Phys.* **31**, 319.
Beckman, J.E. and Ross, J.: 1975, in M. Rowan-Robinson (ed.) *Far Infrared Astronomy*, Pergamon Press, Oxford, p. 79.
Braun, D.C. and Lindsey, C.A.: 1987, *Astrophys. J.* **320**, 898.
Clark, T.A. and Boreiko, R.T.: 1980, *Bull. Amer. Astron. Soc.* **12**, 817.
Clark, T.A. and Boreiko, R.T.: 1982, *Solar Phys.* **76**, 117.
Clark, T.A., Naylor, D.A., Tompkins, G.J. and Duncan, W.D.: 1992, *Solar Phys.* (in press).
Foukal, P.: 1992, *Nature*, **358**, 285.
Gingerich, O., Noyes, R.W., Kalkofen, W. and Cuny, W.: 1971, *Solar Phys.* **18**, 347.
Hermans, L.M. and Lindsey, C.A.: 1986, *Astrophys. J.* **310**, 907.
Horne, K., Hurford, G.J., Zirin, H. and de Graauw, Th.: 1981, *Astrophys. J.* **244**, 340.
Jefferies, J.T. and Lindsey, C.A.: 1988, *Astrophys. J.* **335**, 372.
Labrum, N.R., Archer, J.W. and Smith, C.J.: 1978, *Solar Phys.* **59**, 331.
Lindsey, C.A.: 1987, *Astrophys. J.* **320**, 893.
Lindsey, C.A., Becklin, E.E., Jefferies, J.T., Orrall, F.Q., Werner, M.W. and Gatley, I.: 1983, *Astrophys. J. (Letters)* **264**, L25.
Lindsey, C.A., Becklin, E.E., Orrall, F.Q., Werner, M.W., Jefferies, J.T. and Gatley, I.: 1986, *Astrophys. J.* **308**, 448.
Lindsey, C.A., de Graauw, Th., de Vries, C. and Lidholm, S.: 1984, *Astrophys. J.* **277**, 424.
Lindsey, C.A., Hildebrand, R.H., Keene, J. and Whitcomb, C.E.: 1981, *Astrophys. J.* **248**, 830.
Lindsey, C.A. and Hudson, H.S.: 1976, *Astrophys. J.* **203**, 753.
Lindsey, C.A. and Jefferies, J.T.: 1990, *Astrophys. J.* **349**, 286.
Lindsey, C.A., Jefferies, J.T., Clark, T.A., Harrison, R.A., Carter, M., Watt, G., Becklin, E.E., Roellig, T.L., Braun, D.C., Naylor, D.A. and Tompkins, G.J.: 1992, *Nature* **358**, 308.
Marsh, K.A., Hurford, G.A. and Zirin, H.: 1981, *Astron. Astrophys.* **94**, 67.
Michard, R.: 1974, in R.G.Athay (ed.), *Chromospheric Fine Structure*, Reidel, Dordrecht, p. 7.
Newstead, R.A.:1969, *Solar Phys.* **6**, 56.
Roellig, T.L., Becklin, E.E., Jefferies, J.T., Kopp. G.A., Lindsey, C.A., Orrall, F.Q. and Werner, M.W.: 1991, *Astrophys. J.* **381**, 288.
Shimabukuro, F.I., Wilson, W.J., Mori, T.T. and Smith, P.L.: 1975, *Solar Phys.* **40**, 359.
Simon, M. and Zirin, H.: 1969, *Solar Phys.* **9**, 317.
Vernazza, J.E., Avrett, E.H. and Loeser, R.: 1976, *Astrophys. J. Suppl.* **31**, 1.
Vernazza, J.E., Avrett, E.H. and Loeser, R.: 1981, *Astrophys. J. Suppl.* **45**, 635.
Wannier, P.G., Hurford, G.J. and Seielstad, G.A.: 1983, *Astrophys. J.* **264**, 660.

12-μm OBSERVATIONS AT THE 1991 ECLIPSE

D. E. JENNINGS and D. DEMING

Code 693, NASA/Goddard Space Flight Center, Greenbelt, MD 20771, U.S.A.

G. McCABE

Hughes/STX Corporation, Lanham, MD, U.S.A.

R. NOYES

Harvard-Smithsonian Center for Astrophysics, 60 Garden Street, Cambridge, MA 02138, U.S.A.

G. WIEDEMANN

European Southern Observatory, Karl-Schwarzschildstrasse 2, D-8046 Garching bei München, Germany

and

F. ESPENAK

Code 693, NASA/Goddard Space Flight Center, Greenbelt, MD 20771, U.S.A.

Abstract. The 11 July 1991 total solar eclipse over Mauna Kea was a unique opportunity to study the limb profile of the 12.32 μm MgI emission line. Our observations used the NASA 3-meter Infrared Telescope Facility,[1] and a new Goddard large cryogenic grating spectrometer. Spectra of the line were taken in the slitless mode at second contact. The results show that the emission peaks within ∼ 300 km of the 12-μm continuum limb. This agrees with recent theoretical predictions for this line as a NLTE upper photospheric emission feature. However, the increase in optical depth for this extreme limb-viewing situation means that most of the observed emission arises from above $T_{\min}$, and we find that this emission is extended to altitudes well in excess of the model predictions. The line emission can be traced to altitudes as high as 2000 km above the 12-μm continuum limb, whereas theory predicts it to remain observable no higher than 500 km above the continuum limb. The substantial limb-extension observed in this line is qualitatively consistent with limb-extensions seen by other observers in the far-IR continuum, and may be indicative of departures from gravitational hydrostatic equilibrium in the upper solar atmosphere, and/or may result from temperature and density inhomogeneities. The extended altitude of formation of this line enhances its value as a Zeeman probe of magnetic fields.

Key words: eclipses – infrared: stars – Mg I – Sun: atmosphere

1. Introduction

The Mg I emission lines near 12 μm are the most Zeeman-sensitive lines presently observed in the solar spectrum, and of great potential in studies of solar and stellar magnetic and electric fields – Deming *et al.* 1988, Chang and Schoenfeld (1991). First observed by Brault and Noyes (1983), they were originally believed to be formed in the low chromosphere. Early models of the line formation process showed that the emission lines were, in fact probably photospheric in origin (Lemke and Holweger 1987). Whereas Zirin and Popp (1989) argued for chromospheric formation, recent theoretical work (Hoang-Binh 1991, Chang *et al.* 1991, and Carlsson *et al.* 1991, hereafter CRS) strongly suggests that the lines are formed by departures from LTE in the upper photosphere. Indirect evidence from observations of

---

[1] The Infrared Telescope Facility is operated by the University of Hawaii Institute for Astronomy, under contract with the National Aeronautics and Space Administration. The authors were Visiting Astronomers at this Facility.

the 12.32-μm line – Deming et al. (1988) – puts the region of formation near the temperature minimum, as predicted by the recent theory. It is important, however, to show by direct observation that the line is photospheric. Such observations are not possible under normal observing conditions because the spatial resolution attainable with existing solar telescopes at 12 μm is not adequate. The telescope limitations can be exceeded, however, by observing the line intensity during an eclipse using the motion of the lunar limb to obtain very high spatial resolution. We report such observations here, obtained using slitless IR spectroscopy of the 12.3 μm line at the solar limb during the 11 July 1991 total eclipse – Deming et al. (1992).

2. Observations

The 3-meter NASA Infrared Telescope Facility (IRTF) on Mauna Kea was used to observe the 12.3 μm line at the solar limb. A high-resolution cryogenic array grating spectrometer, developed at Goddard for planetary and astrophysical applications, was adapted for this experiment. This instrument, which we call Celeste, is shown schematically in Figure 1. The spectrometer was mounted at the Cassegrain focus of the telescope. The grating and all other internal optics in the instrument are cooled with LHe. The beam enters through a lens (which is also the dewar window) and is focussed through two filter wheels at an aperture wheel. The beam passes through a hole in the center of the primary mirror (M3) and has its pupil imaged near the secondary (M2). The Cassegrain secondary and primary mirrors expand the beam before it strikes the grating. The diffracted beam from the grating is refocussed through the center of the primary. A fixed sphere (M4) at the focus sends the beam to a tiltable sphere (M5) located next to the secondary. The angle of this final sphere can be adjusted to return the beam to the grating via the fixed sphere and the secondary, and this multi-pass cycle can be repeated up to three times. For the eclipse, Celeste used a 15 × 30 cm^2 grating with 31.6 lines mm^{-1}, which was illuminated by a 12.5-cm diameter beam in 4th order. For the eclipse observations, single-pass was used, and the tiltable mirror imaged the beam at the detector array. The detector array was a 10 × 50 element blocked impurity band (BIB) device. A 386 PC computer controlled the array, stored the data, and communicated with the telescope computer.

Observations were planned for both second and third contact, with telescope motion directed by data computer. Excellent data were obtained at second contact (east limb). Unfortunately, third contact was missed due to a communication failure between the data computer and the telescope control system.

We began preparing Celeste at the observatory one week before the eclipse. However, because of problems with operating the arrays which were solved on the last day, we were not able to mount the instrument on the telescope until approximately six hours before the eclipse. During resolution of the problems, we had to disconnect the internal focus adjustment. The input focus was set to a nominal value based on prior experience in the laboratory. Because we were operating in the slitless mode, where the eclipse edge effectively created our entrance slit, the focus position was not critical. Since the eclipse was an early-morning event, we had only

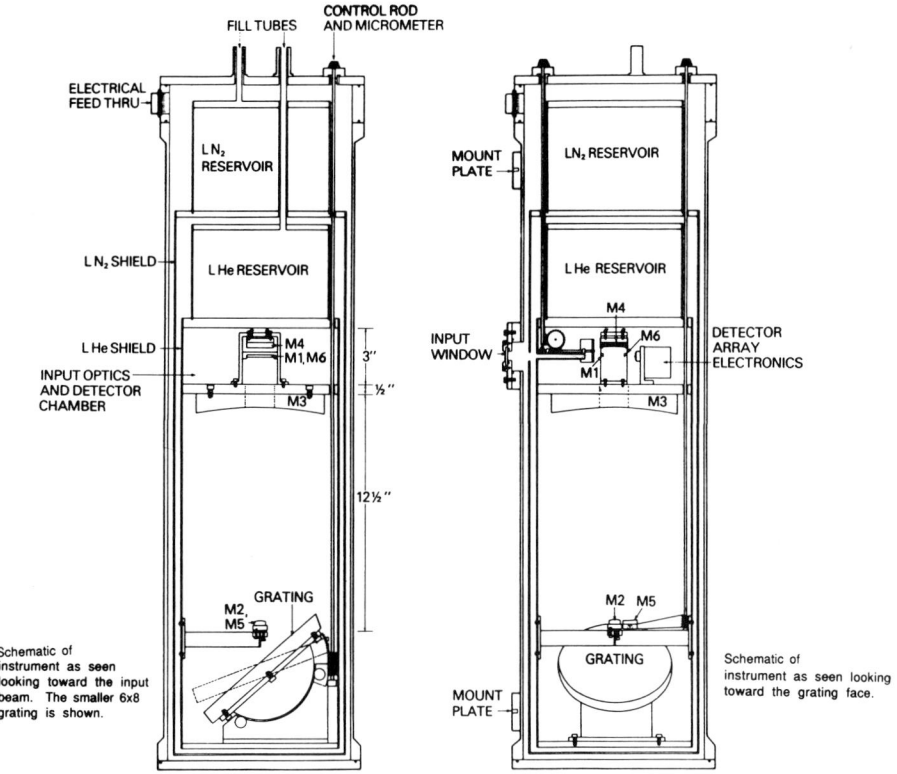

Fig. 1. The cryogenic grating spectrometer, "Celeste", shown schematically. Radiation is focussed by the window/lens, through two filter wheels (fixed and CVF) at an aperture wheel. The beam then is expanded by the Cassegrain system and sent to the 18×33 cm^2 grating (a 15×30 cm^2 grating was used for the eclipse). The diffracted beam is refocussed and transferred by a pair of multi-pass mirrors (used in single-pass) to a final focus at the 10×50 element BIB detector array.

45 minutes available from the time we could first acquire the sun, until second contact at 7:28 AM local time (the eclipse had begun when we first saw the sun!). We set the telescope focus from IRTF documentation, based on the measured position of the Celeste entrance aperture. This left only two adjustments to be made using sunlight: 1) positioning the telescope at the extreme solar limb, and 2) adjusting the grating angle to center the emission line on the detector array. The line was found and centered by first scanning the grating around the previously calibrated position of the line. Limb-brightening of the line was verified by moving the telescope from the limb to a position well on the disk. The resultant grating angle agreed exactly with the prior calibration made using an NH$_3$ gas absorption cell in the laboratory at GSFC. A cold entrance slit was used in Celeste during the line centering and limb positioning, but the aperture wheel was rotated to the slitless position for the eclipse.

The 3-meter telescope aperture was covered with white polypropylene plastic of a variety used for food packaging; we discovered the material by recording the infrared spectrum of a potato chip bag. The polypropylene absorbed and scattered the visible and near-IR radiation enough to prevent damage to the telescope and instrument optics from focussed sunlight. However, at 12 μm the polypropylene was transmissive without significant scattering, allowing the limb to be observed during the setup phase. The polypropylene was removed about 40 seconds before second contact. We observed the second contact limb at the celestial easternmost point of the disk. This point was found by moving east along the bisector of a north-south chord, and we estimate that the uncertainty in the limb positioning was about 5 arcsec. The east-west positioning could be checked in the observed data by measuring the shift of the line peak with respect to the center of the array. This shift was found to be 8 arcsec.

Several cold entrance apertures are selectable in Celeste. We elected to use a slitless mode for the actual observations, choosing a rectangular aperture which projected to 6×20 arcsec^2 on the sky. We oriented the aperture and array with their long axis perpendicular to the solar limb, and the spectral dispersion was along the long (50-pixel) axis of the array. The rationale for slitless operation was to make sure that the limb was captured within the aperture at second contact, to maximize the signal-to-noise ratio by eliminating the losses which occur using an entrance slit, and to possibly sample more than a single location along the limb (this latter advantage was not realized in practice.) Our concern about the signal-to-noise ratio reflects the facts that the solar intensity is much reduced in the 12-μm region, as compared to the visible, and that high temporal and spatial resolution were needed. Even so, it was necessary to read and co-add at the maximum rate to minimize saturation of the array. Each integration was the result of co-adding approximately 50 frames of the array, which was read at 250 Hz, the maximum rate possible with our data system. The observations consisted of a sequence of 150 consecutive measurements, each consisting of 0.2 second integration plus 0.12 second of overhead. Since the lunar limb motion was 0.53 arcsec sec^{-1}, the spatial resolution normal to the solar limb could be as high as 0.15 arcsec. No sky chopping was used. After second contact, the thermal background level on the moon was measured at the center of the lunar disk, and after third contact flat-field calibration of the array response was performed on the dome wall and with the entrance aperture blocked with warm sources (hand, soldering iron).

3. Data Reduction

Each frame of data was processed by subtracting a lunar background frame (a small contribution), and dividing by a flat-field frame. Inspection of the data after this process reveals that the line emission is about an order-of-magnitude broader than the expected 0.03 cm^{-1} diffraction-limited resolution. This is not altogether surprising, since we were running in slitless mode and had had no opportunity to check the telescope focus. Seeing and defocussing probably account for most of the degraded resolution. Resolution was not critical in this experiment, however, since we only needed to be able to distinguish the line from the continuum. An

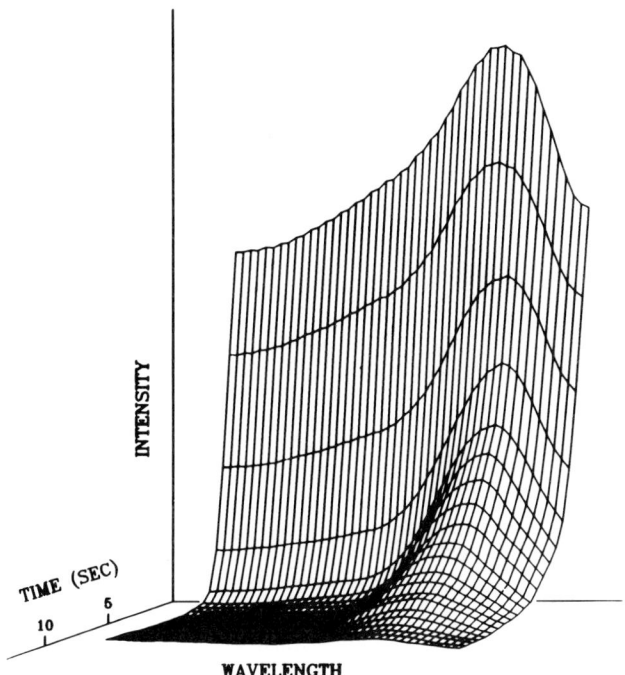

Fig. 2. Spectral data for the 12.32 μm line near second contact, in a 3-D representation. Wavelength increases to the left, frequency increases to the right, with 0.02 cm^{-1} per point. The time scale is equivalent to height in the solar atmosphere, 390 km per second of lunar limb motion), and the vertical axis is relative intensity. The first two detector columns were not operational, so only 48 columns are illustrated.

additional, possibly related, characterisic of the reduced data is that each of the 10 rows of the array gives essentially identical results for the timing of the line and continuum disappearance, consistent with a spatial resolution comparable to the 6 arcsec extent of the array.

Since there seems to be no difference in timing of the eclipse along columns of the array, we averaged along columns to produce a single line spectrum for each frame. The resulting time sequence at second contact is illustrated in Figure 2. The wavelength axis represents the column number of the array, and corresponds to spectral dispersion (0.02 cm^{-1} per column). The intensity axis corresponds to the average in each column of the array. The remaining axis is time, and 21 successive integrations are shown by the lines across the surface at equally spaced time intervals. The continuum radiation from the solar limb in the 1.0 cm^{-1} bandpass of the spectrometer was sufficient to saturate the array until a few seconds before second contact. Figure 2 represents about 7 seconds of data, corresponding to only a few arcseconds of lunar limb motion. The portion of the data near zero on the

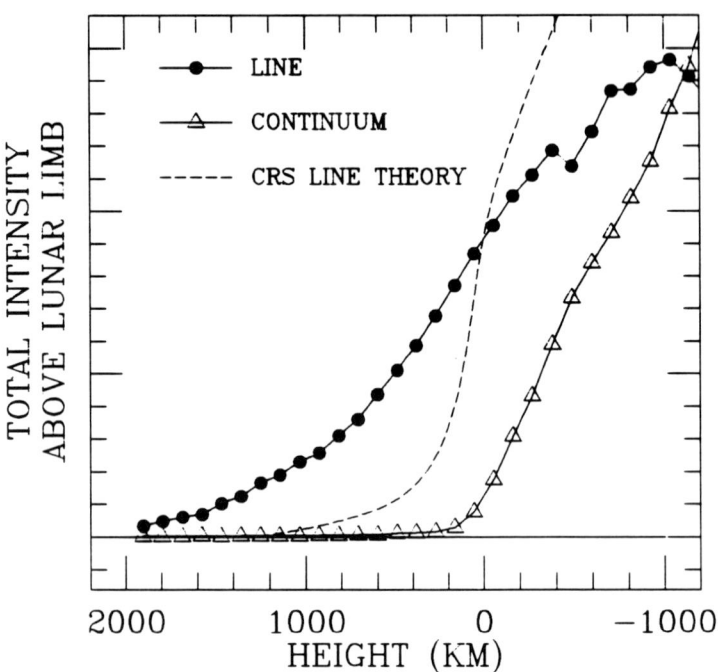

Fig. 3. Observed continuum and 12.32 μm line signal above the height of the lunar limb. The line signal is the total line energy, obtained by integrating the data over wavelength, shown in comparison to the variation predicted by the CRS theory. Height zero is defined to occur at the 12.3-μm continuum limb, and the line and continuum signals are on different relative intensity scales (see text).

time-axis origin is representative of the 12 μm continuum, which drops rapidly to zero at second contact. The emission line is the "ridge" structure apparent in the figure (peaking near column 38); *the line can be observed for up to 5 seconds following the disappearance of the continuum*. Its position on the wavelength scale in Figure 2 is a consequence of the east-west spatial position of the limb within the slitless aperture. In order to derive timing information about the visibility of the line, we averaged the intensity of the line over its profile in each frame. We made no attempt to correct for the fraction of line emission which missed the array on the right in Figure 2. The wavelength-integrated line signal, and the continuum signal, are shown versus time in Figure 3. The line signal was derived by subtracting a continuum contribution from the total flat-fielded array signal. The continuum contribution was calculated by taking column 3 of the flat-fielded array (the leftmost point on Figure 2) as representative of the continuum level, and assuming it to be constant with wavelength. Since the relevant data span only a few seconds of time,

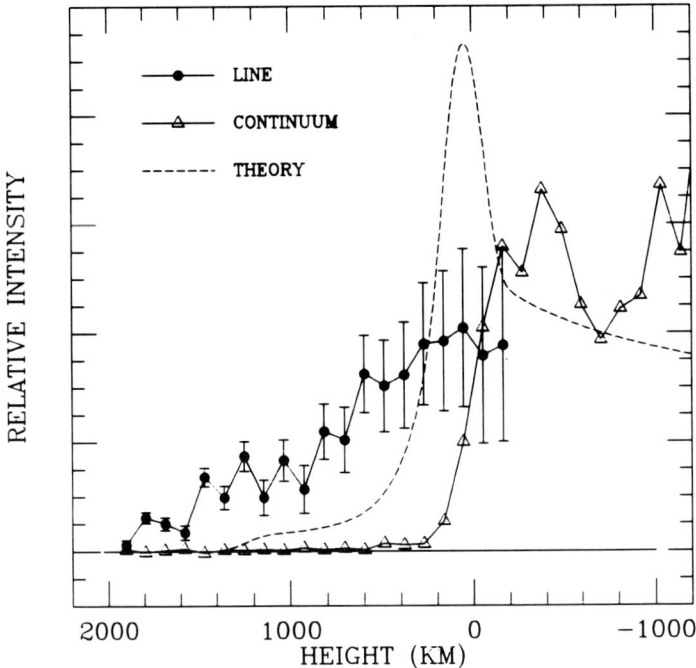

Fig. 4. High resolution limb profiles for the line and continuum, obtained by differencing the Figure 3 data. The dashed line shows the extreme limb profile expected for the total line intensity, computed using the CRS line formation theory, and convolved to the 220 km resolution of the observations. The line and continuum relative intensity scales are different by factor of 10.

the line position does not shift appreciably on the array, and the relative shape of the limb emission profile is unaffected by the fraction of the total line intensity which is missed. The Figure 3 line and continuum data are plotted on different relative intensity scales so that they can be compared more easily; the intensity scale for the line energy is expanded by a factor of 10 from the continuum scale. Also illustrated on Figure 3 is a simulation of the total line energy expected versus time, based on the CRS theory.

The data in Figures 2 and 3 measure the total intensity from regions remaining uncovered by the lunar limb. By taking the difference between successive scans, the intensity due to the portion of the limb covered during the corresponding approximately 0.3 second elapsed time can be derived. The differenced data represents a high spatial resolution scan of the solar limb. The differenced series for our data are shown on Figure 4 for both the line and continuum signals. The height in Figure 4 is defined to be zero at the 12-μm continuum limb; the visible continuum limb is at -400 km on this scale (Chang et al. 1991). Noise due to atmospheric fluctuations is present in the total signals (Figure 3), and becomes more evident in the differenced data (Figure 4). The noise in the original data (Figure 2) can be estimated by the departures of individual values from a smooth (e.g., linear) temporal variation; these fluctuations are typically 3% of the total signal, and occur primarily on a "slow" time scale (typically one to several seconds). Before second contact, when the total signal, dominated by continuum, is largest, its noise fluctuations can result in prohibitively large errors in the differenced line signal. We therefore only plot emission points according to the criterion that the differenced line signal must be larger than a 3% fluctuation in the total signal. This limits us to points at and above the continuum limb (Figure 4). The error bars shown were calculated by assuming that the emission is itself affected by 3% sky-transparency fluctuations, as well as by statistical fluctuations in the thermal background emission. Near the limb, the error bars exceed the fluctuations in the observed points, which is likely due to the fact that the differencing removes much of the relatively slow transparency fluctuations. Well above the limb, the error bars are dominated by statisical background fluctuations, and the fact that the observations show scatter in excess of the errors indicates either that real structure is being observed, or that our estimate of errors is too small in this regime.

In order to estimate the extent of degradation in spatial resolution caused by the irregularity of the moon's limb, we used the charts of Watts (1963) to calculate the range of heights which might be present due to lunar topography, including libration for the summit of Mauna Kea. The relevant topography has a range of heights corresponding to a variation of 0.3 arcsec over our 6-arcsec aperture width. We therefore assume that our height resolution of the solar atmosphere is about 220 km, and this is consistent with the appearance of the 12-μm continuum limb in Figure 4.

The theoretical limb profile of the emission from the CRS theory, accounting for the sphericity of the emitting layers at the extreme limb, is also shown in Figure 4. We extended the CRS model to the smaller optical depths needed for our comparison by reference to the Chang et al. (1991) calculations. We proceeded by adopting the CRS model and extrapolating their line source function to smaller optical depths by assuming that, at line center, $d \log S / d \log \tau = -0.128$. This gradient was taken from Figure 1 of Chang et al. (1991). The actual line profile for these calculations was assumed to be gaussian, with a FWHM of 0.017 cm^{-1}. Because our observations refer to the total intensity in the line, we took the integrated signal in the line. To account for spatial resolution, the resultant limb emission profile was convolved with a rectangular function of 220 km width to simulate the degradation due to lunar topography. The resultant profile is shown as a dashed line on Figure 4. Its sum over height is represented by the dashed line in Figure 3.

4. Results and Discussion

Figure 4 shows that the observed emission profile has a maximum very close to the 12-μm continuum limb, as predicted by theory for an NLTE emission line. The emission peak is not displaced by the approximately 700 km which would be required for an LTE chromospheric emission (Chang et al. 1991). On the other hand, the peak in the emission is not nearly as distinct as predicted by the CRS theory, and the emission extends much higher than calculated from the theory. Using the continuum as a reference, it is possible to compare the magnitude of the observed line energy with theory. The observed line intensity scale on Figure 4 has been expanded by a factor of 10 relative to the continuum scale. Recalling that the continuum refers to the total energy in the 1.0 cm^{-1} bandpass of the array, we see that the observed peak line signal is about $0.7/10 = 7\%$ of the continuum signal, *i.e.*, equal to the energy in 0.07 cm^{-1} of the continuum. Our calculations using the CRS line source function indicate that the line-core-to-continuum specific intensity ratio is ~ 2 just inside the limb, but optical thickness makes the total line energy depend on the form adopted for the line profile. Using a gaussian profile of 0.017 cm^{-1} FWHM we can account for only one-third of the observed line intensity. A better theoretical line shape would place a significant fraction of the opacity in the line wings, thereby giving a greater value for the calculated line intensity. Having no observation of a resolved line profile at the extreme limb, the absolute magnitude of the calculated curves in Figures 3 and 4 is not known, and only their shapes are significant. However, observations on the disk (*e.g.*, Brault and Noyes 1983) imply that the total line intensity above the limb cannot be much greater than the theoretical prediction. The altitude extension we observe does not, therefore, imply that the line contains more emission than predicted, but only that the scale height of the emitting layer is several times greater than in the model.

The good agreement which CRS obtain for the observed profiles of the line at $\mu = 1.0$ and 0.2 shows that their model correctly predicts the dependence of the source function on optical depth, $S(\tau)$. However, the *height* dependence of the emission source function, $S(h)$, is not constrained by comparison with previous observations. The heights from the model are computed under the assumption that the solar atmosphere is in gravitational hydrostatic equilibrium (*e.g.*, Vernazza *et al.* 1981, VAL). Since specification of $S(h)$ is critical to proper interpretation of magnetic data taken using the 12.32 μm line, it would clearly be better to derive it from direct observation. Moreover, an extended infrared limb has been seen before, in the continuum. Eclipse observations in the far-IR continuum have shown that the limb is extended to heights well above the predictions of the VAL model. Lindsey *et al.* (1986) found that the 100- and 200-μm continuum limbs were extended by ~ 1 arcsec above the VAL prediction, implying "large departures from gravitational-hydrostatic equlibrium almost immediately above the chromospheric temperature minimum." Although the 12.32 μm line is photospheric when seen on the disk, the increase in optical depth for the extreme limb-viewing situation means that most of the emission observed here arises from above $T_{\min}$, where such departures occur. This is easily seen by comparing Figure 3 of Chang *et al.* (1991) to the position of $T_{\min}$ in the VAL model. However, the Lindsey *et al.* (1986) results require increases in chromospheric density, and such increases may affect the 12.32-μm emission. In addition, the region above the temperature minimum will probably contain spatial inhomogeneities. Such inhomogeneities will have the effect of extending the emission to greater altitudes when viewed at the extreme limb. This effect is also likely to contribute to the extension which we observe. Full deveveopment of theory incorporating departures from hydrostatic equilibrium and spatial inhomogeneities are needed to fully understand the chromospheric portion of the 12.32-μm emission.

Acknowledgements

We thank the IRTF Director and staff for accommodating this experiment (which placed unusual requirements on the facility), and for their excellent support. We are grateful to Mr. Mark Paules of C. P. Converters for providing the polypropylene plastic used to protect the telescope. Dr. Charles Lindsey made a number of valuable suggestions both before and after the experiment. Dr. Mats Carlsson kindly provided tabulations of the CRS model optical depths and source functions, and Dr. Dennis Reuter assisted in preparing our instrumentation for this experiment. The BIB detector array was supplied by Rockwell International Corporation's Science Center. Celeste was manufactured by IR Systems. The data system and array electronics were fabricated by Electromechanical Design and Fabrication, and Wallace Instruments. This work was supported by the NASA Solar Physics Branch, under RTOP 170-38-53-10, and by the Goddard Director's Discretionary Fund.

References

Brault, J.W., and Noyes, R.W.: 1983, *Astrophys. J. (Letters)* **269**, L61.
Carlsson, M., Rutten, R.J., and Shchukina, N.G. 1992, *Astron. Astrophys.* **253**, 567.
Chang, E.S., and Schoenfeld, W.G. 1991, *Astrophys. J.* **383**, 450.
Chang, E.S., Avrett, E.H., Mauas, P.J., Noyes, R.W., and Loeser, R. 1991, *Astrophys. J.* **379**, 79.
Deming, D., Boyle, R.J., Jennings, D.E., and Wiedemann, G. 1988, *Astrophys. J.* **333**, 978.
Deming, D., Jennings, D.E., McCabe, G., Noyes, R., Wiedemann, G., and Espenak, F. 1992 *Astrophys. J. (Letters)* **396**, L53.
Hoang-Binh, D. 1991 *Astron. Astrophys.* **241**, L13.
Lemke, M., and Holweger, H. 1987 *Astron. Astrophys.* **173**, 375.
Lindsey, C., Becklin, E.E., Orrall, F.Q., Werner, M.W., Jefferies, J.T., and Gatley, I. 1986, *Astrophys. J.* **308**, 448.
Vernazza, J.E., Avrett, E.H., and Loeser, R. 1981 *Astrophys. J. Suppl.* **45**, 635.
Watts, C.B. 1963, *Astron. Papers Amer. Ephem.* **17**, 1.

850 μm OBSERVATIONS OF THE 11 JULY 1991 TOTAL SOLAR ECLIPSE

M. W. EWELL, JR., H. ZIRIN and J. B. JENSEN

Big Bear Solar Observatory, Caltech 264-33, Pasadena, CA 91125, U. S. A.

and

T. S. BASTIAN

National Radio Astronomy Observatory, P. O. Box O, Socorro, NM 87801, U. S. A.

Abstract. We present observations of the 11 July 1991 total solar eclipse made from the Caltech Submillimeter Observatory. The 850 μm limb is extended 3380±140 km above the visible limb, and there is a 10% brightening at the extreme limb. The measured limb height agrees with previous work at shorter and longer wavelengths. The run of limb heights with wavelength is well fit by a single electron density scale height. We argue that there is no need to invoke spicule geometry to explain the observations.

Key words: eclipses – infrared: stars – Sun: chromosphere

1. The Experiment

The beam size of the CSO antenna is too large to make a meaningful measurement of the solar limb profile under normal conditions. However, at second and third contacts of a total eclipse it is possible to measure the brightness of the solar crescent even when it is smaller that the beam. The limb profile is essentially the time derivative of the observed brightness. In principle, the accuracy of such a measurement is limited only by the roughness of the lunar limb. In practice, the imperfect tracking of the CSO antenna precluded simply pointing at the contact locations; it is not possible to untangle brightness changes due to tracking jitter from the measured brightness profile. We decided instead to drive the antenna back and forth across the solar crescent with a 2 second period. The time of maximum signal can then be used to determine the antenna position. A similar method has been employed by Lindsey *et al.* (1983).

To avoid focusing too much near infrared radiation on the secondary and receiver, we covered the antenna with a Griffolyn tent (see Horne *et al.*, 1981, and Clark *et al.*, 1983). We used an SIS tunnel junction receiver tuned to 353 GHz. A description of this device can be found in Ellison *et al.* (1989). At each contact, we obtained 4000 data points at a sampling rate of 50 Hz. The WWV time reference was monitored at the observatory, so the precise time that each data point was obtained is known to one-sixtieth of a second. The expected 850 μm contact time was centered in the 80 second observation. The peak-to-peak amplitude of the antenna motion was 30 arcsec for second contact, and 70 arcsec for third contact.

The raw data for both contacts is shown in Figure 1. The position of the visible limb is also indicated. Several points are immediately apparent. Emission is detected well above the visible limb. The motion of the antenna across the lunar limb is evident when there is no solar signal. The peaks are narrower in the third contact data because the antenna was traversing more quickly, and the alternation between wide and narrow peaks was due to the earth's rotation alternately assisting and

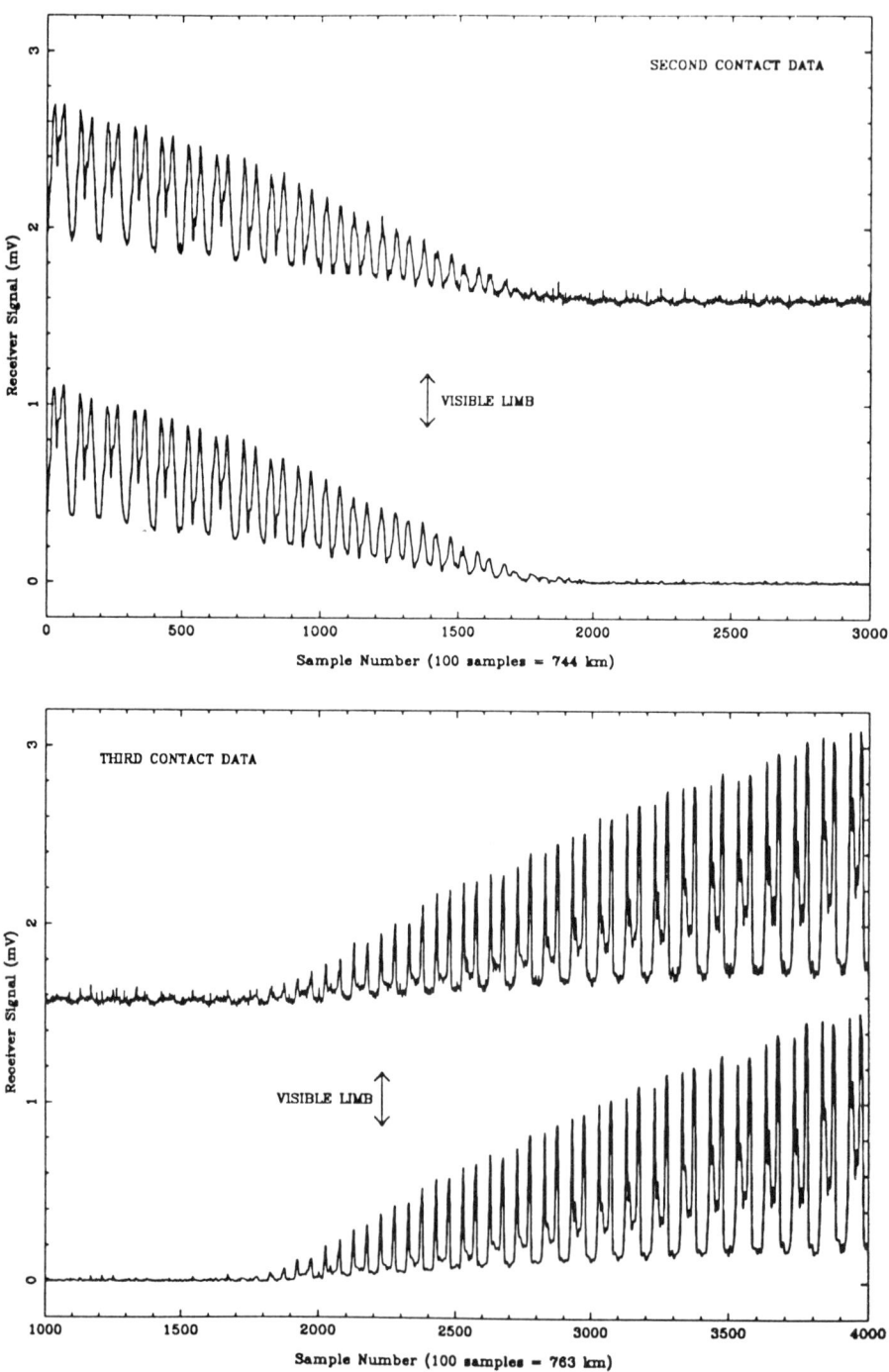

Fig. 1. The raw and smoothed data from second and third contacts.

opposing the telescope motion. The structures between the peaks are due to antenna transients. The heights of the peaks depart from a smooth curve by more than can be accounted for by receiver noise. This effect is due to variations in the atmosphere; the eclipse occured at a 69° zenith angle, and the weather conditions were less than ideal. The sky and moon contributions were subtracted, and the data was then smoothed using a robust non-parametric scheme prescribed by Tukey (1977). The smoothed data is also shown in Figure 1.

The one dimensional beam profile of the CSO antenna was measured by performing a series of drift scans across the lunar limb during the partial phase of the eclipse. The beam profile has also been measured at 290 GHz by Serabyn et al. (1991) using shear interferometry. They found that the beam had a 25" FWHM gaussian core and an 80" extended plateau (20.6" and 66" respectively, when scaled to 353 GHz). Our drift scans were consistent with these values, but we also detected an extended component that was well fit by a 440" FWHM gaussian containing about 17% of the total power. This three component beam was used in the moon subtraction mentioned above, and also in the limb profile determination described in the next section.

2. The Limb Profile

The limb profile is determined from the heights and positions of the peaks in the data. The solar limb is divided into bins whose widths are given by the time separation of the peaks multiplied by the speed of the lunar limb. Each bin is considered in turn, starting with the outermost. A position for the antenna beam is assumed, the brightness of the bin is adjusted to give the observed signal, the position of the beam is moved to the point which would give the maximum signal, and the procedure is iterated until it converges to consistent values for the beam position and bin brightness. Beam deconvolution and differentiation are thus performed simultaneously and self-consistently. If this method is applied directly to the raw data, several of the bins are assigned a negative brightness. This is because the peak heights are not monotonically increasing due to the atmospheric fluctuations. To get around this difficulty, we first fit a quadratic to the peak heights, and used the fitted values in the subsequent analysis. However, for peaks above the limb, we retained the unfitted values. A more complete description of the data reduction can be found in Ewell et al. (1992).

The measured third contact limb is shown in Figure 2. The peak which corresponds to the half-brightness point occured at 17:32:07.8 UT. Visible third contact, with the lunar limb correction applied, occured at 17:32:16.9 UT (Bangert et al., 1989), and the speed of the lunar limb at third contact was 0.50535 arcsec s^{-1}. This gives a 850 μm limb height of 4.6±0.5 arcsec, or 3390±190 km. The error bars are set to one-half of the distance between peaks. The second contact peak at this same brightness level occured at 17:28:18.2 UT, with visible contact at 17:28:09.4 UT and a lunar speed of 0.51843 arcsec s^{-1}, giving a limb height of 3360±190 km. The extremely close agreement between these two values is fortuitous, but it leaves little doubt as to the reliability of the result. Combining the two measurements gives our quoted value of 3380±140 km.

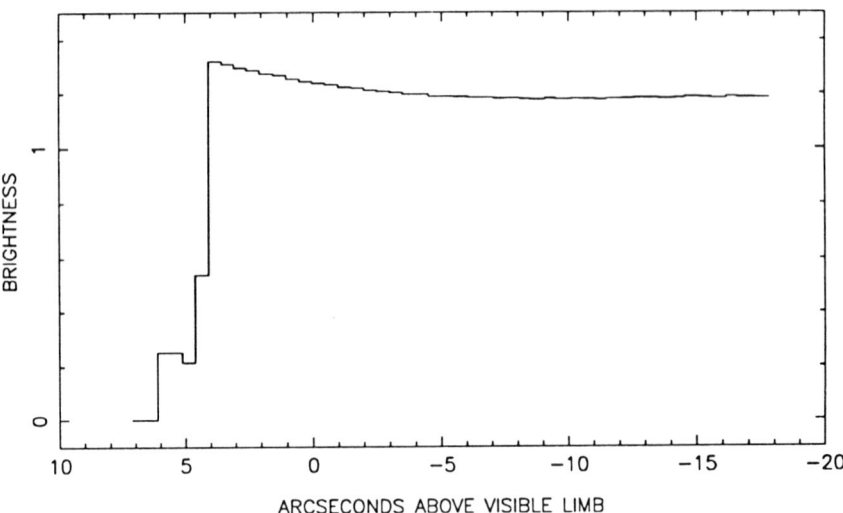

Fig. 2. The third contact limb profile. The half-power point is 4.6″ above the visible limb. The overall normalization is uncertain, but there is clearly a limb brightening of about 10%.

The brightness in Figure 2 is normalized to unit brightness at sun center. There is clearly a 10% brightening in the 5 arcseconds just inside the limb. Of necessity the sun center measurements were taken several minutes before the contact measurements, and because of the atmospheric conditions, the calibration is somewhat uncertain. Therefore, we cannot determine whether the 20% brightening 20 arcseconds inside the limb is real, giving a total limb brightening of 30%, or whether it is simply a calibration error.

3. Discussion

Table 1 shows how our limb height measurement at 850 μm compares with previous measurements by Roellig *et al.* (1991) at 200, 360 and 670 μm and with the value found by Belkora *et al.* (1992) for this eclipse at 3 mm. The data are in mutual agreement, all following a general trend. Table 1 also gives the limb heights predicted by VAL model C (Vernazza *et al.*, 1981), the "standard" model of the solar atmosphere. There is a large and statistically significant discrepency at all listed wavelengths; the far infrared and submillimeter limbs are extended well above where they occur in the VAL. This extension has already been noted by Lindsey *et al.* (1986) as evidence that the chromosphere is not in hydrostatic equilibrium.

We have constructed a very simple model (the Caltech Irreference Chromospheric Model, or CICM) which fits all of the observed limb heights with essentially one adjustable parameter. The opacity source for the conditions and wavelengths of interest is free-free absorption, so we model only temperature and electron density.

TABLE I
Limb Heights, Calculated and Observed

Wavelength (μm)	Observed (km)	VAL limb (km)	CICM limb (km)
200	1450±200	900	1550
360	2300±200	1400	2300
670	3300±200	1700	3100
850	3380±140	1800	3400
3000	5500±400	1800	5400

We use the VAL up to the temperature minimum at 525 km, above which the temperature increases to 6500 K at 1400 km, and to 7500 K at 5000 km. (Model heights are referenced to the photosphere as in the VAL: note that all other heights given in this paper are referenced to the visible limb.) The results are not sensitive to changes of a few hundred kilometers in the 6500 K height; the value adopted is also from the VAL. The temperature at 5000 km was chosen to agree with measurements between 1.4 GHz and 18 GHz by Zirin *et al.* (1991). We take the electron density to fall exponentially from the temperature minimum, and the electron density scale height is our single adjustable parameter. The ionization fraction could be changing rapidly in the modeled region, so the connection between electron density and total density is not clear. The values shown in Table 1 are for a scale height of 1200 km. The agreement with the observations is remarkable; it is not necessary to invoke a rough atmosphere in order to fit the data.

There is an independent line of argument against a rough atmosphere model. Consider the high resolution limb pictures taken at various offsets in the Hα line by R. B. Dunn at Sac Peak (Lynch *et al.*, 1973). The spicules are seen to be sharp and distinct at 0.75 Å and 1 Å from line center, respectively. But at line center, the top of the Hα emission is smooth, and is *higher* than the spicules. Thus it is clear that, at least in Hα, the atmosphere does not have a rough top. Moreover, the height of the Hα limb agrees with the limb height at 3 mm.

Finally, we would like to point out that there is also an independent line of argument showing that the chromosphere is not in hydrostatic equilibrium. Elements with a high first ionization potential are well known to be underabundant in the corona by about a factor of 4 (*c.f.* Breneman and Stone, 1985). The details of the fractionation process are not known, but it clearly involves the acceleration of charged particles. Even if the process was 100% efficient, to achieve the observed abundances it would need to act on 75% of the total mass that reaches the corona, and on an even higher fraction with less than perfect efficiency. Thus most of the mass in the chromosphere is being subjected to accelerations that are not accounted for in the standard atmosphere models, and it is not surprising that the assumption of hydrostatic equilibrium yields predicted limb heights that are unequivocally ruled out by the observations.

Acknowledgements

We would like to thank Prof. Tom Phillips for letting us point his telescope at the sun, Dr. Gordon Hurford for developing the Griffolyn tent, Ms. Michal Peri for help cutting it, Antony Schinckel for his high-altitude acrobatics, and Taco Machilvi for his computer wizardry. We would also like to thank Prof. E. Avrett for several conversations on chromospheric models.

This work was supported by NSF grants AST-9015139, AST-9015755, and by a Caltech SURF.

References

Bangert, J. A., Fiala, A. D., and Harris, W. T.: 1989, *United States Naval Observatory Circular No. 174*.
Belkora, L., Hurford, G. J., Gary, D. E., and Woody, D.: 1992, *Astrophys. J.*, submitted.
Breneman, H. H., and Stone, E. C.: 1985, *Astrophys. J. (Letters)* **299**, L57.
Clark, T. A., Kendall, D. J. W., and Boreiko, R. T.: 1983, *Infrared Phys.* **23**, 289.
Ellison, B. N., Schaffer, P. L., Schaal, W., Vail, D., and Miller, R. E.: 1989, *Int'l J. IR & MM Waves* **10**, 8.
Ewell, M. W., Jr., Zirin, H., Jensen, J. B., and Bastian, T. S.: 1992, *Astrophys. J.*, submitted.
Horne, K., Hurford, G. J., Zirin, H., and de Graauw, Th.: 1981, *Astrophys. J.* **244**, 340.
Lindsey, C., Becklin, E. E., Jefferies, J. T., Orrall, F. Q., Werner, M. W., and Gatley, I.: 1983, *Astrophys. J. (Letters)* **264**, L25.
Lindsey, C., Becklin, E. E., Orrall, F. Q., Werner, M. W., Jefferies, J. T., and Gatley, I.: 1986, *Astrophys. J.* **308**, 448.
Lynch, D. K., Beckers, J. M., and Dunn, R. B.: 1973, *Solar Phys.* **30**, 63.
Roellig, T. L., Becklin, E. E., Jefferies, J. T., Kopp, G. A., Lindsey, C. A., Orrall, F. Q., and Werner, M. W.: 1991, *Astrophys. J.* **381**, 288.
Serabyn, E., Phillips, T. G., and Masson, C. R.: 1991, *Appl. Optics* **30**, 1227.
Tukey, J. W.: 1977, *Exploratory Data Analysis*, Addison-Wesley, Reading, p.523.
Vernazza, J., Avrett, E. H., and Loeser, R.: 1981, *Astrophys. J. Suppl.* **45**, 635.
Zirin, H., Baumert, B. M., and Hurford, G. J.: 1991, *Astrophys. J.* **370**, 779.

OBSERVATIONS OF THE 1991 ECLIPSE AT 3.5 mm WAVELENGTH

S. M. WHITE and M. R. KUNDU

Department of Astronomy, University of Maryland, College Park, MD 20742, U.S.A.

Abstract. We report preliminary results of the partial eclipse seen from the BIMA millimeter-wavelength interferometer in northern California. The use of an interferometer has many advantages over previous single-dish eclipse observations at 3 mm, and in particular it allows a quite different type of measurement of the height of the "3-mm limb" which is direct and precise : we find it to be $11.9'' \pm 0.4''$ above the optical limb.

Key words: eclipses – techniques: interferometric – Sun: radio radiation

1. Introduction

At millimeter wavelengths, as in other wavelength ranges, solar eclipses have long been recognized as providing a means to carry out otherwise-impossible investigations of the Sun's atmosphere. In particular, the motion of the Moon's limb can be used to provide higher spatial resolution than a millimeter telescope can achieve alone. Since the typical resolution of single-dish telescopes operating at 3 mm wavelength is not great ($1.1'$ for a 10-m telescope; $15''$ for the Nobeyama 46-m telescope), it is important to make use of any opportunities for better spatial resolution. During an eclipse we can achieve this by using the fact that the difference between two successive measurements made while the Moon is moving across the Sun represents the flux from a narrow strip whose width equals the distance moved by the Moon in the corresponding interval. Since typically the Moon moves at $0.5''$ per second, then a 1 second integration time can produce subarcsecond resolution in the direction of the Moon's motion.

Within the last few years the intrinsic resolution achievable at millimeter wavelengths has improved enormously through the development of millimeter-wavelength interferometers. Information on arcsecond-scale structures in flares is now readily available (Kundu *et al.*, 1990; White *et al.*, 1992). However, the Sun is so complex and constantly changing that true imaging is still difficult with the 3-element interferometers presently available. Thus the structure of the quiet-Sun atmosphere has not yet been thoroughly studied with arcsecond resolution, and eclipse observations are still important for such studies.

We observed the 1991 July 11 eclipse with the 3-element Berkeley-Illinois-Maryland Array (BIMA) in northern California at 3.5 mm wavelength. Seen from this location, the Moon obscured half the solar disk. However, since the field of view of the telescope is about $2'$ there is little one can do with a total eclipse that cannot also be achieved in a partial eclipse. Here we report the results of a preliminary analysis of the BIMA observations. We focus mainly on the height of the "3-mm limb", since the use of an interferometer provides a quite different means for measuring this quantity, which proves to be very precise and much less dependent on model comparisons.

2. The Observations

BIMA consists of three 6-m dishes operating as an interferometer in the 70–115 GHz frequency range. It is important to bear in mind some aspects of interferometers in interpreting these observations. Each pair of antennas provides an interferometer pair, and the output of each interferometer pair is a one-dimensional spatial Fourier transform of the brightness distribution within the primary beam of each 6-m dish (2.3′). Two sidebands at 86 GHz and 89 GHz were the observing frequencies. The fringe spacing and orientation of the Fourier pattern on the sky are determined by the locations of the two antennas: for these observations the 3 BIMA antennas were at locations 12.2 m W, 12.2 m N and 30.5 m N of the array center. The resulting baselines provided fringe spacings typically of 40″, 40″ and 20″, with differing orientations. The interferometer measures both an amplitude and a phase. If there is a single source dominating the flux in the field of view, the phase is a measure of its location (in one dimension) relative to the center of the field of view; if the source is moving, the phase will change by 360° as the source moves a distance corresponding to one fringe spacing in the appropriate direction. This is important, as we show below. Solar observations with BIMA are described more fully by White and Kundu (1992).

3. First Contact

First contact occurred at 17:18 UT at a position angle of 253° (measured east around the solar circumference from celestial north). The geometry of the eclipse is very similar to the figure shown by Belkora *et al.* (1992) for the Owens Valley observatory, except that Hat Creek is further north and first contact accordingly occurred further south on the Sun's west limb. The Moon's velocity vector relative to the Sun at the time was $(0.48'' \, s^{-1}, -0.09'' \, s^{-1})$ in RA and Dec., respectively. The projected component along the normal to the Moon's surface at the point of contact was $0.44'' \, s^{-1}$. The time resolution of the BIMA observations was 0.4 s, so the intrinsic spatial resolution was 0.2″ in the direction of motion.

The curvature of the Moon's limb across the field of view of the BIMA dishes was only 2.3″, so to a good approximation both it and the Sun's limb can be regarded as straight edges (given that the fringe spacing to be used was 40″). The limbs are nearly vertical on the sky during first contact. For such an arrangement, an east-west baseline gives optimal results because the sharp edge in the brightness distribution at the solar limb is parallel to the (north-south) fringes and produces a strong response in the interferometer. At first contact the baseline formed from antennas 1 and 2 was the most east-west of the baselines, with an effective EW fringe spacing of 51″ and NS fringe spacing of 95″; the position angle of the fringes is therefore 242°, compared with 252° for the limbs.

The observation of first contact on baseline 12 at 89 GHz is shown in Figure 1. The three curves are the correlated amplitude (solid line) and phase (dots) measured by the interferometer pair, as well as the total power (solid line; arbitrary units) measured by one of the telescopes acting as a single dish. The amplitude and total power plots have both been smoothed with a 4-second boxcar; however,

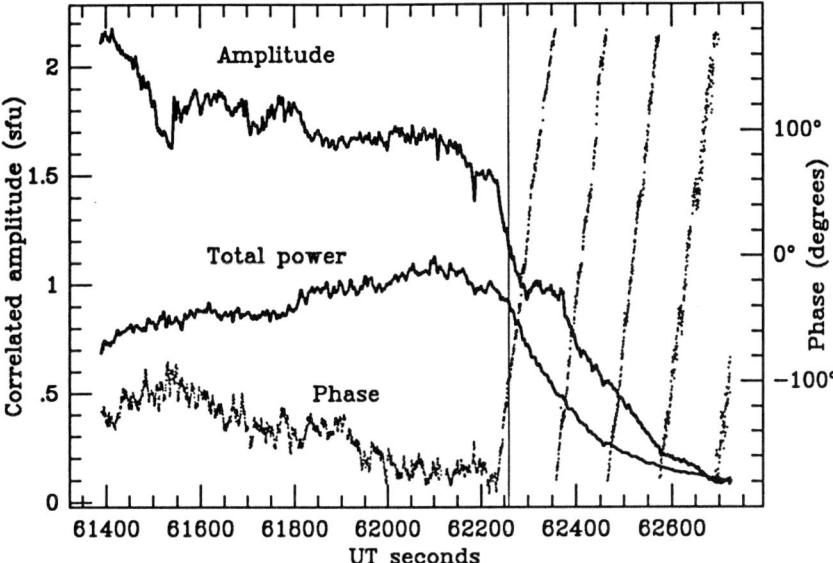

Fig. 1. Data from baseline 12 at BIMA at 89 GHz during first contact. The three curves show the correlated amplitude (solid line), the phase (dots) and the single–dish total power (solid line; arbitrary units). The thin vertical line indicates the time of optical first contact.

no smoothing has been applied to the phase data. The time of optical contact is indicated by a vertical line.

The immediately striking aspect of this figure is the rapid change in the phase properties, which occurs 27 seconds prior to the first contact of the optical limbs. In Figure 2 we show the phase variation at the time of the eclipse on an expanded scale for both sidebands (86 GHz and 89 GHz). Effectively, prior to the arrival of the Moon the phase is a measure of the position of the solar limb where the brightness temperature jumps by 7000 K, and thus the phase is roughly constant with time; variations in the phase prior to the eclipse are primarily due to atmospheric effects. However, once the Moon starts covering the Sun, the effective solar limb begins to move, and the phase begins to change in a linear fashion accordingly (called "phase winding" in radio astronomy). One wind of 360° phase takes about 110 s, corresponding to a distance of 50″, which is consistent with the fringe spacing on baseline 12.

The onset of motion is clearly very sharp. This seems to imply that the Sun's limb at 3 mm wavelength is sharp, although further modelling will be required to determine this. We can estimate the height of the "3-mm limb" from the time between the onset of the eclipse at BIMA, and the ephemeris for the optical eclipse (the latter has been calculated independently by us and by D. Gary, Caltech). From Figure 2, the onset time at BIMA can be determined to better than 1 second; we thus estimate this time difference as 27 ± 1 s, corresponding to a height of 11.9″ ±

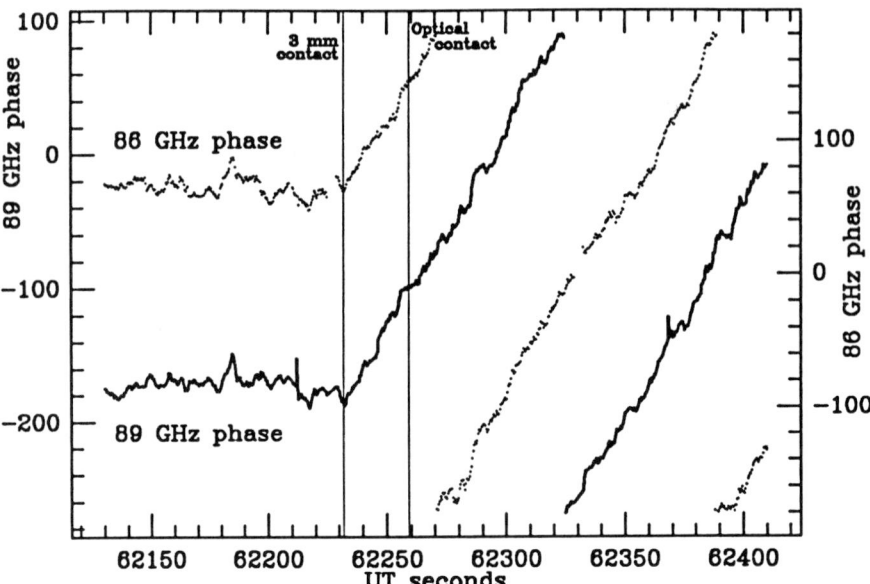

Fig. 2. The variation of the phase with time at both 86 GHz and 89 GHz at the time of first contact. Units are degrees.

0.4″ (8000 km). The onset of the phase winding appears to be very sharp, implying that the contrast at the "3-mm limb" is also sharp. The total power rises somewhat as the eclipse approaches. This is probably due to the warm (220 K) surface of the Moon filling a previously empty region of sky. Note also that the total power curve is typical of previous observations of eclipses at 3 mm: the difference in determining the onset of the eclipse at 3 mm from the phase information and from the total power (or even correlated amplitude) information is readily evident.

4. Chromospheric models

Previous measurements of the height of the limb at similar wavelengths include those of Coates et al. (1958; 10″ ± 5″ at 4.3 mm), Takahashi (1967; half–power height of 16″ at 3.0 mm), Simon (1971; < 10″ at 3.5 mm), Swanson (1973; 13″ ± 5″ at 3.2 mm), Labrum et al. (1978; 7″ ± 1″ at 3.0 mm), Belkora et al. (1992; 7.5″ ± 0.8″ at 3.0 mm) and Horne et al. (1981; 8.6″ ± 1″ at 1.2 mm). We note that these measurements are not all directly comparable because most have used comparison of observations with predictions based on models, and different observers have found that different models fit their data best. The observations by Belkora et al. (1992) are nearly identical to ours in nature, for the same eclipse, and it is surprising that two such similar experiments should produce such different results. There is always the possibility that a localized large-scale structure at the limb will determine the measured height, but in this case we can see no such structure in the Hα image (there is a filament well south of the field of view). There are no features evident on the limb of the Moon at the relevant location which can explain differences larger than 1″. Thus we do not presently understand the reasons for the difference between our measurement and Belkora et al. (1992).

It is well known from previous measurements that the limb height at 3 mm is inconsistent with homogeneous chromospheric models (*e.g.*, those of Vernazza et al. 1981 predict a 3-mm height of at most 2000 km or 3″). The explanation most commonly given for this result is that the observed height is actually the top of the forest of spicules which will dominate lines of sight tangent to the solar surface. Spicules are jets of gas seen at the solar limb to be projecting up from the surface through the chromosphere to heights typically of $\sim 8''$. The height measured here at 3 mm wavelength is somewhat larger than in most previous observations. Although spicule heights greater than 10″ are common, they usually occur for individual isolated spicules which would not cover enough of the solar limb to explain a measurement over a large field of view such as this one. Thus it is not clear that the height of 12″ found here for the "3-mm limb" can be explained by the spicule interpretation.

More detailed modelling is required in order to derive a brightness profile across the limb from these data (this has been carried out for the Owens Valley data by Belkora et al. 1992). It is however clear from the data that no sharp limb spike, such as those reported by Hagen (1957) and Beckman et al. (1975), was observed. We also observed the covering and uncovering of several active regions and a large filament to search for small-scale features, and these data will permit us to obtain some information on the distribution of 3-mm-emitting regions with respect to optical features. Fourth contact was observed in these observations, but no useful data were obtained because none of the baselines were suitably oriented.

Acknowledgements

We thank Rick Forster and the staff of BIMA for help with the observations, and D. Gary, W. Ewell and L. Belkora for helpful discussions. Solar research at Maryland is supported by NSF grant ATM 90-19893 and NASA grants NAG-5-1540, NAG-W-1541 and NAG-W-2172. Scientific research at U. Md. with BIMA is supported by NSF grant AST 91-00306.

References

Beckman, J. E., Lesurf, J. C. G., and Ross., J.: 1975, *Nature* **254**, 38.
Belkora, L., Hurford, G. J., Gary, D. E., and Woody, D. P.: 1992, *Astrophys. J.*, submitted.
Coates, R. J., Gibson, J. E., and Hagen, J. P.: 1958, *Astrophys. J.* **128**, 406.
Hagen, J. P.: 1957, in H. C. van der Hulst (ed.), *Radio Astronomy*, IAU Symp. No. 4, Cambridge Univ. Press, Cambridge, p. 263.
Horne, K., Hurford, G. J., Zirin, H., and de Graauw, T.: 1981, *Astrophys. J.* **244**, 330.
Kundu, M. R., White, S. M., Gopalswamy, N., Bieging, J. H., and Hurford, G. J.: 1990, *Astrophys. J. Lett.* **358**, L69.
Labrum, N. R., Archer, J. W., and Smith, C. J.: 1978, *Solar Phys.* **59**, 331.
Simon, M.: 1971, *Solar Phys.* **21**, 297.
Swanson, P. N.: 1973, *Solar Phys.* **32**, 77.
Takahashi, K.: 1967, *Astrophys. J.* **148**, 497.
Vernazza, J., Avrett, E., and Loeser, R.: 1981, *Astrophys. J. Suppl.* **45**, 635.

White, S. M., Kundu, M. R., Bastian, T. S., Gary, D. E., Hurford, G. J., Kucera, T., and Bieging, J. H.: 1992, *Astrophys. J.* **384**, 656.
White, S. M., and Kundu, M. R.: 1992, *Solar. Phys.*, in press.

NEAR IR OBSERVATIONS OF THE 11 JULY 1991 TOTAL SOLAR ECLIPSE FROM MAUNA KEA, HAWAII

T. A. CLARK

Physics Department, University of Calgary, Calgary, Alberta T2N 1N4, Canada

D. A. NAYLOR and G. J. TOMPKINS

Department of Physics, University of Lethbridge, Lethbridge, Alberta T1K 3M4, Canada

and

C. LINDSEY

Solar Physics Research Corporation, 4720 Calle Desecada, Tucson, AZ 85718, U.S.A.

Abstract. Near IR total eclipse measurements have provided clear evidence during both 2nd and 3rd contacts for a limb extension of about 125 km for wavelengths in the range containing the CO fundamental vibration-rotation bands between 4.3 and 5.5 µm, when compared to the limb at nearby shorter wavelengths. This is interpreted as a "flash" spectrum in the CO lines, with the above extension representing the outer level of the CO emission layer. This height can be compared to the $\tau_{CO} = 1.0$ level incorporated into recent representative atmospheric models (Ayres and Wiedemann, 1989) which is 90 km above the visible limb for a semi-empirical "hot chromosphere" model (Avrett, 1985) and 220 km for a "cool" radiative equilibrium model based upon work by Anderson (1989).

Key words: CO molecule – eclipses – infrared: stars – Sun: chromosphere

1. Introduction

The vibration-rotation bands of CO in the near IR represent a valuable probe of the conditions in the atmospheres of the Sun and other cool stars (e.g. Ayres and Testerman, 1981; Heasley et al., 1976). In the solar atmosphere, limb darkening of lines in these CO bands (Ayres and Testerman, 1981) and the line characteristics of low core brightness temperature and no central emission peak reinforced the conclusion of Noyes and Hall (1972), that the CO does not participate in the "normal" chromospheric heating. The upper photosphere and chromosphere appear to be bifurcated into a hot chromosphere ($T \sim 6000$ K) within the chromospheric network or above UV bright points and a cool ($T \sim 3500$ K) atmosphere above supergranular cells containing the majority of the CO gas (Ayres, 1990). Models of the solar atmosphere have been devised which incorporate both the CO cool regions and the hotter network from which the reversed cores of chromospheric lines arise and which include the expected height distribution of the CO emission. So far, no observation has measured this distribution at the limb.

The present measurement represents a preliminary attempt to determine the vertical extent of the CO layer by monitoring the intensity of the near IR spectrum between 2.8 and 5.4 µm including the fundamental CO V-R band, during the 11 July, 1991 total solar eclipse from Mauna Kea. This experiment was carried out as part of the visible eclipse monitoring in support of the international collaborative millimeter wave eclipse experiment on the 15-meter James Clerk Maxwell Telescope (Lindsey et al., 1992).

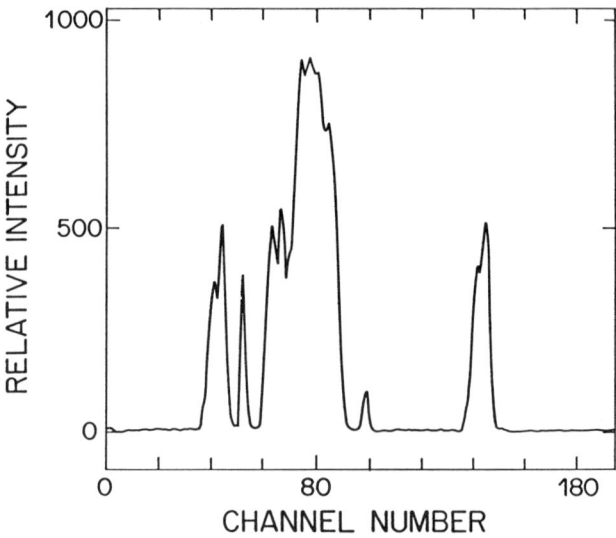

Fig. 1. Typical CVF spectrum of the Sun, taken during the eclipse.

2. Instrumentation

A heliostat and 0.125 meter f/24 telescope mounted on the roof of the JCMT entrance porch on Mauna Kea projected a 28-mm image of the Sun onto a slit of width 0.52 mm and length 4 mm (35 × 270 arcsec^2), behind which was placed a 4-segment continuously-variable filter wheel, rotated at 200 steps per revolution by stepping motor, at 1 revolution per second. Focussing optics re-imaged the slit onto a 1-mm InSb detector, while a multi-blade chopper placed in this beam modulated the resulting light at about 800 Hz. Three of the filter segments were within the pass-band of this detector, a single 2.5-4.5 μm filter (the "near IR" channel) with a single-step resolution of 0.045 μm resolution, and two 4.4-8.0 μm filters for which the detector sensitivity limited the observed spectrum to 5.4 μm. These latter two filters, on opposite sides of the filter wheel, covered the spectral region containing the CO fundamental V-R bands at 4.3-5.5 μm twice per second to a single-step resolution of about 0.08 μm. The resulting signals from the phase-sensitive detection system were digitized and stored in synchronism with the stepping motor drive and were tape-recorded along with standard time signals and displayed upon a high-speed chart recorder for monitoring and record-keeping during the eclipse. A beam-splitter fed a visual image through a 5 × 10 arcmin^2 slit onto an unfiltered Si photocell for visible eclipse monitoring. This signal was also digitized and recorded in synchronism with the millimeter wave signal monitored by the JCMT. A video camera recorded both the visual image and WWV time signals to provide an audio and video log-book of the experiment. A typical spectrum, recorded about 5 minutes before 2nd contact, is shown in terms of channel number in Figure 1, the CO band regions occupying channels 37-48 and 137-148 respectively.

3. Observing Conditions

The eclipse on Mauna Kea began when the Sun was only 8° above the horizon, or just a few degrees above the mountain summit, to the east of the JCMT. Thus, the early calibration spectra were taken at disk center through several air-masses. Totality occurred when the Sun had reached an elevation of 21° while 4th contact was observed at 37° elevation, at an air-mass of 1.66. Localized clouds were seen over the summit in the direction of the Sun soon after 1st contact but cleared through the central eclipse and totality periods, only to return again several times during the recovery beyond 3rd contact. These clouds did not affect the signal around totality. Indeed, the noise level on the signal decreased significantly around totality. Thin high clouds extended across the sky throughout the eclipse but did not appear to affect the IR signal.

The Moon moved across the Sun's disk at 0.519 arcsec per second. Thus, the spectrophotometer provided an angular resolution at the Sun's limb of about 0.25 arcsec in the CO band region and about twice this figure for the near IR band. Several calibration runs on the solar disk center were recorded before and after the eclipse. Digital recording of the period around totality was started 5 minutes before 2nd contact when the width of the remaining arc of the solar disk had become less than about 1/2 of the slit length. The whole photometer head was rotated to align the slit to be perpendicular to the eclipsed limb. This slit alignment for 2nd contact was within a few degrees of optimum alignment for 3rd contact and no attempt was made to re-align the slit during totality.

4. Results

Eclipse curves are shown in Figures 2a and b for the few seconds around 2nd and 3rd contacts respectivley, for a single channel in the 4.0 μm "continuum" region where solar CO absorption lines are absent and where atmospheric absorption is low, for a single channel at 4.6 μm within the region of strong solar CO absorption and for the visible channel at 2nd contact. These data have been normalized to agree at a point some 10 seconds before 2nd contact for Figure 2a and 10 seconds after 3rd contact for Figure 2b, and an equivalent solar-disk distance scale has been placed upon these graphs. It is immediately apparent that the CO-band 2nd contact occurs a small but significant time later than that for the visible or continuum near-IR contact, and that the order is reversed for the 3rd contact data by about the same time difference. This is interpreted to indicate that the CO bands are exhibiting a flash spectrum in showing remanent emission up to about 120 km above the near IR and the visible limbs, or about 460 km above the $\tau_{0.5} = 1$ level in the solar atmosphere.

Figure 3 shows a more detailed 2nd-contact differential eclipse curve using normalized and averaged data from 5 adjacent channels in the near IR continuum,

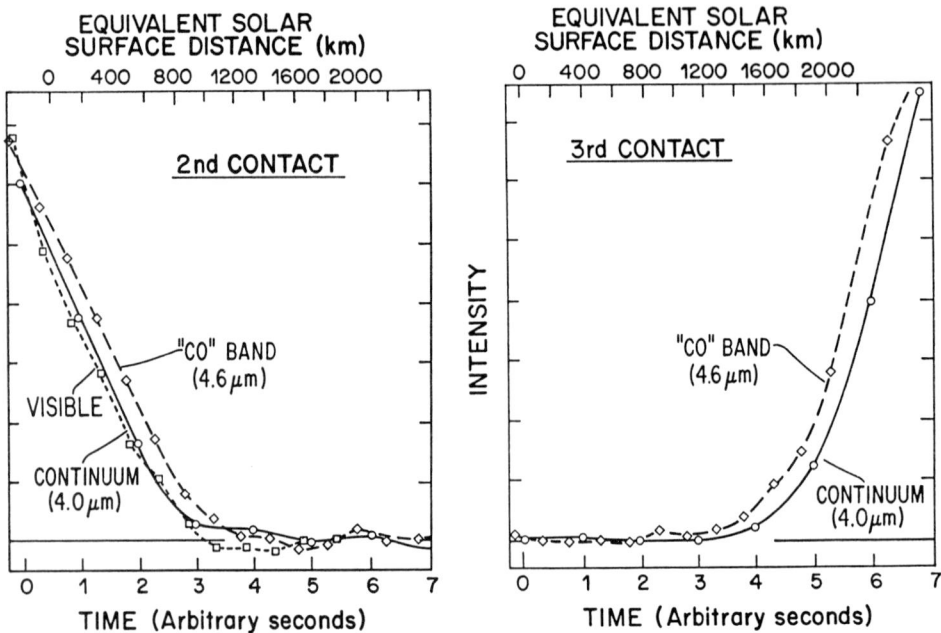

Fig. 2. Panels a and b show respectively the occultation curves for the 2nd and 3rd contacts, for single channels in the visible, near-IR continuum, and CO band regions. One second of time corresponds to 0.519 arcsec.

from 4 adjacent channels in the CO-band region and from a smoothed visible channel record. A first attempt has been made in these data to calibrate the intensity in terms of the fraction of equivalent disk-center intensity per resolution element (0.519 arcsec wide, at 1 spectrum per second) using early and late spectra for this radiometric calibration. This absolute intensity scale is tentative only and is dependent upon the atmospheric-extinction correction over the wide range of airmass encountered during this eclipse. The visible record has been normalized to the expected limb darkening at $\mu = \cos\theta = 0.2$ from standard limb darkening curves and agrees reasonably with the eclipse curve of Weart (unpublished, but shown in Athay (1976)) which is shown for comparison with the visible curve. The equivalent visible eclipse record of Makita (1972), modeled by Noyes and Kalkofen, shows a much steeper final decrease than either the present record or that of Weart. The scale height of the final visible limb decrease for the present data is about 840 km compared to 535 km for the data of Weart and about 100 km for Makita's curve. By comparison, the IR continuum scale height is 175 km while that for the CO band region is 185 km. In this graph, the solar limb position has been placed at the point of inflection of the near IR continuum curve to match these other visible

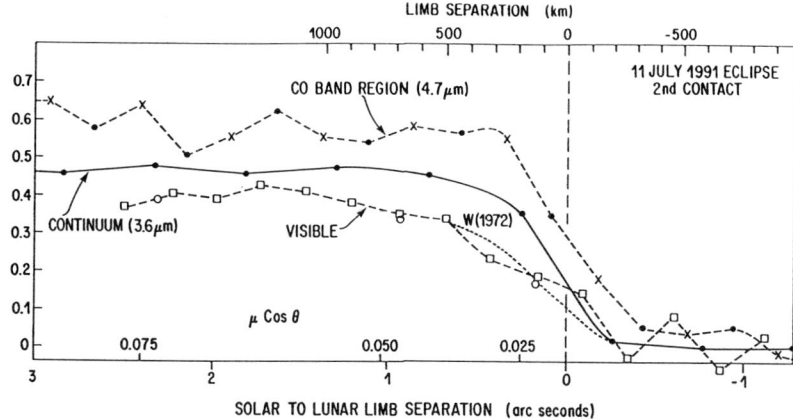

Fig. 3. Differential eclipse curve for 2nd contact, plotted as the fraction of the equivalent disk-center intensity, for averages of 5 channels in the continuum band, 4 channels in the CO band region, and the visible channel. The limb position has been placed at approximately the point of inflection of the final limb decrease. W(1972) refers to the visible eclipse curve of Weart.

eclipse curves. The zero-point values for height and $\mu = \cos\theta$ scales have been fixed at this level. Establishment of the true limb position requires further analysis of the visible eclipse curve, but is unlikely to be inaccurate in an absolute sense by more than 150 km.

The dashed curve in Figure 3 shows the relative extension of the CO-band region beyond the near IR and visible limb by 125 ± 15 km, with maybe a more extensive but low intensity (0.05 of the central intensity) residual out to 640 km above the near-IR limb. There is evidence of limb darkening within the last few arcseconds in all of these eclipse curves. Analysis is still proceeding on the 3rd contact data but these results may be affected by a large prominence noted near this contact point on eclipse photographs.

5. Conclusions

The present near-IR eclipse measurement at wavelengths containing the CO fundamental V-R band shows clear evidence for an extension (of about 125 km) of the solar limb beyond the visible and near-IR continuum limbs. If this is interpreted as the flash spectrum of the composite CO band, then this indicates that the main CO emission arises in a layer slightly below the temperature minimum, at a depth of about 465 km above the photosphere. This result can be compared to the $\tau_{CO} = 1.0$ level predicted by two model atmospheres used in the recent non-LTE investigation of Ayres and Wiedemann (1989). These are the semi-empirical "hot" chromosphere model of Avrett (1985) and radiative equilibrium model of Anderson (1989), which place this level 90 and 220 km above the visible limb, or 430 km and 560 km above the $\tau_{0.5} = 1.0$ level in the photosphere, respectively.

Acknowledgements

It is a pleasure to thank the staff of the James Clerk Maxwell Telescope for their efforts in support of this experiment, particularly in mounting the equipment on the entrance roof. This telescope is operated by the Royal Observatory, Edinburgh on behalf of the Science and Engineering Council of the United Kingdom, the Netherlands Organisation for Scientific Research and the National Research Council of Canada. The work was supported by a travel grant from the National Research Council of Canada to TAC and DAN, and by Natural Sciences and Engineering Research Council operating grants. This support is gratefully acknowledged.

References

Anderson, L.S.: 1989, *Astrophys. J.* **339**, 558.
Athay, R. G.: 1976, *The Solar Chromosphere and Corona: Quiet Sun*, Reidel Publishing Company, Dortrecht, Holland, p. 173.
Avrett, E.H.: 1985, in B. Lites (ed.), *Chromospheric Diagnostics and Modeling*, National Solar Observatory, Sunspot, N.M., p. 67.
Ayres, T.R.: 1990, in J.O.Stenflo (ed.) 'Solar Photosphere: Structure, Convection and Magnetic Fields', *IAU Symp.* **138**, 23.
Ayres, T.R. and Testerman, L.: 1981, *Astrophys. J.* **245**, 1124.
Ayres, T.R. and Wiedemann, G.R.: 1989, *Astrophys. J.* **338**, 1033.
Heasley, J.N., Ridgway, S.T., Carbon, D.F. Milkey, R.W. and Hall, D.N.B.: 1978, *Astrophys. J.* **219**, 790.
Kurucz, R.L.: 1991, in A. N. Cox, W. C. Livingston, and M. S. Matthews (eds.), *Solar Interior and Atmosphere*, University of Arizona Press, Tucson, Arizona, p. 666.
Lindsey, C., Jefferies, J.T., Clark,T.A., Harrison, R.A., Carter, M.K., Watt, G., Becklin, E.E., Roellig, T.L., Braun, D.C., Naylor, D.A., and Tomkins, G.J.: 1992, *Nature* **392**, 739.
Makita, M.: 1972, *Solar Phys.* **24**, 59.
Noyes, R.W. and Hall, D.N.B.: 1972, *Astrophys. J. (Letters)* **176**, L89.

INFRARED IMAGES OF THE SUN DURING THE JULY 11, 1991 SOLAR ECLIPSE

E. V. TOLLESTRUP and G. G. FAZIO

Smithsonian Astrophysical Observatory, Cambridge, MA 02138, U.S.A.

J. WOOLAWAY and J. BLACKWELL

Amber Engineering, Inc., Santa Barbara, CA 93117, U.S.A.

and

K. BRECHER

Department of Astronomy, Boston University, Boston, MA 02215, U.S.A.

Abstract. Infrared images (1.65 μm) of the eclipsed Sun were taken atop Mauna Kea, Hawaii, during the July 11, 1991 total eclipse with an Amber Engineering 128×128 InSb array camera. The camera, mounted on a portable solar tracker, had a 3.8-cm, f/2 objective that produced a 4.9° field of view. The primary objective of the experiment was to search for dust or rocky rings around the Sun, previously detected at about 4 $R_\odot$. High thin clouds, atmospheric dust and aerosols from the June 1991 explosion of Mount Pinatubo in the Philippines, and the overall brightness of the solar corona resulted in a very high infrared background. Despite this, high signal-to-noise radial infrared intensity profiles were obtained of the solar corona from the Moon's limb out to about 10 $R_\odot$. Preliminary analysis shows some evidence for an enhanced surface brightness between 3 to 4 $R_\odot$ along the east-west direction, but much fainter than seen in previous solar eclipses. The transition region between the K-corona and the F-corona clearly shows at 2.5 $R_\odot$, and the surface brightness of the F-corona as a function of radius (from about 2 to 10 $R_\odot$) can be fit by a simple power law.

Key words: eclipses – infrared: stars – interplanetary medium – Sun: corona

1. Introduction

The detection of possible dust rings around the Sun has been an "on again, off again" affair ever since they were first reported in 1967. During the 1966 solar eclipse, Peterson (1967) and MacQueen (1967) independently detected excess 2.2 μm emission along the solar equatorial direction at a radius of about 4 $R_\odot$. MacQueen (1967) verified his results in 1967 by using a stratospheric balloon-borne coronagraph to scan the corona. Peterson found a broad emission peak with a maximum near 3.3 $R_\odot$ and a sharp peak at 3.9 $R_\odot$. MacQueen found two components during the 1966 eclipse, one at 3.5 and one at 4 $R_\odot$. During the 1967 balloon experiment, MacQueen detected peaks at 4, 8.7, and 9.2 $R_\odot$. These early results show clear evidence of enhanced infrared emission. But despite this, the more recent results have been ambiguous. Detections have been reported at 10 μm for the 1973 solar eclipse (Lena *et al.* 1974) and at several wavelengths for a multi-wavelength (1.25, 1.65, 2.25, and 2.8 μm) eclipse experiment in 1983 (Mizutani *et al.* 1984; Isobe *et al.* 1985). However, these results are not consistent with the 1966-67 results. The eclipse experiments in 1970, 1978, and 1980 have all been unsuccessful in finding dust rings (*e.g.*, Mankin *et al.* 1974; Peterson 1971; and Rao *et al.* 1981).

The solar corona is extremely bright and spatially complex, which makes these type of observations (scanned photometers) difficult. Not only do the K- and F-coronae overlap, but they have radial surface brightness profiles that are both steep and varying. More troublesome are the solar streamers. They are very complicated, 3-dimensional structures which can only be viewed in projection as a 2-dimensional image on the sky. Therefore, without a true 2-dimensional image, the comparatively faint signal of any dust ring will be difficult to decipher against the spatial structure created by the K- and F-coronae and the solar streamers. The new generation of large format imaging infrared arrays is ideally suited to untangle the different components, search for the dust rings, and study the spatial structure of the infrared corona.

2. Observations

The observations were made at the NASA Infrared Telescope Facility (IRTF) on Mauna Kea during the July 11, 1991 solar eclipse. The camera, a modified Amber Engineering 'Infrared Imaging System', used a 128×128 InSb array with a 3.8-cm, f/2 lens, which produced a 4.9 degree field of view. The filter set included a 1.65 and 2.2 μm filter; a cold 2.5 μm thermal-blocking filter was also used. The camera, mounted on a portable solar tracker, was located on the eastern loading dock of the IRTF. The orientation of the InSb array was rotated to position the cardinal points along the 45 degree diagonals of the array. This orientation increased the radial distance along the ecliptic plane to 14 $R_\odot$. In addition, this orientation minimized any possible contamination of the solar signal by unknown row- or column-dependent fixed pattern noise sources.

During the four minutes of totality, the observations were divided into two 2-minute long intervals, one for the 1.65 μm filter and the other for the 2.2 μm filter. Only the 1.65 μm observations are presented in this paper. The 1.65 μm observations were divided into one minute of data taken at a frame rate of 1.69 frames/sec and another minute of data taken at data rates varying from 3 to 54 frames/sec. The data at 1.69 frames/sec produced the best images of the outer corona, but saturated most of the K-corona. The faster frame rates where required to obtain unsaturated images.

As a further precaution against contamination by fixed-pattern noise and dead, hot, or excessively noisy pixels, the image of the Sun was allowed to slowly drift southward across the array, and after each 30 second interval, the image of the Sun was quickly stepped by several pixels to the south. By doing so, as many different pixels as possible imaged the same region of the corona.

For each set of frames with different frame rates the data reduction included flat-fielding, bad pixel restoration, image shifting, and co-addition of the individual frames in each set. The absolute calibration was obtained by observing α Lyra on the night of July 11-12 and by laboratory observations of a calibrated blackbody. The two methods agreed to about 5 percent.

3. Results

The eclipsed Sun at 1.65 μm looks very similar to that seen at visible wavelengths. The K- and the F-corona and the solar streamers are clearly seen. But, as is shown in Fig. 1, no obvious dust rings are evident. This figure shows a plot of the surface

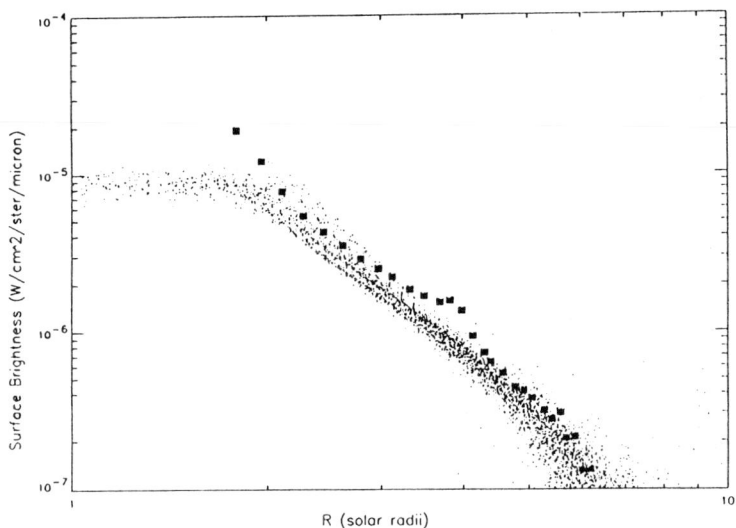

Fig. 1. A plot of surface brightness vs radial distance for each pixel in the co-added 0.6 sec, 1.65 μm image (dots). The large square symbols are the 1.65 μm data points from Mizutani et al. (1984). Below a radius of $2R_\odot$ the data is saturated.

brightness of each pixel versus the radial distance of that pixel. For reference, the 1.65 μm results from the 1983 solar eclipse (Mizutani et al. 1984) are over-plotted as large square symbols. Although the underlying corona from the two eclipses is nearly the same, no emission peak as bright as that seen in 1983 is evident in our 1991 eclipse.

Any faint dust ring signature would be difficult to detect on top of the very steep and bright corona. Therefore, we modeled the surface brightness of the F-corona, using an empirically determined model, and subtracted this signal from the observed signal. The empirical model was $S(R,\theta) = a + b(\theta)(R/R_\odot)^c$, where a compensates for zero-point offsets and b is dependent on polar angle θ. No physics is implied by the choice of this model; its sole purpose is to uniformly subtract off a reasonable approximation for the F-corona so that small scale fluctuations can be detected on the corona.

The model was determined by examining thin radial cuts at many different position angles around the sun (chosen to avoid the solar streamers). An example of how well this model worked is shown in Fig. 2, which shows a radial cut out the north pole and overplots the model fit. Also included is a plot of the residual signal after subtraction. The plots of the residual signal showed 4 features: 1) the K to F-corona transition region, which occurs abruptly at $2.5R_\odot$, 2) the best value of the power law, which was $c = 2.7 \pm 0.3$, 3) the total noise, which is about 2×10^{-7} W/cm²/ster/μm peak-to-peak, and 4) the completeness of the F-coronal subtraction, although some nonzero signal is apparent at a level of 5×10^{-8}

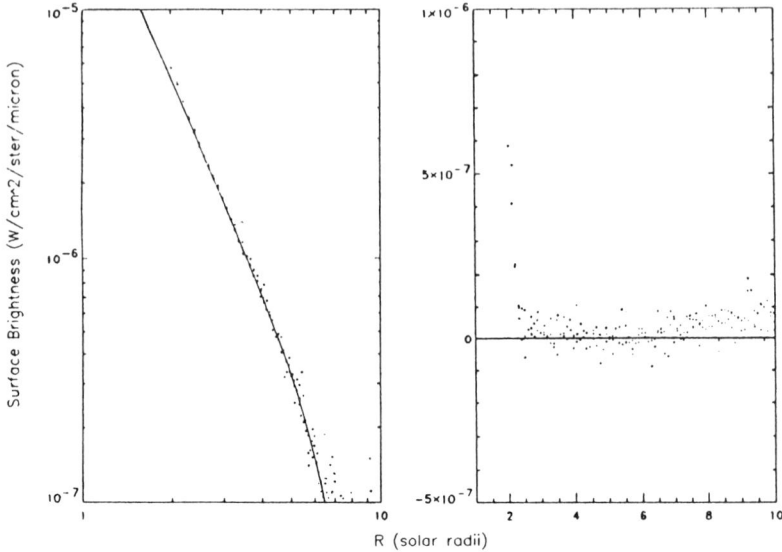

Fig. 2. The left plot compares the model coronal profile (line) to the data (dots) for a radial cut out the north pole. The right-hand plot shows the residuals after the model is subtracted from the real data.

W/cm^2/ster/μm. Another general result is that the equatorial regions are 15% brighter than the polar regions.

After subtracting the model from the real data, an image of the residual signal showed two faint emission peaks, one located on each side of the sun, along the ecliptic, and at about $4R_\odot$. No other similarly shaped peaks are apparent, the only other remaining features are the solar streamers and the saturated inner regions of the K-corona. Fig. 3 shows four plots that compare cuts through the east with cuts through the north and south, and similar comparisons for a western cut. In all four comparisons, excess emission is apparent in both the east and the west cuts (as compared to the north or south cut) for radial distances between 3.5 and $4.5R_\odot$. The excess emissions are too faint to definitively state that dust rings exist, but their presence is apparent, as illustrated in Fig. 3. They are equally spaced, and lie along the east-west plane at a radius of about $4R_\odot$. Only half of the data has been reduced, so any further conclusions must be deferred until after all of the data is completely analyzed.

Acknowledgements

The authors were Visiting Astronomers at the IRTF which is operated by the University of Hawaii under contract from the National Aeronautics and Space Ad-

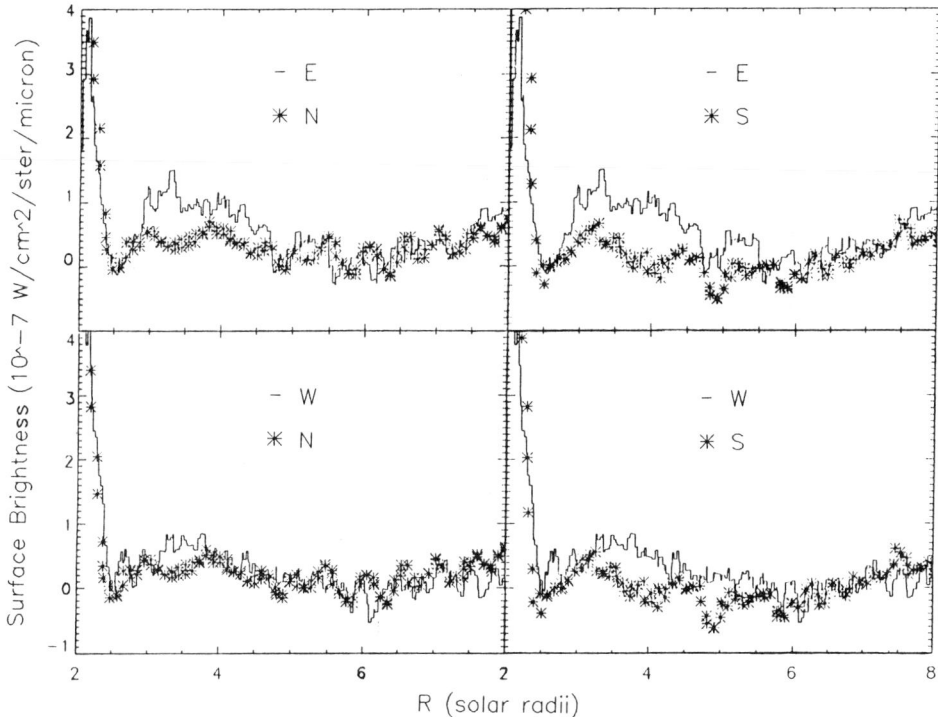

Fig. 3. The four plots compare radial cuts in the east or west direction (lines) to cuts in the north or south direction (stars) of the residual signal that remains after the model signal is subtracted from the real data. The data has been smoothed with a 3 pixel wide box-car filter to dampen the high spatial frequency noise.

ministration. We wish to thank Amber Engineering for their financial and technical support of this experiment, as well as their generous loan of a 128×128 InSb Infrared Imaging System and the National Solar Observatory for the loan of the solar tracker.

References

Isobe, S., Hirayama, T., Baba, N., and Miura, N.: 1985, *Nature* **318**, 644.
Lena, P., Viala, Y., Hall, D., and Soufflot, A.: 1974, *Astron. Astrophys.* **37**, 81.
MacQueen, R. M.: 1968, *Astrophys. J.* **154**, 1059.
Mankin, W. G., MacQueen, R. M., and Lee, R. H.: 1974, *Astron. Astrophys.* **31**, 17.
Mizutani, K., Maihara, T., Hiromoto, N., and Tamaki, H.: 1984, *Nature* **312**, 134.
Peterson, A. W.: 1967, *Astrophys. J. (Letters)* **148**, L37.
Peterson, A. W.: 1969, *Astrophys. J.* **155**, 1009.
Peterson, A. W.: 1971, *Bull. Amer. Astron. Soc.* **3**, 500.
Rao, U. R., Alex, T. K., Iyengar, V. S., Kasturirangan, K., Marar, T. M. K., Mathur, R. S., and Sharma, D. P.: 1981, *Nature*, **289**, 779.

IR OBSERVATIONS OF THE K AND F CORONA DURING THE 1991 ECLIPSE

J. R. KUHN and H. LIN

Michigan State University, East Lansing, MI 48824, U.S.A.

P. LAMY

Laboratoire d'Astronomie Spatiale, F-13012 Marseille, France

S. KOUTCHMY

Institut d'Astrophysique, Boulevard Arago, F-75014 Paris, France

and

R. N. SMARTT

National Solar Observatory, Sunspot, NM 88349, U.S.A.

Abstract. The availability of relatively large format IR array detectors is incentive for reexamining the classic question of whether or not there are "dust rings" around the sun – a problem for which there are conflicting observational answers. The 1991 eclipse path included a high altitude observatory and provided a potentially ideal opportunity to study the infrared properties and dust content of the corona. Here we report results from an experiment conducted from Mauna Kea using a HgCdTe array detector sensitive to wavelengths between 1-2.5 μm. Surface brightness measurements in the H-band and polarization data in the J-band were obtained over a field-of-view of ± 6 $R_\odot$ while K-band images further extend to 15 $R_\odot$ on the western side of the Sun. J-band polarization data and H and K-band surface brightness data clearly show the inhomogeneous structure in the K corona and the ecliptical flattening of the F corona. We see no evidence of a circumsolar, local dust corona (dust rings) out to 15 $R_\odot$.

Key words: eclipses – infrared: stars – interplanetary medium – Sun: corona

1. Introduction

The Fraunhofer or F corona is that component of the surface brightness with absorbtion lines in its spectrum. Unlike the K corona, which has no spectral features and results primarily from photospheric light scattered by hot coronal electrons, the F component results from sunlight scattered off of zodiacal dust. It is unclear to what extent this component results from light scattered by dust in the immediate environment of the sun's corona. The F-corona may simply be "zodiacal" light, or photons scattered by grains far from the sun but close to the line-of-sight to the corona.

Peterson (1963) was one of the first to suggest that it should be possible to look for a "local" coronal contribution to the F component by looking for the thermal emission of hot dust in the outer corona. In this model dust may be deposited in the inner solar system by comets. Small grains that orbit the sun should then experience a drag force as they scatter or reradiate solar radiation (*cf.* Wyatt and Whipple 1950). Thus a grain gradually spirals toward the sun until it reaches its sublimation temperature. At this point the solar radiation force to gravitational force ratio increases and the grain either vaporizes or evolves into an elliptic orbit that takes it away from the sun. The inner circumsolar radius within which there are no grains depends on the grain properties. Silicate grains with radii of a few microns should survive for a few hundred orbits near 4 solar radii (*cf.* Mukai 1985) and grains that start out much smaller will not get close to the sun. Several authors have reported observations of the thermal and large-angle scattered light from such dust rings, starting with Peterson (1963) and MacQueen (1968). None of these or later reports have been entirely consistent (*cf.* Koutchmy and Lamy 1985) and others have been unable to detect any local F-coronal light contribution (Mampaso *et al.* 1982). While these are difficult observations it seems clear that a definitive observation of the local F component will significantly constrain our understanding of the solar environment.

Since the sublimation temperature of the grains is expected to be near 2000 K, the thermal emission should peak at a wavelength between 1 and 2 μm and the thermal emission throughout the circumsolar dust region should be redder than the Thomson-scattered K coronal flux component. In addition a local F corona component resulting from large-angle scattering from dust grains should show some linear polarization. Clearly an experiment designed to detect this light should incorporate polarization and infrared surface brightness observations. Until the 1991 eclipse, all prior IR experiments used scanning single-element detectors to look for excess surface brightness "bumps" along the ecliptic. During a typical eclipse experiment, where there are only a few seconds or minutes of totality, a large angular region to be mapped, and under conditions of a changing background sky brightness – these are difficult observations. It is evident that a sensitive array detector experiment may make the identification of dust-related surface brightness enhancements considerably easier since each pixel records the coronal surface brightness simultaneously.

2. The Experiment

The detector is a 128×128 HgCdTe array device, produced by Rockwell International (part TCM-1000C). Each pixel has a large photoelectron well capacity (3×10^7) with a read noise of about 2000 electrons. Our detector has an average quantum efficiency of about 70% for light with wavelengths between 1 and 2.5 μm. The array operates at liquid nitrogen temperature, behind cold filters, polarizers, and objective lens. The device is electronically shuttered so it does not require a fast mechanical shutter to define the exposure time. Figure 1 shows a schematic of the eclipse experiment. Standard (Barr Associates Inc.) infrared H and K filters are mounted on a cold filter wheel. Three J band filters with type HR (Polaroid, Inc.) sheet polarizers are also mounted on the cooled wheel assembly. The filter wheel is controlled by the experiment computer and a motor outside the dewar. The 110mm f.l., f/2 (stopped down to f/4 during the experiment) doublet objective was mounted in the dewar. Because of the large wavelength range of the observations we used an internally mounted stepper motor to make focus adjustments. The final image scale ranged between 0.094 and 0.098 $R_\odot$ per pixel. The detector system is described more fully elsewhere (McPherson *et al.* 1992).

The field-of-view of the detector is about 3.2 degrees or ±6 $R_\odot$ so that the corona will be about 10^3 times brighter near the limb of the moon than it is at the edge of our field. Laboratory measurements show that there is scattered light in the optics, probably due to multiple reflections from the lens, at a level near

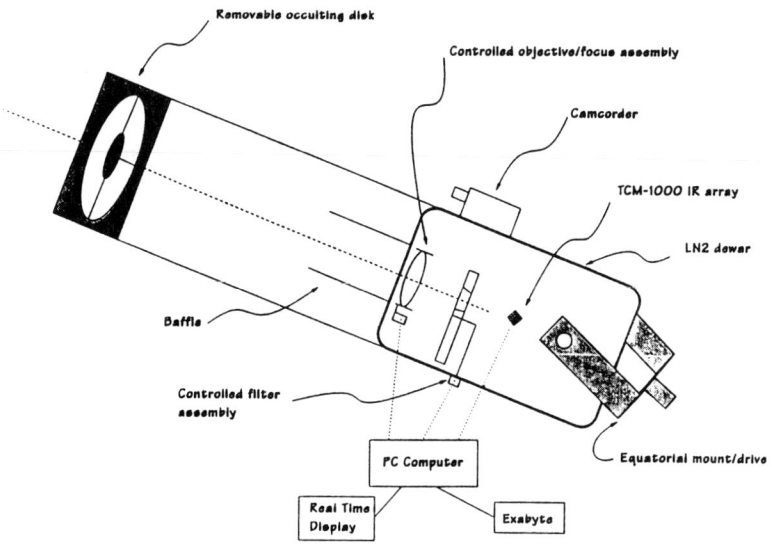

Fig. 1. Schematic of the principle components of the eclipse experiment.

10^{-3} of the surface brightness at the center of the field. Unfortunately this residual signal changes with the illumination pattern, filter, and focus, and so it cannot easily be modeled. Our solution to this problem was to externally occult the bright inner corona. We accomplish this with a mask approximately 1.2 m in front of the focal plane. The mask supports a removable 3-cm radius occulting disk that can be positioned in a 6-cm radius opening. With the central disk removed the telescope may also be tilted away from the sun to see the outer corona – so that the bright inner corona does not illuminate the telescope optics. The thermal emission of the occulting disk at the K-band was less than the sky background signal and is not a problem.

The dewar and occulter assembly were mounted directly to a simple equatorial sun-tracker. The equatorial mount was driven at a fixed rate, but allowed for rapid manual offsets in RA and Dec. During the eclipse we manually shifted the telescope to the west to see the outer corona. Since the moon's limb was not in the field of the IR array during the offset observations we mounted a camcorder with a large field of view on the tracker. Thus we have a video record that could be digitized to measure the precise spatial offset of the image when the limb was not visible on the array.

The filter/polarizer wheel, focus mechanism, and array were controlled by a PC. Images could be read, flat-fielded and displayed in real time, while reading and recording them to Exabyte tape required about 1 s. We had an ambitious observing run planned for the 4 minutes of totality which required a computer script to run smoothly. During this time we would observe J polarization and H and K flux both with, and without the occulter. In addition we would offset the telescope to observe the K flux beyond 14 $R_\odot$. In order to observe the inner corona without saturating the detector we also needed exposures as short as 3.8 ms, and for measuring the outer corona our longest exposure was 10 s. The largest "overhead" in the observations was the time spent manually shifting the telescope and the time required by the internal stepper-motor to change the focus for the K-band data (about 15 s). Table I shows a log of the eclipse data – the sequence number, filter bandwidth, exposure duration, time, and occulter position.

3. Analysis

The IR array images were calibrated like most optical CCD data. Flat-fields were obtained separately for each of the polarization angles, occulter positions and wavelengths from blank sky observations. Because the Mauna Kea sky clouded over soon after the eclipse we used flat-field data that were obtained the previous day. Amplifier bias and dark field corrections were obtained from "overscan" data and dewar "cold stop" observations. The field-flattened flux data were calibrated using observations of Vega and Altair obtained on the night of July 9. Unfortunately the eclipse sky conditions were far from photometric and we estimate that our overall flux calibration may be in error by as much as 60%.

An image of the "sky" can be produced by looking at the difference between two eclipse images obtained in rapid succession near second or third contact – when the overall illumination level is changing rapidly. Since the unscattered coronal light does not vary (except at the limb of the moon!) on a timescale of a second or less, a difference image will show primarily the change in the scattered light caused by both clouds and our optics. The difference of externally occulted images shows the change in scattered light caused by clouds. Thus the changing lunar limb position gives an independent check on the scattered light in our measurements. Figure 2 shows a difference image taken using H-band data obtained at 43 and 46 seconds after 2nd contact. A similar wavy pattern, near the edges of the frame, also appears in occulted images and so is apparently due to atmospheric scattered light. It has an rms amplitude of about 1.7×10^{-3} of the change in the H brightness at the lunar limb. Figure 2 shows the relative scattered light in the H and K band data, normalized by the brightness change observed near the lunar limb. Within 1.7 $R_\odot$ the long exposure data is saturated – requiring that they be normalized (indirectly) by the product of the measured time rate of change in the integrated coronal brightness, the peak limb brightness, and the time between exposures. Thus, Figure 3 is a direct measure of the scattered-light solar aureole in the H and K bands, normalized by the average surface brightness near the lunar limb.

The calibrated, circularly averaged, H and K surface brightness is plotted in Figure 4. In these units solar color implies a K band flux which is a factor of about 3 less than the H flux (which is what we observe near the limb). Between 1 and 7 $R_\odot$ the H and K brightness decline by a factor of about 300 and 500 respectively. Using the measured limb brightness and scattering function profile from Figure 3 we can estimate the scattered light component in the data, which is also plotted on Figure 4. Figure 3 shows that the scattered light declines more rapidly in the longer wavelength band, *i.e.*, that the scattered light gets bluer further

TABLE I
IR Observing Log for July 11, 1991 Eclipse

Seq. No.	Filter	Occulter	Exp. Time	Start Time
01	K	NO	0.0038	17:28:19
02	K	NO	0.5	17:28:21
03	H	NO	0.0038	17:28:48
04	H	NO	0.0038	17:28:50
05	H	NO	0.5	17:28:52
06	H	NO	0.5	17:28:55
07	J1	NO	0.0038	17:29:05
08	J1	NO	0.0038	17:29:07
09	J1	NO	0.5	17:29:09
10	J1	NO	0.5	17:29:11
11	J2	NO	0.0038	17:29:18
12	J2	NO	0.0038	17:29:20
13	J2	NO	0.5	17:29:22
14	J2	NO	0.5	17:29:24
15	J3	NO	0.0038	17:29:31
16	J3	NO	0.0038	17:29:32
17	J3	NO	0.5	17:29:34
18	J3	NO	0.5	17:29:37
19	J3	YES	2.5	17:29:44
20	J3	YES	2.5	17:29:48
21	J3	YES	2.5	17:29:53
22	J2	YES	2.5	17:30:01
23	J2	YES	2.5	17:30:05
24	J2	YES	2.5	17:30:10
25	J1	YES	2.5	17:30:18
26	J1	YES	2.5	17:30:22
27	J1	YES	2.5	17:30:27
28	H	YES	2.5	17:30:39
29	H	YES	2.5	17:30:43
30	H	YES	2.5	17:30:48
31	K	YES	2.5	17:31:17
32	K	YES	2.5	17:31:22
33	K	YES	2.5	17:31:25
34	K	NO	0.0038	17:31:32
35	K	NO	0.0038	17:31:34
36	K	NO	0.5	17:31:36
37	K	NO	0.5	17:31:39
38	K	OFFSET	1.5	17:31:53
39	K	OFFSET	1.5	17:31:56
40	K	OFFSET	1.5	17:32:00
41	K	OFFSET	10.5	17:32:03

Column 1 identifies the sequence number; column 2 shows the filter bandwidth; column 3 indicates whether the occulter was in place or if the telescope was offset to the west; column 4 is the exposure time in seconds; column 5 shows the start time (UT) of the exposure.

190   J. R. KUHN ET AL.

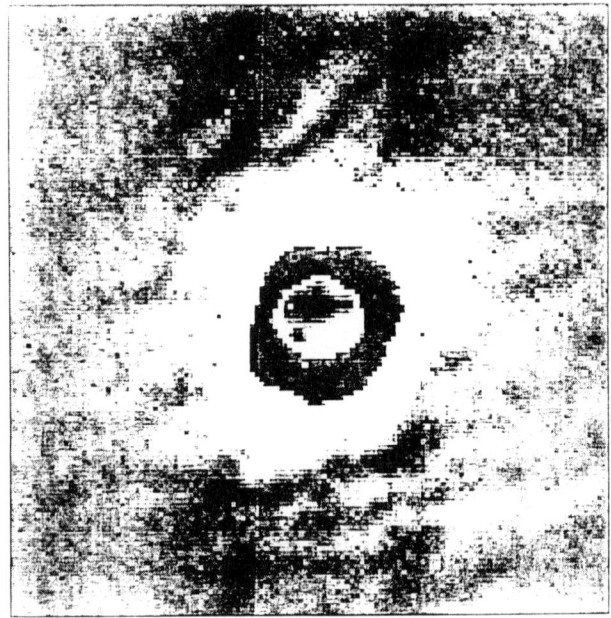

Fig. 2. The residual image obtained as the difference of two consequtive H-band observations. The change in the scattered light component is visible as a wavy pattern in this picture. Its amplitude is about 0.2% of the change in H brightness near the limb of the moon during this period.

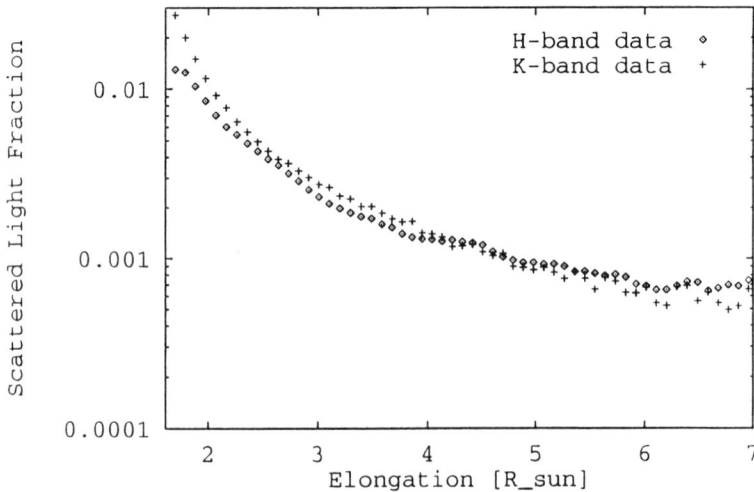

Fig. 3. This shows an estimate of the scattered light contribution to our observations. The curve is obtained from the circular average of the Fig. 2 data and has been normalized by the change in brightness near the limb.

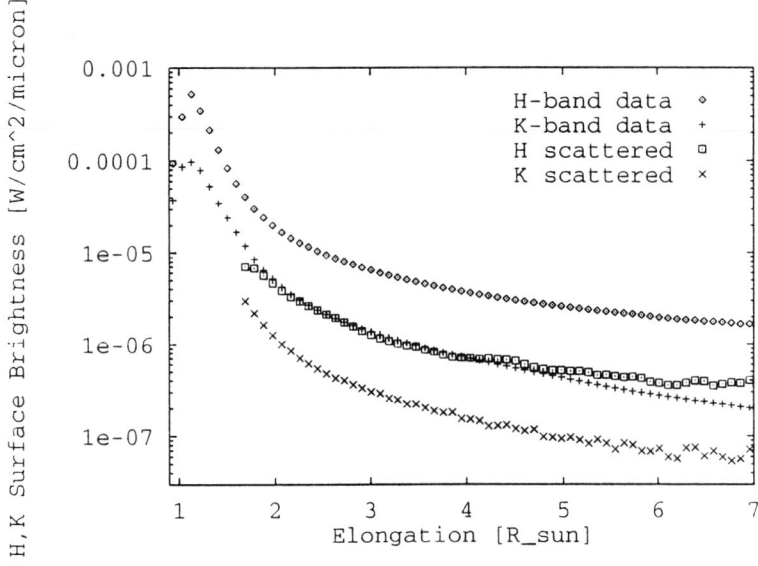

Fig. 4. The calibrated H and K-band surface brightness. The scattered light contribution (estimated from the data in Fig. 3) is also plotted here.

from the limb. This may explain the apparent trend to bluer colors away from the limb in Figure 4. Unfortunately the atmospheric scattered light contribution is difficult to subtract from the figure 4 data – there is probably a factor of 2 normalization uncertainty in the scattered light function in addition to evidence of atmospheric variability. Nevertheless, there is no evidence of a "redder-than-solar" coronal component. Mukai (1985) also found the H and K corona to be "bluer" than solar color in a balloon observation.

We have combined the centered short and long exposure K-band data with the offset frames in order to construct a composite image that shows the coronal surface brightness from the limb of the moon out to about 14 $R_\odot$. The offset images were obtained by pointing the instrument to the west by approximately 9 $R_\odot$. The angular offset was accurately determined after the eclipse from digitized frames obtained with the camcorder. The IR frames were merged by finding the best linear regression between corresponding pixels in the overlapped region of each frame. The intensity scale and offset derived in this way were consistent with our uncertainty in the expected scale and sky background variations between frames. The scale change resulted from exposure time differences but was also accurately determined by this regression procedure. The resulting image is shown in Figure 5 using a logarithmic intensity scale. The mean K-band brightness and the K and F coronal model of Koutchmy and Lamy (1985) are plotted in Figure 6. Several points should be noted: First, neither the model nor data have been rescaled before plotting these

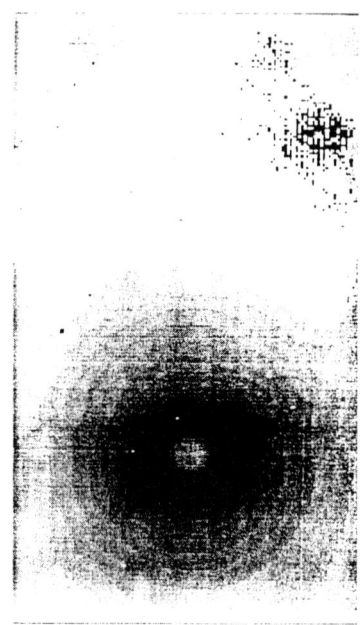

Fig. 5. The offset K band data has been combined with the centered data to construct this composite image which shows the corona out to 14 $R_\odot$ in the western direction. With a rotational orientation accuracy of about 1 degree, the north direction is to the left. The greyscale of this image is logarithmic and the darkest shade is approximately 10^4 brighter than the lightest shade.

curves. Second, both the data and model show a "break" in the surface brightness near 2.5 $R_\odot$ beyond which the F corona dominates the K corona surface brightness. Lastly, there is an apparent scale difference between the model and data, in the sense that the data is everywhere brighter than the model predictions. That this is simply a scale change is demonstrated by Figure 7 which shows the data scaled by 0.5. Note that data beyond 8 $R_\odot$ are easily affected by small changes in the assumed background or zero level and have not been plotted here. Nevertheless, it is clear that the IR data are consistent with the visible wavelength coronal model and show no anomalous surface brightness bumps out to at least 8 $R_\odot$.

It is clear from Figure 5 that coronal streamers can be observed in the infrared, much the same as in visible wavelengths. Figure 8a,b shows images of the fractional deviation in H and K-band light from the mean radial brightness profile. These "residual" images clearly show the aspherical structure in the corona. Since this coronal component results from "colorless" Thomson scattering of photospheric light from hot electrons, the visible structure should be independent of wavelength (as it is).

While the K corona should be linearly polarized, in a direction tangent to the nearest limb point, a local F coronal component may also be polarized. Thus, a search for anomalous polarization "rings" might also provide evidence of a local

IR OBSERVATIONS OF THE K AND F CORONA DURING THE 1991 ECLIPSE 193

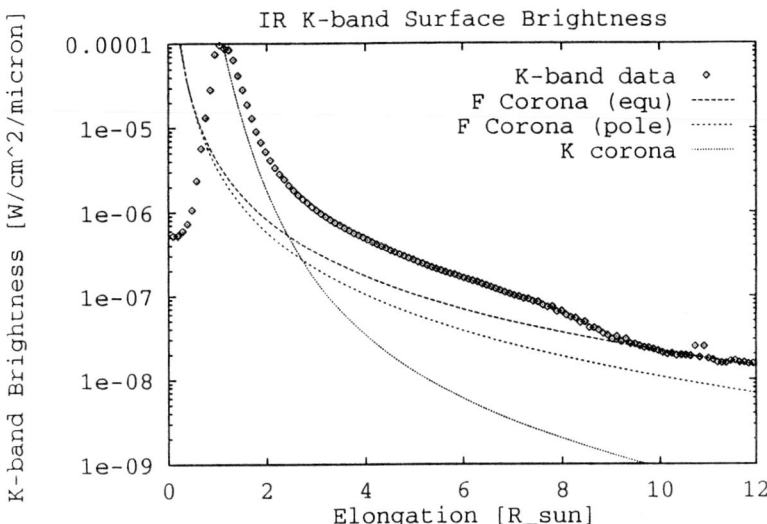

Fig. 6. The mean K-band brightness and the Koutchmy-Lamy K and F coronal model are plotted here.

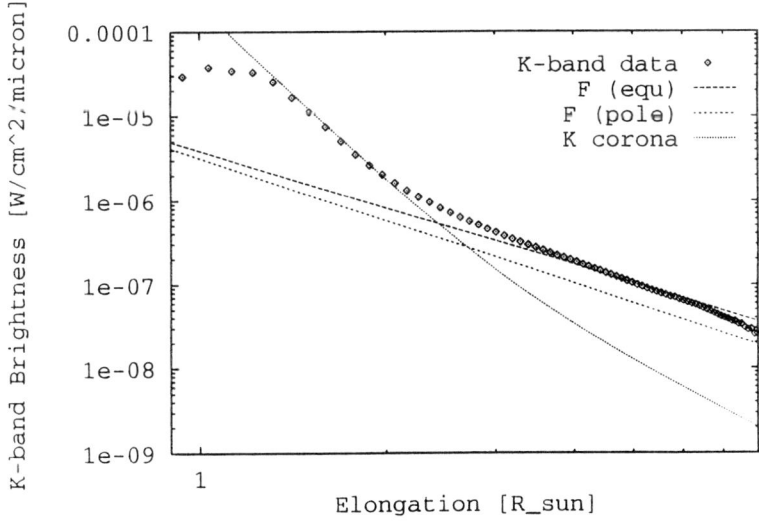

Fig. 7. This shows a comparison of the K data and the coronal model after allowing for a scale or calibration uncertainty in the data.

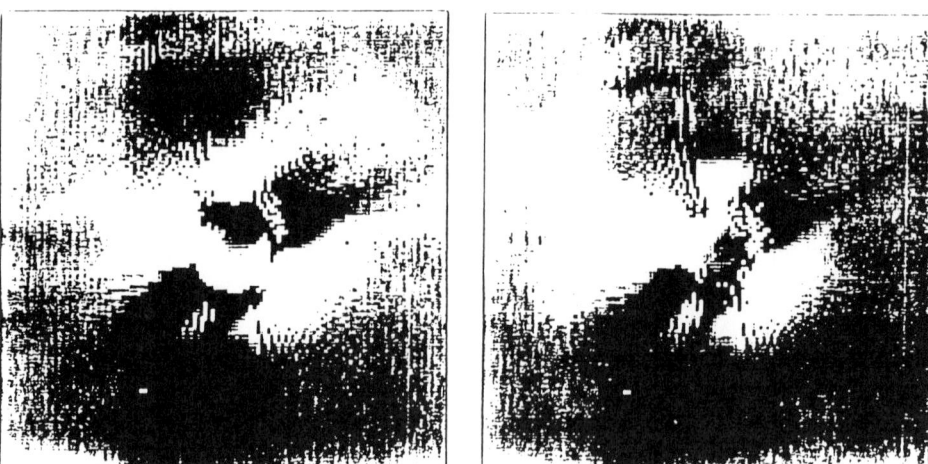

Fig. 8. These are images that show the residual structure in the H (Fig. 8a) and K-band (Fig. 8b) data. They were obtained by subtracting the circularly averaged surface brightness profile.

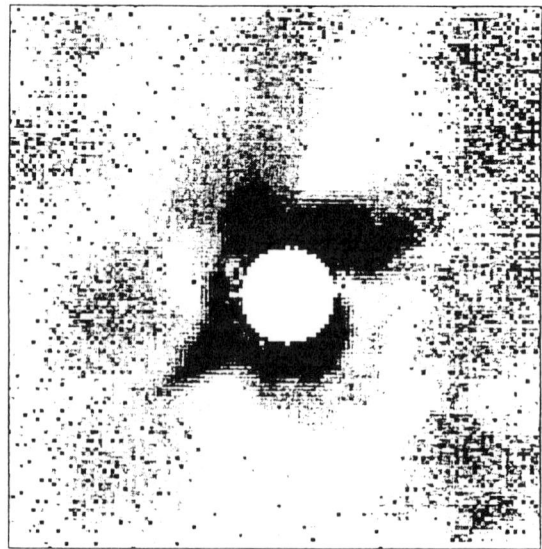

Fig. 9. This shows the mean J-band polarization fraction. The greyscale is linear in polarization where the darkest shade represents polarization of 0.4 and the lightest is a value of 0.05.

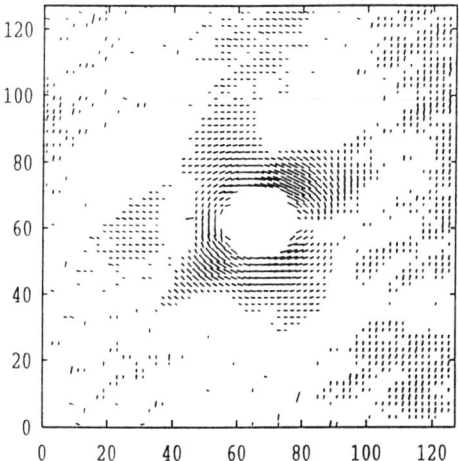

Fig. 10. The direction and magnitude of the polarization is indicated graphically in this diagram. The length of each line segment is proportional to the polarization fraction, while its direction shows the polarization plane.

F corona. Thus, J-band data were obtained through three different fixed linear polarizers, oriented at three fixed angles. Knowing these angles and the measured J-band surface brightness at each pixel we calculated the polarization fraction and direction in the image. Figure 9 is a map of the J-band polarization while Figure 10 shows the corresponding polarization direction in those pixels with polarization fraction greater than 0.1. Figure 11 shows the mean polarization versus distance from the sun.

The rapid drop in polarization near 2 $R_\odot$ is consistent with the break in K-band light we saw in Figure 6 which we interpreted as the "boundary" of the K corona. The mean polarization results are in good agreement with the data reviewed by Koutchmy and Lamy (1985), although it is clear from Figure 9 that the mean polarization is dominated by the complex aspherical shape of the streamers. The polarization direction is tangent to the limb, as we expect for Thomson scattering, and where it deviates the polarization is small and probably insignificant.

The plane of the ecliptic is inclined about 9 degrees CCW of vertical in these images. By subtracting the average radial light profile from the data we can more easily see the elliptical "zodiacal" light. Figure 12 shows such a residual image. A smooth, but flattened, coronal component is apparent beyond the K corona. This is not a local "F corona" since it not redder than solar color and lacks any significant polarization. Furthermore, although patchy scattered light variations limit our measurements, the apparent equatorial excess is consistent with the "zodiacal" flattening of the F corona described by the Koutchmy-Lamy model. According to their model the difference between the polar and equatorial K-band surface bright-

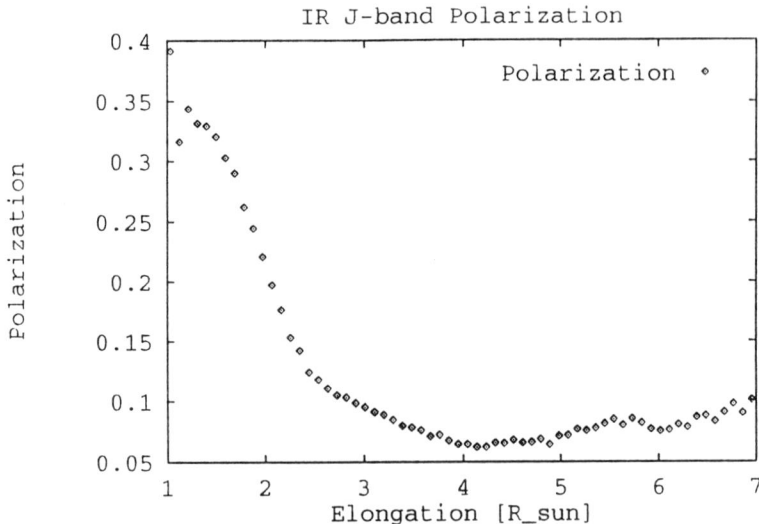

Fig. 11. The average polarization is plotted versus distance from the sun in this graph.

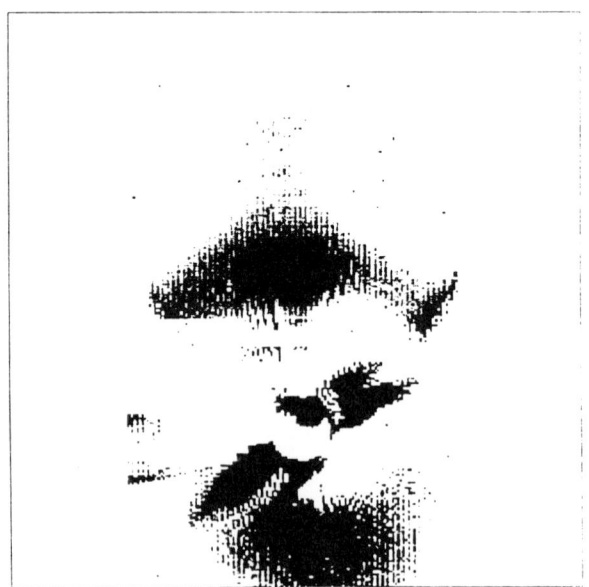

Fig. 12. A residual K-band image from the data in Figure 5 shows the ecliptical excess associated with the "zodiacal" light. This image was obtained by subtracting the circularly averaged data from Figure 5. The greyscale is linear in intensity. The range between the brightest (dark shades) and the dimmest pixel is 3.7×10^{-8} W cm^{-2} ster^{-1} μm^{-1}.

ness near 8 $R_\odot$ should be about 10^{-8} W cm^{-2} ster^{-1} μm^{-1}, which is consistent with the ecliptical brightening seen in Figure 12.

4. Conclusions

Although sky conditions at Mauna Kea during the 1991 eclipse were not ideal, we succeeded in observing the K and F corona using a new infrared array system. Our observations through broad-band J,H and K filters reveal an infrared corona with no surprises. J-band polarization data and H and K surface brightness observations clearly show the inhomogeneous structure in the K corona and the ecliptical flattening of the "zodiacal" light. While we detect the flattening of the non-local F corona with a surface brightness amplitude of about 10^{-8} W cm^{-2} ster^{-1} μm^{-1}, we see no evidence of a circumsolar, local F corona with a brightness amplitude near 10^{-7} W cm^{-2} ster^{-1} μm^{-1} (MacQueen 1968). If there is a local dust component to the corona it must now be considerably smaller than it was in the past.

Acknowledgements

This experiment was supported by the Laboratoire d'Astrophysique Spatial, the Canada-France Hawaii consortium, and Michigan State University. We are grateful to David Hoadley and Nathan Blair for help with the camera software and polarization analysis. The IR camera system was developed in collaboration with the University of Wyoming and Haverford College through the Astrophysical Research Consortium.

References

Koutchmy, S. and Lamy, P.: 1985, in R.H. Giese and P. Lamy (eds.), *Properties and Interactions of Interplanetary Dust*, p. 63.
McPherson, M., Lin, H. and Kuhn, J.R.: 1992, *Solar Phys.* **139**, 255.
MacQueen, R.M.: 1968, *Astrophys. J.* **154**, 1059.
Mampaso, A., Sanchez-Magro, C., and Buitrago, J.: 1982, in W. Fricke and G. Tedeki (eds.), *Sun and Planetary Systems*, p. 257.
Mukai, T.: 1985, in R.H. Giese and P. Lamy (eds.), *Properties and Interactions of Interplanetary Dust*, p. 59.
Peterson, A.W.: 1963, *Astrophys. J.* **138**, 1218.
Wyatt, S., Whipple, F.L.: 1950, *Astrophys. J.* **111**, 134.

INFRARED CORONAL OBSERVATIONS AT THE 1991 SOLAR ECLIPSE

R. M. MACQUEEN

Rhodes College, 2000 North Parkway, Memphis, TN 38112, U.S.A.

and

K-W. HODAPP and D. N. B. HALL

University of Hawaii, 2680 Woodland Drive, Honolulu, HI 96822, U.S.A.

Abstract. Observations of the solar corona in a narrow band filter at wavelength 2.12 μm during the total solar eclipse of 11 July 1991 are described, and compared with results obtained by two observers during the eclipse of November 1966, and shortly thereafter. The lack of any observable signature of thermal emission in the 1991 results suggests that during 1966/67, the near-solar environs were subjected to a locally enhanced dust population, supplied by one or more sungrazer comets. Possible conditions which match the observational circumstances are discussed.

Key words: eclipses – infrared: stars – interplanetary medium – Sun: corona

1. Introduction

We have recently reported the initial analysis of results of near infrared observations of the solar corona obtained at the total solar eclipse of 1991 (Hodapp, MacQueen and Hall, 1992; hereafter referred to as Paper I). In that report, we present observations obtained with a narrow band filter at wavelength 2.12 μm from among observations carried out in the J, H, and K bands.

In this report, we elaborate further on the above observations, compare these current results with some prior results, and discuss possible causes for the differences between observations at the 1966 and 1991 total solar eclipses.

2. Observations

Details of the instrument are presented in Paper I. Particular attention was paid to eliminating any possibility of spurious artifacts which might be interpreted as signatures of thermal emission from circumsolar dust. To address this potential problem we placed an occulting disk in front of the 1-cm diameter CaF_2 objective lens of the *NICMOS* camera. The serrated occulter subtended an angular diameter of 90 arcmin, or 2.5 $R_\odot$ from sun center. Tests of this system were carried out at the time of nearly-full moon, on June 27, 1991 and the results indicated that in the K band, the radiance of diffracted light on the disk periphery was less than 10^{-3} that of the moon. In addition, no artifacts were present in the instrument field to at least a factor of 10^{-2} times the level of the diffracted light. These results held for pointing accuracies of up to 0.4 $R_\odot$; only at the point when extreme offset pointings of about 0.8 $R_\odot$ were carried out could artifacts be identified. These precautions ensured that no spurious instrumental artifacts would be identified as thermal emission features.

Sky conditions during totality, and throughout the day of eclipse, were substandard for the Mauna Kea site. The presence of both Mt. Pinatubo volcanic dust and cirrus clouds limited the quality of observations during totality and quality and quantity of calibrations. The cirrus clouds present during totality drifted rather rapidly through the field of view of the instrument, gradually clearing over this period. As a result, data obtained during the latter portion of totality was less influenced by the effects of the cirrus than that obtained earlier. Thus, initial data reduction has emphasized frames obtained late in the period of totality, especially a 10- second narrow band 2.12 μm filter exposure obtained just prior to third contact.

The 2.12 μm image presented in Paper I exhibits the usual features of the solar K- and F-coronal components. The structured K-corona is dominated by prominent coronal streamers whose bases lie close to the solar east and west limbs, as determined by comparison of eclipse images and High Altitude Observatory K-coronameter maps (Sime and Streete, 1992).

The unstructured F-coronal and background K-coronal contribution is concentrated in (or near) the ecliptic plane. Figure 1 presents equatorial and polar radial scans of the 2.12 μm coronal radiance observed at the 1991 eclipse and, for comparison, near-equatorial radial scans at wavelength 2.2 μm reported from independent observations at the 1966 eclipse (Peterson, 1967; MacQueen, 1968). The latter measurements differ in their absolute radiances by a factor of three. The current 1991 observations more closely agree with the radiances obtained by Peterson, while those of MacQueen (1968) are lower by about a factor of three. As noted in Paper I, the current results have been calibrated by direct observation of α Lyra and α Boo in the 2.12 μm filter, employing the results of Strecker, et al. (1979). This straightforward procedure is inherently more reliable than the calibration procedures required in 1966, and despite sky variations, we adopt the current values.

The comparison of the 1966 and 1991 radial falloffs in Figure 1 vividly shows the relative smoothness of the current results compared to the earlier observations. There is no hint of a signature of thermal emission in the current results. As we have reported, we estimate that the upper limit for such a signature is roughly a factor of twenty smaller than the signature observed in 1966 by MacQueen, and thus about a factor of fifty smaller than that reported by Peterson.

3. Discussion

Although the 1966/67 observations were carried out with single element detectors, it is difficult to picture how the radial scans carried out by the two investigators could have intercepted any extant K-coronal feature whose signature could have been attributed to thermal emission. White light images of the corona at the 1966 eclipse confirm that there were no visible coronal mass ejections (undiscovered until the late 1960's) or anomalous K-coronal features present. And, the presence of a similar (but not exactly identical) feature about 8 weeks after the total eclipse, observed with an infrared balloon-borne coronagraph, lends additional credence to the reality of the feature. Likewise, K-coronameter scans from early 1967 revealed only typical coronal features, and even no bright streamers on the relevant solar limb.

In Paper I we summarize efforts prior to 1991 to observe a coronal dust emission signature. Since the 1966/67 observations remain the most unambiguous detection,

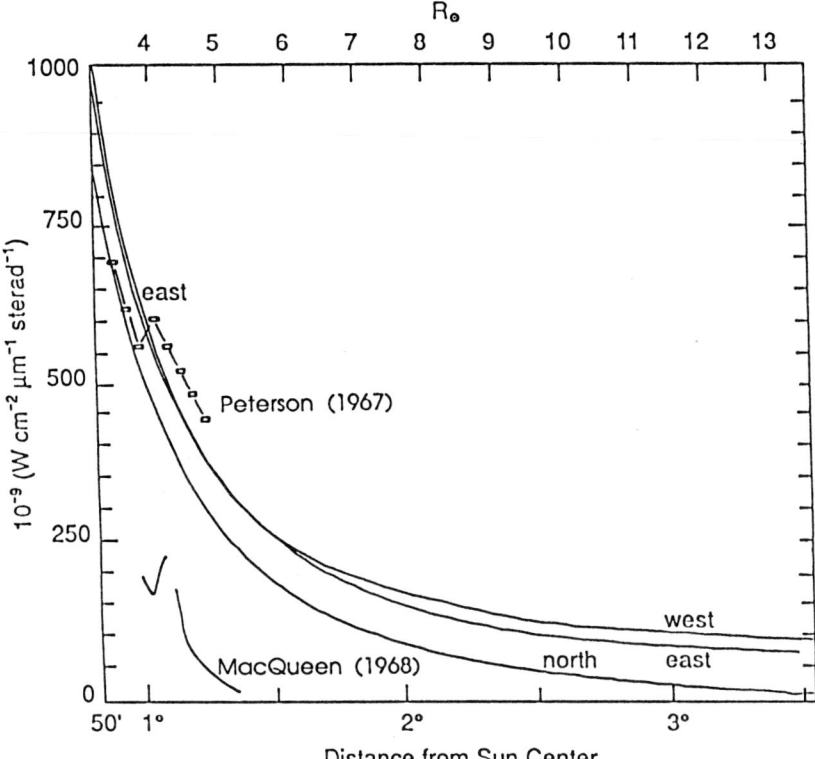

Fig. 1. Radial scans of the near infrared coronal radiance observed on the east, west and north solar limbs at the 1991 eclipse, and equatorial scans by observers at the 1966 eclipse.

we address the question of what variable input to the interplanetary dust population could have been responsible for features at that time, but not later (certainly not in 1991).

The time scale for capture of interplanetary dust is generally far in excess of the several-decade time scale required to explain the variation between 1966/1967 and 1991. However, one possibility of short-term injection of dust into the near-solar environment involves that from sungrazing comets, and we briefly explore this possibility in the following paragraphs.

The appearance of smaller members of the so-called Kreutz group of sungrazing comets is now known to be highly episodic on an annual basis (MacQueen and St. Cyr, 1991). Space-borne coronagraphs have been responsible for the detection of the majority of this class of comet; during a little less than a decade, sixteen sungrazer comets were so observed. To date, no ground-based observing methods have revealed these smaller members, either near the sun, or deeper in space.

The appearance of a Kreutz group member near the sun is similar to that of more usual comets whose perihelion passage is not near the solar surface. However, sungrazer comets apparently do not survive perihelion passage: no such comet has been observed to emerge from the near-solar region occulted by the space-borne coronagraphs. However, using the orbital elements of the Kreutz group, it was shown that these small comets apparently survive perihelion approach down to a distance of about 3 $R_\odot$ without discernable effects (MacQueen and St. Cyr, 1991). With one spectacular exception (Michels, et al., 1982), coronagraph observations do not reveal any visible light effect of the disappearance of the comets near perihelion.

Could a sungrazing comet (or comets) be responsible for the near infrared signatures of 1966/67? This possibility requires at least that (a) such small members of the Kreutz group contain sufficient material to account for the required local enhancement of near-solar dust, based upon thermal emission models of the corona, and (b) such a density enhancement not be apparent in the visible wavelength region, dominated by scattered radiation.

Mukai and his collaborators (Mukai, et al., 1974; Mukai and Yamamoto, 1979; Mukai, 1985) have investigated the dust density enhancement required to approximately match the infrared signatures observed in the 1966/67 period. Following Belton (1966), they have suggested that the circumsolar dust orbits, initially determined by the action of Poynting-Robertson drag, may, upon the onset of sublimation, become stabilized and a locally-enhanced dust region result. Mukai and Yamamoto (1979) suggested that a local enhancement of a factor of 5-10 would be required to match the observed 4 $R_\odot$ signature.

To estimate the F-coronal dust mass requires that we specify the dust particle size distribution, space distribution and dust mass density. For the former, we assume the distribution given by Lamy and Perrin (1986): a two-component distribution (Population I and Population II) of large and small particles, each separately with power law falloffs. For the space distribution of dust, we assume that the number of particles per cm^3 varies with distance from the sun as r^{-q}, with the index q equal to 1.0 or 1.3 (Lamy and Perrin, 1986). Finally, we assume a mass density $\rho = 2.5$ gm cm^{-3}, (Giese, 1961). If we consider a local coronal volume of approximate dimensions 0.2 $R_\odot$ wide, centered on a distance 4 $R_\odot$ from sun center, 4 $R_\odot$ along the line of sight and extending 2 $R_\odot$ in heliographic latitude, that volume would contain approximately 10^6 grams of dust.

This total mass could be easily supplied by even a small member of the Kreutz sungrazer comet group. For example, MacQueen and St. Cyr (1991) estimated that, on the assumption that these comets fully sublimed during perihelion passage, the smallest sungrazing comet was only about 30 meters in diameter. Even such a small member of the Kreutz group could provide roughly 10^{10} grams of material – more than adequate to achieve the above local enhancement of the F-corona, or to enhance a larger volume.

The visibility of such a near-solar dust enhancement depends upon the nature of the scattering properties of the material. Generally, however, dust far from the sun contributes significantly to the F-coronal scattered radiance (forward scattering), while that close to the sun less so (scattering at nearly 90°). To assess the potential contribution of a locally enhanced region of the solar F-corona, we have computed the line-of-sight brightness of an element 4 $R_\odot$ along the line of sight, placed at a distance of about 4 $R_\odot$ from sun center, relative to the total line of sight contribution of all dust. For this computation, we have employed the material properties of pyrrhotite (Perrin and Lamy, 1989) as approximating that relevant to Population II zodiacal cloud material. Population I material has been represented by the complex index of refraction of chondrite (Perrin and Lamy, 1989). The volume scattering functions and line of sight radiance have been computed using the Mie scattering intensities, from a newly developed code (MacQueen and Greeley, 1992).

We find that the 4 $R_\odot$ volume element contributes between 2-3 $\times$ 10^{-3} that of the full line of sight F-coronal radiance, and thus a dust enhancement of a factor of 5 would contribute at most roughly 1% to the visible wavelength radiance of the $K + F$ coronal radiance–at or below the limit of detectability of visible wavelength eclipse observations.

As a result, we conclude that it is possible to explain the appearance of an infrared signature of thermal emission of dust at the eclipse of 1966 by injection of an enhanced dust population from a sungrazing comet. However, to explain the presence of the observed features in early 1967, about two months after the eclipse, is more difficult; it is surely unlikely that additional sungrazing comets were responsible for a second transient appearance of a thermal emission signature. Alternatively, the injection of dust provided by a single comet may reside near the sun for an extended period, but as noted in Paper I, we are unaware of any calculations of the dynamics of such dust.

References

Belton, M.J.: 1966, *Science*, **151**, 35-44.
Giese, R.H.: 1961, *Zeits. fur Astrophys.*, **51**, 119-147.
Hodapp, K-W., MacQueen, R.M. and Hall, D.N.B.: 1992, *Nature*, **355**, 707-710.
Lamy Ph. and Perrin, J-M.: 1986, *Astron. Astrophys.*, **163**, 269- 286.
MacQueen, R. M.: 1968, *Astrophys. J.*, **154**, 1059-1076.
MacQueen, R.M. and St. Cyr, O.C.: 1991, *Icarus*, **90**, 96-106.
Michels, D.J., Sheeley, N., Howard, R., and Koomen, M.J., 1982: *Science*, **215**, 1097-1102.
Mukai, T., 1985: in R. Giese and Ph. Lamy (eds.), *Properties and Interactions of Interplanetary Dust*, pp. 59-62.
Mukai, T., Yamamoto, T., Hasegawa, H., Fujiwara, A. and Koike, C.: 1975, *Pubs. Astron. Soc. Japan*, **26**, 445-458.
Mukai, T. and Yamamoto, T.: 1979, *Pubs. Astron. Soc. Japan*, **31**, 585-595.
Perrin J-M. and Lamy, Ph.: 1989, *Astron. Astrophys.*, **226**, 288- 296.
Peterson, A.W.: 1967, *Astrophys. J.*, **148**, L37-L39.
Sime, D. and Streete, J.L.: 1992, (in preparation).
Strecher, D.W., Erickson, E.F., and Witteborn, F.C.: 1979, *Astrophys. J. Supp. Ser.*, **41**, 501-512.

ON THE CORONAL AND PROMINENCE STRUCTURES OBSERVED AT THE TOTAL SOLAR ECLIPSE OF 11 JULY 1991

Y. SUEMATSU, H. FUKUSHIMA and Y. NISHINO

National Astronomical Observatory, Mitaka, Tokyo 181, Japan

Abstract. Coronal images were taken in the light of the He I 10830 Å line, the 10000 Å continuum, and the Fe XIV 5303 Å line, with the aim of studying the thermal structure of the corona. In addition, spectroscopic observations were made in the violet wavelength region (3760–4060 Å) and near-infrared (10745–10835 Å), to obtain details of physical conditions of the corona, especially of its cool part. The data obtained do not show any distinct cool structures other than ordinary prominences. Some preliminary results concerning the corona and prominence structures are given.

Key words: eclipses – He I 10830 Å – infrared: stars – Sun: corona – Sun: prominences

1. Introduction

The solar corona is generally known to have temperatures in the million-degree range. However, several types of features of relatively low temperature exist in the low corona such as spicules, prominences ($T = 10^4$ K) and EUV-jets ($T = 10^4$ K, Dere *et al.*, 1983; $T = 10^5$ K, Brueckner and Bartoe, 1983). Since some of these cool features are transient, they should at least in part interact with the hot corona through mass and energy transfers. For instance, spicules ascend from the chromosphere into the corona, and then some of them fade out, others fall back to the chromosphere (Beckers, 1972). Wagner *et al.* (1983) showed that a coronal rain prominence forms underneath a coronal void, using eclipse white-light pictures. Hence, it is probable that material having temperatures between 10^4 and 10^6 K exists in the vicinity of the well-known cool features or could even spread over the entire corona.

Some authors have reported the existence of cool material other than the well-known features in the corona. From eclipse observations, Deutsch and Righini (1964) obtained a surprising Ca II H and K emission spectrum up to one solar radius above the limb; the intensity ratio of K to H was much larger than 2. They drew the conclusion that material with a characteristic temperature of 10^5 K existed over a wide region of the corona (Cavallini and Righini, 1975). Bappu *et al.* (1972) also obtained a strange metallic-line emission spectrum, which seemed not to arise from normal prominences, although they did not carry out an analysis of the data.

Outside of eclipses, Leroy (1972) found many small-scale faint features in Hα filtergrams from coronagraph observations (*cf.* Öhman, 1972), while Gnevishev and Gnevisheva (1963) reported such features in He I D$_3$ filtergrams. Furthermore, it is interesting to note that the weakening of the EUV line emission with wavelengths shorter than 912 Å could be caused by Lyman continuum absorption due to unknown cool material in the corona (*e.g.*, Kanno and Suematsu, 1982).

However, there has been criticism directed at the validity of the evidence for cool material mentioned above, because of its strange properties. Some of the emissions might come from atmospheric or instrumental scattering of the chromospheric or prominence radiation (*e.g.*, Caccin *et al.*, 1971). Our main purpose in these eclipse observations was to confirm the existence or non-existence of cool material other than the ordinary cool features, and further, to study the origin of such material if it exists.

We used the He I 10830 Å line for this purpose because this line suffers less atmospheric scattering than visual or violet lines. In order to know the spatial distribution of cool material, He I 10830 Å filtergrams were taken with a CCD camera. Moreover, He I 10830 Å spectra, including the well-known Fe XIII 10747/10798 Å lines, were taken to understand the detailed physical conditions of both cool and hot material, although this did not yield a useful result. In addition, spectra of metallic lines in the violet region (3760–4060 Å), including the Ca II H and K lines, were taken because these lines give much more information on the physical conditions of cool material. Fe XIV 5303 Å filtergrams were taken in order to study the hot corona and the relationship between the cool and hot structures. As a matter of course, we were also interested in the thermal structure of the corona, especially the regions surrounding prominences, and in the physical conditions in the faint parts of prominences.

The observations were carried out under good sky conditions, at the campus of Universidad Autonoma de Baja California Sur (UABCS), La Paz, Mexico.

2. Instrumentation and Observing Procedure

A 3-channel telescope was used consisting of three tubes mounted on a single equatorial mount and was designed to obtain three different monochromatic images of the corona. Two of the three tubes had apochromatic lenses of 76 mm diameter ($f/7.9$), telecentric lens systems, filter boxes, camera lenses, and CCD cameras (TAKENAKA TM-840N). A He I 10830 filter (passband of 6 Å) was used in one filter box and a 10000 Å-continuum filter (passband of 200 Å) in another. The solar image diameter at the CCD camera was about 5 mm. The image data from the CCD were stored on a VCR with S-VHS mode, as well as on a hard disk of a personal computer through a 512×512 pixel image-processor with 8-bit precision. The image processor automatically integrated video images up to the saturation level.

The third tube consisted of an apochromatic lens of 100-mm diameter ($f/8$), a relay lens system, a Fe XIV 5303 Å filter (passband of 3.5 Å), and a 35-mm camera (Nikon F3, with 250 exposure capability). The solar image diameter on the film was about 25 mm. Kodak TMAX400 emulsion was used. Camera exposures were controlled by a note-type computer; the exposure time was changed between 1/2000 second and 64 second during the eclipse observation. The telescope was pointed to the east solar limb before mid-totality and to the west limb after mid-totality.

A spectrograph was designed to observe the violet (3760–4060 Å, dispersion = 8.5 Å mm^{-1}) and near-infrared (10745–10835 Å, dispersion = 10.3 Å mm^{-1}) wavelength regions simultaneously, consisting of a straight mirror slit, a collimator (off-axis paraboloid), grating (50×50 mm^2, 1200 grooves mm^{-1}), and two camera lenses. The solar image was focused on the mirror slit by a Cassegrain-type telescope (objective of 100 mm diameter; $f/12.4$ system). The spectrograph and the feed telescope were equatorially mounted.

The width and the length of the slit were 30 μm and 15 mm, respectively, for the chromosphere and low corona, while the width was changed to 200 μm during the observation of the high corona. The direction of slit axis was always north-south in the sky: the slit was positioned tangentially to the solar east or west limb. In order to examine linear polarization of the light, a Gran-Taylor prism, which is effective for both infrared and violet wavelengths, was placed in front of the mirror slit. The prism was rotated around its axis in 60-degree steps.

An improved CCD camera (National CD-55) was used with an evacuating and cooling device for the infrared observation. A film-camera (Nikon F-3, with 250 exposure capability) was used for the violet region. Another personal computer was employed to control the exposure of both the CCD and film cameras, the rotation of the polarizer, and the fine settings of the spectrograph slit. The telescope-spectrograph was moved so that the slit was set to the extreme east solar limb around the time of second contact, and moved higher above the east limb (up to 100 arcseconds) by 90 seconds before third contact. Finally, the telescope was manually pointed to the west just above the lunar limb until third contact.

We used Kodak TMAX400 emulsion for the violet observations and exposure times of 1/15, 1, 4, 8, and 180 s, synchronizing with the rotations of the polarizer, for the east corona, and a fixed value of 8 s for the west corona. For the cooled CCD camera, we selected exposure times of 1/15 and 180 s for the east corona, storing the data on a VCR. This CCD did not operate for the west corona.

White-light slit-jaw images, which were also linearly polarized because of the Gran-Taylor prism, were also recorded on a VCR through a CCD camera (Sony XC-77RR) with a neutral density filter and f55 mm camera lens (Micro Nikkor).

Finally, large-scale coronal images in white-light were taken with a film-camera (Kodak TMAX100 emulsion and Fuji Color HRII 100) and a video camera.

3. Preliminary Results and Discussion

During this eclipse, we observed three outstanding streamers: two extended in the northeast direction and the other in the southwest. Also, prominent coronal plumes (ray-like structures), *not always polar plumes* in this case, were observed in the north-to-west region. Two large prominences were evident: one was at the east limb and another at the west.

White-light coronal pictures were analyzed to study the width-variation of the plumes as a function of distance along them. Although we can see both bright plumes and dark extended voids, the result presented here is for seventeen plumes. It is found that most of the plumes expand linearly with distance: the expansion rate is 10–70 arcsec per solar radius. The widest and most prominent plume seems to connect to the large west prominence and hence to be abnormal. When the widths are extrapolated to the solar surface, we obtain values of 10–40 arcsecond. These values suggest that the narrow coronal plumes may be related to the enhanced unipolar magnetic fields emanating from the supergranulation network boundaries, whose typical cell diameter is about 40 arcsecond (*cf.* Newkirk and Harvey, 1968).

In the Fe XIV 5303 filtergrams (Figure 1) we can see many fine-scale structures such as loops, threads and rays. In the long exposures filtergrams, we can even

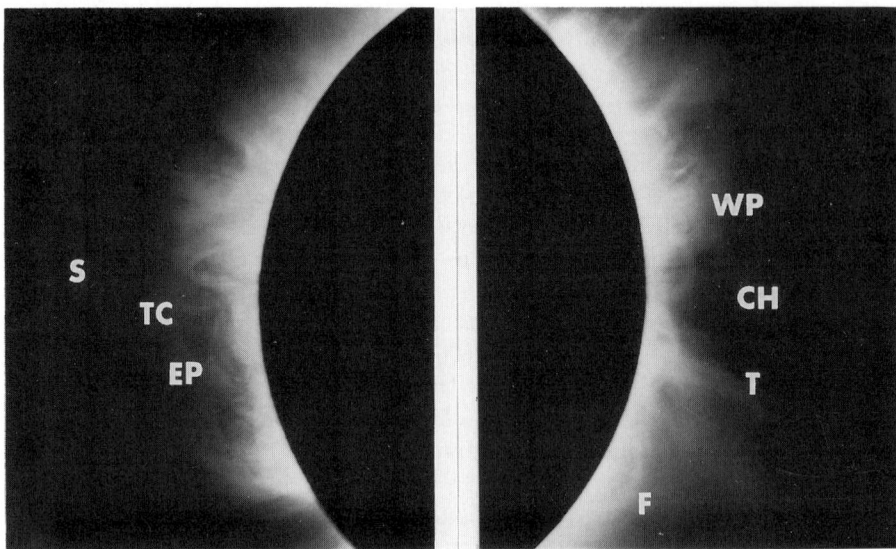

Fig. 1. The east (left) and west (right) corona in Fe XIV 5303 Å line. The exposure times were 32 s. North is at the top. The symbol TC indicates the twin cavity, S the east streamers, EP the location of the east prominence, CH the coronal hole, T the thread, WP the location of the west prominence, and F the location of the faintest prominence (see Figure 2).

perceive the coronal streamers (the symbol S in Figure 1) seen in the white-light pictures, although they are very faint. Of special interest in the east corona is a twin cavity structure (TC) underneath the east streamer – two cavities that are in close proximity. The large east prominence (EP) is sitting inside one of these cavities. The 1966 eclipse showed such twin cavity structure too: Two adjacent cavities were enclosed by a helmet-streamer structure and each contained a prominence (Tandberg-Hanssen, 1979). In the west corona, a coronal hole (CH) is seen and a thread (T) at the south edge of this coronal hole seems to be twisted. We cannot see a distinct cavity above the large west prominence (WP).

We cannot detect any cool structures other than the ordinary prominences for the He I 10830 line and the Ca II H and K lines to the limit of the detector sensitivity which is about 10^{-2} of the ordinary prominence intensity for He I 10830 filtergrams (Figure 2), and 10^{-4} for Ca II H and K line spectra. Unfortunately, the coronal parts of the Ca II H and K line spectra are contaminated by atmospheric scattered light from the prominences: False emission lines are seen in the regions neighboring the prominences. This negative result might not necessarily mean that cool material other than that of ordinary prominences does not exist in the corona at all. Our

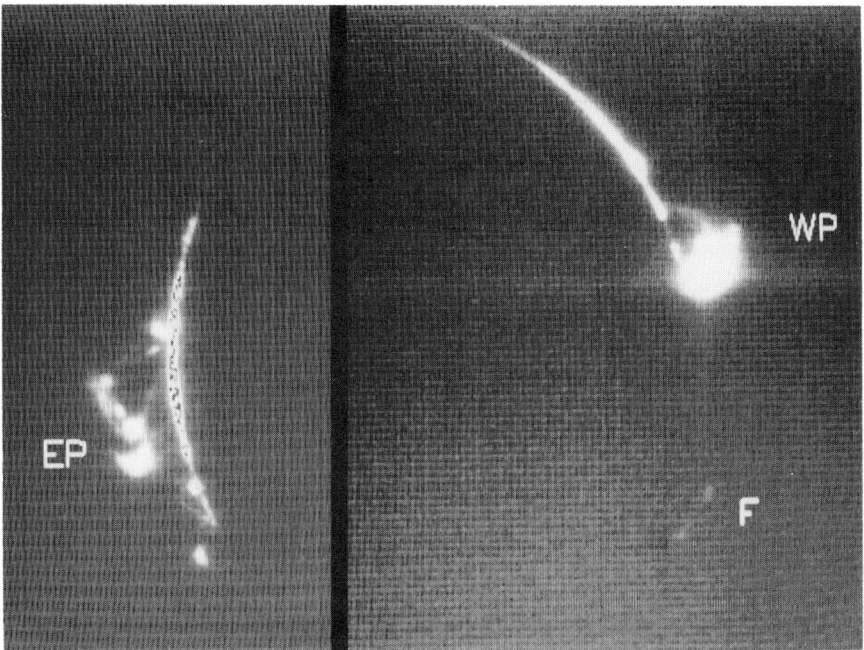

Fig. 2. The east (left) and west (right) corona/prominence in He I 10830 line. A total of six images and nine images were summed up for the east and the west, respectively. In order to enhance faint regions, bright regions are saturated. North is at the top. The symbol EP and WP indicate the large east and west prominences, respectively, and the symbol F indicates the faintest prominence we could detect.

detector sensitivities were insufficient for this purpose, and one might find such cool material in other eclipses. This problem will need further investigation in the future.

The faintest prominence (F in Figure 2) in our observations is in the southwest quadrant. Its intensity is a few hundredths that of a normal prominence in He I 10830. This is a nearly horizontal bar shaped prominence and shows highly-shifted line spectra in Ca II H and K, corresponding to a velocity of 100 km s^{-1}. However, the prominence is barely seen in higher Balmer or metallic lines. This might imply that the temperature of this prominence is higher than that of more typical ones (see, *e.g.*, Alikayeva, 1975).

We can see tilted line spectra in the violet at the extreme top of the large east prominence. This tilting probably indicates the rotational motion of the prominence material. At the leg part of this prominence, between the prominence body and the chromosphere, we have highly-shifted Ca II H and K line spectra, corresponding to a line-of-sight velocity of several tens of km s^{-1}.

In the violet spectra, in addition to the chromospheric/prominence lines, we can detect three coronal lines; Co XII 3800.7, Fe XI 3987.1, and Cr XI 3998.0, which are formed in plasma of about one million degrees. With these lines, as well as the Fe XIV 5303, we can estimate the temperature structure in the corona in the future.

Acknowledgements

We are grateful to Dr. H. Kurokawa, Kyoto University, the chief of the Japanese Eclipse Observation Team (JEOT), for his kind and helpful coordination between CUPOE (Comite Universitario Para la Observacion del Eclipse) /COMES (Comite Mexicano Eclipse Solar) and our team. We wish to express our hearty appreciation to CUPOE/COMES, especially, Drs. M. Oseguera (CUPOE, UABCS) and J. Bohigas (COMES, UNAM), and to the staff of the Japanese Embassy of Mexico, for their kind and helpful cooperation in Mexico: we could not have succeeded with our observations without their practical help. We are obliged to Profs. Y. Yamashita, T. Hirayama, and E. Hiei, National Astronomical Observatory of Japan (NAOJ), for their useful suggestions and encouragement. Finally, but not least, our thanks go to all members of Solar Physics Division, NAOJ, especially Messrs. N. Tanaka, K. Sano and H. Miyazaki, Mechanical Factory Division, NAOJ, and all members of JEOT and National Committee for Solar Eclipse, Science Council of Japan, for their useful assistances during this eclipse project. This project was carried out under the support of a Grant in Aid of Overseas Scientific Research No. 02041094, Ministry of Education, Science and Culture, Japan. This manuscript was written while one of us (Y.S.) was staying at the National Solar Observatory at Sacramento Peak. Y.S. is grateful to the staff of the Observatory for their warm hospitality.

References

Alikayeva, K.V.: 1975, *Solar Phys.* **41**, 89.
Beckers, J.M.: 1972, *Ann. Rev. Astron. Astrophys.* **10**, 73
Bappu, M.K.V., Bhattacharyya, J.C., and Sivaraman, K.R.: 1972, *Solar Phys.* **26**, 366.
Brueckner, G.E. and Bartoe, J.-D.F.: 1983, *Astrophys. J.* **272**, 329.
Caccin, B., Moschi, G., Rigutti, M. and Falciani, R.: 1971, *Solar Phys.* **17**, 89.
Cavallini, F. and Righini, A.: 1975, *Solar Phys.* **45**, 291.
Dere, K.P., Bartoe, J.-D.,F., and Brueckner, G.E.: 1983, *Astrophys. J. (Letters)* **267**, L65.
Deutsch, A.J. and Righini, G.: 1964, *Astrophys. J.* **140**, 313.
Gnevishev, M.N. and Gnevisheva, R.S.: 1963, in J.W. Evans (ed.) *The Solar Corona*, New York, Academic Press, p. 241.
Kanno, M. and Suematsu, Y.: 1982, *Pub. Astron. Soc. Japan* **34**, 449.
Leroy, J.L.: 1972, *Solar Phys.* **25**, 413.
Newkirk, G.Jr. and Harvey, J.: 1968, *Solar Phys.* **3**, 321.
Öhman, Y.: 1972, *Solar Phys.* **28**, 399.
Tandberg-Hanssen, E.: 1979, in E. Jensen, P. Maltby, and F.Q. Orrall (eds.), 'Physics of Solar Prominences', *IAU Colloq.* **44**, 139.
Wagner, W.J., Newkirk, G., and Schmidt, H.U.: 1983, *Solar Phys.* **83**, 115.

THE WHITE-LIGHT, FAR-RED (600-700 nm) AND EMISSION CORONAE AT THE JULY 11, 1991 ECLIPSE

V. RUŠIN, M. RYBANSKÝ and M. MINAROVJECH

Astronomical Institute, Slovak Academy of Sciences,
059 60 Tatranská Lomnica, Czecho-Slovakia

and

T. PINTÉR

The Slovak Centre of Amateur Astronomy, 947 01 Hurbanovo, Czecho-Slovakia

Abstract. Preliminary results of the analysis of the white-light, emission (green and red), and far red (600-700 nm) corona during the July 11, 1991 eclipse are given. Even though the corona is of nearly-maximum type, four different principal coronal structures are seen, combined with faint, small-scale structures (loops, arches, cavities, voids or plasmoids). Scattered light is seen up to 10 $R_\odot$ in helmet streamers. The Ludendorff index of the corona shape turns out to be $a + b = -0.02$, and the estimated brightness of $J_K = 1.47 \times 10^{-6}\ B_\odot$. Some aspects of multiwavelength observations are discussed.

Key words: eclipses – Sun: corona

1. Introduction

Total solar eclipses provide a unique opportunity not only to observe all parts of the solar corona $(E,\ K,\ F,\ T\ (?))$ under optimum conditions, but to study other aspects of solar activity in a wide range of wavelengths, and heights above the solar surface. The combination of eclipse and non-eclipse observations, both ground-based and from space, should provide one of the best opportunities for developing multispectral models of the solar corona. This opportunity occurred on July 11, 1991, when many observers located around the world were able to observe the Sun and its surroundings both in the eclipse path and outside it. In this article we present some preliminary results of the white-light, far-red, and emission coronae as observed on July 11, 1991.

2. Observations

2.1. THE WHITE-LIGHT AND FAR-RED (600-700 NM) CORONAE

Classical pictures of the white-light corona were obtained in observations made at La Paz (Baja California Sur, Mexico) with two telescopes: a 3-meter focal length telescope as a part of the MICE experiment (Zirker et al., 1992) and a 1-meter focal length telescope . To find coronal streamers or F-corona properties very far from the Sun in the interplanetary space, an 80-mm camera was used (with Fujichrome film). The far-red corona was imaged with video recording (Panasonic) during the whole eclipse (more than 6 min). The blue and green components of the camcoder record were separated electronically from the video record and digitized for storage.

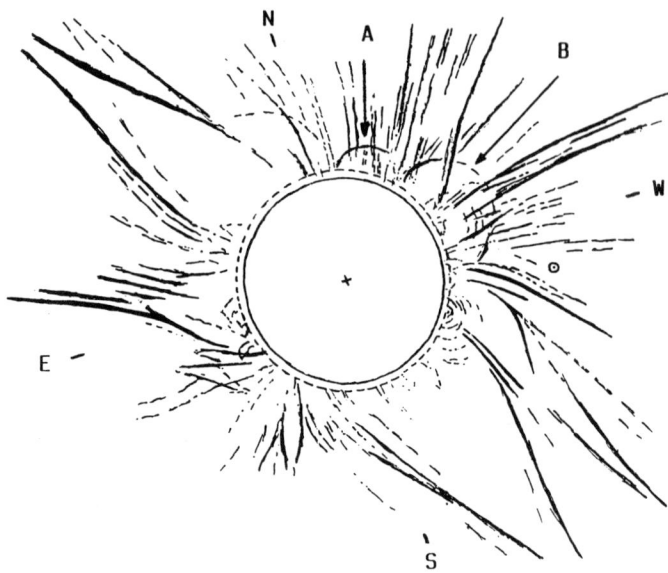

Fig. 1. Large-scale stuctures of the white-light corona (Zirker et al., 1992). Notable features in the NW-quadrant are denoted by the letters A, B.

2.2. The green (530.3 nm) and red (637.4 nm) corona, and prominences (non-eclipse observations)

Standard limb observations (of about 40″ above the solar limb with a lag of 5° from the north solar pole towards east) of the green and red corona during the eclipse day and around it, were made at Lomnický Peak coronal station.

3. Results and Short Discussion

The eclipse occurred after the maximum of cycle 22 – mid-1989 according to the sunspot number, however increased activity on the Sun was observed for three months, from June 1991 to August 1991, when it reached values comparable with the maximum values (see Solar Geophysical Data for 1992). Four different principal large-scale types of coronal structures were found in the white-light corona (Figure 1) during the eclipse.

(a) Helmet streamers are located in the NE-quadrant in P.A. (at the moon's limb) 14°–60°, and in the SW-quadrant in P.A. 155°-260°. These streamers represent multiple systems located above the quiet filamentary channels of both north and south hemispheres (some small prominences in the NE-region were observed at P.A. 37° and 41° on July 9, 1991). A separation of helmet streamers with height above the Sun enables us to reach the conclusion that they have different bases above the photosphere with regard to heliographic longitude. Nevertheless, some of them are seen up to 9-10 $R_\odot$ (Figure 2),and their orientation above the Sun in

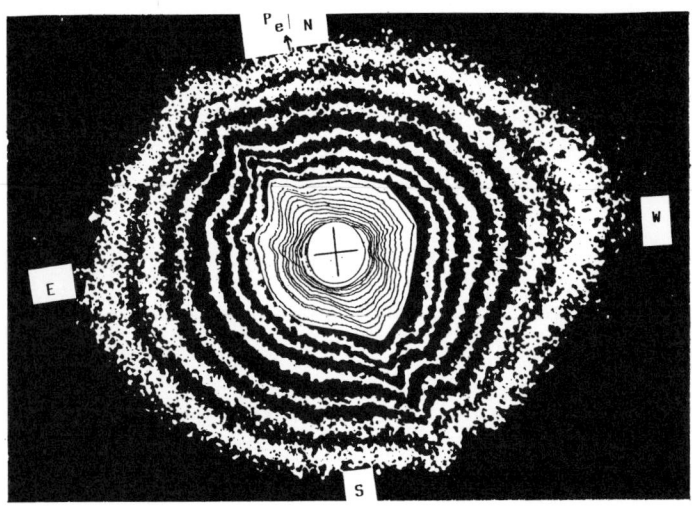

Fig. 2. Isophotes (arbitrary units) of the inner and outer solar corona.

nearly radial with height.

(b) Narrow polar rays are mainly extended above the north solar pole, and should be seen up to 3-4 $R_\odot$. Such narrow rays are not seen above the south solar pole. Different types of coronal structures above both poles confirm a temporally-shifted development of solar activity between individual hemispheres over the solar cycle as has been observed earlier.

(c) A typical coronal hole was localized above the north solar pole, between P.A. 315°-15°, where a decrease in coronal brightness (Figure 2), is clearly seen. Similar cases are found in the NW-quadrant at P.A. 265°-278° and in the SW-quadrant at P.A. 130°-155°. It may be supposed that both these regions should be good candidates for coronal holes too (equatorial and high altitude), and a good location for searching for the F-corona or T-corona. We note that the angular size of these coronal holes is smaller than for the comparable north coronal hole, and faint, narrow streamers are localized there.

(d) Thin coronal rays which were mostly non-radial (especially very close to the solar limb) are found to be located above the active regions, which should be centered above the P.A. 70°-120° and 280°-310°.

Thin coronal loops, arches, cavities or voids (e.g., Koutchmy, 1977) may be found at nearly all the principal coronal structures mentioned above. The ellipticity is one of the basic parameters of the coronal shape for a given eclipse (the coronal shape is closely connected with the prominence distribution). The ellipticity of the isophotes (as derived from Fig.2) amounts, even in the innermost corona, to 0.1 only and decreases outward to 0.01. The Ludendorff index of the coronal shape turns out to be $a + b = -0.02$ ($a = 0.144$, $b = -0.164$), and the white-light corona is a maximum type. Using a relation between the total brightness of the white-light corona and ellipticity (Rušin and Rybanský, 1985), we found that the

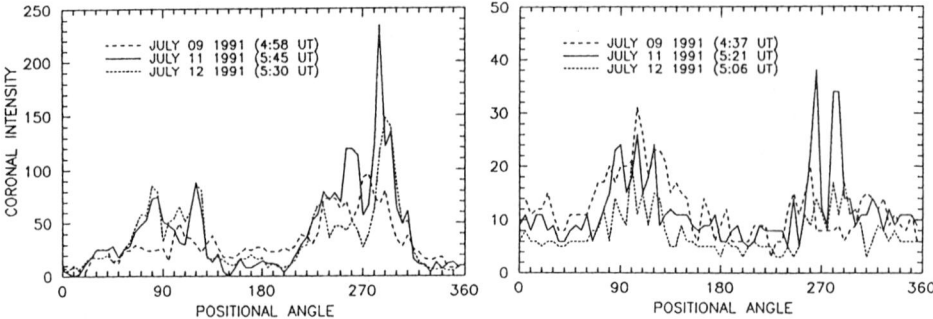

Fig. 3. The green (left) and red (right) spectral limb intensities at Lomnický Peak coronal station (40″ above the solar limb). Intensities are expressed in absolute coronal units ($\times 10^{-6} B_\odot$). The day of observation is shown in the Figure.

integral brightness reached $J_K = 1.47 \pm 0.15 \times 10^{-6} B_\odot$, and the white-light corona was one of the brightest observed at an eclipse at this phase of a cycle $\Phi = 0.41$ (preliminary value for cycle 22). We note that a strong excess in favour of the southern hemisphere is clearly seen in the inner-most corona $(K + F)$. Moreover, one of the SW-streamers overlaps the south solar pole. Both these features may perturb the Ludendorff index. Nevertheless, the 11 July corona is of the maximum type. We also note that, during the July 22, 1990 eclipse, the corona was also of the maximum type, even if a remarkable coronal hole was observed above the south solar pole (Marková et al., 1992). The video recording and small camera images extended the coronal isophotes up to $9 - 10 R_\odot$. We see $3 - 4$ helmet streamers (created by scattered light) in the NE- and SW-quadrants (see Figure 2). The distribution of the $K + F$ corona above $3 R_\odot$, with an excess of the F-corona, is non-spherically symmetry around the Sun. Its maximum values do not lie in the E-W solar direction, but are tilted to the solar equator by 10° (the direction in P.A. 100°-280°) – that is, along the ecliptic plane. A similar result was obtained for the infrared corona by MacQueen et al. (1993). The green and red line coronal intensities are shown in Figure 3. Rapid changes in their intensities were seen in the equatorial regions above both the E and W hemispheres.

These changes are too great to be caused by solar rotation over 4 days, and must result from rapid changes connected with photospheric activity in the active regions. In Figure 4 we present a comparison between the green corona and the NIXT experiment for the eclipse day (Golub, 1992). No typical coronal mass ejection (CME) was observed during the eclipse passage from Hawaii to Brazil, even if such could not be excluded for the days before or after the eclipse day. Prominences above the active region showed high dynamical activity over these days. Moreover, two very interesting features, which resemble a head of CMEs or a plasmoid in the solar corona were found in the Hawaii and Mexico pictures in the NW-limb (their central positions are at P.A. 290° and 330°). However, they remain at nearly the same height above the solar surface over the 1.5 hours between the Hawaii and

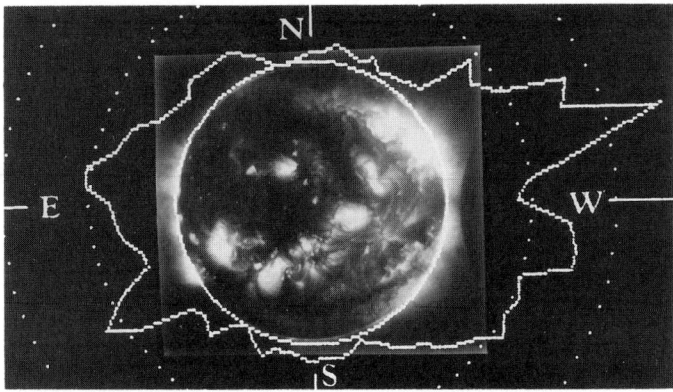

Fig. 4. Comparison of the July 11 1991 green coronal intensity with the NIXT (4-6 nm) corona (Courtesy L. Golub). Inner circle: 100×10^{-6} $B_\odot$; outer circle: 200×10^{-6} $B_\odot$.

Mexico totalities. No notable changes were found between the far-red and white-light coronal images but we do not exclude differences showing up in a detailed study. The intensity distribution of the F-coronal spectrum is generally assumed to be the same of that of the photosphere both in the visual and the infrared. A brief comparison between the results given by Macqueen et al. (1993), and our white-light, F-coronal data, showed nearly the same picture of the intensity distribution, which may be taken as a partial confirmation of that assumption. The F-coronal intensity distribution around the Sun does not depend on solar activity, except for transitory events (CME's), but mat well be connected with the distribution of asteroids or sun-grazing comets, see, e.g., Chocol et al. (1983).

Acknowledgements

The Astronomical Institute Expedition to Mexico was sponsored by Grant 496/91 of the Slovak Academy of Sciences, Bratislava, National Geographic Society; the private sponsors of Mr.J. Mencak and Mr.S. Tomas (CSFR); and the Institut d'astrophysique, Paris. One of us (V.R.) would like to thank the IAU and NSO Tucson for providing him with financial support to the meeting. We would also to thank to Mr. P. Bendík, and Mr.P. Gašpar for preparing the Figures.

References

Chochol, D., Rušin, V., Kulčár, L., and Vanýsek, V.: 1983, *Astrophys. Space Sci.* **91**, 71.
Koutchmy, S.: 1977, in C. J. Durrant and A. Bruzek (eds.), *Illustrated Glossary for Solar and Solar-Terrestrial Physics*, Reidel, Dordrecht, p. 39.
MacQueen, R.M., Hodapp, K-W., and Hall, D.N.B.: 1993, these proceedings.
Marková, E., Vyskočil, L., Rušin, V., and Rybanský, M.: 1992, Solnechnyje Dannyje, in press.
Rušin, V., and Rybanský, M.: 1985, Bull. Astron. Inst. Czechosl., 77, 81.
Zirker, J.B., Koutchmy, S., Nitschelm, C., Stellmacher, G., Zimmermann, J.P., Martinez, P., Kim, I., Dzjubenko ,N., Kurochka, L., Makarov, V., Fatianov, M., Rusin, V., Klocok, L., and Matsuura, O.T.: 1992, *Astron. Astrophys.*, in press.

THE STRUCTURE OF THE WHITE-LIGHT CORONA AT THE 1991 ECLIPSE

A. SANCHEZ-IBARRA

Centro de Investigación en Física, Universidad de Sonora, Sonora 83190, México

M. CISNEROS-MOLINA and G. HINOJOSA-PALAFOX

Departamento de Matemáticas, Universidad de Sonora, Sonora 83190, México

F. CISNEROS-PEÑA and J. GUERRERO DE LA TORRE

Departamento de Física, Universidad de Sonora, Sonora 83190, México

M. NORZAGARAY-COSÍO

Departamento de Matemáticas, Universidad de Sonora, Sonora 83190, México

and

C. TAPIA-FONLLEM

Departamento de Física, Universidad de Sonora, Sonora 83190, México

Abstract. The total solar eclipse of July 11, 1991 was observed from "La Matanza", Baja California Sur, México, only 5 km south of the center line of totality, with several small instruments intended to obtain images of the corona during totality, and using a range of exposure times which allowed us to detect both the inner and outer corona. Relations between large and fine scale structures of the corona, the photospheric and chromospheric activity, and the presence of coronal holes are presented.

Key words: eclipses – Sun: corona

1. Observations

As one of several projects of our team to observe the total solar eclipse of July 11, 1991, we obtained a sequence of photographic images of the corona, on Ektachrome 100 emulsion, and using a Schmidt-Cassegrain telescope of 125 mm aperture and f/10, and a 135-mm camera with a 200-mm telephoto lens. We obtained a sequence of pictures with exposures ranging from 1/1000 to 1/4 s with the telescope, and from 1/500 to 2 s with the camera. The longest exposures were made at the center of totality. Seventy images were obtained in the 6m 49s of totality.

The shorter exposures gave images of the smaller streamers. The longer exposures provide images of the maximum extension of the larger streamers and contain information on the presence of fine structure in the corona.

2. Analysis of Data

With resolutions of about 15 arcsec for the telescope images, and 5 arcmin for the camera images, it was possible to identify 48 coronal streamers. Figure 1 shows the identified streamers; ten of these were identified in both sets of pictures and were classified with a letter, while the streamers resolved only in the images taken with the telescope were classified with consecutive numbers.

Three parameters were measured for each streamer in every image: position-angle, inclination-angle, and extension. The position-angle was determined from

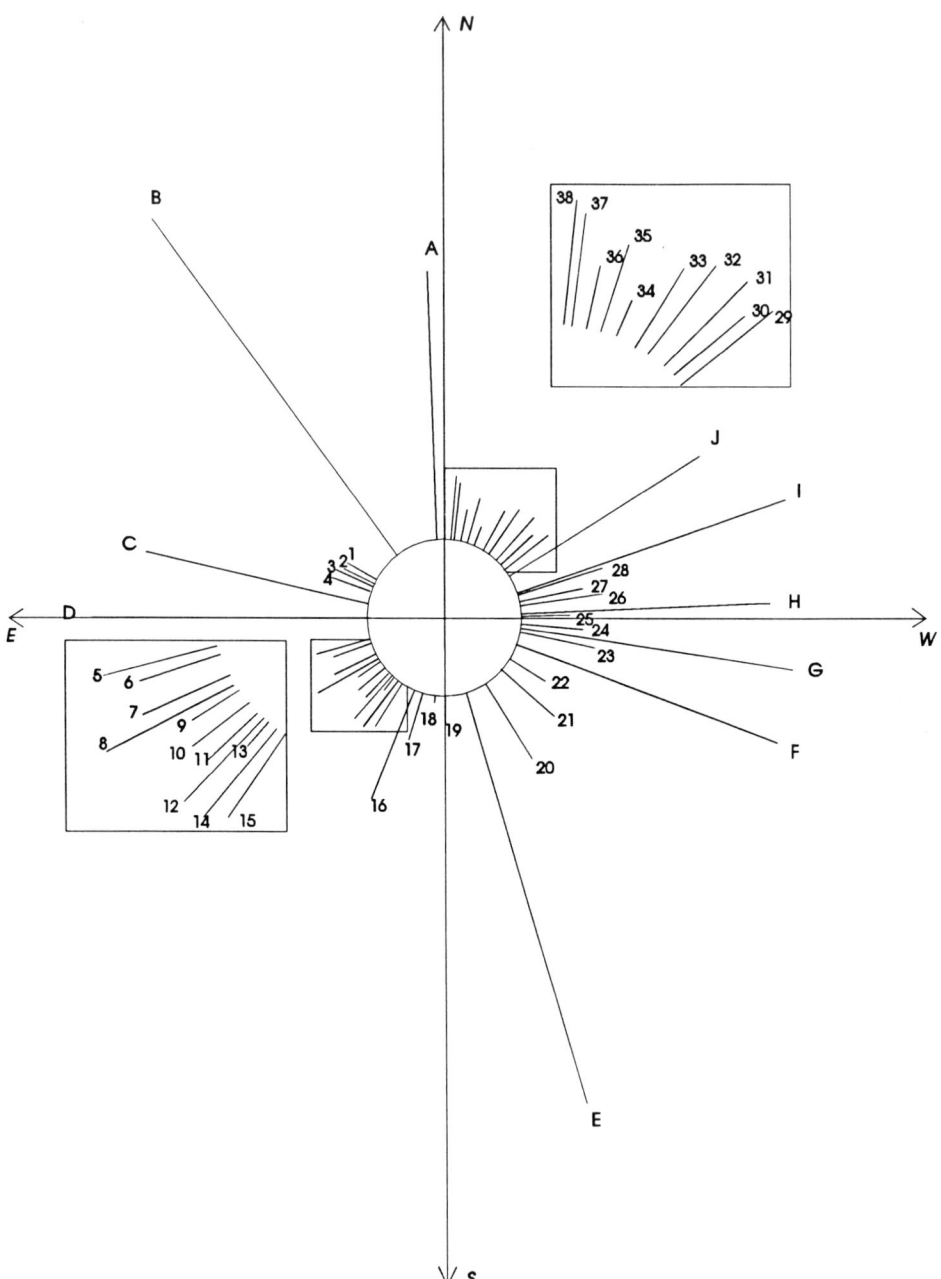

Fig. 1. Streamers identified at the total solar eclipse of July 11 1991, with maximum extension.

the north solar pole in the eastward direction. The angle was measured at the midpoint of the streamer base on the lunar limb. The inclination angle was measured from the normal to the position angle, and the extension is measured in $R_\odot$.

For comparison with the coronal streamers observed during the eclipse, the solar activity was reviewed on visible and Hα observations made by ourselves both one Carrington Rotation before and after the eclipse. In addition, the Solar Geophysical Data tabulations were used to check the large-scale magnetic fields, and He I 10830 Å data from the National Solar Observatory-Kitt Peak for coronal-hole activity.

The main problem was to infer the activity present on the solar hemisphere which is not visible from the earth. The persistence of sunspot groups and Hα features was noted from daily observations spanning the three Carrington rotations. However some uncertainty exists in the possible disappearance of active regions or the emergence of ephemeral regions (ER). Based on this persistence, and the activity present in the visible hemisphere, we selected 16 solar features as possible sources of the coronal streamers.

To study the relationships, a radial vector was extended from the center of the solar disk to the limb, passing through the source selected in accordance with its position at the time of the eclipse. The position angle of the vector was also measured from 0 to 360 degrees in an eastward direction from the north solar pole.

Comparing the position angles projected from the sources with the position angles of the streamers, it was possible to relate fourteen of the 48 streamers with fifteen of the 16 sources selected. These are listed in Table I with their identification, position angles, and maximum extension in $R_\odot$. The sources were identified with a roman number.

3. Discussion and Conclusion

Streamer B, one of the most prominent during the eclipse, was not related with any special solar feature. We consider as a possible explanation the drift of sectors of different magnetic polarity near to the north solar pole – the first north polar coronal hole that appeared after the reversal of polarity of cycle 22 (which occurred during 1990) was in development there (Sanchez-Ibarra and Barraza-Paredes, 1992). Streamer B in fact consisted of two streamers superposed, and streamer A was at the limit of the north polar coronal hole, perhaps having the same origin as streamer B.

The other streamers could be related with ephemeral regions and the large-scale magnetic field of the photosphere.

Acknowledgements

We are very grateful for the help kindly give us by Don Ignacio Castro, Comisario of La Matanza, and the people of this ejido. Also, the support given by the CIAS-La Paz was fundamental, as also the help of many other friends. This project was supported by CONACYT under contract D111-904459.

TABLE I
Identified Streamers and their Relation to the Sources

Streamer	PA Degrees	Max. Extension (L)	Source
A	1.90	3.40	
B	36.73	5.36	
1	61.13	0.46	I
3	67.30	0.56	II
4	71.17	0.53	II
C	77.57	3.00	
D	91.03	3.60	
6	109.30	0.50	III
7	116.00	0.67	IV
18	173.50	0.06	V
E	196.42	5.36	VI
22	242.10	1.16	VII
F	249.88	3.64	VII
G	261.61	3.60	
H	272.34	3.24	
26	278.80	1.08	VIII
27	281.90	0.85	IX
28	287.50	1.14	X
I	288.60	3.76	XI
J	301.78	2.88	XII
29	308.00	0.73	XIII

Listed below is the identification of each source with the NOAA/USAF number if it was a sunspot group; with a position angle if a prominence; and with its Carrington heliographic coordinates at the meridian passage if a filament. The sources were:

Source I.   6685; region with moderate activity that was in the invisible hemisphere.
Source II. λ=N15°, L=95°; filament present in a quiet region.
Source III. 6686,6729; two regions in a sector with moderate activity.
Source IV. 6728; another sunspot group in the sector of 6729.
Source V.   PA=171°; small quiet prominence visible the day of the eclipse.
Source VI. PA=196/199 ; a prominence with two peaks. Not defined as one or two prominences.
Source VII. PA=240°; small quiet prominence.
Source VIII. PA=279°; the west big prominence called "seahorse".
Source IX. 6714; small sunspot group in a region with high activity one rotation before the eclipse date.
Source X.   6709; sunspot group in the same region of 6714.
Source XI. 6711; sunspot group in the same region of 6714.
Source XII. 6701; sunspot group disappeared but active region persisted.
Source XIII. λ=N35° L=308°; filament associated with region 6701.

References

Sanchez-Ibarra, A., and Barraza-Paredes, M.: 1992, *Catalogue of Coronal Holes*, submitted for publication.
Solar-Geophysical Data: 1991, NOAA World Data Center A for Solar-Terrestrial Physics.

PART 3

INFRARED PERSPECTIVES ON ATMOSPHERIC DYNAMICS

SUBPHOTOSPHERIC CONVECTION

ROBERT F. STEIN
Physics and Astronomy Department, Michigan State University,
East Lansing, MI 48824, U.S.A.

and

ÅKE NORDLUND
Copenhagen University Obs., Øster Voldgade 3, Dk-1350 Copenhagen K, Denmark

Abstract. Three-dimensional simulations of solar convection are described. The simulations show that viewing convection as a hierarchy of eddies does not properly represent the large scale topology. A better picture is to view convection as a broad warm upflow with embedded cool, narrow, downdrafts. These downdrafts penetrate many scale heights through the convection zone and carry most of the net convective flux. Near the solar surface there are extremely large fluctuations in the temperature (5000–11000 K), entropy and pressure (factor of four). Radiation temperature does not provide an accurate measure of the gas temperature at a given geometric depth, because the opacity is very temperature sensitive. The emergent intensity in the infrared is smaller and has a smaller contrast than in the visible. However, in terms of radiation temperature the infrared is hotter and has a higher contrast than the visible.

Key words: convection – infrared: stars – Sun: granulation – Sun: interior

1. The Simulation

We simulate the upper portion of the solar convection zone by solving the equations of mass, momentum and energy conservation. The equation of state includes the effects of ionization and excitation of hydrogen, helium and other abundant elements. Three-dimensional radiative transfer at the surface is included assuming LTE and using a 4-bin opacity distribution function. The boundary conditions at the top and bottom are transmitting and in the horizontal directions are taken to be periodic (Nordlund and Stein 1990, Nordlund 1982). We have made two simulations: They both have the upper boundary at the temperature minimum, about 4 pressure scale heights above the top of the convection zone. One extends down about 2.5 Mm and covers a range of about 6 pressure scale heights in the convection zone. The other extends down 9 Mm and covers a range of 10 pressure scale heights through the convection zone.

The mean (horizontally and temporally averaged) atmosphere in these simulations is shown in Figure 1. We first present the topology of the convective flow, then the relation between the flow and the thermodynamic variables. Next we discuss the energetics and finally describe what can be observed.

2. Topology

On a large scale looking at convection as a hierarchy of eddies is not accurate. A better picture is warm, diverging, slow upflows with embedded cool, filamentary, fast downdrafts. Kinetic energy flux isosurfaces (Figure 2) reveal the regions of high velocity. They stop at the visible surface because the velocity decreases rapidly in the photosphere. Long thin downdrafts which penetrate through the entire simulation domain are clearly seen. Another way to visualize the fluid motion is to

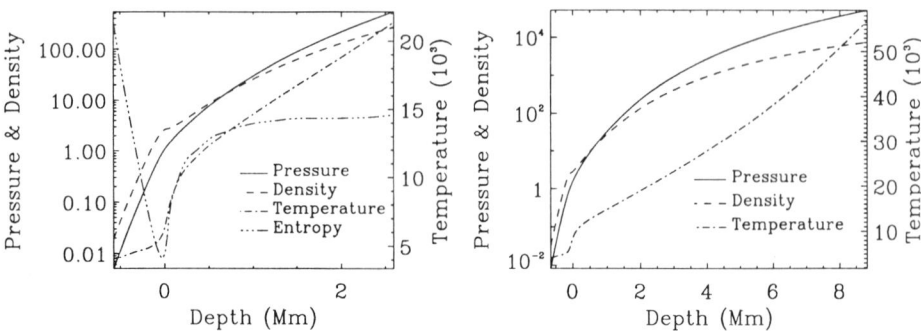

Fig. 1. Mean structure of atmosphere in the numerical simulations. The horizontal and temporal averages are plotted. (a) 2.5 Mm deep atmosphere. (b) 9 Mm deep atmosphere.

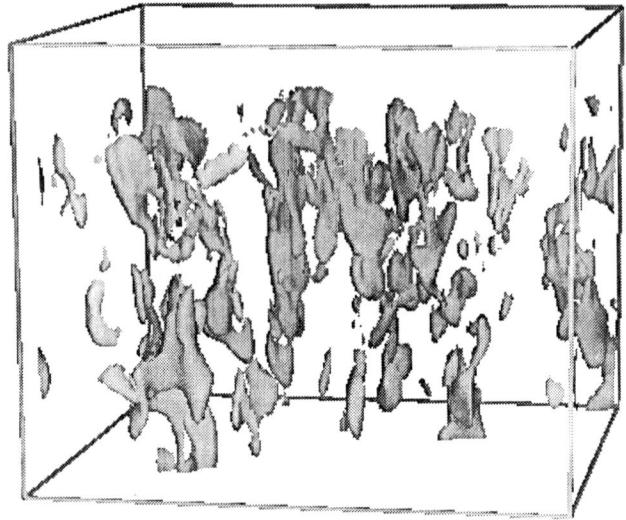

Fig. 2. Kinetic energy flux isosurfaces. The flux level shown is 10^{10} ergs cm^{-2} s^{-1}

follow fluid parcels in time. Figure 3 shows fluid parcels which are moving upward through the visible surface at a given time, where they come from and where they move to. These are the fluid parcels that form the centers of the granules. Nine minutes earlier, most of these parcels have come from about the same depth below the surface, but from a much smaller region of the horizontal plane, because the upflow has to diverge in order to conserve mass. Nine minutes later, these fluid parcels are heading downward. They have again concentrated into very small horizontal regions. Downward moving filaments are clearly revealed. They have much larger velocity than the upflow, since in 9 minutes they reach the bottom at 2.5 Mm, whereas in the same amount of time the upflows only moved from about 1 Mm

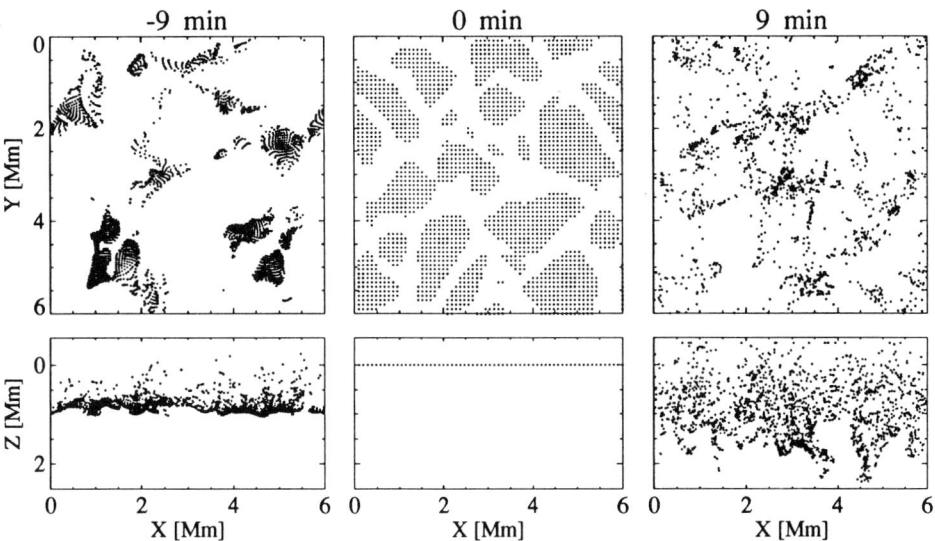

Fig. 3. The origin and destination of fluid parcels ascending through the visible surface at time $t = 0$. Nine minutes earlier, most of these parcels originated from a small source volume – vertically (because they have nearly the same vertical velocity) and horizontally (because the upflow diverges). Nine minutes later, most of the fluid has descended a substantial distance, concentrating into a few filamentary downdrafts.

depth to the surface.

Broad upflows and filamentary downdrafts can also be clearly seen in vertical slices through the simulation domain. Figure 4 shows the velocity in the xz plane and the temperature fluctuation in two such slices from different locations. In Figure 4a, a downdraft extends through the entire depth (10 pressure scale heights). Figure 4b shows two upflows, which diverge and turn over below the surface. Most of the fluid has to turn over before reaching the surface because of mass conservation. Note that the upflow region is very broad compared to the downdraft.

3. Thermodynamics and Flow

Below about 1 Mm the flow is nearly adiabatic. The entropy is nearly constant and its fluctuations are small (Figure 5). At the surface the entropy gradient is superadiabatic and very large. At different locations on the surface this steep jump occurs at slightly different depths ranging from about -50 km, up in the photosphere, to 250 km below the surface. Note that the zero of the height scale lies near the surface, but is arbitrary. $\langle T \rangle = 5800$ K at $z = -30$ km and $\langle \tau_{0.63\mu m} \rangle = 1$ at $z = -72$ km. There is a very tight correlation between the entropy and the temperature (Figure 6a). Most of the upward moving plasma is hot (11000 K) and most of the cool plasma (6000–7000 K) is moving down (Figure 6b). The hot upflowing plasma has

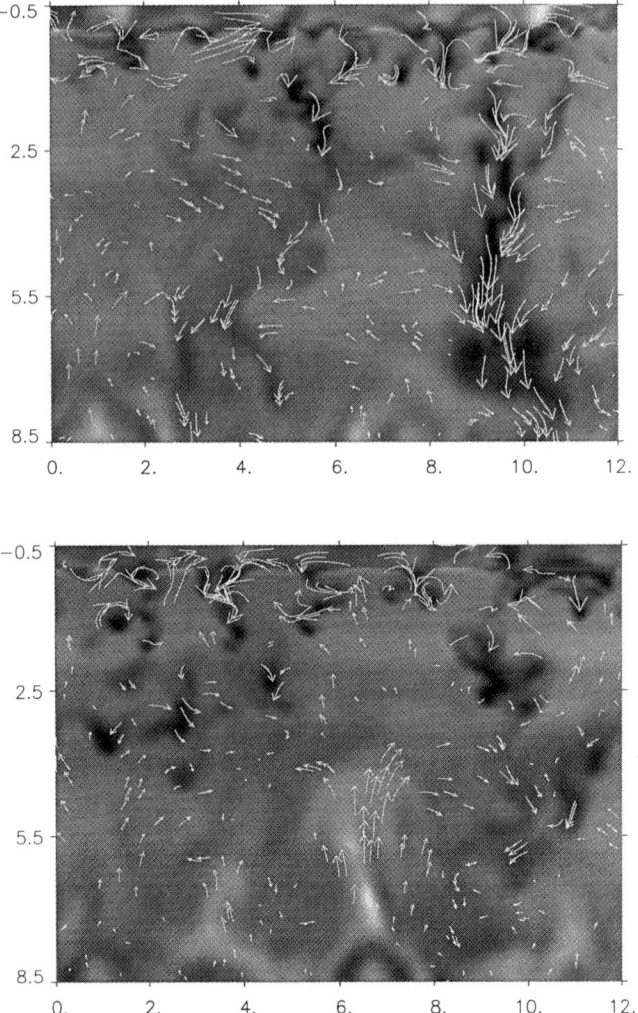

Fig. 4. Velocity and temperature fluctuation $(T-\overline{T})$, scaled by $T_{\max}-T_{\min}$ at each depth, in two vertical slices through the computational domain. Dark is cooler and light is hotter than average. Tick marks indicate horizontal and vertical scale in Mm. The surface is at $z=0$. (a) Note the filamentary downdraft near the right edge that penetrates through the entire 10 scale heights of the convective region. (b) Note the broad diverging upflows near the center and the left side, with significant overturning several Mm below the surface.

high entropy which gets radiated away when it reaches the surface. Figure 6c shows that this high-entropy, hot, upflowing plasma has low density, while the low entropy, cool, downflowing plasma has higher density. The resulting pressure is fairly constant from point to point at a given depth (Figure 6d).

Near the surface, at a given geometric depth, there is a huge range in tem-

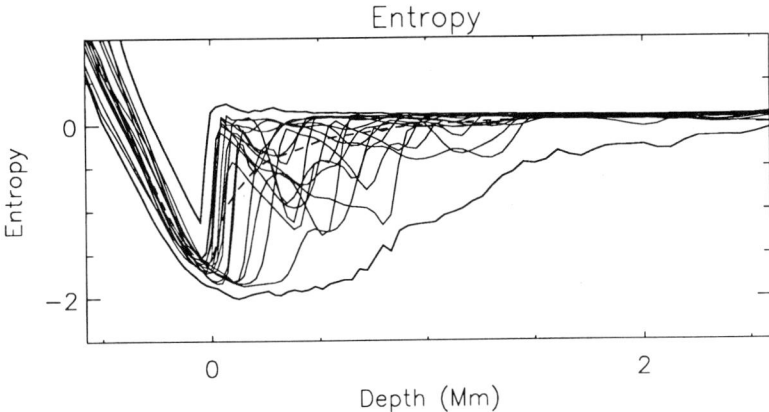

Fig. 5. Entropy as a function of depth is shown for several horizontal locations. Bounding curves are the minimum and maximum entropy. Dashed curve is the mean entropy. Near the surface the entropy gradient is huge and superadiabatic. At depth the entropy becomes nearly uniform.

perature (5000-11000 K) from point to point (Figure 6b). The rms temperature fluctuation is almost 2000 K. In this superadiabatic region the mean temperature gradient peaks at 30 K/km. With increasing depth, the magnitude of the temperature fluctuations decreases rapidly, becoming less than 100 K below 1 Mm.

The rms up, down and horizontal velocities all peak at slightly less than 2 km/sec near the visible surface. The upward rms velocity peaks about 50 km below, the downward rms velocity peaks about 300 km below and the horizontal rms velocity peaks about 100 km above the surface. The downward moving fluid reaches a maximum of about 13 km/s, at 600 km depth where the minimum temperature is 10000 K. The horizontal velocity has maxima of about 10 km/s near the visible surface. Where the horizontal velocity is high in cool intergranular lanes this flow is supersonic and shocks develop. These are nearly vertical standing shocks, like walls around some of the granules.

There is a horizontal cellular structure, of course. Figure 7b shows the temperature distribution at the surface with superimposed horizontal velocities. Note that the regions of high temperature, which are the centers (sometimes the edges) of the granules, are fairly small and the low temperature lanes are quite broad. They are much broader than the intergranular lanes seen in the intensity or the areas of downflow, which correspond closely with the intergranular lanes. Temperature images 50 km further down have larger granules but are still quite different from intensity images. In particular, the hot areas have nearly uniform temperatures and hence appear "burnt out".

Above the surface (Figure 7a), the temperature distribution reverses. Granules become cool due to adiabatic expansion of the diverging upflowing gas in a subadiabatic region, and the intergranular lanes become warm due to adiabatic compression heating. Below the surface (Figure 7c), the hot regions broaden and

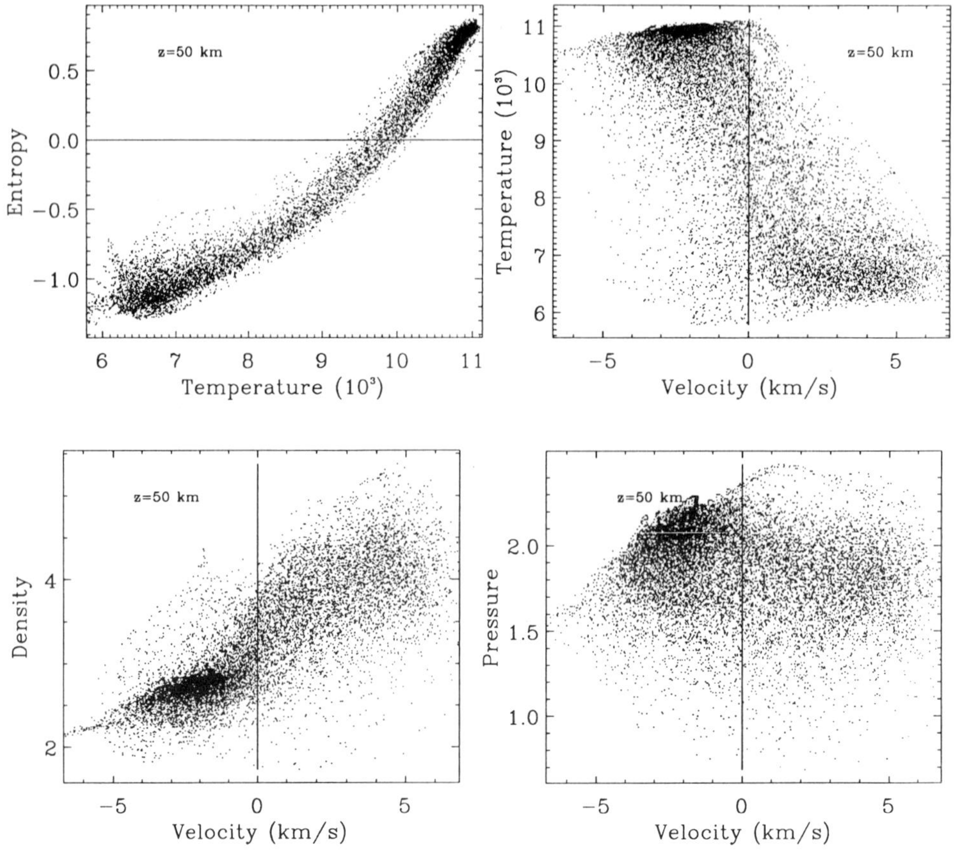

Fig. 6. Atmospheric structure at 50 km depth. Shown are the correlation of entropy with temperature and the correlation of temperature, density and pressure with velocity.

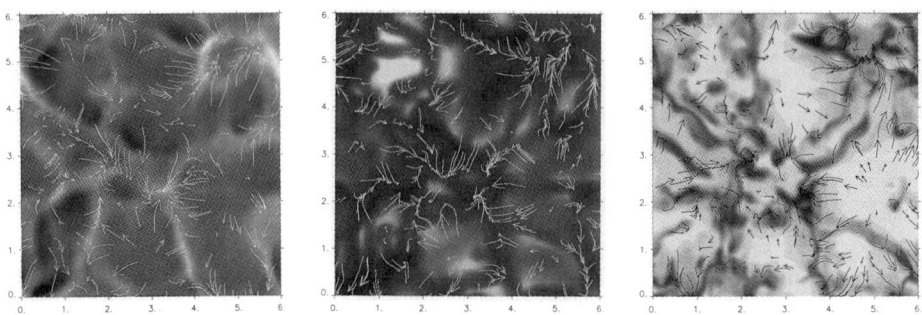

Fig. 7. Temperature and horizontal velocity at (a) -250 km, (b) 0 km (visible surface) and (c) 260 km.

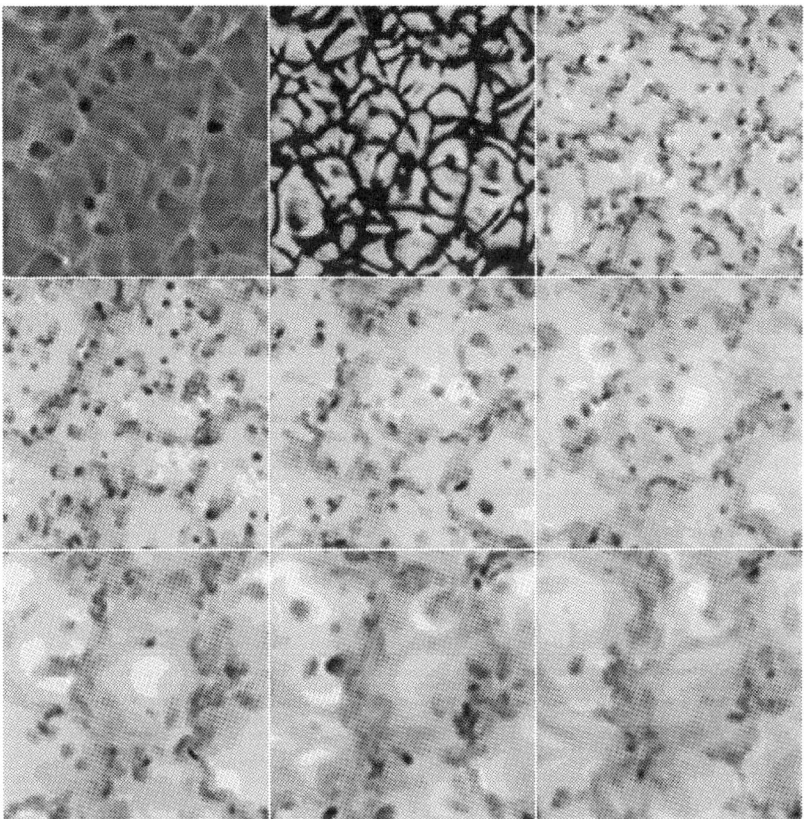

Fig. 8. Temperature on horizontal planes at intervals of 0.5 Mm from -0.5 Mm (temperature minimum) to 3.5 Mm. Light regions are hot and dark areas are cool. Note the increasing size of the hot regions with increasing depth.

the cool intergranular lanes become quite narrow. The flow converges into these cool regions which become the isolated downdrafts. The beginning of this can already be seen at 260 km below the surface. A montage of horizontal slices showing the temperature in steps of 0.5 Mm from -0.5 Mm above the surface to 3.5 Mm below it (Figure 8) shows the cool downflows embedded in the hot upflows. Fine scale structure develops a little below the surface. Then the intergranular lanes break up into isolated cool downdrafts in the generally hot upflow. The size of these hot upflow cells increases with depth.

4. Energetics

Above the surface energy is transported outward by the radiative flux, below the surface by the enthalpy (convective) flux (Figure 9). In our model, we take the

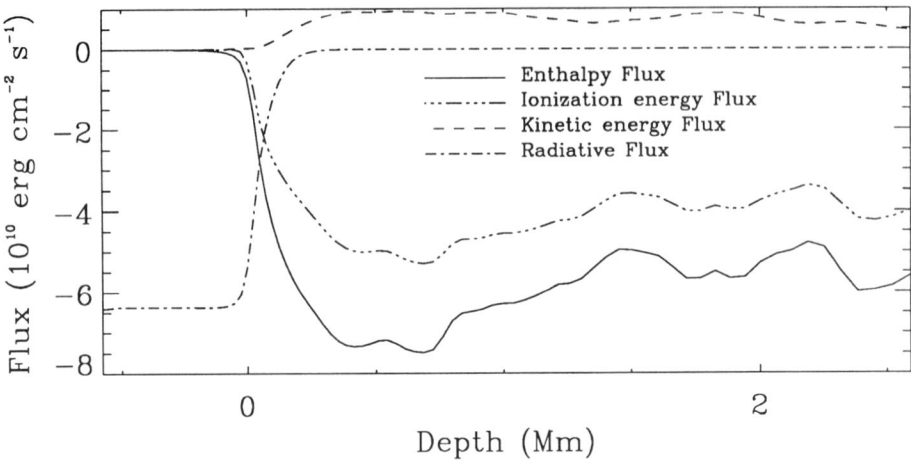

Fig. 9. Radiative, enthalpy (convective), ionization energy (latent heat) and kinetic energy fluxes as a function of depth.

z pointing downward, so upward fluxes are negative. The net upward flux is the sum of the enthalpy flux and the kinetic energy flux, which is always downward. However, the kinetic energy flux is always small compared to the enthalpy flux and so has only a minor impact. Note the large contribution to the enthalpy flux made by the latent heat of ionization. About 2/3 of the heat is carried to the surface as ionization energy.

The various contributions to the flux, summed over the surface area ordered by the vertical velocity, is shown in Figure 10, which reveals the contributions of the up and downflowing plasma. Two depth are shown: 50 km and 1 Mm below the surface. For instance, consider the net flux. Near the surface about half the net flux is carried by the upflows and half by the downflows. Down deeper, most of the net flux, about 70%, is carried by the downflows and only 30% is carried by the upflows. Similarly, near the surface about half the buoyancy work is done in the upflows and half in the downflows, while deeper down most (about 70%) of the work is done in the downflows. It is the cool, low entropy plasma which is doing most of the driving of the convective motions in the interior. However, the amount of driving decreases with depth because the entropy fluctuations decrease with depth (Figure 5).

Several other groups have also been making convective simulations (*e.g.*, Chan and Sofia 1986 and Cattaneo *et al.* 1991). They have simulated inefficient convection of an ideal gas with most of the flux carried by conduction and their results are somewhat different. Cattaneo *et al.* find that in their downflows the downward kinetic energy flux nearly cancels the upward enthalpy flux, so that the net flux is carried almost entirely in the upflows. We have run such a case starting from a snapshot provided by Cattaneo and Malagoli, and find that indeed for inefficient convection of an ideal fluid, most of the flux is carried by the upflows. Hence, there is something different in the physics when one uses an equation of state including

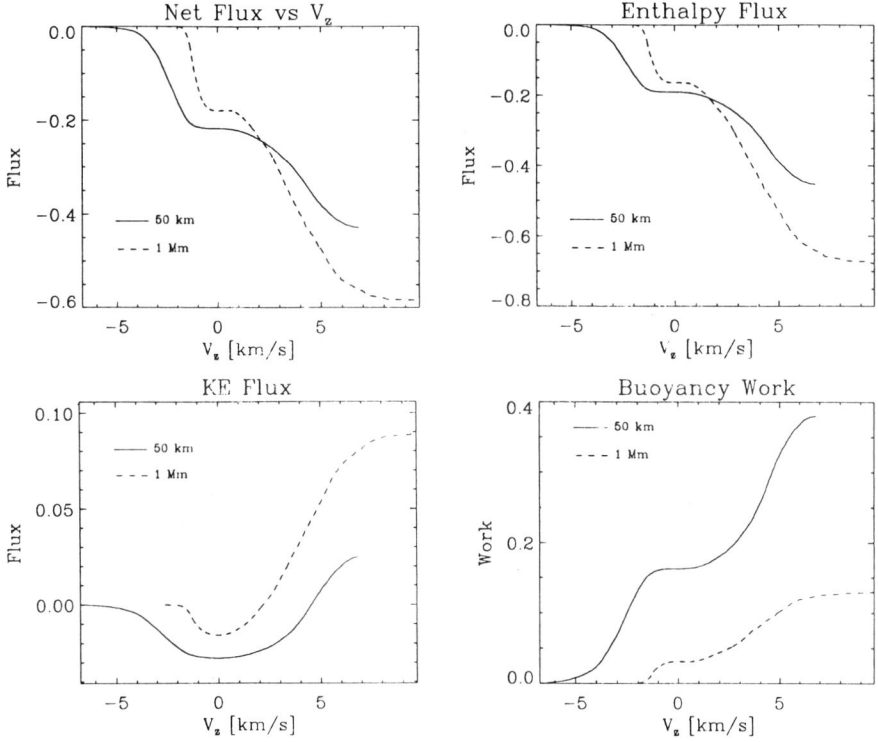

Fig. 10. Net flux, enthalpy flux, kinetic energy flux and buoyancy work at 50 km and 1 Mm depth as a function of fluid velocity. The fluxes and work are summed over area for fluid velocities ranging from maximum upflow to maximum downflow.

ionization and considers a situation of efficient convection.

5. Observables

The emergent intensity generated in the simulation, when smoothed with a point spread function to account for the effects of a telescope and seeing, is very similar to the observed granulation intensity pattern (Figure 11) (see also Lites et al. 1989). There are some interesting similarities and differences. In the simulation, many granules are brightest along the edges. Smoothing removes most of these bright edges. The observed image also has granules with bright edges. More recent observations, with exceptional seeing, confirm that granules are often brightest along their edges (Keller and von der Lühe 1992). A comparison of the size spectrum of the simulated and observed granulation is shown in Figure 12. The simulation has less small scale structure, but the large scale structure is reproduced quite well.

We now compare the intensity in the infrared and the visible. To calculate this we used monochromatic Planck functions as the source functions, one at 0.63 μm and the other at 1.6 μm. We used our normal opacity for the 0.63 μm calculation

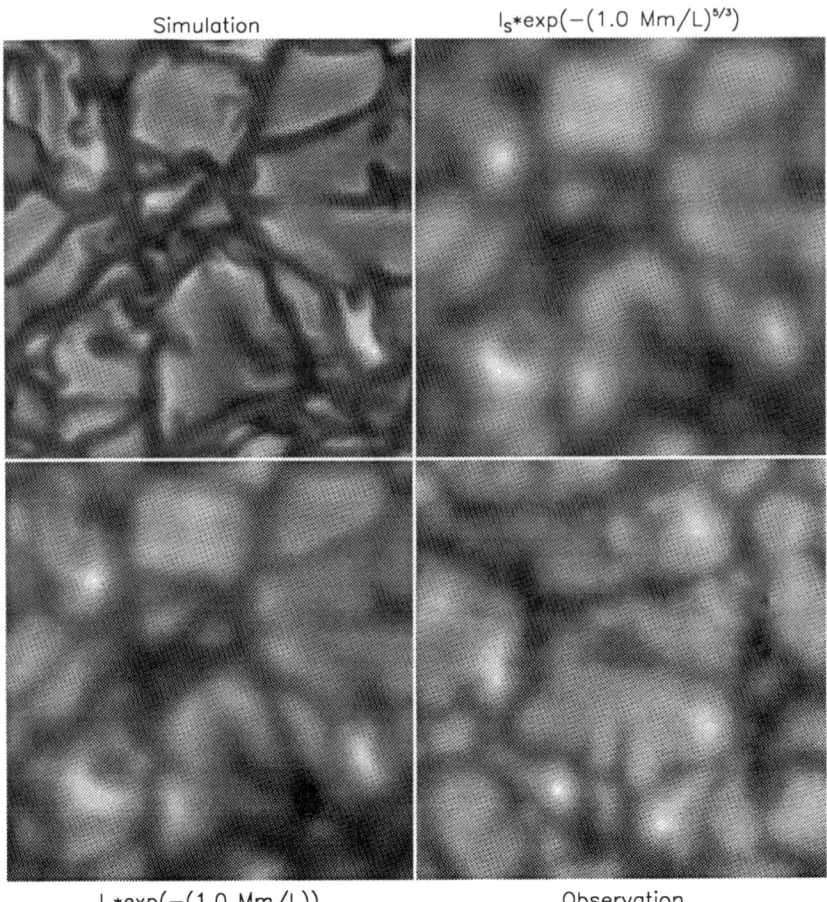

Fig. 11. Comparison of the intensity pattern from the numerical simulation with observations from the Swedish Solar Observatory on La Palma. From left to right and top to bottom the images are (upper left) the emergent intensity from the simulation; (upper right) the same intensity, smoothed with a $\exp(-(k/k_0)^{5/3})$ point spread function representative of a finite instrumental resolution and atmospheric seeing; and (lower left) the same intensity smoothed with an $\exp(-k/k_0)$ point spread function. For both point spread functions $k_0 = 2\pi/1$ Mm. An area of the same size (6×6 Mm) from a slit jaw image obtained at the Swedish Solar Observatory by Bruce Lites is shown at lower right (cf. Lites et al. 1989).

and reduced the opacity by a factor of 1.6 for the 1.6 μm calculation. Data from Bob Kurucz (private communication) shows that is about the ratio near optical depth one. The emergent intensity at the the two different wavelengths is shown in Figure 13. The two images differ only slightly. Observations at the two wavelengths are likely to differ more for reasons of different seeing and telescope resolution, and

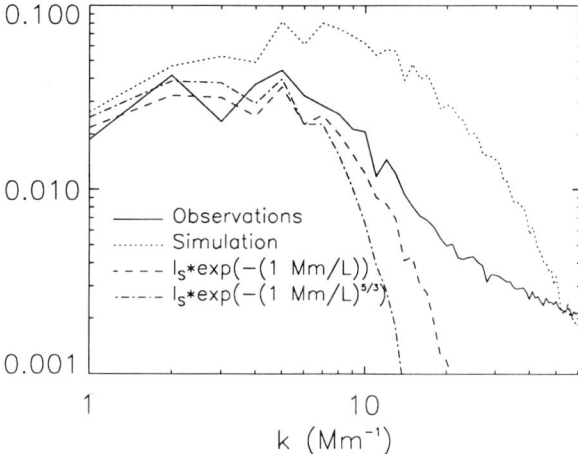

Fig. 12. Horizontal size spectrum of observed and simulated granules including effects of smoothing by two different point spread functions.

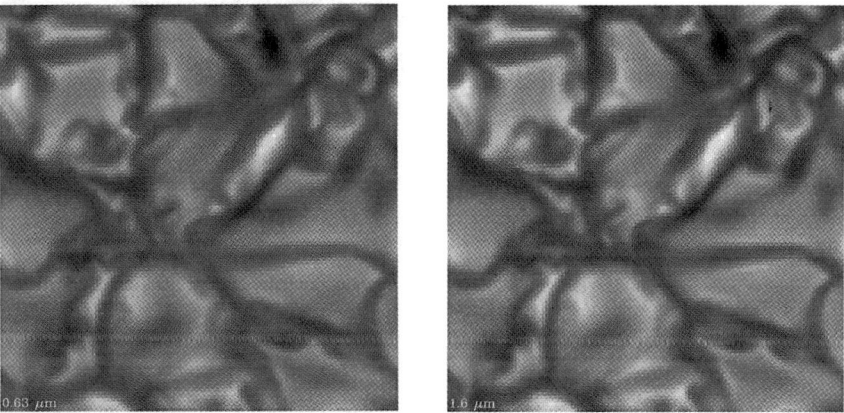

Fig. 13. Emergent intensity at 0.63 μm (visible) and 1.6 μm (IR) from the same snapshot. Each image is scaled so that its intensity range covers the full grey scale.

have to be of very high quality to reveal true differences. Because of the wavelength dependence of the Planck function, the intensity is smaller and its contrast is less in the infrared compared to the visible. The rms relative intensity fluctuation in the visible is 0.14 and in the infrared is 0.08, so their ratio is 1.7. In terms of the radiation temperature, the infrared is about 300 K hotter and has a 20% greater contrast than the visible (Figure 14).

The radiation temperature (even in LTE) does not correlate well with the gas temperature at a given geometric depth. At $\langle \tau_{1.6\mu m} \rangle = 1$ the gas temperature varies between 5000 K and 10300 K. The radiation and gas temperatures are well correlated in the cool regions, but the radiation temperature exhibits a much smaller

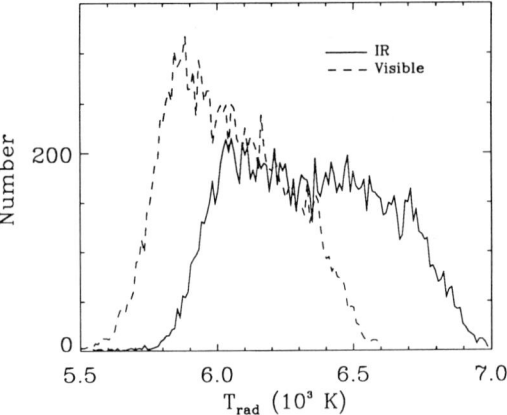

Fig. 14. Histogram of the radiation temperature at 0.63 and 1.6 μm. The average radiation temperature is 6047 K at 0.63 μm and 6352 K at 1.6 μm. The rms relative radiation temperature fluctuation is 0.036 at 0.63 μm and 0.043 at 1.6 μm.

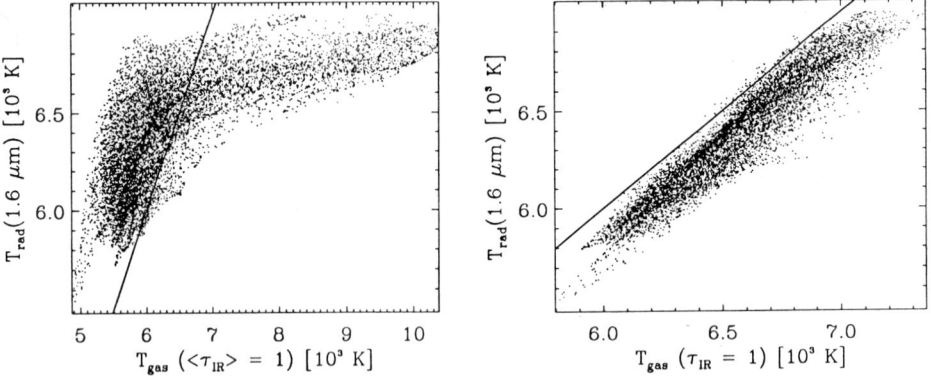

Fig. 15. a) Correlation of radiation temperature at 1.6 μm with gas temperature at depth where $\langle \tau_{1.6\mu m} \rangle = 1$. b) Correlation of radiation temperature with temperature at the depth where $\tau_{1.6\mu m} = 1$ locally.

range than the gas temperature (Figure 15a). This is a result of the very rapid increase in opacity with temperature, so one looks into shallower depths in hotter regions and doesn't see the very high temperature gas. In accordance with the Eddington-Barbier relations, the radiation temperature is approximately equal to the gas temperature at optical depth one (Figure 15b). The geometrical height where this occurs varies from place to place and has an rms excursion of 34 km.

Acknowledgements

This work was supported in part by NASA grant NAGW-1695 (RFS) and the Danish Natural Science Research Council and the Danish Space Board (ÅN). The calculations were performed on the Michigan State University Convex 240 and the National Center for Supercomputing Applications Cray 2. The authors appreciate the support of these organizations.

References

Cattaneo, F., Brummell, N. H., Toomre, J., Malagoli, A. and Hurlburt, N. E.: 1991 *Astrophys. J.* **370**, 282.
Chan, K. L. and Sofia, S.: 1986 *Astrophys. J.* **307**, 222.
Keller, C. U. and von der Lühe, O.: 1992, in J. M. Beckers and F. Merkle (eds.), *High Resolution Imaging by Interferometry II*, ESO Conference, in press.
Lites, B. W., Nordlund, Å. and Scharmer, G. B.: 1989 in R. J. Rutten and G. Severino (eds.), *Solar and Stellar Granulation*, NATO ASI Series **263**, Kluwer Academic Publishers, Dordrecht, pp. 381-399.
Nordlund, Å.: 1982, *Astron. Astrophys.* **107**, 1.
Nordlund, Å. and Stein, R. F.: 1990, *Comp. Phys. Comm.* **59**, 119.

THE INFRARED GRANULATION: OBSERVATIONS

SERGE KOUTCHMY

Institut d'Astrophysique, CNRS, 98 bis, Boulevard Arago, F-75014 Paris, France
and
National Solar Observatory, Sunspot, NM 88349, U.S.A.

Abstract. High spatial resolution observations of the solar granulation up to 3 μm are possible on existing vacuum solar telescopes. They permit the analysis of the deepest photospheric layers near 1.7 μm. A high photometric accuracy is achieved using scanning techniques with a pinhole photometer; imaging methods are used for 2D-analysis. Statistically significant results on granulation, including power spectrum and histogram analysis, center-limb variations and lifetime analysis, are presented. Temperature fluctuations of periods near 5 minute are considered at 1.7 μm, as well as large-scale variations at the scale of the meso- and, especially, the super-granulation. We also discuss the "abnormal" granulation in magnetic regions and the umbral granulation in the cores of sunspots.

Key words: infrared: stars – Sun: granulation – Sun: photosphere – sunspots

1. Introductory Remarks

1.1. Present View of Solar Granulation

Several reviews recently appeared on the subject, coming from observers, see Muller (1989), Karpinsky (1990), Title *et al.* (1990), Topka and Title (1991) and from theoreticians, see Chan *et al.* (1991). The book of Bray *et al.* (1984) is still a good introduction to the subject. Magneto-convection is considered by Weiss (1990). None of these works specifically considered the IR granulation, although a series of results were already presented, see Turon and Léna (1973), Albregtsen and Lynne Hansen, (1977), Worden (1975) and more recently Koutchmy (1990), hereafter referenced K90. The mentioned reviews by observers concentrate on the excellent optical observations and imaging of Pic du Midi and La Palma Observatories, results from the analysis of the Soviet stratoscope experiment of Pulkovo Observatory or of the SOUP experiment (seeing-free) on SpaceLab 2. They raise problems such as the following: Is granulation "turbulent" on small scales (R. Muller)? What is the true time variation of granules when the oscillatory component near 5 minutes is removed (Title *et al.*)? Is granulation well described by the network of dark lanes (V. Karpinsky) and what is the role of the (weak?) magnetic field? Where are shocks predicted (Cattaneo *et al.*, 1989) in numerical simulations? What are the temperature fluctuations corresponding to the observed meso- and super-granulation? Optical granulation was also extensively observed at the VTT at Sacramento Peak Observatory by J. Evans, J. Beckers, R. Dunn, S. Keil and many others; here we shall concentrate on quantitative results obtained there in the IR, over the last 17 years. Many IR images, and even processed movies, have been obtained and are reported elsewhere in the present proceedings.

1.2. Why Observe the IR Granulation? – A Theoretical Point of View

The radiative opacity of the solar atmosphere reaches its absolute minimum in the 1.65 μm spectral region of the continuum, see Figure 1. Obviously the 1.7 μm

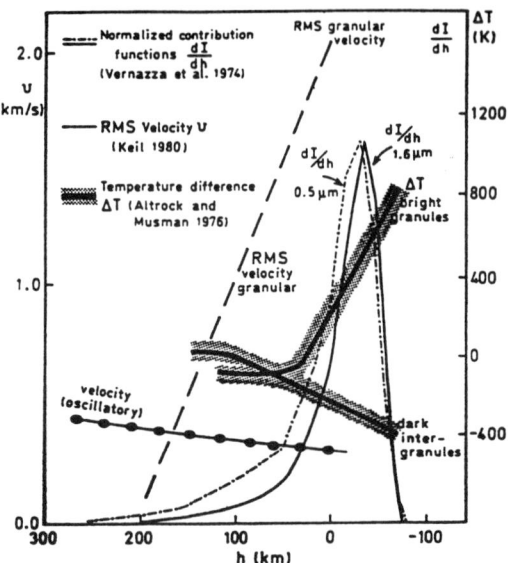

Fig. 1. Contribution functions for the wavelength regions 0.5 μm and 1.6 μm, after Vernazza et al. (1974). On the same height scale are shown the behaviour of the temperature fluctuations from the granulation model of Altrock and Musman (1976) model and the RMS granular velocity and oscillations from Keil's (1980) analysis.

region of the opacity minimum region (OMR) is the most interesting for studying the layers which are most affected by the convection: Not only do we see deeper, we also partially avoid layers affected by 5-minute oscillations, and layers where the temperature contrast shows an inversion, see Figure 1. Granular convective velocities seem to increase drastically when interpolated toward deeper regions – Keil (1980) – so granules are expected to be seen in a more dynamical regime: shorter lifetimes and/or smaller size. Note also that the amplitude of 5-minute oscillations decreases with depth and, accordingly, convective motions are better measured in the OMR. Finally, because of the spectral behaviour of the Planck function, the interpretation of measured intensity fluctuations is made more easily in IR than in optical regions, when layers of different temperatures are considered, see Figure 2. Not only does a mixing of hot and cool gas along the line of sight occur in the optical regions, but the phenomenon is non-linear in terms of temperature, the main thermodynamical parameter; this is especially misleading when images are considered. Direct images of granulation in the OMR are published in K90 ; another example is given on Figure 2.

1.3. Advantages and Disadvantages of Studying IR Granulation

From the beginning, IR measurements have been made photoelectrically (or thermoelectrically), with strictly linear detectors of large dynamic range, so mistakes made in the visible through the use of the photographic films, which are often poorly

Fig. 2. Images of the solar granulation simultaneously observed in IR at 1.6 μm in the OMR (top-left), and in the optical continuum at 525.93 nm (bottom-left) and the upper atmosphere at 518.41 nm, with the UBF of the VTT of NSO/SPO. Note the good correlation between the IR and the optical granulation (separated by 77 seconds of time) and the lack of obvious correlation with magnetic elements shown in Mg b1 + 0.4 (lower-right) evidenced by their bright emissions. The photographs were processed by F. Stauffer and T. Darvann at NSO/SPO.

calibrated, were avoided. Even the recently-performed observations in the visible with CCD cameras suffered from a lack of dynamic range (important for measuring the MTF of the telescope) and a limited field of view. We believe the combination of imaging techniques with scanning techniques yields a powerful and needed tool to extract good quantitative data for analyzing granulation.

Good photometric work requires the use of stable, well-controlled, observing conditions and the availability of good windows. In the IR, both the Earth's atmospheric transparency and the seeing are improved, sometimes dramatically, (at Sacramento Peak the transparency near 1.7 μm was measured to be as high as 98% !) Windows available for observation of the solar continuum are also better than in the visible; line-blocking effects, which are especially severe in the blue region of the spectrum, are considerably reduced in IR. Moreover, the effect of the Earth's atmospheric differential refraction is also drastically decreased so that large spectral passbands, giving high signal-to-noise ratios, can be used for the fast imaging or scanning without losing spatial resolution (see Fig. 3). This advantage can be used to compensate for the disadvantage of losing the multiplex gain when scanning techniques are used with a single detector of large dynamical range – Koutchmy et al. (1977).

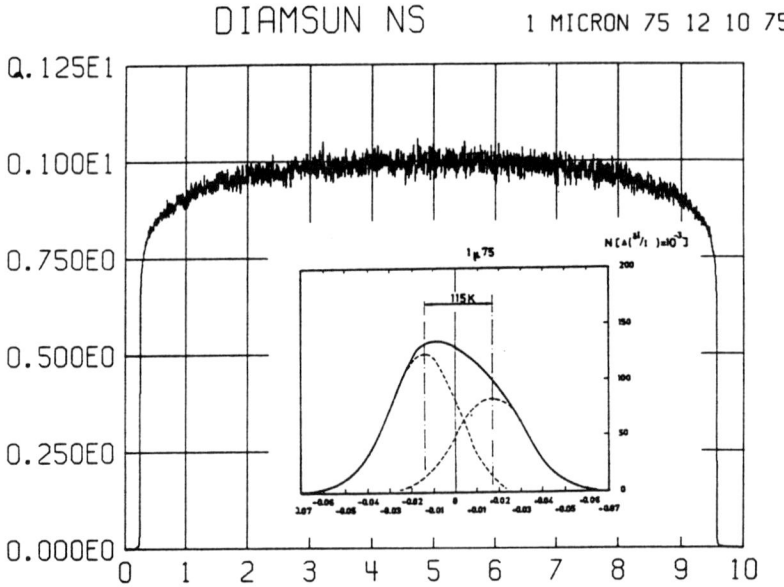

Fig. 3. A typical solar scan used to measure the RMS intensity variations of the IR granulation. The whole scan contains 11000 points. Note the level of noise, barely seen outside the limb, and the center-limb variation of the amplitude of the RMS fluctuations. The measured spatial resolution closely corresponds to the theoretical, when the combined effects of instrumental diffraction and the digitization step, with the 0.75 arcsec-diameter pin-hole, were taken into account. At the bottom, we have inserted the result of an histogram analysis of the central part of selected scans over an observing period of more than 2 hours. The histogram shows the largely skewed distribution in the intensities of IR granulation. Dark areas surpass bright areas by a ratio 1 : 0.69.

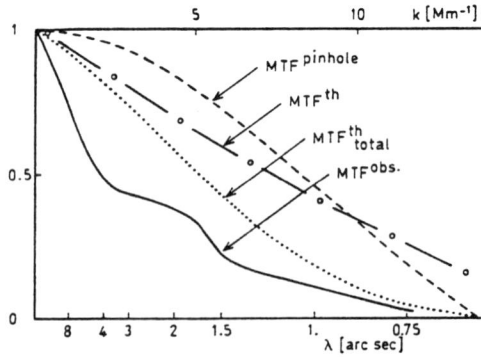

Fig. 4. Modulation Transfer Functions observed for different conditions, at the NSO/SP VTT near 1.7 μm. The total observed MTF was deduced from the analysis of many fast scans of the extreme-limb of the Sun, assuming a perfect step function for the external part of the solar limb. Theoretically computed MTF's are also shown.

Up to now only telescopes designed for the optical region have been used to study the IR granulation; at 1.7 μm, such telescopes become diffraction-limited, because all aberrations are reduced by a factor 3 (see below). With respect to temporal variations, we find once again that working in the IR is more interesting. At a good, high-altitude, dry site, Earth-atmospheric induced effects are reduced, and indeed can easily be identified, because they are related to variations in microbarographic and water-vapor-content values.

The main disadvantage of working in the IR has been the lack of good 2D-detectors; granulation images were made by scanning the Sun or by using very low quantum efficiency IR vidicons (see K90). In recent years the situation has completely changed, thanks to the appearance of new (not classified) highly-efficient 2D-detectors based on cooled Hg-Cd-Te chips (see these proceedings), so we anticipate that many new results on IR granulation will soon appear. Finally, we note that intensity modulations produced by temperature fluctuations are considerably smaller at 1.7 μm than in the visible. A 100 K temperature fluctuation corresponds to a 8.5% variation at 500 nm and only to 3.5% at 1.7 μm. This makes sunspot cores easier to study in the IR.

2. Statistical Analysis of the IR Granulation

2.1. AMPLITUDE OF SMALL-SCALE GRANULATION INTENSITY FLUCTUATIONS

In 1975-76, we used a specially designed PbS pin-hole photometer to accurately measure intensity fluctuations over the solar disc observed at the prime focus of the VTT of NSO-SPO. The photometer used a chopper at 1100 Hz frequency and phase lock-in amplification, with a digitization rate of 100 point sec^{-1}. A 14 bit/point precision A/D converter was used to record the data. Spectral selection was performed with a small IR f/60 monochromator with a 28 × 28 mm^2 grating blazed at 2 μm, and broad-band IR interference filters. For "fast" analysis we used only interference filters near 1.75 μm and 2.2 μm; then the typical S/N ratio over the granulation scans was 2000 when a circular 0.75 arcsec diameter pin-hole was used (see Fig. 3). Only the very quiet Sun was considered in this analysis.

In this way, precise measurements of center-limb variations measurements were collected (see Koutchmy *et al.*, 1977); the MTF of the telescope, including effects of scattered light, was accurately calibrated (see Fig. 4) by the use of fast scans of the extreme-limb. We computed the derivative of each scan to define the position of the inflection point and used only the external part of each scan to get the edge-like smearing function; taking the FFT of its derivative, we readily deduced the MTF, making the assumption that the true solar limb is far narrower than the FWHM of the line spread function. We found this method better than using the lunar limb at solar eclipses, because the lunar profile provides a rather poor knife edge.

An easy way to look at the observed FWHM of the smearing function is to analyze the auto-correlation function of the best granulation scans performed near the center of the Sun (see Fig. 5a); the half-width found from our best scans was of the order of 0.48 arcsec, which is the lowest value ever reported in the literature, including the visible region. This low value seems really to correspond to what

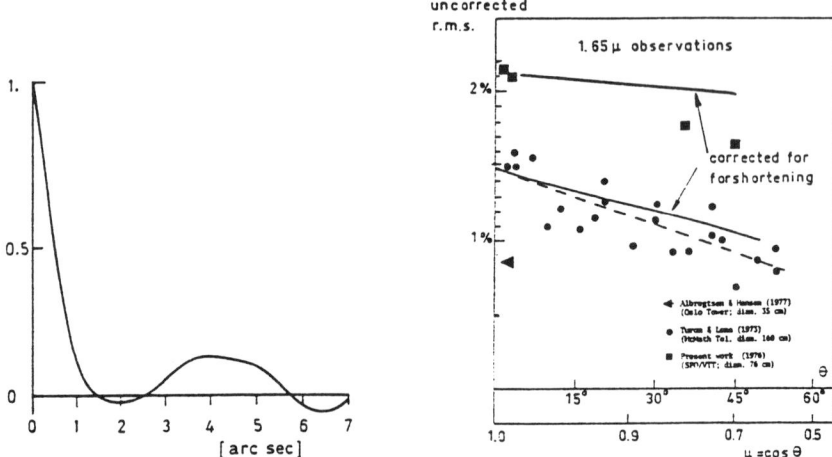

Fig. 5. a) Auto-correlation function computed for a selected 200 arcsec long scan taken near the center of the Sun, at 1.75 μm. The corresponding RMS was found equal to 2.17% and its width is 342 km or 0.476 arcsec. Note the minimum at 2 arcsec and the broad maximum around 4 arcsec. b) Center-limb variations of the RMS, calculated without corrections for the smearing. Data collected at NSO–SPO with high S/N ratio are the best.

we observe on the Sun in the OMR; it cannot be explained by the noise (S/N $\geq$ 2000) nor the value of the sampling resolution (0.19 arcsec). It is an indication of the high quality of the data and suggests that the granulation in the OMR is possibly finer. The first indication that something is definitely different was given by the histogram analysis – see the inset at the bottom of Figure 3. The skewness of the distribution is remarkable; the bi-distribution shows that areas occupied by darker than the average pixels dominate those of bright pixels in the ratio 1:0.69. In the visible, Keil (1977) obtained 0.88, and Karpinsky (1990) found 0.80 from high resolution stratoscope images. However, the bi-distribution shows an average difference of temperatures between dark and bright area of only 115 K. This value should not be confused with that which could be deduced from the RMS of intensity fluctuations; Figure 5b shows our results compared to other published results, for different positions over the disk ($\cos\theta$). The center-limb decrease is well apparent (after correction for foreshortening). These values are not corrected for the smearing shown by the measured MTF, Figure 4. After considering the spectral distribution of the IR granulation intensities (see further) we determined a correction factor of the order of 4.2, so the corrected RMS at the disc center reaches 8.6% (instead of 2.05%) at 1.75 μm; we should, however, confess that this value is affected by a large error bar due to the uncertainty of the correction at very small scales.

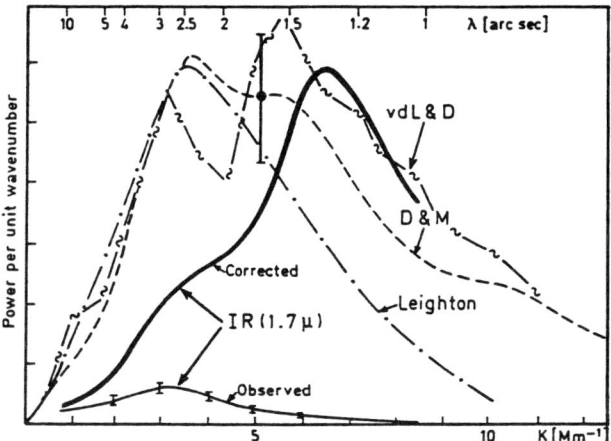

Fig. 6. Observed and corrected 2D spatial power spectra obtained in the IR (full lines). For comparison, we show results reported in visible light by different authors, over a 12-year interval, with improved resolution and correction methods: Leighton (1963), Deubner and Mattig (1975), and von der Lühe and Dunn (1987). Note that the power determinations in the visible extend to higher frequencies with improved methods; the relative lack of power at low frequencies in the OMR IR spectrum is striking.

2.2. POWER-SPECTRUM ANALYSIS

We selected the best scans performed at 1.75 μm during a 3-hour run, to perform a 1D power-spectrum analysis of very low noise, superposing several tens of spectra. The 1D average spectrum was readily transformed to a 2D power-spectrum by inverting the Abel transform (we also used the Hankel transform of auto-correlation functions like the one of Fig. 5a); spectra were smoothed to increase the level of confidence. A 2D power-spectrum per unit surface was published, without correction for the MTF, in K90, to show the difference in behaviour of raw data, between the visible light and the IR power spectra. In Figure 6 we show our best 2D spectrum before and after correction using the "observed" MTF of Figure 4. For comparison, Figure 6 also shows 2D power spectrum obtained by different authors at different epochs, in the visible. There is a clear tendency to report more and more power at higher frequencies; in the IR the dominance of power at high frequencies is more pronounced, bringing more arguments in favor of turbulent convection, see Muller (1989).

2.3. TEMPORAL ANALYSIS: LIFETIMES; TEMPORAL FLUCTUATIONS

Using the scanning method (see Fig. 3), we analyzed the time variations of intensity fluctuations at granular size scales near the center of the Sun. Power spectrum analysis of a temporal sequence 46 minutes in length showed relative power in the

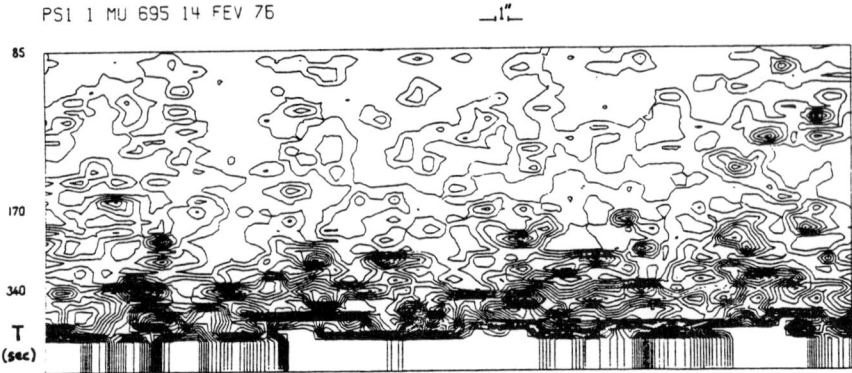

Fig. 7. Distribution of the power in temporal IR intensity variations spectra computed along a single direction near the center of the Sun. The cut-off frequency corresponds to 11.7 mHz (T = 85 second) and the time sequence is of 47 minute length.

5-minute range at high spatial resolution but considerably less power when a 200 arcsec average is used (see K90). In Figure 7, curves of iso-power obtained at high spatial resolution are shown along one direction near the center of the Sun. Peaks of power are identified at the scale of the granulation from 3 to 6 minutes period. It is, however, difficult to distinguish between true recurring convection phenomena and oscillatory cells, so we should keep this in mind when looking at the problem of the lifetime of granules.

We tried to compute the lifetime of IR granules using 1D scans and cross-correlations; looking at the decrease of the maximum amplitude of the cross-correlation function computed with different time lags we found a rather short lifetime, of the order of 3.3 minute for the decrement of the best fit of the data to an exponential function (lifetime should be considered equal to at least twice the value of the decrement). Figure 8a illustrates the typical cross-correlation function we obtained with a 160-second time lag. Note the shift of the bisector of the function, which corresponds to the shift produced by solar rotation, seen by the displacements of granules in the E-W direction. The rather short lifetime found in this analysis is attributed to the dominance of very small granules, partially to fragments of granules, which cause a loss of correlation using this method; oscillations could possibly also influence the results. In 1988-90 several movies of IR granulation were obtained (see K90). A movie of 45 minutes of continuous observations was processed by T. Darvann (these proceedings). Using the filtered and selected images of this movie, correlation coefficients between images were computed – see Figure 8b. Now the decrement of the average curve corresponds to 7.5 minute (lifetime 15 minute), which seems rather long. The processing which has been applied to permit a precise tracking, to compute the flow map, seems to favor longer-ifetime granules and possibly agglomerates of granules and cell-like structures with with a

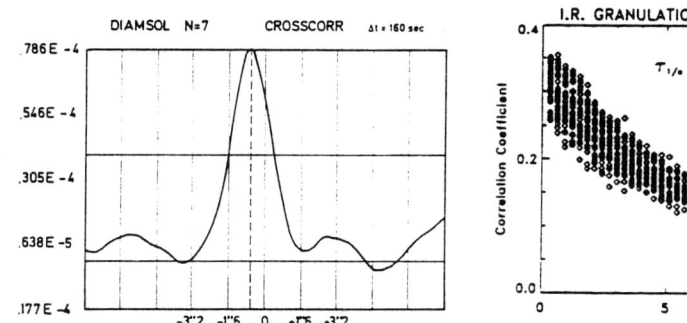
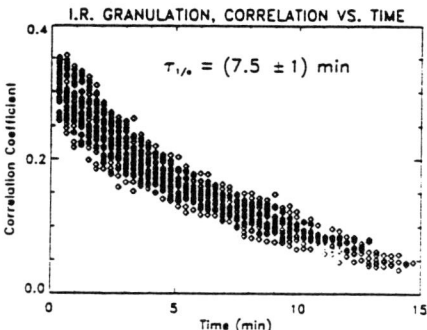

Fig. 8. a) Typical cross-correlation function obtained using two 1D scans made near the equator at 1.75 μm. The time lag is 160 second; note the shift due to solar rotation. b) Lifetime of granules computed by T. Darvann from a processed 45-minute movie at 1.6 μm.

longer lifetime. We conclude that both 6- and 15-minute lifetimes are statistically significant, depending of the size of granules.

3. Large-Scale Granulation

3.1. Statistical Analysis

Sizes of convective cells are not limited to the granulation size, from about 1 to 3 arcsec. In the visible, both Doppler shifts and intensity modulations reveal cells at a meso-granulation size, between 5 and 40 arcsec, and also at the supergranulation (SG) size near 42 arcsec. A temperature contrast of the order of 20 K was found by Koutchmy and Lebecq (1986) for meso-granules, after eliminating 5-minute oscillations. In the OMR, at 1.7 μm, Worden (1975) and Koutchmy (1978) gave the first results on the SG. We analysed a 940 × 940 arcsec^2 matrix of intensity fluctuations, see Figure 3, taken over the central part of a very quiet Sun (Oct.11, 1975) at 1.7 μm. After accurately removing the center-to-limb variations of normalized scans, we performed a 1D-autocorrelation analysis 10^3 times and looked at spacings near the size of SG – see the resulting curve in Figure 9a. For sizes near 27 Mm, the autocorrelation function is definitively positive; only at 36 Mm a slightly negative value is found; again near 55 Mm a large maximum is observed. From this analysis we conclude that the boundaries of SG are definitely bright, statistically speaking – see Figure 9a – contrary to the first results reported by Worden (1975), which were made with a rather low S/N ratio through the large dispersion spectrograph of the 80-cm Kitt Peak auxiliary heliostat. A positive contrast at SG size is also reported in Lin and Kuhn (1992); it is obviously related to the magnetic nature of the boundaries of the SG. To estimate the temperature modulation at the bound-

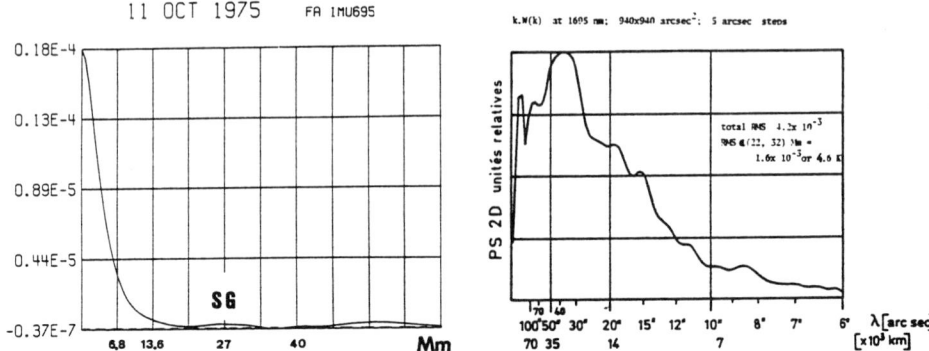

Fig. 9. a) Superposed autocorrelation function of the central part of the quiet Sun to show the positive temperature contrast observed at the super-granulation size. b) The corresponding 2D-Power spectra per unit of wavenumber. The main peak shows a SG size of around 42 arcsec.

ary of SG, we used the 2D power-spectrum of Figure 9b showing a total RMS of 4.2×10^{-3}. The power measured between 22 and 32 Mm, which includes the main peak corresponding to the SG, gives an RMS temperature contrast of 4.6 K; because 5-minute oscillations are not fully eliminated in this analysis, the true contrast is probably smaller but still positive.

3.2. Mesogranulation Flow

Using visible granules as tracers, L. November (1986) showed how it is possible to compute a flow map which reveals the mesogranules. The 45-minute movie obtained in 1988 at 1.6 μm, (see K90), has been processed with this method. T. Darvann, (1992), performed a full analysis of the movie with very interesting results. The mesogranulation flow, and probably the SG flow, are well evidenced, see Figure 10; boundaries are well represented by the positions of "corks" which follow the flow after more than 8 hours. A very interesting deep sink appears in the field of view and boundaries show a kind of sheared flow. Several meso-scale cells are seen on the divergence map – see the contours in Figure 10. Contrary to the divergence, a vorticity-map apparently shows large changes over a time lag of 20 minutes, reversing its sign in several locations; this behaviour suggests the presence of torsional waves at meso-scales with periods of 20 to 40 minutes. More data are needed to confirm this very interesting finding coming from the analysis of the 1.6 μm OMR granulation movie. An additional analysis, which is needed in the future, is a comparison with the map of magnetic elements of the photosphere, which are presumably located where the flow converges. The presence of meso-scale torsional

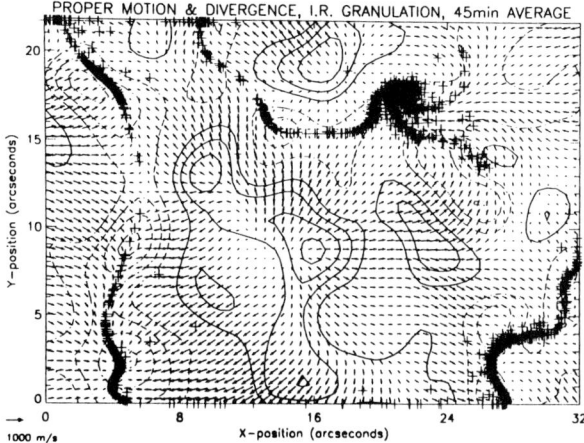

Fig. 10. Map of the flow deduced from the 45-minute granulation movie of 1988 obtained in the OMR. Contours show divergence of the average flow, solid line positive, dashed line negative (downflow). Cork positions after 8 hours are superposed to show the boundaries of a SG cell; note also several meso-scale cells inside the SG.

waves, with 20 to 40 minute periods, would be of great importance in enhancing magnetic-field effects in higher layers, up to the transition region.

4. Discusion and Conclusions

4.1. IR Granulation and the Magnetic Field

Examples of what granulation phenomena observed in the OMR can bring to the analysis of the origin of small magnetic elements are shown in Figs. 2 and 10; Figure 9a shows statistically the effect of concentration of magnetic elements at the boundaries of SG of the quiet Sun: these produce an excess of temperature (network bright point) and they occur at the locations of converging large scale flows. Note that we are referring to magnetic elements outside active regions, avoiding pores and pore-like structures which, obviously, show a negative contrast (see Fig. 2 and the paper of Darvann and Koutchmy in the present proceedings).

In regions of large concentration of magnetic field, namely the cores of sunspots, umbral granulation or umbral dots are observed. At 1.6 and 2.2 μm, we observed some sunspots by scanning, and also by imaging with diffraction-limited resolution at video speed. Umbral dots are easily seen; the images are stable and because the average contrast of the core is of the order of 0.6 or 0.7 – see Kotov and Koutchmy (1992) – there is no problem of scattered light as in the visible. We believe that the bright IR umbral dots will be more seriously analysed in the future, permitting new insights into the problems of magneto-convection and the origin of sunspots. Not only is the problem of spurious scattered light avoided, but magnetic effects are better separated from thermal and velocity effects; and the seeing is improved.

4.2. CONCLUDING REMARKS

Observing in the OMR has provided new results on granulation:

i) a large spectrum of cells, including the smallest at sub-arcsecond size, is observed;

ii) Convective phenomena are well observed with different spatial and temporal scales; there is now a need for line-profile analysis with high spatial resolution;

iii) The best spatial resolution is achieved in the OMR; simultaneous observations at two or more wavelengths are entirely feasible for a better coverage in heights.

Existing and planned solar telescopes should be adapted to IR needs and new ground-based facilities can be envisaged in the near future, based on new technology designs for building good IR telescopes.

References

Albregtsen, F. and Lynne Hansen, T.: 1977, *Solar Phys.* **54**, 31.
Altrock, R.C. and Musman, S.: 1976, *Astrophys. J.* **203**, 533.
Bray, R.J., Loughhead, R.E. and Durrant, C.J.: 1984, *The Solar Granulation*, Cambridge University Press.
Cattaneo, F., Hurlburt, N.E. and Toomre, J.: 1989, in R. J. Rutten and G. Severino (eds.), *Solar and Stellar Granulation*, Kluwer Academic Publishers, Dordrecht, p. 415.
Chan, K.L., Norland, A., Steffen, M. and Stein, R.F.: 1991, in A. Cox, W. Livingston, and M. Matthews (eds.), *Solar Interior and Atmosphere*, University of Arizona Press, Tucson, Arizona, p. 223.
Darvann, T. and Koutchmy S.: 1993, these proceedings.
Deubner, F.L. and Mattig, W.: 1975, *Astron. Astrophys.* **45**, 167.
Foukal, P., Little, R., Graves, J., Rabin, D. and Lynch, D.: 1990, *Astrophys. J.* **353**, 712.
Karpinsky, V.N.: 1990, in J. Stenflo (ed.), 'Solar Photosphere: Structure, Convection and Magnetic Fields', *IAU Symp.* **138**, 67.
Keil, S.L.: 1977, *Solar Phys.* **53**, 359.
Keil, S.L.: 1980, *Astron. Astrophys.* **82**, 144.
Kotov, V. and Koutchmy, S.: 1993, these proceedings.
Koutchmy, S.: 1978, in "Pleins Feux sur la Physique Solaire", *Proceedings of the 2nd Europ. Ass. Solar Phys. Meeting*, ed. CNRS, p. 155.
Koutchmy, S.: 1990, in J. Stenflo (ed.), 'Solar Photosphere: Structure, Convection and Magnetic Fields', *IAU Symp.* **138**, 81.
Koutchmy, S., Koutchmy, O. and Kotov, V.: 1977, *Astron. Astrophys.* **59**, 189.
Koutchmy, S. and Lebecq, C.: 1986, *Astron. Astrophys.* **169**, 323.
Leighton, R.B.: 1983, *Ann. Rev. Astron. Astrophys.* **1**, 69.
Lin, K. and Kuhn, J.R.: 1992, *Solar Phys.* , in press.
Muller, R.: 1989, in R. J. Rutten and G. Severino (eds.), *Solar and Stellar Granulation*, Kluwer Academic Publishers, Dordrecht, p. 101.
November, L.J.: 1986, *Appl. Optics*, **25**, 3, 392.
Title, A.M., Shine, R.A., Tarbell, T.D., Topka, K.P. and Scharmer, G.B.: 1990, in J. Stenflo (ed.), 'Solar Photosphere: Structure, Convection and Magnetic Fields', *IAU Symp.* **138**, 49.
Topka, K.P. and Title, A.M.: 1991, in A. Cox, W. Livingston, and M. Matthews (eds.), *Solar Interior and Atmosphere*, University of Arizona Press, Tucson, Arizona, p. 727.
Turon, P.J.: 1975, *Solar Phys.* **41**, 271.
Turon, P.J. and Léna, P.: 1973, *Solar Phys.* **30**, 3.
Vernazza, J.E., Avrett, E.H. and Loeser, R.: 1981, *Astrophys. J. Suppl.* **45**, 635.
von der Lühe, O. and Dunn, R.B.: 1987, *Astron. Astrophys.* **177**, 265.
Weiss, N.O.: 1990, *Proc. IAU Symp.* **142**, 139.
Worden, S.P.: 1975, *Solar Phys.* **45**, 521.

SIMULTANEOUS IR AND VISIBLE LIGHT MEASUREMENTS OF THE SOLAR GRANULATION

S. KEIL

U. S. Air Force Phillips Laboratory, Sunspot, NM 88349, U.S.A.

J. KUHN and H. LIN

Michigan State University, East Lansing, MI 48824, U.S.A.

and

K. REARDON

Department of Astronomy, Williams College, Williamstown, MA 01267, U.S.A.

Abstract. Movies of the solar granulation were made simultaneously at 5575 Å and 1.64 μm using the Vacuum Tower Telescope at NSO/SP. A 128 × 128 HgCdTe array was used in the infrared and an RCA 504 CCD in the visible. From the movies, we determine and compare statistical properties of the granulation and seeing conditions.

Key words: convection – infrared: stars – Sun: granulation

1. Introduction

The near infrared spectrum at 1.64 μm offers several advantages for observing the solar granulation. The effects of atmospheric seeing are reduced (the Fried parameter, r_0, which scales as $\lambda^{6/5}$, increases by a factor of 3.5), and since the solar opacity reaches a minimum near 1.64 μm, we see deeper into the solar atmosphere where the granulation is a more dominant effect. (Keil, 1980, Koutchmy, 1988). One difficulty of observing at the longer wavelength is that the granular contrast is reduced. Computations show this reduction to be a factor of ~2 at 1.64 μm (Stein, 1992). Another difficulty is that the angular resolution of the telescope is reduced by a factor of ~3. Although 1.64 μm radiation is formed only 30-40 km below the 5575 Å radiation, the convective heat flux decreases very rapidly with height between these two layers (Bray et. al., 1984) and thus, IR observations can provide a more useful upper boundary condition for models of small-scale solar convection than do observations at visible wavelengths. The effects of the 5-minute oscillations are also greatly reduced at the deeper layer (Koutchmy, 1988) and the granular field is more easily observed.

Turon and Léna (1973) made one of the earliest attempts to measure the solar granulation in the opacity minimum region. They used a linear array and obtained two dimensional images by scanning the sun. In addition, they made center-to-limb photometric scans. While the photometric scans produced results on the center to limb variation of the granular contrast, the imaging was not of sufficient quality to derive properties of the granules. Koutchmy (1988) used a two dimensional IR vidicon equipped with a video digitizer to make movies of the granulation at 1.75 μm. He recorded white light images using a conventional video camera and a VCR at the same time. Because of the non-uniform response of the IR vidicon and the eight bit limitation of the A/D in the video digitizer, he made only qualitative studies of the granulation. He found that the IR granules exhibited a bi-modal intensity distribution and that their lifetimes are shorter in the IR (3.5 min vs. 6 min).

We take advantage of improvements in IR arrays and CCD cameras to simultaneously measure the granules at IR and visible wavelengths and obtain more quantitative properties. In Section 2 we describe the observations. In Section 3 statistical properties of the seeing are discussed and in Section 4 properties of the granules are compared as observed in the two wavelength bands.

2. Observations

Observations were made at the center of the solar disk using the NSO/SP Vacuum Tower Telescope (VTT) on 15 Oct 1990. The beam from the VTT was split between the Universal Birefringent Filter (UBF), which was alternately tuned to 5575 Å (continuum) and 5576.09 Å (core of the Fe I 5576 line) with a 1/8-Å bandpass, and the IR array, which was fed using a 2800Å FWHM filter centered at 1.64 μm. The CCD camera used with the UBF had 300 × 240 pixels, each 16 × 20 μm^2, corresponding to 0.09" × 0.11" pixels on the sun, and a field of view of 27" × 26". The exposure time was 90 ms and the cadence was one continuum and one line center image every 8 s. The IR array has 128 × 128 pixels, each 60 × 60 μm^2, corresponding to 0.2" × 0.2" pixels and a field of view of 26" × 26". The exposure time was 1.5 s and the cadance was one image every 4 s. The results presented below were obtained from the first 55 minutes of data taken from a three hour observing sequence. This time period corresponded to the most stable seeing conditions, thus alleviating some of the differences caused by exposure time.

From the raw observations, we produced a 400 frame movie in the continuum at 5575 Å and an 800 frame movie at 1.64 μm. Each movie was then correlation tracked to remove gross motions, destretched to remove differential motions, and finally, passed through a sub-sonic filter to remove the effects of the 5 minute oscillations.

A side by side movie was produced to visually assess the quality of the data. Watching the movie, one gets a distinct impression that seeing conditions are more stable in the IR. Granular evolution is easier to follow in the IR, although when you freeze the movie on any particular frame, it is easy to draw a one to one correspondence between features in the visible and IR. We quantify these impression in the next section.

3. Atmospheric Seeing at 5575 Å and 1.64 μm

We have looked at several properties of the data that are related to atmospheric seeing. These include the amount of image motion, the rms variance of the intensity contrast, and the spatial scale of the data determined from the spatial autocorrelation function of individual images.

As the images were correlation tracked, the values of the x and y displacements needed to align the images were stored. The maximum displacements at 5575 Å were $\Delta x = \pm 0.5''$ and $\Delta y = \pm 0.6''$, with a variance of $\sigma_x = 0.15''$ and $\sigma_y = 0.20''$, while at 1.64 μm, $\Delta x = \pm 0.35''$ and $\Delta y = \pm 0.40''$, $\sigma_x = 0.10''$, and $\sigma_y = 0.14''$. Dispersion plots between the fluctuations at 5575 Å and 1.64 μm showed the displacements were not well correlated, probably owing to the difference in

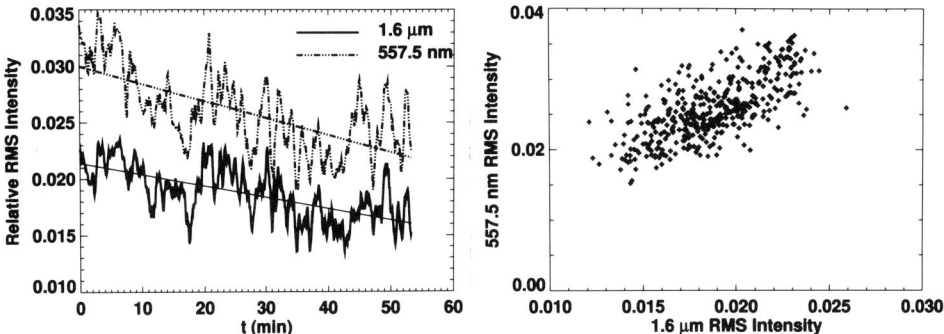

Fig. 1. The panel on the left shows the normalized rms intensity fluctuations observed at 5575 Å and at 1.64 μm. The right hand panel shows the dispersion between the rms values at 5575 Å and 1.64 μm.

exposure times.

The images were destretched by local correlation tracking on a grid of ~2″ boxes against a reference image that was obtained from a one minute temporal running mean of the data. For each image we stored the displacement at each grid point required to destretch the images. From these displacements, an average displacement was generated for each image and then for the entire time sequence. The mean displacements at 5575 Å and 1.64 μm were $\Delta r = 0.065''$ and $\Delta r = 0.035''$ respectively.

Figure 1a plots the normalized rms intensity fluctuations of each image over the 55 min observing run. The quality of the seeing decreases with time over the 55 minutes. The rms fluctuations observed in the visible and IR track fairly well. Figure 1b shows the dispersion relationship between the two wavelengths. The data has not been corrected for optical or atmospheric transfer functions. If the IR data is corrected for the expected factor of two reduction in contrast at 1.64 μm, we are seeing greater contrast in the IR than in the visible, before correction for atmospheric and instrumental effects. Figure 2 shows spatial autocorrelation functions at the two wavelengths averaged over the 55 min time sequence. The half width at half maximum of the autocorrelation functions gives an estimate of the features sizes being resolved. These widths are 390 km at 1.64 μm compared to the 316 km cutoff of the VTT and 460 km at 5575 Å compared to the 110 km (diffraction) cutoff of the VTT. Thus we are resolving features nearer the telescope cutoff at 1.64 μm than at 5575 Å, in spite of the longer exposures and reduced telescope resolution. This would indicate substantially better seeing conditions at 1.64 μm.

The results for the image motion and correlation tracking displacements can be partially explained by the difference in exposure times. The longer exposures in the IR tend to smear out the atmospherically induced large scale and differential image motions. Thus on the average, we find smaller displacements in the IR. However, these same effects would tend to smear the granules and reduce their contrast in

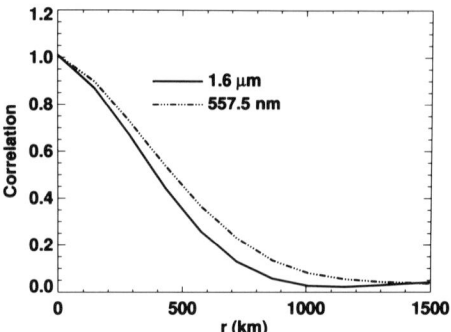

Fig. 2. Mean spatial autocorrelation functions plotted for 1.64 μm and 5575 Å. The autocorrelation of each image was computed as a function of x and y, azimuthally averaged to obtain the radial dependence, and then averaged over time to get the mean functions shown here.

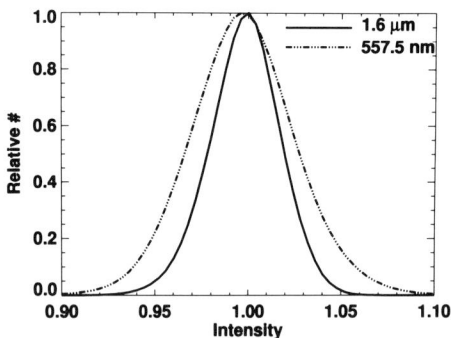

Fig. 3. Histograms of the intensity fluctuations, averaged over the first 15 minutes of data.

the IR and increase the width of the autocorrelation function. Since we are seeing near diffraction limited data in the IR, we conclude that the seeing has less effect on the IR observations.

4. Properties of the Granules at 1.64 μm and 5575 Å

Figure 3 shows intensity histograms, at the two wavelengths, obtained by computing individual histograms for each image in the first 15 minutes of data (the period exhibiting the highest rms intensity fluctuations, see Figure 1) and then averaging the histograms. The distributions are both skewed very slightly to darker values (in

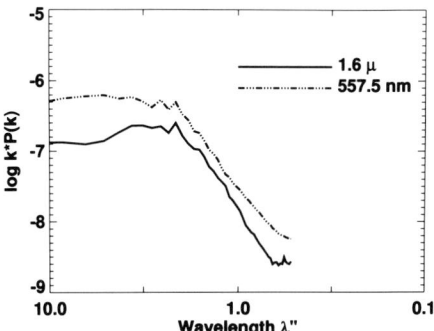

Fig. 4. Radial spatial power spectra, averaged over the 55 min observing sequence.

agreement with the findings of Keil, 1977 and Pravdjuk et al., 1974. We do not see strong evidence for a bimodal distribution as reported by Koutchmy (1988). Seeing could play a role in causing our distribution to be more symmetrical, although our observed rms at 1.64 μm is slightly higher than his observed values at 1.75 μm. The distributions give a ratio of bright to dark elements of 0.88 in the visible and 0.89 in the IR.

Measuring the size and spatial distribution of the solar granulation is extremely difficult and algorithm dependent (Title et al., 1989, Roudier and Muller, 1987). Some of the algorithms that have been used include spatial power spectra and autocorrelations (Deubner and Mattig, 1975), fractal dimensions (Roudier and Muller, 1987), and granule center and lane finding algorithms (cf. Title et al.). Rather than evaluating the techniques, we compare differences between the IR and visible data. We again emphasize that no corrections for transfer functions have been applied to the data.

Figure 4 compares the spatial power of the intensity fluctuations (as a function of radial wavenumber) at the two wavelengths. The spectra have been averaged over the 55 minute observing run. The power spectra in the IR and visible behave similarly at high wavenumber, corresponding to wavelengths from about 3″ down to the telescope cutoff. This portion of the spectra appears to follow a power law. At low wavenumbers, the IR spectra shows a decrease in power, probably due to the diminished fluctuations associated with the 5 min oscillations discussed below. The IR power is smaller at all wavelengths, in agreement with the observed intensity fluctuations. We have also computed the fractal dimension of the granules using the granular locating algorithm described by Newbury and Keil (1990). We find a smooth increase in the fractal dimension from about 1.3 for small granules to 1.9 for the largest granules in both the visible and IR. This is similar to the result Newbury and Keil find for granules observed in the visible.

The temporal power spectrum was computed at each pixel in the array and then averaged over the array to obtain Figure 5a. The 5 min oscillations show up clearly

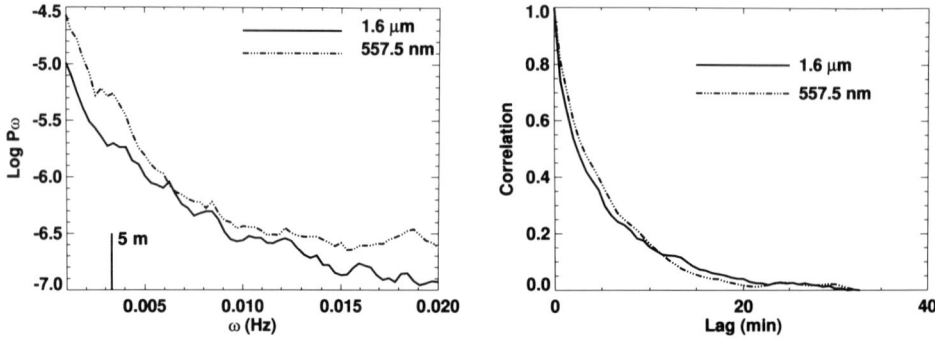

Fig. 5. The left hand panel shows the temporal power spectra averaged over all spatial points. On the right is shown the cross-correlation between the reference frame and subsequent frames. A reference frame was chosen at 40 second intervals and then cross correlated with the subsequent images. These cross correlation functions where then averaged to obtain the mean curves shown in the figure.

in the 5575 Å data but are greatly reduced in the 1.64 μm data. The noise level is substantially reduced in the IR observations, corresponding to a reduction in seeing induced fluctuations. The temporal cross correlation functions are shown in Figure 5b. The correlation between time steps is seen to decrease more rapidly for the IR granulation during the first 10 minutes. This lends support to Koutchmy's (1988) argument that the granules are shorter lived in the opacity minimum region. However, the difference we observe is much smaller than he measures. In addition to the cross correlation functions, we tracked individual granules, starting from a reference frame and then following the granule forward and backward in time until it disappeared. Granules were defined by an algorithm that searched for local maxima. Histograms of the granular lifetimes determined in this manner yield a mean lifetime near 5 minutes in both the IR and visible. We were able to track a few granules for periods near 40 minutes in the IR. None of the visible granules could be tracked longer than 32 minutes.

Acknowledgements

Our thanks to the NSO/SP VTT staff (Dick Mann, Steve Hegwer, and Roy Coulter) for helping obtain the data. The IR array was developed in collaboration with the Astrophysical Research Consortium, with assistance from Michigan State University, the University of Wyoming, and Haverford College.

References

Bray, R. Loughhead, R. and Durrant, C.: 1984, *The Solar Granulation*, Cambridge University Press, 2nd Edition.
Deubner, F. and Mattig, W.: 1975, *Astron. Astrophys.* **45**, 167.

Keil, S.: 1977, *Solar Phys.* **53**, 359.
Keil, S.: 1980, *Astron. Astrophys.* **82**, 144.
Koutchmy, S.: 1988, in J. Stenflo (ed.), 'Solar Photosphere: Structure, Convection and Magnetic Fields', *Proc. IAU Symp.* **138**, 81.
Newbury, J., Keil, S.: 1991, *Bull. Amer. Astron. Soc.* **23**, 1048.
Pravdjuk, L., Karpinsky, V., and Andreiko, A.: 1974, *Soln. Dann.* **2**, 70.
Roudier, T., Muller R.: 1987, *Solar Phys.* **107**, 11.
Stein, R.: 1992, private communication
Title, A., Tarbell, T., Topka, K., Ferguson, S., and Shine, R.: 1987, *Astrophys. J.* **336**, 475.
Turon, P. J. and Léna, P.: 1973, *Solar Phys.* **30**, 3.

MEASUREMENTS OF HORIZONTAL FLOWS
IN 1.6 μm GRANULATION

TRON A. DARVANN

Institute of Theoretical Astrophysics, University of Oslo,
P. O.Box 1029 Blindern, N-0315 Oslo, Norway
and
National Solar Observatory, Sunspot, NM 88349, U.S.A.*

Abstract. We report on a first analysis of the horizontal motions in a 45 minute (32 × 22 arcsec^2 field of view) granulation time series (movie presented at the present IAU Symposium) obtained at 1.6 μm with the Vacuum Tower Telescope (VTT) of the National Solar Observatory at Sacramento Peak (NSO/SP). High signal/noise flow maps are obtained by use of the local cross correlation technique (November 1986) which incorporates efficient attenuation of seeing and 5-min oscillations. The flow pattern, showing a ($\approx$ 30 arcsec diameter) supergranule with ($\approx$ 8–15 arcsec) mesogranules superposed, is long lived compared to the 45 min of observations. The computed flows (velocity, divergence, vorticity) resemble the ones obtained at visible wavelengths (*e.g.*, by Brandt *et al.* 1991, November 1989, November and Simon 1988, Simon *et al.* 1988). The high quality of the flow maps (due to a large number of selected images (1500), and (supposedly) smaller 5-min oscillations and better seeing conditions at 1.6 μm) allows us to study time evolution (resolution $\approx$ 15 min) of the details of the flow (spatial resolution $\approx$ 3 arcsec). An interesting new finding is the short lifetime (< 45 min) of vorticity as opposed to the long lived ($\gg$ 45 min) divergence of the flow. Our study demonstrates the possibility of using the 1.6 μm window to the opacity minimum region to study the horizontal flows at these deep layers of the photosphere.

Key words: infrared: stars – Sun: granulation – Sun: photosphere

1. Introduction

The technique of local cross correlation of granulation has been developed during the last few years to the point where it is a trusted and highly accurate tool for measuring horizontal flow patterns in the photosphere (November 1986, Simon *et al.* 1988). The method has been used in several investigations of the solar convection (Brandt *et al.* 1991, Müller *et al.* 1990, November 1989, Simon *et al.* 1988, Title *et al.* 1986, 1987, 1989). Flow patterns can be studied in detail, and *e.g.*, the flow divergence gives a proxy for vertical flow, while test particle ("cork") simulations demonstrate the redistribution of magnetic flux by the convection. However, only presently, with greatly increasing data acquisition rates and computer capacity, the full power of the technique is starting to be seen, when applied together with other diagnostics *e.g.*, obtained from simultaneous multi-wavelength data (*e.g.*, Yi 1992). Particularly interesting is a *simultaneous* measurement of both small and large scale (from granular to global scale) flows for a study of interaction between different convective scales, and between convection and differential rotation (Darvann 1991), by observing granulation proper motion over a very large field of view. Noise limitations have been studied by November and Simon (1988), demonstrating that noise ($\approx$ 100 m s^{-1}) from random granular motions dominate in a typical 1-hour average flow map, while seeing and 5-min oscillations contribute negligibly. Darvann (1991) showed that the noise level may be significantly further reduced by averaging over

---

* Operated by the Association of Universities for Research in Astronomy, Inc. (AURA) under cooperative agreement with the National Science Foundation.

a much longer time than a couple of hours. While all the work up to now has been carried out at visible wavelengths, some gain in signal to noise ratio for the flow measurement might possibly be achieved at 1.6 μm, thereby allowing smaller velocities to be detected. Granular velocities are larger at this wavelength (Keil 1980), and contrast is lower, but seeing and 5-min oscillations both have smaller amplitude. We present here a first computation of the topology of the flow at this wavelength. Keil *et al.* (1993, these proceedings) have obtained high resolution time series obtained in the visible and IR *strictly simultaneously*, and such a data set should be able to show the detailed differences of the flow topology, in addition to the signal/noise properties, between the two wavelengths.

2. Observations and Preprocessing

The observations were carried out on April 19, 1988 with the Vacuum Tower Telescope (VTT) at NSO/SP. The CHIRP image processing system was used in order to digitize and flatfield the granulation frames from a Vidicon IR video camera at a 1 Hz rate with subsequent recording on video tape. Visual image selection was thereafter carried out on a time series of 45 min duration showing variable, fair seeing conditions. The selection resulted in 1500 images approximately evenly spaced in time ($\Delta t \approx 1.8$ s). The time series was thereafter correlation tracked by an FFT method (von der Lühe 1983) in order to remove image motion. Images were also "brushed" (high gradient "bad" pixels removed, November 1986) and Fourier filtered to remove bad video lines that sometimes moved across the images. Figure 1b shows one of the images used for the measurement of horizontal flows.

3. Proper Motion Maps of Infrared Granulation

Proper motion maps were computed by local cross correlation (November 1986, Darvann 1991) of the 1500 image pairs (Fig 1b) and were noise-optimized through the technique of temporal summation of cross correlation functions (November and Simon 1988). Noise was further reduced by averaging results from 3 different computations applying three different correlation time lags (= the time difference between the correlated images in a pair); 50, 52 and 54 s, respectively. Figure 2 shows the average proper motion map (represented by the flow vectors) for the full 45 minute time series. The maximum speed is 900 m s^{-1}, RMS speed is 300 m s^{-1}, estimated accuracy $\pm$ 100 m s^{-1}. The effective spatial resolution of the map is determined by the Gaussian correlation window of 3.8 arcsec FWHM used in the computation.

A strong outflow with spatial scale corresponding to a little less than the typical size of a supergranule is evident near the center of the 32 $\times$ 22 arcsec^2 field of view. The outline of the "supergranule boundary" becomes more evident to the eye by computing the paths of evenly distributed "test particles" or "corks" moving with the flow (Simon *et al.* 1988). In Figure 2 we have allowed the corks (+ signs in the plot) to move for 8 hours in order to get an impression of the location of the "supergranulation network" for the 45 min average. The corks avoid regions of positive divergence ("outflow"), as shown in Figure 10 in the review paper by Koutchmy (elsewhere in these proceedings), and concentrates in the regions of neg-

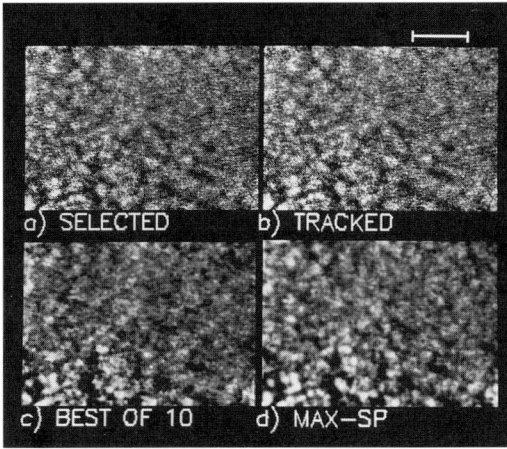

Fig. 1. Example of a single frame of the movie comparing image quality after 4 steps of processing have been applied. Length of the bar showing the scale is 7 arcsec. Difference in contrast and amount of detail between upper right and lower left of each frame is due to variations in sensitivity of the IR chip. *a)* Images after visual selection, filtering and "brushing." *b)* Images after correlation tracking. This is the type of image used for the computation of horizontal motions. *c)* Best (highest RMS intensity) images out of every 10 images, subsequently smoothed by 3 × 3 pixel bilinear interpolation. *d)* Images reconstructed from groups of 10 sequential images by use of spatial power maximization (MaxSP) (Koutchmy and Koutchmy 1989), smoothed by 3 × 3 pixel bilinear interpolation. A movie shows a large improvement by MaxSP in terms of sharpness and stability.

ative divergence ("downflow"). The local extrema of divergence (typically $\pm 3 \times 10^{-4}$ s^{-1}) may be interpreted to represent mesogranules (November and Simon 1988). The contours in Figure 2 show the vorticity averaged over the same 45 minute observing period.

A very useful evaluation of the quality of a flow map (complementary to noise estimation by temporal power spectral analysis of flow maps, November and Simon 1988), is to compare maps computed from different independent subsets of the data. For our data set, we computed a time series of 16 independent 2.8-min maps (90 image pairs contributing to each). From these maps we formed an "ODD" average map by averaging the 8 *odd* "interlaced" maps, and an "EVEN" average by averaging the 8 *even* ones. The ODD and EVEN maps (the divergences of which are shown in the bottom two figures of Figure 3) in this way become independent in terms of seeing (no common image pairs), and also almost independent in terms of granulation noise (2.8 min is a large fraction of the granulation lifetime, so that different granulation "realizations" contribute to the two maps). The correlation coefficients are high (0.7–0.8) (Table I), demonstrating the high quality of the computation. In a similar fashion, we averaged the 8 *first* 2.8 min maps to form a "FIRST" 22.5 min average map, and the 8 *last* 2.8 min maps to form a "SECOND" (later in time) 22.5 min map (the divergence of these are shown in the top two figures of Figure 3). Table I shows that correlation coefficients are high also in this case (0.5–0.6),

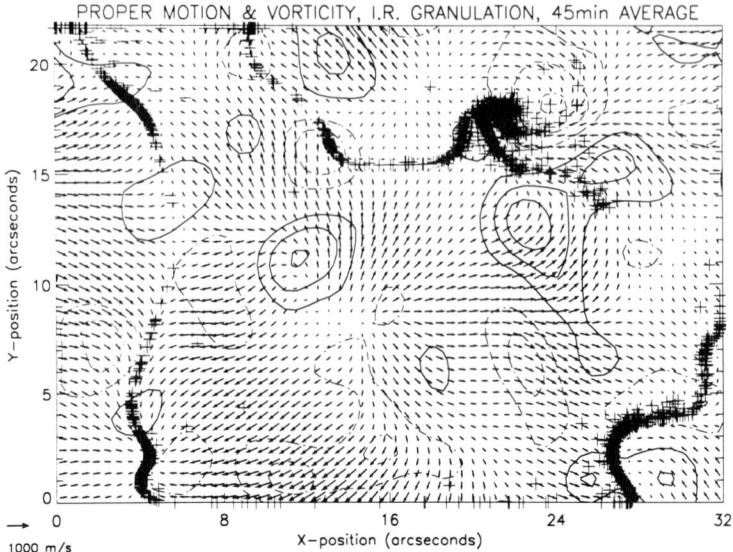

Fig. 2. A 45 min average proper motion map (flow vectors). Note the 1 km s^{-1} calibration flow vector to the lower left. Contours show flow vorticity (solid contour is positive (counterclockwise) vorticity, dashed contour negative (clockwise)), +-signs locate "corks" that have followed the flow vectors for 8 hours after initiating as an evenly distributed cork field at time 0. The same flow map, but with contours of divergence, is shown in Figure 10 of the review paper by Koutchmy elsewhere in these proceedings. Contour interval is 1×10^{-4} s^{-1} for both figures; the zero-contour is not shown.

TABLE I

Correlation coefficient for comparison between the horizontal flow in the first (subscript *1*) and second (subscript *2*) 22.5 min average maps, and between the even (subscript EV) and odd (subscript ODD) maps (see the text). The correlation is given separately for the x- and y-component of the flow (v_x, v_y), and the correlated divergence (DIV) maps are the ones shown in Figure 3. VOR denotes vorticity maps.

	$v_{x,1}$	$v_{y,1}$	DIV_1	VOR_1		$v_{x,EV}$	$v_{y,EV}$	DIV_{EV}	VOR_{EV}
$v_{x,2}$	0.65				$v_{x,ODD}$	0.82			
$v_{y,2}$		0.56			$v_{y,ODD}$		0.72		
DIV_2			0.50		DIV_{ODD}			0.70	
VOR_2				0.07	VOR_{ODD}				0.44

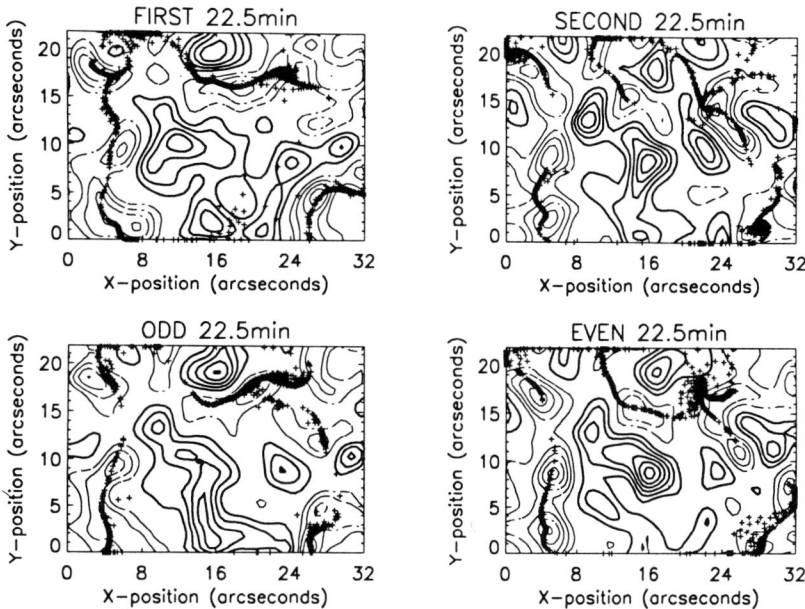

Fig. 3. Four divergence maps overlayed with cork positions after 8 hours, solid contour is positive divergence ("upflow"), dashed contours represent negative divergence ("downflow"). Upper two maps show the first and last 22.5 hr divergence maps making up the 45 min average shown in Fig 2. Lower two maps are 45 min maps composed by averaging odd and even ("interlaced") 3 min maps. The large similarity (see Table I) between the four maps demonstrates the low noise level of the measurement.

but slightly lower than for the ODD/EVEN comparison, and we interpret this to be due to a slow evolution of a long lived flow pattern. The corresponding drop in correlation coefficient for the divergence is similar (Table I). For the vorticity, however, there is a dramatic decrease in correlation, and apparently the "lifetime" of vorticity must be very short compared to the "lifetime" of the divergence. Flow maps influenced by seeing has a tendency to show large values of vorticity, but the relatively large correlation (0.44) between the ODD and EVEN vorticity maps, rules out seeing noise as an explanation. Previous (white light) data sets that we have worked with also show a slight tendency for somewhat shorter "lifetime" of vorticity compared to divergence, but not at all as dramatic as in the present case. A time series of three 15 min average vorticity maps reveals large changes, and in some locations we see reversal of sign in the course of 45 min, leading one to think of torsional waves (see Koutchmy's review paper in these proceedings).

We can also note here that the best way of qualitatively checking a flow map is to study an accelerated movie of the granulation. We were able to draw a map by hand that shows quite similar large scale features as the computed one.

4. Conclusions

We have demonstrated that it is now possible to obtain high quality time series of granulation proper motion maps using local cross correlation of 1.6 μm granulation images. Our preliminary study shows flow properties (flow amplitude, size scale of flow patterns etc.) that are within the range found by previous measurements in the visible. One possible exception is the apparent short lifetime of the flow vorticity (on a mesogranular scale); this needs to be investigated by temporal power spectral analysis. In order to detect smaller differences between the flows in the visible and IR, it will be necessary to carry out a statistical analysis of data obtained *strictly simultaneously* like the observations by Keil *et al.* (1993, these proceedings). Also, a larger field of view than presently obtained, would be of great value.

Acknowledgements

Travel support from ITA, Oslo, and Institute d'Astrophysique, Paris, is gratefully acknowledged. I would like to thank NSO/SP for the hospitality and outstanding working conditions offered to me during several visits, and for financial support in connection with attending the IAU Symposium. The present work has benefitted greatly from the never ending supply of inspiration and excellent observations provided by Dr. Serge Koutchmy. The software for proper motion measurements has been developed by Dr. Larry November. Fritz Stauffer programmed CHIRP, and Larry Wilkins provided electronics for the IR camera.

References

Brandt, P.N., Ferguson, S., Scharmer, G.B., Shine, R.A., Tarbell, T.D., Title, A.M., Topka, K.: 1991, *Astron. Astrophys.* **241**, 219.
Darvann, T.A.: 1991, Cand. Scient. Thesis, Univ. Oslo.
Keil, S.L.: 1980, *Ap.J.* **237**, 1024.
Keil, S.L., Kuhn, J.R., Lin, H., Reardon, K.: 1993, these proceedings.
Koutchmy, S.: 1993, these proceedings.
Koutchmy, O., Koutchmy, S.: 1989, in O. von der Lühe (ed.), *High Resolution Solar Observations*, National Solar Observatory, p. 217.
Muller, R., Roudier, Th., Vigneau, J., Frank, Z., Shine, R., Tarbell, T., Title, A., Simon, G.W.: 1990, in L. Dezso (ed.), *The Dynamic Sun*, Debrecen, p. 44.
November, L.J.: 1986, *Appl. Opt.* **25**, 391.
November, L.J.: 1989, *Ap. J.* **344**, 494.
November, L.J., Simon, G.W.: 1988, *Ap. J.* **333**, 427.
Simon, G.W., Title, A.M., Topka, K.P., Tarbell, T.D., Shine R.A., Ferguson, S.H., Zirin, H., and The SOUP Team: 1988, *Ap. J.* **327**, 964.
Title, A.M., Tarbell, T.D., Simon, G.W., and the SOUP Team: 1986, *Adv. Space Res.* **6**, 253.
Title, A.M., Tarbell, T.D., Topka, K.P.: 1987, *Ap. J.* **317**, 892.
Title, A.M., Tarbell, T.D., Topka, K.P., Ferguson, S.H., Shine, R.A., and the SOUP Team: 1989, *Ap. J.* **336**, 475.
von der Lühe, O.: 1983, *Astron. Astrophys.* **119**, 85.
Yi, Z.: 1992, *Quiescent Filaments, Magnetic Field, and Flows in the Photosphere*, Chapter 4 of Ph.D. Thesis, Univ. Oslo.

ON SUNSPOT AND FACULAR CONTRAST VARIATIONS NEAR 2 μm AND 4 μm

V. A. KOTOV

CSSA, Stanford University, Stanford, CA 94305, U.S.A.
and
Crimean Astrophysical Observatory, Nauchny, Crimea 334413, USSR

and

S. KOUTCHMY

Institut d'Astrophysique, CNRS, 98 bis, Boulevard Arago, F-75014 Paris, France

Abstract. Observations of the Sun at 2.2 and 3.75 μm have been made at the Pic-du-Midi and Crimean observatories. High resolution ($\sim 1''$) records of the sunspot and facular contrasts at various heliocentric distances as well as those at the extreme solar limb are presented. We find substantial variations in the sunspot-core relative intensity caused by magnetic activity of the spot. The extreme-limb contrasts of faculae at 3.75 μm are strong thus support flux-tube models which result in an enhancement of the extreme-limb facular brightness.

Key words: flux tubes – infrared: stars – Sun: faculae, plages – sunspots

1. Introduction

Observations of the Sun at 2.2 and 3.75 μm were made in 1974 at the Pic-du-Midi and Crimean observatories but were never reported. Because of possible contamination by absorption lines within the filter passband the 2.2 μm records were not suitable for accurate determination of spot contrasts. The basic observations were carried out at the CrAO Solar Tower (with the 90-cm main mirror) using the liquid-nitrogen cooled PbS cell combined with a Ge-filter and a 0.34-μm interference filter centered at 3.75 μm (Koutchmy et al., 1977). We used the main optical system of the Tower producing a 50-cm image of the Sun with resolution $\sim 1''$ at 3.75 μm. The detector had a circular entrance aperture, chosen to sample at the $\sim 1''$ diffraction limit of the telescope; the sampling frequency was 400 Hz.

2. Extreme Limb Profile

Observations of the extreme limb made at 2.2 and 3.75 μm showed that, in the first approximation, the blurring function may be represented by $S(r) = (\pi b^2)^{-1} e^{-(r/b)^2}$ with $b \approx 1.4''$. Since the aureole at $r \approx 7''$ outside the limb did not exceed 0.5%, the influence of blurring on the contrast measurements (for umbral diameters $\geq 15''$) is of small significance. The total contribution of stray light was estimated to be $\leq 1\%$.

The 3.75 μm records of the extreme limb differ appreciably from those obtained at 2.2 μm: the former reveal the existence of intensity enhancement $\approx 1.4\%$ at $r \approx 13''$. This enhancement should be probably attributed to an excess due to faint "invisible" IR faculae located near the limb. The intensities near the limb, corrected for seeing, together with the Allen's (1976) values, are given in Table 1. Our data significantly improve the limb profile for $r < 10''$, due to the higher spatial resolution of the present observations.

TABLE I
The extreme limb intensities at 3.75 μm (outside faculae).

r (")	44	19	5	4	3	2.0	1.3
Observed	-	0.858	0.800	0.790	0.768	0.735	0.690
Corrected	-	0.860	0.819	0.814	0.800	0.790	0.780
Allen (1976)	0.889	0.856	0.803	-	-	-	0.704

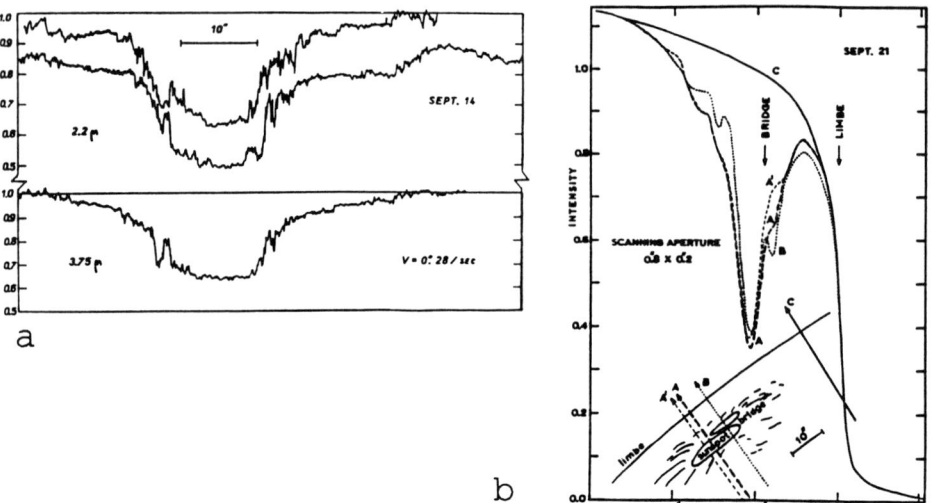

Fig. 1. (a) Typical IR photometric scans across the same location of a sunspot at 2.2 μm (top) and 3.75 μm (bottom); and (b) the continuum intensity near the Hα line of the spot shown in Figure 2 (21 September; 3.75 μm).

3. Sunspot Contrasts

Figure 1a presents records across a large sunspot (the spot group N 225, *Solar Geophysical Data*) on 14 September 1974 where the bright bridges are clearly seen at both wavelengths. The 3.75 μm records across the same spot made on 19–21 September are shown in Figure 2. One can readily see a *prominent facular area*, accociated with the "Secci" ring, with contrasts up to ≈ 4% and ≈ 1" fine structure. The maximum contrast substantially *increased*, from 2-3 to 3-4%, as the area approached the limb.

On 21 September the spot was photographed almost simultaneously in the continuum near $H\alpha$ with a 2 Å passband interference filter (Fig. 1b). A bright bridge ≈ 1.2" in width was clearly seen in both passbands, near Hα and at 3.75 μm. The bridge (and faculae) had larger contrasts in the IR than near Hα (bridge contrasts ≈ 8.2 and 4.4% respectively). The Hα faculae clearly seen in the penumbra were

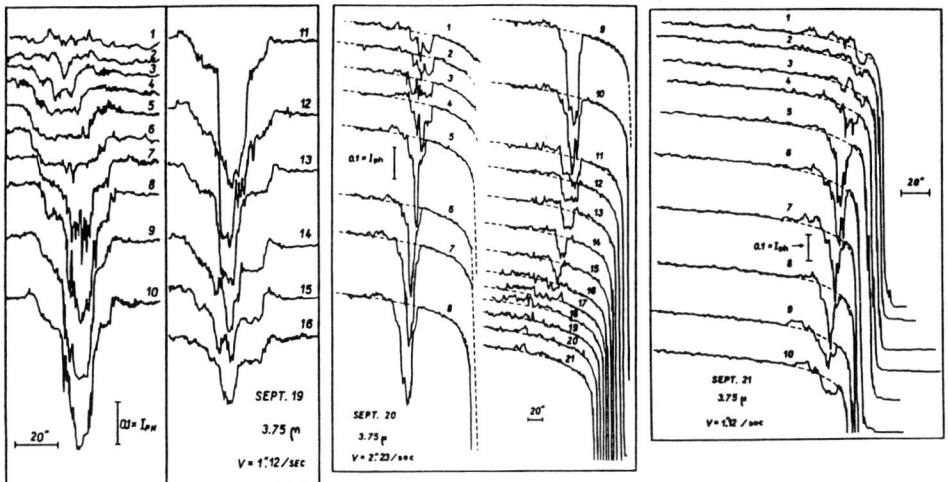

Fig. 2. The 3.75 μm photometric scans across a single spot obtained on 19 (left), 20 (middle) and 21 (right) September; the spacings between subsequent scans is 2.6″.

practically invisible outside the spot. Near Hα, we also did not observe any significant facular enhancements for $r > 22''$, contrary to the IR data, which explicitly showed facular regions with contrasts up to $\approx 4\%$.

The average contrasts for the umbra I_u and penumbra I_p at various heliocentric distances R are listed in Table 2. (The August observations of the same spot were performed by Koutchmy at Pic-du-Midi.) We see significant changes of I_u, presumably caused by the spot evolution and/or variation in the spot magnetic field (Fig. 3). The mean values $I_u = 0.65$ and $I_p = 0.94$ fairly agree with Allen's (1976) data for 3.75 μm. Our September average contrast I_p coincides also with the Maltby's (1972) result 0.936 ± 0.008 for 3.8 μm. The noticable tendency of I_u to increase slightly toward the limb is compatible with Wittmann and Schroter's (1969) conclusion based on observations at 0.468 and 0.790 μm. (However it contradicts the results of Makita and Morimoto (1960) obtained at similar wavelengths by averaging contrasts of various spots. Mattig (1969) argued that these latter measurements actually showed an increase of I_u toward the limb if one considers individual spots separately.) We mighty therefore conclude that the temperature gradient in the spot was a bit smaller than in the photosphere; however another interpretation can also be valid: a time evolution of the spot itself.

Ekmann and Maltby (1974) stressed the necessity of IR observations of spots. They propose that the scatter in IR intensities for different spots be considered as real. This could easily explain the variation in spot contrasts obtained by different observers. We note that the contribution of umbral dots in sunspot core measurements is often ignored; it is therefore difficult to discuss further the discrepancies in results reported by different authors.

The intensity gradient across the penumbra at the W-side of the spot (Fig. 2) appears to be larger than that at the E-side. This supports a similar observation

TABLE II

Sunspot contrast variations at 3.75 μm (uncorrected for a stray light).

Date (1974)	$R = \sin\theta$	Umbra	Penumbra
8 August	0.527	0.553	-
9 - " -	0.415	0.527	-
10 - " -	0.341	0.770	-
11 - " -	0.368	0.680	-
12 - " -	0.475	0.675	-
14 - " -	0.807	0.692	-
15 - " -	0.944	0.695	-
14 September	0.201	0.622	0.930
19 - " -	0.810	0.652	0.933
20 - " -	0.922	0.658	0.938
21 - " -	0.986	0.653	0.958

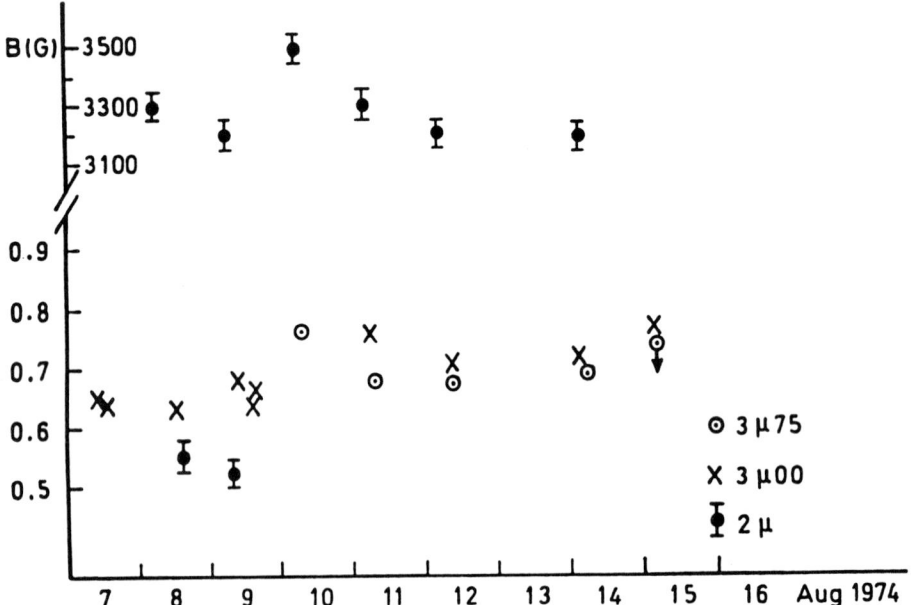

Fig. 3. Time variation of the magnetic field B (as measured by N.N.Stepanyan and colleagues at CrAO) and of the umbral core intensity I_u recorded in August 1974. If one ignores a plausible center-limb effect, a striking correlation emerges between B(top) and I_u (bottom).

TABLE III
Mean facular contrasts at 2.2 and 3.75 μm.

$R = \sin \theta$	0.402	0.200	0.810	0.922	0.986
Contrast, 2.2 μm	< 0.01	-	0.020	0.014	-
Contrast, 3.75 μm	-	< 0.01	≈ 0.012	0.027	0.042

by Maltby (1972) (though he suggested that the difference might be caused by an observational selection). We suppose that the asymmetry is real and indicates more regular structure of the leading side of the penumbra. In general, our data do not show a substantial decrease of I_u or I_p toward the limb found earlier by Albregtsen et al. (1984) for the 0.387–2.35 μm range. This may be due to several circumstances: (a) the difference in wavelength; (b) that we studied only a single spot at various radial distances instead of averaging contrasts of different spots; (c) temporal variations in the spot, see Figure 3.

4. Facular Contrast

The deduction of true facular contrasts near the limb is of a paramount importance for the modelling of filigree (magnetic flux tubes), especially in view of the controversy between different authors: models with a Wilson depression with "hot walls", as opposed to "hot hill" or "cloud" models. We believe that both models coexist, and predominance of one or other depends on the magnetic field configuration (Koutchmy and Stellmacher, 1978).

The most recent summary of monochromatic measurements of facular contrasts for the range 0.33 to 1.00 μm at various $\cos \theta$ values has been compiled by Foukal et al., (1991). Our near-IR contrasts (Table 3) agree well with their values (Figure 4) and may presumably represent true IR contrasts due to high (≈ 1″) spatial resolution of the observation. We note also that our contrasts appear to be significantly different from those obtained by Lindsey and Heasley (1981) for 10-25 μm. This discrepancy might be attributed to the differencies in wavelengths and especially in spatial resolutions. It confirms our poor understanding of the geometry and temperature distribution of real solar faculae as corrections for the filling factor cannot, at this stage, be properly introduced.

5. Conclusion

High resolution photometry of sunspots and faculae in the 2 to 4 μm region shows important properties needed for modelling these magnetic structures and their dynamics. More detailed measurements are now possible due to the availability of new IR imaging technology (see these proceedings). This makes precise identification of the origin of these variations a possibility. It is important that telescopes now be adapted for such observations.

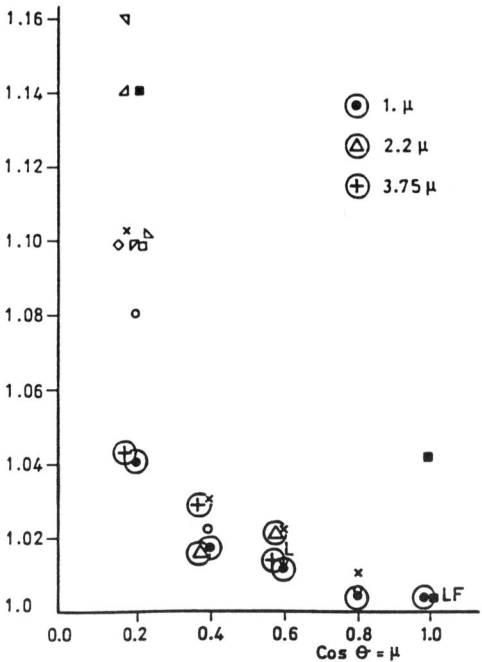

Fig. 4. Observed facular contrast vs $\mu = \cos\theta$ as adopted from Foukal *et al.* (1991) but plotted with our values for 2.2 μm (triangles) and 3.75 μm (crosses). Contrasts at 1.00 μm, taken from Foukal *et al.*, 1991, are represented by filled dots. All IR contrasts are shown inside circles; other symbols correspond to the 0.33-0.87 μm range.

References

Albregtsen, F., Joras, P.B., and Maltby, P.: 1984, *Solar Phys.* **90**, 17.
Allen, C.W.: 1976, *Astrophysical Quantities*, Athlone Press, London.
Ekmann, G., and Maltby, P.: 1974, *Solar Phys.* **35**, 317.
Foukal, P., Harvey, K., and Hill, F.: 1991, *Astrophys. J.* **383**, L89.
Koutchmy, S., Koutchmy, O., and Kotov, V.: 1977, *Astron. Astrophys.* **59**, 189.
Koutchmy, S., and Stellmacher, G.: 1978, *Astron. Astrophys.* **67**, 93.
Lindsey, C., and Heasley, J.N.: 1981, *Astrophys. J.* **247**, 348.
Makita, M., and Morimoto, M.: 1960, *Pub. Astron. Soc. Japan* **12**, 63.
Maltby, P.: 1972, *Solar Phys.* **26**, 76.
Mattig, W.: 1969, *Solar Phys.* **6**, 413.
Wittmann, A., and Schröter, E.H.: 1969, *Solar Phys.* **10**, 357.

SOLAR 5-MINUTE OSCILLATIONS AT 2.23 μm

TORBEN LEIFSEN

Institute of Theoretical Astrophysics, University of Oslo,
P. O. Box 1029, Blindern N-0315, Oslo 3, Norway[*]

Abstract. Large amplitude solar 5-min intensity oscillations have recently been detected at 2.23 μm using broad band (650 Å FWHM) photometry (Leifsen and Maltby, 1990). Large intensity amplitudes in a broad range in the near infrared was unexpected, and several questions concerning the source of the high amplitudes were raised. In an attempt to study the nature of these oscillations, time series of spectra have been obtained with the Fourier Transform Spectrometer (FTS) of the McMath telescope at National Solar Observatory at Kitt Peak. We present preliminary results from a 10 day long run in May 1991 in support for the suggestion that the results may be useful in both helio- and asteroseismological investigations.

Key words: infrared: stars – stars: oscillations – Sun: oscillations

1. Introduction

The interpretation of the observed solar 5-min oscillations as the evanescent tails of standing acoustic eigenmodes trapped in a sub-photospheric cavity (Ulrich, 1970; Leibacher and Stein, 1971) is presently widely accepted. Deubner (1975) confirmed this interpretation when the predicted ridges in the $k - \omega$ diagram was found from velocity measurements. The techniques have later been refined (e.g., Harvey and Duvall, 1984), and lately Nishikawa and Hirayama (1986) detected the predicted ridges in the $k - \omega$ diagram in broad band intensity measurements. Ground based observations of global luminosity oscillations have proven to be very difficult (Andersen and Domingo, 1986; Jimenez et al., 1988), and most ground based observations of global solar oscillations are velocity measurements (e.g., Libbrecht et al., 1990). High quality measurements of oscillations in the solar irradiance have been obtained from space based instruments (Woodard and Hudson, 1983).

Large amplitude infrared 2.23 μm solar intensity oscillations were detected in photometer observations obtained at Oslo Solar Observatory in 1987 and 1988. Five wavelength regions ranging from 0.67 μm to 2.23 μm and 7 circular entrance apertures ranging from 0.5 to 4.3 arcmin were observed simultaneously at all wavelengths with the same detector. The 2.23 μm region showed remarkably higher amplitudes than the other wavelength regions. The observed power was concentrated to the 2.5–3.5 mHz region, suggesting that we observe the well known 5-min oscillations. A comparison of the integrated power in the 2.5–3.5 mHz range as a function of spatial resolution shows that the power drops of toward lower resolution.

Full disk observations from June 1988 show the familiar 5-min oscillations near 3 mHz but also a strong feature near 4 mHz. This feature varies considerably from day to day, and it coincides in frequency with the fundamental p-mode resonance of the chromosphere (Deming et al., 1986). However, using a second order Fourier Transform and an autocorrelation analysis it was possible to identify a 135 μHz

---

[*] Observations obtained as a Visiting Astronomer, National Solar Observatory, National Optical Astronomy Observatories, operated by the Association of Universities for Research in Astronomy, Inc., under cooperative agreement with the National Science Foundation.

spacing between peaks in the 4 mHz feature. This suggests a high frequency tail of the 5-min oscillations.

2. Observations and Data Reduction

Two main questions arise from the photometer observations; one relates to the high power observed at 2.23 μm and the other concerns the nature of the power spectrum. We have choosen to address these questions by using the McMath telescope with its Fourier Transform Spectrometer at the National Solar Observatory on Kitt Peak. Two observing modes were used: A 50 arcsec circular area at the solar disk center was observed using the McMath East Auxilliary Telescope. Spatially integrated light was observed using the McMath main heliostat and two flat mirrors. Active guiding was used in the imaged mode. Two time series were obtained in May 1991; one in the imaged mode at the solar disk center over a 4 day long period and one with integrated light lasting 6 days.

These observations consist of time series of spectra separated in time by 1 minute and with a spectral resolution of 0.014 Å in the spectral range 2.0–2.5 μm. Each spectrum contains 825 Kilobytes of data. Twelve hours of observations result in 580 Megabytes of data. In order to handle these long time series each spectrum was divided in shorter spectral intervals (typically 630 Å for a time series of four days duration) and stored on several optical disks. Each series of intensity measurements in time at a given wavelength element was then treated as a separate time series. Each time series was normalized by dividing by a 20 min running mean, thus removing all long period variations, and cosine bells were added at the ends and before and after interruptions in the time series. Then power spectrum analysis was applied to each time series.

One of the main aims of the project and the main reason for using the Fourier Transform Spectrometer is to identify the sources of high intensity amplitudes in the 2.0–2.5 μm region of the solar spectrum and distinguish between solar oscillations and possible oscillations originating from the terrestrial atmosphere. Line identifications were taken from *An atlas of the Solar Spectrum from 1850 to 9000* cm^{-1} (Livingston and Wallace, 1991) and the Hall (1973) atlas for solar lines, and from the HITRAN molecular database (Rothmann et al., 1992) for terrestrial lines.

In order to study the variation of the power spectra as a function of wavelength, each powerspectrum was plotted in a 3-dimensional shaded-surface image (Figs. 1 and 2). The x-z plane represents power spectra at given wavelength elements. In this way we can study the variation of the power spectra by comparing the results obtained for solar atomic and molecular lines, the continuum and for lines formed in the Earths atmosphere.

3. Results

A study of the four days of disk center observations revealed the following: It was immediately apparent that a large number of spectral lines in the observed wavelength region show large intensity oscillations. A majority of these lines show a frequency distribution in the power spectrum that resembles the well known solar *p*-

mode oscillations. A few lines show a distinctly different frequency distribution, with some power in the 2–4 mHz region but with the power increasing strongly towards lower frequencies. These lines were indentified as water vapor lines originating in the terrestrial atmosphere. All the other lines showing high power were identified as solar lines.

There are several hundred solar spectral lines in the observed wavelength region, including the 2-0 and 3-1 bands of CO and several atomic lines. Figure 1 shows a power spectrum of an extract of the solar spectrum containing several lines from the CO 3-1 band. The ridges parallel to the y-axis correspond in wavenumber to CO lines. The familiar distribution around 3 mHz from solar p-modes are clearly seen. Figure 2 shows similar ridges in power spectra from a region in the solar spectrum containing several atomic lines.

All terrestrial molecular lines (expect water vapor), i.e., mainly CH_4, show no intensity oscillations above noise. This excludes the possibility of bands of terrestrial lines as the source of the observed high amplitude oscillations. The wavelength region shown in Figure 2 includes several CH_4 lines. None of these show intensity oscillations.

Even though water-vapor lines in the region show intensity oscillations, they are far to few and have a distinctly different power distribution to account for the observed intensity oscillations. A comparison of time series integrated over wavenumber both including and excluding the water-vapor lines show that the water vapor lines have little or no influence on the observed power spectra. The main reason for this is that relatively few water vapor lines contribute to the power and they tend to be concentrated to the edges of the 2.23 μm window.

Both molecular and atomic solar lines also show velocity oscillations with the familiar p-mode power distribution. We are currently analysing the velocity data and the phase between the velocity and line intensity oscillations. There is no doubt that the observed intensity and velocity oscillations in the solar lines are of solar origin.

There are indications of power in the 2.0–3.5 mHz region in the continuum region of the spectrum. However, the signal is comparable with noise, and no firm conclusions can be drawn yet as the noise level approaches the signal level in the photometer observations. At this point further observations are needed.

A six day long time series in spatially integrated light mode gives no firm conclusions, mainly because of telescope tracking problems during observations, introducing low-frequency power (around 1 mHz) with a severe leakage into the 3 mHz region.

4. Conclusion

The new observations obtained, with different equipment at an other site show intensity oscillations of solar origin in the 2.23 μm region. This conclusion is strengthened by the fact that terrestrial spectral lines (except water vapor) show no sign of oscillations above noise. The possible influence of water vapor absorption in the Earth's atmosphere is of minor importance due to the fact that the wavelength region in question contains only a few lines. The disk center observations show both

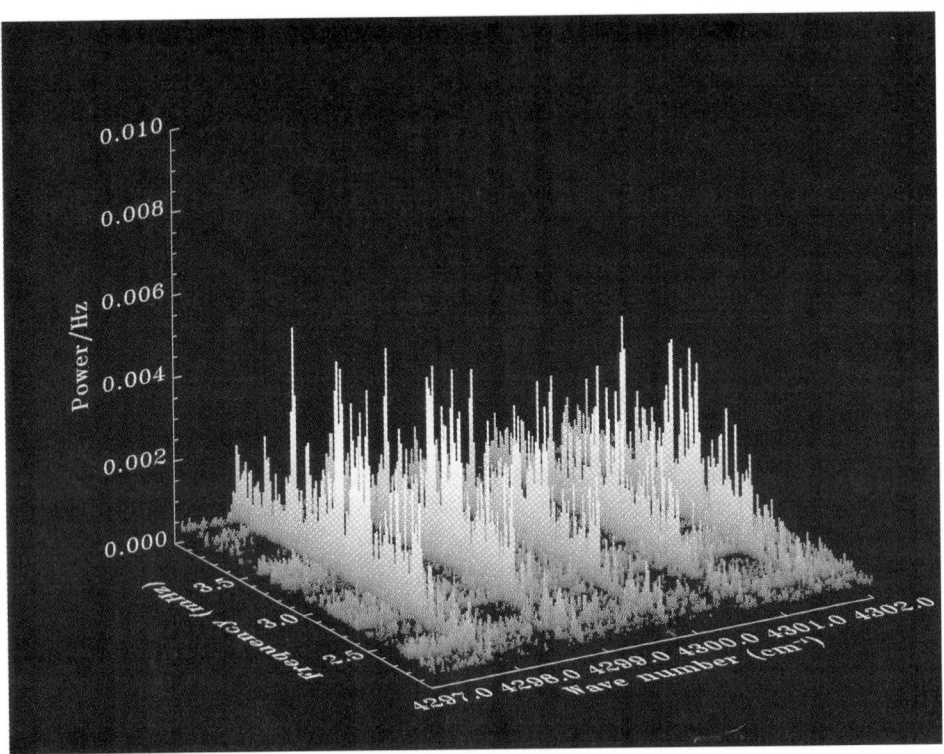

Fig. 1. Power as a function of wavelength (x-axis) and oscillation frequency (y-axis) in a time series lasting four days observed at the solar disk center. A 50 arcsec entrance aperture was used. The ridges with maximum around 3 mHz are solar intensity oscillations in the solar R36-R40 lines of the CO 3-1 band.

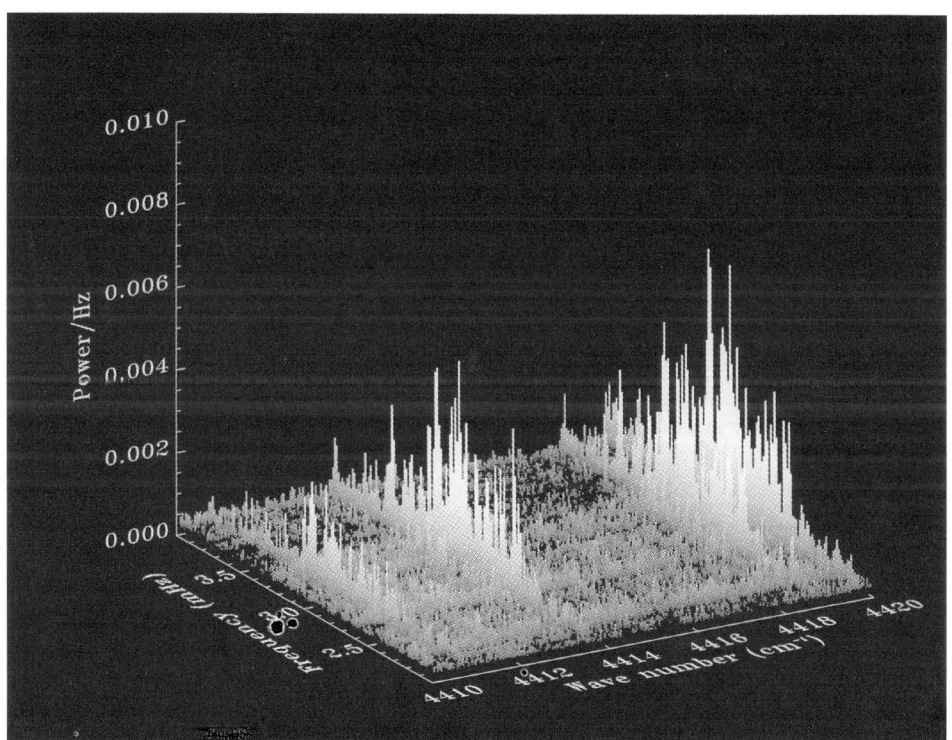

Fig. 2. Power as a function of wavelength (x-axis) and oscillation frequency (y-axis) in a four day long time series at the solar disk center with a 50 arcsec entrance aperture. The ridges are solar intensity oscillations in solar Si (4410.7 cm^{-1}), CaI (4413.1 cm^{-1}), CaI (4413.6 cm^{-1}) CaI (4418.4 cm^{-1}), CaI (4418.7 cm^{-1}) and Fe (4419.7 cm^{-1}) lines. Note that terrestrial CH$_4$ lines in this spectral region show no intensity oscillations.

intensity and velocity oscillations in solar lines from the CO 2-0 and 3-1 bands as well as in atomic lines in the wavelength region. We are currently working on velocity–intensity phase studies. In the case of continuum oscillations and the spatially integrated light observations an improved observing procedure including a better normalization of the FTS spectra, an improved signal/noise ratio and better telescope tracking is needed in order to obtain a signal level comparable to that of the photometer observations. This will be attempted on the McMath/FTS in a run scheduled for June 1992. In addition simultaneous photometer observations in five wavelength bands are planned with the same telescope.

Acknowledgements

I wish to thank the Telescope Allocation Committee at the National Solar Observatory for allocating observing time to the project, Paul Hartmann, Claude Plymate and Jeremy Wagner for enthusiastic observing support, Dr. Jim Brault and Greg Ladd for valuable help, and Professor Per Maltby for helpful discussions and suggestions.

References

Andersen, B. N. and Domingo, V.: 1986, in J. Christensen-Dalsgaard and S. Frandsen (eds.), 'Advances in Helio- and Asteroseismology', *IAU Symp.* **123**, 67.
Deming, D., Glenar, D., Käufl, H. U., Hill, A. A., and Espenak, F.: 1986, *Nature* **282**, 591.
Deubner, F. L.: 1975, *Astron. Astrophys.* **44**, 371.
Hall, D. N. B.: 1973, *An Atlas of Infrared Spectra of the Solar Photosphere and of Sunspot Umbrae*, Kitt Peak National Observatory, Tucson, Arizona.
Harvey, J. W. and Duvall, T. L.: 1984, *Jet Propulsion Lab. Publ.* **84-84**, 165.
Jiménez, A., Pallé, P. L., Roca Cortés, T., Andersen, B. N., Domingo, V., Alvarez, M. and Ledezma, E.: 1988, *Seismology of the Sun and Sun-like Stars*, ESA SP-286, p. 163.
Leibacher, J. W. and Stein, R. F.: 1971, *Astrophys. J. (Letters)* **7**, 191.
Leifsen, T. and Maltby, P: 1990, *Solar Phys.* **125**, 241.
Libbrecht, K. G., Woodard, M. F., and Kaufmann, J. M.: 1990, *Astrophys. J. Suppl.* **74**, 1129.
Livingston, W., and Wallace, L.: 1991, *An Atlas of the Solar Spectrum in the Infrared from 1850 to 9000* cm^{-1} *(1.1 to 5.4 μm)*, National Solar Observatory Technical Report no. 91-001.
Nishikawa, J. and Hirayama, T.: 1986, in Y. Osaki (ed.), *Hydrodynamic and Magnetohydrodynamic Problems in the Sun and Stars*, University of Tokyo, p. 215.
Rothmann, L. S., Gamache, R. R., Tipping, R. H., Rinsland, C. P., Smith, M. A. H., Chris Benner, C., Malathy Devi, V., Flaud, J.-M., Camy-Peyret, C., Perrin, A., Goldman, A., Massie, S. T., Brown, L. R., and Toth, R. A.: 1992, *J. Quant. Spect. and Rad. Transf.*, to appear in special edition.
Ulrich, R. K.: 1970, *Astrophys. J.* **162**, 993
Woodard, M. and Hudson, H. S.: 1983, *Nature* **305**, 589.

GROUND-BASED NEAR-INFRARED OBSERVATIONS OF GLOBAL SOLAR OSCILLATIONS

L. V. DIDKOVSKY

Crimean Astrophysical Observatory, Nauchny, Crimea 334413, USSR

and

V. A. KOTOV

Crimean Astrophysical Observatory, Nauchny, Crimea 334413, USSR
and
Center for Space Science and Astrophysics, Stanford University,
Stanford, CA 94305, U.S.A.

Abstract. Near-infrared (0.7-1.0 μm) observations of solar brightness oscillations were performed in 1983-1991 at the Crimean Solar Tower with the use of two (16 × 16 and 32 × 32) photodiode arrays. In 1991 new observations of p-modes were made simultaneously in two spectral ranges, near 0.7 and 1.0 μm. The data is analysed to check for the presence of solar variation with the 160-min period. It is found that the mean relative amplitude for the 160-min solar irradiance variation at 0.73 - 1.65 μm wavelengths, $\sim 2 \times 10^{-6}$, is much lower than the upper limits set by the ACRIM and IPHIR space experiments.

Key words: infrared: stars – Sun: oscillations

1. Introduction

The most valuable information about the deep solar interior can be attained by observing long-period oscillations (g-modes) of the Sun. Yet it is widely believed that the most detailed seismic data on the Sun's interior might come entirely from Doppler measurements of p-modes. There are some indications however that acoustic frequencies do not fit the standard model of the Sun the with required accuracy (Gough and Toomre, 1991). Measurements of the solar brightness may provide valuable data to improve helioseismic inferences. Of particulart concern is the 160-min periodicity found in the Sun by several groups of observers (Kotov et al., 1983), which so far lacks a reasonable explanation.

Infrared observations (especially those made in atmospheric transparency windows) are to a large extent free from the Earth's atmosphere influences. The remaining errors connected with the atmosphere and instrument can be substantially reduced by the use of differential (center-to-limb) techniques primarily intended for detection of temporal variations of the limb-darkening profile.

The first infrared study of solar global oscillations was made in 1977-1978 by Koutchmy et al. (1980) at 1.65 μm wavelength with the use of a linear mechanical monitor of the solar brightness. As a result, the presence of 160-min oscillations with a mean differential amplitude $\sim 2 \times 10^{-5}$ was firmly established (Kotov et al., 1983). Later this IR monitor was replaced by modern devices based on the use of two photodiode arrays (PA; 16 × 16 and 32 × 32 pixels). The spectral windows of the new instruments are determined by the spectral sensitivities of the PA's and glass filters.

2. Observations

The PA-1 (16 × 16), 3.83 × 3.85 mm in size, is illuminated by a 5.9 mm pin-hole image of the Sun. The brightness signals were collected from 12 pixels corresponding to the N, S, E, and W limbs at heliocentric distances r $\sim$ 0.64 and to the solar disk center (C) (Didkovsky and Kotov, 1986), with spatial resolution $\sim$ $0.18D_\odot$ and effective wavelength 0.82 ± 0.13 μm. The high S/N was ensured by a low ratio, $\sim 2 \times 10^{-5}$, of thermal electrons to to photoelectrically gennerated ones. The registration frequency, $\sim$ 500 Hz, makes it possible to obtain brightness readings from all 5 areas monitored on the Sun's disk every 24 ms; the data were integrated for 1-min intervals and stored.

The analysis consisted of calculation of the differential signals for 4 limb regions on the disk (i = N, S, E or W): $d_i = (I_c - I_i)/I_c)$, where I_c, I_i are the mean intensities of the center-disk and limb regions. The averages of the four values of d_i were also computed to get "average limb intensity" signal. Due to low spatial resolution this instrument was devoted exclusively to measurements of low-degree global oscillations ($l \lesssim 5$).

Another, more powerful, device, using a 32 × 32 PA and a lens to form a solar image on it, was installed at the Solar Tower in 1986 (Didkovsky and Kotov, 1988). The PA is illuminated by a 9-mm image of the Sun and allows us, under $\sim 0.027 D_\odot$ resolution, to study oscillations of low and intermediate degree ($l \lesssim 40$). The spectral passbands correspond to 0.73 ± 0.10 and 0.98 ± 0.15 μm, with $\sim$ 38 seconds integration time. This new instrument records brightness signals from all 1024 pixels and permits an accurate orientation of the PA with respect to solar equator and, accordingly, permits us to obtain the two-dimensional brightness pattern of the photosphere. Thorough tests have been made to determine the degree of array homogeneity and slow temporal trends of pixel sensitivity. The spatial response function has been calculated for both arrays. The slow daily trends, caused primarily by the Earth's atmospheric transparency variations, have been removed by 2nd order polynomials fits.

3. Results

An example of the power spectrum (PS) of residuals (observations minus trend) is shown in Figure 1. It clearly shows the presence of major peaks corresponding to solar p-modes within the range 2.5 to 3.4 mHz, with average relative amplitudes $\sim$ 4×10^{-4}. The most interesting point is a good correlation between the frequencies of dominant peaks in Figure 1 with those inferred by Woodard and Hudson (1983) for p-modes of low degree from bolometric measurements on board the SMM satellite. The mean spacing we find between major peaks in Figure 2 corresponds to 68 μm, exactly that of Woodard and Hudson (1983), supporting our interpretation that the signal analyzed is of pure soalr origin.

The use of PA-2 (32 × 32) has significantly extended our capability to study space-time characteristics of solar oscillations. After removing 10-min running-mean values, the residuals (observations minus "quiet Sun"; see Didkovsky and Kotov, 1988) exhibit 5-min oscillations of solar brightness (Fig. 2).

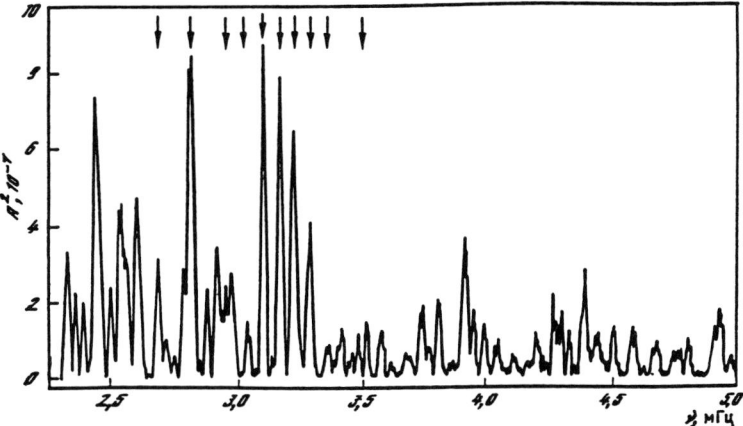

Fig. 1. Power spectrum of the brightness signal I_c (0.82 µm, August 6, 1983). The arrows show 10 frequencies of degree $l = 0$-2 inferred from bolometric ACRIM measurements.

In 1991 we observed solar brightness oscillations simultaneously in two spectral windows, 0.73 ± 0.04 and 1.00 ± 0.10 µm, with the use of the PA-2 device and two sets of glass filters covering separately N and S hemispheres of the Sun. The analysis of the observations carried out on 18 and 19 July showed appreciable differences between the power spectra computed (with a spatial filter for the sectorial harmonic degree $l = 8$, $m = 0$) for the N (0.73 µm) and S (1.00 µm) brightness oscillations, with concentration of power near 3.4 and 3.1 mHz respectively (Fig. 3). This study is now in progress, with the main goal of searching for the asymmetry in global oscillations over the solar disk and for plausible differences between the two spectral ranges.

The photodiode-array data obtained in 1983-1988 at different wavelengths (from 0.7 to 1.0 µm) allow us to check the authenticity of the phase-coherent oscillation with a period of $P_0 = 160.0101$ min (Brookes et al., 1976; Severny et al., 1976). Earlier IR data obtained by Koutchmy et al., 1980 in 1977-1978 at 1.65 µm clearly showed the presence of p_0 oscillation with the mean differential amplitude $\sim 2.5 \times 10^{-5}$. This finding was later supported by the 1981-1982 data obtained also at 1.65 µm (Kotov et al., 1983). Further evidence in favour of a true solar origin of the 160-min oscillation was given by careful consideration of all conceivable potentially spurious sources of a potential 160-min periodicity (Didkovsky and Kotov, 1987). The analysis of all data obtained in 1983-1988 with both arrays within the range 0.7-1.0 µm supports the conclusion by Koutchmy et al., 1980 about phase stability, in average, of the P_0 oscillation of the Sun. The average amplitude, however, is found to be several times smaller than that in the previous years, 1977-1982, (a similar decrease of the P_0-amplitude during a few late years was reported also by Kotov et al., 1992 on the basis of velocity observations).

We now present a superposed-epoch analysis of the total set of brightness mea-

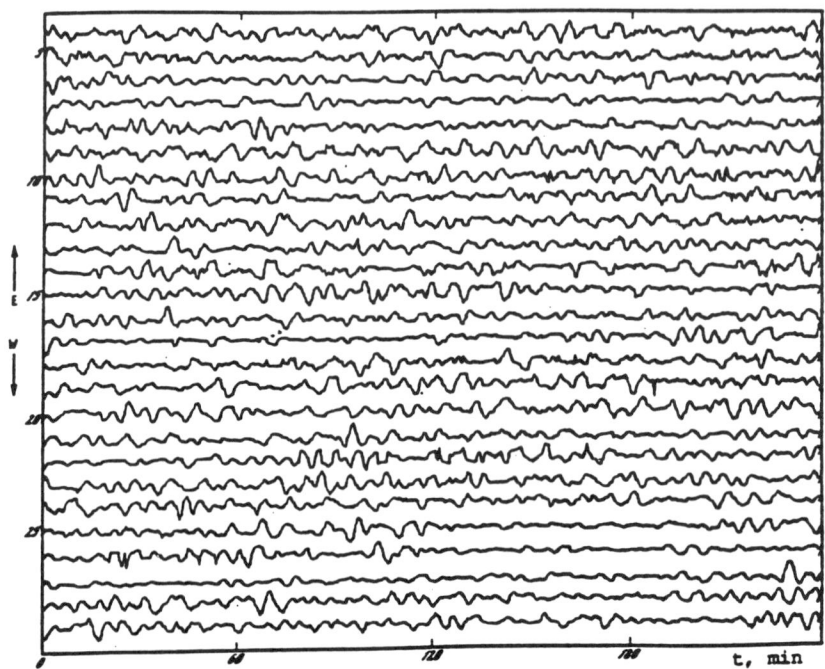

Fig. 2. Two-dimensional time-spatial distribution of 5-min brightness oscillations of the Sun as recorded by pixels 4-29 parallel to the solar equator. Spatial resolution corresponds to 53″; maximum amplitude of brightness oscillations ~ 0.25%.

surements made in the Crimea from 1976 through 1988. This data is mostly in the near IR [at 0.7–1.0 μm – with the use of two PA's, and at 1.65 μm – using the IR monitor (Koutchmy et al., 1980)]. In all there are 3102 hours of observations over 565 days (the total number N of 5-min periods is 37227). The resultant brightness curve for the folding period, P_0 is shown in Figure 4. The presence of P_0 periodicity in this 13-year set of observations appears to be highly significant (at about 4 σ confidence level). The most interesting is a perfect coincidence of the maximum of the near-IR brightness curve with the maximum of flare occurrences on the Sun. The latter curve was obtained by Kotov et al. (1992) from the analysis of $N_f = 24410$ chromospheric flares of importance $B \geq 1$ observed in 1935-1980.

The average harmonic amplitude of the differential brightness P_0 variation is ~ 1.5 × 10^{-5}. It corresponds to ~ 2 × 10^{-6} for an amplitude of solar irradiance changes in the 0.73–1.65 μm spectral range (Didkovsky and Kotov, 1987), and is therefore much smaller than the upper limits inferred from bolometric and irradiance measurements performed in 1980 and 1988 on board the SMM and PHOBOS space missions (see, e.g., Woodard and Hudson, 1983).

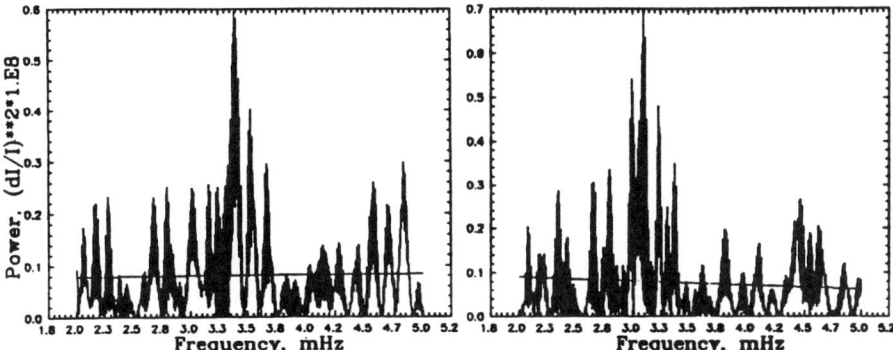

Fig. 3. The power spectra of solar brightness oscillations with spatial filter $l = 8$, $m = 0$ for the spectral windows 0.73 ± 0.10 μm (left) and 1.00 ± 0.04 μm (right).

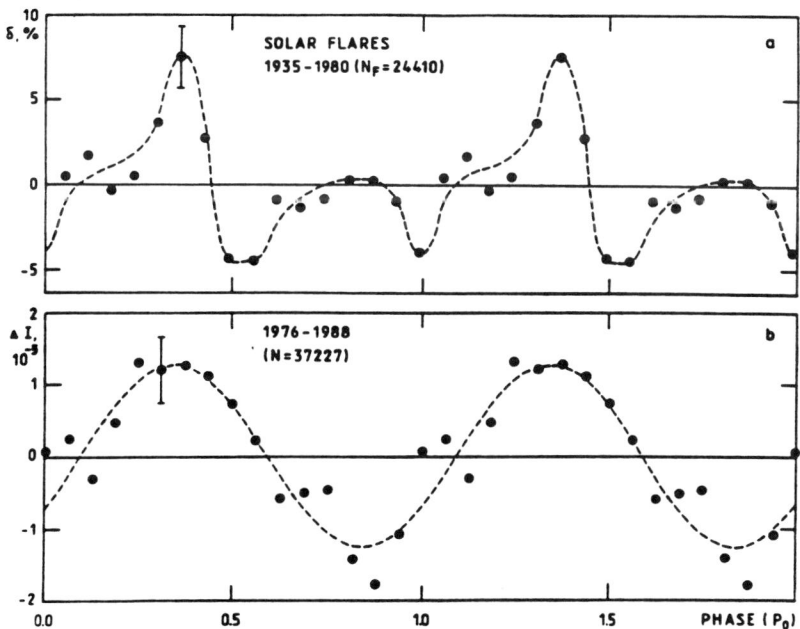

Fig. 4. The average solar brightness P_0 curve (b), compared with the average profile of frequency modulation of solar flares, 1935-1980, with a period $P_0 = 160.0101$ min, (a).

References

Brookes, J.R., Isaak, G.R., and van der Raay, H.B.: 1976, *Nature* 259, 92.
Didkovsky, L.V., and Kotov, V.A.: 1986, *Izv. Krym. Astrofiz. Obs.* **74**, 132.
Didkovsky, L.V., and Kotov, V.A.: 1987, *Izv. Krym. Astrofiz. Obs.* **76**, 119.
Didkovsky, L.V., and Kotov, V.A.: 1988, *Izv. Krym. Astrofiz. Obs.* **80**, 118.
Gough, D., and Toomre, J.: 1991, *Ann. Rev. Astron. Astrophys.* **29**, 627.
Kotov, V.A., Haneychuk, V.I., and Tsap, T.T.: 1992, *Crimean J. Astrophys.*, in press.
Kotov, V.A., Koutchmy, S., and Koutchmy, O.: 1983, *Solar Phys.* **82**, 21.
Koutchmy, S., Koutchmy, O., and Kotov, V.A.: 1980, *Astron. Astrophys.* **90**, 372.
Severny, A.B., Kotov, V.A., and Tsap, T.T.: 1976, *Nature* **259**, 87.
Woodard, M., and Hudson, H.: 1983, *Solar Phys.* **82**, 67.

SOLAR OSCILLATIONS INSTRUMENT AT AN INFRARED WAVELENGTH OF 1.6 μm AT YUNNAN OBSERVATORY

LI RUFENG, YE BINXUM, CHEN HAILIN, LIU SHAOHUA, DENG BAILIAN and MA JAGU

Yunnan Observatory, Kunming 650011, PRC

and

H. A. HILL and P. H. OGLESBY

Department of Physics, University of Arizona, Tucson, AZ 85721, U.S.A.

Abstract. A photometric solar seismograph, as part of an international network, was installed at Yunnan Observatory, the Chinese Academy of Sciences, and put into operation in the spring of 1991. This instrument is used to detect solar oscillations by measuring the continuum radiation intensity on the solar disk with a spatial resolution of 50 arcsec, in heliocentric coordinates, for research on p-modes and g-modes when $l < 50$, where l is the angular degree of the eigenfunction. The solar oscillations can be observed simultaneously at two wavelengths of the detector with a sensitivity of 10 of the average intensity. This paper reports on the optical system of the instrument. Also introduced in this paper is the compensation system for the noise signals produced by the changes in the transparency of the Earth's atmosphere. This is based on solar photometry at wavelengths of 0.55 mm and 1.6 mm.

Key words: infrared: stars – instrumentation: photometers – Sun: oscillations

1. Introduction

The 5-minute solar oscillation was first discovered by Leighton and collaborators in 1960 (Leighton, Noyes, and Simon, 1962). Ten years later, the prediction that there should be an acoustic resonant wave cavity below the solar surface was proposed by Ulrich et al. (1977) and by Leibacher and Stein (1971) to explain the 5-minute oscillations. Ulrich made an important prediction that there should be a dispersion relationship between k and l. This characteristic relation was respectively demonstrated in 1975 by Deubner (1975) while observing the Sun, and by Grec and Fossat (1983) while observing the low p-modes at the South Pole. Hill and Stebbins (1976) found a series of oscillations of the Sun's diameter with periods ranging from ten minutes to one hour. The observational method for the study of solar oscillations has been improved along with progress made in other aspects of research on solar oscillations. In order to research the structure and dynamics of the solar interior and the long-term period variation of the Sun, a cooperative observatory network has been set up by SCLERA of the University of Arizona, and Yunnan Observatory. The first solar seismograph of this network was installed at Yunnan Observatory in 1990-1991 and was put into operation in March 1991. In the near future, a network of these instruments will be set up to remedy the defect that arises when the Sun cannot be continuously observed at a single observatory. This instrument can be used to observe the Sun simultaneously at the two wavelengths of 0.55 mm and 1.6 mm, with the sensitivity of the detector being 10^{-6} of the average intensity of the Sun to monitor oscillations with $l < 50$.

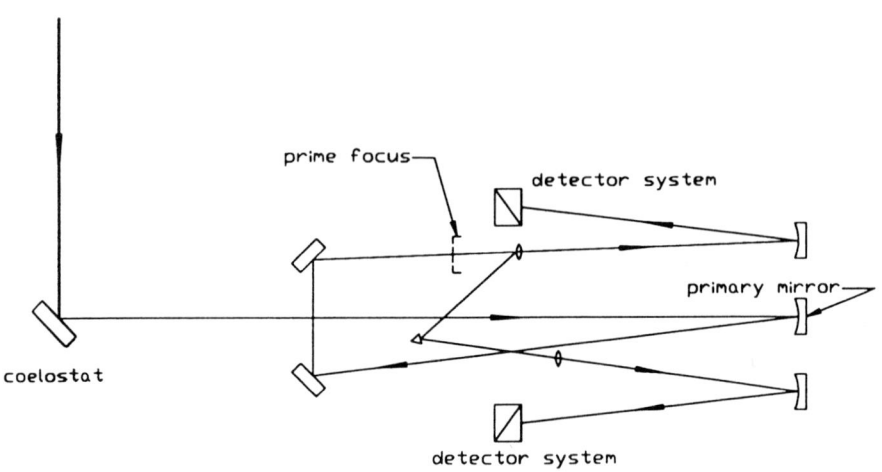

Fig. 1. The optical system

2. Optical System of the Solar Seismograph

The solar seismograph at Yunnan Observatory is used as a part of the network to observe solar oscillations by determining changes in the intensity of the continuum on the solar disk with a spatial resolution of ~ 50 arcsec in heliocentric coordinates. The optical system of the instrument consists of a coelostat, a primary mirror, and a detector system, as shown in Figure 1.

Sunlight passes through a window to the 20-cm-diameter coelostat mirror and then to the primary mirror with an aperture of 7.0 cm and a focal length of 200 cm. After the sunlight passes through the prime focus, two solar images are formed in the optical system. One of them has a diameter of 19 mm for measurements at disk center; the other has a 28 mm diameter for measurements at the solar limb. The solar images are scanned across three Ge 1×16 diode arrays at 0.55 μm, and across a second set of 3 Ge arrays at 1.6 μm. In front of each linear array is a mask which is designed to yield a spatial resolution of approximately 3° in solar latitude and longitude (viewed normally at the solar equator) per pixel. Scanning is implemented to give a coverage of $\pm 90°$ in longitude and $\pm 60°$ in latitude.

3. Data Acquisition

The operating system and data acquisition system of the telescope are controlled with a Masscomp 5500 on-line computer. The signals received by the detectors are sampled every 50 ms with a 12-bit A/D converter. Signals for a given location on the solar disc are filtered by a digital triangle filter with a width of 16 s. The output of the triangle filter is sampled every 4 s and recorded. This data-acquisition system can produce 31 Mbytes of data in 10 hours, and the data is recorded on

1/4-inch cartridge tape. The data are exchanged between Yunnan Observatory and SCLERA.

4. Compensation Role of Simultaneous Observations at Wavelengths of 0.55 and 1.6 μm.

Since the original solar signals produced by oscillations are measured through the Earth's atmosphere, the measured results include both of the solar signals and signals produced by temporal variations in the transparency of the Earth's atmosphere. Care must be taken to minimize the transparency effects due to water vapor. In this regard, there are no absorption bands of water vapor at the two wavelengths of 0.55×0.01 μm and 1.6×0.015 μm (Oglesby, 1987).

However, the variation in the solar intensity depends on wavelength and satisfies the radiation transfer equation

$$\mu \frac{dI_\lambda(\tau_\lambda, \theta, \phi)}{d\tau_\lambda} = I_\lambda(\tau_\lambda, \theta, \phi) - S(\tau_\lambda) \tag{1}$$

where $d\tau_\lambda = -\rho \kappa_\lambda \, dz$, κ_λ is the opacity at wavelength λ, ρ the density, S the source function, and μ is the cosine of the angle between the line of sight and the normal to the surface.

If solar oscillations are regarded as small perturbations, the basic equation for the Eulerian perturbation I' can be expressed as follows:

$$I'(r,t) = I(r,t) - I_0(r) \tag{2}$$

Under the linearized perturbation, equation (1) can be written as

$$\mu \frac{dI'_\lambda}{d\tau_\lambda} = I' - S' + \frac{(\rho \kappa)'}{\rho \kappa}(I - S) \tag{3}$$

where the superscript prime denotes the Eulerian perturbation. The solution of this equation is:

$$I'_\lambda = \int_0^\infty [S' - \frac{(\rho \kappa)'}{\rho \kappa}(I - S)] e^{-\tau_\lambda/\mu} \, d\tau_\lambda/\mu \tag{4}$$

The properties of I'/I were studied by Hill and Rosenwald (1986) of SCLERA, with the following results:

$$\frac{I'_{0.55}}{I} \approx 7.4 \times 10^{-5} \tag{5}$$

and

$$\frac{I'_{1.6}}{I} \approx -5.5 \times 10^{-5}. \tag{6}$$

The characteristics of I' in the above results at the two wavelengths are opposite in sign to each other, and in particular are different from the wavelength dependence of changes in I due to transparency changes in the Earth's atmosphere. Therefore, this kind of characteristic can be used to discriminate true solar variations from changes in transparency in the Earth's atmosphere.

References

Deubner, F. L.: 1975, *Astron. Astrophys.* **44**, 371.
Grec, G., Fossat, E., and Pomerants, M. A.: 1983, *Solar Phys.* **82**, 55.
Hill, H. A., and Rosenwald, R.: 1986, *NATO Advanced Workshop, Logan, Utah Proceedings*.
Hill, H. A., Stebbins, R. T., and Brown, T. M.: in J. H. Sanders and A. H. Wapstra (eds.), *Atomic Masses and Fundamental Constants*, **5**, Plenum, New York, p. 622.
Leibacher, J. W., and Stein, R. F.: 1971, *Astrophys. J. (Letters)* **7**, 191.
Leighton, R. B., Noyes, R. W., and Simon, G. W.: 1962, *Astrophys. J.* **135**, 474.
Oglesby, P. H.: 1987, Ph.D. dissertation, University of Arizona.
Ulrich, R. K., and Rhodes, E. J.: 1977, *Astrophys. J.* **218**, 521.

MAGNETIC FIELDS, OSCILLATIONS, AND HEATING IN THE QUIET SUN TEMPERATURE MINIMUM REGION FROM ULTRAVIOLET OBSERVATIONS AT 1600 Å

J. W. COOK

*E.O. Hulburt Center for Space Research, Naval Research Laboratory,
Washington, DC 20375, U.S.A.*

Abstract. The High Resolution Telescope and Spectrograph (HRTS) instrument has obtained broadband spectroheliograph images at 1600 Å of the solar temperature minimum region. I discuss HRTS observations of quiet areas and their relation with magnetic fields, five minute oscillations, and heating. The brightness temperature of solar fine structure elements composing the supergranular network is found to be linearly proportional to the local absolute value of magnetic field strength. There is evidence for a 250-s period oscillation occurring in 10-arcsec scale patches, which however is energetically unimportant to the local heating budget. A general nonmagnetic background heating and five minute oscillations occur globally, while the network bright points occur in magnetic regions, heated perhaps from partial dissipation of Alfvén waves (whose energy flux is linearly proportional to B) in individual elemental 1500-G (at the photosphere) flux tubes which expand to form the temperature minimum fine structure bright points.

Key words: flux tubes – infrared:stars – Sun: magnetic fields – Sun: UV radiation

1. Introduction

The temperature minimum region of the solar atmosphere can be observed at infrared wavelengths both in the continuum around 350 μm, and in absorption lines such as the vibration-rotation lines of CO, for example the 3-2 R(14) line at 4.67 μm. This region can also be observed in the far ultraviolet, and observations as far back as 1978 from sounding rocket flights have been made with arcsecond spatial resolution. A comparison with such observations can lead to a fuller understanding of the temperature minimum.

The High Resolution Telescope and Spectrograph (HRTS) instrument consists of a 30-cm diameter telescope, a broadband spectroheliograph, a slit spectrograph which can cover a wavelength range from 1175–1710 Å with 0.05 Å spectral resolution, and an Hα system which can both display images using a TV camera and record them on film. Both the spectroheliograph and Hα systems use reflected images from the spectrograph slit jaw mirrors. Slit spectra and spectroheliograph images are recorded on film. The spatial resolution, while potentially an arcsecond or better, has typically been 1–2 arcsec.

The spectroheliograph has been tuned on a number of sounding rocket flights to cover a passband centered on 1600 Å, where the predominant flux contributor is continuum emission from the temperature minimum region (approximately 70% of the integrated intensity over the passband in disk quiet regions; the remainder is chromospheric line emission). I discuss HRTS observations from quiet areas of bright point fine structure, its relation to magnetic fields seen in photospheric magnetograms, the evidence for five minute oscillations, and estimates of required heating. Where available, I compare results from the far ultraviolet observations with work in the infrared, noting both similarities and also several puzzling differences.

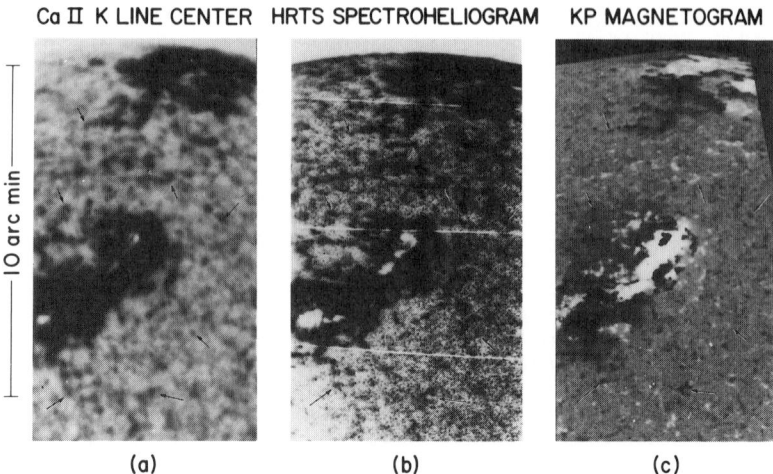

Fig. 1. Comparison of Sac Peak Ca II image, HRTS 1600 Å spectroheliogram, and Kitt Peak magnetogram from the day of the HRTS II flight.

2. Structure of Quiet Regions

The central problem of the physics of the solar atmosphere is to understand the physical processes which reverse the temperature decline in the upper photosphere, as consistent with radiative equilibrium, and instead lead to the temperature rise from the temperature minimum at 4400 K, through the chromosphere and transition region, up to the several times 10^6 K corona. This input of additional energy is clearly connected with the magnetic field, either as a channeling agency, such as for magneto-acoustic waves, or more directly through input of energy from changes in the magnetic field such as reconnection.

The observed structure of the solar photosphere is a granular pattern with individual granules of order 1–2 arcsec in size which rise and fall vertically, have sideways motions of order 1 km s^{-1}, and last around 8 minutes. But by the temperature minimum, where the atmosphere already departs from a radiative equilibrium model and additional heating is occuring (Anderson and Athay 1990), the observed structure in quiet areas is the supergranular network.

Small network elements composed of clumped arcsec scale bright points occur at edges of the supergranulation, coincident with stronger field regions in photospheric magnetograms, while the cell centers are filled with the order of 20 individual bright points evolving on a 1 minute timescale (Cook, Brueckner, and Bartoe 1983). Figure 1 illustrates a Ca II image, HRTS 1600 Å spectroheliogram, and photospheric magnetogram from the day of the HRTS II flight on 13 February 1978, showing the similarity of structure in Ca II and the 1600 Å image, and the relationship of bright areas (dark in these negative images) with strong field regions in the magnetogram (Cook, Brueckner, and Bartoe 1983). In comparison, a new, unpublished image at 350 μm obtained by the JCMT was shown at this meeting, with a spatial resolution of approximately 10 arcsec, which only begins to suggest the network element supergranular structure of quiet areas (Lindsey 1993).

This supergranular network structure of quiet regions can be observed throughout the chromosphere and into the transition region, with a good degree of correspondence in location of network elements in images from a range of temperatures. This basic network structure persists up to at least 500,000 K, as can be seen in Skylab data from the NRL SO82A instrument (Sheeley et al. 1975; Cook 1991). Images in Ne VII, formed at 500,000 K, show that this temperature represents something of a transitional range for solar morphology. The quiet disk is still organized in the network pattern, but loop structures are beginning to appear in active regions. At temperatures of 10^6 K (seen in Mg IX) and higher (Fe XV), the disk and limb show large complete loop systems, typically connecting separated active regions, and the supergranular pattern has disappeared. The basic quiet Sun structure up to at least 500,000 K is closely related to structures of lower temperature plasmas, instead of to the corona.

3. Relationship of Brightness and Magnetic Flux

Using HRTS 1600 Å spectroheliograms and Kitt Peak magnetograms, Cook and Ewing (1990) examined the quantitative relationship of brightness temperature at 1600 Å to photospheric magnetic field strength in the quiet Sun. We used a technique which obtained the best-fit relationship of a given functional form between two histogram distributions, in our case for brightness temperature at 1600 Å and for the distribution of absolute magnetic flux, in a 486 × 452 arcsec^2 sample quiet area from the HRTS V flight on 11 December 1987.

Figure 2 illustrates the distribution of brightness temperature at 1600 Å which we found for our sample quiet area (the solid and dashed lines were obtained from two different exposures of the same sample area). The functional form used for mapping one distribution to the other is very general, and was originally meant to reproduce the S-curve shape of film characteristic curves in photometric photometry (see Cook, Ewing, and Sutton 1988). The observations alone determined the fit, which was not imposed at the start. In Figure 3 we show the functional relationship which we found (the solid and dashed lines from two different 1600 Å exposures of the same sample area are again in good agreement), which is essentially linear for field strengths above the ~7 G noise level of the Kitt Peak magnetograms. We found that even magnetic bipoles, unlike in the corona, are no brighter than implied by their absolute flux magnitude.

We estimated the additional energy flux necessary to maintain the network bright points by integrating the LTE approximation $\Delta E = 16\sigma T_0^3 \Delta T \tau$, where ΔT is the observed brightness temperature excess in network bright points above an average temperature T_0, and τ is the optical depth (at 5000 Å) at the temperature minimum level, over the observed distribution of brightness temperature above 4600 K, which represents the bright point population. This average additional energy flux, 7.5×10^6 ergs cm^{-2} s^{-1}, is directly comparable to the flux required to heat the normal average global background at the $T_{\min}$ level, 8×10^6 ergs cm^{-2} s^{-1}, from the prescribed heating models of Anderson and Athay (1990).

This linear result is consistent with heating by Alfvén waves, whose flux $F_A =$

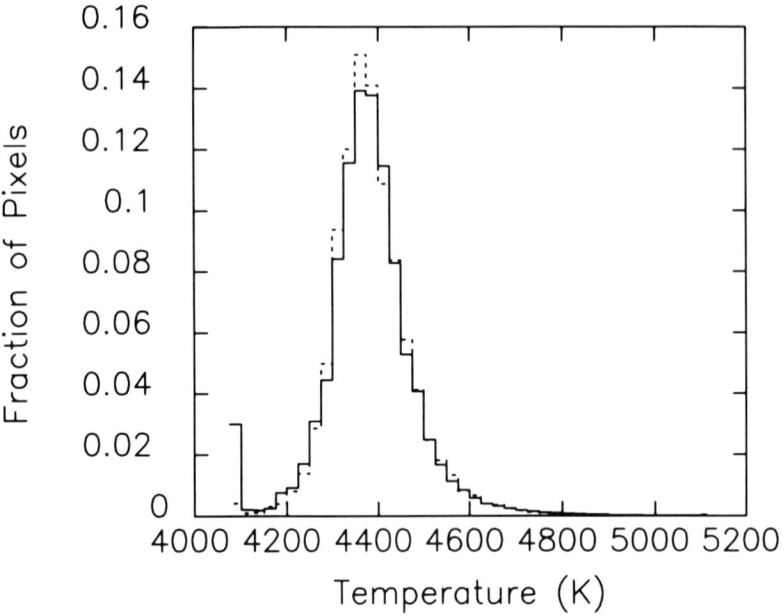

Fig. 2. Histogram distribution of brightness temperature in a sample quiet area.

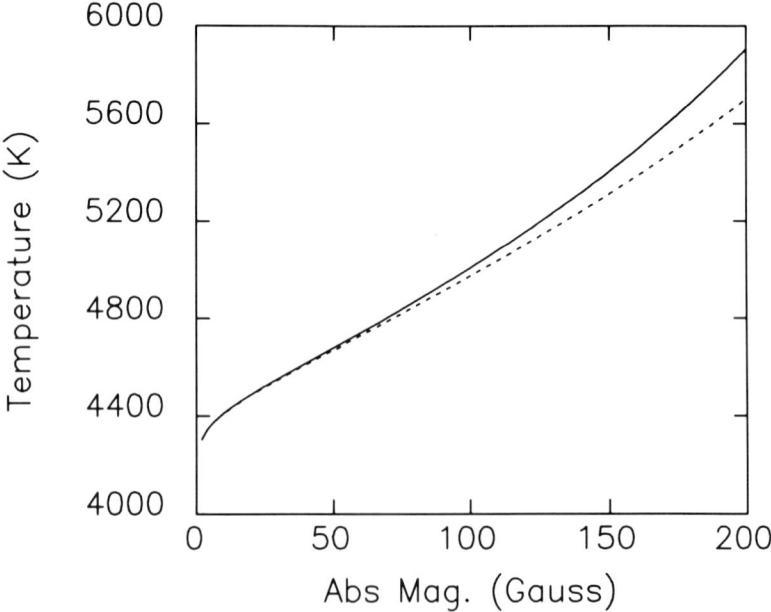

Fig. 3. Relationship of 1600 Å brightness temperature vs. absolute magnetic field strength in a sample quiet area.

$\rho v^2 V_A$, where $V_A = B/(4\pi\rho)^{1/2}$ is the Alfvén velocity, is linear in B. We estimated the potential flux at photospheric levels produced from granular buffeting in magnetized areas. Using $v \sim 1$ km s^{-1} as the typical horizontal granular velocity (Title et al. 1989) and $\rho = 2.7 \times 10^{-7}$ g cm^{-3} as the density at the $\tau = 1$ level of the photosphere (Vernazza, Avrett, and Loeser 1981), there is 10 times the flux required for heating the $T_{\min}$ network element bright points potentially available in Alfvén waves generated from photospheric granular buffeting. The problem still remains, however, to find a viable dissipation mechanism for the $T_{\min}$ region.

Our comparison with the energy available in Alfvén waves was suggested by the observed linear relationship of excess heating to magnetic flux, but we did not mean to rule out other mechanisms which could yield such a result.

Cook and Ewing (1990) suggested that the individual network bright points of ~ 1 arcsec diameter seen at 1600 Å, associated with typical Kitt Peak observed field strengths of 50 G obtained for 2 arcsec seeing, were consistent with individual flux tubes with true, unresolved photospheric field strengths of around 1800 G and diameters of $\sim 1/3$ arcsec. They used a simple scaling law for the diameter d of a flux tube based on conservation of flux and equipartition of magnetic pressure and gas pressure p, $(d_2/d_1) = (p_1/p_2)^{1/4}$. The VAL model C atmosphere (Vernazza, Avrett, and Loeser 1981) gives the ratio of gas pressures at $\tau = 1$ and the $T_{\min}$ as 88, and so $d(T_{\min}) \sim 3.1 d(photosphere)$. A field of 50 G at 2 arcsec resolution then corresponds to $50 \times 36 = 1800$ G at 1/3 arcsec. Magnetograms from the Lockheed SOUP instrument (Title et al. 1990) have reached 0.5 arcsec spatial resolution, and found that field strengths associated with Ni I bright points are of the order of 600 G. Spatial resolution below 0.5 arcsec should show even higher field strengths for elementary photospheric flux tubes. The $T_{\min}$ is the highest level where network and active region features still keep a resolvable bright point structure. At greater heights the decreasing gas pressure and expanding fields cause the pointlike network features to spread out to more continuous patches of emission.

4. Brightness Oscillations at 1600 Å

Cook and Ewing (1991) also examined the brightness variations occuring in HRTS V 1600 Å data in a series of 56 images over 330 s of the sounding rocket flight. We found variations in small 10 arcsec patches in cell centers, which were immediately reminiscent of the five minute oscillations in the photosphere, and the oscillations observed by Noyes and Hall (1972) in the infrared line of CO at 4.67 μm which is formed in the $T_{\min}$. I show in Figure 4 a sample result obtained by fitting the brightness temperature variations with time in individual 5 or 10 arcsec boxes filling a 350×350 arcsec^2 field from our time series of 1600 Å images. We used sine functions with individual average temperatures, amplitudes, periods, and phases for each box.

With only 330 s temporal coverage from our sounding rocket observations, a Fourier analysis ($\omega - \kappa$ diagram) was inappropriate. Instead we looked at the histogram distributions of the periods, amplitudes, etc., found for the individual 5 or 10 arcsec boxes into which we divided our field. Figures 5 and 6 show the distribution of the periods and the amplitudes found from our sine fits to the variation

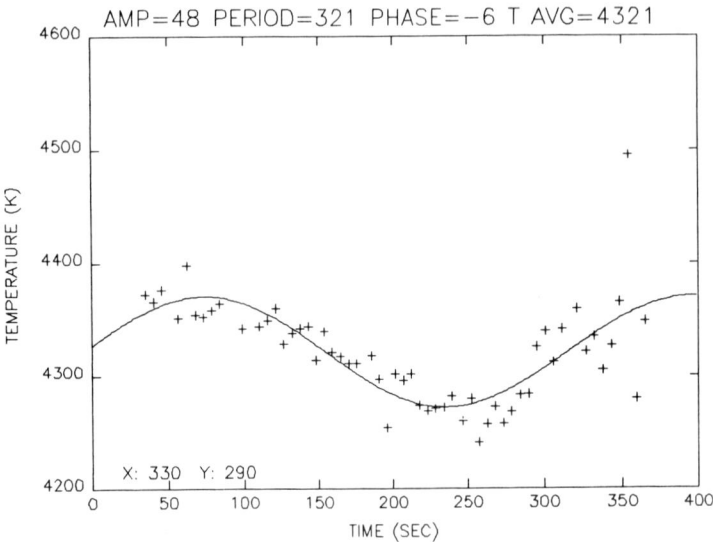

Fig. 4. Sine wave fit to brightness temperature variation for an individual 10 arcsec square box. The scatter to the fit illustrates the noise level of the data.

of brightness temperature within individual boxes. The periods show a broad peak running from 150 to 400 s, centered at 250 s. A range of brightness temperature amplitudes was found, with the average of the 5 arcsec box distribution at 50 K.

These results from our 1600 Å observations can be compared with T_{min} observations in the infrared, and also with Ca II K line observations. Observations of CO infrared absorption lines by Noyes and Hall (1972) and Ayres and Brault (1990) give periods near 250–300 s for oscillations, but significantly greater values of 100–150 K for the amplitudes. Observations by Lindsey and Roellig (1987) at 350 μm also give periods near 300 s; their amplitudes are not really directly comparable because of their large field of view (beam size). In contrast, observations using the Ca II K line (for example Dame, Gouttebroze, and Malherbe 1984) typically give periods in the three minute range in cell centers, which is usually interpreted as a chromospheric period. It is unclear how to directly reconcile the 1600 Å, infrared, and Ca II results into a single picture. The resolution of these differences may come from a model for the temperature minimum and low chromosphere which is highly structured and possibly dynamic.

We estimated the average energy flux $(1/2)\rho v^2 c_s$, where the sound speed $c_s = (\gamma kT/\mu m_H)^{1/2}$, for a simple propagating, adiabatic, undamped wave (however, it is not clear that these waves are actually propagating!). Other regions of the atmosphere depart from adiabaticity, but in the T_{min} the radiative relaxation time for radiative damping of compressional waves is usually estimated to be long compared with the compression time of an acoustic wave; in addition γ should be close to the monatomic ideal value of 5/3 because of the low degree of ionization. In this case

QUIET SUN TEMPERATURE MINIMUM REGION   293

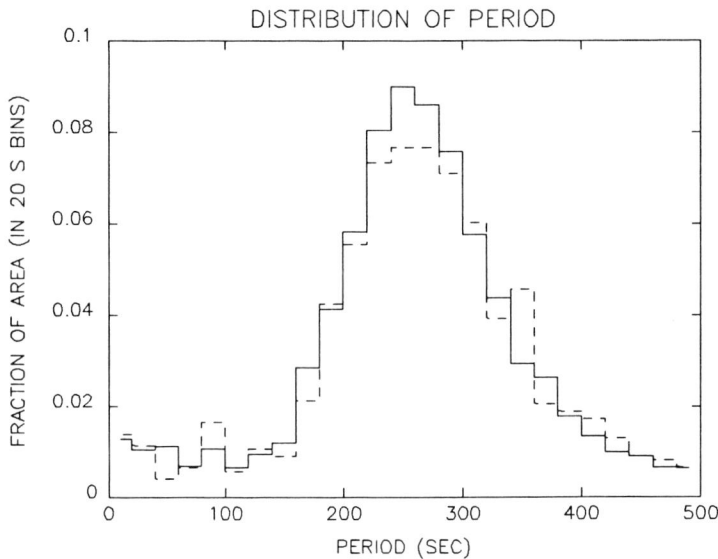

Fig. 5. Histogram distribution of period for 5 arcsec boxes (*solid line*) and for 10 arcsec boxes (*dashed line*).

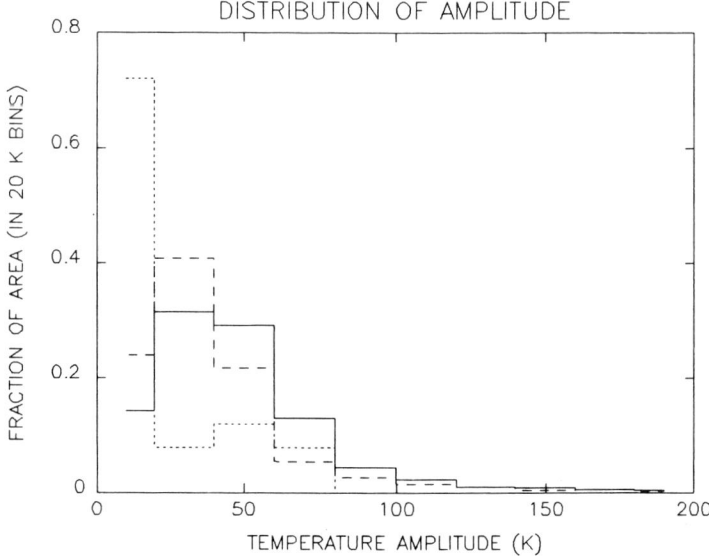

Fig. 6. Histogram distribution of amplitude for 5 arcsec boxes (*solid line*), 10 arcsec boxes (*dashed line*), and 70 arcsec boxes (*fine dashed line*).

$(v/c_s) = (\gamma - 1)^{-1}(\Delta T/T)$, and using $\Delta T = 50$ K gives an acoustic flux of 1.8×10^5 ergs cm^{-2} s^{-1}. This flux is small compared to an average global requirement of 8×10^6 ergs cm^{-2} s^{-1} to heat the $T_{\min}$ level (Anderson and Athay 1990), and so the 250-s oscillations are energetically unimportant in the local heating budget.

To summarize these results, we suggested a picture for the quiet Sun in which a general nonmagnetic background heating and 250 s oscillations occur globally, while additional heating (perhaps from partial dissipation of the flux in Alfvén waves produced from photospheric granular buffeting of individual flux tubes) produces the network bright points. These bright points may be individual elemental ∼1800 G (at the photosphere) flux tubes which expand to arcsec diameter with decreasing gas pressure by the temperature minimum level. The average additional energy flux required to maintain the network bright points, representing only around 2% of the total surface area, is directly comparable to the flux required to heat the normal global average background at the temperature minimum level.

The relationship of brightness at 1600 Å and magnetic flux in the photosphere must eventually flatten out above 50 G, as finally active region field strengths are reached. Already the brightest few elements in quiet areas are as bright as any features seen at 1600 Å except for flares, and active regions simply contain a higher fractional area of these brighter elements. When field strengths characteristic of sunspots are reached, a presumably completely different physical mechanism takes over to actually suppress heating.

Acknowledgements

This work was supported by NASA under DPR W-14,541 and by the Office of Naval Research.

References

Anderson, L.S., and Athay, R.G.: 1990, *Astrophys. J.* **346**, 1010.
Ayres, T.R., and Brault, J.W.: 1990, *Astrophys. J.* **363**, 705.
Cook, J.W.: 1991, in P. Ulmschneider, E.R. Priest, and R. Rosner (eds.), *Mechanisms of Chromospheric and Coronal Heating*, Springer, Berlin, p. 83.
Cook, J.W., Brueckner, G.E., and Bartoe, J.-D.F.: 1983, *Astrophys. J. (Letters)* **270**, L89.
Cook, J.W., and Ewing, J.A.: 1990, *Astrophys. J.* **355**, 719.
Cook, J.W., and Ewing, J.A.: 1991, *Astrophys. J.* **371**, 804.
Cook, J.W., Ewing, J.A., and Sutton, C.S.: 1988, *Pub. A.S.P.* **100**, 402.
Dame, L., Gouttebroze, P., and Malherbe, J.-M.: 1984, *Astron. Astrophys.* **130**, 331.
Lindsey, C.: 1993, these proceedings and private communication.
Lindsey, C., and Roellig, T.: 1987, *Astrophys. J.* **313**, 877.
Noyes, R.W., and Hall, D.N.: 1972, *Astrophys. J. (Letters)* **176**, L89.
Sheeley, Jr., N.R., Bohlin, J.D., Brueckner, G.E., Purcell, J.D., Scherrer, V., and Tousey, R.: 1975, *Solar Phys.* **40**, 103.
Title, A., Tarbell, T., Topka, K., Cauffman, D., Balke, C., and Scharmer, G.: 1990, in C.T. Russell, E.R. Priest, and L.C. Lee (eds.), *Physics of Magnetic Flux Ropes*, American Geophysical Union, Washington, p. 171.
Title, A.M., Tarbell, T.D., Topka, K.P., Ferguson, S.H., Shine, R.A., and the SOUP Team: 1989, *Astrophys. J.* **336**, 475.
Vernazza, J.E., Avrett, E.H., and Loeser, R.: 1981, *Astrophys. J. Suppl.* **45**, 635.

PART 4

INFRARED ATOMIC PHYSICS AND LINE FORMATION

ATOMIC PHYSICS OF THE 12-μm AND RELATED LINES

EDWARD S. CHANG

Department of Physics and Astronomy, University of Massachusetts,
Amherst, MA 01003, U.S.A.

Abstract. The 12 μm emission lines were unexpectedly detected about a decade ago. Great progress has been made in understanding the atomic physics underlying these high-l Rydberg transitions in Mg I and other atoms. In a magnetic field, their Landé g factor is shown to be unity. At disk center, the shift of the absorption trough relative to the emission peak is demonstrated to be due to the quadratic Stark Effect, permitting measurement of the photospheric electric field strengths. Other related lines of Mg I require accurate atomic fine structure data to interpret properly their complex line profiles. Related lines are found in the *ATMOS* spectra for C I, Na I, Al I, Si I, Ca I, and Fe I, in addition to H I.

Key words: atomic processes – electric fields – infrared: stars – magnetic fields

1. Introduction

The solar 12 μm emission lines were apparently first recorded in 1976 by Goldman *et al.* (1980) in the New Atlas of IR Solar Spectra. However the lines were so startling that they were handmasked in the Atlas, being suspected to be instrumental artifacts. Subsequent observations with a higher resolution instrument at the South Pole, and in a balloon flight, established the authenticity of two 12 μm lines at 811.575 and 818.058 cm^{-1} and one 7 μm line at 1356.182 cm^{-1} (Murcray *et al.*, 1981).

Detailed studies of the 12 μm lines were undertaken by Brault and Noyes (1983) at the McMath telescope. Among the many puzzling results were the strong limb brightening, the shift of the absorption trough relative to the emission peak, the Zeeman splitting patterns and the identity of these and of 41 other weaker emission lines. In 1989 the ground-based observations were augmented by the satellite *ATMOS* spectra, which covered the entire 2 to 16 μm range at disk center (Norton and Farmer 1989, Kurucz 1990), adding many more unidentified lines. Most of the stronger lines turned out to be high-l Rydberg lines. The goal of this article is to describe the atomic physics of these lines and to show how to extract from these lines information on the physical properties of the sun.

The 12 μm lines were identified as high-l Rydberg transitions in Mg I, Al I, and Si I (Chang and Noyes,1983, Chang, 1984), using theory to be reviewed in Section 2. Laboratory verification of the Mg I lines was accomplished by Lemoine, Demuynck, and Destombes (1988). Further they verified that the Landé factor g for the two 12 μm lines was unity, as predicted. Magnetic properties of these lines will be explored in Section 3. Deming and collaborators have cleverly exploited the 12 μm lines for vector magnetometry, and he will present his results in these proceedings. Recall that the Mg I 12 μm lines display distinct trough-to-peak shifts. Examination of the *ATMOS* spectra revealed other lines, *e.g.*, at 11 μm, with greater shifts – often in the opposite sense. These shifts were interpreted (Chang and Schoenfeld 1991) as quadratic Stark shifts. In Section 4, the electric properties of these lines will be elucidated. The modeling and formation of these lines in the solar atmosphere have been successfully explained by Chang *et al.* (1991) and by Carlsson *et al.*

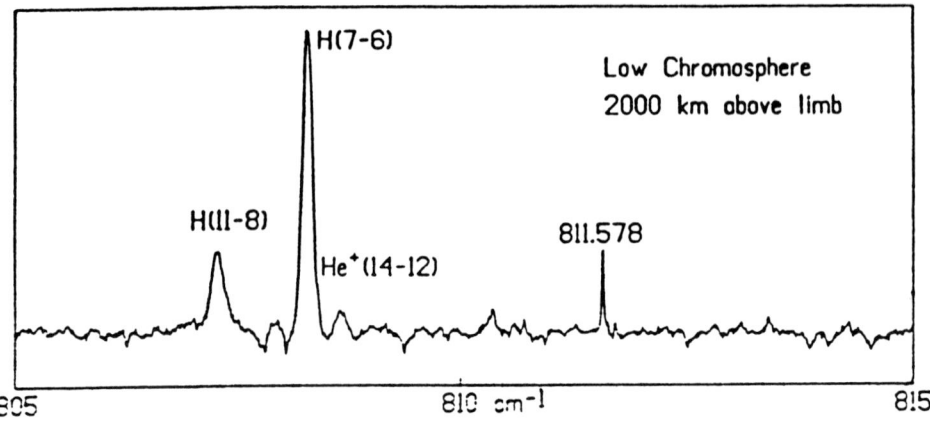

Fig. 1. The 12 μm and the hydrogen lines from Brault and Noyes (1983).

(1992), and are presented separately by Avrett, and by Rutten and Carlsson, in these proceedings.

In addition to the Mg I 12 μm lines, other high-l Rydberg lines are observed at 7 and 11 μm, which also display emission peaks on absorption troughs. Other Mg I lines with intermediate l and resolvable fine structure have been addressed by Jefferies (1991) as perplexing. All these constitute the related lines of Mg I to be discussed in Section 5. Section 6 delves into the similar lines in other elements. In addition to the fore-mentioned lines of Al I and Si I, newly discovered lines in Fe I are the subject of two poster papers (Johansson et al. and Schoenfeld et al., these proceedings). Further lines of C I, Na I, and Ca I have also been identified, but these assignments remain unpublished. In Section 7, I present the conclusions, and prospects for the future.

2. High-l Rydberg Atomic Lines

The most important clue to identifying the 12 μm lines is that they lie just on the higher frequency side of the hydrogen $n = 7$ to 6 line at 808.283 cm^{-1} as shown in Figure 1. This immediately suggests that they belong to the 7-6 transition of some other atom, where the levels have very small quantum defects. Such a requirement is automatically met by high-l Rydberg states (where l is the orbital, and n the principal, quantum numbers). For then the large centrifugal barrier prevents the Rydberg electron from penetrating the atomic core, which gives rise to the usual quantum defect from short-range interactions. Thus the only contribution comes from the weak long-range interaction which may be regarded as a perturbation on the pure Coulomb potential, responsible for the hydrogen levels. The Rydberg electron's electric field E induces a dipole moment in the atomic core $\mu = \alpha E$, where α stands for the core's polarizability. The resultant perturbation energy is $-\mu E$, which is approximately proportional to $-\alpha n^{-3} l^{-5}$ (Chang and Noyes 1983). Since the lower level is perturbed more than the upper, the transition frequency must exceed that of the corresponding hydrogen transition. However, unlike hydrogen, each n manifold splits into several distinct levels designated by l, with decreasing energy as l decreases. The selection rule limits allowed transitions to $\Delta l = 1$ and -1, but the former are typically 2 orders of magnitude weaker than the latter, and hence usually unobservable. Recall that the number of radial nodes in the hydrogenic wavefunction is $n_r = n - l - 1$. In the $\Delta l = -1$ case, n_r is very different for the initial and the final states, resulting in large cancellations in the transition probability. On the other hand, n_r is identical for the Δl (and Δn) $= -1$ transitions, resulting in little cancellation. Indeed for the circular orbits ($l = n - 1$), the nodeless wavefunctions suffer no cancellations at all. Then the transition probability is proportional to n^4, which is 3 orders of magnitude larger than the typical value for the resonance lines. For the same values of n but successively lower values of l, the transition probabilities are weaker by about a factor of two.

These considerations led Chang and Noyes (1983) to identify the 811 and the 818 cm^{-1} lines as the $7i$–$6h$ and the $7h$–$6g$ transitions in a neutral atom. The identification was suggested by the large separation which required a value for α two orders of magnitude greater than for helium. A good candidate was the alkali-like core which, with its lone loose electron, was known for its large polarizability. The first abundant element with such a core is magnesium. Using its known value of α, we calculated those transition frequencies and found that they agreed well with the observed ones. Additional lines of Mg I were also identified from theory for $l > 3$, and from experiment for lower l levels. Improved theoretical values were given later (Chang 1987).

Laboratory confirmation of the high-l Rydberg lines also came about serendipitously. In measuring the diode-laser spectra of MgH and MgD, Lemoine et al. (1988) noted the presence of some atomic Mg I lines. Further studies of the magnesium discharge under high resolution revealed some 30 absorption lines, including the two 12 μm lines. Comparison of laboratory, solar, and theoretical line positions for the high-l Rydberg lines are given in Table 1. Note that the excellent agreement between the positions of the emission peaks and the laboratory data implies that the peaks are essentially unshifted, while the absorption troughs have substantial shifts as will be interpreted in Section 4.

The identification of the emission lines as high-l Rydberg transitions has enormous implications for solar physics. Since these levels are hydrogen-like, their atomic properties are mostly accurately known. This fact greatly facilitates radiative transfer calculations in model atmospheres (Chang et al., 1991; see article by Avrett et al., these proceedings; Carlsson et al., 1992; see article by Rutten and Carlsson, these proceedings). As alluded to earlier, the large opacity of the 12 μm lines is responsible for the greater height of formation, where a small departure from LTE is possible. Thus the rapidly rising line source function in the upper photosphere is responsible for the emission peak and limb brightening. In fact the chromospheric temperature rise has nothing whatsoever to do with the emission lines. Other hydrogen-like properties of these lines will be exploited to measure the magnetic and the electric fields in the solar photosphere next.

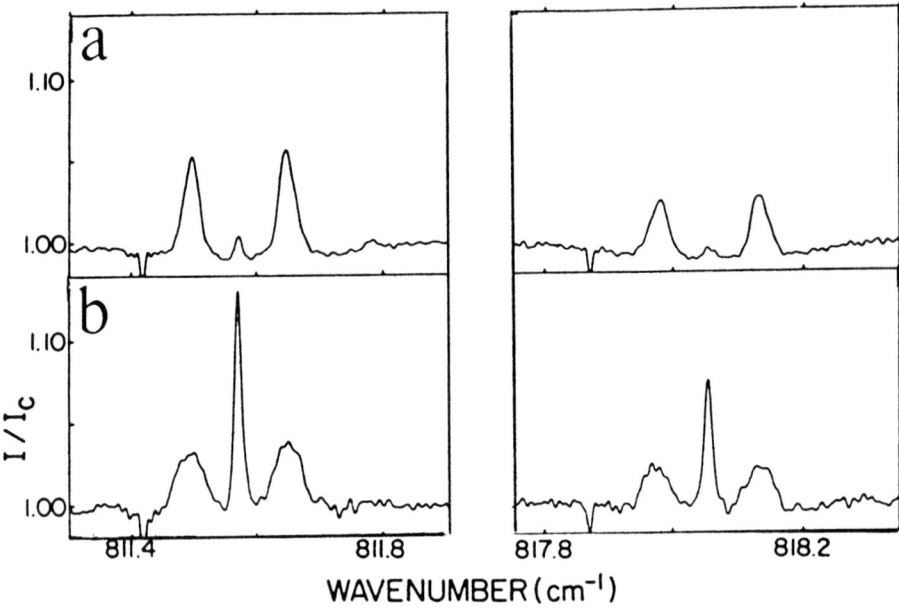

Fig. 2. Penumbral spectrum of the Mg I 12 μm lines from Brault and Noyes (1983): (a) the σ components in the longitudinal direction; (b) the π component, with twice the strength of the σ, in the transverse direction.

3. Magnetic properties

From a simplistic point of view, the Landé g factor of any high-l Rydberg level is predicted to be unity. First in the absence of a magnetic field, fine structure splittings in the 12 μm lines have not been observed even in the high resolution diode-laser measurement. Theoretical estimates indicate that they are less than 0.001 cm^{-1}, and therefore the spin-orbit coupling must be very weak. Second in the presence of a magnetic field B, the "classic" interaction with the Rydberg electron exceeds interactions involving the atomic core. Indeed, the g-factor prediction has been confirmed by diode-laser measurements of the σ components from 200 to 1000 G. The measured g-factors for the 811 and the 818 cm^{-1} lines were 0.992(10) and 1.002(10) respectively (Lemoine et al., 1988) – with one standard deviation error in parenthesis.

The penumbral observations of Brault and Noyes (1983) are shown in Figure 2; Figure 2a shows the longitudinal, and Figure 2b the transverse, Zeeman pattern. Without the benefit of line identification, these authors inferred $g \sim 1$ from some sodium lines, and correctly deduced a measured magnetic field of 1600 G. Recently, exciting measurements of vector magnetic fields using polarimetry of the 12 μm lines have been carried out by Deming and his collaborators (1988 and these proceedings). For a complete understanding of the magnetic properties of high-l Rydberg

TABLE I
Observed Mg I Spectral Lines (cm^{-1})

Transition	Solar	Theory[a]	Laboratory[a]
8h–7g	530.986	.983	–
7i–6h	811.575	.589	.5749(10)
7g–6f	848.010	.007	.0109(5)
			.2141(8)
	848.060	.057	.0610(10)
			.0698(7)
9k–7i	885.524	.521	.5292(2)
9i–7h	886.869	.863	.8717(2)
6h–5g	1356.182	.186	–

[a] Integer in column 2 deleted.

lines, one must simultaneously consider internal (magnetic fine structure) and external field effects, as Chang (1987) has done. As long as one of these dominates over the other, the splitting will vary linearly with B. The case of a strong internal field gives the familiar anomalous Zeeman effect with the splitting proportional to $g\mu_0 B$, where the Bohr magneton μ_0 has the value 4.67×10^{-5} cm^{-1} (an erroneous value was given in Chang 1987). For the 12 μm lines, even an external field of 100 G will overwhelm the internal field, resulting in the Paschen-Back effect. Expressions for the splittings and line strengths can be found in Chang (1987). Unfortunately his equation (10b) contained a typographical error in that one of the factors in the numerator should read $(l \pm m + 2)$. Thus the relative strength of each σ line when summed over all m values should be $1/3$ – the same as for the π line.

4. Electrical Properties

At disk center, the 12 μm lines display an emission peak which agrees with the laboratory line position, and an absorption trough which is shifted. We have undertaken a study of these and some related lines using the *ATMOS* spectra. The troughs for the 811 and 1356 cm^{-1} lines are shifted to the higher frequency while the 818, 885, and 886 cm^{-1} lines (see Table 1) are shifted to the lower frequency side. This perculiarity is accounted for by the quadratic Stark effect. In an electric field of strength E, the energy-level shift is given by $\delta E_{nl} = -\alpha_{nl} E^2$, where α_{nl} stands for the electric dipole polarizability of the Rydberg atom in the state nl. Its values have been calculated for the transitions of interest by Chang and Schoenfeld (1991). Note that these Rydberg atomic polarizabilities are 4 to 6 orders of magnitude larger than the magnesium core polarizability α.

To understand the opposite direction of the trough shifts, let us consider the upper levels of the 811 and the 818 cm^{-1} lines. The 7i ($l = 6$) and the 7h ($l = 5$) levels are separated by less than 2 cm^{-1}, and may be regarded as an isolated two-level pair. In the presence of an electric field they admix, and repel each other.

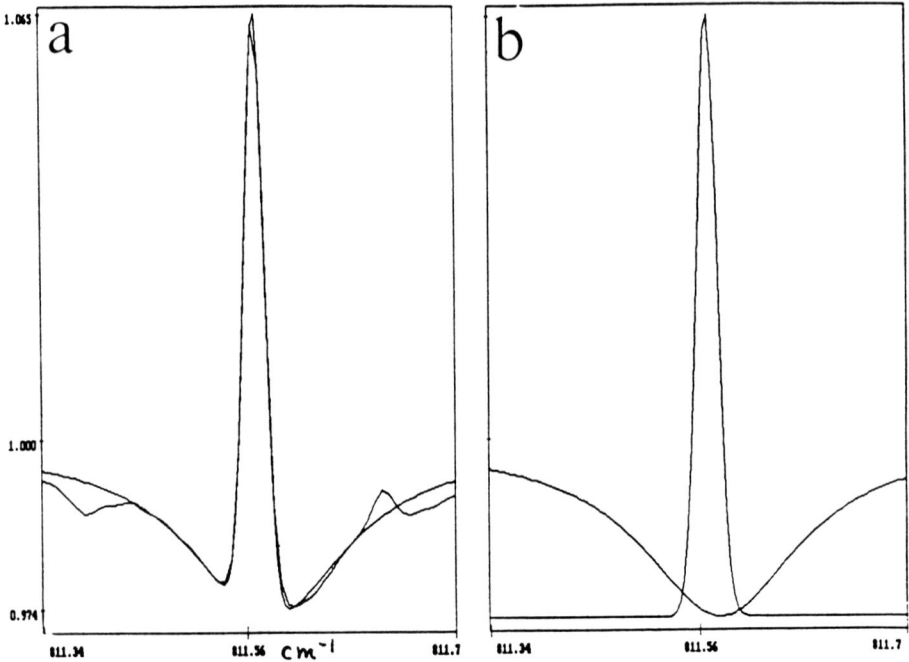

Fig. 3. The 811 cm^{-1} line profile from Chang and Schoenfeld (1991): (a) The computer-fitted (broken line) and the observed (full line) profiles; (b) The emission peak and the absorption trough.

Therefore the $7i$ level is shifted upwards ($\alpha_{7i} < 0$) and the $7h$ level downwards ($\alpha_{7h} > 0$), in agreement with observations. Actually there are two complications which merit further discussion. One is that the $7h$ level is also repelled upwards by the $7g$ level. However the $7g$ level is 3 times further away than the $7i$ (with about the same off-diagonal matrix element), so the net shift of the $7h$ level is downwards. The other is that similar shifts occur for the lower levels of the transition. However it can be shown that α_{nl} varies as $n^7 l^5$, so the upper level always dominates the shift. Thus we are led to the conclusion that the 811 and the 1356 cm^{-1} lines, whose upper levels are circular states, have absorption troughs shifted towards higher frequency, while the 818, 885, and 886 cm^{-1} lines, whose upper levels are not circular, have absorption troughs shifted towards lower frequencies.

Figure 3 shows the 811 cm^{-1} line at disk center from the *ATMOS* spectra. Panel b shows that the peak has a Gaussian form, and the shifted trough a Lorentzian – as predicted by Chang and Schoenfeld (1991). From the shift of this, and other, lines, we determine electric field strengths of 100 to 200 V cm^{-1} except for the 1356 cm^{-1} line which is formed deeper in a stronger electric field.

5. Related Lines in Magnesium

So far we have seen that Rydberg lines with $l > 3$ have negligible fine structure splittings. However lines with $l = 3$ (or 2) reveal resolvable splittings of a few hundredth cm^{-1}. For example, the $7g$–$6f$ line (see Table 1) was first observed by Brault and Noyes (1983) at the solar limb as a doublet whose components were of nearly equal intensity and had a spacing of 0.050 cm^{-1}. The identification was suspect since the existing atomic data then revealed that the $6f$ ^1F and ^3F were only 0.03 cm^{-1} apart and that the intensity ratio should be 1:3. As indicated by the last column of Table 1, the controversy was resolved when high-resolution laboratory data became available. Figure 4 shows that each peak in the solar spectrum was a blend of 2 fine-structure transitions with the $^3F_2$ blending into the $^1F_3$ component, thereby yielding nearly equal peaks. Solar modeling of the line showed very good agreement with the observed profiles both at the disk center and at the limb (Chang et al., 1992).

Other related lines involving F levels in the *ATMOS* data were recently reported by Jefferies (1991). Using the then-current atomic data, he was led to the enigmatic conclusion that the triplets were in emission, while the singlet was in absorption. Using more-recent atomic data, we have computed line profiles from our solar model and have concluded that there is, in fact, no difference between the behavior of triplet and singlet components. Comparison with the observed profiles revealed some subtleties. In the $6g$–$5f$ line profile shown in Figure 5, an opacity minimum appeared between the triplet and the singlet emission peaks, all residing in a deep absorption trough. From the observations alone, it is impossible to distinguish the opacity minimum from the fine-structure peaks. In the $5g$–$4f$ line at 2586 cm^{-1}, our computed profile showed a weak peak for the singlet as for the triplet components. However, the spacing between this peak and the adjacent troughs was less than the *ATMOS* instrumental resolution. Therefore the weak peak disappeared in the observed profile.

The importance of accurate atomic data to the interpretation of solar spectra has just been clearly demonstrated. Fortunately, in Mg I, the recent diode-laser data (Lemoine et al., 1990) and the FTS data (Biémont and Brault, 1986), are available. The high ($l > 3$) Rydberg levels appear to be reliably calculated by theory. Theoretical energies are accurate to within a few thousandths of a wavenumber, and their atomic properties are well understood.

6. Related Lines in Other Atoms

Since the first identification of the Mg I 12 μm lines, similar high-l Rydberg transitions in other elements were sought and found in the 12 μm window line list. More recently the search has expanded to include the more extensive *ATMOS* data. Hydrogen lines are easily distinguished by their large broadening, due primarily to the linear Stark Effect. These lines become broader and shallower as n increases. At $n \sim 10$, the broadened troughs become indistinguishable from the continuum.

Lines of other atoms are conveniently divided into two categories. One consists of those elements whose atomic cores have an isotropic S state. Like Mg I, their

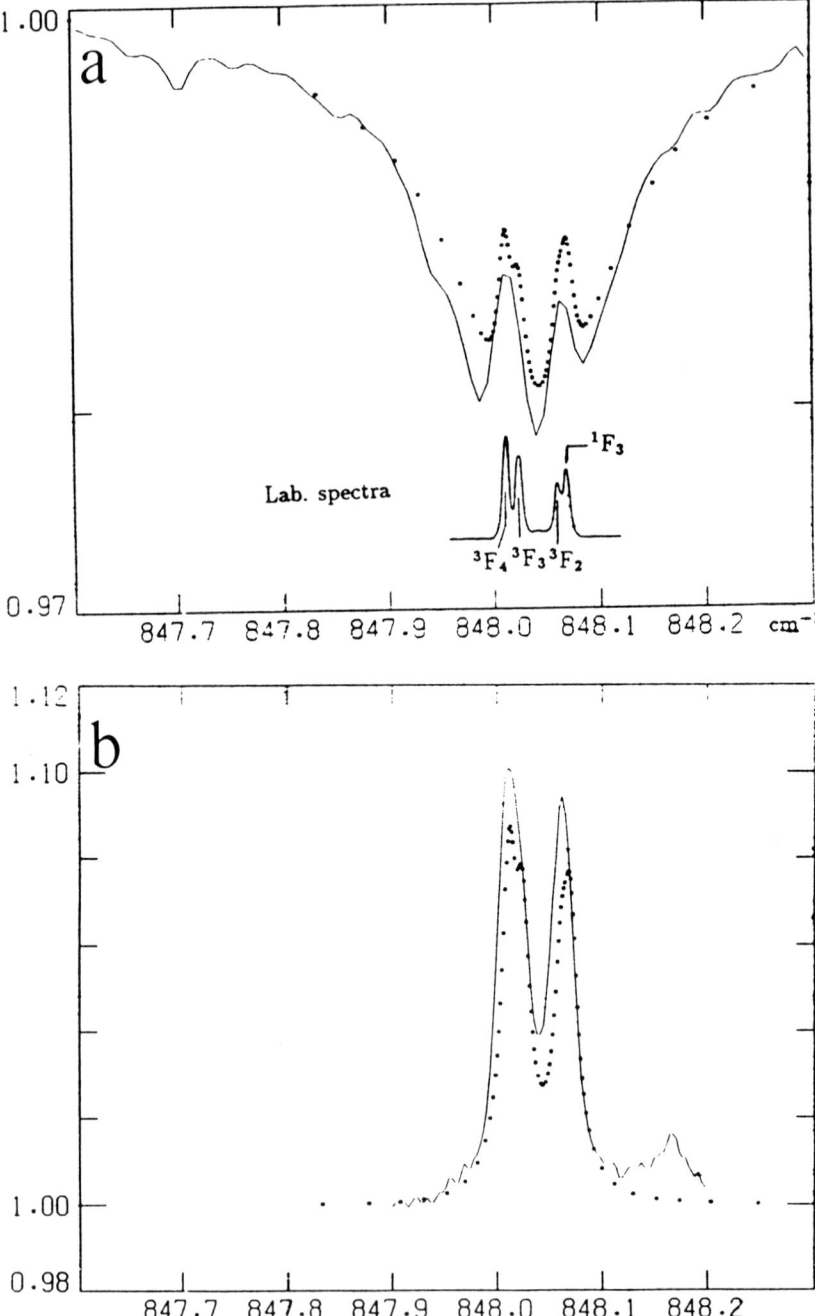

Fig. 4. The Mg I $7g$–$6f$ line from Chang et al., (1992). The solid line shows the observations, and the dotted line is the computed profile: (a) at disk center (inset at bottom shows the laboratory spectra); and (b) at the limb.

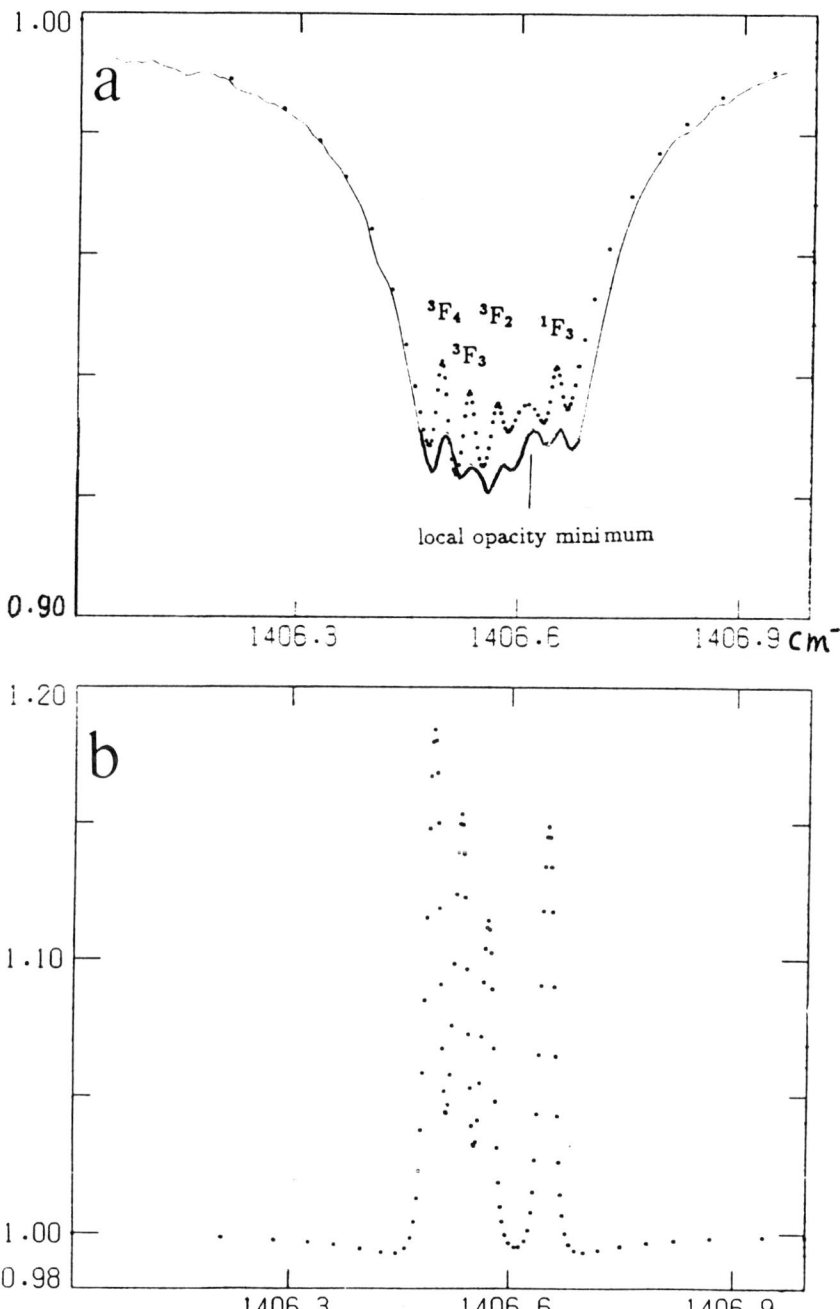

Fig. 5. The Mg I $6g$–$5f$ line. The solid line shows the observed, and the dotted line the computed, profiles. Panel (a) refers to the disk center; (b) to the limb.

energy levels are determined by just one parameter, the core polarizability, aside from the trivial reduced-mass Rydberg-constant correction. Some high-l Rydberg lines have been identified for Na I, Al I, and Ca I. They are observed in absorption, with the exception of a few Al I lines which also have weak emission peaks. The other category of lines arise in those atoms whose anisotropic cores have a permanent electric quadrupole moment Q, in addition to the core polarizability. The electrostatic interaction between Q and the orbiting Rydberg electron gives rise to a further splitting of each nl level into sub-levels designated by the K quantum number. Thus a single Rydberg line in the previous category may fractionate into many lines in these atoms. High-l Rydberg lines in this group have been identified in C I, Si I (Chang 1984), and Fe I (Johansson et al.; Schoenfeld et al.; these proceedings). Our analysis of the $ATMOS$ spectra in the 4, 7, and 12 μm regions reveals no unidentified lines of significant strength. Therefore we do not expect to see high-l Rydberg lines of any other elements.

7. Conclusions

In the decade following the discovery of the 12 μm emission lines, great advances have been made in understanding their underlying atomic physics. The hydrogen-like properties of these lines greatly simplify their analysis in the solar spectrum, and enhance the accuracy of the extracted solar properties. As valuable as the 12 μm lines have proven to be for measurements of the solar electric and magnetic fields, it is desirable to explore similar lines in the far infrared. They are formed higher up, reaching into the low chromosphere where little is known about these fields, and in addition the sensitivity increases as the wavelength increases. Consider, for example, the 20 μm line of Mg I which was detected by Brault and Noyes (1983) at the limb. Measurements of this, and other lines, both at the limb and at disk center (and preferably from a satellite) will greatly enhance our knowledge of the astrophysics of the sun.

Acknowledgements

I thank E. H. Avrett for useful discussions and W. G. Schoenfeld for discussions and assistance in preparation of this manuscript.

References

Avrett, E. H., Chang, E. S., and Loeser, R.: 1993, these proceedings.
Biémont, E. and Brault, J. W.: 1986, *Physica Scripta* **34**, 751.
Brault, J. W. and Noyes, R. W.: 1983, *Astrophys. J. (Letters)* **269**, 61.
Carlsson, M., Rutten, R. J., and Shchukina, N. G.: 1992, *Astron. Astrophys.* **253**, 567.
Chang, E. S. and Noyes, R.W.: 1983, *Astrophys. J. (Letters)* **275**, 11.
Chang, E. S.: 1984, *J. Phys. B* **17**, 11.
Chang, E. S.: 1987, *Physica Scripta* **35**, 792.
Chang, E. S., Avrett, E. H., Mauas, P. J., Noyes, R. W., and Loeser, R.: 1991, *Astrophys. J. (Letters)* **379**, 79.
Chang, E. S., Avrett, E. H., Mauas, P. J., Noyes, R. W., and Loeser, R.: 1992, in M. Giampapa, and J. A. Bookbinder (eds.), *Seventh Cambridge Workshop on Cool Stars, Stellar Systems, and the Sun*, ASP Conference Series, Vol. 26, p. 521.

Chang, E. S., and Schoenfeld, W. G.: 1991, *Astrophys. J.* **383**, 450.
Deming, D., Boyle, R. J., Jennings, D. E., and Wiedermann, G. R., : 1988, *Astrophys. J.* **333**, 978.
Deming D., Hewagama T., and Jennings D. E., McCabe, G., and Wiedemann, G.: 1993, these proceedings.
Farmer, C. B. and Norton, R. H.: 1989, *A High-Resolution Atlas of the Infrared Spectrum of the Sun and the Earth's Atmoshpere from Space*, NASA Ref. Pub. 1224, Vol. 1.
Goldman, A., Blatherwick, R. D., Murcray, F. H., Van Allen, J. W., Bradford, C. M., Cook G.R., and Murcray D. H.: 1980, *New Atlas of IR Solar Spectra*, Vol. 1, Line Positions and Identifications, Vol.2, The Spectra, Department of Physics, University of Denver.
Jefferies, J. T.: 1991 *Astrophys. J.* **377**, 337.
Johansson, S., Nave, G., Geller, M., Sauval, A. J., Grevesse, N.: 1993, these proceedings.
Kurucz, R.: 1990, in A. N. Cox, W. C. Livingston, and M. Mathews (eds.), *The Solar Atmosphere and Interior*, Tucson, University of Arizona Press, p. 663.
Lemoine, B., Demuynck, C., Destombes, J. L., and Davis, P. B.: 1988, *J. Chem. Phys.* **89**, 673.
Lemoine, B., Demuynck, C., and Destombes, J. L.: 1988, *Astron. Astrophys.* **191**, L4.
Lemoine, B., Petitprez, D., Destombes, J. L., and Chang, E. S.: 1990, *J. Phys. B* **23**, 2217S.
Murcray, F. J., Goldman, A. Murcray, F. H., Bradford, C. M., Murcray, D. G., Coffey, M. T., and Mankin, W. G.: 1981, *Astrophys. J. (Letters)* **247**, 97.
Rutten, R., and Carlsson, M.: 1993, these proceedings.
Schoenfeld, W. G., Chang, E. S., and Geller, M.: 1993, these proceedings.

THE FORMATION OF INFRARED RYDBERG LINES

ROBERT J. RUTTEN

Sterrekundig Instituut, Postbus 80 000, NL–3508 TA, Utrecht, The Netherlands

and

MATS CARLSSON

Institute of Theoretical Astrophysics, P.O. Box 1029, Blindern, N–0315 Oslo 3, Norway

Abstract. We review the formation of infrared solar spectral lines from highly excited levels in neutral atoms. The lines of Mg I and H I are the most interesting ones. We explain the NLTE processes by which they are affected and we study the sensitivity of the Mg I 12 μm lines to granulation and to fluxtubes.

Key words: atomic processes – hydrogen – infrared: stars – line: formation – line: profiles – magnesium – Sun: atmosphere

1. Introduction

In his concluding remarks to the Sixth Cambridge Cool Star workshop, Linsky (1990) listed the formation of the Mg I 12 μm emission features in the solar spectrum as one of the "major unanswered questions for the next workshop". By the next Cool Star workshop this enigma was indeed solved, by Carlsson et al. (1990, 1992a, 1992b) and Chang et al. (1991, 1992). In this review, we expand on our detailed analysis in Carlsson et al. (1992a, henceforth Paper I) and its Cool Star summary (Carlsson et al., 1992b) with further explanation of the Mg I 12 μm emission, additional Mg I infrared absorption line modeling, and a parallel discussion of H I infrared line formation. The older literature is reviewed in Paper I. Relevant atomic physics is discussed by Chang (1993) in the preceding paper in this volume; in the next one, Avrett (1993) displays additional Mg I and H I computations. Our H I modeling is presented in Carlsson and Rutten (1992, henceforth Paper II) and elsewhere in this volume (Carlsson and Rutten, 1993).

All the above modeling is "one-dimensional", assuming lateral homogeneity. In view of the obvious inhomogeneity of the real solar photosphere, Linsky (1990) also put the apparent validity of such plane-parallel modeling for solar Fe I and other lines in the visible, as reviewed by Rutten (1990), on his list of major unanswered questions for the future, and suggested that things may be different in the infrared. We show below that the infrared transitions of Mg I, both the emission and the absorption lines, are actually better reproduced by our one-dimensional computations than any other Fraunhofer species sofar. We try to answer the question why this modeling is so good by computing Mg I 12 μm profiles from granules.

More generally, using the infrared high-n lines to study the spatial structure of the solar atmosphere becomes possible now that their NLTE formation is basically understood. The strongest H I lines provide diagnostics of the chromosphere while the strongest Mg I lines are useful for upper-photosphere magnetometry. We lay a basis for the latter by computing Mg I 12 μm profiles from magnetic fluxtubes.

2. NLTE Rydberg Line Formation

2.1. NLTE Populations

Figure 1 shows the principal ingredients of Rydberg (high-n) line formation in the Sun. The lefthand panel is from Paper I and specifies LTE Rydberg populations N^* for the MACKKL quiet-Sun model atmosphere (Maltby et al., 1986). It illustrates that infrared lines of neutral metals (Mg I, Si I, Al I etc.) come from the upper photosphere, with Mg I supplying the strongest ones. In contrast, the stronger hydrogen lines have split formation. The lower chromosphere and lower photosphere both contribute opacity to them, with a wide gap at the temperature minimum (located at $\lg \tau_{500} = -4$). Such behavior is well known for Hα (e.g., Schoolman, 1972); it is shared by the higher H I $n\alpha$ lines (see the contribution functions in Carlsson and Rutten, 1993).

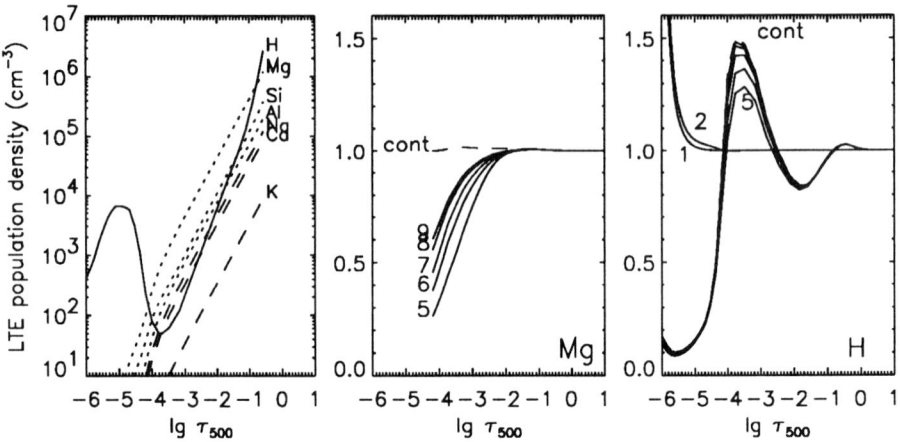

Fig. 1. *Left:* Saha-Boltzmann populations N^* of comparable high-n levels in various atoms. *Middle and right:* NLTE departure coefficients β_n of levels $n = 5 - 9$ and the ion for Mg I and H I, respectively; levels 1 and 2 added for H I. *Abscissae:* logarithm of continuum optical depth at $\lambda = 500$ nm.

The curves in the lefthand panel shift upward for increasing n, but the pattern remains the same. Since the quantum defects of the neutral-atom levels decrease with increasing n, the corresponding Rydberg transitions (split in l) approach each other towards longer wavelengths and merge with the hydrogen lines from the short wavelength side. The Mg I and H I lines dominate the resulting features in the spectrum. The Mg I lines are stronger at 12 μm (Brault and Noyes, 1983), whereas the 11–12 and 12–13 H I lines are about as strong as their Mg I counterparts. H I dominates at higher n, above $\lambda = 100$ μm (Boreiko and Clark, 1986).

The other panels of Figure 1 are from Papers I and II, respectively. They show NLTE population departure coefficients $\beta_n = N/N^*$, with N the actual NLTE population of level n. Note that these Zwaan coefficients β_n differ from the more commonly used Menzel coefficients b_n. The latter are normalized by the N/N^* ratio of the next ion and smaller by $1/\beta_{\text{cont}} = N^*_{\text{cont}}/N_{\text{cont}}$ (Menzel and Cillié, 1937;

Wijbenga and Zwaan, 1972; cf. Vernazza et al., 1981 p. 663). The LTE populations N^* of H I and Mg I in the lefthand panel should be multiplied by β values from the middle and righthand panels to obtain actual MACKKL populations. The main change is a tenfold reduction of the chromospheric H I population peak. The patterns remain similar, however.

2.2. NLTE MECHANISMS

The divergences between adjacent curves in the β panels of Figure 1 cause NLTE emission in the corresponding infrared Mg I and H I transitions. How do these departure divergences come about? We refer to Papers I and II for details; here, we illustrate the pertinent properties of high-n levels more generally. These are the following:

(i)—Rydberg lines have long wavelengths. In the infrared, the correction factor $1 - (\beta_u/\beta_l)\exp(-h\nu/kT)$ for stimulated emission reaches zero already for small excess of the upper-level departure β_u over the lower-level departure β_l, *i.e.*, for slight departure divergence. For example, for $T = 5000$ K and $\lambda = 12$ μm, $\exp(-h\nu/kT) = 0.79$, so that a departure ratio $\beta_u/\beta_l > 1.27$ already produces a negative correction factor (lasering). For a smaller offset of β_u/β_l from unity the line extinction decreases by the correction factor, while the line source function increases by its inverse. These changes differ qualitatively. The opacity decreases rapidly with height but the source function does not, especially in the upper photosphere (where the temperature gradient is small) and in the infrared (where the temperature sensitivity of the Planck function is slight). The effect of a small $\beta_u > \beta_l$ departure divergence is therefore much larger for the source function than for the opacity. Figure 11 of Paper I demonstrates that a departure excess of only 5% ($\beta_u/\beta_l = 1.05$) already produces emission in the 12 μm lines. Thus, as small as the departure divergences in Figure 1 are, their effects are large.

(ii)—Rydberg levels are hydrogenic. The fact that classical theory applies well to them not only facilitates model-atom building for NLTE computations, but also results in regular patterns in collision cross-sections. Figure 5 of Paper I shows the bound-bound collision matrix of our Mg I model atom. Close-lying levels with the same n are coupled strongly through "l-changing" collisions with electrons and neutral hydrogen atoms. The resulting high-l same-n manifolds are primarily coupled by $\Delta n = 1$ collisions. Together, these constitute a "Rydberg ladder" along which collisional coupling is strongest.

(iii)—Rydberg transitions have small energy. Their energy separations are much smaller than the 1 eV kinetic energy typically available in collisions. In addition, Rydberg atoms are very large. Therefore, collision rates dominate over radiative rates. The collisional coupling between Rydberg levels is very strong, as is their collisional coupling to the continuum. NLTE departure divergences can therefore *only* exist between Rydberg levels if they are driven by processes elsewhere in the Grotrian diagram.

Indeed, there are also non-Rydberg ingredients to the realization of Rydberg statistical equilibrium. These are:

(i)—NLTE photon pumping. Non-local photons that are sufficiently energetic to

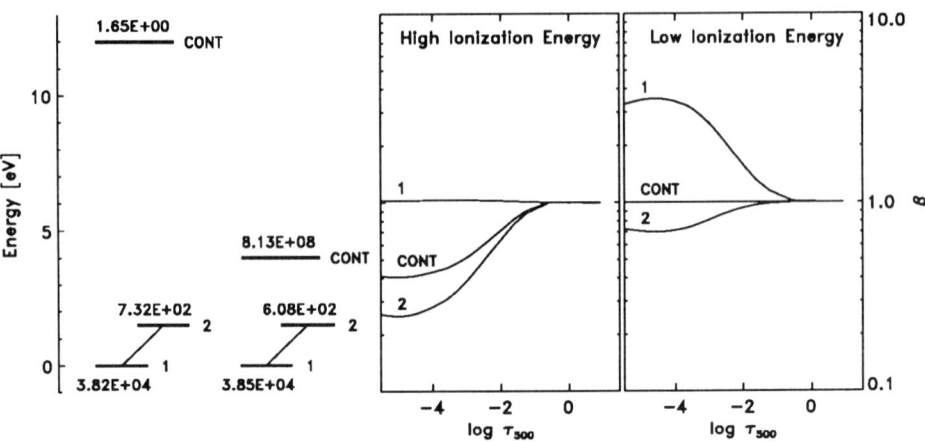

Fig. 2. Didactic computation of photon suction. *Left:* two-level-plus-continuum model atoms. The lefthand atom has 12 eV ionization energy, the righthand atom 4 eV. The numbers specify NLTE population densities in cm^{-3} at $\log \tau_{500} = -3$. The abundances are scaled to obtain similar extinction in the single line, which resembles K I 769.9 nm. In the lefthand case, the neutral stage contains most of the element; at right, the element is ionized. *Right:* corresponding NLTE departure coefficients against continuum optical depth. The split between β_1 and β_2 is set by two-level resonance scattering. Photon suction occurs for the case of low ionization energy at right; it produces overpopulation of the neutral stage (β_1). Graph taken from Bruls (1992).

impart non-local information on local populations are either incoming or outgoing. In the first case, they cause optical pumping; in the second, photon suction. Photospheric photon pumping is best known in the form of ultraviolet overionization, for example of Fe I (Lites, 1972; Athay and Lites, 1972; *cf.* Rutten, 1988, 1990), but line pumping occurs also, for example in Fe II (Cram *et al.*, 1980). Photon pumping employs ultraviolet radiation fields from the deeper photosphere, which are characterized by superthermal radiation temperatures because their mean intensity J_ν drops less steeply with height than the Planck function B_ν. The corresponding excess in available photons causes larger upward than downward radiative rates and depletes the lower-state populations until these are small enough to attract a balancing downward net collisional rate.

(ii)—NLTE photon suction. This process follows from photon losses and operates in minority species such as the photospheric alkalis. It is demonstrated in Figure 2 after Bruls (1992), for two-level-plus-continuum model atoms following Bruls *et al.* (1992). Radiative deexcitations transfer electrons out of the upper level 2 into the lower level 1. Loss of some of the resulting photons to outer space results in underpopulation of level 2 and overpopulation of level 1. The resulting split between β_1 and β_2 starts deeply, due to strong resonance scattering; it produces the characteristic line source function deficit $S^l < B_\nu$ of a resonance line. In the case of high ionization energy (lefthand Grotrian diagram and lefthand β panel),

β_1 departs much less from unity than β_2 because the population of the ground state is much larger, as shown by the population densities in the Grotrian diagrams. However, β_1 rises well above unity in the case of low ionization energy; this overpopulation arises through bound-free coupling. The ion (CONT) is coupled to levels 1 and 2 by three-body recombination, with stronger coupling to level 2 than to level 1. In the lefthand case of large ionization energy, the ion has very small population and follows level 2 into depopulation. However, if the ion population is much larger than the neutral-atom population, it constrains the population of level 2 rather than be wagged around. Photon losses from level 2 are then compensated collisionally from the ion. This is photon suction. It increases the population of the ground state until the neutral-stage overpopulation (β_1) is sufficiently large to balance the ionization equilibrium with a net collision rate up from level 1. Whereas the photon-loss divergence between β_1 and β_2 is the same for the two cases, producing similar NLTE $S^l < B_\nu$ source function behavior, the photon suction produces triple overpopulation of the neutral stage in the case of low ionization energy. Bruls et al. (1992) supply further demonstrations of the suction process using realistic K I model atoms of increasing size.

(iii)—LTE reservoirs. For low ionization energy, the ion population acts as a large reservoir which is impervious to the neutral atom and its photon losses. It is an LTE reservoir, with $\beta_\text{cont} \approx 1$, because it contains most of the species. For high ionization energy, the ground state of the neutral atom constitutes the LTE reservoir. In that case, the continuum population follows the photon losses.

(iv)—NLTE replenishment flows. In the case of low ionization energy, the photon losses and the continuum reservoir supply boundary conditions to the establishment of net recombination into level 2. This constitutes a replenishment flow, driven by the resonance line. In real atoms, intermediate levels at the bottom of the Rydberg ladder and the continuum at the top supply similar boundary conditions. Departure divergence between these is then accommodated along the ladder, preferentially in small $\Delta n = 1$ steps. A radiative-collisional *departure diffusion flow* results. It is similar to the Rydberg population diffusion occurring in optically-thin recombination (*e.g.*, Seaton, 1964; Section 5.4 of Sobelman et al., 1981; Biberman et al., 1987). For sufficiently high levels, it is fully dominated by collisions, with $\Delta n = 1$ collisions carrying most of the net downward flow.

2.3. NLTE IN Mg I INFRARED LINES

Figure 3 has a schematic representation of the Mg I term diagram at left. Magnesium is predominantly singly ionized in the photosphere, so that the Mg II ground state constitutes the Mg LTE population reservoir. The lowest Mg I levels are too far from the continuum for ultraviolet overionization, which is most effective at 3–4 eV ionization energies ($\lambda = 400 - 300$ nm). The resonance transitions and other strong lines such as 3s–3p and the b lines are opaque and close to detailed balance. The main NLTE driver consists of a sequence of $\Delta n = 1$ and $\Delta n = 2$ lines at high l. These are strong in the laboratory sense, having large transition probabilities. The lines are sufficiently weak in the solar spectrum, due to their small Boltzmann populations, that they are thin in the middle photosphere. Their combined radiative losses cause photon suction, replenishing photon losses from the continuum just as in the neutral alkalis. The resulting overpopulation of levels 3 (comparable to level 1 in the two-level case, split in l) and the superthermal radiation in their bound-free continua balance the ionization equilibrium.

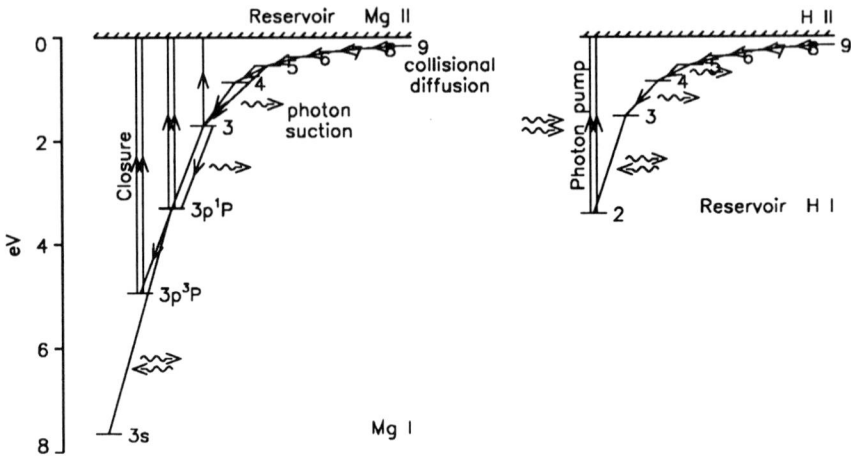

Fig. 3. Schematic Grotrian diagrams of Mg I (left) and H I (right), with the continua aligned. In Mg I, the NLTE driving is primarily by photon losses in infrared lines with large transition probability. These are replenished from the LTE continuum reservoir along a Rydberg ladder of $\Delta n = 1$ steps *because* the latter levels have very strong collisional coupling. In H I, the lower levels constitute the reservoir and the Balmer continuum is the main NLTE driver in the temperature minimum region. It pumps the continuum into overpopulation and so produces departure diffusion along the Rydberg ladder.

The replenishment flow comes down primarily along the Rydberg ladder. It is collisionally dominated along its top. This is visualized in Figure 4. The lefthand waterfalls illustrate that most of the replenishment flow enters at the top of our model ($n = 9$). The Mg II—9 waterfall would undoubtedly be split into yet higher steps if our atomic model accommodated these, but such splitting would not affect the 7-6 step which harbors the 12 μm lines. The cascade at right shows the split between net collisional and radiative deexcitation along $\Delta n = 1$ steps. Collisions already dominate the 7-6 transition, and gain even more higher up.

In Paper I and in Carlsson et al. (1992b) we have shown results of various experiments and of a "multi-MULTI" sensitivity analysis. In the latter the code (named MULTI; Carlsson, 1986) was run 2525 times, doubling each input parameter one by one. The results demonstrate that the NLTE driving is indeed by infrared photon losses with some help from violet overionization, that the replenishment flow takes $\Delta n = 1$ ladder steps preferentially, and that the flow is made up by collisions along this ladder. Therefore, the 12 μm lines have NLTE emission peaks by virtue of the strong collisional coupling along the ladder steps. Only if this coupling is strong

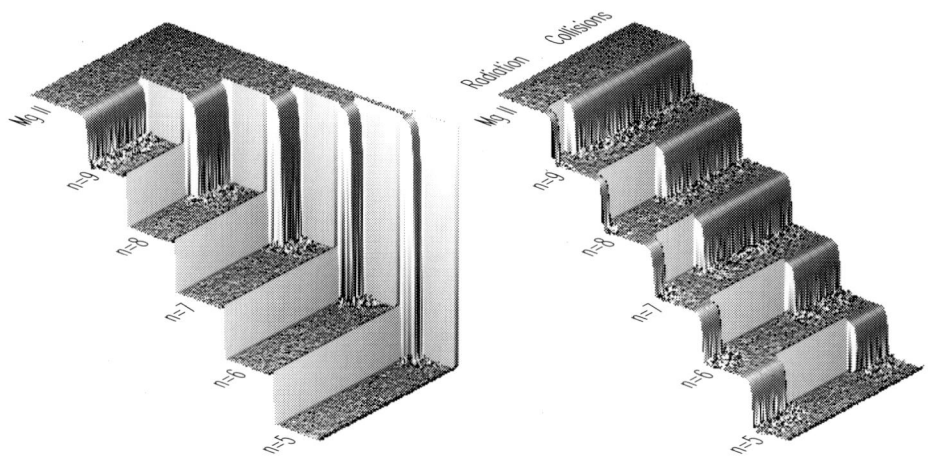

Fig. 4. Kayaker's views of the population departure diffusion flow along the Rydberg ladder in Mg I. *Left:* distribution of the total net recombination over levels 5–9. *Right:* division between collisions and radiation of the net downward rates along $\Delta n = 1$ steps.

enough does the replenishment flow come down along the laddder, rather than using other channels such as $\Delta n = 2$ level skipping or direct recombination. The step-by-step adjustments to the boundary conditions along the ladder maintain level-to-level departure divergences. *The Mg I 12 µm lines have NLTE emission peaks because of strong collisional coupling between the pertinent levels!*

In addition, the sensitivity experiments show that our computation is very robust. The largest changes to the computed 12 µm peaks amount to only 20% when doubling transition probabilities (bound-bound or bound-free), 10% for doubling collision cross-sections.

2.4. NLTE IN H I INFRARED LINES

Hydrogen is neutral in the photosphere. Levels 1 and 2 (with Ly α in detailed balance) constitute the LTE reservoir. The Balmer continuum has 3.4 eV ionization energy and feeds strongly on superthermal radiation with $J_\nu > B_\nu$ from the deep photosphere. This pumping is a major NLTE driver. Photon losses in the principal Paschen and Brackett transitions probably contribute as well.

The resulting H I β pattern in Figure 1 differs from the Mg I pattern primarily in that the ion population is out of LTE. The initial hump and dip of $\beta_{\rm cont}$ stem from reversed $B_\nu > J_\nu$ excesses in infrared continua and photon exchange between infrared lines and their background continua, while the peak in the temperature minimum region is caused by Balmer continuum pumping. The wide minimum at large height results from the chromospheric temperature rise.

Note that the H I β panel differs from the corresponding Menzel b departure plot in Figure 30 of Vernazza *et al.* (1981), which has reversed peaks and dips and $b \approx 1$ for high-n levels in the low chromosphere. Note also that much of the H I recombination literature (references in paper II) adopted $\beta_{\text{cont}} = 1$ implicitly. This assumption is invalid in the low chromosphere where hydrogen is predominantly neutral and $\beta_{\text{cont}} \approx 0.1$.

In the temperature minimum region, the peak of β_{cont} again acts as upper boundary condition to the high-n ladder to produce level-to-level departure divergence and NLTE source function enhancements. These are less computationally robust than for Mg I, because they depend more strongly on the $J_\nu > B_\nu$ excesses in the ultraviolet which are difficult to quantify due to the enormous line crowding in that region. Experiments show that the peak of β_{cont} is indeed very sensitive to the details of the line haze representation. It is not sensitive, however, to change of the model atom size, an experiment easier accomplished for H I than for Mg I. We computed H I lines with $n < 10$ for an 11-level-plus-continuum model in addition to the 19-level-plus-continuum model of Paper II. There is no difference in the resulting profiles.

2.5. Mg I Rydberg Line Profiles

Figure 5 is assembled from Paper I and shows our results for the strongest infrared emission line, Mg I 12.32 μm. The fit is quite good. Note that we applied no parameter fitting in our computation. Standard choices were made for microturbulence (1.0 km s^{-1}), macroturbulence (1.5 km s^{-1}), and van der Waals damping enhancement (times 2.5). Other choices were made regarding neutral hydrogen collisions (assumed sufficiently important in l-changing transitions to produce detailed balancing of close-lying levels with equal n — *cf.* Omont, 1977) and for forbidden transitions (smaller collision strengths than comparable permitted transitions). As noted above, the resulting line profiles are not very sensitive to the cross-sections. What counts is the regularity of the patterns in the collision matrix, and the corresponding selection of the Rydberg ladder as primary replenishment channel.

The righthand panel of Figure 5 shows the computed source function with superimposed formation heights. Comparison with the lefthand panel illustrates the mapping of source function structure into line profile. The peaks originate in the upper photosphere even when observed near the limb.

The modeling of Chang *et al.* (1991) reproduces the observed 12 μm profiles less well than ours. Chang *et al.* (1992) added computations of other infrared Mg I lines, observed by ATMOS (Farmer and Norton, 1989) and examined by Jefferies (1991). Their new computations also do not reproduce the observations very closely. At Avrett's request, we have computed the $7g - 6f$ and $6h - 5g$ lines using the setup of Paper I without change, except that l-splitting was taken into account by adopting detailed balance between the corresponding fine structure levels. Figure 6 shows results (actually computed during the meeting, and kindly displayed by Avrett in his talk). These fits are good too. It appears that we have a sound rendering of solar Mg I atoms inside our computer. We suspect that the major difference from Chang *et al.* lies in collision matrices and the treatment of photoionization. The atom-size experiment with H I indicates that our Mg I atom (complete to $n = 9$) is

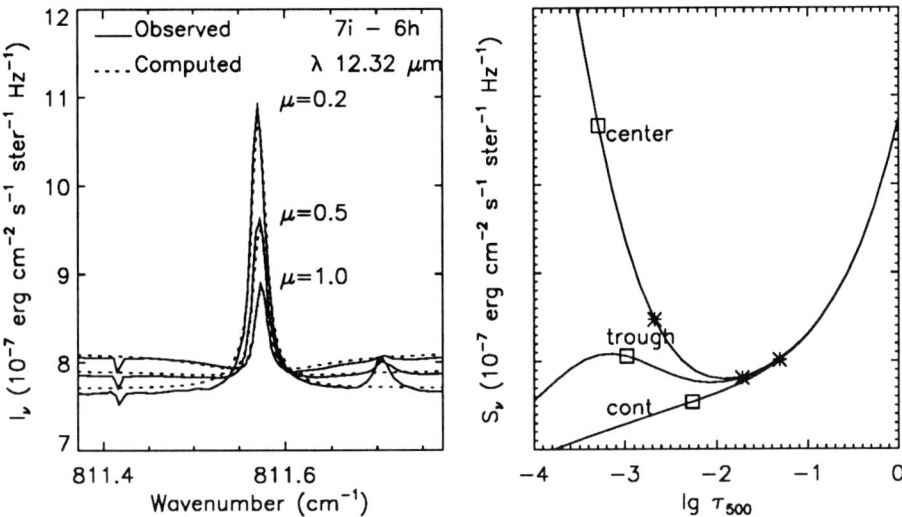

Fig. 5. Profiles and source functions of the Mg I 12.32 μm line ($7i - 6h$). *Left:* observed and computed profiles on absolute intensity scales, for the three viewing angles θ indicated by $\mu = \cos\theta$. Observed profiles are based on Brault and Noyes (1983). *Right:* computed source functions, for line center, an inner-wing wavelength ("trough") and the continuum. Average formation heights are specified by squares for $\mu = 0.2$, stars for $\mu = 1$.

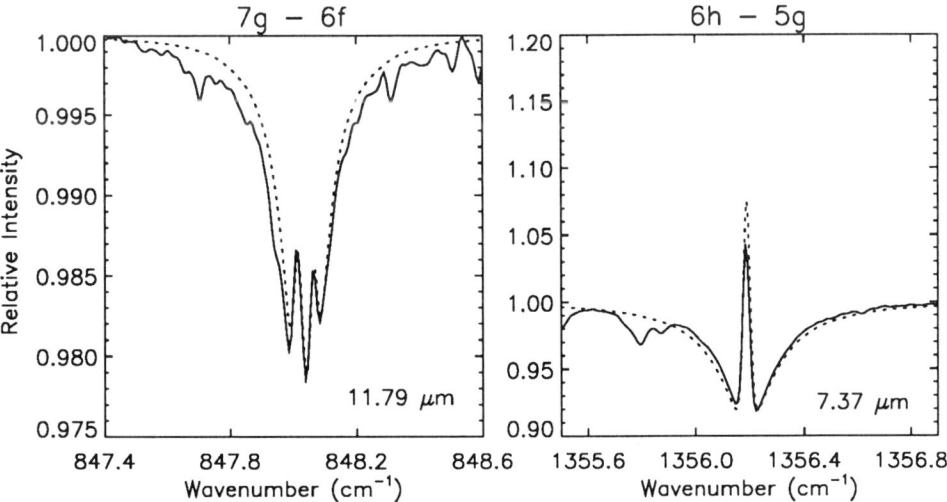

Fig. 6. Modeling (dashed) of two Mg I absorption lines using the setup of Paper I. Observed profiles (solid) are from ATMOS.

sufficiently large, as does the waterfall visualization in Figure 4.

There are also differences between the H I profiles displayed in our contribution elsewhere in this volume and Avrett's H I profiles in the next paper. These are likely to arise from different line-haze formalisms which affect the Balmer continuum (opacity distribution functions versus opacity sampling functions, though both from Kurucz's line tables; LTE versus scattering).

3. The Real Sun: Granules

The good fits which our Mg I computations produce without any parameter adjustment rather seem too good, in the sense that plane-parallel modeling has never managed to reproduce center-to-limb profile behavior of multiple lines in any spectral species without resorting to adjustment of the many fitting parameters represented by height and angle-dependent micro- and macro-turbulence and damping enhancements. The most successful example to date is the analysis of the Na I spectrum by Caccin et al. (1980), which indeed relied on such fitting.

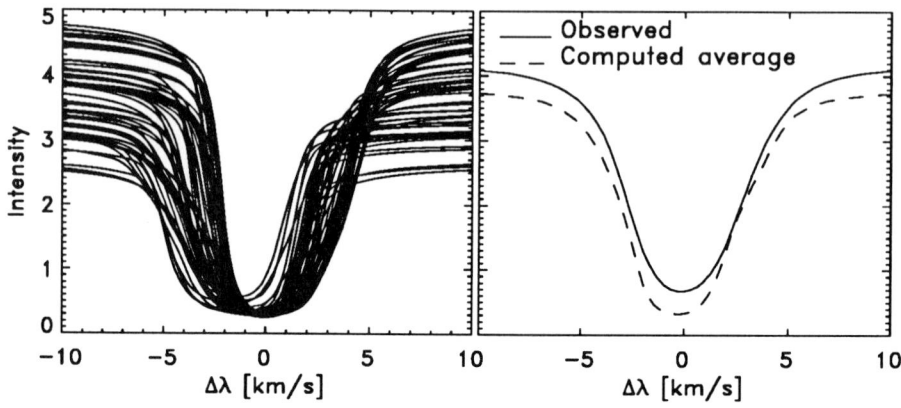

Fig. 7. Profiles of K I 769.9 nm computed from a numerical simulation of the solar granulation by Nordlund and Stein, after Bruls and Rutten (1992). *Left:* profiles per resolved surface element. *Right:* spatially averaged profiles. *Dashed:* average of the computed profiles. *Solid:* observed. *Abscissae:* wavelength expressed as Dopplershift, with upward velocity (blueshift) positive. *Ordinates:* absolute intensity in 10^{-5} erg s^{-1}cm^{-2}Hz^{-1}ster^{-1}.

Why is our Mg I modeling so good? We suspect that the success of these plane-parallel computations of infrared Rydberg lines is enhanced by lessened response to photospheric inhomogeneities. In a parallel analysis, Bruls and Rutten (1992) computed alkali line profiles using output from a Nordlund & Stein granulation simulation as input (*cf.* Stein and Nordlund, 1989; Nordlund and Stein, 1990). Results are shown in Figure 7. The lefthand panel shows profiles of the K I 769.9 nm resonance line per simulation surface element. No turbulent broadening or damping enhancement was employed. The granulation produces appreciable broadening of the mean profile (righthand panel), which will be hard to match precisely with one-dimensional micro- and macroturbulence or collisional damping enhancements.

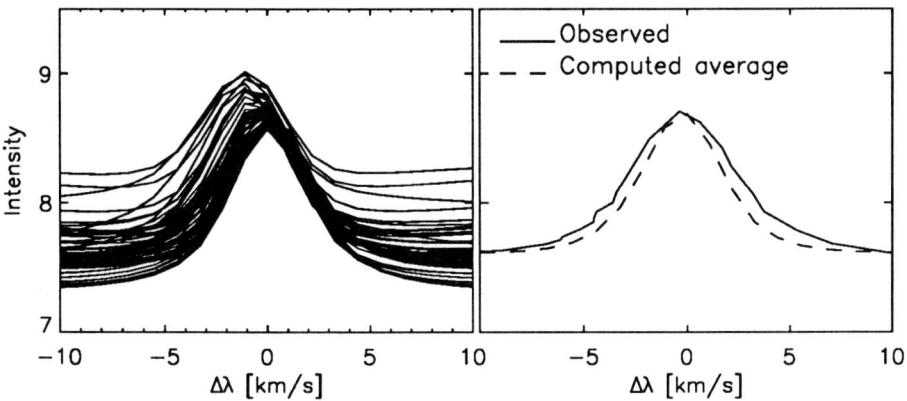

Fig. 8. Same as Figure 7, but for the Mg I 12.32 μm line. *Ordinates:* absolute intensity in 10^{-7} erg s^{-1}cm^{-2}Hz^{-1}ster^{-1}. Note the offset from zero.

The Mg I 12.32 μm line originates from the upper photosphere, as does the core of K I 769.9 nm. Figure 8 shows results for the Mg I line from a similar experiment, using the same Nordlund & Stein simulation data. Again, there is much variation in the left panel whereas the computed average is close to the observed profile. However, there are significant differences between the two tests. The K I background continuum originates much deeper in the photosphere and therefore displays much larger granular modulation, up to 50%. The granular contrast in the Mg I background continuum ranges over less than 10%. In K I, the granular correlation between upward motion and large continuum brightness produces distinct asymmetry in the resolved profiles, both per profile and in their occurrence patterns. The resolved Mg I profiles display more symmetry and less or even reversed Dopplershift-brightness correlation. Note that these profiles are presented on velocity scales to make them comparable. The line cores have similar widths (about 10 km s^{-1} at their base) but the Mg I wings, which map source function structure, are more extended. In summary, the Mg I 12 μm line suffers much less from granulation-induced asymmetries than the K I resonance line because its background continuum originates higher. Gaussian broadening, as prescribed by micro- and macro-turbulence, may therefore indeed work better as simulacrum.

4. The Real Sun: Fluxtubes

The Mg I 12 μm emission features are promising magnetometers, although their Landé factors g_{eff} are only unity. The widths of their emission peaks, set by the NLTE source function structure, are reasonably small (Figure 8). This NLTE sharpness and their clean hydrogen-like splitting pattern gives the Mg I peaks their value. On the other hand, their NLTE origin requires Detailed line formation modeling for any application beyond simple splitting measurement. We display initial NLTE magnetic-profile computations here.

Figure 9 shows profiles and contribution functions for the 12.32 μm line. We

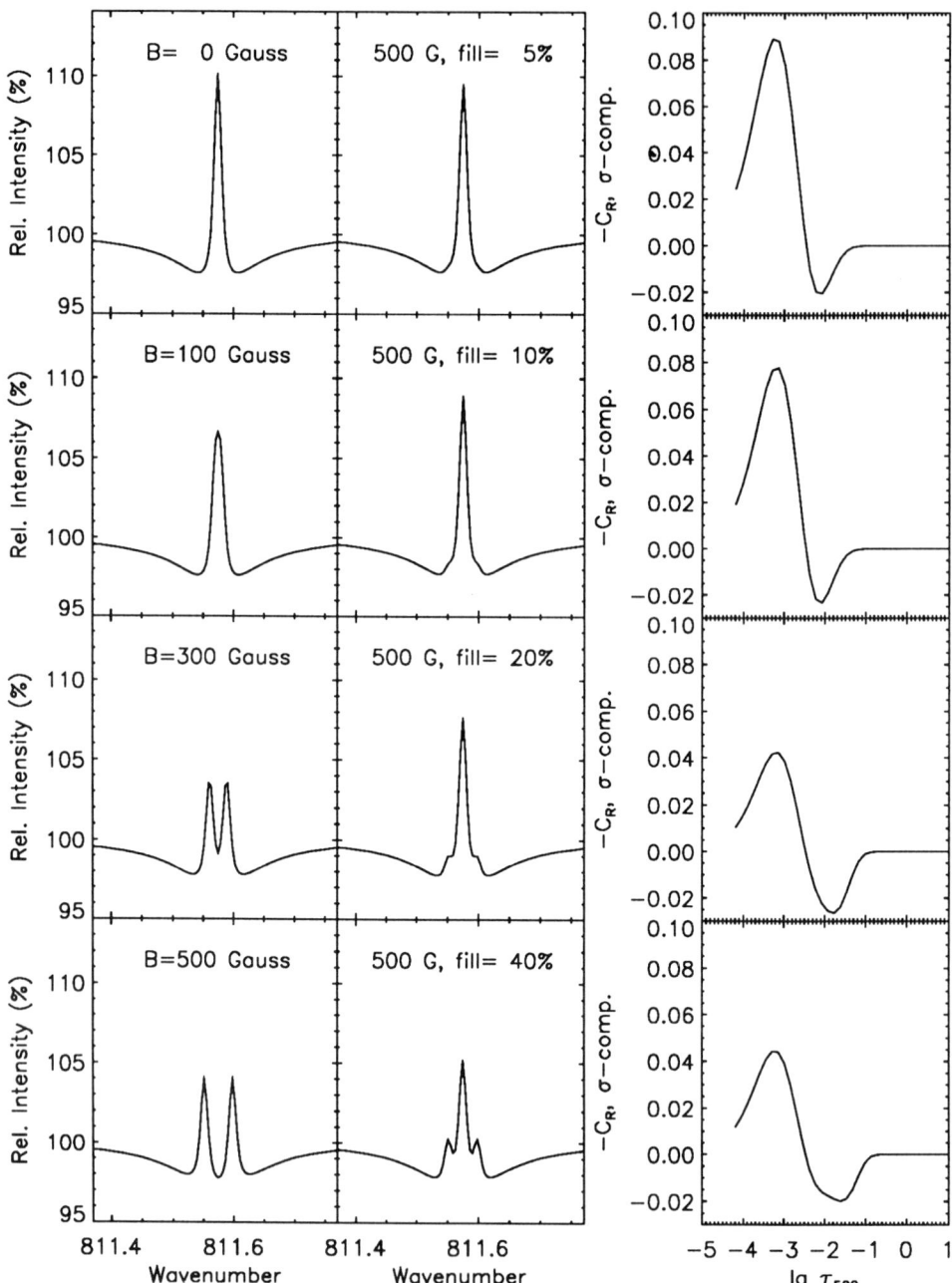

Fig. 9. Magnetic-field modeling of the Mg I 12.32 μm line. *Left:* fluxtube profiles for the indicated field strengths. *Middle:* non-magnetic profiles mixed with 500 G profiles to the specified filling factors. *Right:* σ-component contribution functions, corresponding to the field strengths in the lefthand column.

have adopted the field-free approximation (Rees, 1969, 1987), $g_{\text{eff}} = 1$, the T5780 model atmosphere (Paper I), and height-independent vertical fields of the specified strengths. The lefthand column shows intensity profiles from such first approximations to a solar fluxtube. The σ components are well separated above $B = 300$ G. The middle column shows profiles computed for a mix of 500 G and non-magnetic plasma specified by the filling factor. The filling factor must be appreciable to deliver noticeable σ components. The quiet Sun does not contain sufficient field to upset our plane-parallel reproduction of its spatially-averaged profiles.

Finally, the righthand column displays relative-intensity contribution functions to a σ component, corresponding to the field strengths in the lefthand column. These are line-depression contribution functions as defined by Magain (1986), plotted negatively to let emission be contributed by an upward peak. They are similar to the non-magnetic line center contribution function plotted in Figure 7 of Paper I, at half the magnitude for a fully split profile.

5. Conclusion

The previous reviews of infrared line formation were the Liège ones of de Jager (1964, 1975). In 1964, he remarked that for $\lambda > 2$ μm the solar spectrum was disappointingly void of spectral lines, none having been identified above $\lambda = 5$ μm. In 1975 he had to quote his 1964 review, because nothing had changed.

In 1992, we at last have at least infrared Mg I and H I lines of interest. Neither species is valuable for granulation research. High-resolution narrow-band images taken in Mg I 12.32 μm will be remarkably bland, except for regions of close-packed strong magnetic fields (Figures 8 and 9). This confirms that the primary Mg I interest lies in upper-photosphere plage and penumbra magnetometry. The H I lines appear useful as diagnostics of the temperature and density structure of the low chromosphere (Paper II — see also our poster contribution and Avrett's review in this volume).

Using Mg I and H I emission features to measure properties of the solar atmosphere requires detailed non-LTE modeling. Complete model atoms are then needed, as well as proper accounting for Stark broadening and Stark shifts and correct treatment of the ultraviolet line haze. These tasks are not trivial.

References

Avrett, E. H.: 1993, in these proceedings.
Athay, R. G. and Lites, B. W.: 1972, *Astrophys. J.* **176**, 809.
Biberman, L. M., Vorob'ev, V. S., and Yakubov, I. T.: 1987, *Kinetics of Nonequilibrium Low-Temperature Plasma*, Plenum, New York.
Boreiko, R. T. and Clark, T. A.: 1986, *Astron. Astrophys.* **157**, 353.
Brault, J. and Noyes, R.: 1983, *Astrophys. J.* **269**, L61.
Bruls, J. H. M. J.: 1992, *Formation of diagnostic lines in the solar spectrum*, PhD Thesis, Utrecht University.
Bruls, J. H. M. J. and Rutten, R. J.: 1992, *Astron. Astrophys.* **265**, 257.
Bruls, J. H. M. J., Rutten, R. J., and Shchukina, N. G.: 1992, *Astron. Astrophys.* **265**, 237.
Caccin, B., Gomez, M. T., and Roberti, G.: 1980, *Astron. Astrophys.* **92**, 63.

Carlsson, M.: 1986, *A Computer Program for Solving Multi-Level Non-LTE Radiative Transfer Problems in Moving or Static Atmospheres*, Report No. 33, Uppsala Astronomical Observatory.
Carlsson, M. and Rutten, R. J.: 1992, *Astron. Astrophys.* **259**, L53 (Paper II).
Carlsson, M. and Rutten, R. J.: 1993, in these proceedings.
Carlsson, M., Rutten, R. J., and Shchukina, N. G.: 1990, in L. Deszö (Ed.), *The Dynamic Sun*, Proc. EPS 6th European Solar Meeting, Publ. Debrecen Heliophysical Observatory **7**, Debrecen, p. 260.
Carlsson, M., Rutten, R. J., and Shchukina, N. G.: 1992a, *Astron. Astrophys.* **253**, 567 (Paper I).
Carlsson, M., Rutten, R. J., and Shchukina, N. G.: 1992b, in M. S. Giampapa and J. A. Bookbinder (Eds.), *Cool Stars, Stellar Systems, and the Sun*, Proc. Seventh Cambridge Workshop, ASP Conf. Series, **26**, p. 518.
Chang, E. S.: 1993, in these proceedings.
Chang, E. S., Avrett, E. H., Mauas, P. J., Noyes, R. W., and Loeser, R.: 1991, *Astrophys. J.* **379**, L79.
Chang, E. S., Avrett, E. H., Mauas, P. J., Noyes, R. W., and Loeser, R.: 1992, in M. S. Giampapa and J. A. Bookbinder (Eds.), *Cool Stars, Stellar Systems, and the Sun*, Proc. Seventh Cambridge Workshop, ASP Conf. Series, **26**, p. 521.
Cram, L. E., Rutten, R. J., and Lites, B. W.: 1980, *Astrophys. J.* **241**, 374.
de Jager, C.: 1964, in M. Migeotte (Ed.), *Les Spectres Infrarouges des Astres*, Coll. Int. Liège, Mém. 8° Soc. Roy. Sci., Liège, p. 151.
de Jager, C.: 1975, *Space Sci. Rev.* **17**, 645.
Farmer, C. B. and Norton, R. H.: 1989, *A High-Resolution Atlas of the Infrared Spectrum of the Sun and the Earth Atmosphere from Space*, NASA Ref. Publ. 1224, Vol. 1.
Jefferies, J. T.: 1991, *Astrophys. J.* **377**, 337.
Linsky, J. L.: 1990, in G. Wallerstein (Ed.), *Cool Stars, Stellar Systems and the Sun*, Proc. Sixth Cambridge Workshop, Astron. Soc. Pac. Conf. Series **9**, p. 500.
Lites, B. W.: 1972, *Observation and Analysis of the Solar Neutral Iron Spectrum*, NCAR Cooperative Thesis No. 28, High Altitude Observatory, Boulder.
Magain, P.: 1986, *Astron. Astrophys.* **163**, 135.
Maltby, P., Avrett, E. H., Carlsson, M., Kjeldseth-Moe, O., Kurucz, R. L., and Loeser, R.: 1986, *Astrophys. J.* **306**, 284.
Menzel, D. H. and Cillié, G.: 1937, *Astrophys. J.* **85**, 88.
Nordlund, Å. and Stein, R. F.: 1990, *Comp. Phys. Comm.* **59**, 119.
Omont, A.: 1977, *Journal de Physique* **38**, 1343.
Rees, D. E.: 1969, *Solar Phys.* **10**, 268.
Rees, D. E.: 1987, in W. Kalkofen (Ed.), *Numerical Radiative Transfer*, Cambridge University Press, Cambridge, Great Britain, p. 213.
Rutten, R. J.: 1988, in R. Viotti, A. Vittone, and M. Friedjung (Eds.), *Physics of Formation of FeII Lines Outside LTE*, IAU Colloquium 94, Reidel, Dordrecht, p. 185.
Rutten, R. J.: 1990, in G. Wallerstein (Ed.), *Cool Stars, Stellar Systems and the Sun*, Proc. Sixth Cambridge Workshop, Astron. Soc. Pac. Conf. Series, Volume 9, p. 91.
Schoolman, S. A.: 1972, *Solar Phys.* **22**, 344.
Seaton, M. J.: 1964, *Mon. Not. R. Astron. Soc.* **127**, 177.
Sobelman, I. I., Vainshtein, L. A., and Yukov, E. A.: 1981, *Excitation of Atoms and Broadening of Specral Lines*, Springer, Berlin.
Stein, R. F. and Nordlund, Å.: 1989, *Astrophys. J. Lett.* **342**, L95.
Vernazza, J. E., Avrett, E. H., and Loeser, R.: 1981, *Astrophys. J. Suppl. Ser.* **45**, 635.
Wijbenga, J. W. and Zwaan, C.: 1972, *Solar Phys.* **23**, 265.

MODELING THE INFRARED MAGNESIUM AND HYDROGEN LINES FROM QUIET AND ACTIVE SOLAR REGIONS

E. H. AVRETT

*Harvard-Smithsonian Center for Astrophysics, 60 Garden Street,
Cambridge, MA 02138, U.S.A.*

E. S. CHANG

*Department of Physics and Astronomy, University of Massachusetts,
Amherst, MA 01003, U.S.A.*

and

R. LOESER

*Harvard-Smithsonian Center for Astrophysics, 60 Garden Street,
Cambridge, MA 02138, U.S.A.*

Abstract. The emission lines of Mg I at 7.4, 12.2, and 12.3 μm are now known to be formed in the upper photosphere; the line emission is due to collisional coupling of higher levels with the continuum together with radiative depopulation of lower levels. These combined effects cause the line source functions of high-lying transitions to exceed the corresponding Planck functions. However, there are uncertainties in a) the relevant atomic data, particularly the collisional rates and ultraviolet photoionization rates, and b) the sensitivity of the calculated results to changes in atmospheric temperature and density. These uncertainties are examined by comparing twelve calculated Mg I line profiles in the range 2.1–12.3 μm with ATMOS satellite observations. We show results based on different rates, and using different atmospheric models representing a range of dark and bright spatial features. The calculated Mg profiles are found to be relatively insensitive to atmospheric model changes, and to depend critically on the choice of collisional and photoionization rates. We find better agreement with the observations using collision rates from van Regemorter (1962) rather than from Seaton (1962). We also compare twelve calculated hydrogen profiles in the range 2.2–12.4 μm with ATMOS observations. The available rates and cross sections for hydrogen seem adequate to account for the observed profiles, while the calculated lines are highly sensitive to atmospheric model changes. These lines are perhaps the best available diagnostics of the temperature and density structure of the photosphere and low chromosphere. Further calculations based on these infrared hydrogen lines should lead to greatly improved models of the solar atmosphere.

Key words: atomic processes – infrared: stars – line: formation – Sun: atmosphere

1. Introduction

The Spacelab-3 ATMOS experiment of Farmer and Norton (1989) has obtained high quality disk-center solar spectra in the infrared wavelength range $\lambda = 2.1$–16.5 μm. These spectra include profiles of many atomic and molecular lines that originate from a wide range of depths in the photosphere, and provide the best observations available for determining the structure of the photosphere.

In this paper we compare calculated infrared profiles for both magnesium and hydrogen lines with the ATMOS observed profiles. The line profile calculations for these atoms must account for departures from LTE and are consequently more complex than LTE calculations of molecular lines. However, the Mg and H lines are formed over a greater range of depths than strong lines of CO and other molecules. Also, the H and CO lines have opposite temperature sensitivities, as shown in Sections 5 and 6.

The magnesium results given here represent a continuation of the results given in (1) our two previous papers, Chang et al. (1991, 1992); and (2) the extensive paper of Carlsson, Rutten, and Shchukina (1992a, hereafter CRS). See also Carlsson, Rutten, and Shchukina (1992b) and the paper of Rutten and Carlsson (1993) in these proceedings.

The hydrogen calculations reported here may be compared with the recent results of Carlsson and Rutten (1992, 1993).

2. Mg I Atomic Data

CRS have tested the sensitivity of their computed line profiles to changes in the various atomic data used in their calculation. They find that the results are most sensitive to changes in the Einstein A coefficients, but these are relatively well-established values. For high-l Rydberg lines the dipole matrix elements are essentially hydrogenic (see Hoang-Binh, 1993), while for low-l transitions the A coefficients have been independently calculated by several groups, giving values that are in very good agreement with each other.

The atomic processes that are next in importance are the bound-bound transition rates due to collisions with electrons. We and CRS both adopt the collision rates compiled by Mauas et al. (1988) for transitions between the low-lying levels.

For the remaining high-lying permitted transitions CRS use the impact parameter approximation of Seaton (1962). They assume no intersystem collisional coupling except for the intersystem rates given by Mauas et al. For each of the forbidden transitions within the singlet and triplet systems CRS assume "a collision strength of 5% of the closest permitted transition."

The procedure we use for the remaining high-lying permitted transitions is a choice of either 1) the same Seaton impact parameter formulae, or 2) the approximation by van Regemorter (1962) which gives collision rates that are larger by a factor of 3 or so. The rate that is calculated in either of these cases is proportional to the oscillator strength, f, of the transition. We treat all forbidden transitions in the same way as the remaining permitted transitions, but assume that $f = 0.1$. This way of treating the collision rates for forbidden transitions differs from that of CRS, and may partly explain why our computed profiles shown in Section 3 agree only approximately with those of CRS when we use the Seaton formulae. Another possible difference is that CRS included levels up to $n = 9$ while we have included additional hydrogenic levels up to $n = 15$.

Inelastic collisions with hydrogen atoms appear to have a much smaller effect than those with electrons, as discussed by CRS. For completeness, however, we have included the Mg-H collision rates from Kaulakys (1985) for transitions between the high-l (hydrogenic) levels of Mg I.

The calculated line profiles are also sensitive to the choice of photoionization rates. We find that different ways of treating photoionization from the $3p\,^3P^0$ level can significantly affect the solution. This rate depends on the photoionization cross section and on the radiation intensity in the $\lambda < 251$ nm wavelength region. We and CRS use similar cross section data. We use the results of Ueda, Karasawa, and Fukuda (1982) and CRS use the slightly higher values from Moccia and Spizzo

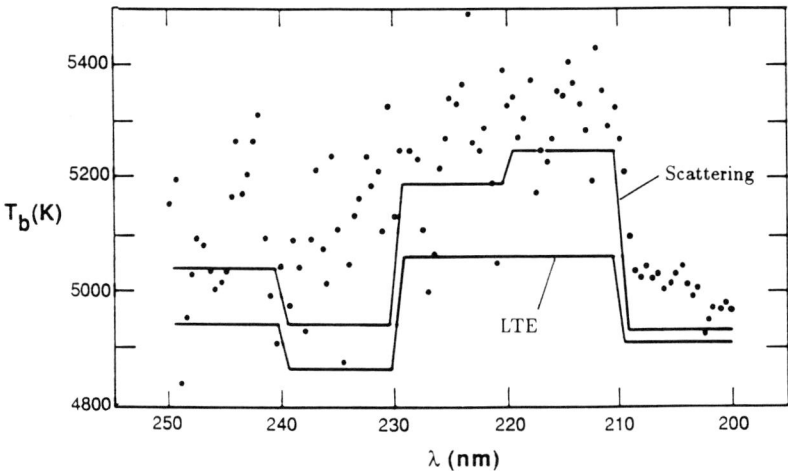

Fig. 1. Observed disk-center brightness temperature of the quiet Sun averaged over 0.5-nm intervals (dots), compared with calculated brightness temperatures averaged over 10-nm intervals.

(1988). The $\lambda < 251$ nm spectrum is filled with absorption lines, and both calculations use line opacity sampling based on the line lists of Kurucz (1991). CRS assume LTE in computing the $\lambda < 251$ nm radiation intensity. We show the results of assuming either a) approximate LTE, or b) that photons absorbed at line wavelengths are scattered so that the resulting mean intensity in the photoionization rate is less strongly coupled to the local Planck function than in LTE.

From a study of non-LTE line blanketing effects in the Sun, Anderson (1989) suggested the following expression for the depth-dependent scattering albedo σ for a given line: $\sigma = 1/(1 + C/A)$, where $C/A = Q \times (n_e/1.1 \times 10^{14})(\lambda/500)^3$. Here n_e is the electron number density at the given atmospheric depth, λ is the wavelength of the line in nm, and the parameter Q has the value 1 for ions, and 0.12 for atoms. An atomic line at $\lambda = 250$ nm formed at a depth where $n_e = 10^{13}$ cm^{-3} would have a scattering albedo $\sigma = 0.9986$. The scattering results we give in this paper use the value $Q = 0.12$. Our results assuming approximate LTE for the photoionizing lines are based on the value $Q = 10^4$, for which $\sigma = 0.0009$ in the above example.

We need to determine which of these two approximations ($Q = 0.12$ or 10^4 for scattering in the photoionizing lines) gives calculated mean intensities vs. depth in the range $\lambda < 251$ nm in better agreement with those in the solar atmosphere. The only comparison that can be made is between the calculated and observed emergent intensities in this wavelength range.

Figure 1 shows the brightness temperature corresponding to the observed diskcenter intensity of the average quiet Sun plotted vs. wavelength. These are rocket observations by Kohl, Parkinson, and Reeves as tabulated by Vernazza, Avrett, and Loeser (1976). Figure 1 also shows the brightness temperatures corresponding to our calculated emergent intensities, averaged over 10 nm intervals, for the scattering ($Q = 0.12$) and LTE ($Q = 10^4$) line approximations discussed above. This comparison suggests that the $Q = 0.12$ scattering approximation for the photoionizing lines is better than assuming LTE for these lines.

3. Effect of Different Atomic Parameters

Scattering in the photoionizing lines is a non-LTE effect that enhances the emission in the Mg I 12 μm line. We find that using the Seaton collision rates with scattering in the $\lambda < 251$ nm lines gives excessive 12 μm emission, and that we obtain better agreement with the observations by using the larger van Regemorter collision rates. We will refer to these two cases as Se-Sct and VR-Sct, respectively.

Assuming LTE for the photoionizing lines tends to decrease the 12 μm emission so that using the Seaton rates gives better agreement with the observations in this case. Assuming LTE and the larger van Regemorter collision rates gives much less emission than is observed. These two cases are designated as Se-LTE and VR-LTE.

Figure 2 shows the observed ATMOS profiles for two Mg lines, and the computed profiles corresponding to the four cases referred to above. The two lines are the $7i$–$6h$ line at 811.6 cm^{-1} (12.32 μm) and the $6g$–$5f$ line at 1406.6 cm^{-1} (7.11 μm). The solid curve is the observed line in all four panels. The dotted profiles in panels a and c are the VR-Sct results, while the Se-LTE results are plotted with long-short dashes. These are the two cases in closest agreement with the observations. Panels b and d show the Se-Sct and VR-LTE results as long-dash and short-dash profiles, respectively. These results seem clearly inconsistent with the observations.

The profiles in Figure 2 correspond to disk center. Figure 3 shows the $7g$–$6f$ line at 848 cm^{-1} (11.79 μm) at $\mu = 0.14$ close to the solar limb. Here we compare the profile observed by Brault and Noyes (1983) with the computed VR-Sct and Se-LTE profiles (dots and long-short dashes, respectively). The VR-Sct results here are closer to the observations.

Given the results shown in Figures 1–3, we adopt the VR-Sct case for the purpose of further comparisons with the ATMOS profiles. The Se-LTE results in Figures 2 and 3 represent our attempt to duplicate the CRS calculations. However, CRS match the observed 12 μm line almost exactly while our Se-LTE profile is not a close fit. The results shown above therefore should not be regarded as a comparison of our results with those of CRS, but rather a comparison of the effects of using van Regemorter *vs.* Seaton collision rates, and of using scattering *vs.* LTE for the photoionizing lines.

4. Atmospheric Models

The computed profiles shown in Figures 2 and 3 were based on the atmospheric model for the average quiet Sun given by Maltby *et al.* (1986), hereafter called model C. Figure 4 shows the temperature as a function of height for this model and two others, A and F, discussed by Avrett (1985) and tabulated by Fontenla, Avrett, and Loeser (1992). In the upper photosphere and low chromosphere these three models have been constructed to account for the range of quiet-Sun Ca II H

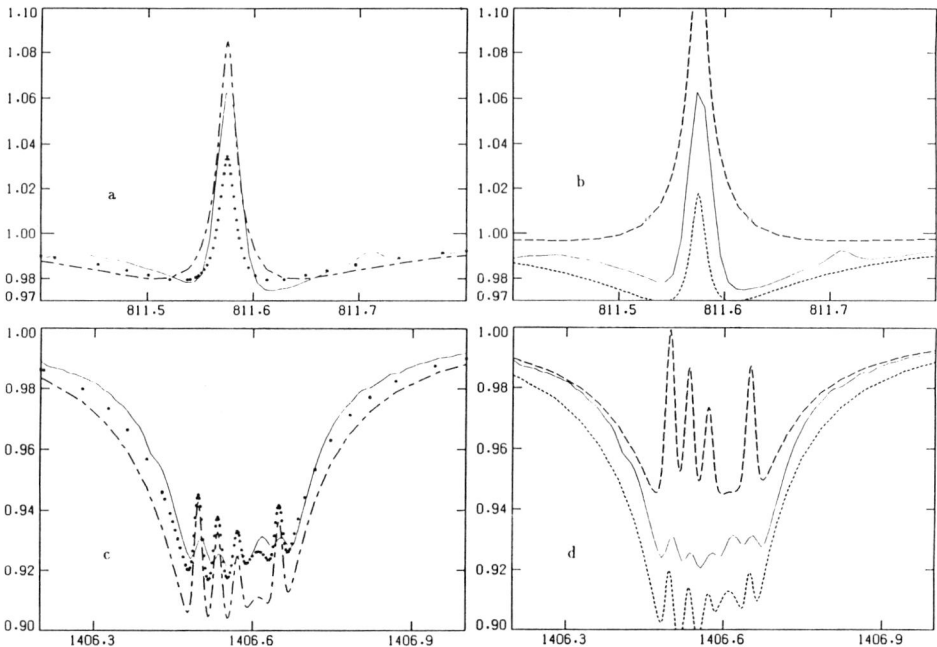

Fig. 2. Disk-center ATMOS profiles (solid lines) for the 811.6 and 1406.6 cm^{-1} Mg lines compared with the calculated results in four cases: Se-LTE (long-short dashes), VR-Sct (dots), Se-Sct (long dashes), and VR-LTE (short dashes).

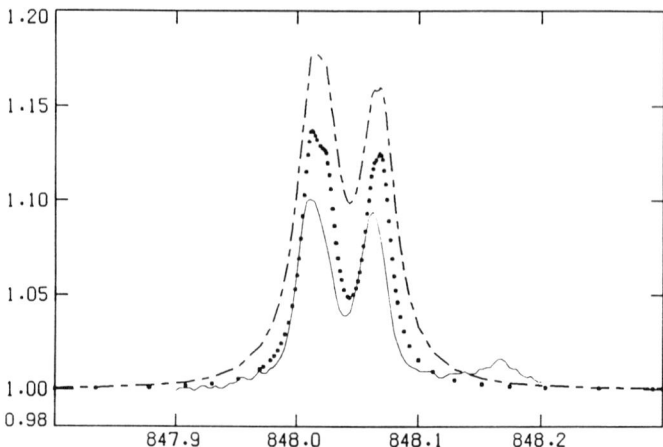

Fig. 3. Profile of the 848 cm^{-1} line at $\mu = 0.14$ observed by Brault and Noyes (solid line) and the calculated profiles in the two cases: Se-LTE (long-short dashes) and VR-Sct (dots).

line profiles observed at high spatial resolution by Cram and Damé (1983). Models A and F give H-line profiles that roughly match the profiles observed in the faintest 10% and the brightest 10% of quiet areas, while model C gives a computed H-line profile that roughly agrees with the observed spatially-averaged profile. Deeper in the photosphere, model C accounts for a range of infrared observations, as discussed by Maltby et al. (1986), and agrees with the model of Holweger and Müller (1974) derived from an analysis of a large number of lines throughout the visible spectrum. The models have an increase in temperature in the low chromosphere to account for the emission in the cores of the Ca II H and K lines and to account for the increase in the observed brightness temperature for $\lambda > 150$ μm in the far infrared and for $\lambda < 160$ nm in the ultraviolet. Models based on the CO lines, however, do not show a chromospheric temperature rise – see Ayres, Testerman, and Brault (1986). Note that the differences between models F and C occur deeper in the atmosphere than the differences between models A and C. The variations of models A and F relative to C are not well established. Very few unambiguous diagnostics are available to determine the properties of these colder and hotter component models. As shown in the next section the calculated infrared Mg lines have very little sensitivity to the differences between models A and F. However, a number of the hydrogen profiles shown later in Section 6 are highly sensitive to the differences between the three atmospheric models, and should be useful in obtaining improved models.

5. Calculated Mg I Lines

The Mg I term diagram in Figure 5 shows the lines considered in this section. These are the prominent Mg I lines that appear in the Farmer and Norton (1989) atlas.

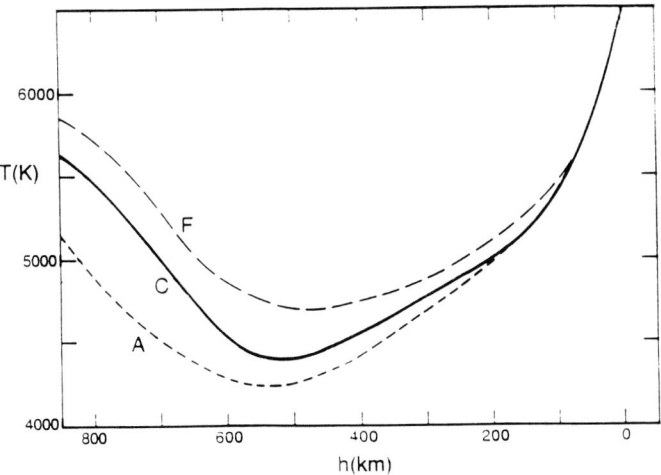

Fig. 4. Temperature *vs.* height for models A, C, and F.

The solid lines in panels *a* through *l* of Figure 6 are the observed profiles. The profiles computed from models A, C, and F are those plotted with short dashes, dots, and long dashes, respectively. All of these lines except those in panels *h* and *i* were discussed earlier by Chang *et al.* (1991, 1992).

From a theoretical viewpoint, the 958 cm^{-1} $7f$–$6\,^3D$ line in panel *i* is similar to the 848 cm^{-1} $7g$–$6f$ line in panel *j*, except that the 958 line is weaker, with three weak emission components instead of two blended pairs of stronger components in the case of the 848 line. The computed emission peaks in the 958 line are evidently too large. The 812, 818, 848, and 958 lines are successively weaker multiplet transitions between the $n = 7$ and $n = 6$ levels. As noted below, the emission we calculate is too small for the 812 line and too large for the 848 line. Correcting this disparity may also reduce the difference between the computed and observed 958 profiles.

The $7h$–$5g$ magnesium line in panel *h* is evident only as a broad depression. The two narrower features at 2166.03 and 2166.1 cm^{-1} are CO lines. Only in this panel of Figure 6 are the model A, C, and F results well separated, showing that the CO lines are much more sensitive to model changes than the Mg I lines.

In our calculations the groups of hydrogenic sublevels $5f$, g, $6f$, g, h, and $7f$, g, h, i in Figure 5 are each treated as single levels for the purpose of computing the non-LTE departure coefficients, b_n. Thus, the $n' - n$ line source function ($n' > n$) is

$$S_\nu = \frac{2h\nu^3/c^2}{(b_n/b'_n)\exp(h\nu/kT) - 1} \quad (1)$$

for the 1407 and 1356 lines in panels *f* and *g*, and for the 848, 818, and 812 cm^{-1} lines in panels *j*, *k*, and *l*. We account for the ν dependence of S_ν in equation (1) between the different multiplet lines. Note that $\exp(h\nu/kT)$ varies by the factor 1.01 between 848 and 812 cm^{-1} for $T = 5000$ K. Nevertheless, the central emission we calculate for the 812 line is less than observed, the calculated emission peaks

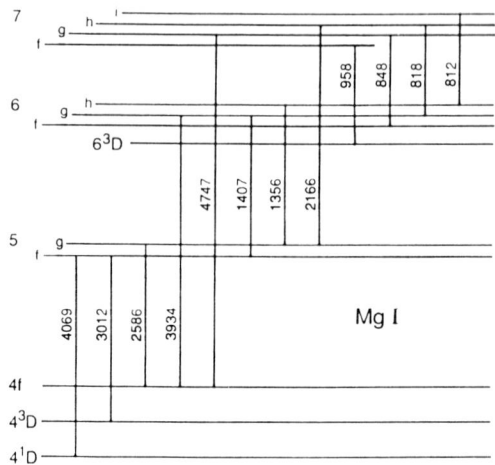

Fig. 5. Mg I term diagram showing the lines discussed in this paper.

for the 848 line tend to be higher than observed, while we obtain good agreement with the intermediate 818 line. CRS treat these sublevels as distinct levels, but they assume very large collision rates for $nl'-nl$ transitions to obtain essentially a common b_n for each group. Thus we would expect the relative behavior of their calculated 812 and 848 lines to be similar to ours. However, they report very close agreement with the observations for both lines (see Rutten and Carlsson, 1993). The differences between our calculation and theirs need to be examined further.

We conclude from the present study that the infrared Mg lines are relatively insensitive to the choice of atmospheric parameters, but are highly sensitive to some of the atomic parameters used in the calculation, particularly the bound-bound collision rates and the degree of scattering in photoionizing lines. When we adopt a high degree of scattering for these lines, to be consistent with UV observations, we find that the van Regemorter collision rates give better agreement with infrared Mg observations than do the smaller Seaton values.

6. Hydrogen Lines

We have also computed the profiles of the infrared hydrogen lines that appear in the Farmer and Norton atlas. Because space is limited here we do not include a description of the various rates and cross sections used in our 15-level hydrogen calculation. These details will be provided in a forthcoming paper by Chang, Avrett, and Loeser (in preparation). Here we simply note that our rates and cross sections for hydrogen are basically the same as those described by Carlsson and Rutten (1992).

The hydrogen and magnesium calculations differ in that the hydrogen results show much less sensitivity to changes in the rates and cross sections. The profiles

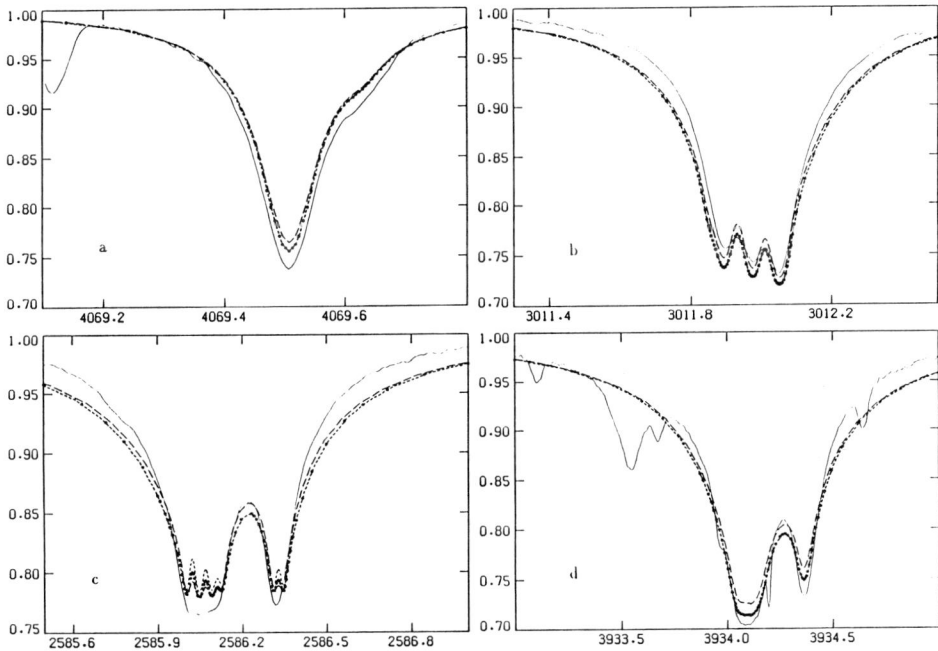

Fig. 6. Profiles of the 12 Mg I lines indicated in Figure 5. ATMOS observations (solid lines) compared with the profiles calculated from model A (short dashes), model C (dots), and model F (long dashes).

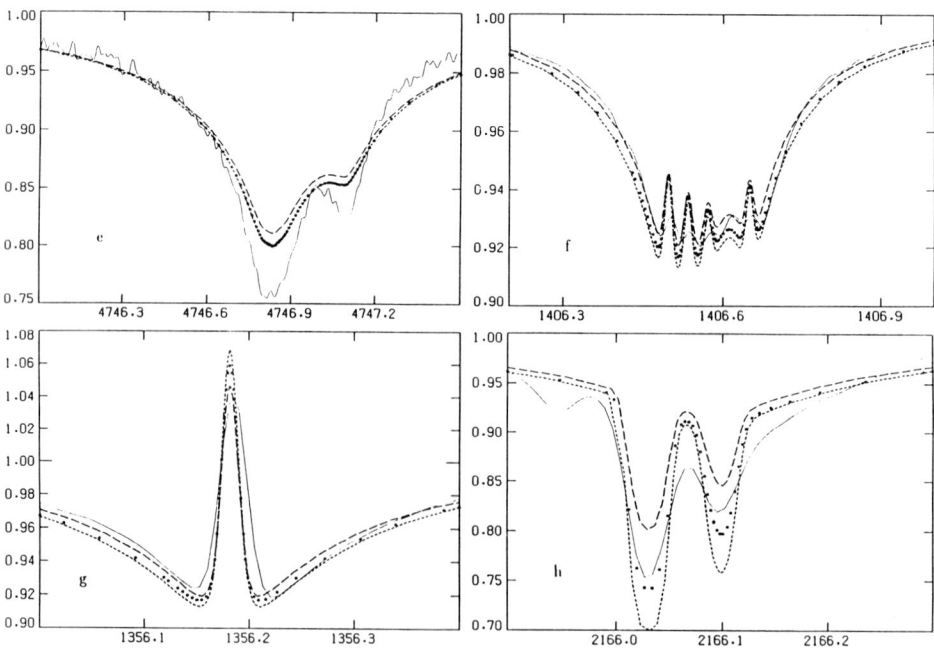

Fig. 6. *Continued.*

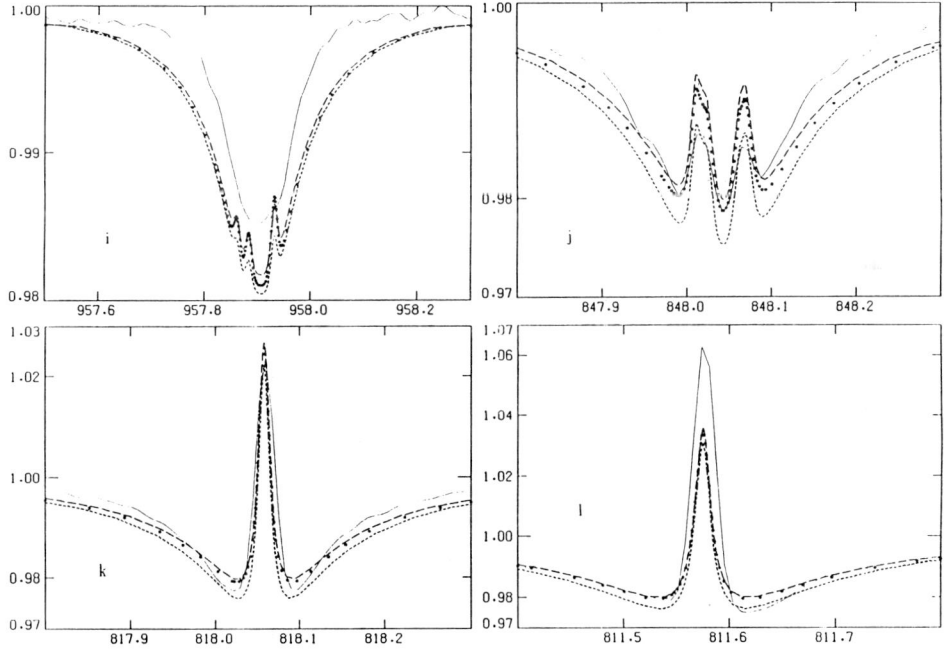

Fig. 6. *Continued.*

we show in Figure 7 below are essentially unaffected by changes in the bound-bound collision rates by factors of 2 to 4 (which occur when we use either the van Regemorter or Seaton rates instead of the values from Johnson 1972 that we use otherwise). Adopting LTE instead of scattering for the photoionizing lines also causes only small changes in the calculated hydrogen lines.

We show profiles for the hydrogen line transitions between levels with the following $n'-n$ principal quantum numbers: 5–4, 6–4, 7–4; 6–5, 7–5, 8–5, 9–5, 10–5; 7–6, 8–6, 9–6; and 9–7. In each case we compare the disk-center profiles for models A, C, and F with the observed ATMOS profile. The results are shown in panels a through i of Figure 7.

The striking difference between these hydrogen profiles and those for Mg I in Figure 6 is the sensitivity of the hydrogen lines to the different atmospheric models. The model F profiles differ substantially from those calculated from models A and C. Note that except for the 7–6 and 9–7 lines, the model F central intensities are smaller than those for models A and C even though the model F temperatures are larger. In contrast, the CO-line model F intensities in Figure 6h are larger. The hydrogen lines behave in this way because the upper level number densities are highly sensitive to temperature. At 500 km, for example (see Fig. 4), the calculated number densities of the $n = 4$ level are 3.2×10^1 (A), 1.3×10^2 (C), and 1.1×10^3 (F) cm^{-3} for models A, C, and F, respectively, and the values of $\tau_0(5-4)$, the 5–4 line center optical depth, are 0.023 (A), 0.21 (C), and 1.5 (F) at this height.

The 2586 cm^{-1} line of Mg I in Figure 6c is also a 5–4 transition (see Fig. 5), but the lower-level number density at 500 km in this case is about 1.7×10^3 for all three models while the central optical depth of the Mg I 2586 line is 0.041 (A), 0.044 (C), and 0.047 (F) in the three cases.

For both H and Mg I, the $n = 4$ number density is closely related to the ion number density, because this level is so close to the continuum (0.85 eV in both cases). The proton number density is highly sensitive to the temperature in this region where hydrogen is almost completely neutral. In contrast, almost all magnesium is ionized so that the Mg II number density varies little with temperature.

In Figure 7d the 6–5 calculated profile for model C (and for model A) has a weak central emission feature which is due to $b_6 > b_5$ (see equation 2) in the upper photosphere and temperature minimum region, and not due to the chromospheric temperature rise. This calculated emission feature is analogous to the central peaks of the Mg I $6g$–$5f$ line at 1407 cm^{-1} (Fig. 6f). The central optical depth $\tau_0(6-5)$ of this hydrogen line has the value unity near 200 km in model C. The observed profile clearly does not show the central emission that we calculate in this case.

The deep absorption in the 6–5 model F profile is due to the much larger values of n_5, causing $\tau_0(6-5) = 1$ to occur near 800 km in the low chromosphere, and causing $b_6 \approx b_5 \approx 1$ in the upper photosphere. Furthermore, we find that $b_6 < b_5$ in the low chromosphere, which prevents the line source function from increasing as T increases in the 500–800 km region. We calculate a residual intensity at line center for model F of about 0.90, which is too low compared to the observed value of about 0.925. (The narrow central absorption feature in Figure 7d is not part of the solar line, but is due to H$_2$O absorption within the observing instrument.) We stated earlier that $\tau_0(5-4) = 1.5$ for model F at 500 km. $\tau_0(5-4) = 1$ occurs near

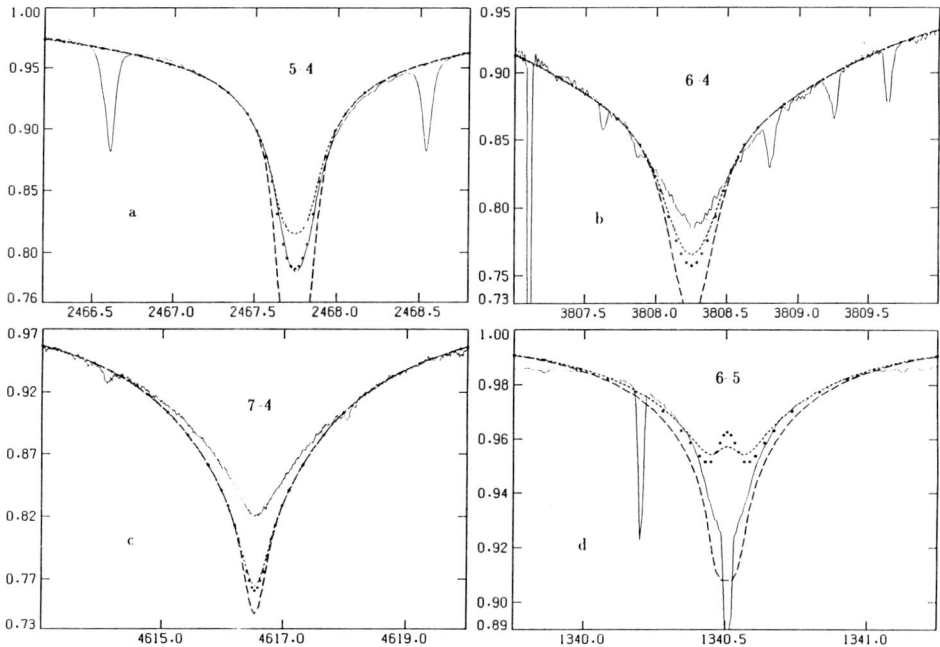

Fig. 7. Profiles of 12 hydrogen lines. ATMOS observations (solid lines) compared with the profiles calculated from model A (short dashes), model C (dots), and model F (long dashes).

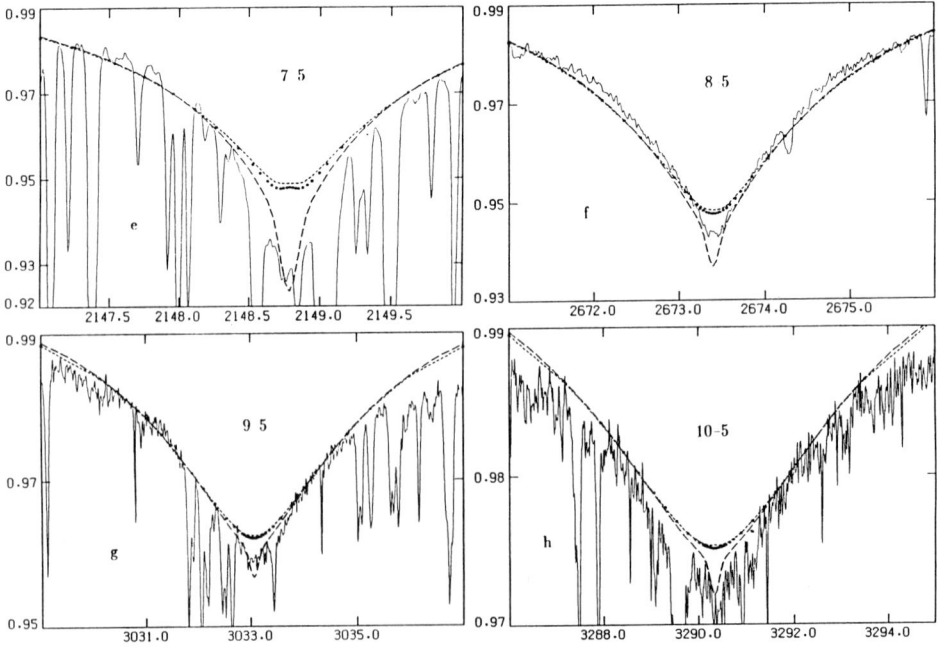

Fig. 7. *Continued.*

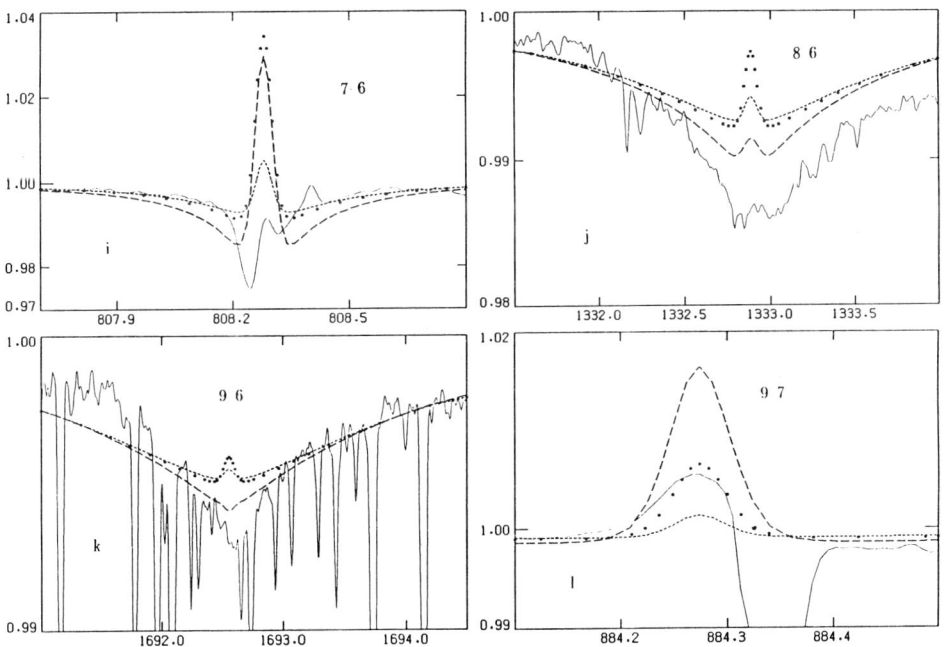

Fig. 7. *Continued.*

850 km, where the temperature is about 1200 K above the minimum value. The 5–4 model F profile (Fig. 7a) does not have a central emission feature because $b_5/b_4 \approx$ 0.8 at 850 km. The $\tau_0(7-6)$ values at 500 km are 0.030 (A), 0.22 (C), and 1.3 (F) which are similar to the $\tau_0(5-4)$ values of 0.023 (A), 0.21 (C), and 1.5 (F) at the same height. However, the 7–6 profiles in Figure 7i bear no resemblance to the 5–4 profiles in Figure 7a. These differences are due to two effects: 1) the line opacity is a much smaller fraction of the continuum opacity at 808 cm^{-1} than at 2468 cm^{-1}, and 2) at and just above the temperature minimum region the model calculations show that $b_5/b_4 < 1$ while $b_7/b_6 \geq 1$. The model F profiles of the 5–4 and 6–5 lines do not show a central emission feature because of the $b'_n/b_n < 1$ effect. All three computed 7–6 profiles show central emission due to both the chromospheric temperature rise and $b'_n/b_n > 1$.

The observed profile in Figure 7i shows a central H(7–6) emission feature together with a strong Mg I ($6d\,^1D$–$6p\,^1P$) absorption line just redward of the H(7–6) line center, and an emission line due to Si I ($7i[7\frac{1}{2}]$–$6h[6\frac{1}{2}]$) in the blue wing. The H(7–6) central emission we calculate is too large except perhaps for model A.

The 6–4 and 7–4 lines in Figures 7b and 7c are weaker than the 5–4 line but are less sensitive to atmospheric model changes. The 7–5, 8–5, 9–5, and 10–5 lines are weaker versions of the 6–5 line. The 8–6 and 9–6 lines are not only weaker than the 7–6 line but are less affected by the chromospheric temperature rise; the non-LTE effects giving central emission for models A and C are reduced in model F due to the higher densities.

Finally we consider the 9–7 line in Figure 7l. At 500 km the $\tau_0(9-7)$ values are 0.007 (A), 0.04 (C), and 0.2 (F), but $b_9/b_7 = 1.05$ (A), 1.03 (C), and 1.00 (F) in the three cases. The model A profile shows weak non-LTE emission while the model F emission is due to the chromospheric temperature rise. Both of these effects influence our model C profile, which matches the observed H(9–7) line rather well. The absorption feature at 884.35 cm^{-1} is a blend of two solar OH lines.

The 8–7, 9–8, 10–9, ... lines at wavenumbers 524.6, 359.7, 257.3, ... are emission lines, mainly due to the temperature rise. Brault and Noyes (1983) measured the profile of the 8–7 line at $\mu = 0.14$ near the limb and found the central emission to be 1.15 times the continuum. We compute 1.29 for model C, which is substantially larger. Boreiko et al. (1993) have detected the disk-center 8–7 line in emission from ground-based observations and have obtained profiles of the 12–11, 13–12, 14–13, and 16–15 emission lines from measurements at balloon altitudes. They find the central emission of the 14–13 line relative to the continuum to be about 1.10. Our calculated 14–13 profiles have central values 1.03 (A), 1.13 (C), and 1.30 (F), showing that only the model C result is consistent with the Boreiko et al. measurement.

7. Conclusions

We have compared ATMOS profiles of the Mg I and H lines in the 2.2–16.7 μm range with theoretical profiles calculated from three given atmospheric models corresponding to faint, average, and bright regions of the quiet Sun. The calculated Mg I profiles have only minor differences due to these three models, but are quite sensitive to the atomic rates and cross sections on which the calculation depends. The results suggest the possibility of using these detailed comparisons to accept or reject various alternative rates and cross sections.

The hydrogen profiles are less affected by uncertainties in the atomic data but are quite sensitive to changes in the atmospheric model parameters. We have interpreted the calculated results but have not made any changes in the atmospheric models to get better agreement with the observations.

The results shown here represent only the first step in utilizing the ATMOS infrared spectra to redetermine the structure of the solar photosphere and low chromosphere.

Acknowledgements

We are very grateful to Dr. R. L. Kurucz for providing the calibrated ATMOS profile data shown here. This work has been supported by NASA Grant NSG-7054.

References

Anderson, L. S.: 1989, *Astrophys. J.* **339**, 558.
Avrett, E. H.: 1985, in B. W. Lites (ed.), *Chromospheric Diagnostics and Modelling*, National Solar Observatory, Sunspot, NM, p. 67.
Ayres, T. R., Testerman, L., and Brault, J. W.: 1986, *Astrophys. J.* **304**, 542.
Boreiko, R. T., Clark, T. A., Naylor, D. A., and Busler, J.: 1993, these proceedings.
Brault, J. and Noyes, R.: 1983, *Astrophys. J.* **269**, L61.
Carlsson, M., and Rutten, R. J.: 1992, *Astron. Astrophys.* **259**, L53.
Carlsson, M., and Rutten, R. J.: 1993, these proceedings.
Carlsson, M., Rutten, R. J., and Shchukina, N. G.: 1992a, *Astron. Astrophys.* **253**, 567 (CRS).
Carlsson, M., Rutten, R. J., and Shchukina, N. G.: 1992b, in M. S. Giampapa and J. A. Bookbinder (eds.), *Cool Stars, Stellar Systems, and the Sun, Seventh Cambridge Workshop*, Astron. Soc. Pacific, San Francisco, p. 518.
Chang, E. S., Avrett, E. H., Mauas, P. J., Noyes, R. W., and Loeser, R.: 1991, *Astrophys. J.* **379**, L79.
Chang, E. S., Avrett, E. H., Mauas, P. J., Noyes, R. W., and Loeser, R.: 1992, in M. S. Giampapa and J. A. Bookbinder (eds.), *Cool Stars, Steller Systems, and the Sun, Seventh Cambridge Workshop*, Astron. Soc. Pacific, San Francisco, p. 521.
Cram, L. E., and Damé, L.: 1983, *Astrophys. J.* **272**, 355.
Farmer, C. B., and Norton, R. H.: 1989, *A High-Resolution Atlas of the Infrared Spectrum of the Sun and the Earth Atmosphere from Space*, NASA Ref. Pub. 1224, Vol. 1.
Fontenla, J. M., Avrett, E. H., and Loeser, R.: 1992, *Astrophys. J.*, in press.
Hoang-Binh, D.: 1993, these proceedings.
Holweger, H., and Müller, E. A.: 1974, *Solar Phys.* **39**, 19.
Johnson, L. C.: 1972, *Astrophys. J.* **174**, 227.
Kaulakys, B.: 1985, *J. Phys. B.*, **18**, L167.
Kurucz, R. L.: 1991, in L. Crivellari, I. Hubeny, and D. G. Hummer (eds.), *Stellar Atmospheres: Beyond Classical Models*, p. 441.
Maltby, P., Avrett, E. H., Carlsson, M., Kjeldseth-Moe, O., Kurucz, R. L., and Loeser, R.: 1986, *Astrophys. J.* **306**, 284.
Mauas, P. J., Avrett, E. J., and Loeser, R.: 1988, *Astrophys. J.* **330**, 1008.
Moccia, R., and Spizzo, P.: 1988, *J. Phys. B.*, **21**, 1133.
Rutten, R. J., and Carlsson, M.: 1993, these proceedings.
Seaton, M. J.: 1962, *Proc. Phys. Soc. London*, **79**, 1105.
Ueda, K., Karasawa, M., and Fukuda, K.: 1982, *J. Phys. Soc. Japan*, **51**, 2267.
van Regemorter, H.: 1962, *Astrophys. J.* **136**, 906.
Vernazza, J. E., Avrett, E. H., and Loeser, R.: 1976, *Astrophys. J. Suppl.* **30**, 1.

COMPUTATION OF INFRARED HYDROGEN LINES

MATS CARLSSON

Institute of Theoretical Astrophysics, P.O. Box 1029, Blindern, N-0315 Oslo 3, Norway

and

ROBERT J. RUTTEN

Sterrekundig Instituut, Postbus 80 000, NL-3508 TA, Utrecht, The Netherlands

Abstract. We compare infrared hydrogen lines observed with ATMOS with computations for two models of the solar atmosphere, one without and one with a chromosphere. The weaker H I lines are formed in the photosphere. Proper evaluation of Stark broadening is required to reproduce their profiles; the heavy ion contribution is most important. The cores of the stronger lines are sensitive to the structure of the chromosphere, but detailed NLTE modeling is needed for diagnostic applications.

Key words: atomic processes – hydrogen – infrared: stars – line: formation – line: profiles – Sun: atmosphere

1. Introduction

In an early review of the infrared solar spectrum, de Jager (1964) summarized his pioneering H I Paschen and Brackett studies (de Jager and Neven, 1950; de Jager *et al.*, 1956). He suggested that it would be worthwhile to compare the infrared hydrogen lines with modern broadening theory and noted that Pfund α and the next α line (6–7) should be observable from 20 km or higher altitude. A quarter century later, the ATMOS space interferometer (Farmer and Norton, 1989) has finally yielded data which allow us to follow his suggestion. We do that here, using an ATMOS atlas kindly forwarded by R. Kurucz. More detail is given in Carlsson and Rutten (1992) and elsewhere in these proceedings (Rutten and Carlsson, 1993).

2. Collisional Broadening

Figure 1 shows relative H I broadening contributions against $\lg \tau_{500}$ for the MACKKL quiet-Sun model atmosphere (Maltby *et al.*, 1986). In the deepest photosphere ($\lg \tau_{500} > 0$), the broadening is dominated by Stark broadening due to protons. This contribution falls off quickly with height and then peaks again in the low chromosphere, above the temperature minimum, which is at $\lg \tau_{500} = -4$. Electron broadening shows a similar pattern, with an outward rise from increasing metal ionization. Throughout the photosphere, where the damping wings of the H I absorption lines are formed, Stark broadening by metal ions outweighs all other contributions except for the van der Waals broadening of Brackett α (top left panel).

3. Line Profiles

Figure 2 shows ATMOS data and computed H I line profiles, in the form of radiation temperatures to cancel variations in Planck function sensitivity with wavelength. All panels have the same temperature scale to facilitate comparison; similarly, each panel spans ± 100 km s^{-1} in Dopplershift from line center.

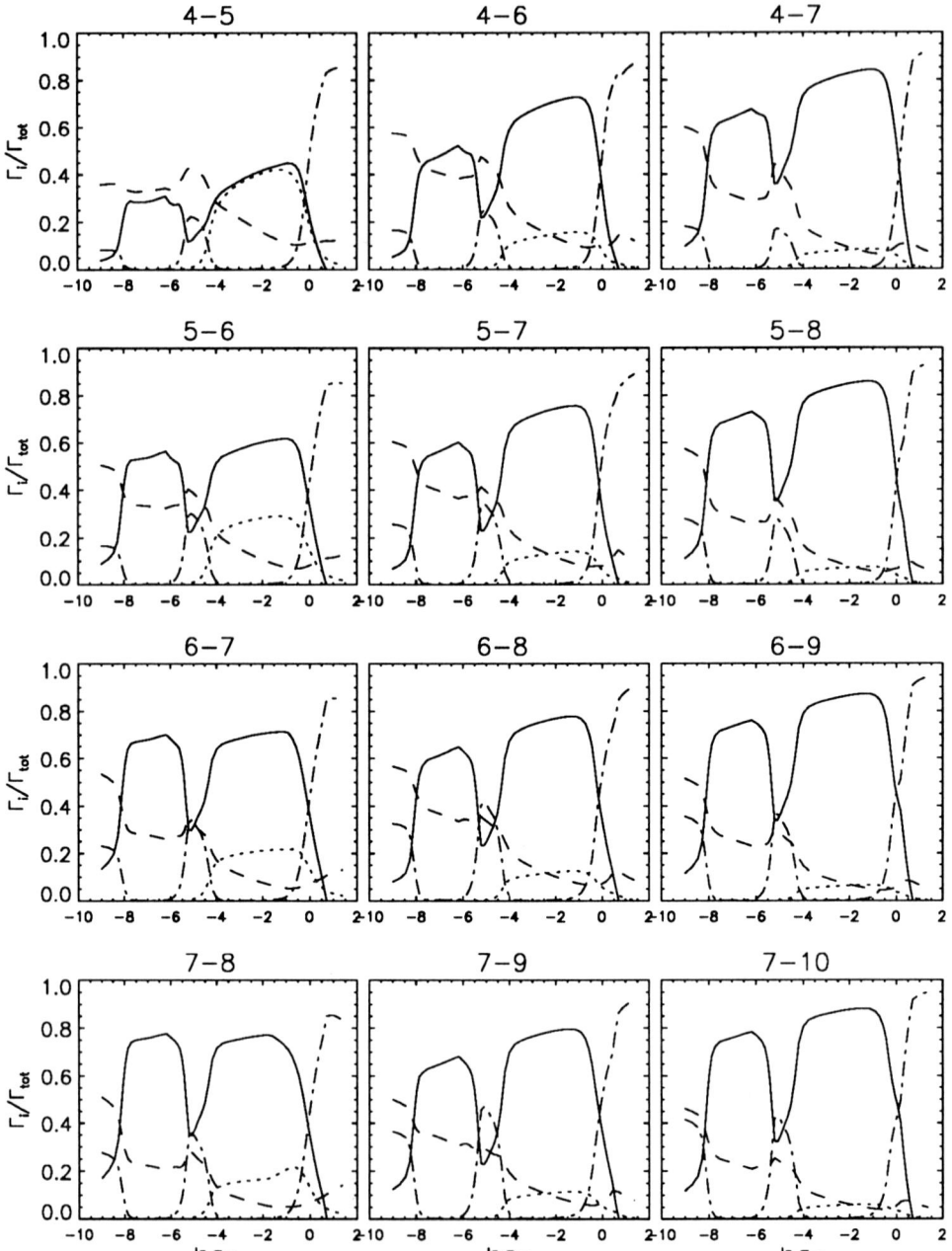

Fig. 1. Broadening contributions for infrared H I lines against continuum optical depth. *Solid:* Stark broadening by ions. *Dot-dashed:* Stark broadening by protons. *Dashed:* Stark broadening by electrons. *Dotted:* van der Waals broadening.

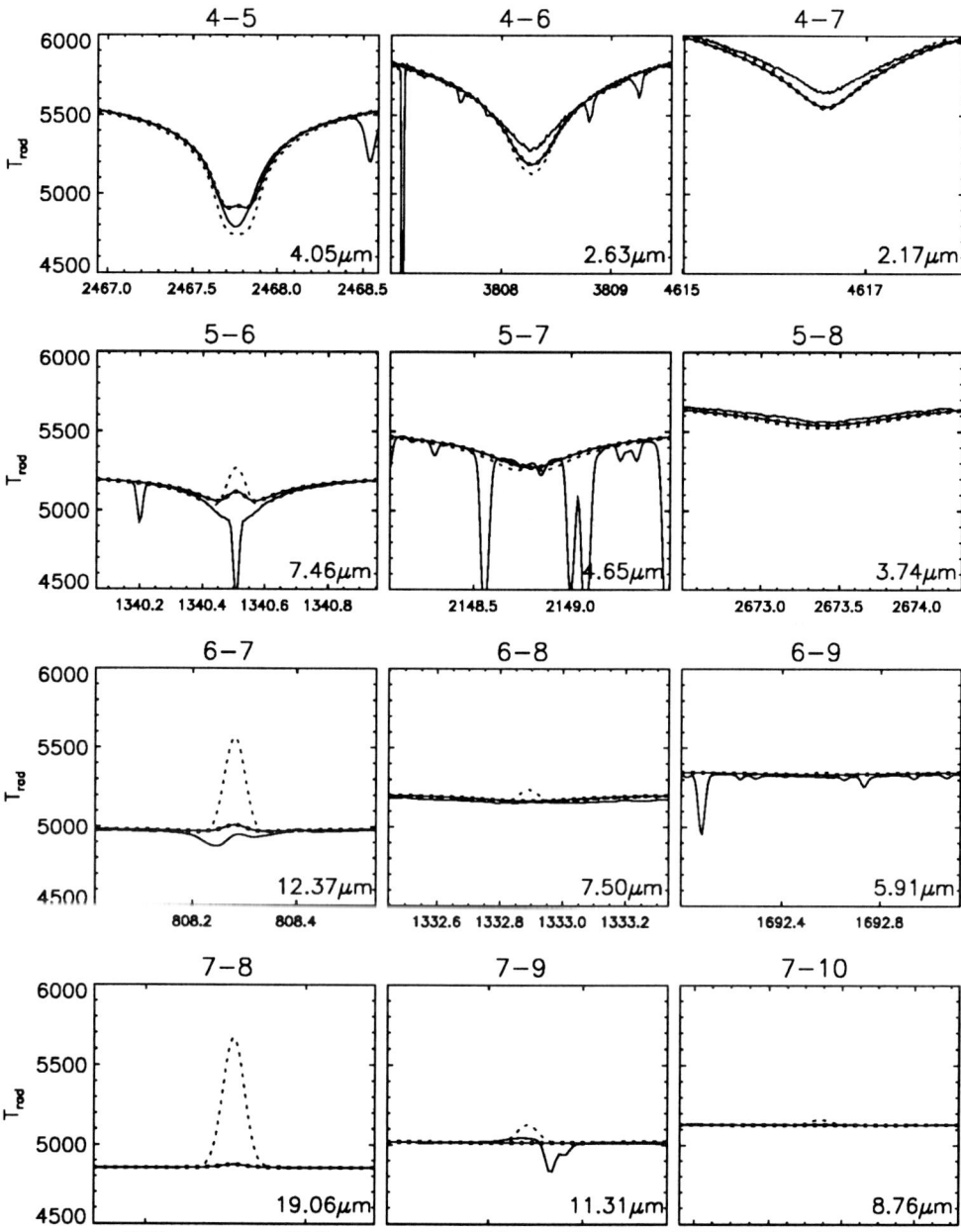

Fig. 2. Comparison of observed spectra from ATMOS (solid) with H I line profiles computed from MACKKL (dashed) and T5780 (solid+dashed). All profiles are normalized to the continuum intensity computed from the T5780 model. *Ordinates:* radiation temperature in K. *Abscissae:* wavenumber in cm^{-1}. The line-center wavelength in μm is given in each panel.

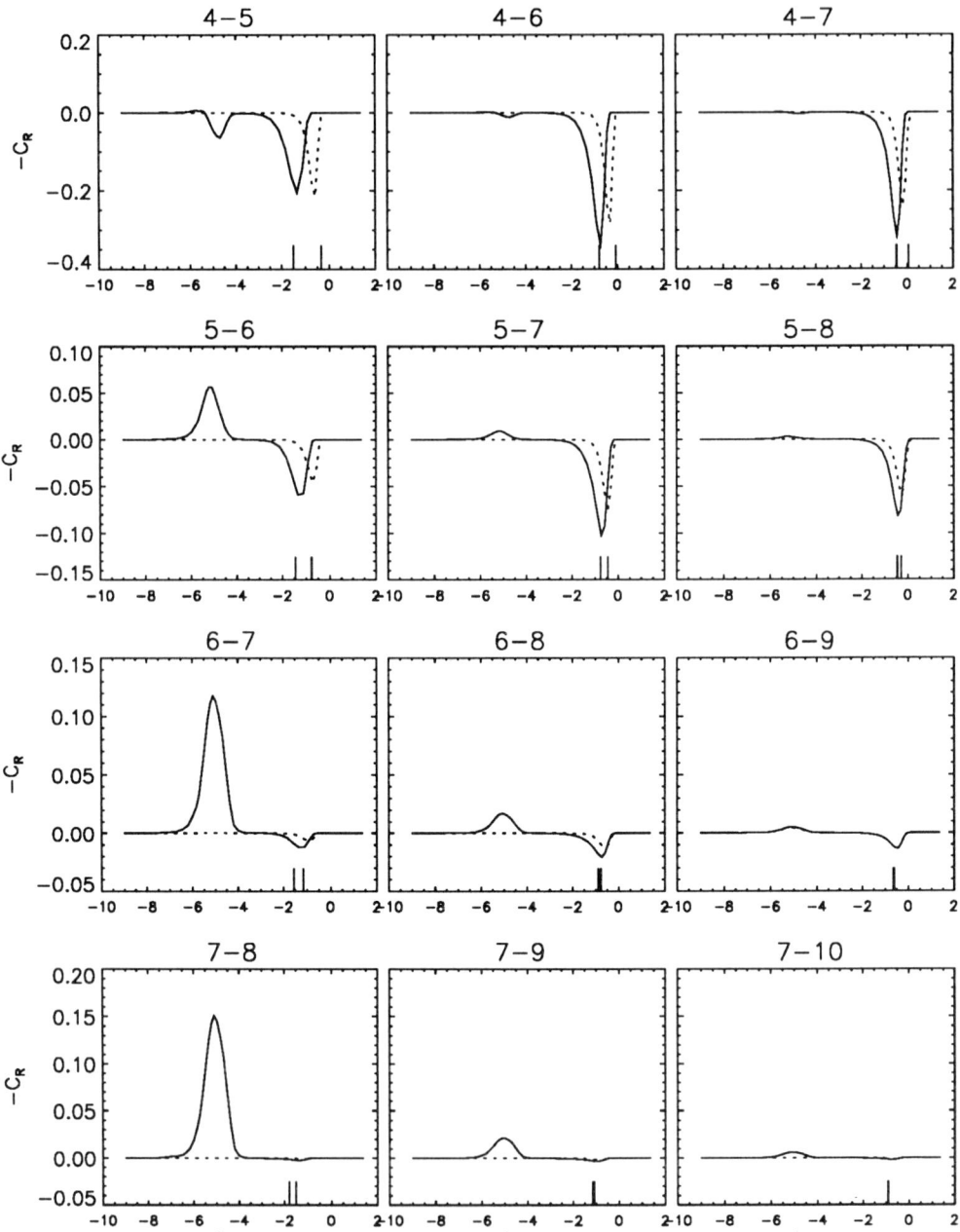

Fig. 3. H I relative-emission contribution functions for MACKKL against lg τ_{500}. *Solid:* at line center. *Dashed:* for $\Delta\lambda$ corresponding to the half-maximum value of the emergent line depression. The ticks on the abscissae mark the $\tau_\nu = 1$ locations at line center and in the adjacent continuum.

The dashed profiles are for the MACKKL model, the solid-plus-dashed for the T5780 model. The latter is a radiative-equilibrium model without chromosphere from Uppsala (Edvardsson *et al.*, 1993). The wide wings of the photospheric profiles are reproduced very well. In the line cores, there are noticeable differences between computations and observations, and also between the two models.

The extinction coefficient of the α lines increases with the lower-level statistical weight, downward along the lefthand column of Fig. 2. Nevertheless, the lines weaken in equivalent width due to the steep increase of the H^- free-free continuum opacity with wavelength (Figure 1 of Carlsson and Rutten, 1992). The 4-5 line is shallowest for the T5780 computation, in which the core is filled in by NLTE departure divergence. The MACKKL computation has a deeper core, slightly deeper than the observed profile. The other α lines show a reversed pattern. The T5780 computations have weak emission reversals due to NLTE effects, while the MACKKL produces additional emission from its chromosphere. LTE computations with the MACKKL give peaks of about half the NLTE values (Figure 3 of Carlsson and Rutten, 1992). The absorption spike in the 5-6 line is due to gas within the ATMOS interferometer. In the observed 6-7 spectrum, H I seems to cause the slight upward hump, not the dip at left. The 7-8 line was not observed by ATMOS. Its LTE MACKKL peak height is half the NLTE value shown here. The corresponding LTE line-center intensity is $I_0/I_c = 1.07$, which corresponds better to the observation of Boreiko and Clark (1986) than the NLTE value.

The line and the continuum extinction decrease both to the right along the rows in Figure 2. Both models produce slightly too deep 4-6 and 4-7 lines. The 5-7 and 5-8 lines are reproduced very well. The higher β and γ lines (6-8 to 7-10) are better modeled by the T5780 atmosphere than by the MACKKL model. The latter again predicts too much line-center emission.

4. Contribution Functions

Figure 3 shows relative-emission contribution functions for MACKKL. These are line-depression contribution functions as defined by Magain (1986), plotted on reversed scales to let positive contributions correspond to line emission above the continuum. The $\tau_\nu = 1$ locations for line center and the continuum are given by vertical tick marks on the abscissae. Their convergence with increasing wavelength (downward along the columns) illustrates that the continuum gains more opacity than the lines.

The $\tau_\nu = 1$ locations all lie within the photosphere. The strongest, highest-formed line (7-8) reaches unit opacity below $\lg \tau_{500} = -2$. Figure 1 of Carlsson and Rutten (1992) shows that even the H I 18-19 line reaches $\tau_\nu = 1$ only at the temperature minimum.

Nevertheless, the H I lines are sensitive to the chromosphere far above these $\tau_\nu = 1$ locations. The H I Rydberg populations have a dip at the temperature minimum (Figure 1 of Rutten and Carlsson, 1993). The line-to-continuum extinction ratio increases rapidly with height above the temperature minimum (Figure 1 of Carlsson and Rutten, 1992). Plots of the variation of τ_ν with τ_{500} show plateaus at the temperature minimum. The lines therefore reach substantial optical depth above the temperature minimum, even though their $\tau_\nu = 1$ location is in the photosphere. It causes the appreciable chromospheric relative-emission contributions in Figure 3.

5. Discussion

Various improvements in modeling infrared H I lines are desirable. First, Stark broadening should be computed with better accuracy. Ion densities should be quantified per species; the Holtsmark distribution should be used; and Stark shifts should be accounted for in detail. Unfortunately, atomic physics theory does not provide definitive recipes yet; the deeper-formed H I lines may help in formulating these.

Second, the presence of NLTE population departure divergences in the upper photosphere requires more investigation (cf. Rutten and Carlsson, 1993). Overionization in the Balmer continuum is probably the main driving process, but experimentation should bear that out. A "multi-MULTI" sensitivity analysis (cf. Carlsson et al. 1992a, 1992b) may identify the role of various population and depopulation processes.

Third, the cores of the H I α lines are sensitive to the structure of the chromosphere. Zirker (1985) showed that their off-limb emission strengths, as observed by Brault and Noyes (1983), fit characteristic prominence properties; their disk-center and center-to-limb profile shapes may constrain empirical models of the solar atmosphere. This sensitivity makes the infrared H I lines interesting, and makes detailed NLTE modeling of their formation worthwhile. The differences between profiles with and without a chromosphere in Figure 2 and between NLTE and LTE peak heights in Figure 3 of Carlsson and Rutten (1992) represent a first step. Other comparisons are discussed by Avrett (1993) in this volume.

References

Avrett, E. H.: 1993, in these proceedings.
Boreiko, R. T. and Clark, T. A.: 1986, *Astron. Astrophys.* **157**, 353.
Brault, J. and Noyes, R.: 1983, *Astrophys. J.* **269**, L61.
Carlsson, M. and Rutten, R. J.: 1992, *Astron. Astrophys.* **259**, L53.
Carlsson, M., Rutten, R. J., and Shchukina, N. G.: 1992a, *Astron. Astrophys.* **253**, 567.
Carlsson, M., Rutten, R. J., and Shchukina, N. G.: 1992b, in M. S. Giampapa and J. A. Bookbinder (Eds.), *Cool Stars, Stellar Systems, and the Sun*, Proc. Seventh Cambridge Workshop, ASP Conf. Series, 26, p. 518.
de Jager, C.: 1964, in M. Migeotte (Ed.), *Les Spectres Infrarouges des Astres*, Coll. Int. Liège, Mém. 8º Soc. Roy. Sci., Liège, p. 151.
de Jager, C., Migeotte, M., and Neven, L.: 1956, *Ann d'Astrophys.* **19**, 9.
de Jager, C. and Neven, L.: 1950, *Proc. Kon. Ned. Akad. Wetensch.* **53**, 1578.
Edvardsson, B., Gustafsson, B., Lambert, D. L., Nissen, P. E., Tomkin, J., and Andersen, J.: 1993, *Astron. Astrophys.*, in press.
Farmer, C. B. and Norton, R. H.: 1989, *A High-Resolution Atlas of the Infrared Spectrum of the Sun and the Earth Atmosphere from Space*, NASA Ref. Publ. 1224, Vol. 1.
Magain, P.: 1986, *Astron. Astrophys.* **163**, 135.
Maltby, P., Avrett, E. H., Carlsson, M., Kjeldseth-Moe, O., Kurucz, R. L., and Loeser, R.: 1986, *Astrophys. J.* **306**, 284.
Rutten, R. J. and Carlsson, M.: 1993, in these proceedings.
Zirker, J. B.: 1985, *Solar Phys.* **102**, 33

NEW ATOMIC DATA FOR Mg I LINES

D. HOANG-BINH

LAM, Observatoire de Paris, F-92195 Meudon Cédex, France

Abstract. Theoretical oscillator strengths are reported for transitions between excited terms of the singlet and triplet S, P, D, F and G manifolds of atomic magnesium. The principal quantum numbers of the jumping electron range from $n = 4$ to 100. The calculations are based on the Coulomb approximation. Low-n results, for which other data are available previously, agree reasonably well with the sophisticated calculations of Moccia and Spizzo (1988). Data for $^{1,3}F$–$^{1,3}G$ transitions may deviate from hydrogenic values by as much as 50%.

Key words: atomic data – infrared: stars – Mg I

1. Introduction

Absorption oscillator strengths for transitions in Mg I have been calculated by many authors, e.g., Froese-Fisher (1975 a, b), Victor et al. (1976), Moccia and Spizzo (1988, hereafter MS). In these works, the principal quantum numbers of the jumping electron are smaller than 10. Over the last decade, several lines of Mg I have been discovered in the far-infrared spectrum of the sun (e.g., Brault and Noyes 1983; Chang and Noyes 1983; Jefferies 1991), and their study has become an important topic in solar physics. As the recent works of Hoang-Binh (1991) and Chang et al. (1991) have shown, investigations on the formation of these Rydberg lines require a large body of atomic data, in particular for principal quantum numbers $n \geq 10$. We report new calculations of multiplet oscillator strengths for transitions between excited terms of Mg I, for n up to 100. These data are required to compute radiative and collisional transition rates entering the statistical equilibrium equations, and are also relevant to the study of the line broadening (Hoang-Binh et al. 1987).

2. Theory

The multiplet absorption oscillator strength, for a transition between a lower term i and an upper term k, is given by (Wiese et al. 1969):

$$f(i,k) = (303.7/g_i\lambda)S_{i,k}, \qquad (1)$$

where λ is the wavelength of the transition in Å, $S_{i,k}$ is the multiplet strength in atomic units (a.u.), and g_i is the statistical weight of the lower term. The strength of a transition $3s\,nl\,^{2S+1}L$ to $3s\,n'\,l'\,^{2S+1}L'$, involving no equivalent electrons, is, for $l' = l - 1$ (Goldberg 1939),

$$S(i,k) = (2S + 1)(2L + 1)l(2l - 1)\sigma^2, \qquad (2)$$

where

$$\sigma^2 = a_0^2\, e^2/(4l^2 - 1)\, R^2, \qquad (3)$$

$$R = \int_0^\infty P_{n,l}(r)\, P_{n',l'}(r)\, dr, \qquad (4)$$

and n, n' and l, l' are the principal and azimuthal quantum numbers respectively of the jumping electron; S is the total spin and L, L' the total azimuthal quantum numbers, a_0 is the Bohr radius (1 in a.u.), and $P_{n,l}(r)$ and $P_{n',l'}(r)$ are normalized radial eigenfunctions, in atomic units, of the jumping electron in the relevant configurations.

Thus, the calculation of $f(i, k)$ reduces to that of the integral R. For $n, n' \geq 10$, it is justified to use the Coulomb approximation, for which (Hoang-Binh et al. 1979)

$$P_{n,l}(r) = z^{1/2} K(\nu, l) W_{\nu, l+1/2}(2\rho/\nu), \tag{5}$$

where z is the effective charge, ν is the effective quantum number, $\rho = zr$, and $W_{\nu, l+1/2}$ is the Whittaker function. The normalizing factor is (Seaton 1958)

$$K(\nu, l) = [\zeta(\nu) \nu^2 \Gamma(\nu + l + 1) \Gamma(\nu - l)]^{-1/2}. \tag{6}$$

Provided ν is not too small, $\zeta(\nu) \approx 1$; here we shall take $\zeta(\nu) = 1$ and use the analytical formula given by Hoang-Binh et al. (1979) to calculate R. The quantum defects have been calculated, using either the term positions of Bashkin and Stoner (1975) (for n, n' up to ≈ 8), or the extended Ritz formulae given by Risberg (1965) for larger values of n, n'. For ^1D terms, values for large n have been obtained by extrapolation of Risberg's values; $^{1,3}G$ terms are taken to be hydrogenic.

3. Results and Discussion

Calculations have been performed for transitions between the $^{1,3}S, ^{1,3}P^0, ^{1,3}D$, and $^{1,3}G$ manifolds. The principal quantum numbers of the jumping electron range from 4 or 5 to 100. Although we are mainly interested in the domain $n, n' \geq 10$, where data are generally not available, lower terms have been considered for the sake of self-consistency and for comparison with other theoretical results. The full results will be published elsewhere. Here, we show only the trends of f-values for large n and Δn. Let the principal quantum numbers of the upper and lower terms be denoted by m and n respectively; and $\Delta n = m - n$. Figure 1 displays an example of the typical behavior of the oscillator strength with respect to Δn (for n fixed), and indicates that interpolation or extrapolation are possible. An exception concerns the transitions $np\ ^1P - md\ ^1D$, for which the variation of f as a function of Δn is not monotonic for $n < 10$. In general, it is found that $f \sim 1/(\Delta n^\gamma)$, with $\gamma \approx 2.7$ in the limit of large Δn, while f is approximately linear with respect to n in the limit of large n.

To our knowledge, the most extensive body of theoretical data on Mg I has been given by MS, who considered only $n, n' \leq 9$. The agreement of our results with these sophisticated calculations is excellent in some instances, e.g., for the transitions $np\ ^1P - ms\ ^1S$ (see Table 1). This is somewhat surprising, and probably fortuitous, since we do not expect the Coulomb approximation to be very good for transitions involving low n and strongly core-penetrating orbits. One the other hand, a few large discrepancies may be also noted (e.g., $np\ ^1P^0 - md\ ^1D, \Delta n \geq 1$). On the whole, most of our low-n results agree reasonably well with those of MS. Our high-n results should be more reliable, since the radial wave function is rather well represented by the Whittaker function at large radial distances. For Mg I, Wiese

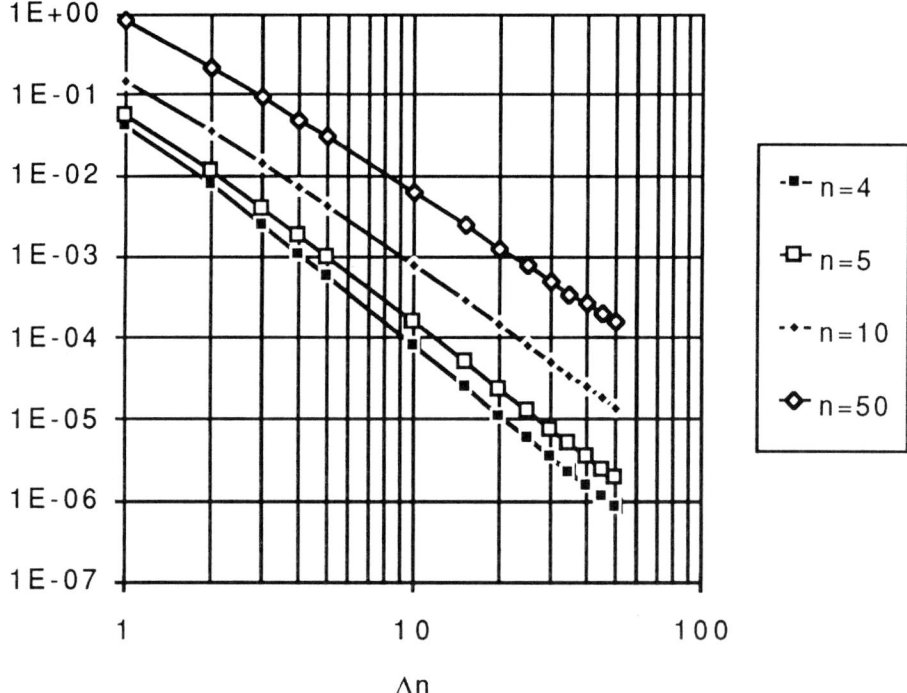

Fig. 1. $f(ns\,^1S, mp\,^1P^0)$ as a function of Δn, for selected values of n and m ($= n + \Delta n$).

et al. (1969) estimated the uncertainty of the Coulomb approximation to be about 50%, but they felt that this was quite conservative. Our results support this view; in any case, an uncertainty of $\sim 50\%$ seems quite acceptable in many astrophysical problems. Further, they may be extrapolated to obtain threshold photoionization cross-sections for magnesium Rydberg states, as shown by Hoang-Binh (1983 and unpublished) for the case of hydrogen. In this connection, it may be noted that an efficient analytical method (also based on the Coulomb approximation), which allows one to deal easily with $n > 10$, has been proposed by Hoang-Binh and Van Regemorter (1979).

It is interesting to see how far the f-values for transitions between Mg I Rydberg states deviate from hydrogenic values. Table 2 lists f-values for $nf\,^1F^0 - mg\,^1G$ transitions in Mg I together with the corresponding hydrogenic values (quantum defects set equal to 0). It can be seen that the f-values still differ significantly from hydrogenic ones for $nf\,^1F^0 - mg\,^1G$ transitions, which involve high angular momenta ($l = 3, 4$). This should be borne in mind when interpreting solar spectra (Jefferies 1991). Thus, it is only for transitions between $l \geq 4$ Rydberg states that

TABLE I

Values of $f(np\,^1P^0, ms\,^1S)$. The last column shows the percentage difference between this work and the results of Moccia and Spizzo (1988).

Δn	n	This Work	MS	Difference(%)
1	4	2.97E-01	2.99E-01	-0.53
	5	4.31E-01	4.31E-01	-0.02
	6	5.65E-01	5.64E-01	+0.12
	7	6.98E-01	6.97E-01	+0.24
	8	8.32E-01	8.29E-01	+0.38
2	4	1.98E-02	1.98E-02	-0.14
	5	2.96E-02	2.97E-02	-0.36
	6	3.89E-02	3.91E-02	-0.57
	7	4.78E-02	4.79E-02	-0.11
3	4	6.11E-03	6.10E-03	+0.21
	5	9.19E-03	9.20E-03	-0.16
	6	1.20E-02	1.20E-02	+0.33

TABLE II

Values of $f(nf\,^1F^0, mg\,^1G)$

Δn	n	H	Mg I	Difference (%)
1	5	1.18E+00	1.05E+00	13.25
	10	1.08E+00	8.19E-01	31.31
	20	1.36E+00	9.48E-01	43.45
	50	2.51E+00	1.65E+00	51.80
2	5	2.29E-01	2.28E-01	0.47
	10	2.72E-01	2.34E-01	16.25
	20	3.46E-01	2.71E-01	27.70
	50	6.17E-01	4.55E-01	35.85
5	5	2.39E-02	2.61E-02	-8.74
	10	3.99E-02	3.78E-02	5.38
	20	5.28E-02	4.53E-02	16.55
	50	8.10E-02	7.17E-02	25.48

the use of hydrogenic values is fully justified. In this connection, we note that a very efficient method for calculating exact hydrogenic radial integrals, for principal quantum numbers up to 1000, has been proposed (Hoang-Binh et al. 1979; Hoang-Binh 1990).

References

Bashkin, S., Stoner, J.O.: 1975, *Atomic Energy Levels and Grotrian Diagrams 1: Hydrogen-Phosphorus XV*, North Holland, Amsterdam.
Brault, J., Noyes, R. W.: 1983,*Astrophys. J. (Letters)* **269**, L61.
Chang, E.S., Avrett, E.H., Mauas, P.J., Noyes, R. W., Loeser, R.: 1991,*Astrophys. J. (Letters)* **379**, L79.
Chang, E.S., Noyes, R.W.: 1983, *Astrophys. J. (Letters)* **275**, L11.
Froese-Fisher, C.: 1975a, *Can. J. Phys.* **53**, 184.
Froese-Fisher, C.: 1975b, *Can. J. Phys.* **53**, 338.
Goldberg, L.: 1939, *Astrophys. J.* **90**, 414.
Hoang-Binh, D.: 1983, *Astron. Astrophys.* **121**, L19.
Hoang-Binh, D.: 1990, *Astron. Astrophys.* **238**, 449.
Hoang-Binh, D.: 1991, *Astron. Astrophys.* **241**, L13.
Hoang-Binh, D., Van Regemorter, H.: 1979, *J. Phys. B* **14**, L329.
Hoang-Binh, D., Prud'homme, M., Van Regemorter, H.: 1979, *Astrophys. J. (Letters)* **230**, L127.
Hoang-Binh, D., Brault, P., Picart, J., Tran-Minh, N., Vallee, O.: 1987, *Astron. Astrophys.* **181**, 134.
Jefferies, J.T.: 1991, *Astrophys. J.* **377**, 337.
Moccia, R., Spizzo, P.: 1988, *J. Phys. B* **21**, 1133.
Risberg, G.: 1965, *Ark. Fys.* **28**, 381.
Victor, G.A., Stewart, R. F., Laughlin, C.: 1976, *Astrophys. J. Suppl.* **31**, 237.
Wiese, W.L., Smith, M.W., Miles, B.M.: 1969, *Atomic Transitions Probabilities II, Sodium through Calcium*, NSRDS-NBS 22, U.S. Government Printing Office, Washington, D.C.

ON THE ION BROADENING OF THE 12 μm LINES OF ATOMIC MAGNESIUM

D. HOANG-BINH and H. VAN REGEMORTER

LAM, Observatoire de Paris, F-92195 Meudon Cédex, France

Abstract. We have investigated the impact broadening of the 12 μm lines. Ion broadening is found to follow the adiabatic theory, whereas electron broadening needs a non-adiabatic treatment. This relaxes the fixed width/shift ratio, as found by Chang and Schoenfeld (1991), using a pure adiabatic analysis.

Key words: atomic processes – infrared: stars – Mg I

1. Introduction

Electron broadening of the 12 μm lines has been shown to be important by Chang and Schoenfeld (1991). Ion broadening of the far-infrared lines of hydrogen has been shown to be more important than electron broadening by Hoang-Binh (1982), and Hoang-Binh et al. (1987). Hoang-Binh (1982) has estimated the half-width at half-maximum (HWHM), γ, of the far-infrared lines of hydrogen, broadened by elastic collisions with protons, and found that it is much larger than the HWHM due to inelastic collisions with electrons. This is due to the complete degeneracy of the (n, l) and $(n, l \pm 1)$ levels, which gives transitions rates scaling as the square root of the perturber mass, transition rates for $n \rightarrow n' \neq n$ being negligible. Now, in the case of the 12 μm $(6h-7i)$ and $(6g-7h)$ lines, the broadening is due to collisional transitions between nearly degenerate levels, and it is of interest to find out whether collisions with ions are still predominant in the broadening process.

2. Theory

The number, P, of perturbers in the sphere of Weisskopf radius, for a dipole interaction between degenerate H levels, is

$$P \approx N_p \left[Z_p\, n(n-1)\, e^2 (2\pi a_0)/hv \right]^3 \tag{1}$$

where n is the principal quantum number, v is the relative velocity of the perturber and emitter, and $Z_p e$ is the charge of the perturber.

In the following, we will set $v = <v> = (2kT/\mu)^{1/2}$, where μ is the reduced mass if the interacting particles and the kinetic temperature is taken to be 5,000 K. Let N_1 be the value of N_p corresponding to $P = 1$. The usual condition for validity of the impact approximation, applicable to H, is $P < 1$, or $N_p < N_1$. Thus, using this condition for the 12 μm lines, the impact approximation is applicable for $N_p(\text{Mg}^+) < 1.57 \times 10^{11}$ cm^{-3}, $N_p(\text{H}^+) < 6.93 \times 10^{12}$ cm^{-3}, and electron density $N_e < 5.13 \times 10^{17}$ cm^{-3} (Table 1). But the use of this criterion for the non-hydrogenic case has been much criticized (see *e.g.*, Sahal-Brechot 1969). Although often reliable for electrons, it breaks down for protons and ions. It can be shown that, for the correct effective adiabatic polarization interaction examined below, the impact approximation is valid for $N_p(\text{Mg}^+) < 5 \times 10^{14}$ cm^{-3} and $T = 5,000$ K.

TABLE I

Electron and ion (H$^+$ and Mg$^+$) densities (cm^{-3}), below which the impact approximation is applicable ($P < 1$), for the case of degenerate levels: $T = 5,000$ K.

n	$N_1(e)$	$N_1(H^+)$	$N_1(Mg^+)$
5	4.75E+18	6.42E+13	1.45E+12
6	1.41E+18	1.90E+13	4.30E+11
7	5.13E+17	6.93E+12	1.57E+11
8	2.17E+17	2.92E+12	6.62E+10
9	1.02E+17	1.38E+12	3.11E+10
10	5.22E+16	7.04E+11	1.59E+10

For a line $a \to b$, the collision width, in angular frequency units, is given by the Baranger formula

$$\gamma = N_p < [\sum_i Q(a \to i) + \sum_j Q(b \to j)] > + \gamma \text{ (elastic)}, \tag{2}$$

where the Q are collisional cross-sections, and, for our case, $a = 6g(6h), b = 7h(7i)$, and the summations are over levels $i \neq a, j \neq b$.

The elastic part is given in terms of the phase shifts ϕ_a and ϕ_b, by the equation

$$\gamma \text{(elastic)} = N_p < v [\int |f_a(\Omega) - f_b(\Omega)|^2 \, d\Omega >$$

$$= N_p < v \int 8\pi \rho \, d\rho \sin^2[\phi_a - \phi_b] > \tag{3}$$

where $f_a(\Omega)$, $f_b(\Omega)$ are scattering amplitudes, and

$$\phi_a = (I_H/kT)(\mu/m) \sum_i f_{a,i} (I_H/\Delta E_{a,i})(1/\rho^2) B(\beta_{a,i} \, a_0^2), \tag{4}$$

and ϕ_b is given by the same expression with b, j replacing a, i. The quantities $f_{a,i}$ and $f_{b,j}$ are oscillator strengths, while $\beta_{a,i} = 2\pi \rho \Delta E_{a,i}/hv$, and a corresponding expression holds for $\beta_{b,j}$.

Tables of the well-known functions $A(\beta)$ and $\zeta(\beta)$ (see below) and $B(\beta)$ may be found in Griem's (1974) book. The corresponding shift is

$$d = N_p < v \int 2\pi \rho \, d\rho \sin[2(\phi_a - \phi_b)] > \tag{5}$$

Let us consider a collisional transition $(n, l) \to (n, l+1)$ corresponding to a line of angular frequency $\omega(l, l+1)$. In the process of line broadening by collisions, adiabatic theory will be applicable if

$$\omega(l, l+1) \rho_\epsilon / v \geq 1, \tag{6}$$

TABLE II
Energy of triplet levels of Mg I: T is 5,000 K.

n	l	$E(\text{cm}^{-1})$	$\Delta E(\text{cm}^{-1})$	$\Delta E/kT$
6	3	58575.46	3.534E+01	1.017E-02
6	4	58610.80	8.140E+00	2.343E-03
6	5	58618.94		
7	3	59400.77	2.277E+01	6.552E-03
7	4	59423.54	5.316E+00	1.530E-03
7	5	59428.85	1.664E+00	4.789E-04
7	6	59430.52		

where ρ_ϵ is the effective impact parameter. Let E_l and E_{l+1} be the energies of levels (n, l) and $(n, l+1)$, respectively, and writing $\Delta E_{l,l+1} = (E_{l+1} - E_l)$, expression (6) is equivalent to

$$\Delta E_{l,l+1}/kT \geq 2/l_p \quad (7)$$

where $l_p = \mu v \rho_\epsilon/(h/2\pi)$ is the angular momentum of the perturber.

3. Electron impact widths

For electron impacts, μ is the electron mass, the important values of $l_p (= l_e)$ are 1, 2, 3, 4; thus adiabaticity obtains for $\Delta E_{l,l+1}/kT \geq 0.5$ to 2. Now, Table 2 shows that for the transitions of interest, $\Delta E_{l,l+1}/kT \ll 1$; thus, electronic collisions are strongly non-adiabatic. As equation (2) shows, the calculation of $\gamma (= \gamma_e)$ requires knowledge of cross-sections of collisional transitions between levels a, b and other atomic levels. We will consider only the most important ($\Delta n = 0$) transitions ($l \rightarrow l\pm1$)). Using the impact parameter theory in the dipole approximation (Seaton 1962, Sahal-Brechot 1969), the cross-section for $a \rightarrow i$ is

$$Q(a \rightarrow i) = 8 (I_H/kT) (I_H f_{a,i}/\Delta E_{a,i})[\zeta(\beta_1)/2 + \phi(\beta_1)] \pi a_0^2, \quad (8)$$

where I_H is the Rybderg energy and β_1 and $\phi(\beta_1)$ are defined and tabulated by Seaton (1962).

In the case of electrons, when $kT \gg \Delta E$, the relevant values of $\beta_{a,i}$ and $\beta_{b,j}$ are very small, and $B(\beta) = -\pi \beta^2[1/2 + \ln(1.78 \beta/2)]$. It is easy to show that the phase shifts ϕ_a and ϕ_b are also very small, and γ_e(elastic) and d_e(elastic) are negligible. Using equation (2) and values listed in Table 3, we have calculated the electron widths γ_e listed in Table 4.

4. Ion impact widths

For ions, $\mu \gg m_e$, $l_p = l_i \approx 10^3$ to $10^4 \times l_e$, and adiabaticity obtains for $\Delta E_{l,l+1}/kT$ greater than about 10^{-3} to 10^{-4}. Table 2 shows that for the lines

TABLE III

Cross Sections (Q) in cm^2 and Rates (K) in cm^3 s^{-1} for a number of transitions in a 5,000 K gas.

n	l	$Q(l \to l+1)$	$K(l \to l+1)$	$K(l+1 \to l)$
6	3	1.328E-11	5.169E-04	4.020E-04
6	4	1.380E-11	5.371E-04	4.395E-04
7	3	3.037E-11	1.182E-03	9.195E-04
7	4	3.967E-11	1.544E-03	1.264E-03
7	5	2.970E-11	1.156E-03	9.785E-04

TABLE IV

Comparison of γ_e/N_e with the adiabatic contribution γ_i/Ni given by equation (9) for Mg$^+$. Units are cm^3 s^{-1}; T = 5,000 K.

Transition	γ_e/N_e	γ_i/N_i	d_i/N_i	C_4
$6g$–$7h$	3.36E-03	4.64E-04	3.99E-04	8.07E-11
$6h$–$7i$	1.42E-03	6.50E-04	-5.60E-04	-1.34E-10

in question, this is indeed the case. Thus, ion broadening is adiabatic. Owing to the small ion velocity and the large ρ of interest, $\beta(\text{ion}) \approx (\mu/m_e) \beta(\text{electron})$, for a given ΔE. When $\beta \gg 1$, $B(\beta) = \pi/4\beta$ in (4), and it is a simple matter to derive the well-known Lindholm formulae (Sobelman et al. 1981) for the width $\gamma = \gamma_i$, and shift $d = d_i$, namely,

$$\gamma_i = 38.8 \, C_4^{2/3} \, v_i^{1/3} \, N_i, \qquad (9)$$

$$d_i = 33.4 \, C_4^{2/3} \, v_i^{1/3} \, N_i, \qquad (10)$$

and,

$$C_4 = (1/4\pi) \left[e^2/(h/2\pi) \right] (\alpha_b - \alpha_a) \, a_0^3 \qquad (11)$$

where α_a, α_b are the polarizabilities in atomic units, and a_0 is the Bohr radius. For a level (n, l), we have

$$\alpha_{n,l} = 4 \sum_{n',l'} [f_{(n,l;n',l')} (I_H/\Delta E)^2] \qquad (12)$$

where ΔE is the energy difference (in Rydberg units) between (n, l) and (n', l'). It is sufficient to consider only $n' = n$. Values of $C_4, \gamma_i/N_i$, and d_i/N_i for the lines of interest are listed in Table 4. It can be seen that ion impact broadening, though smaller than electron impact broadening, is not negligible.

5. Discussion

The main conclusion of our study is that, in the impact regime, adiabatic theory applies only to collision with ions, not with electrons. This is in contrast to the treatment given by Chang and Schoenfeld (1991), who use adiabatic theory to describe the broadening and shift by electron impact (ions are ignored). As expressions (9) and (10) show, adiabatic theory predicts a constant width/shift ratio, in disagreement with observations. Our results show that the line shifts are due to ions, not to electrons; hence the width/shift ratio, $(\gamma_e + \gamma_i)/d_i$, is not constant, in better qualitative agreement with observations. However, a close comparison of our results with observations is not warranted, because γ_i and d_i must refer to all ions, not just Mg^+. Further, owing to the high abundance of neutral hydrogen, broadening and shift by collisions with H atoms may be important; unfortunately, no simple reliable theory is available. The simple Fermi model overestimates l-mixing transitions rates for $n < 10$, possibly by more than one order of magnitude. The present work shows that electronic collisional l-mixing rates ($\approx 10^9$ s^{-1} for $7h \rightarrow 7i$ at $N_e = 10^{12}$ cm^{-3}) are among the largest collisional rates entering the statistical equilibrium equations (see Hoang-Binh 1991). Thus, it is not necessary to invoke collisions with H atoms to achieve equal-n-level balancing (Carlsson et al. 1992).

References

Carlsson, M., Rutten, R.J., Shchukina, N.G.: 1992, *Astron. Astrophys.* **253**, 567.
Chang, E.S., Schoenfeld, W.G.: 1991, *Astrophys. J.* **383**, 450.
Griem, H.R.: 1974, *Spectral Line Broadening by Plasmas*, Academic Press, New York.
Hoang-Binh, D.: 1982, *Astron. Astrophys.* **112**, L3.
Hoang-Binh, D: 1991, *Astron. Astrophys.* **241**, L13.
Hoang-Binh, D., Brault, P., Picart, J., Tran-Minh, N., Vallee, O.: 1987, *Astron. Astrophys.* **181**, 134.
Sahal-Brechot, S.: 1969, *Astron. Astrophys.* **1**, 91.
Seaton, M.J.: 1962, *Proc. Phys. Soc.* **79**, 1105.
Sobelman, I.I., Vainshtein, L.A., Yukov, E.A.: 1981, *Excitation of Atoms and Broadening of Spectral Lines*, Springer, Berlin.

HIGH-l RYDBERG LINES OF Fe I IN THE *ATMOS* SPECTRA:

$4f-5g$, $5g-6h$ ...

WILLIAM G. SCHOENFELD and EDWARD S. CHANG

*Department of Physics and Astronomy, University of Massachusetts,
Amherst, MA 01003, U.S.A.*

and

MURRAY GELLER

*Atmospheric and Cometary Sciences Section, Jet Propulsion Laboratory,
Pasadena, CA 91109, U.S.A.*

Abstract. We have identified the Fe I $4f-5g$ lines at 4 μm in the *ATMOS* solar spectra. Using the polarization model as previously applied to silicon, we predict and identify the $5g-6h$ lines at 7 μm. Additional absorption features at 2.5 and 12 μm are also shown to be due to high-l Rydberg transitions in Fe I.

Key words: atomic processes – Fe I – infrared: stars – line: identification

1. Solar Rydberg Lines

The polarization theory for a non-penetrating Rydberg atom (Edlén, 1964) implies that every atom has highly excited levels whose energies are nearly hydrogenic. As is well known, transitions between Rydberg levels of neutral atoms are found in the infrared; and fortunately the *ATMOS* spectrum spans most of the region of interest. Thus, any atom with sufficiently high stellar abundance and a relatively low ionization potential should produce Rydberg lines strong enough to be observed in the uncontaminated *ATMOS* spectra. Previously identified hydrogen-like transitions in the solar spectrum have been attributed to Al I, Si I (Chang, 1984), as well as the extensively-studied 12 μm lines of magnesium (Chang and Noyes, 1983). By virtue of its solar abundance and the relative complexity of its core configuration, iron is expected to produce infrared lines that are approximately as strong and far more numerous than those of the simple core atoms, whose Rydberg lines have already been identified.

2. Rydberg Levels in Fe I

The Rydberg electron in Fe I is loosely coupled to the core ($1s^2 \ldots 3d^6 4s^1$), whose lowest term is ^6D. This term gives rise to five distinct core fine structure levels whose angular momentum (j_c) couples to the Rydberg electron's orbital angular momentum in the jl coupling scheme. For non-penetrating Rydberg states ($l \geq 4$), the quantum numbers j_c, n and l, along with K ($\vec{K} = \vec{j_c} + \vec{l}$), are sufficient to completely specify the Rydberg levels. For partially penetrating f electrons ($l=3$), the fine structure due to the spin of the Rydberg electron is large enough to necessitate the use of the JK coupling scheme (Cowan and Andrew, 1965). The addition of the spin of the Rydberg electron to form the total angular momentum J ($\vec{J} = \vec{K} + \vec{s}$), produces pairs of closely spaced levels.

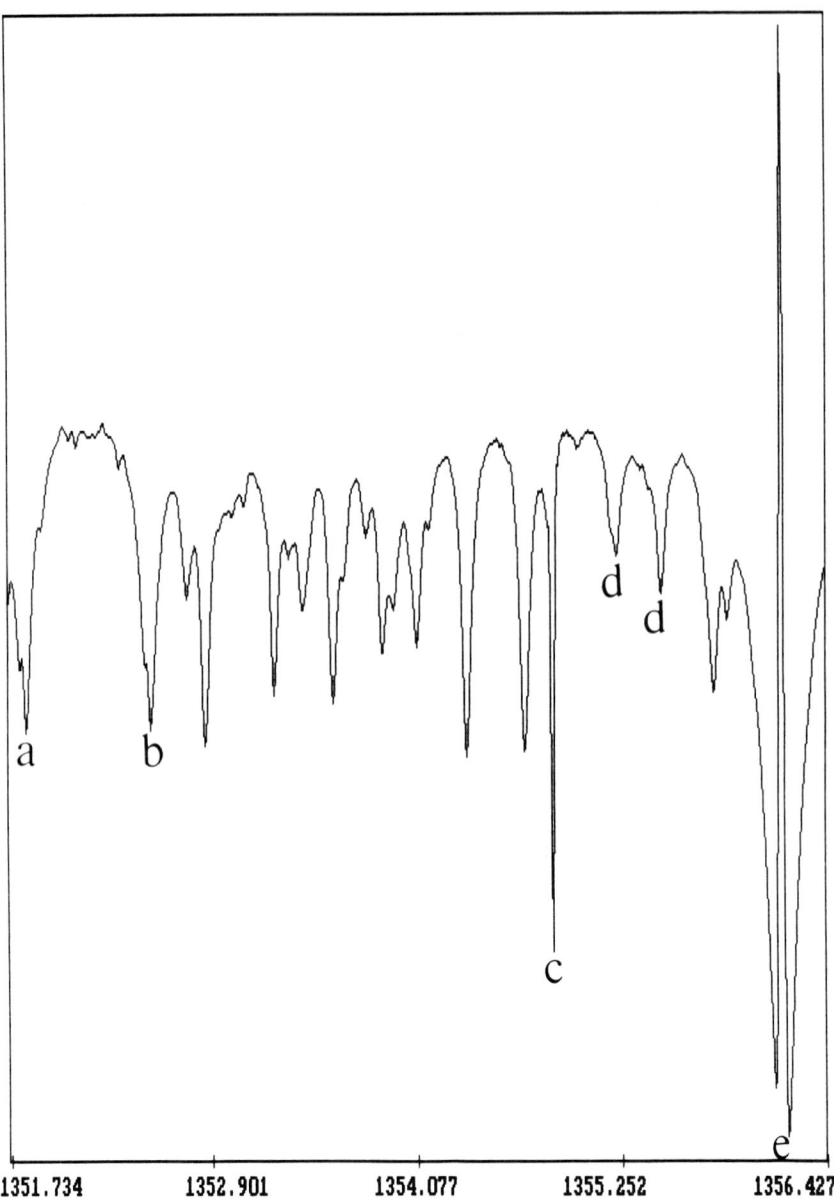

Fig. 1. 5g–6h Rydberg lines in the *ATMOS* solar spectra. a) aluminum, b) silicon, c) satellite gas, d) carbon, e) magnesium, all other lines are attributed to iron.

In the simple version of the polarization model for non–spherical cores (Chang, 1984), the energy of the Rydberg level is

$$E_{nlj_cK} = IP - \frac{\text{Ryd}}{n^2} + E_{j_c}^{\text{core}} - \alpha \langle \frac{1}{r^4} \rangle_{nl} - 2Q \langle j_c lK \| C^2 \| j_c lK \rangle \langle \frac{1}{r^3} \rangle_{nl}.$$

In this expression IP is the ionization potential of the atom, while $E_{j_c}^{\text{core}}$ represents the fine-structure energy of the core. The isotropic dipole polarizability α, is the first multipole polarizability that is induced in the core by the Rydberg electron. This distortion of the electron cloud lowers the overall energy of the level. The last term represents the electrostatic interaction between the permanent quadrupole moment of the core Q, and the Rydberg electron. Both the sign and magnitude of this interaction depend explicitly on j_c, l and K via the angular portion of the matrix element. All core fine structure levels with $j_c > 1/2$ will have a non-zero quadrupole interaction. As usual,

$$\langle \frac{1}{r^3} \rangle_{nl} \text{ and } \langle \frac{1}{r^4} \rangle_{nl}$$

are the hydrogenic radial expectation values over the Rydberg electron wavefunction. These expectation values are simple functions of n and l. The simplicity of the polarization formula allows α and Q to be regarded as fitting parameters used to model the interaction of the Rydberg electron with the atomic core.

3. Line Identification

Since the $4f$ levels in iron have recently been determined (Johansson and Learner, 1990), and the $4f$–$5g$ transitions are known to occur around 4 μm, we searched for pairs of lines of nearly equal strengths, whose splittings matched the splittings of the $4f$ pairs. With tentative identifications for the strongest lines, we were able to fit the model to the observed transition wavenumbers in order to determine the values of the polarization parameters. With reliable values for α and Q, we were able to predict and identify all the remaining Rydberg lines.

The laboratory confirmation of the $4f$–$5g$ lines of iron are presented by Johansson et al. (1993), in this volume, so we shall only mention that we have identified 83 absorption lines between 2543 cm^{-1} and 2582 cm^{-1}. Turning to the $5g$–$6h$ array, we have calculated the positions of all the allowed transitions, and located those that can be observed in the ATMOS spectrum, 31 in all. Most of these are visible in Figure 1, along with Rydberg transitions of other elements. Table 1 identifies the $5g$–$6h$ transitions in Fe I by providing the theoretically predicted line positions along with those actually observed. With a few exceptions, the differences between the two are less than 0.07 cm^{-1}. Many of the weaker Rydberg transitions are blended with other, stronger $5g$–$6h$ Fe I lines. Also included in Table 1 are the calculated hydrogenic line strengths, scaled to facilitate comparison with the observed depths (arbitrary scale). Agreement is excellent, and all observed lines in the region now have identifications except for three very weak features.

As n and l increase, the radial expectation value of the Rydberg electron wavefunction increases. Therefore, the polarization energy that is associated with α

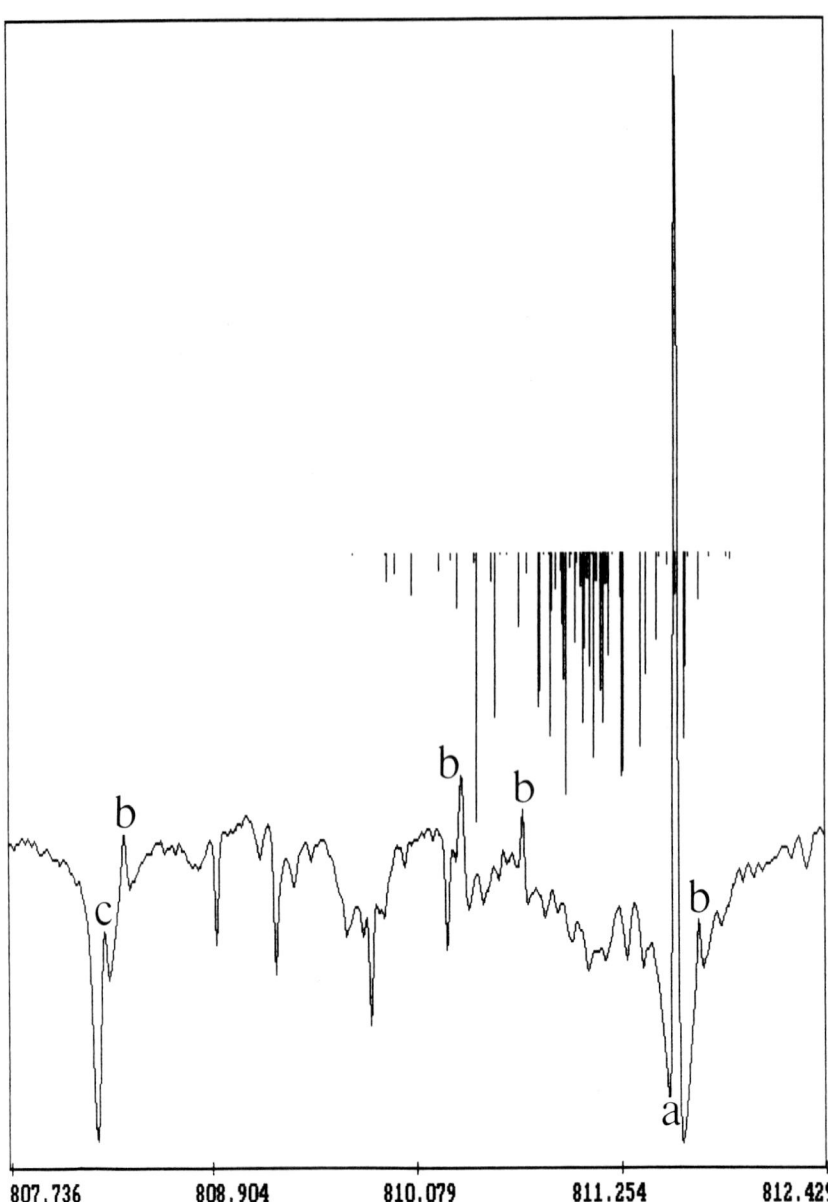

Fig. 2. Rydberg lines at 12μm. The positions and the relative strengths of the $6h$–$7i$ lines of Fe I are superimposed above the *ATMOS* spectra in order to interpret the unknown irregular trough next to the 811 cm^{-1} of Mg I (a). Other lines of interest include b) $6h$–$7i$ emission lines of Si, as well as c) the $n = 6 \rightarrow 7$ line of hydrogen.

TABLE I
5g–6h TRANSITIONS IN Fe I

IDENTIFICATION			ATMOS	THEORY	LINE	DEPTH	
j_c	K_{lower}	K_{upper}	cm^{-1}	cm^{-1}	STRENGTH		
9/2	0.5	0.5	1349.731	1349.827	0.16	0.1	
9/2	1.5	0.5	1350.170	1350.229	0.23	0.3	
9/2	8.5	8.5	1350.279	1350.339	0.21		bl Si
9/2	1.5	1.5	1350.525	1350.689	0.26		bl Si
9/2	2.5	3.5	1351.400	1351.417	0.39	0.5	
9/2	8.5	9.5	1351.521	1351.648	2.24	2.2	
3/2	4.5	5.5	1351.728	1351.828	0.99	1.4	
9/2	2.5	2.5	1351.937	1351.980	0.35	0.3	
5/2	4.5	4.5	1352.577	1352.606	0.17		bl Si
5/2	4.5	5.5	1352.778	1352.822	0.95	0.9	
9/2	3.5	4.5		1352.895	0.58		bl
3/2	3.5	4.5		1352.941	0.80		bl
5/2	5.5	6.5	1352.885	1352.955	1.20	2.3	bl
7/2	0.5	1.5		1352.995	0.25		bl
5/2	3.5	3.5		1353.060	0.15		bl
5/2	5.5	5.5	1353.044	1353.083	0.15	0.1	bl
7/2	7.5	7.5		1353.115	0.15		bl
7/2	1.5	2.5	1353.107	1353.149	0.35	0.2	
7/2	1.5	1.5		1353.255	0.14		bl
7/2	7.5	8.5	1353.276	1353.366	1.81	2.0	
7/2	2.5	3.5	1353.358	1353.404	0.50		sh
5/2	3.5	4.5	1353.440	1353.458	0.75	1.0	bl
9/2	3.5	3.5		1353.480	0.43		bl
7/2	2.5	2.5		1353.530	0.22		bl
1/2	4.5	5.5	1353.614	1353.710	1.02	2.0	bl
1/2	3.5	4.5		1353.710	0.83		bl
7/2	3.5	4.5	1353.679	1353.705	0.68	1.0	
7/2	3.5	3.5	1353.802	1353.828	0.28	0.5	bl
7/2	6.5	6.5		1353.854	0.25		bl
5/2	2.5	2.5		1353.938	0.10		bl
7/2	6.5	7.5	1353.90	1353.944	1.46	1.4	
7/2	4.5	5.5	1353.962	1353.970	0.89	0.9	
7/2	4.5	4.5		1354.060	0.30		bl
7/2	5.5	6.5	1354.096	1354.093	1.15	1.3	bl
7/2	5.5	5.5		1354.115	0.30		bl
9/2	7.5	7.5	1354.167	1354.138	0.36	0.2	
9/2	4.5	5.5		1354.379	0.82		bl
5/2	2.5	3.5		1354.386	0.58		bl
5/2	6.5	7.5	1354.381	1354.446	1.48	2.8	bl
9/2	7.5	8.5	1354.714	1354.723	1.81	2.5	bl
3/2	5.5	6.5		1354.740	1.23		bl
9/2	4.5	4.5	1354.85	1354.884	0.47		sh
5/2	6.5	6.5	1355.017	1355.110	0.10	0.1	
5/2	1.5	2.5	1355.20	1355.241	0.45		bl C
9/2	5.5	6.5	1355.497	1355.499	1.10		bl C
9/2	6.5	6.5		1355.739	0.45		bl
9/2	6.5	7.5	1355.795	1355.796	1.42	1.6	bl
9/2	5.5	5.5		1355.798	0.48		bl
3/2	2.5	3.5	1355.871	1355.854	0.64	0.8	

bl– blended
sh– shoulder

decreases. Accordingly, the quadrupole interaction energy that is associated with the non-spherical core atoms decreases in magnitude, thereby decreasing the separation between states of the same n, l, j_c but different K. The net effect is to push all the lines within a transition array closer together, and also closer to the corresponding line in hydrogen. This is clearly seen in Figure 2, which shows Rydberg lines from a number of different atoms in close proximity to the $n = 6 \rightarrow 7$ line in hydrogen. In Fe I, the $4f$–$5g$ transition array is spread out over 40 cm^{-1}, the whole $5g$–$6h$ array is separated by only 6 cm^{-1}, and the $6h$–$7i$ array in Fe I is essentially packed into a 2 cm^{-1} band. While no specific absorption lines can be associated with the $6h - 7i$ transition array of Fe I, it is evident that the closely spaced weak lines combine to form the irregular depression centered at 811 cm^{-1}. The only other possible $\Delta n = 1$ transitions expected to be seen in the *ATMOS* spectra are the $5f$–$6g$ lines. We have examined the region where they should be located, and have found only weak lines ($< 1\%$). Positive identification remains difficult because the partially penetrating $5f$ levels are unknown experimentally.

In addition, we have looked for and found some of the weaker $\Delta n = 2$ transitions between the $4f$–$6g$ levels at 2.5 μm. More than fifty identifications have been made, although many of these are blended with stronger CO lines. Other $\Delta n = 2$ lines are possible ($n = 5 \rightarrow 7$, $6 \rightarrow 8$ ...), but these are also probably too weak to be seen.

References

Chang, E.S.: 1984, *J.Phys.B* **17**, L11.
Chang, E.S., and Noyes, R.W.: 1983, *Physica Scripta* **35**, 792.
Cowan, R.D., and Andrew, K.L.: 1965, *J. Opt. Soc. Am.* **55**, 502.
Edlén, B.: 1964, *Encyclopedia of Physics vol 27*, Springer, Heidelberg.
Johansson, S., and Learner, R.C.M.: 1990, *Astrophys. J.* **354**, 775.
Johansson, S., Nave, G., Geller, M., Sauval, A.J., Grevesse, N.: 1993, these proceedings.

HIGH-n HYDROGEN LINES IN SOLAR INFRARED SPECTRA FROM BALLOON-BORNE, MAUNA KEA, AND *ATMOS* OBSERVATIONS

R. T. BOREIKO

*Center for Astrophysics and Space Astronomy, University of Colorado,
Campus Box 389, Boulder, CO 80309, U.S.A.*

T. A. CLARK

Physics Department, University of Calgary, Calgary, Alberta T2N 1N4, Canada

D. A. NAYLOR

Department of Physics, University of Lethbridge, Lethbridge, Alberta T1K 3M4, Canada

and

J. R. BUSLER

Physics Department, University of Calgary, Calgary, Alberta T2N 1N4, Canada

Abstract. This paper reports the observation of high-n lines in emission from $n =$ 12–11, 13–12, 14–13 and 16–15 Rydberg transitions in H, Mg and Si in solar far IR spectra taken from balloon altitudes, in which the H I line intensities are found to exceed those from the heavier elements. Tentative identification is also made of the $n =$ 8–7 hydrogen line in emission on 20 μm spectra taken from Mauna Kea. The characteristics of the hydrogen lines are compared with lower-n transitions seen in the Space Shuttle *ATMOS* spectra, in which Brackett, Pfund and $n = 6$ lines with $\Delta n =$ 1, 2, 3 and 4 are seen as broad absorption features, while the $n =$ 7–6 line shows a small emission peak within a broader absorption line and the $n =$ 9–7, and possibly the 11–8, transitions appear as weak emission lines. These results indicate that the transformation from absorption to emission occurs at longer wavelengths for hydrogen lines than for those of heavier elements.

Key words: H I – infrared: stars – line: formation – line: identification – Sun: atmosphere

1. Background

The discovery in the early 1980's of intense emission lines in the 10 μm solar spectrum (Murcray *et al.*, 1982, Brault and Noyes, 1983) and their subsequent identification as lines from $n =$ 7–6 Rydberg transitions in Mg I, Si I and Al I (Chang and Noyes, 1983; Chang, 1984) has opened up a new and important avenue in the study of the sun's high photosphere and chromosphere. These initial discoveries prompted a search for equivalent lines in far IR spectra taken from balloon altitudes with the University of Calgary balloon-borne telescope and the tentative identification of complexes of emission lines from $n =$ 16–15 and 14–13 (Boreiko and Clark, 1986).

In the meantime, the *ATMOS* spectrometer was flown on the Space Shuttle in 1985 to study the Earth's high atmosphere by rapid absorption spectroscopy during occultation of the Sun by the Earth's atmosphere. During these experiments, *ATMOS* also produced a series of superb high-resolution near-IR spectra of the Sun taken at high elevation angles which are almost devoid of atmospheric absorption features but which show a wealth of solar spectral lines. Among these, a series of hydrogen lines from various transitions are clearly seen to follow a trend from strong absorption at $n =$ 5–4 to weak emission at $n =$ 9–7.

In view of this trend within the limited spectral range of *ATMOS* and the obvious importance to solar atmospheric modeling of further and more extended spectral information of this kind, a search for lines from the $n = 8\text{--}7$ transitions in the solar spectrum was also undertaken through the highly structured and variable 20 μm atmospheric window from Mauna Kea.

2. Experimental Details

The 0.23 m University of Calgary balloon-borne solar telescope (Boreiko, 1985; Boreiko and Clark, 1987) was flown to altitudes of 30-35 km from Gimli, Manitoba, Canada in 1982 and 1985 to measure the far IR (50 and 160 cm^{-1}) spectrum of the central 8 arc minute region of the Sun. A rapid-scanning Michelson interferometer with an apodized resolution of 0.015 cm^{-1} was used to make these radiometric measurements. Absolute calibration was performed with reference to a 1160 K black-body source, under the same thermal conditions as for the solar measurements. Direct measurements of the background were obtained by offsetting the telescope by several solar diameters.

A Michelson interferometer (Naylor and Clark, 1986), fed by a heliostat and 0.15-m optics, was used on Mauna Kea (4200 m) in Dec 1989 to explore the 20 μm (490–540 cm^{-1}) solar and atmospheric spectra to an apodized resolution of 0.01 cm^{-1} and to search for solar atomic lines from $n = 8\text{--}7$ transitions.

The *ATMOS* instrument and operating mode are described in Farmer and Norton (1989). Individual spectra between 600 and 4700 cm^{-1} (2.3–16 μm), were measured to 0.01 cm^{-1} apodized resolution with a rapid-scanning Fourier transform spectrometer from the Space Shuttle, at both high and low tangent angles through the Earth's atmosphere.

3. The Spectra

3.1. Balloon Measurements

The balloon-borne spectra were analyzed using standard Fourier spectroscopy techniques. Notable features of the spectra were several emission line complexes along with the expected absorption lines from stratospheric molecules, particularly H_2O and O_3. Sections of these spectra have been analyzed, corrected for instrument transmission and calibrated in terms of solar continuum intensity to yield the emission complexes shown in Figures 1a–d. Stratospheric absorption lines are seen to dominate the $n = 14\text{--}13$ and $16\text{--}15$ regions but are absent or weak in the other two regions. The expected positions of H, Mg and Si lines from $\Delta n = 1$ transitions as well as positions and relative heights of the stratospheric absorption lines are shown on these graphs. It can be seen that the strongest feature in each complex is a peak at the H line position. At the longer wavenumbers, there is evidence of both the Mg line and the 6-fold emission from Si, although the positions of these features do not appear to correspond precisely with the expected line positions.

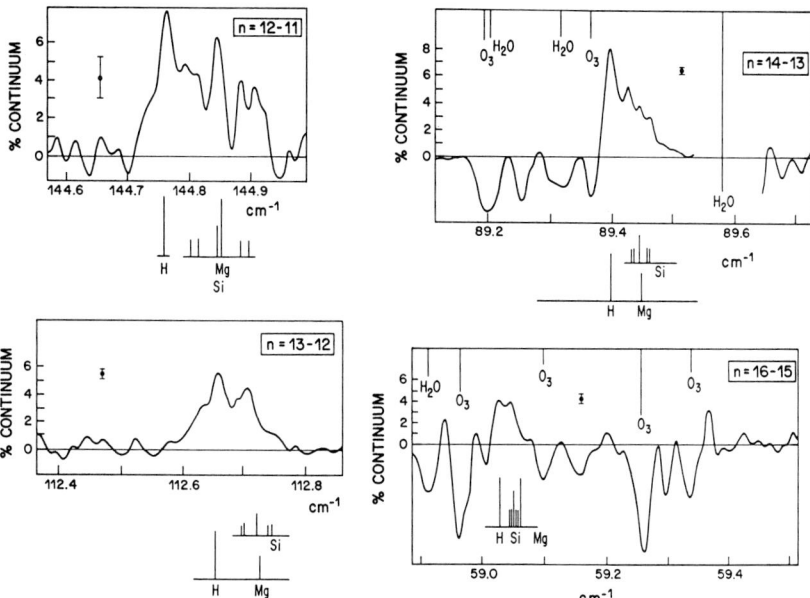

Fig. 1. a–d: Emission line complexes from transitions between high-n states of H, Mg and Si. Error bars represent ±1 standard deviation on the mean continuum intensity.

3.2. 20 μm Spectra from Mauna Kea

Figure 2 shows sections of two representative solar spectra between 523 and 526 cm^{-1}, taken under very dry conditions, along with the FASCOD/HITRAN synthetic transmission spectrum, which indicates the positions and strengths of major absorption features. An emission peak is seen on each of these spectra, of intensity about 4 ±1.5% of continuum, at the position of the $n = 8\text{--}7$ transition of H I. Equivalent Mg I and Si I lines from this transition are predicted to occur on the steep side or in the center of the major H$_2$O line at 526 cm^{-1}. No such features have yet been identified. The large uncertainty in measured H I line intensity reflects the lack of accurate knowledge of the underlying spectral envelope. Thus, this result is regarded as preliminary only and a final conclusion must await further calibration.

3.3. *ATMOS* Spectra

The sequence of H I lines from high level transitions between $n = 5\text{--}4$ to $9\text{--}7$ have been measured on the *ATMOS* spectra. Of particular interest are the 2 sections of the spectrum shown in Figure 3 for lines showing the transition between absorption and emission. The $n = 7\text{--}6$ line shows a small emission peak superimposed upon a broader, offset, absorption feature while the $n = 9\text{--}7$ line is a weak, broad, emission peak upon which two solar OH absorption lines are superimposed.

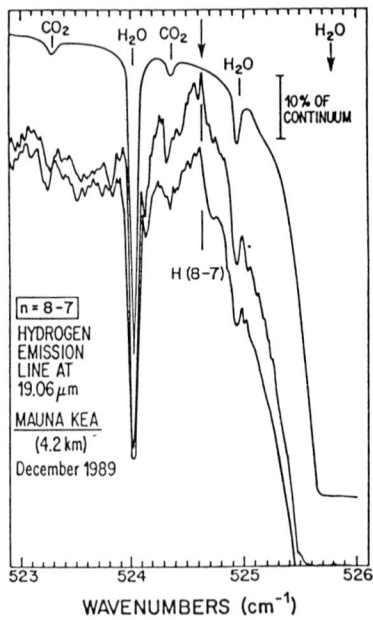

Fig. 2. Two representative 20 μm solar spectra taken from Mauna Kea, Hawaii.

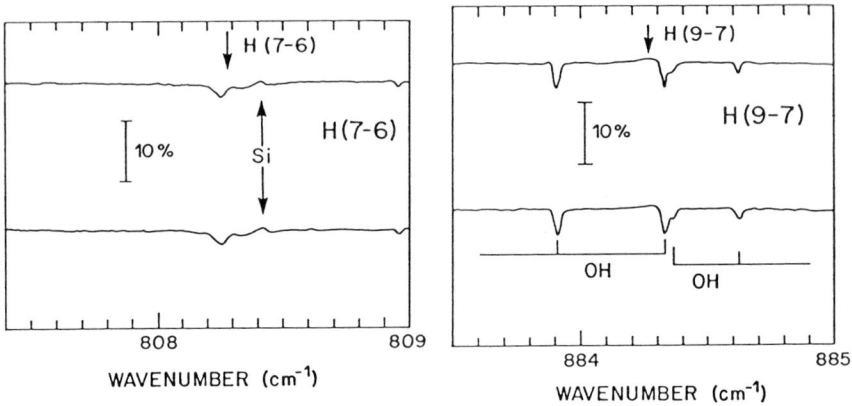

Fig. 3. Two sections of the *ATMOS* spectra, showing emission lines from $n = 7-6$ and $9-7$ transitions in H I.

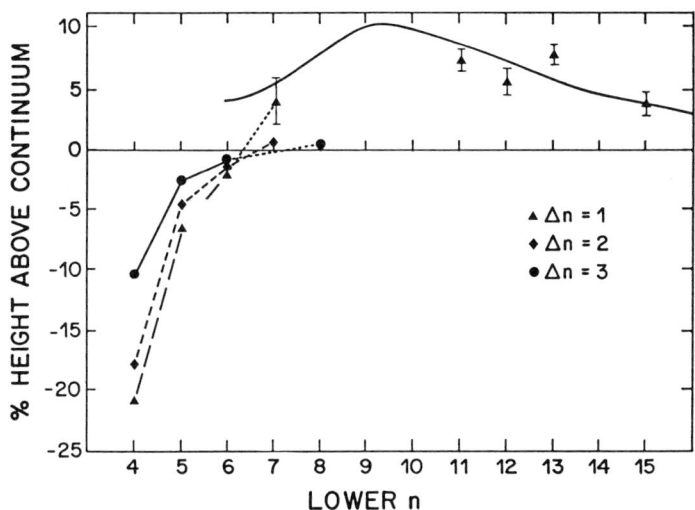

Fig. 4. Summary of the H I line peak intensities as a function of lower n value. The solid line above $n = 7$ shows the prediction of Hoang-Binh (1982).

4. Discussion and Conclusions

Figure 4 summarizes the trend of line depth or height of all of the observed H I recombination lines as a function of lower n value. The higher n-value lines are compared with the model prediction of Hoang-Binh (1982). The uncertainties in these line intensities arise from the correction for atmospheric and instrumental background emissions. There is a clear trend from absorption to emission in this H line sequence where the transition occurs at a higher n than for the Mg I lines, a fact which has been successfully explained by the modeling of Carlsson et al. (1992). Figure 5 shows the equivalent trend in line width, from very broad to relatively narrow H I lines as n increases, in the ATMOS data. The high-n emission lines are also relatively narrow, with FWHM of about 0.05 cm^{-1}, but with relatively wide outer wings, particularly on the $n = 12$–11 and 13–12 H I lines. This trend, from deep, wide absorption features to relatively narrow emission peaks, with the cross-over for $\Delta n = 1$ between $n = 7$–6 and 8–7, and the relative behavior of the H I and heavier element lines, has been explained, at least qualitatively, by Carlsson et al., (1992) and, in these proceedings, by Rutten and Carlsson and by Avrett. The further refinement of line intensity and width measurements of these and other high-n lines should serve to tighten the solar atmospheric model parameters within the line source region.

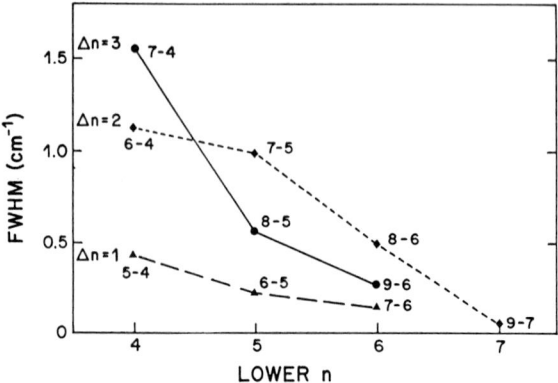

Fig. 5. Line width as a function of lower n for lines in the *ATMOS* spectrum.

Acknowledgements

It is a pleasure to thank the National Research Council of Canada and their contractor, ADGA International, for their support of the balloon program. The hospitality of the staff of the NASA Infra Red Telescope Facility on Hawaii during the 1989 campaign is also acknowledged. This research was supported by NSERC of Canada research grants to TAC and DAN. RTB was a Killam Scholar during part of this research program, while JB held a UC-NSERC Silver Jubilee Scholarship during the summer of 1991.

References

Boreiko, R.T.: 1985, Ph.D. thesis, University of Calgary.
Boreiko, R.T. and Clark, T.A.: 1986, *Astron. Astrophys.* **157**, 353.
Boreiko, R.T. and Clark, T.A.: 1987, *Astrophys. J.* **318**, 445.
Brault, J. and Noyes, R.W.: 1983, *Astrophys. J.* **269**, L61,
Carlsson, M., Rutten, R.J. and Shchukina, N.G.: 1992, *Astron. Astrophys.* **253**, 567.
Chang, E.S.: 1984, *J.Phys.B.: At. Mol. Phys.* **17**, L11.
Chang E.S. and Noyes, R.W.: 1983, *Astrophys. J.* **275**, 111.
Farmer C.B. and Norton, R.H.: 1989, NASA Ref. Publ. 1224, Vol. 1.
Hoang-Binh, D.: 1982, *Astron. Astrophys.* **112**, L3.
Murcray, F.J. et al.: 1982, *Astron. Astrophys.* **247**, L97.
Naylor, D.A. and Clark, T.A.: 1986, *SPIE* **627**, 482.

SOLAR SUBMILLIMETER AND MILLIMETER SPECTROSCOPY BETWEEN 7 AND 30 cm^{-1} FROM THE JAMES CLERK MAXWELL TELESCOPE

D. A. NAYLOR and G. J. TOMPKINS

Department of Physics, University of Lethbridge, Lethbridge, Alberta T1K 3M4, Canada

T. A. CLARK

Physics Department, University of Calgary, Calgary, Alberta T2N 1N4, Canada

G. R. DAVIS

University of Saskatchewan, Saskatoon, Saskatchewan S7N 0W0, Canada

and

W. D. DUNCAN

Joint Astronomy Centre, University Park, Hilo, HI 96720, U.S.A.

Abstract. A two-beam Martin-Puplett polarizing interferometer has been used in the rapid-scan mode on the 15 meter JCMT in conjunction with the facility detector, UKT14, to survey the solar sub-millimeter and millimeter spectrum in the four wavebands at 7–11, 11–15, 21–24 and 27–30 cm^{-1} to a spectral resolution of 0.01 cm^{-1} and at spatial resolutions of 19″, 16″, 7″ and 6″, respectively. Overall atmospheric transmission through these windows has been evaluated by comparison with synthetic spectra generated with FASCOD/HITRAN. A search has been made for contributions to these spectra from high-n transitions of H and heavier elements by several methods, including the comparison of solar with lunar and limb with disk center spectra.

Key words: infrared: stars – instrumentation: interferometric – Sun: radio radiation

1. Introduction

Solar spectra have been measured through the sub-millimeter and millimeter atmospheric windows from Mauna Kea under dry conditions in February 1991 with a 2-beam polarizing spectrometer mounted on the James Clerk Maxwell Telescope (JCMT). These spectra have been analyzed to evaluate the influence of both the telescope-detector system and variable atmospheric transmission upon the overall quality of such data and to explore the possibility of detecting and observing high-n recombination lines of strength 1% or less. Measurement of the intensities of these lines, extending the measurements already made, from space by *ATMOS* (Farmer and Norton 1989), from the ground in the near IR at 12 μm (*eg.*, Brault and Noyes 1983) and from balloon altitudes in the far IR (Boreiko and Clark 1986), will provide further verification of the modeling of their source regions (see Rutten and Carlsson 1993, and Avrett, Chang and Loeser 1993, in this volume).

2. Instrumentation, Observations and Data Processing

The present data were obtained on February 27, 1991 with a Martin-Puplett polarizing interferometer (Martin and Puplett 1970) mounted at the Nasmyth $f/35$ focus of the JCMT in the double-input port, single output port mode. UKT14, the broad-band bolometric detector (Duncan *et al.* 1990) operating at 0.36 K was used at the output port, with wide-band filters to define the sub-millimeter and millimeter windows at 350, 450, 850 and 1,100 μm. Focal plane apertures between 21 and 65 mm were selected to match the 6″–19″ range in diffraction beam-widths of the telescope over these wavelengths. In practice, this beam pattern was modified to a greater or lesser extent, depending on wavelength, by the small amplitude but significant beam-pattern extension detected by solar limb scanning (Lindsey and Roellig 1991, Clark et al. 1992). The spectrometer uses a metal-grid polarizer/analyzer and beam-splitters on thin Mylar substrates. With the narrow $f/35$ beam of the JCMT Nasmyth focus, it was possible to achieve the design resolution of 0.01 cm^{-1} with no collimation in the spectrometer.

This instrument provides several significant advantages over the conventional IR Michelson interferometer. Modulation efficiency of the metal-grid beamsplitter is constant and much higher over a wide wavelength range than that of thin-film beamsplitters. The complete separation of radiation from the two input ports removes uncertainty about the source of signal modulation.

The highest quality data was obtained in the "rapid-scan" staring mode with an ambient temperature black-body reference source in the second input beam and the telescope secondary mirror held stationary while the scanning roof-top mirror was moved through a complete scan in 45 s. Sets of 10 spectra were taken alternately, on Sun center and on a point 2,000″ from Sun center (or about one solar radius outside of the solar limb) at the same air mass. Spectra were also taken in a slow-scan mode by chopping the JCMT secondary mirror 120″ across the solar limb, using synchronous rectification to take the difference signal and remove atmospheric emission fluctuations during the 15 to 40 minute scans. These data proved to be significantly inferior in quality and resolution to those taken in the rapid-scan mode.

Data analysis followed standard procedure for Fourier transform spectroscopy and used a linear phase correction, since there was little dispersion in the optical elements of the instrument at these wavelengths.

3. Synthetic Spectra

The synthetic spectra used for comparison with the measured spectra were derived from the multi-layer FASCOD2 program (Clough et al. 1981), using the HITRAN database (Rothman et al. 1987) and the 1976 US Standard Atmosphere for the season and latitude of these observations. This synthesis included the major absorbers H_2O, O_2, O_3, N_2O, CO, CO_2 and CH_4 and the FASCOD continuum absorption term discussed by Clough et al. (1981), which contributes significant opacity at these wavelengths. Since the major absorber, H_2O, is usually highly variable, the amount of this constituent was derived from measurements of atmospheric opacity at 225 GHz, (monitored continuously at the Caltech Submillimeter Observatory (CSO), located adjacent to the JCMT) and incorporated into the synthesis. In practice, the atmosphere was found to be remarkably stable through the run and was estimated to contain 0.3 mm of precipitable water vapor. The final synthetic spectrum also incorporated the expected shape of the solar continuum, the UKT14 filter transmissions to low resolution and the airmass relevant to each spectrum.

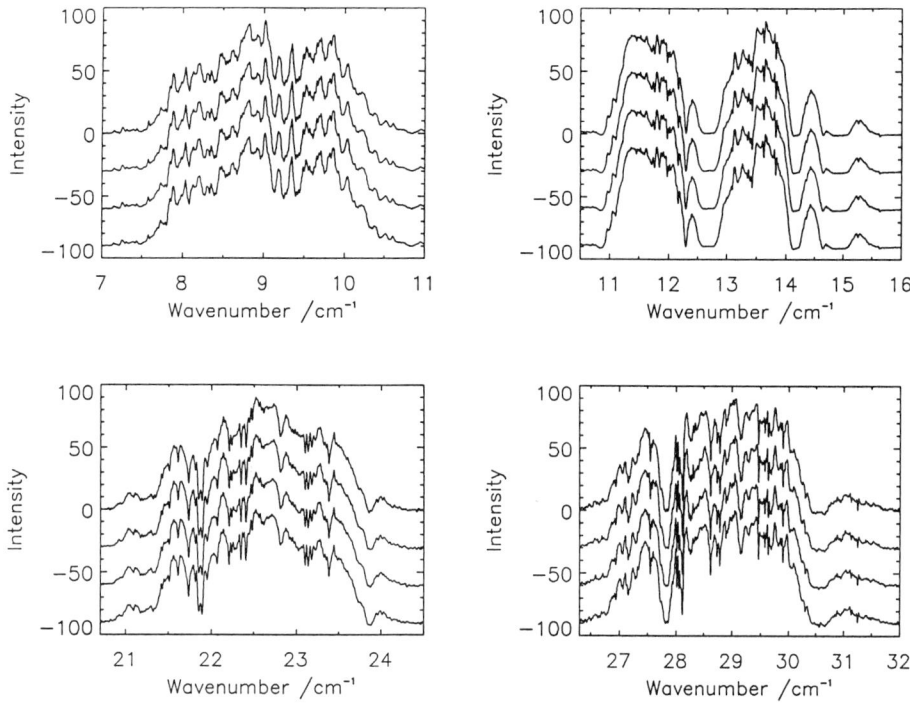

Fig. 1. Sets of 4 individual spectra through the 4 windows at 9, 12.5, 22.5 and 29 cm^{-1}.

4. Discussion of the Spectra

Representative sets of 4 individual sequential spectra at each wavelength, centered upon 9, 12.5, 22.5 and 29 cm^{-1} (270, 375, 675 and 870 GHz, respectively), each taken in 45 s, are shown in Figure 1. These spectra demonstrate excellent repeatability all wavelengths, with the greatest variability in the lowest-transmission, 29 cm^{-1} window, as expected. Overall noise estimates for the instrument in this configuration, (S/N between 80 and 1,000) agree very well with the noise equivalent flux density quoted for UKT14 for wide-band photometry on the JCMT. This indicates that the instrument added no significant noise to the spectra and that the atmospheric transmission fluctuations were small during this period of very dry atmospheric conditions above Mauna Kea.

Figure 2 compares the mean of each set of spectra to its respective synthetic spectrum, generated as described above, and to an equivalent black-body calibration spectrum from ambient-temperature and liquid-nitrogen-cooled cones in the two beams of the interferometer. The synthesis is a good fit to the measured atmospheric transmission in the 26–31 cm^{-1} and 20–25 cm^{-1} intervals, with narrow O_3 lines playing a significant role. A major feature of the 11–16 and the 7–11 cm^{-1}

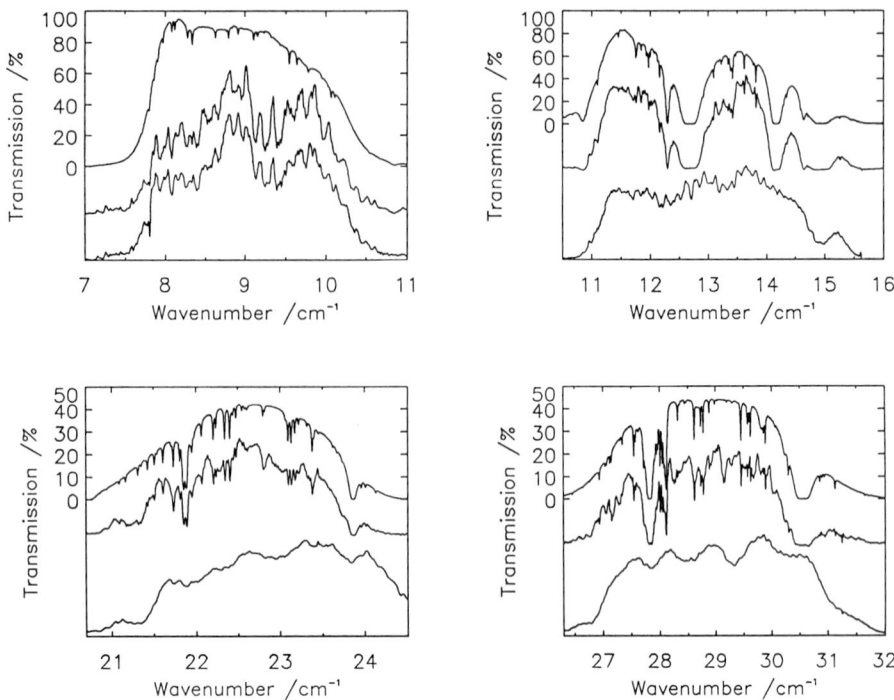

Fig. 2. Comparison of average solar spectra (middle trace) with normalized synthetic transmission spectra (top trace) and calibration spectra (bottom trace) in the 4 windows.

spectra is the rapidly varying structure on the continuum which, since it is reproduced in the calibration spectra, is thought to originate from detector, probably as a result of inadvertent optical cavities within the UKT14 cryostat. This "channel fringe" structure, large enough to mask all but the deepest O_3 lines in the 7–11 cm^{-1} interval, appears to have several characteristic "frequencies" and is fairly close to that in the calibration spectrum. This is not true, however, in the 20–25 and 26–31 cm^{-1} intervals, where the calibration spectra show slowly varying fluctuations with wavenumber and not the more rapid changes apparent in the solar and lunar spectra. It is also apparent in this comparison that the shape of the measured spectral envelope is dominated by the wings of the strong H_2O lines and that this shape will be very dependent upon airmass. This in turn means that calibration sources should be measured at very similar air masses wherever possible to avoid errors in the correction for atmospheric transmission shape. The Moon is a suitable calibration source for these solar measurements, but its position is rarely ideal, and the low overall signal and variations in emission across its surface may provide less than ideal calibration.

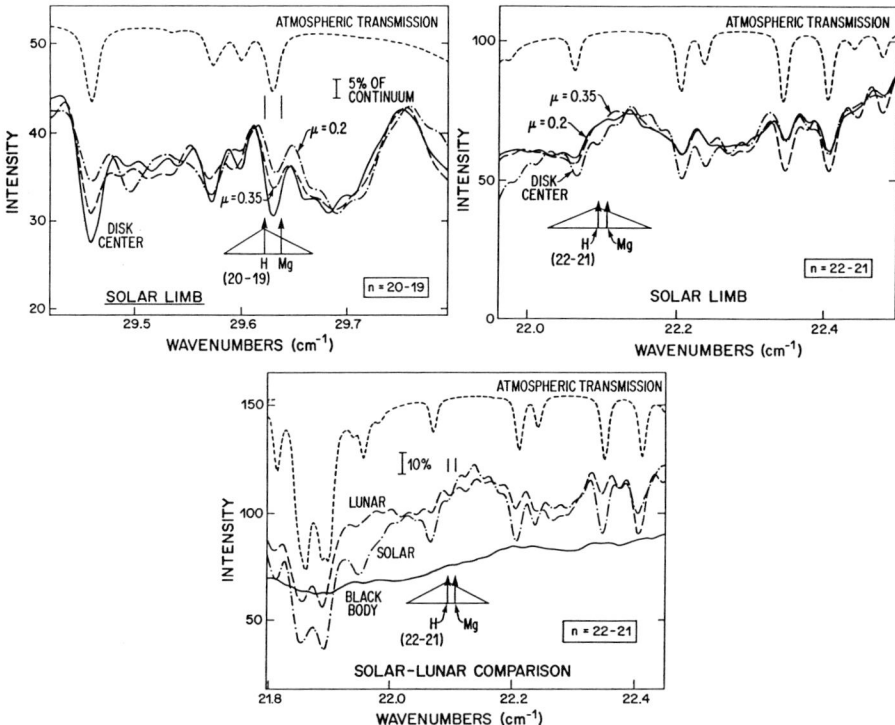

Fig. 3. (a) and (b) Comparison of normalized sections of spectra around the $n = 20\text{--}19$ and $22\text{--}21$ recombination line positions at 29.622 and 22.095 cm^{-1} at disk center and at $\mu = 0.35$ and 0.2 and (c) comparison of solar and lunar spectra at 22.095 cm^{-1}, along with the calculated atmospheric transmission.

5. Line Search

Several techniques were used in this initial search for high-n solar emission lines with the polarizing interferometer. Rapid-scan spectra at solar disk center were compared to lunar spectra from another time and airmass. Slow-scan spectra were taken in 2 positions, chopping the secondary mirror 120″ across the solar limb. The first was taken with one beam 60″ inside the limb ($\mu = 0.35$), and the second only 20″ inside the limb ($\mu = 0.20$) to utilize any limb brightening effects. Spectral quality, was significantly poorer in the slow-scan spectra, probably due to atmospheric transmission fluctuations over the scan time. Figure 3 shows normalized spectra near expected line positions to compare the solar-disk-center and lunar spectra, and solar-limb and disk-center spectra, for $n = 20\text{--}19$ and $22\text{--}21$. Continuum intensity and line widths predicted by Hoang-Binh (1982), are shown for each line. There is little evidence of excess emission at these positions. Better calibration with reduced channel-fringe structure will be needed in future searches.

6. Conclusions

The present spectra of the Sun show excellent reproducibility and signal-to-noise, but contain significant structure due to detector "sensitivity" variations across the filter band-passes, which appear to be beam-width dependent and somewhat difficult to calibrate out at the present time. Determination of the instrumental transmission function using both ambient-temperature and cooled black-body cones in the two beams of the Martin-Puplett interferometer is shown to reproduce the general overall shape at the longer wavelengths but provides only a hint of the rapid fluctuations in instrument function seen in the solar and lunar spectra at the shorter wavelengths, perhaps because of different input beam geometry into the detector optics. This structure, without adequate calibration, makes it difficult to identify weak, relatively broad line features such as the expected emission lines from high-n states of H and other elements. Modification of the optical system of UKT14 to remove the major cavity effects has already been accomplished but is unlikely to eliminate these effects completely. The calibration of spectra from the Sun, planets and other sources will pose difficulties, in view of the requirement for beam-width matching, particularly in the accurate representation of the continuum envelopes of sources.

Acknowledgements

It is a pleasure to thank the staff of the James Clerk Maxwell Telescope for their efforts in mounting the interferometer and in supporting these experiments. This telescope is operated by the Royal Observatory Edinburgh on behalf of the Science and Engineering Research Council of the United Kingdom, the Netherlands Organization for Scientific Research and the National Research Council of Canada. This work has been supported by NSERC and NRC of Canada research and travel grants and this support is gratefully acknowledged.

References

Avrett, E. H., Chang, E. S. and Loeser, R.: 1993, these proceedings.
Boreiko, R.T. and Clark, T.A.: 1986, *Astron. Astrophys.*, **157**, 353.
Brault, J.W. and Noyes, R.W.: 1983, *Astrophys. J. (Letters)*, **269**, L61.
Clark, T.A., Naylor, D.A., Tompkins, J.G. and Duncan, W.D.: 1993, *Solar Phys.*, in press.
Clough, S.A., Kneisys, F.X., Rothman, L.S., Gallery, W.O.: 1981, *SPIE*, **277**, 152.
Duncan, W.D., Robson, E.I., Ade, P.A.R., Griffin, M.J. and Sandell, G.: 1990, *Mon. Not. Roy. Astron. Soc.*, **243**, 126.
Farmer, C.B. and Norton, R.H.: 1989, *A High Resolution Atlas of the Infrared Spectrum of the Sun and the Earth Atmosphere from Space*, Vol. I: The Sun, NASA Ref. Pub. 1224, Washington, D.C.
Hoang-Binh, D.: 1982, *Astron. Astrophys.*, **112**, L3.
Lindsey, C.A. and Roellig, T.L.: 1991, *Astrophys. J.*, **375**, 414.
Martin, D.H. and Puplett, E.F.: 1970, *Infrared Phys.*, **10**, 105.
Rothman, L.S., Gramache, R. R., Goldman, A., Brown, L. R., Toth, R. A., Pickett, H. M., Poynter, R. L., Flaud, J.-M., Camy-Peyret, C., Barbe, A., Husson, N., Rinsland, C. P. and Smith, M. A. H.: 1987, *Applied Opt.*, **26**, 4058.

PART 5

MAGNETIC FIELDS AND INFRARED MAGNETOMETRY

VECTOR MAGNETOMETRY USING THE 12-μm EMISSION LINES

D. DEMING

Code 693, NASA/Goddard Space Flight Center, Greenbelt, MD 20771, U.S.A.

T. HEWAGAMA

Hughes/STX Corporation, 4400 Forbes Boulevard, Lanham, MD 20706, U.S.A.

D. E. JENNINGS

Code 693, NASA/Goddard Space Flight Center, Greenbelt, MD 20771, U.S.A.

G. McCABE

Hughes/STX Corporation, 4400 Forbes Boulevard, Lanham, MD 20706, U.S.A.

and

G. WIEDEMANN

*European Southern Observatory, Karl-Schwarzschildstrasse 2,
D-8046 Garching bei München, Germany*

Abstract. Recent polarimetric observations of the 12.32-μm emission line have provided the observational basis for deriving vector magnetic fields in the upper photosphere with great sensitivity. We use a line source function from the non-LTE model of Carlsson, Rutten and Shchukina, and calculate the radiative transfer of the Stokes I, Q, U, and V profiles. The results show that the profiles are not significantly affected by magneto-optical effects or by saturation, and reliable vector fields can be extracted by simply fitting the Seares relations to the Stokes profiles. Vector field observations for sunspots have shown that the field extends well beyond the photometric boundary of the sunspot, but that the field strength at the penumbral/photospheric boundary is less than half of the sunspot-center value. Within a mature sunspot, the 12-μm line profiles contain essentially no unpolarized radiation, indicating that the field is not intermittent in the sense of containing discrete flux tubes separated by field-free regions. We describe the design of a 12-μm Stokes polarimeter incorporating a high-resolution Fabry-Perot etalon and a 128 × 128 infrared array detector.

Key words: infrared: stars – instrumentation: polarimeters – line formation – Sun: magnetic fields

1. Introduction

The Mg I emission lines near 12 μm are the most magnetically-sensitive lines which can currently be observed in the solar spectrum, and are of great interest in studies of solar magnetic and electric fields (Deming *et al.* 1988; Chang and Schoenfeld 1991). The well-understood nature of the atomic physics (Chang 1993) and the recent development of convincing NLTE formation theories for these lines (Chang *et al.* 1991, Carlsson, Rutten and Shchukina 1992, hereafter CRS), in combination with limb measurements at the 1991 total eclipse (Jennings *et al.* 1993), lay a firm foundation for their diagnostic use.

We now understand that the lines are formed in the upper photosphere, several pressure scale heights above the regions sensed in visible magnetograms. Most magnetic observations have utilized the Mg I line at 12.32 μm. CRS have given the line-center contribution function for this transition, showing that it peaks near $\log(\tau_{500}) = -2.7$, whereas a typical visible-region photospheric line senses the region near $\log(\tau_{500}) = -1.0$, (see, *e.g.*, Rees, Murphy and Durrant 1989). While this difference in formation level is quite modest, it occurs over an important regime.

Consulting model atmospheres (*e.g.*, Maltby *et al.* 1986), we find that the ratio of external gas pressure to the magnetic pressure of a 1,000-Gauss field is 0.1 at the 12-μm height, versus a value near unity for a typical visible photospheric line. This change from the regime where gas and magnetic pressures are comparable, to the regime where magnetic pressure dominates, is of considerable physical significance. Simultaneous magnetic observations at 1.6 μm (Rabin 1992) and 12 μm would be especially interesting, because the significant difference in the gas-to-magnetic pressure ratio would be accompanied by the large magnetic splitting which is characteristic of infrared (IR) lines.

Vector magnetometry using the 12-μm lines requires both polarimetric observations and the means to interpret them and extract the magnetic vector. The capability to make polarimetric 12-μm observations using the McMath Fourier transform spectrometer (FTS) was developed by Hewagama (1991), and Hewagama *et al.* (1992). The polarization properties of the McMath telescope are quite favorable at this long wavelength and the observations do not require correction for telescope polarization (Hewagama 1991; Deming *et al.* 1991). The greatest observational limitation at present is the single-detector nature of the FTS observations. We have recently begun to use array detectors for 12-μm observations, and we will here describe the methodology of the observations and progress toward constructing a 12-μm Stokes polarimeter using a 2-D IR array detector.

As concerns the extraction of vector fields, we need to evaluate the impact of the NLTE models. Hewagama (1991) assumed that the emission was optically thin and derived vector fields by fitting the Seares (1913) relations to the observed Stokes profiles. The assumption of optical thinness should be re-examined in the light of the new models, which show line-center optical depths reaching values greater than unity. We have therefore calculated theoretical Stokes profiles for the 12.32-μm line based on the CRS model results, and we will show that the Hewagama (1991) fits to the Seares (1913) relations are quite adequate for the purpose of extracting the field vector.

2. Quantum Mechanics of the Transition

The 12-μm lines were identified by Chang and Noyes (1983), and Chang (1987, 1992) has discussed the quantum mechanics of the transitions. We here reiterate and expand on a few aspects of the 12.32-μm line which are especially relevant to vector field determinations. The upper state has $n = 7$, $l = 6$ and the lower state $n = 6$, $l = 5$. Fine structure due to spin-orbit coupling is quite small, of order 0.0005 cm^{-1}, much less than the quiet-Sun line width of 0.017 cm^{-1}.

Figure 1 (from Hewagama 1991) shows the level configurations for the line calculated using the Chang (1987) formulae and assuming that the atom is immersed in a 300 Gauss external magnetic field. In the presence of even this relatively weak magnetic field, the interaction energy of the total angular momentum vector with the magnetic field is larger than the spin-orbit fine structure. The total angular momentum in the field direction is m in Figure 1, and spin-orbit interaction produces the fine structure at each m. The transitions $\Delta m = 0$ produce the π component in the "Zeeman" pattern, while $\Delta m = \pm 1$ produce the σ components. Figure 2

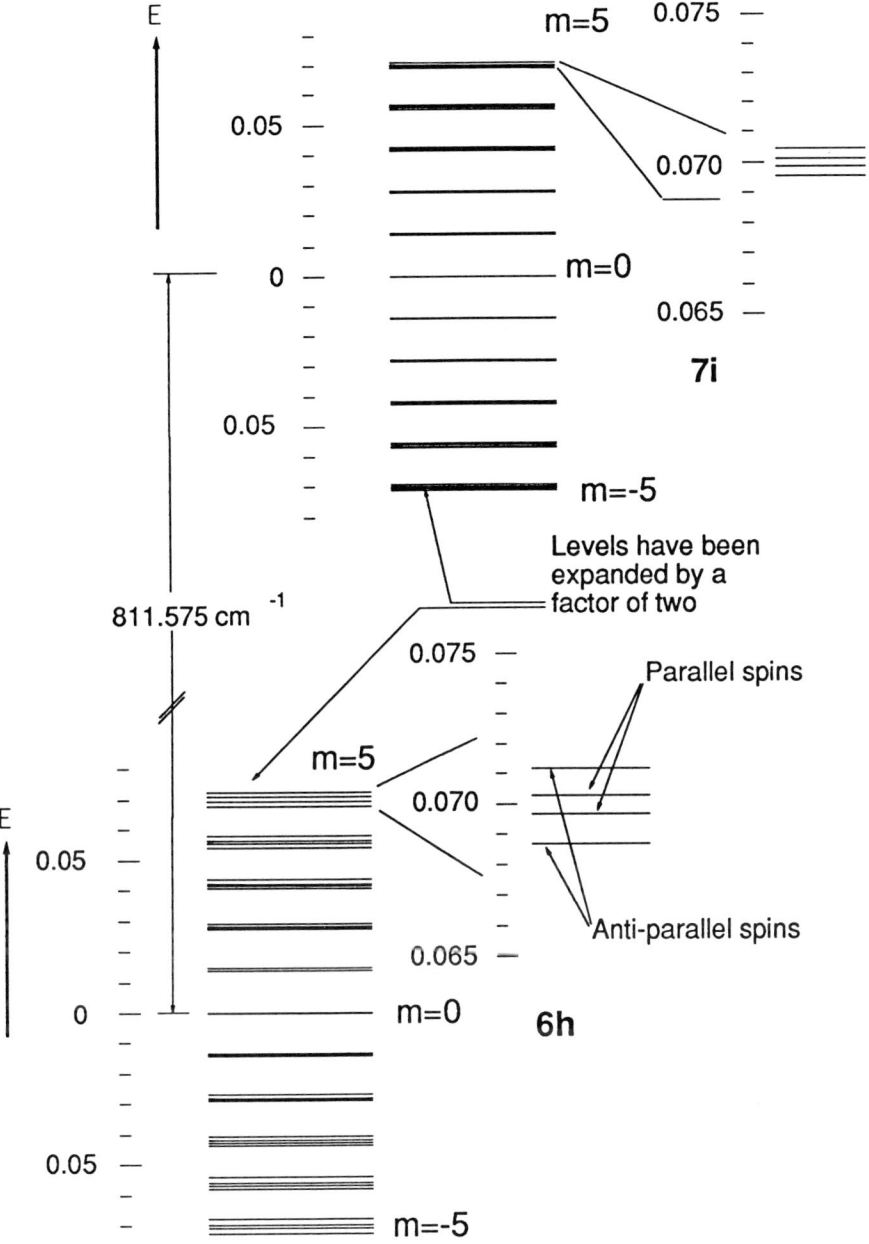

Fig. 1. Energy level diagram for the 12.32-μm line. The $m = 6$ states in the upper level are not illustrated.

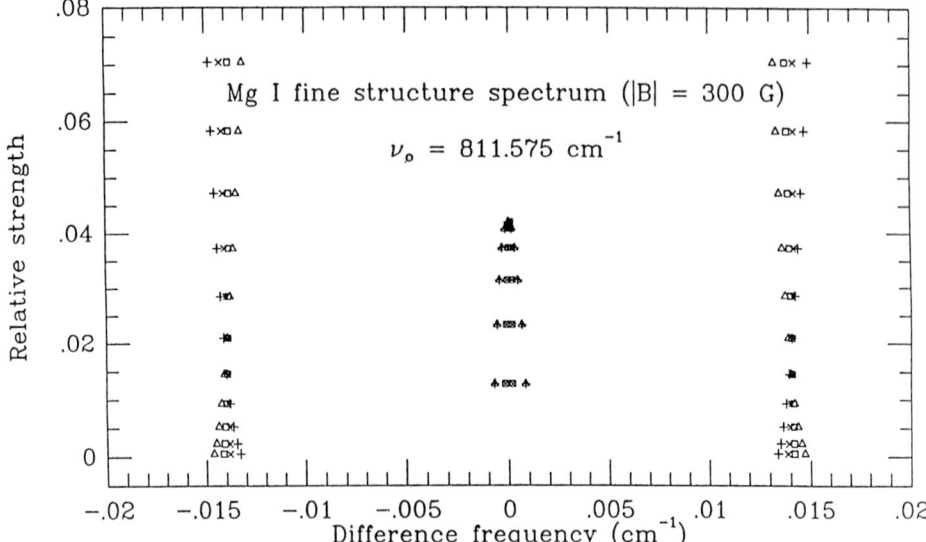

Fig. 2. Positions and relative strengths of "Zeeman" components in the 12.32-μm line spectrum for a 300-Gauss magnetic field strength.

shows the resultant "Zeeman" pattern, with fine structure in the components being significantly less than the splitting. Solar active-region magnetic fields observed in this line to date are seldom much weaker than 300 Gauss. Consequently, the line can be well approximated as a normal Zeeman triplet having Landé-g factor of unity (Chang 1987), and the fine structure in each component can be ignored. For truly weak fields (tens of Gauss or less) the fine structure in the magnetic pattern would have to be accounted for.

3. Radiative Transfer and the Stokes Profiles

Figure 3 shows an example of Stokes profiles observed for this line by Hewagama *et al.* (1992), in comparison with profiles calculated using the Seares relations. The Seares fit assumes that the line components are Gaussian emissions having widths equal to the observed profiles. They were fit to the observations using non-linear least squares, resulting in the field strength, $|\mathbf{B}|$, line-of-sight inclination, γ, and azimuth, α, shown in the I profile (upper left panel) in the Figure. The assumption underlying the Seares fit is that the emission components are optically thin, *i.e.*, that photons emitted upward from the upper state progress without significant re-absorption. Departures from optical thinness could, in principle, affect the fits in two ways. First, the ratio of amplitudes in the π and σ components, which indicates the inclination to the line-of-sight, could be affected by saturation. In the case where the line is very optically thick, the central intensities in the π and σ components will all approach the value of the source function in the line-forming

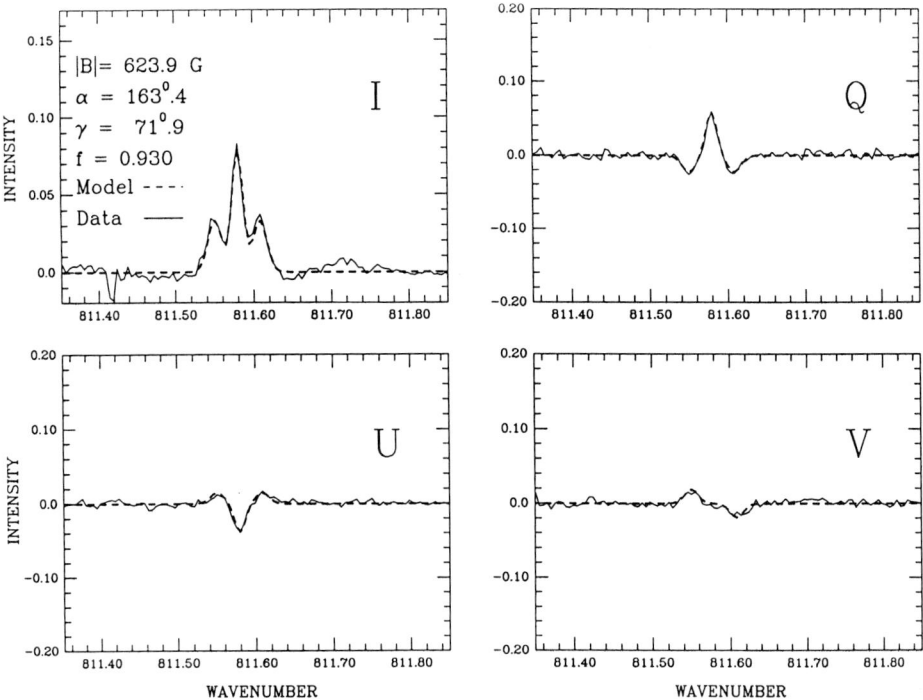

Fig. 3. Observed Stokes profiles of the 12.32 μm lines in a sunspot penumbra (solid line), in comparison to a fit of the Seares formulae (dashed line). The derived field strength and orientation are noted, as well as the fractional polarization, *i.e.*, the fraction (f) of the emission in Stokes I which is accounted for by the polarized light intensity, $\sqrt{Q^2 + U^2 + V^2}$.

region of the photosphere, relatively independent of the field inclination. Second, the Stokes profiles can, in principle, exhibit magneto-optical effects, where line absorption results in conversion of one polarization state to another (*e.g.*, from linear to circular).

To test whether these two possibilities are important in practice, we have calculated Stokes I, Q, U and V profiles for this line using a more rigorous radiative transfer approach. We adopted the CRS line-center source function, and used a Voigt profile for the line. The Doppler width for the line was taken to be 0.017 cm^{-1} FWHM, the same as the observed line. The damping parameter was taken to be proportional to gas pressure, and the constant of proportionality was adjusted to reproduce the absorption wings which are present on the observed quiet-Sun line. We used the DELO method described by Rees, Murphy and Durrant (1989), and we checked our implementation of the method by comparing to the results of a direct quadrature integration of the Stokes radiative transfer equations. The DELO calculations included magneto-optic effects.

Figure 4 (left) shows DELO profiles computed with and without magneto-optics.

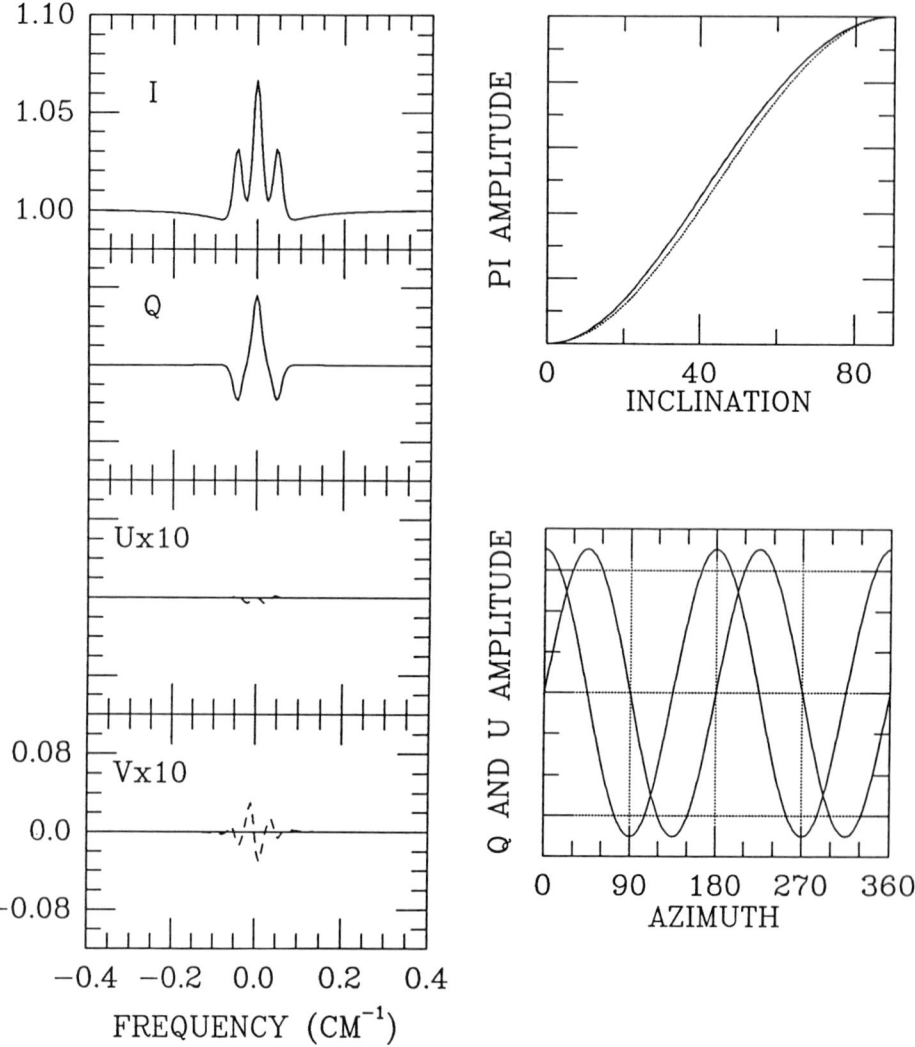

Fig. 4. Results of radiative transfer calculations for the Stokes profiles of the 12.32-μm lines. The $I, Q, U,$ and V profiles on the left were calculated for a field inclination of 90° (*i.e.*, horizontal) and an azimuth of 0°. The dashed line shows the result of including magneto-optical effects (note the expanded scales for U and V). The π-component amplitude in the upper right includes the effects of magneto-optics, and is shown in comparison to a sin-squared relation (dotted line). The Q and U amplitudes in the lower right panel also include the effect of magneto-optics, which does not cause a significant perturbation in their azimuth dependence. All of the calculations were made for a field strength of 1,000 Gauss.

The field inclination was taken to be 90° and the azimuth was 0°, putting all of the polarization signal into Stokes Q. The inclusion of magneto-optics (dashed lines) puts a small fraction of the polarization into Stokes V and Q, but the amount is so small that the effect is unimportant for vector field determinations at present. We also calculated the amplitude of the π component in Stokes Q and U as a function of field azimuth for 90° inclination (lower right), in order to test whether magneto-optics causes an apparent rotation of the azimuth. As is evident from the Figure, the angles of maxima and zeros of Q and U for the π component do not differ significantly from the values expected from the Seares relations. As a further test, we calculated the line-center amplitude of the π component as a function of field inclination (upper right), including magneto-optical effects. This relation is shown in comparison to the sin-squared dependence predicted from the Seares relation (dotted line). The maximum deviation from the sin-squared dependence amounts to less than 2°. We conclude from these calculations that the Hewagama (1991) Seares relation fits are adequate for the derivation of vector fields from the observed Stokes profiles.

4. Magnetic Configuration of Sunspots

Sunspots are the most prominent magnetic regions on the solar surface, so it is logical to apply the 12-μm lines first to sunspot studies. Several sunspots have now been observed at 12 μm (Deming et al. 1988; Hewagama et al. 1992), and some general conclusions have emerged. The dependence of magnetic field strength with distance from sunspot center has been observed in the visible by many authors (e.g., Beckers and Schröter 1969), and this functional dependence is important as a test of magnetostatic sunspot theories (e.g., Osherovich and Garcia 1989). Figure 5 shows this relation for two sunspots observed at 12 μm (Deming et al. 1988, Hewagama et al. 1992).

The first conclusion from these measurements is that azimuthal symmetry is a poor approximation for sunspot field strengths. The 12-μm splittings typically determine the field strength with an error of less than 50 Gauss, whereas the scatter in field strengths at a constant radius is several hundred Gauss, indicating that real azimuthal structure is present. Another point of interest is the value of the field strength at the penumbra/quiet Sun boundary, $R = R_p$. Beckers and Schröter (1969) concluded that the field at this boundary was half of the sunspot-center field strength, whereas Figure 5 shows that it is significantly less. The incomplete splitting exhibited in visible lines, in combination with the small field strength, and potential scattered light contamination at this boundary, have historically made this determination problematical. Indeed, some authors (Adam 1990; Wiehr and Balthasar 1989) have concluded that the magnetic field strength approaches *zero* a few arcseconds beyond this photometric boundary. The 12-μm data, however, show that the sunspot magnetic field extends to (typically) half a sunpot radius beyond the boundary with the surrounding photosphere ($R \sim 1.5R_p$). This conclusion is unequivocal, because the 12-μm splittings are resolved down to field strengths of 300 Gauss. The penumbral field at $R = R_p$ being 600 Gauss, scattered light cannot cause this to be confused with the 200–300 Gauss fields seen at $1.5R_p$. The

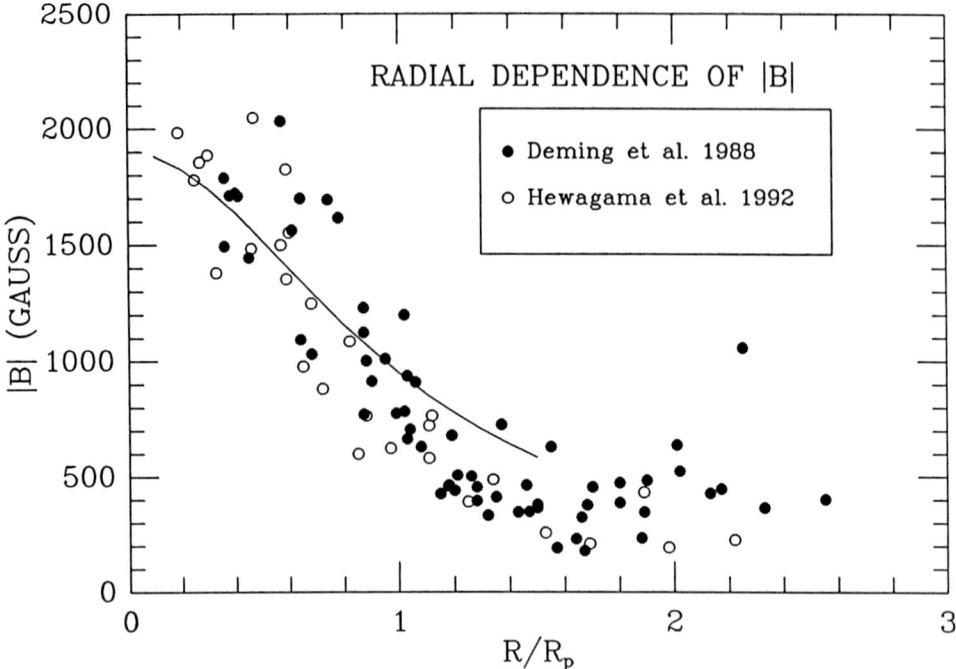

Fig. 5. Magnetic field strength versus distance from sunspot center in units of the penumbral radius, based on 12-μm measurements by Deming et al. (1988) and Hewagama et al. (1992). The solid line is the Beckers and Schröter (1969) relation.

conclusion that the field at $R = R_p$ is weaker than the Beckers and Schröter (1969) relation, but extends to a large fraction of a sunspot radius, is consistent with recent Stokes polarimetry in the visible (Lites and Skumanich 1990).

Parker (1979) has suggested that the sub-surface structure of sunspots in made up of a collection of discrete flux tubes separated by field-free gas. Since it is well known that sunspot penumbrae show filamentation in high-spatial-resolution images, it is natural to associate the filamentation seen in high-resolution images with magnetic filamentation. The 12-μm data can give us valuable insight into the possible intermittent nature of the sunspot field. Because the splitting is resolved, unpolarized radiation in the 12-μm π component can only arise as emission from field-free gas, not as the cancellation of orthogonal polarizations from incompletely-split σ components. Magnetic filling factors, based on the fraction of the 12 μm emission which is polarized, were derived by Hewagama et al. (1992) and are shown in Figure 6.

Within a mature sunspot, the 12-μm emission is essentially 100% polarized, decreasing to 90% at the boundary with the surrounding photosphere. This implies that field-free regions do not contribute significantly to the 12-μm sunspot emission, so the field at this altitude does not appear to be filamentary in the sense of being composed of discrete flux tubes separated by field-free gas. This conclusion is not

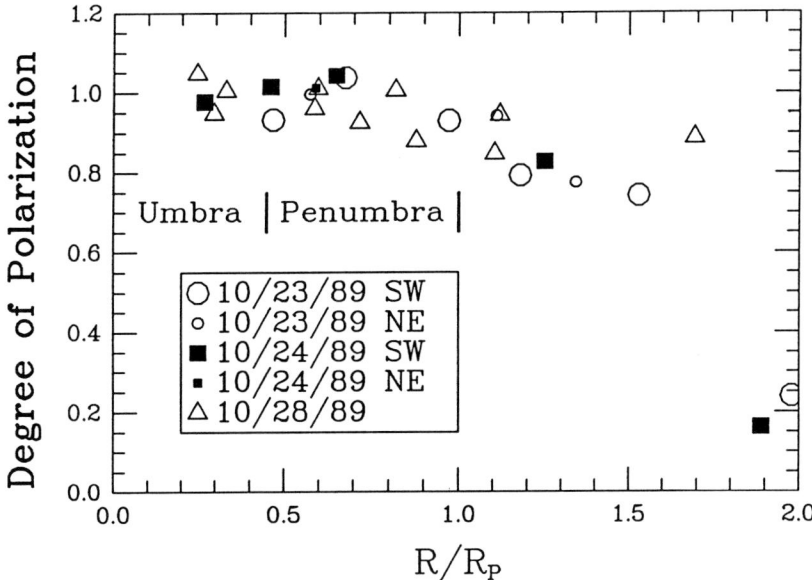

Fig. 6. Degree of polarization in the 12.32-μm line versus distance from sunspot center in units of the penumbral radius, from Hewagama et al. (1992).

surprising, since the lower gas pressure at the 12-μm altitude will allow discrete flux tubes to expand and merge.

The 12-μm data have also been used to deduce the inclination of the field to the surface normal, and Figure 7 shows this for a sunspot observed by Hewagama et al. (1992). The 12-μm inclinations in this instance differ considerably from the average values seen in the visible (see e.g., Lites and Skumanich 1990). The smaller data symbols in Figure 7 represent a region of the sunspot which may have been affected by the presence of adjacent magnetic regions, and a filament. The inclinations to the surface normal are dramatically smaller in this region of the penumbra, which Hewagama et al. (1992) attribute to perturbation by the adjoining magnetic structure. While this seems plausible, visible observations have not shown such anomalies. Additional observations are needed in order to see whether inclination anomalies are a common feature in 12-μm data.

5. Progress Toward a 12-Micron Stokes Polarimeter

To date, most 12-μm magnetic observations have been made using the single-detector McMath FTS. Much could be learned about the magnetic configurations of solar active regions if 12-μm polarimetry were possible using a 2-D array detector. In addition to the magnetic configuration of sunspots discussed above, other outstanding problems in solar physics could benefit from such instrumentation. We note two examples of such problems:

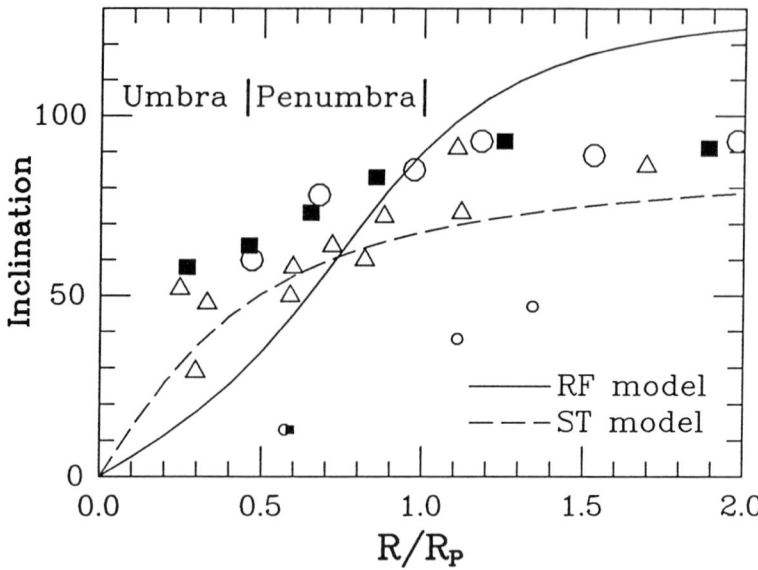

Fig. 7. The inclination of the magnetic field versus distance from the sunspot center in units of the penumbral radius. The inclinations are in heliographic coordinates, measured from a normal to the solar surface, in the 12.32-μm line by Hewagama et al. (1992).

(i) The Expansion of Plage and Network Flux Tubes. It has long been believed that most solar magnetic flux is present in strong-field form and that the relatively low field strengths seen by longitudinal magnetographs simply reflect the fact that the signal is diluted by the small filling factor for the magnetic elements. Rabin (1992) has used observations in the very Zeeman-sensitive 1.56-μm line to confirm the strong-field nature of plage flux tubes. His results also show spatially-resolved patterns in the field strength and filling factor. The field strength in an isolated thin flux tube varies with height as $\exp(-h/2H)$, where H is the gas pressure scale height. The factor $2H$ arises from the fact that the gas pressure difference between the flux tube and the surrounding photosphere is balanced by the magnetic pressure in the flux tube, the latter being proportional to $|\mathbf{B}|^2$. Over the 2.4 gas pressure scale heights which separate the 1.6- and 12-μm observations (125 to 400 km levels), the field strength should drop by a factor of 3.3. Plage field strengths at 12 μm are approximately 600 Gauss (Brault and Noyes 1983; Zirin and Popp 1989), whereas Rabin (1992) typically observes field strengths of 1400 Gauss at 1.6 μm. The observed field strength ratio of $1400/600 = 2.3$ is less than expected by an amount which exceeds the errors in the IR observations. However, the 1.6- and 12-μm observations have, to date, not been simultaneous. Moreover, since Rabin observes a large-scale pattern in the field strengths, as well as filling factors as large as 30%, the real situation is doubtless much more complex than the thin-flux-tube model. Simultaneous vector field obser-

vations over extended spatial scales at both 1.6 and 12 μm could tell us much about the manner in which plage fields are organized and their expansion with height.

(ii) Magnetic Field Relaxation Associated with Solar Fields. It is almost universally believed that the energy released in solar flares is extracted from that stored in non-potential magnetic field configurations (Haisch, Strong and Rodono 1991). However, observational proof of this belief has been difficult to obtain. It has been observationally established that flare occurrence is often associated with the existence of sheared photospheric fields (Hagyard et al. 1984). However, since the magnetic energy release occurs at coronal altitudes, its signature in photospheric magnetic observations is likely to be rather subtle. The approach which must be adopted is to calculate the magnetic free energy of an active region configuration, using photospheric observations as a boundary condition, and look for changes in this free energy which can be associated with a flare (Gary et al. 1987). Since this involves essentially an extrapolation of vector field measurements to coronal altitudes (Gary and Musielak 1992), the results are likely to be more robust if the vector fields are measured as high as possible. The upper photospheric nature of the 12-μm line, and its great magnetic sensitivity, suggest that 12-μm vector magnetograms would be excellent observational material with which to attack the flare problem.

We are developing a Stokes polarimeter for the 12.32-μm line, using IR array detectors. Array technology at this thermal IR wavelength is developing very rapidly. We are currently testing a 128 × 128 Si:As array from Rockwell as the detector for this polarimeter. The first decision which we made was that using a relatively large format array detector was impractical in an FTS, because the data rates would be impossibly high. Instead, the choice was between a grating spectrometer or a Fabry-Perot approach. There are really three dimensions to be observed, two spatial dimensions and one spectral dimension. Hence, we would ideally like to have a 3-D detector so as to observe all spatial positions and wavelengths simultaneously. Since we have at most a 2-D detector, we have to decide how to use it optimally. We could use one detector dimension spectrally, and the other spatially. This is the grating spectrometer approach with an input slit, and the slit has to be scanned in the remaining spatial dimension. The Near Infrared Magnetograph (NIM) being developed at NSO uses this method. Alternatively we could use both detector dimensions spatially, and scan the spectrum using a Fabry-Perot etalon. Since the Sun is significantly fainter at 12 μm than at 1.6 μm, the 12-μm observations take longer. Consequently, the efficiency with which we use the detector array is a greater concern. Measurements at approximately 50 points across a line are needed for adequate definition of Stokes profiles. The 128 × 128 array is already larger than this, so we gain in total efficiency by using both array dimensions spatially, and scanning an etalon through the spectrum.

The design of the 12-μm polarimeter is shown in Figure 8. Spectral resolution of 0.007 cm^{-1} is provided by a high-resolution etalon. It is scanned while maintaining parallelism using capacitance sensing of the plate separation and correction using

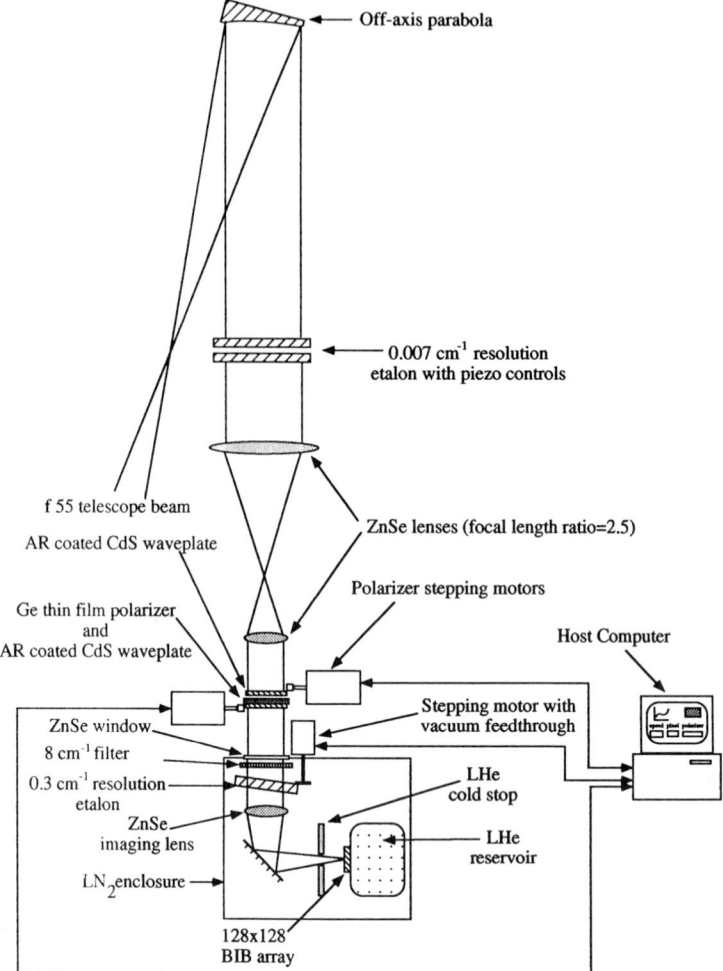

Fig. 8. Design of a 12-μm Stokes polarimeter using a 128 × 128 Si:As infrared array detector and Fabry-Perot etalons to provide spectral resolution.

piezoelectric actuators in a servo-loop. It was developed by Queensgate, Inc. and has a free spectral range of 0.4 cm^{-1}. A cold (LN$_2$) constant-gap etalon having a free spectral range of 8 cm^{-1} and resolution of 0.4 cm^{-1} is used to isolate a single high-resolution fringe. The lower-resolution etalon is mechanically tilt-tuned to follow the piezoelectric scanning of the high-resolution etalon. One low-resolution fringe is isolated using a narrow-band filter.

There are potential pitfalls involved in scanning the spectrum. The principal problem is likely to be changes in atmospheric transmission and instrumental sensitivity which are presumably $1/f$ in nature. The maximum read rate of the 128 × 128 array is 30 Hz, so almost two seconds are required to scan the spectrum. Figure 9 shows spectra taken using a very preliminary version of the 12-μm polarimeter,

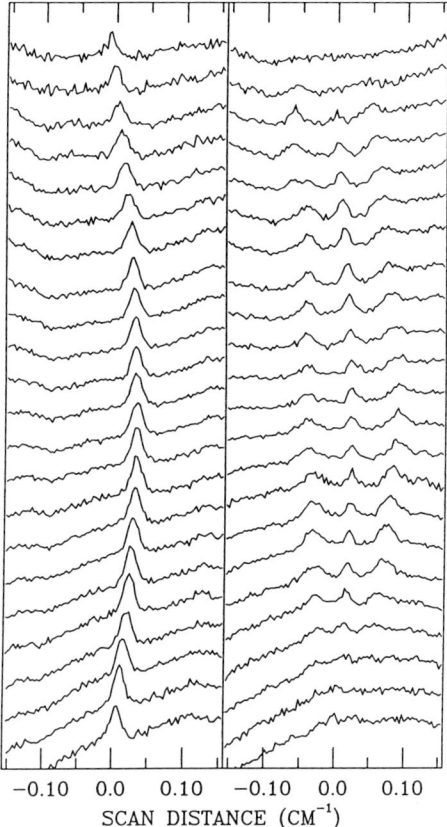

Fig. 9. Fabry-Perot-etalon spectra of the 12.32-μm line obtained using a very preliminary version of the Figure 8 polarimeter and a 24-element linear array detector. Quiet Sun spectra are illustrated on the left, and sunspot spectra on the right. Each spectrum corresponds to a single pixel of the 24-element array, viewing 3″ × 3″ on the solar disk.

scanning the spectrum even more slowly (about 30 seconds). This version used a 24-element linear array, rather than the 128 × 128. It also used a low resolution etalon which was uncooled, and its limiting noise level was determined by thermal background radiation from the warm etalon. The Figure 9 spectra show both a quiet-Sun region (left), and a sunspot region (right) where the line shows large splitting. The noise level in these test observations is considerably higher than will be possible when the low-resolution etalon is used cold. Nevertheless, it is encouraging that even the very slow scanning rate used for Figure 9 does not produce noise in excess of the thermal background from the warm etalon. When this thermal background noise is reduced, we may find that atmospheric and instrumental changes limit the observations. If so, the high-resolution etalon can be stepped to a reference continuum point on every other sample. This would double the observation time needed for the observations, but it would allow us to remove the effect of atmospheric and instrumental changes. We plan to generate Stokes profiles by subtracting successive measurements of $I + V$ and $I - V$, etc., as described by Hewagama et al. (1992) for FTS data. The polarizers are rotated, under control of the data acquisition computer, using stepping motors.

The spectra shown in Figure 9 cover about 72" on the solar disk (about 3" per pixel). There is loss of finesse in the etalon for the pixels at each end of the linear array. The Figure 8 design will correct this by using a larger collimated beam through the high resolution etalon. The spatial resolution using the 128 × 128 array will be 0.8" per pixel. This matches a single pixel to the diffraction-limited resolution of the proposed 4-meter McMath-Pierce telescope (Livingston 1991), and provides several-pixel sampling of its current resolution.

Acknowledgements

This work was supported by the NASA Solar Physics Branch, under RTOP 170-38-53-10. We thank Rockwell International for providing the 128 × 128 array, and Rob Rutten and Mats Carlsson for sending results of their calculations of source functions for the 12-μm lines.

References

Adam, M.G.: 1990, *Solar Phys.* **125**, 37.
Beckers, J.M., and Schröter, E.H.: 1969, *Solar Phys.* **10**, 384.
Brault, J.W., and Noyes, R.W.: 1983, *Astrophys. J. (Letters)* **269**, L61.
Carlsson, M., Rutten, R.J., and Shchukina, N.G.: 1991, *Astron. Astrophys.* **253**, 567 (CRS).
Chang, E.S.: 1987, *Phys. Scripta* **35**, 792.
Chang, E.S.: 1993, these proceedings.
Chang, E.S., and Noyes, R.W.: 1983, *Astrophys. J. (Letters)* **275**, L11.
Chang, E.S., and Schoenfeld,W.G.: 1991, *Astrophys. J.* **383**, 450.
Chang, E.S., Avrett, E.H., Mauas, P.J., Noyes, R.W., and Loeser, R.: 1991, *Astrophys. J. (Letters)* **379**, L79.
Deming, D., Boyle, R.J., Jennings, D.E., and Wiedemann, G.: 1988, *Astrophys. J.* **333**, 978.
Gary, G.A., and Musielak, Z.E.: 1992, *Astrophys. J.* **392**, 722.
Gary, G.A., Moore, R.L., Hagyard, M.J., and Haisch, B.M.: 1987, *Astrophys. J.* **314**, 782.
Hewagama, T.: 1991, Ph.D. Thesis, University of Maryland at College Park.
Hewagama, T., Deming, D., Jennings, D.E., Osherovich, V., Wiedemann, G., Zipoy, D., Mickey, D.L., and Garcia, H.: 1992, *Astrophys. J. Suppl.* in press.
Jennings, D.E., Deming, D., McCabe, G., Noyes, R., Wiedemann, G., and Espenak, F.: 1993, these proceedings.
Hagyard, M.J., Smith, J.B.Jr., Teuber, D., and West, E.A.: 1984, *Solar Phys.* **91**, 115.
Haisch, B., Strong, K.T., and Rodono, M.: 1991, *Ann. Rev. Astron. Astrophys.* **29**, 275.
Lites, B.W., and Skumanich, A.: 1990, *Astrophys. J.* **348**, 747.
Livingston, W.: 1991, *Proc. SPIE* **1494**, 50.
Maltby, P., Avrett, E.H., Carlsson, M., Kjeldseth-Moe, O., Kurucz, R.L., and Loeser, R.: 1986, *Astrophys. J.* **306**, 284.
Osherovich, V.A., and Garcia, H.A.: 1989, *Astrophys. J.* **336**, 468.
Parker, E.N.: 1979, *Astrophys. J.* **230**, 905.
Rabin, D.: 1992, *Astrophys. J. (Letters)* **390**, L103.
Rees, D.E., Murphy, G.A., and Durrant, C.J.: 1989, *Astrophys. J.* **339**, 1093.
Wiehr, E., and Balthasar, H.: 1989, *Astron. Astrophys.* **208**, 303.
Zirin, H., and Popp, B.: 1989, *Astrophys. J.* **340**, 571.

PROPERTIES OF MAGNETIC FEATURES FROM THE ANALYSIS OF NEAR-INFRARED SPECTRAL LINES

SAMI K. SOLANKI

Institute of Astronomy, ETH-Zentrum, CH-8092 Zürich, Switzerland

Abstract. An overview is given of the structure and the physics of magnetic features in solar plages, as derived from observations of near-infrared lines. First, the diagnostic potential of near-infrared lines is compared with that of lines in the visible and at 12 μm. Then, the results on the magnetic and velocity structure of magnetic features obtained from 1.5 μm lines are described, discussed and compared with results of observations in the visible and with theoretical predictions. Finally, the past and present achievements of near-infrared investigations of Zeeman-split lines are summarized.

Key words: infrared: stars – Sun: faculae, plages – Sun: magnetic fields

1. Introduction

Of the many branches of solar research that have been enriched by investigations of infrared radiation, none has been transformed to the same extent as the measurement of magnetic fields. The opening of the infrared has qualitatively enhanced our capability of studying magnetic features. It is, therefore, a pleasant task to review results obtained in solar plages using near-infrared lines, which, in the context of the present review, implies lines with wavelengths between 1.5 and 1.8 μm (magnetic fields in solar plages have very rarely been observed in other near-infrared wavelength ranges, but see, *e.g.*, Harvey and Hall 1971). I shall concentrate on describing and discussing the information which has been derived from the observations using simple models. Thus, the present review concentrates on the interface between observation and theory, each of which is reviewed very competently elsewhere in the present volume (Rabin 1993 and Steiner 1993, respectively).

2. Comparison Between Visible, 1.5 μm and 12 μm Spectral Lines

The most obvious difference between the three wavelength ranges, as far as diagnostics based on the Zeeman-effect are concerned, is in the sensitivity of the spectral lines to the magnetic field (Zeeman sensitivity). The ratio of Zeeman splitting, $\Delta\lambda_H$, to non-magnetic line width, $\Delta\lambda_D$, determines the Zeeman sensitivity: $\Delta\lambda_H/\Delta\lambda_D$. Since $\Delta\lambda_D$ is roughly proportional to the central wavelength of the line, λ, while $\Delta\lambda_H \sim g\lambda^2$, where g is the Landé factor, we have approximately $\Delta\lambda_H/\Delta\lambda_D \sim g\lambda$. The largest g value, $g_{\max}$ and $g_{\max}\lambda$ are listed in Table 1 for each of the three wavelength regions. The Zeeman sensitivity translates directly into the smallest field strength, $B_{\min}$, measurable using different techniques (Table 1). The line-ratio technique mentioned in the table is based on the ratio between the V profiles of two lines that are almost identical except for their g values (Stenflo 1973). Solanki *et al.* (1992a) showed that by a judicious choice of lines at 1.5 μm, the smallest field strength measurable with the line-ratio technique can be lowered to a value otherwise only achievable with the 12 μm lines. Unfortunately, all the strong emission lines at 12 μm have $g = 1$, so that their ratios have no value as

TABLE I
Diagnostic properties of Zeeman-sensitive lines

Property	Visible	1.5 µm	12 µm
g_{max}	3	3	1
$g_{max} \cdot \lambda$	≈ 1.6	≈ 4.7	≈ 12.3
B_{min} from:			
a) complete splitting	1500–2000 G	400–600 G	150–200 G
b) profile fits	800–1000 G	250–300 G	≲ 100 G
c) line ratios	300–500 G	≈ 100 G	—
τ_{5000} of formation	10^{-2}	$\gtrsim 10^{-1}$	10^{-3}–10^{-4}
$\Delta B/B$ in umbra	3–5%	2–4%	—
$\Delta B/B$ in penumbra	5–25%	2–5%	0.5–2%
$\Delta B/B$ in kG flux tubes	10–15%	1–4%	5–10%
line formation	mainly LTE	LTE	NLTE

field-strength diagnostics. The three wavelength ranges also differ in the formation heights of their Zeeman-sensitive lines. Representative values are given in Table 1.

Rough estimates of the best achievable accuracy, $\Delta B/B$, in each wavelength range are given in Table 1 as well. For small-scale kG fields the *relative* accuracy at 1.5 µm is better than at 12 µm, since the field at the low height of formation of the 1.5 µm line is 4–8 times stronger than at the formation height of the 12 µm lines. Table 1 also lists whether the lines are formed in LTE or NLTE (more on the formation of the 12 µm lines can be found in the reviews by Avrett 1993 and Rutten and Carlsson 1993). Lines formed in LTE are generally simpler to interpret. Finally, let me mention two points not listed in Table 1.

1. The temperature sensitivity of the continuum intensity decreases rapidly with increasing wavelength, so that the problems posed by stray light in sunspots and by the unknown continuum intensity of magnetic elements are greatly reduced in the infrared.
2. The visible is rich in spectral lines with different temperature and velocity sensitivities. Thus these quantities can be reliably diagnosed in conjunction with the magnetic field. At 1.5 µm most of the atomic lines are temperature insensitive. Although this improves the accuracy of the measured magnetic vector, it is a substantial disadvantage for the study of the thermodynamics within magnetic features. At 12 µm the choice of lines is even smaller. Although the Mg I emission lines are known to be temperature sensitive, their temperature behavior has not yet been studied in sufficient detail.

3. Field Strengths from 1.5 µm Spectra

In a truly pioneering piece of work, Harvey and Hall (1975) made the first solar field-strength measurement using an infrared line (*cf.* Harvey 1977). They used Stokes V profiles (*i.e.*, spectra in net circular polarization) of the $g = 3$, Fe I 1.5648 μm line (in the following, referred to simply as "the $g = 3$ line"). The large Zeeman sensitivity of this line allowed the first measurement of B outside sunspots directly from the Zeeman splitting. Harvey and Hall obtained field strengths between 1.2 and 1.7 kG and thus confirmed the dominance of kG fields, suggested by earlier painstaking analysis of visible spectra (*e.g.*, Howard and Stenflo 1972, Stenflo 1973, *cf.* Beckers and Schröter 1968, Harvey *et al.* 1972).

For over a decade no further investigations using the near infrared for solar magnetic measurements were reported, although stellar investigators were not idle in the interval (see Saar 1993). Then, in 1986 Sun *et al.*, using only spectra in unpolarized light (Stokes I), confirmed the kG fields in sunspots and plages. They did find one plage region with 600 G, but it is unclear whether this low value is real or is due to uncertainties introduced into the technique due to the unavailability of Stokes Q, U and V. Sun *et al.* (1986) were mainly interested in testing the viability of 1.5 μm lines for stellar magnetic measurements, for which only Stokes I can be used (Saar 1993).

Stenflo *et al.* (1987b) analyzed the center-to-limb variation (CLV) of Stokes V of the $g = 3$ line. They found that the Zeeman splitting decreases measurably towards the limb, from which they concluded that B decreases with height, z, but did not determine $B(z)$ quantitatively. Note that lines in the visible are too Zeeman insensitive to allow magnetic gradients to be obtained from their CLV (Solanki *et al.* 1987).

Zayer *et al.* (1989) fitted observed Stokes V profiles of the $g = 3$ line and of Fe I 1.5822 μm (effective Landé factor $g_{\text{eff}} = 0.75$) using synthetic profiles formed in flux-tube models. By simultaneously fitting these two lines, which differ mainly in their Landé factors, they were able to separate the influence of the thermodynamics from that of the magnetic field and thereby could diagnose the range of field-strengths in the spatial resolution element. Furthermore, by combining these lines with Zeeman-sensitive lines in the visible, which are formed higher in the atmosphere, they could also distinguish between horizontal and vertical magnetic gradients and thus deduce the magnetic stratification in small magnetic elements. Their results confirmed the basic theoretical picture (*e.g.*, Knölker *et al.* 1988, Steiner and Pizzo 1989) that small-scale fields are concentrated in flux tubes, whose kG fields are confined by gas pressure. They also found evidence for a weak field (400–800 G), carrying 3–7% of the flux, but, unfortunately, their analysis was restricted to only two infrared spectra.

Muglach and Solanki (1992) carried out a statistical analysis involving all unblended Fe I lines in the wavelength range 1.5–1.8 μm and confirmed the results of Zayer *et al.* (1989). They also found that for the kG fields of magnetic elements, lines with $g_{\text{eff}} \gtrsim 1.5$ are completely split. To illustrate this, the wavelength separations between the red and blue Stokes V peaks, $\lambda_r - \lambda_b$, of all the analyzed lines are plotted in Figure 1 *vs.* the normalized Zeeman splitting v_H/B (*i.e.*, basically *vs.* g_{eff}). The diamond in the upper right-hand corner is the $g = 3$ line. Only a single infrared spectrum was analyzed.

The investigations by Livingston (1991), Rabin *et al.* (1991) and Rabin (1992a,b) heralded a new era of plage magnetic field measurements. Mainly due to improved detectors, profiles could now be easily and reliably measured at many positions

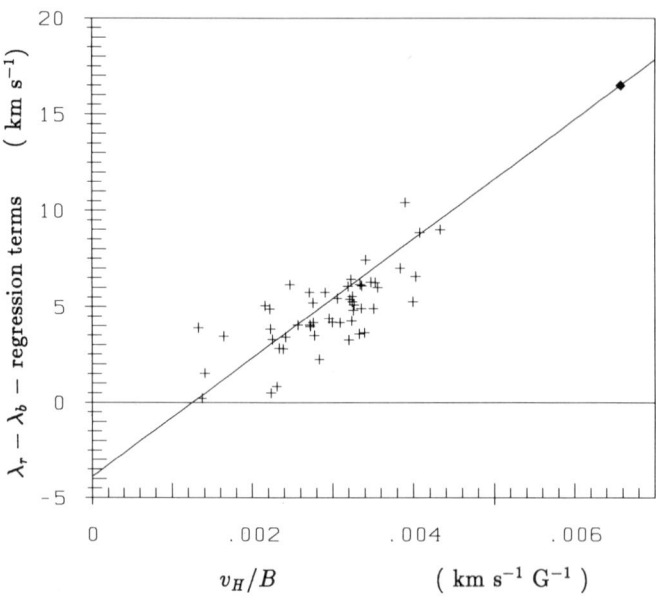

Fig. 1. Difference between the wavelengths of the red and blue Stokes V maxima, $\lambda_r - \lambda_b$ (in km s^{-1}), of all unblended FeI lines in the 1.5–1.8 μm band vs. v_H/B. The non-magnetic influences on $\lambda_r - \lambda_b$ have been removed with a multivariate regression. The slope of the diagonal line corresponds to the splitting induced by 1,550 G. The diamond in the upper-right corner represents the $g = 3$ line at 1.5648 μm.

on the Sun and the variation of B from one position to the next analyzed. Most profiles gave kG fields, but smaller field strengths were also observed. In addition, many (in the case of Livingston's data), or at least some (for Rabin's data), of the observed V profiles of the $g = 3$ line had highly anomalous shapes. The solid curve in Figure 3 (left frame) shows an example of an anomalous V profile from the data set of Livingston (1991), while a normal V profile is plotted for comparison in Figure 2. Before further progress could be made, the nature of these anomalously shaped profiles had to be explained.

To find such an explanation was one of the aims of Rüedi et al. (1992a). Like Zayer et al. (1989) they used numerical solutions of the Unno-Rachkovsky equations obtained in flux-tube models to fit 27 V spectra of λ 1.5648 μm ($g = 3$) and λ 1.5652 μm ($g_{\text{eff}} = 1.53$). They found that all V profiles (both the normally and the anomalously shaped ones) are well reproduced by standard flux-tube models. A single magnetic flux-tube component is generally sufficient to fit the normal profiles, while the anomalously shaped profiles require two separate flux-tube components. Examples of the fits to normal and anomalous profiles are shown in Figures 2 and 3, respectively. Note that in Figure 2 the adopted flux-tube model, with $B(z = 0) = 1,520$ G, not only reproduces the larger splitting of the $g = 3$ line, but also its larger σ-component width without requiring any ad hoc broadening. The σ-

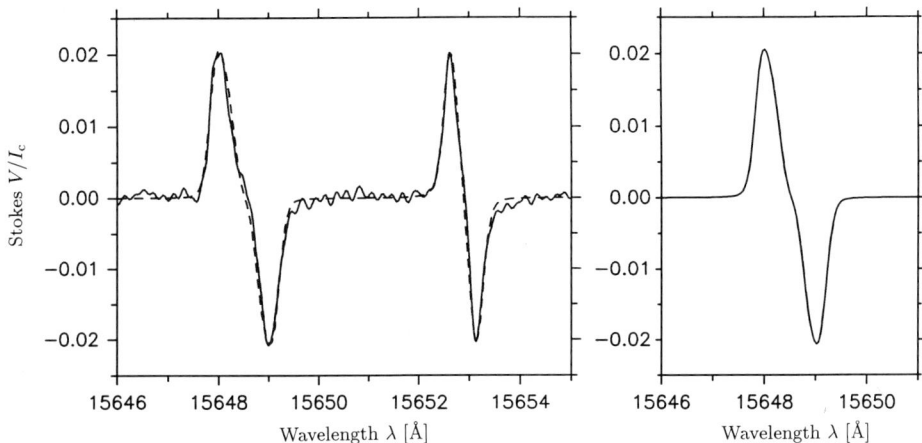

Fig. 2. a) Observed (solid) and synthetic (dashed) "normal" Stokes V profiles of FeI 1.5648 μm ($g = 3$) and FeI 1.5652 μm ($g_{\text{eff}} = 1.53$). The synthetic profile results from a model composed of a single thin magnetic flux tube with field strength $B(z = 0) = 1,520$ G. $z = 0$ refers to unit optical depth at $0.5\,\mu$m in the quiet Sun. b) Synthetic profile of the $g = 3$ line.

component is broadened by the vertical field-strength gradient naturally introduced into the model by horizontal pressure balance. The observed profiles in Figure 3 are best fit by two flux-tube components having $B(z = 0) = 1,700$ G and $-1,080$ G, respectively, and possessing no relative Doppler shift.

Rüedi et al. (1992a) also found that approximately 90% of the magnetic flux visible in Stokes V of the $g = 3$ line is in strong-field form, i.e., with $B(z = 0) \gtrsim 1,250$ G, or, equivalently, $\beta(z = 0) \leq 1$ (plasma $\beta = 8\pi P/B^2$, where P is the gas pressure). This result agrees well with the lower limit of 90% on strong-field flux set by Howard and Stenflo (1972) and Frazier and Stenflo (1972). A new feature is the definite detection of weak fields: 10% of the magnetic flux is found to have 400 G $\lesssim B(z = 0) \lesssim 1,250$s G, i.e., $\beta(z = 0) > 1$. Note that $B(z = 0) \approx 400$ G corresponds to the smallest B reliably measurable with the observed line pair. Thus available observations cannot rule out the existence of still weaker fields. Figure 4 shows $B(z = 0)$ vs. the spatially averaged field strength. The results of fits to simple profiles are represented by circles, complex profiles by plusses (each magnetic component is represented by a separate '+'). The solid line is a regression through the strong fields.

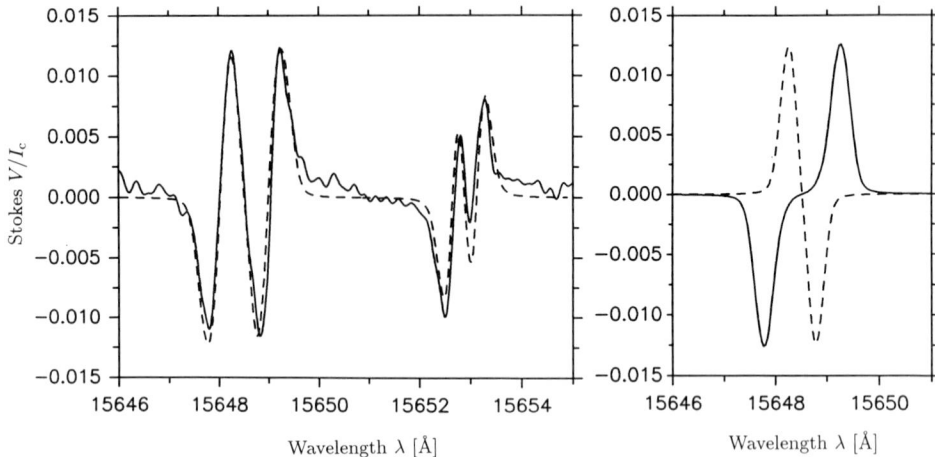

Fig. 3. a) Same as Figure 2, but for "anomalous" Stokes V profiles with strong inversions in their cores. The synthetic (dashed) curve results from two flux-tube components having opposite polarities [$B_1(z = 0) = 1,700$ G, $B_2(z = 0) = -1,050$ G] and no relative wavelength shift. b) Synthetic Stokes V profiles of each individual flux-tube component. Solid curve: V profile due to the B_1 component; dashed curve: V profile resulting from the B_2 component.

Figure 4 is in good agreement with the results of Rabin (1992 a, b): strong fields are found at all fluxes, but intrinsically weak fields are limited to small fluxes. Thus all measurements suggest a large empty region in the lower right part of the B vs. $\langle B \rangle$ diagram.

How do the near-infrared results compare with theory and with observations in the visible? The solid vertical bar to the right of Figure 4 indicates the range of $B(z = 0)$ found by Zayer et al. (1990), based on the line ratio between $\lambda 5250.2$ Å and $\lambda 5247.1$ Å. The agreement between the infrared and visible results is surprisingly good for the strong fields. However, the Zeeman sensitivity of the visible lines is insufficient to measure the strength of the intrinsically weak fields. The dashed vertical bar in Figure 4 represents the theoretical predictions of Spruit (1979). He calculated the convective collapse of an initially weak field into a final, convectively stable strong field. The good agreement of the theoretical predictions with the measured strong fields suggests that kG flux tubes are indeed formed by the convective collapse of weaker fields. Convective instability and collapse have been investigated or reviewed by, e.g., Parker (1978), Webb and Roberts (1978), Hasan (1984, 1985), Schüssler (1990) and Thomas (1990).

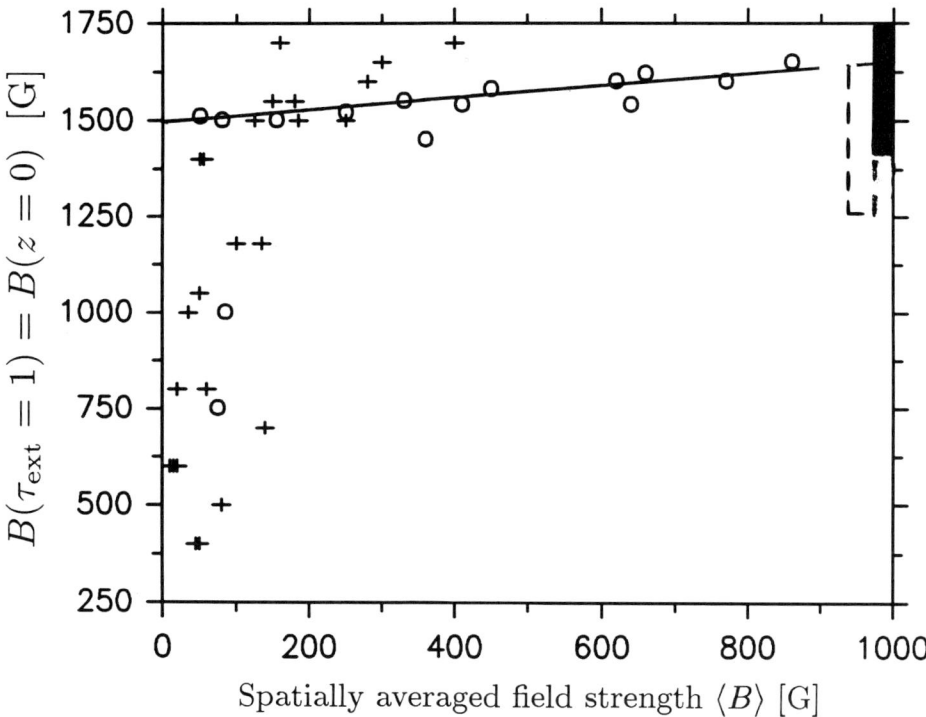

Fig. 4. $B(z = 0)$ vs. $\langle B \rangle$, the field strength averaged over the spatial resolution element, for all the regions analyzed by Rüedi et al. (1992a). If two magnetic components are needed to fit the observed profile, then $B(z = 0)$ for each component is plotted (denoted by a '+'). $B(z = 0)$ values obtained from single magnetic components are marked by a 'o'.

One of the major questions raised by the 1.5 μm observations concerns the nature of the weak fields. Three possible answers come to mind.

1. The weak fields may be attached to the strong fields in the form of return flux, i.e., field lines that bend over and return to the solar interior close to the parent flux tube. Return-flux models of flux tubes were proposed by Frazier and Stenflo (1972) and first constructed by Osherovich (1982). Observational support for some form of return flux associated with plage flux tubes has been claimed by Koutchmy (1991) and Koutchmy et al. (1991). However, this interpretation of the weak fields can be ruled out by the following arguments: a) The observed weak-field component has the same polarity as the closest strong field just as often as it has the opposite polarity. This is in contrast to the return-flux model, which predicts that the weak-field component must always have the opposite polarity to the strong field. b) A weak-field component is sometimes seen without any nearby strong field.

2. At least some of the weak fields seen in the infrared data may be connected to sunspots. Solanki et al. (1992b) have shown that, near sunspots, the *super*penumbral canopy (i.e., the almost horizontal magnetic field lines forming the sunspot super-

penumbra and overlying field-free gas, Giovanelli 1980) can produce a signal similar to the weak fields studied by Rüedi et al. (1992a).

3. A straightforward interpretation is that the weak fields form discrete magnetic features. Although the present observations cannot definitely confirm the existence of such "weak-field flux tubes", they do allow us to constrain their possible properties. From the field strength and the smallest measured flux an upper limit of 350–500 km can be set on the diameters of the smallest such features. Since for the weak-field features $\beta > 1$, their internal energetics are dominated by the gas (in contrast to the strong-field features, for which on average $\langle\beta\rangle \approx 0.32$, so that the magnetic energy density dominates over the internal energy density of the gas). In addition, for most weak fields $\beta > 1.8$, so that according to Spruit and Zweibel (1979) they are convectively unstable and ought to be undergoing a convective collapse, which should be visible as a redshift of the weak-field V profiles. However, no such redshifts are seen in at least nine out of twelve cases analyzed by Rüedi et al. (1992a). This implies that the weak fields must be relatively long lived and relatively stable against the convective instability. U-shaped loops (Spruit et al. 1987) are expected to satisfy these observational constraints and may well be responsible for the observed weak fields.

4. Velocity from 1.5 μm Spectra

4.1. The 1.5 μm Downflow Problem and its Resolution

Newer Stokes V observations in the visible show no downflows $\gtrsim 0.20$ km s^{-1} in small-scale magnetic features (e.g., Stenflo and Harvey 1985, Solanki 1986, Stenflo et al. 1987a, Wiehr 1987, Solanki and Pahlke 1988, Fleck 1991). The sizable downflows suggested by older observations (e.g., Giovanelli and Slaughter 1978, Wiehr 1985) turned out to be an artifact, produced by the blue-red asymmetry of the V profiles, of the low spectral resolution of these observations (Solanki and Stenflo 1986).

In the near infrared the situation has been clarified only very recently. The earliest measurements of the $g = 3$ line at 1.56 μm suggested an average downflow of 1.6 km s^{-1} in magnetic elements (Harvey 1977), although Harvey later repeatedly pointed out that deficiencies in the instrumentation may well have been responsible for the large observed wavelength shifts. A remeasurement of this line using improved instrumentation (the FTS polarimeter) resulted in a lower downflow velocity of 0.6 km s^{-1} (Stenflo et al. 1987b), which, however, still lies well outside the limit set in the visible. Recently Muglach and Solanki (1992) reanalyzed the data set of Stenflo et al. (1987b), but instead of considering only the $g = 3$ line, they investigated all the unblended Fe lines in the 1.5–1.8 μm spectral range of the data. They found no sign of a stationary flow in the observed magnetic features (Fig. 5). They also found that the uncertainty in the zero-crossing wavelength of the $g = 3$ line is of the same order as its measured shift (0.5 km s^{-1}). The large uncertainty is an indirect result of its large Zeeman splitting.

A combination of 1.5 μm and visible observations suggests that in general no stationary flows $\gtrsim 200$ m s^{-1} are present in all the photospheric layers of magnetic elements.

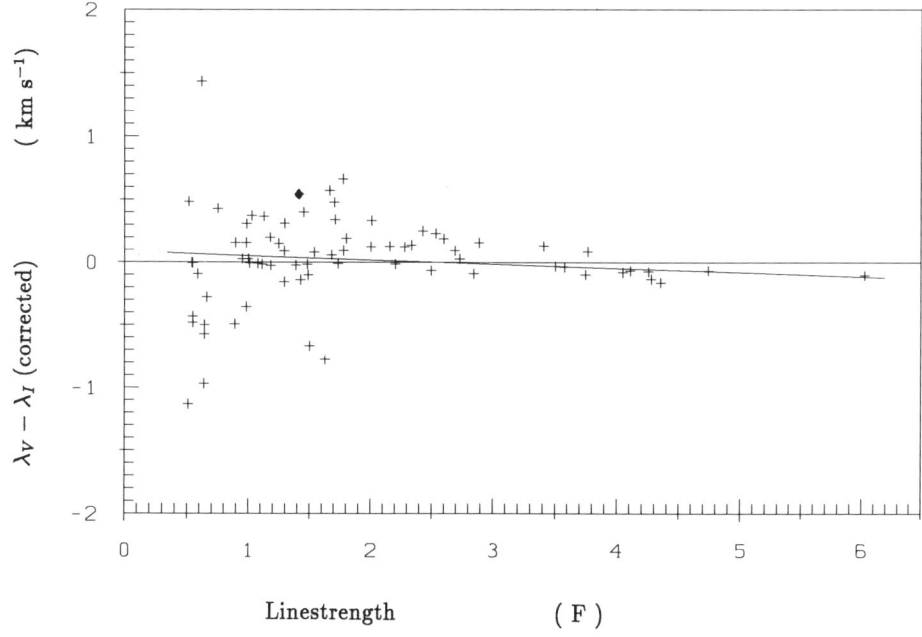

Fig. 5. Difference between the Stokes V zero-crossing wavelength and Stokes I core wavelength, $\lambda_V - \lambda_I$, vs. line strength of the I profile, S_I. The blueshift of λ_I due to the solar granulation has been compensated (Nadeau 1988). Therefore the plotted velocities represent vertical flows in the magnetic elements relative to the solar center of gravity. A linear regression through the data points is also plotted.

4.2. Siphon Flows

Stokes V profiles of $\lambda 1.5648$ μm and $\lambda 1.5652$ μm, observed with the Kitt Peak infrared array in the vicinity of a neutral line, as well as the field strengths and velocities derived from them are plotted in Figure 6 (Rüedi et al. 1992b). Near the neutral line, which intersects the spectrograph slit close to spectrum No. 8, the profiles have an anomalous, highly asymmetric shape. According to Figure 6 the two magnetic components contributing to these anomalous profiles have different field strengths and line-of-sight velocities. The negative polarity has a higher field strength $[B(z = 0) = 1,500 \text{ G}]$ and a *downflow* of $\approx 1 \text{ km s}^{-1}$, while the positive polarity has $B(z = 0) = 1,200$ G and an *upflow*. This correlation between field strength and velocity corresponds to exactly the theoretically predicted spectral signature of siphon flows. Although siphon flows along solar magnetic loops have been the subject of extensive theoretical study (*e.g.*, Meyer and Schmidt 1968, Cargill and Priest 1980, 1982, Thomas 1988, Montesinos and Thomas 1989, Thomas and Montesinos 1990, 1991, 1993, Degenhardt 1989, 1991), only 1.5 μm observations have so far provided convincing evidence for their existence in the solar atmosphere.

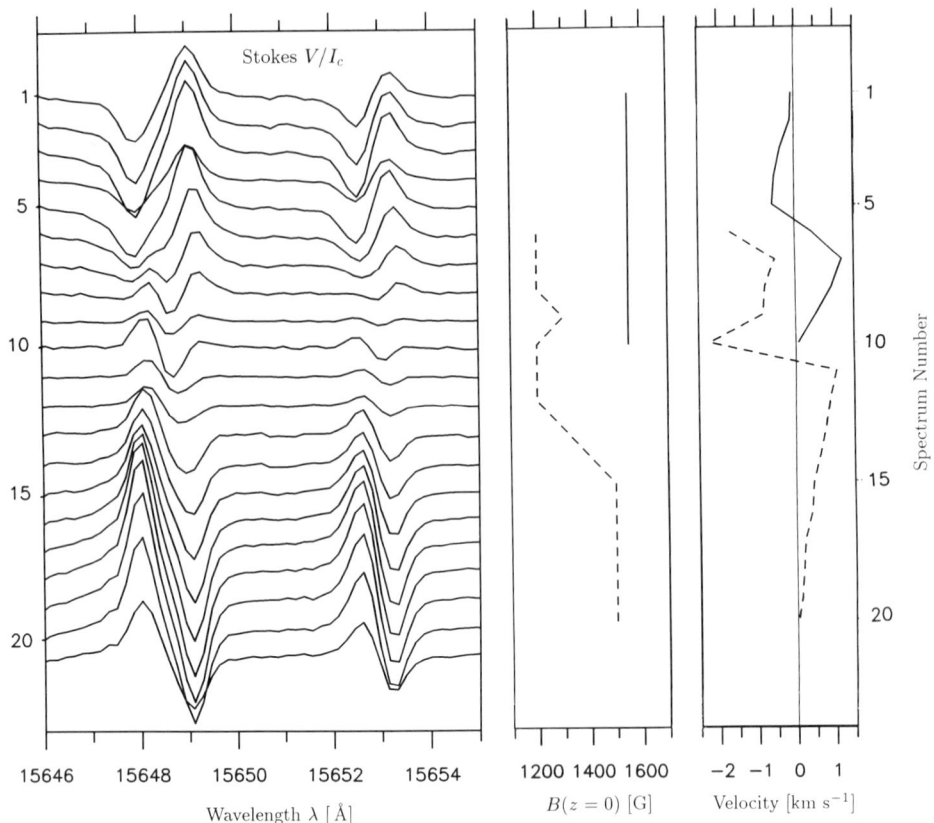

Fig. 6. a) Stokes V spectra at different positions along the slit (each spectrum is offset by 2.5″ with respect to its neighbors). The spectra are numbered at the left of the frame. b) Field strength B along the slit. Dashed curve: positive-polarity fields, solid curve: negative-polarity. c) Line-of-sight flow velocities. Positive velocities are directed downward in the atmosphere.

The infrared is particularly suited to the study of siphon flows, since their detection requires the simultaneous measurement of velocity and magnetic field strengths with great accuracy.

Note that peculiar V profiles, similar to the ones discussed above, have been seen in visible spectra of Ap stars (*e.g.*, Babcock 1951, Mathys 1988) and of neutral lines, generally in sunspot penumbrae (*e.g.*, Kjeldseth Moe 1967, Grigojev and Katz 1972, Golovko 1974, Skumanich and Lites 1991, Sánchez Almeida and Lites 1992). The presence of such peculiar Stokes V profiles has sometimes been referred to as the crossover effect. They have generally been explained by superposing mutually shifted profiles of opposite polarity, although Sánchez Almeida and Lites have been able to reproduce such profiles using vertical gradients of velocity and magnetic inclination.

5. Summary and Outlook

It is my hope that this review has provided a flavor of the magnetic-field related science possible with near-infrared lines. Even the limited effort spent so far on the interpretation and analysis of near-infrared observations of Zeeman-split lines has returned rich scientific dividends. For example, such investigations have

- enabled accurate, reliable and simple magnetic-field measurements (Livingston 1991, Rabin 1992a,b, Solanki et al. 1992a, etc.);
- shown how the field strength is spatially distributed within solar plages (Rabin 1992a,b);
- confirmed the prediction that magnetic fields are principally confined by gas pressure (Zayer et al. 1989);
- confirmed that approximately 90% of the magnetic flux outside of sunspots is in strong-field form ($\beta(z = 0) \leq 1$, Rüedi et al. 1992a);
- provided the first unequivocal detection of intrinsically weak fields in plages [$\beta(z = 0) > 1$, Livingston 1991, Rabin 1992a,b, Rüedi et al. 1992a];
- given estimates of the velocity structure in the deepest layers of magnetic elements: No stationary flows are seen, but a non-stationary RMS velocity of ≈ 1.5 km s^{-1} is detected (Muglach and Solanki 1992);
- set limits on the continuum brightness of magnetic elements (Muglach and Solanki 1992);
- provided the first positive detection of a siphon flow (Rüedi et al. 1992b);
- measured accurate values of the magnetic field in all parts of sunspots, including the outer edge of the penumbra (McPherson et al. 1992, Kopp and Rabin 1992, Solanki et al. 1992b);
- ruled out return-flux models of sunspots (Solanki et al. 1992b, Solanki and Schmidt, 1992);
- measured the Evershed flow in the deepest photospheric layers (McPherson et al. 1992);
- extended the correlation between temperature and field strength to the whole sunspot (Kopp and Rabin 1992) and provided an estimate of the Wilson depression throughout the sunspot (Solanki et al., in preparation);
- shown that sunspot penumbrae are deep (Solanki et al. 1993, Solanki and Schmidt 1992),
- suggested that magnetic elements (Zayer et al. 1989) and sunspots (Solanki and Schmidt, 1992) are bounded by current sheets.

Although the above list is already quite long, the near-infrared Zeeman-split lines are only beginning to fulfil their considerable promise. Some of the results that I hope will eventually emerge from this line of research are (in no particular order):

- the detection of a convective collapse;
- an accurate determination of the relationship between the field strength and the temperature in solar magnetic elements;

- identification of the nature of the weak-field magnetic component seen by e.g., Rüedi et al. (1992a);
- estimates of, or at least limits on, the field strength of intranetwork fields;
- improved limits on (or an outright detection of) a possible turbulent magnetic field (whose presence is suggested by Hanle effect measurements, Stenflo 1982, Faurobert-Scholl 1992);
- a better knowledge of the properties and the frequency of siphon flows;
- accurate (i.e., stray-light independent) field strengths of pores;
- an estimate of the strongest field strengths possible in sunspot photospheres;
- an improved understanding of the nature of the Evershed effect (e.g., is the flow visible in the superpenumbral canopy?);
- many surprises!

References

Avrett E.H., 1993, these proceedings.
Babcock H.W.: 1951, *Astrophys. J.* **114**, 1.
Beckers J.M., Schröter E.H.: 1968, *Solar Phys.* **4**, 142.
Cargill P.J., Priest E.R.: 1980, *Solar Phys.* **65**, 251.
Cargill P.J., Priest E.R.: 1982, *Geophys. Astrophys. Fluid Dyn.* **20**, 227.
Degenhardt D.: 1989, *Astron. Astrophys.* **222**, 297.
Degenhardt D.: 1991, *Astron. Astrophys.* **248**, 637.
Faurobert-Scholl M.: 1992, *Astron. Astrophys.* **258**, 521.
Fleck B.: 1991, *Rev. Modern Astron.* **4**, 90.
Frazier E.N., Stenflo J.O.: 1972, *Solar Phys.* **27**, 330.
Giovanelli R.G.: 1980, *Solar Phys.* **68**, 49.
Giovanelli R.G., Slaughter C.: 1978, *Solar Phys.* **57**, 255.
Golovko A.A.: 1974, *Solar Phys.* **37**, 113.
Grigorjev V.M., Katz J.M.: 1972, *Solar Phys.* **22**, 119.
Harvey J.W: 1977, in *Highlights of Astronomy*, E.A. Müller (ed.), Vol. 4, Part II, p. 223.
Harvey J.W., Hall D.: 1971, in *Solar Magnetic Fields*, R. Howard (ed.), Reidel, Dordrecht, p. 279.
Harvey J.W., Hall D.: 1975, *Bull. Amer. Astron. Soc.* **7**, 459.
Harvey J.W., Livingston W., Slaughter C.: 1972, in *Line Formation in the Presence of Magnetic Fields*, High Altitude Obs., NCAR, Boulder, CO, p. 227.
Hasan S.S.: 1984, *Astrophys. J.* **285**, 851.
Hasan S.S.: 1985, *Astron. Astrophys.* **143**, 39.
Howard R.W., Stenflo J.O.: 1972, *Solar Phys.* **22**, 402.
Kjeldseth Moe O.: 1967, in 'Structure and Development of Active Regions', K.O. Kiepenheuer (ed.), , *IAU Symp.* **35**, 202.
Knölker M., Schüssler M.: 1988, *Astron. Astrophys.* **202**, 275.
Kopp G., Rabin D.: 1992, *Solar Phys.* **141**, 253.
Koutchmy S.: 1991, in *Solar Polarimetry*, L. November (ed.), National Solar Obs., Sunspot, NM, p. 237.
Koutchmy S., Zirker J.B., Darvann T., Koutchmy O., Stauffer F., Mann R., Coulter R., Hegwer S.: 1991, in *Solar Polarimetry*, L. November (ed.), National Solar Observatory, Sunspot, NM, p. 263.
Livingston W.: 1991, in *Solar Polarimetry*, L. November (ed.), National Solar Obs., Sunspot, NM, p. 356.
Mathys, G.: 1988, *Astron. Astrophys.* **189**, 179.
McPherson M.R., Lin H., Kuhn J.R.: 1992, *Solar Phys.* **139**, 255.
Meyer F., Schmidt H.U.: 1968, *Zs. Angew. Math. Mech.* **48**, 218.
Montesinos B., Thomas J.H.: 1989, *Astrophys. J.* **337**, 977.
Muglach K., Solanki S.K.: 1992, *Astron. Astrophys.* **263**, 301.
Nadeau D.: 1988, *Astrophys. J.* **325**, 480.

Osherovich V.A.: 1982, *Solar Phys.* **77**, 63.
Parker E.N.: 1978, *Astrophys. J.* **221**, 368.
Rabin D.: 1992a, *Astrophys. J.* **390**, L103.
Rabin D.: 1992b, *Astrophys. J.* **391**, 832.
Rabin D.: 1993, these proceedings.
Rabin D., Jaksha D., Plymate C., Wagner J., Iwata K.: 1991, in *Solar Polarimetry*, L. November (ed.), National Solar Observatory, Sunspot, NM, p. 361.
Rüedi I., Solanki S.K., Livingston W., Stenflo J.O.: 1992a, *Astron. Astrophys.* **263**, 323.
Rüedi I., Solanki S.K., Rabin D.: 1992b, *Astron. Astrophys.* **261**, L21.
Rutten R. and Carlsson, M.: 1993, these proceedings.
Saar S.H.: 1993, these proceedings.
Sánchez Almeida J., Lites B.W.: 1992, *Astrophys. J.* **398**, 359.
Schüssler M.: 1990, in 'Solar Photosphere: Structure, Convection and Magnetic Fields', J.O. Stenflo (ed.), , *IAU Symp.* **138**, 161.
Skumanich, A., Lites, B.W.: 1991, in *Solar Polarimetry*, L. November (ed.), National Solar Observatory, Sunspot, NM, p. 307.
Solanki S.K.: 1986, *Astron. Astrophys.* **168**, 311.
Solanki S.K., Pahlke K.D.: 1988, *Astron. Astrophys.* **201**, 143.
Solanki S.K., Schmidt H.U.: 1992, *Astron. Astrophys.* submitted.
Solanki S.K., Stenflo J.O.: 1986, *Astron. Astrophys.* **170**, 120.
Solanki S.K., Keller C., Stenflo J.O.: 1987, *Astron. Astrophys.* **188**, 183.
Solanki S.K., Rüedi I., Livingston W.: 1992a, *Astron. Astrophys.* **263**, 312.
Solanki S.K., Rüedi I., Livingston W.: 1992b, *Astron. Astrophys.* **263**, 339.
Solanki S.K., Rüedi I., Livingston W., Schmidt H.U.: 1993, these proceedings.
Spruit H.C.: 1979, *Solar Phys.* **61**, 363.
Spruit H.C., Zweibel E.G.: 1979, *Solar Phys.* **62**, 15.
Spruit H.C., Van Ballegooijen A.A., Title A.M.: 1987, *Solar Phys.* **110**, 115.
Steiner O.: 1993, these proceedings.
Steiner O., Pizzo V.J.: 1989, *Astron. Astrophys.* **211**, 447.
Stenflo J.O.: 1973, *Solar Phys.* **32**, 41.
Stenflo, J.O.: 1982, *Solar Phys.* **80**, 209.
Stenflo J.O., Harvey J.W.: 1985, *Solar Phys.* **95**, 99.
Stenflo J.O., Solanki S.K., Harvey J.W.: 1987a, *Astron. Astrophys.* **171**, 305.
Stenflo J.O., Solanki S.K., Harvey J.W.: 1987b, *Astron. Astrophys.* **173**, 167.
Sun W.-H., Giampapa M.S., Worden S.P.: 1987, *Astrophys. J.* **312**, 930.
Thomas J.H.: 1988, *Astrophys. J.* **333**, 407.
Thomas J.H.: 1990, in *Physics of Magnetic Flux Ropes*, C.T. Russell, E.R. Priest, L.C. Lee (eds.), Geophysical Monograph 58, American Geophysical Union, Washington, DC, p. 133.
Thomas J.H., Montesinos B.: 1990, *Astrophys. J.* **359**, 550.
Thomas J.H., Montesinos B.: 1991, *Astrophys. J.* **375**, 404.
Thomas J.H., Montesinos B.: 1993, *Astrophys. J.* in press.
Webb A.R., Roberts B.: 1978, *Solar Phys.* **59**, 249.
Wiehr E.: 1985, *Astron. Astrophys.* **149**, 217.
Wiehr E.: 1987, in *Cool Stars, Stellar Systems, and the Sun*, V., J.L. Linsky, R.E. Stencel (eds.), Lecture Notes in Physics Vol. 291, Springer-Verlag, Berlin, p. 54.
Zayer I., Solanki S.K., Stenflo J.O.: 1989, *Astron. Astrophys.* **211**, 463.
Zayer I., Solanki S.K., Stenflo J.O., Keller C.U.: 1990, *Astron. Astrophys.* **239**, 356.

THEORETICAL MODELS OF MAGNETIC FLUX TUBES: STRUCTURE AND DYNAMICS

O. STEINER

Kiepenheuer-Institut für Sonnenphysik, Schöneckstrasse 6, D-7800 Freiburg, FRG

Abstract. Two types of model calculations for small scale magnetic flux tubes in the solar atmosphere are reviewed. In the first kind, one follows the temporal evolution governed by the complete set of the MHD and radiative transfer equations to a (quasi) stationary solution. From such a solution the continuum contrasts of a photospheric flux tube in the visible and in the infrared continuum at 1.6 μm have been computed and are briefly discussed. The second, more empirical type of method assumes the flux tubes to be in magnetohydrostatic equilibrium. It is computationally faster and more flexible and allows us to explore a wide range of parameters. Models and insights obtained from such parameter studies are discussed in some detail. These include an explanation for the peculiar variation of the area asymmetry of Stokes V profiles across the solar disk in terms of mass motions in the surroundings of magnetic flux tubes.

Furthermore, a two-dimensional model of the lower chromosphere that has been developed is presented. Emphasis is laid on the effect of thermal bifurcation of the lower chromosphere on the structure of the chromospheric magnetic field. If the cool carbon monoxide clouds, observed in the infrared, occupy the non-magnetic regions, the flux tubes expand very strongly and form a magnetic canopy with an almost horizontal base. This has consequences for the spatial distribution of the Ca II K spectral line emission.

Finally, some consideration is given to the formation and destruction of intense magnetic flux tubes in the solar photosphere. The formation is described as a consequence of the flux expulsion process that leads to a convective instability. A possible observational signature of this mechanism is proposed.

Key words: infrared: stars – MHD – Sun: atmosphere – Sun: magnetic fields

1. Introduction

This paper concentrates on the structure of *stationary, quasi-stationary* and *static* flux tubes in the solar photosphere. A description of the dynamical formation and destruction process is given in Section 6. Emphasis is laid on observable properties of model tubes and, despite the title, formal exposition is avoided.

So far no model calculation includes the complete sequence of formation, life, and destruction of intense magnetic flux tubes. All model calculations of quasi-stationary or static magnetic flux tubes assume the existence of them at the very beginning of the calculation. This is done, for example, by the prescription of a density reduction factor which is the ratio between the density within the tube and outside it at a specified height in the atmosphere.

There exist now basically two approaches to numerical models of intense photospheric flux tubes. A first kind of calculation (Deinzer et al., 1984a,b) start from a given magnetic field/plasma configuration and solve the complete set of MHD equations at subsequent time steps together with a realistic equation of state which takes partial ionization into account. By solving the equations at subsequent time steps, the solution is advanced in time and evolves away from the initial configuration to a quasi-stationary solution. This kind of calculation can be called a detailed simulation in the sense of Oran and Boris (1987). A second kind of model calculations (*e.g.*, Pizzo, 1986; Steiner *et al.*, 1986) presumes the flux tube to be in magnetohydrostatic equilibrium thereby making the time stepping unnecessary. Chapter 3 and after will be devoted to this type of models.

The terms (magnetic) flux tube and flux sheet refer to models of *magnetic elements*. Magnetic elements are the tiny magnetic field concentrations which are inferred from polarimetric observations of the network and active region plages (Stenflo, 1989). They are called magnetic elements because they all show very similar physical properties, and, in a wider sense, because we think that they are the basic elements for the understanding of more global phenomena such as the heating of the chromosphere and corona, the luminosity variation with solar cycle, the thermal bifurcation in the lower chromosphere, the modification of the p-mode eigenfrequencies by magnetic fields, and activity in stellar atmospheres.

2. A Detailed Simulation

Figure 1 is the result of a detailed simulation obtained with the code originally described in Deinzer *et al.* (1984a). It shows contour plots of the density and temperature (a, c), magnetic lines of force (b), and the velocity field (d) of a quasi-stationary solution, resulting after having advanced the calculation through many time steps, starting from an initial flux tube configuration. The input parameters are (1) the geometrical dimension of the computational box, of which Figure 1 shows only a part, (2) the initial density reduction factor, (3) the total magnetic flux across the boundaries, and (4) the temperature at the bottom and the temperature gradient at the top boundary. The left boundary is an axis of symmetry.

The boundary conditions are chosen such that the magnetic field lines cross the upper and lower boundaries at right angles and that no magnetic flux enters nor leaves the box sideways. No material enters or leaves the box. The energy equation takes radiative transfer perpendicular to the 2-D plane of Figure 1 into account, assuming a grey medium, and incorporates a mixing length formalism for small-scale turbulent energy transport in the layers susceptible to convective instability. Since 2-D Cartesian coordinates are used, the term flux sheet is appropriate for the magnetic structure shown in Figure 1.

Since the strong magnetic pressure within the flux sheet must be balanced by the gas pressure outside it, the density within the tube is substantially lower than in the quiet atmosphere at the same geometrical height (Fig. 1a). This, and the reduced temperature at equal geometrical heights, have the effect that the atmosphere within the sheet is optically more transparent than the surroundings. Hence, the intense magnetic flux sheet appears as a tiny crevice in the solar atmosphere, offering a glimpse into deeper layers.

2.1. Energy Budget and Continuum Contrast

The energy budget is determined by two competing effects: suppression of convective energy transport by the strong magnetic field and lateral radiative energy influx into the tube because of the reduced opacity. The lateral radiative energy influx enforces on the external atmosphere a convective pattern as shown in Figure 1d. A narrow strong downflow with velocity up to 6 km s^{-1} forms in the close vicinity

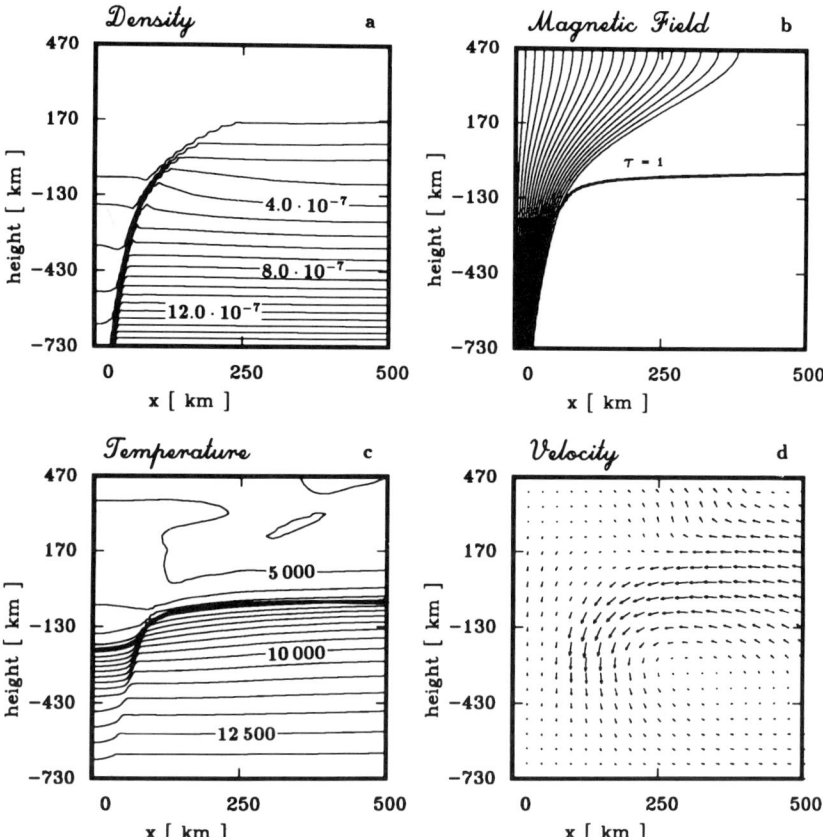

Fig. 1. Density (a) and temperature (c) contours, labeled in units of g cm^{-3} and K, magnetic field lines (b), and velocity field (d). The maximum velocity is about 6 km s^{-1}. Courtesy M. Knölker and M. Schüssler.

of the magnetic field concentration. This mass flow is fed by a broad upflow and a horizontal flow, which, at the same time, advects the energy radiated horizontally into the flux sheet. This particular interaction of convective and radiative energy flux results in the temperature structure shown in Figure 1c. Due to the inhibition of convection by the strong magnetic field, the magnetic flux sheet is distinctly cooler than the surrounding atmosphere at the same *geometrical* height.

The thick curve in Figure 1c indicates the location at which the continuum optical depth reaches unity for an observer peering vertically down onto the structure. A close inspection shows that, if moving along this curve from the middle plane of the flux sheet outwards, the temperature is first higher, and in the close surroundings of the flux sheet boundary lower, than the temperature far away from the flux sheet. From this results the continuum appearance of the magnetic element shown in Figure 2a (solid curve). The "bottom" of the sheet appears bright, the peak

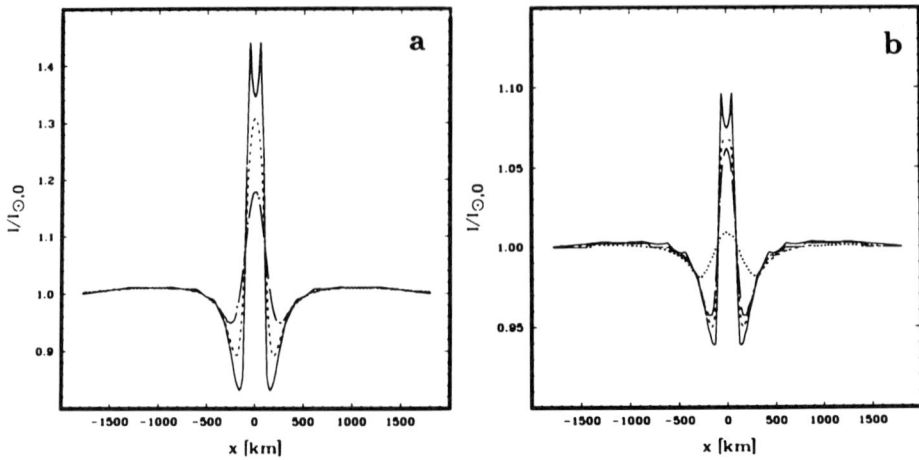

Fig. 2. a) Disk center continuum contrast of the flux sheet of Figure 1 at $\lambda = 5000$ Å: theoretical profile (solid), convoluted with a point spread MTF for a 0.7 m telescope (dashed), additional convolution with a seeing MTF (dot-dashed). b) Continuum contrast at 1.6 μm. The curves represent the theoretical profile (——), convolution with a point spread MTF for a 4 m (Super McMath) (— — ——), 2.4 m (LEST) (— · — · —), and 0.7 m (VTT) (·····) aperture. Note the different scales in Panels a and b. Courtesy M. Knölker.

values stemming from the even hotter "corners" of the depression. The magnetic element is surrounded by a dark well. The dashed curve of Figure 2a represents the disk center continuum contrast when a point spread modulation transfer function (MTF) for a 0.7 m telescope is applied to the theoretical contrast curve. The bright peaks then completely disappear. The dot-dashed curve results with the additional application of a seeing MTF appropriate for the German Vacuum Tower Telescope (VTT) on Tenerife.

While the positions of the isotherms in Figure 1c are determined by the energy budget, the $\tau = 1$ level shifts downwards when changing from the visible to the infrared continuum at 1.6 μm. The shift is different within flux sheet from the shift outside, mainly because of the difference in temperature gradient and temperature, causing a contrast profile which is different from Figure 2a. The theoretical disk center contrast profile at 1.6 μm is the solid curve in Figure 2b. If integrating over a width of 5″ a negative contrast of about −0.1% is left over compared to +0.9% in the visible. A higher spatial resolution would cut away parts of the dark well, thus leading to a positive contrast in the infrared, too (compare observations by Foukal et al.1990; Lin and Kuhn 1992; and Foukal and Moran 1993, in these proceedings). Figure 2b also shows the curves resulting after the application of a point spread MTF on the theoretical contrast curve for a 4 m (Super McMath), a 2.4 m (LEST), and a 0.7 m (VTT) aperture.

Further detailed simulations with different initial values including center-to-limb continuum contrast curves, have been published by Knölker et al. (1988a,b; 1991) and Grossmann-Doerth et al. (1989) (see also the review article by Schüssler, 1990).

3. Hydrostatic Model Flux Tubes

3.1. Basic Equations and Model Parameters

A second kind of numerical model calculation for solar magnetic flux tubes assumes hydrostatic equilibrium from the beginning. Such model calculations have been carried out for different purposes. Pizzo (1986, 1989) and Jahn (1989) have computed models of sunspots; Cally (1990) has developed a code for studies of the transition zone; and Steiner et al. (1986) have used this method for small scale photospheric flux tubes. The equations being solved are the force balance equation,

$$0 = -\nabla p + \rho \mathbf{g} + \mathbf{j} \times \mathbf{B} ,$$

Ampères law,

$$\nabla \times \mathbf{B} = 4\pi \mathbf{j} ,$$

and

$$\nabla \cdot \mathbf{B} = 0 .$$

Because the model is static, the continuity and the induction equations are absent. These equations are solved for a rotationally symmetric, untwisted, vertical flux tube.

As an example, Figure 6 shows a typical model magnetic flux tube. Within the cylindrical computational domain the nearly vertical, curving magnetic field lines delineate the flux tube, which is separated from the surrounding non-magnetic photosphere by a thin layer at which a sheet current travels in azimuthal direction. The thickness of this layer has been estimated by Schüssler (1986) and, recently, by Hirayama (1992) to 2 to 10 km. Due to the stratification of the atmosphere, the confining gas pressure, and hence the magnetic field strength, decrease with height, so that the tubes must expand to guarantee flux conservation. Ultimately, the field spreads out to fill the computational region, simulating the merging with neighboring flux tubes which prevent the tube from further expansion. Far above their merging height the flux tubes form a more or less uniform magnetic field. Inserted into the figure at several heights are curves which show the vertical and the radial magnetic field components as functions of radius.

The following parameters characterize the model: the filling factor, which is the square of the ratio of the tube radius to the width of the computational domain, the magnetic field strength, and the radius or total magnetic flux of the tube, all values specified at the height $z = 0$, where $\tau_{5000} = 1$ in the external atmosphere.

Usually the energy equation is replaced by prescribing the atmospheric structure within and outside the tube, making use of published 1-D atmospheres. This lack of self-consistency makes it possible to rapidly explore a wide range of parameters and atmospheric combinations, possibly not accessible to a detailed simulation. In addition, the computational efficiency of the present method makes it possible to derive the atmospheric structure of magnetic elements from observations of the Stokes V profiles of spectral lines.

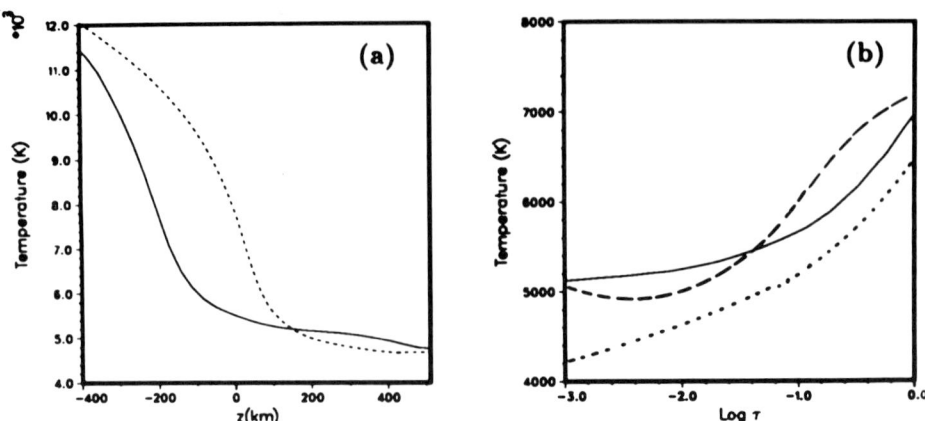

Fig. 3. Temperature as a function of geometrical height (a) and optical depth (b). The dashed curve is the result of an inversion of network Stokes V profiles, while the solid curves are the temperature along the flux sheet axis for the simulation in Figure 1. The dotted curves show the quiet-Sun model of Vernazza, Avrett and Loeser (1981). From Knölker et al. (1991).

3.2. FLUX TUBE STRUCTURE FROM INVERSION OF STOKES SPECTRA

Keller et al. (1990) made a least-squares fit to a number of characteristics of eight Fe I and two Fe II Stokes V profiles obtained from Fourier transform spectrometer observations of network and plage magnetic elements by Stenflo et al. (1984). By adjusting the flux-tube model parameters, they obtained the temperature as a function of $\log \tau_{5000}$, shown as the dashed curve in Figure 3b. The solid curve corresponds to the temperature along the axis of symmetry of the detailed model computation of Figure 1. They both are hotter than the quiet Sun model (dotted curve) at the same *optical depth*. In contrast to the optical picture, the flux tube is distinctly cooler than the quiet Sun model at the same *geometrical* height (Fig. 3a), with the exception of the radiatively heated upper layers. A theoretical 2-D thermal structure for the photospheric layers of magnetic flux tubes using non-grey radiative transfer has been calculated by Steiner and Stenflo (1990).

4. Stokes V Area Asymmetry

Near disk center one observes in active region plages and the quiet solar network that the area and amplitude of the blue wing of Stokes V profiles exceed those of the red wing by several percent (Solanki and Stenflo, 1984; Wiehr, 1985). At the same time, the shift of the zero-crossing wavelength is always close to zero (Solanki and Stenflo, 1986). Towards the limb the area asymmetry shows a sign reversal, e.g., at $\mu = 0.4$ for the Fe I 5250.22 line (Stenflo et al., 1987).

In Figure 4, a static flux tube is surrounded by a non-magnetic plasma in stationary motion. The velocity field is obtained by matching two potential flows for the upflow and the downflow region. It shows a downflow close to the flux tube

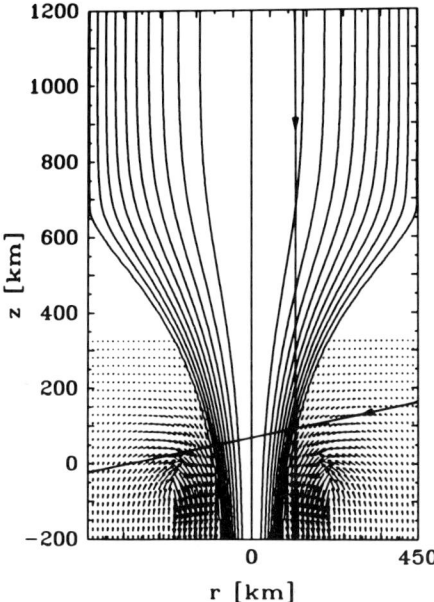

Fig. 4. Static flux tube containing the plage model atmosphere of Solanki (1986) surrounded by a non-magnetic plasma in stationary motion. The velocity field is obtained by matching two potential flows for the upflow and the downflow region. The plasma within the tube is at rest. The vertical line of sight, passes from the magnetic into the non-magnetic, down-drafting atmosphere, along which a Stokes V profile with strong area asymmetry and without shift is produced. The contributions to the area asymmetry at the two magnetic/non-magnetic interfaces of the inclined line of sight are of opposite sign.

surface (similar to Fig. 1d) and a granular upflow further away, crudely mimicking a simulation of granular flow (e.g., Steffen, 1991). The plasma within the tube is at rest. The vertical line of sight, indicated in the Figure, passes from the magnetic into the non-magnetic, down drafting, atmosphere. This situation produces a strong Stokes V area asymmetry without shifting the zero-crossing of the profile. It can be explained by an analysis of the radiative transfer equations for polarized light (Grossmann-Doerth et al., 1988, 1989). Averaging over a large number of lines of sight, weighted with the area they represent, results in Stokes V profiles may be compared with spatially unresolved spectra.

The sign of the area asymmetry depends on the sign of the expression

$$d|B|/d\tau \cdot dv(\tau)/d\tau,$$

where τ is the continuum optical depth and v the mass velocity, both along a line of sight. This expression is clearly positive for the vertical line of sight shown in Figure 4. The inclined line of sight (vertical tube close to the limb), passes two magnetic/non-magnetic interfaces for which the above expression, and hence the contributions to the area asymmetry, are of opposing sign. It therefore is not obvious

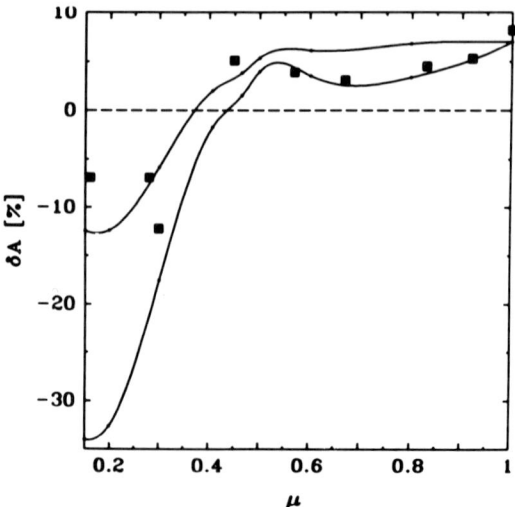

Fig. 5. Observed (squares) and calculated (curves) center-to-limb variation of the relative area asymmetry of Stokes V profiles of the Fe I 5250.22 Å line. The computed values have been obtained from the model of Figure 4 (upper curve). The lower curve results when the radial velocities are arbitrarily increased by a factor of 2.0.

what the center to limb behavior of the area asymmetry will be.

Since flux tubes cluster in network points or active region plages, Bünte et al. (1991) take a periodic arrangement of model flux tubes for the diagnostic procedure. Hence, close to the limb the lines of sight may pass through several flux tubes. They consider flux tubes of a field strength of 1,600 G, a diameter of 200 km, and a filling factor of 5%, all values referring to $\tau = 1$ for the external atmosphere. The diagnostic procedure is carried out for a wide range of velocity fields and temperature stratifications. These include purely vertical or purely radial motion, radial inflow and downflow, and upward and downward flow as in the example shown in Figure 4, as well as a temperature-velocity correlation to approximate a warm upflow and a cool downflow.

As the number of properties exhibited by detailed model calculations increases, such as shown in Figure 1, the center-to-limb variation (CLV) of the area asymmetry approaches the observed behavior. Observed values for the iron line 5250.22 Å as a function of $\mu = \cos\theta$ are shown in Figure 5 (filled squares), together with the results of the model data. Best agreement is obtained with the model of Figure 4, with $v_{z\,\min} = -4.4$ km s^{-1}, $v_{r\,\min} = -2.2$ km s^{-1}, and a maximum temperature excess for the upflow of 1,000 K (upper solid curve). The lower solid curve is obtained if the radial inflow is arbitrarily increased by a factor of 2.0, which demonstrates the high sensitivity of the Stokes V area asymmetry to the precise velocity field surrounding a flux tube. However, similar calculations with the Fe 5083.3 line are less satisfactory, although the sign reversal is reproduced (Bünte et al., 1992).

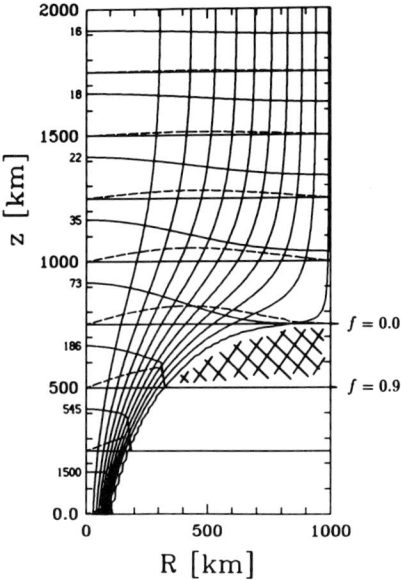

Fig. 6. Model magnetic flux tube containing the FLUXT atmosphere of Ayres et al. (1986) embedded in the radiative-equilibrium atmosphere of Anderson (1989). Superimposed on the field lines are horizontal curves showing the radial variation of the vertical, B_z (solid lines), and radial, B_r (dashed lines), magnetic field components at eight different heights marked by horizontal straight lines. Both, $B_z(r)$ and $B_r(r)$ are normalized to B_z at the axis, the value in Gauss of which is indicated to the left of each curve. The radius is 100 km, the field strength 1,500 G, and the filling factor 0.01; all values are specified at $z = 0$, which corresponds to $\tau_{5000} = 1$ in the external atmosphere. The hatched region indicates the location of a cool carbon monoxide cloud with a strongly height-dependent filling factor, indicated in the right margin.

5. 2-D Models for the Lower Chromosphere

Model calculations over a wide range of parameters (Steiner and Pizzo, 1989) show that, in general, the spreading of flux tubes with height is too small to form a sizable radial field component in photospheric or chromospheric layers. These are inconsistent with extensive observational investigations of Giovanelli and Jones (Giovanelli, 1980; Giovanelli and Jones, 1982; Jones and Giovanelli, 1983; Jones, 1985), which invariably show magnetic canopies extending as low as in the upper photosphere. We have found that canopy fields originating from intense photospheric flux tubes can only be obtained if the flux tubes are embedded in a cool atmosphere, while the chromospheric temperature rise is present only within the flux tubes (Solanki and Steiner, 1990). The crucial point is that the gas pressure in the cool, non-magnetic atmosphere decreases rapidly to a value close to the pressure within the tube, at which height the field is forced to spread nearly horizontally, thereby forming a magnetic canopy.

An example of such a model is shown in Figure 6. The flux tube contains the

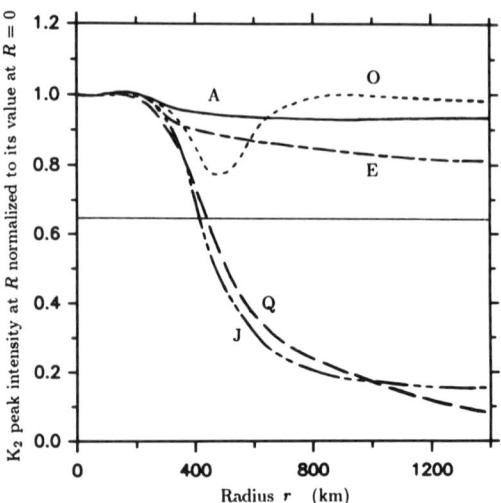

Fig. 7. Intensity of the Ca II K_2 peak with radius r, normalized to the intensity at the flux tube axis ($r = 0$) for different atmospheric model combinations. The horizontal line marks the value to which the K_2 intensity drops at large r according to observations of Grossmann-Doerth et al. (1974). From Solanki et al. (1991).

plage flux tube atmosphere FLUXT of Ayres et al. (1986) and is embedded in the radiative equilibrium model of Anderson (1989). The Anderson atmosphere shows a dramatic temperature drop above the classical temperature minimum, due to the formation of carbon monoxide. The tube shows a sizable radial field component and merges with neighboring flux tubes at a height of only 800 km above $\tau = 1$ of the external atmosphere. Other atmospheric combinations, including the empirical cool atmosphere of Ayres et al. (1986) result in merging heights between 800 to 1,200 km.

In this picture the much debated carbon monoxide cloud (Ayres 1991 and Ayres 1993, in these proceedings) occupies the hatched region of Figure 6. A shortcoming of the present model is the temperature discontinuity at the flux tube boundary which sharply delineates the CO cloud. Full 2-D radiative transfer is needed to obtain a smooth transition. Note, however, the strongly height-dependent filling factor of the CO cloud, which must have a distinct effect on the center-to-limb behavior of the CO lines. This model suggests that a high-resolution filtergram in the 4.7 μm CO lines would depict the CO clouds in quiet supergranular centers. However, according to Figure 6, they are rather shallow, so that disk center observations are probably not suitable. Furthermore, the lines would not be affected by any Zeeman splitting or broadening because they would be formed in field free regions.

The present model has other advantages besides the ability to incorporate the CO cloud and to form a relatively low lying magnetic canopy. Solanki et al. (1991) showed that it also may explain the spatial scale and amplitude of the horizontal Ca II K_2 intensity variation in the quiet Sun. These authors computed synthetic Ca II K spectral lines along many vertical lines of sight across models with different flux tube atmospheres surrounded by Ayres's (1986) empirical cool atmosphere. It turned out that the source function for this line is very sensitive to the height extent of the cool atmosphere, which in turn is limited by the base field strength of the magnetic canopy. Strong Ca II K core emission is always present for rays close to the axis of symmetry of the tube, while rays passing through the canopy may fail to show any core emission at all. This situation is reflected in Figure 7, which shows the Ca II K_2 peak intensity across the radius of flux tubes with a moderate chromospheric temperature rise (curves J and Q) and a strong temperature rise (A and E). Models J and Q show a Ca II K grain size of about 1", in agreement with the observational values of Cram and Damé (1983). If the temperature profiles of the upper layers of the flux-tube atmosphere lie between those of the plage models VALP of Ayres *et al.* (1986) (model E) and VALF of Vernazza *et al.* (1981) (model J), the curve can be expected to drop to the indicated observed value and, at the same time, the average profile over the radius would then best reproduce the Ca II K spectral line of the average quiet Sun.

6. Formation and Destruction of Intense Magnetic Flux Tubes and a Possible Infrared Signature

6.1. Flux Expulsion and Convective Collapse

Two mechanisms have been proposed for the concentration of magnetic flux into structures of high field strength: flux expulsion and convective collapse. Flux expulsion is the process of field advection by the granular and supergranular velocity field, which expels the magnetic flux from the cell interiors into the narrow intergranular lanes. The order of magnitude of the field strength resulting from this process is the equipartition field strength

$$\frac{B_{eq}^2}{8\pi} = \frac{\rho}{2} v^2 ,$$

where ρ is the density and v the typical granular velocity, both at the solar surface. It results in an equipartition field strength of $B_{eq} \approx 400$ G close to the solar surface.

Flux expulsion by convective flows have been investigated numerically in the kinematic as well as in the dynamic case (*e.g.*, Proctor and Weiss, 1982; Hurlburt and Toomre, 1988). Figure 8 shows four snapshots of a flux expulsion simulation in a compressible, stratified medium, similar to an example of Hurlburt and Toomre, 1988. The computation has been started with a homogeneous vertical field. The temperature at the top and the temperature gradient at the bottom are prescribed. The ratio of the initial density at the bottom to that at the top is 5. The Rayleigh number is far above critical, so that magnetoconvection sets in. The flux expulsion process leads to particularly strong field concentration in the downflow region of the cells.

However, field concentration by the expulsion process is insufficient to explain the kG fields of magnetic elements. As a consequence of the flux expulsion process the growing magnetic field concentration causes the retardation of the horizontal flow towards the downflow region. This leads to a cooling of the magnetic region,

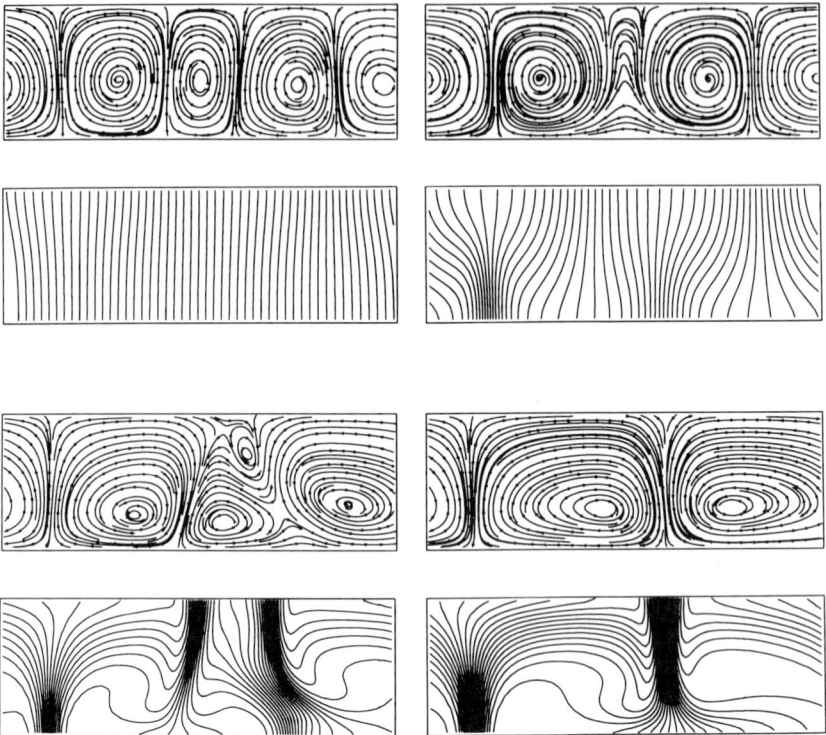

Fig. 8. Four snapshots of the numerical simulation of flux expulsion in a compressible, stratified medium, similar to Hurlburt and Toomre (1988). The top panel shows streak lines of the mass motion, the bottom panel magnetic lines of force. Starting with a homogeneous, vertical magnetic field the flow evolves to a stationary pattern with the field concentrated to equipartition strength in the downflow region.

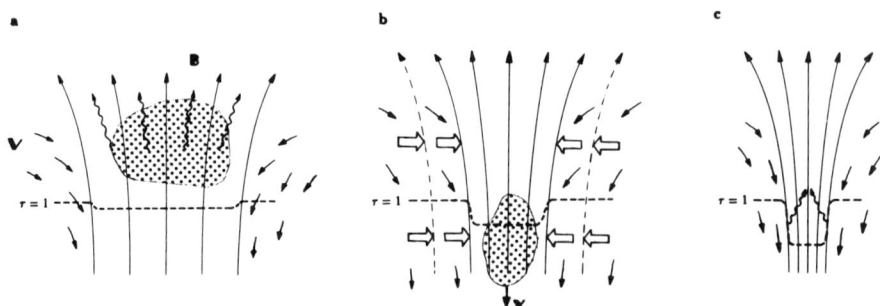

Fig. 9. Schematic sketch of the formation of an intense magnetic flux tube by means of the convective collapse. The dotted cloud represents an arbitrary material package. Short thin arrows indicate plasma motion, whereas the long, vertically oriented arrows represent field lines. The wavy arrows signify radiative energy losses and the bold arrows in the middle sketch stand for the fast magnetic field concentration by the convective collapse.

because the radiative losses can no longer be replenished by the throttled horizontal flow (Fig. 9). This cooling effect causes an increase in the magnetic field, since the gas pressure decreases and leads to a downflow. Provided that an adiabatic downflow in the thermally isolated flux tube takes place in a superadiabatically stratified environment, an instability ensues, known as the convective collapse. This causes a partial evacuation of the tube and leads to a magnetic field which is confined by the external gas pressure and hence of the order of $B \approx (8\pi p_{\text{ext}})^{1/2}$. This formation scenario is described in greater detail by Schüssler (1992) with a complete list of references on the subject.

6.2. CAN CONVECTIVE COLLAPSE BE OBSERVED?

While the convective collapse phenomenon has been studied in detail with analytical and numerical methods (see Schüssler, 1992 and references therein), there have been no direct or indirect observations of it, heretofore.

I propose that the key to a possible detection of a convective collapse is the strong splitting of Zeeman sensitive lines in the infrared (*e.g.*, Fe I 1.5648 μm, $g = 3$). This seems to be the only means to directly observe intrinsically weak photospheric fields of the order of the equipartition field strength (Rüedi *et al.*, 1992; Stenflo, 1992; Solanki 1993, in these proceedings). In the visible, the detection of weak fields is difficult, since the Zeeman broadening is dominated by velocity broadening.

The convective collapse lasts at most a few minutes. Therefore, a time resolution of about ten seconds is needed. The shortness of the process dictates the selection of a location on the Sun with high probability for the occurrence of a collapse, such as a flux-emergent region. Because of the fast field concentration in the course of the convective collapse, we expect to observe a sudden rise of field strength at a location where a magnetic flux density of the order of equipartition strength has been measured previously. More precisely, a quick increase of the peak separation of the infrared Stokes V profiles should be observed. At the same time, due to the adiabatic downflow within the forming flux tube, the Stokes profiles are shifted to the red. However, this shift should disappear in the final, stationary state of the flux tube. A variation in Stokes V amplitude may occur rather because of the increasing temperature of the downshifting material than because of the increasing field strength, since the equipartition field already almost saturates the Stokes V profiles in the infrared (Solanki, private communication).

In real observations, however, the magnetic flux density within the finite spatial resolution element may vary strongly or change sign and not all magnetic flux may become involved in the convective collapse. A mixture of weak and strong field components is expected to be present after the collapse. This gives rise to peculiar Stokes V profiles (Rüedi *et al.*, 1992), so that the interpretation of the observation may not be as straightforward as described above. The magnetohydrodynamic simulation of the convective collapse, together with the computation of synthetic infrared Stokes profiles, will probably be most useful for the understanding and interpretation of such observations.

6.3. A Destruction Mechanism for Magnetic Flux Tubes

Small scale magnetic flux tubes with a total magnetic flux of typically a few times 10^{18} Mx are susceptible to the interchange (flute) instability. Schüssler (1984) has shown that thin flux tubes can be stable if they are surrounded by a whirl flow. Bünte et al. (1992) and Bünte et al. (1993), in these proceedings) found that an azimuthal velocity component of up to 4 km s^{-1} is needed, depending on the atmospheric structure of the flux tube and its close surroundings, to prevent interchange instability. This value refers to the flux tube surface at a depth of about 100 km below the $\tau_{5000} = 1$ level.

Whirl flows are expected to arise from the "bathtub effect" in granular downdrafts which are a natural location of magnetic flux tubes. Moreover, as seen from Figure 1, the mere existence of a magnetic flux tube leads to the formation of converging downflows in its close surroundings which may survive even if the granular pattern changes. However, a strong perturbation may cause the disruption of the whirl flow and hence give way to the flute instability. This MHD instability induces a shredding of the flux tube possibly leading to its disappearance by a flux rearrangement within a few minutes.

Acknowledgements

I am very grateful to M. Schüssler for critically commenting on and carefully reading the manuscript. I also wish to thank M. Knölker for Figures 1 and 2 and M. Bünte for Figures 4 and 5. This work has been supported by the Deutsche Forschungsgemeinschaft under grant SCHU 500/6-1.

References

Anderson, L. S.: 1989, *Astrophys. J.* **339**, 558.
Ayres, T. R.: 1991, in P. Ulmschneider et al. (eds.), *Mechanisms of Chromospheric and Coronal Heating*, Springer-Verlag, p. 228.
Ayres, T. R.: 1993, these proceedings.
Ayres, T. R., Testerman, L., Brault, J. W.: 1986, *Astrophys. J.* **304**, 542.
Bünte, M., Solanki, S. K., Steiner, O.: 1992, *Astron. Astrophys.*, submitted.
Bünte, M., Steiner, O., Solanki, S. K.: 1991, in L. J. November (ed.), *Solar Polarimetry*, National Solar Observatory, Sunspot, New Mexico, p. 468.
Bünte, M., Steiner, O., Pizzo, V. J.: 1992, *Astron. Astrophys.*, submitted.
Bünte, M., Steiner, O., Solanki, S. K., Pizzo, V. J.: 1993, these proceedings.
Cally, P. S.: 1990 *J. Comput. Phys.* **93**, 411.
Cram, L. E., Damé, L.: 1983, *Astrophys. J.* **272**, 355.
Deinzer, W., Hensler, G., Schüssler, M., Weisshaar, E.: 1984a, *Astron. Astrophys.* **139**, 426.
Deinzer, W., Hensler, G., Schüssler, M., Weisshaar, E.: 1984b, *Astron. Astrophys.* **139**, 435.
Foukal, P., and Moran, T.: 1993, these proceedings.
Foukal, P. Little, R., Graves, J., Rabin, D., Lynch, D.: 1990, *Astrophys. J.* **353**, 712.
Giovanelli, R. G.: 1980, *Solar Phys.* **68**, 49.
Giovanelli, R. G., Jones, H. P.: 1982, *Solar Phys.* **79**, 267.
Grossmann-Doerth, U., Kneer, F., von Uexküll, M.: 1974, *Solar Phys.* **37**, 58.
Grossmann-Doerth, U., Schüssler, M., Solanki, S. K.: 1988, *Astron. Astrophys.* **206**, L37.
Grossmann-Doerth, U., Knölker, M., Schüssler, M., Weisshaar, E.: 1989, in R. J. Rutten and G. Severino (eds.), *Solar and Stellar Granulation*, NATO ASI Series, Vol. **263**, Kluwer, Dordrecht, p. 481.
Grossmann-Doerth, U., Schüssler, M., Solanki, S. K.: 1989, *Astron. Astrophys.* **221**, 338.

Hirayama, T.: 1992, *Solar Phys.* **137**, 33.
Hurlburt, N. E., Toomre, J.: 1988, *Astrophys. J.* **327**, 920.
Jahn, K.: 1989, *Astron. Astrophys.* **222**, 264.
Jones, H. P.: 1985, in B. W. Lites (ed.), *Chromospheric Diagnostics and Modelling*, National Solar Observatory, Sunspot, New Mexico, p. 175.
Jones, H. P., Giovanelli, R. G.: 1983, *Solar Phys.* **87**, 37.
Keller, C. U., Solanki, S. K., Steiner, O., Stenflo, J. O.: 1990, *Astron. Astrophys.* **233**, 583.
Knölker, M., Schüssler, M.: 1988, *Astron. Astrophys.* **202**, 275.
Knölker, M., Schüssler, M., Weisshaar, E.: 1988, *Astron. Astrophys.* **194**, 257.
Knölker, M., Grossmann-Doerth, U., Schüssler, M., Weisshaar, E.: 1991, *Adv. Space Res.* **11**, (5)285.
Lin, H., Kuhn, R.: 1992, *Solar Phys.* **141**, 1.
Oran, E. S., Boris, J. P.: 1987, *Numerical Simulation of Reactive Flow*, Elsevier, New York.
Pizzo, V. J.: 1986, *Astrophys. J.* **302**, 785.
Pizzo, V. J.: 1990, *Astrophys. J.* **365**, 764.
Proctor, M. R. E., Weiss, N. O.: 1982, *Rep. Prog. Phys.*, **45**, 1317.
Rüedi, I., Solanki, S. K., Livingston, W. C., Stenflo, J. O.: 1992, *Astron. Astrophys.* **263**, 323.
Schüssler, M.: 1984, *Astrophys. J.* **140**, 453.
Schüssler, M.: 1986, in Deinzer et al. (eds.), *Small Scale Magnetic Flux Concentrations in the Solar Photosphere*, Vandenhoeck and Ruprecht, Göttingen, p. 127.
Schüssler, M.: 1990, in J. O. Stenflo (ed.), 'Solar Photosphere: Structure, Convection, and Magnetic Fields', *Proc. IAU Symp.* **138**, 161.
Schüssler, M.: 1992, in J. T. Schmelz and J. C. Brown (eds.), *The Sun – a Laboratory for Astrophysics*, NATO Advanced Study Institute, Kluwer, Dordrecht, in press.
Solanki, S. K.: 1986, *Astron. Astrophys.* **168**, 311.
Solanki, S. K.: 1987, *Ph.D. Thesis*, ETH Zürich.
Solanki, S. K.: 1993, these proceedings.
Solanki, S. K., Stenflo, J. O.: 1984, *Astron. Astrophys.* **140**, 185.
Solanki, S. K., Steiner, O.: 1990, *Astron. Astrophys.* **234**, 519.
Solanki, S. K., Steiner, O., Uitenbroek, H.: 1991, *Astron. Astrophys.* **250**, 220.
Steffen, M.: 1991, in L. Crivellari et al. (eds.), *Stellar Atmospheres: Beyond Classical Models*, NATO ASI Series, Vol. 341, Kluwer, Dordrecht, p. 247.
Steiner, O., Pizzo, V. J.: 1989, *Astron. Astrophys.* **211**, 447.
Steiner, O., Stenflo, J. O.: 1990, in J. O. Stenflo (ed.), 'Solar Photosphere: Structure, Convection, and Magnetic Fields', *Proc. IAU Symp.* **138**, 181.
Steiner, O., Pneuman, G. W., Stenflo, J. O.: 1985, *Astron. Astrophys.* **170**, 126.
Stenflo, J. O.: 1989, *Astron. Astrophys. Rev.* **1**, 3.
Stenflo, J. O.: 1992, in D. S. Spicer (ed.), *Electromechanical Coupling of the Solar Atmosphere*, Proc. OSL Workshop, Capri, Italy, Amer. Inst. Phys., in press.
Stenflo, J. O., Harvey, J. W., Brault, J. W., Solanki, S. K.: 1984, *Astron. Astrophys.* **131**, 33.
Stenflo, J. O., Solanki, S. K., Harvey, J. W.: 1987, *Astron. Astrophys.* **171**, 305.
Vernazza, J. E., Avrett, E. H., Loeser, R.: 1981, *Astrophys. J. Suppl.* **45**, 635.
Wiehr, E.: 1985, *Astron. Astrophys.* **149**, 217.

THE THERMAL AND MAGNETIC STRUCTURE OF SUNSPOTS

P. MALTBY

Institute of Theoretical Astrophysics, University of Oslo,
P. O. Box 1029, Blindern N-0315, Oslo 3, Norway

Abstract. The continuum intensity observations of sunspot umbrae and penumbrae in the visible and infrared are reviewed. The intensity in the darkest part of the umbra and the average penumbral intensity are known with relatively high accuracy in *large* sunspots. The importance of including infrared observations in the construction of semi-empirical sunspot models is emphasized.

Magnetic field measurements are discussed. Special attention is given to recent high-spatial-resolution observations that show large fluctuations in magnetic field inclination, suggesting that the sunspot magnetic field changes its inclination – but not its magnitude – between bright and dark penumbral features.

Key words: infrared: stars – line: formation – Sun: magnetic fields – sunspots

1. Introduction

Our empirical knowledge of sunspots is mainly based on observations in the visible part of the spectrum (*e.g.* Bray and Loughhead, 1964; Cram and Thomas, 1981; Obridko, 1985; Cox *et al.*, 1992). In this review we will emphasize the importance of the infrared observations for the determination of the thermal structure of sunspots, including the construction of semi-empirical sunspot models, *i.e.*, models that are in hydrostatic equilibrium and are adjusted to fit the observations.

In solar physics it is useful to make clear the nomenclature that will be used. The darkest part of the sunspot will be referred to as the umbral core. Sunspot umbrae are not uniformly dark but contain extensions of penumbral filaments and small, bright regions called umbral dots. For larger sunspots the bright umbral dots are usually located close to the rim of the umbra and may be called peripheral umbral dots (Krat *et al*, 1972). In umbral cores the umbral dots show less contrast (Loughhead *et al.*, 1979) and to some observers the core gives the impression of being a featureless void (Livingston, 1991; Ewell, 1992). One should be aware, however, that examples of sunspots showing umbral dots distributed over nearly the whole umbra have been presented (Lites *et al.*, 1991). The penumbral structure has been studied in detail (see Muller, 1992). Here we shall limit the discussion to regarding the penumbra as consisting of dark and bright filaments, except when we need to underline that the filament appears to be broken up in fibrils.

Studies of the thermal structure of sunspots should be able to give information about the energy transport at different depths. In this paper we review the continuum intensity observations and show that the radiative flux decreases as a function of depth. Since the sunspot magnetic field will prevent ordinary convection, the problem is to explain why sunspots are as bright as observed. This brings us to the basic problem of how energy is transported to the surface in sunspots. If the energy is transported by oscillatory convection (*e.g.*, Weiss *et al.*, 1990), the monolithic flux tube picture of the sunspot may be used. One difficulty with this approach is that the convection cells will have to be very narrow in order to transfer energy laterally by radiation. This problem, combined with the observation that the umbral brightness is practically independent of the umbral diameter, led Parker (1979) to suggest that the sunspot consists of a cluster of narrow fluxtubes. In his model the energy is transported to the surface in field-free intrusions by ordinary convection. The cluster model requires a downdraft to keep the cluster of flux tubes together. There is, however, no observational support for this downdraft. In both models bright umbral dots are regarded as evidence for the way energy is transported to the surface. The possibility that the umbral core contains a region free of umbral dots, as suggested by some observers, may accordingly have important implications for our understanding of the energy flow in sunspots. In the following we shall concentrate on the continuum observations and the deepest observable layers.

Magnetic field observations in the infrared will be discussed. It is, however, necessary to include recent findings in the visible part of the spectrum to update our knowledge regarding the magnetic structure of sunspots. In particular, we shall draw attention to the importance of the fine structure in the magnetic field with spatial scale of the penumbral fibrils (*e.g.*, Title *et al.*, 1992). These new observations appear to solve the problem of understanding the conflicting results derived earlier from magnetic-field and from velocity-field measurements.

2. Continuum Observations

In the 1960's it was generally accepted that large sunspots were darker than small sunspots. However, this view was based on observations that were insufficiently corrected for scattered light (Zwaan, 1965). The stray light originates in the Earth's atmosphere and in the instrument and causes small amounts of light to be scattered from each position in the solar image to other positions. Since the sunspot is darker than the surroundings, the corrections for stray light become critical. In order to measure the amount of stray light, nearly simultaneous observations outside the solar disk have to be carried out. Observers who carefully correct their observations for scattered light have hitherto not found any evidence for a decrease in umbral brightness with increasing umbral diameter (*e.g.*, Rossbach and Schröter, 1970; Albregtsen *et al.*, 1984). This does not exclude the possibility that a systematic change in umbral brightness with umbral size may exist.

We emphasize the importance of a proper correction method for stay light (*e.g.*, Birkle and Mattig, 1965; Iuell and Staveland, 1975). One should be aware, however, that this correction method will not produce a complete restoration of the sunspot image. The method gives the corrected intensity in the center of the umbra as well as the corrected penumbral intensity as averaged over penumbral filaments. In observations of penumbral filaments image restoration methods have been applied (Grossmann-Doerth and Schmidt, 1981; Collados *et al.*, 1987). It seems likely that new algorithms like those for maximum entropy image restoration (Skilling, 1984) will be applied to sunspot observations in the near future. The observed fine structures in the form of bright umbral dots show dimensions comparable to the scale height, and knowledge about their thermal stratification may be important for understanding the energy transfer in sunspot atmospheres.

We note that an extension of the work on infrared imaging of sunspots (*e.g.*, Ewell, 1992) may be an important contribution for constructing two-component models with one hot and one cold component. A two-dimensional study of the sunspot atmosphere will require simultaneous observations with a series of two-dimensional detectors covering the wavelength region from 0.4 to 2.5 μm, properly corrected for stray light through the use of image restoration techniques.

2.1. UMBRAL CONTINUUM OBSERVATIONS

As relatively few intensity observations have been published recently, we have to rely on earlier observations, such as the sunspot intensity observations carried out for 10 months each year at the Oslo Solar Observatory between 1967 and 1986. The broad band pinhole photometers recorded intensities in a total of 11 wavelength regions in the spectral range 0.387–3.8 μm. Only measurements with an accuracy of 0.015 in the umbra/photosphere[1] intensity ratio were retained.

Let us consider the spectral distribution of the umbra/photosphere continuum intensity ratio in umbral cores, for sunspots with umbral radii larger than 4″. The results presented by van Ballegooijen (1984), Albregtsen et al. (1984) and Sobotka (1988, four largest sunspots) are in general agreement with the intensity values suggested by Albregtsen and Maltby (1981), after an evaluation of eleven papers published by different observers during the period 1968 to 1981. The extreme darkness observed in one sunspot by Bumba et al. (1990) is probably related to the fact that no correction for spectral line haze has been applied to that result.

Albregtsen et al. (1984) examined their data for connections between umbral temperature and sunspot parameters like the magnetic induction, size, age and the type of sunspot. No connection was apparent. Although the data do not show any systematic change of brightness temperature with sunspot size for large sunspots, such a relation could exist for smaller sunspots and has been suggested (see Sobotka, 1988).

Older data, not properly corrected for stray light, showed the umbral limb-darkening to be less than that of the photosphere, whereas more recent observers were unable to detect any center-limb variation in the umbra/photosphere intensity ratio. By observing the very same sunspot on several days during its passage over the solar disk, Albregtsen et al. (1984) found that all sunspots studied showed a real and significant decrease in the umbra/photosphere intensity ratio towards the limb. We note that the umbral limb darkening is easiest to detect in the infrared. Hitherto differences in limb-darkening between sunspots have not been detected.

A solar cycle variation in the umbra/photosphere intensity ratio was first detected in the infrared (Albregtsen and Maltby, 1978). Let the phase be measured as the time t elapsed since the last minimum in the cycle, the duration of which is t_0. After correction for limb-darkening the umbra/photosphere intensity ratio ϕ_u may be expressed as a linear function of the phase t/t_0 in the solar cycle, i.e.,

$$\phi_u(\mu = 1, \lambda, t/t_0) = c(\lambda) + d(\lambda)t/t_0.$$

At 1.67 μm the correlation coefficient between the umbra/photosphere intensity ratio and t/t_0 is 0.86 (−0.12, +0.07). The variation is caused by differences in umbral

---
[1] Here "photosphere" refers to the quiet Sun.

intensity; the observed change in the intensity ratio is too large to be explained by a variation in the photospheric intensity.

It is well known that the average heliographic latitude of sunspots varies throughout the solar cycle. The correlation coefficient between the umbra/photosphere intensity ratio and the heliographic latitude is -0.63 (-0.12, $+0.16$) at 1.67 μm. Hence, the umbral intensity is better correlated with t/t_0 than with the sunspot heliographic latitude.

The hypothesis (e.g., Adjabshirzadeh and Koutchmy, 1983) that a variation in the relative number of umbral dots from one sunspot to another may account for the observed difference in the umbra/photosphere intensity ratios has been investigated. Whereas the hypothesis would give the largest change in the visible, the observations show the largest change in the infrared. Suggestions as to the nature of the observed relation between the umbral temperature and the solar cycle have been presented (Schüssler, 1980; Yoshimura, 1983; Nordlund and Stein, 1990). The point to note here is that the observed relation may have implications for the relationship between sunspots and the solar dynamo.

According to Kusoffsky and Lundstedt (1986), the half-life time of "normal" umbral dots is about 60 min. It is often assumed that the umbral dots have almost photospheric temperature and sizes of 100–200 km. However, considerably lower dot temperatures are observed in the core of the umbra (Loughhead et al., 1979; Grossmann-Doerth et al., 1986) and some sunspots show cores without umbral dots (Livingston, 1991; Ewell, 1992). Possibly one may unite these conflicting results by introducing a hypothesis that only umbrae with radii larger than a certain value, say 4″, have an umbral core without umbral dots (Maltby, 1992).

2.2. PENUMBRAL CONTINUUM OBSERVATIONS

Observations by Grossmann-Doerth and Schmidt (1981) and Collados et al. (1987) agree well with the results given by Maltby (1972) for the wavelength dependence of the average continuum penumbra/photosphere intensity ratio, i.e., as averaged over penumbral filaments. No corrections for spectral line differences between the penumbra and the photosphere have been applied to these observations. Within the accuracy of the observations it has not been possible to detect any center-limb variation in the penumbra/photosphere intensity ratio.

Although the average penumbral intensity only changes slightly from one sunspot to another, differences in penumbral intensities between different sunspots are observed and found to be real (Ekmann, 1974). Sunspots with a darker than average umbra usually have a relatively dark penumbra. The correlation coefficient between penumbral and umbral intensities is 0.90 at a wavelength of 1.67 μm.

A description of the morphology of the penumbra is given by Muller (1992). Here it is sufficient to regard the penumbra as consisting of bright and dark filaments. Grossmann-Doerth and Schmidt (1981) and Collados et al. (1987) have studied the intensities in bright and dark penumbral filaments. It is apparent that the bright (dark) filaments within the very same sunspot are not equally bright (dark). This led Grossmann-Doerth and Schmidt (1981) to argue that their results were in conflict with those of Muller (1973). Collados et al. (1987) confirm the spread in brightness between the filaments, but find that their results are in general agreement with those of Muller (1973).

3. Semi-Empirical Models

In semi-empirical models the atmosphere is assumed to be either in hydrostatic or in hydrodynamical equilibrium and the temperature-optical depth relation is adjusted until the calculated intensities account for the observations.

In this paper the emphasis will be on the core model that fits the observations for an *average* large sunspot. Next, we shall take into account the presence of umbral dots and mention a few aspects of two-component umbral models. In penumbral semi-empirical models one has to consider that a non-vertical magnetic field may introduce a lifting force. Furthermore, the marked filamentary structure of the penumbra calls for two or more components in the horizontal plane.

3.1. Umbral Core Model

In semi-empirical umbral core models the atmosphere is assumed to be in hydrostatic equilibrium along a vertical magnetic field such that the field does not influence the stratification. The easiest way to construct a model for the umbral core is by scaling the quiet-Sun model. This approach has been used by several authors, most recently by Sobotka (1988) who sets $\Delta\theta =$ constant, where

$$\Delta\theta = 5,040 \, (1/T_\mathrm{u} - 1/T_\mathrm{ph})$$

and T_u and T_ph are respectively the temperature of the umbra and the photosphere. The advantage of this method is that the $\Delta\theta$ value characterizes the entire model. The disadvantage is that the model does not fit the observations (see below).

We shall consider umbral core models presented by van Ballegooijen (1984), Maltby *et al.* (1986), Obridko and Staude (1988) and Sobotka (1988). Since these models have been calculated with different computer codes and input parameters, differences other than those caused by the different temperature-optical depth relations may occur. In order to minimize these other effects the models have been recalculated using the same computer code.

A comparison between calculated continuum intensities and the observed continuum intensities at the center of the solar disk is presented in Figure 1. It is apparent that all four models may account for the observations in the visible and near infrared part of the spectrum up to approximately 1 μm. Hence, observations in the infrared are required to differentiate between the models. We note that the model by Sobotka (1988) with $\Delta\theta = 0.45$ shows considerably lower intensities than those observed in the far infrared. The model by Obridko and Staude (1988) predicts intensities that are too low at wavelengths above 1.6 μm. Both the models by van Ballegooijen (1984) and by Maltby *et al.* (1986) may account for the observations shown in Figure 1. A recent model by Ming-de and Cheng (1990) was not included in Figure 1; the model is quite similar to the model by Sobotka (1988).

A reasonable fit to the center-limb observations is obtained with the models by van Ballegooijen (1984) and by Maltby *et al.* (1986). The deviations between

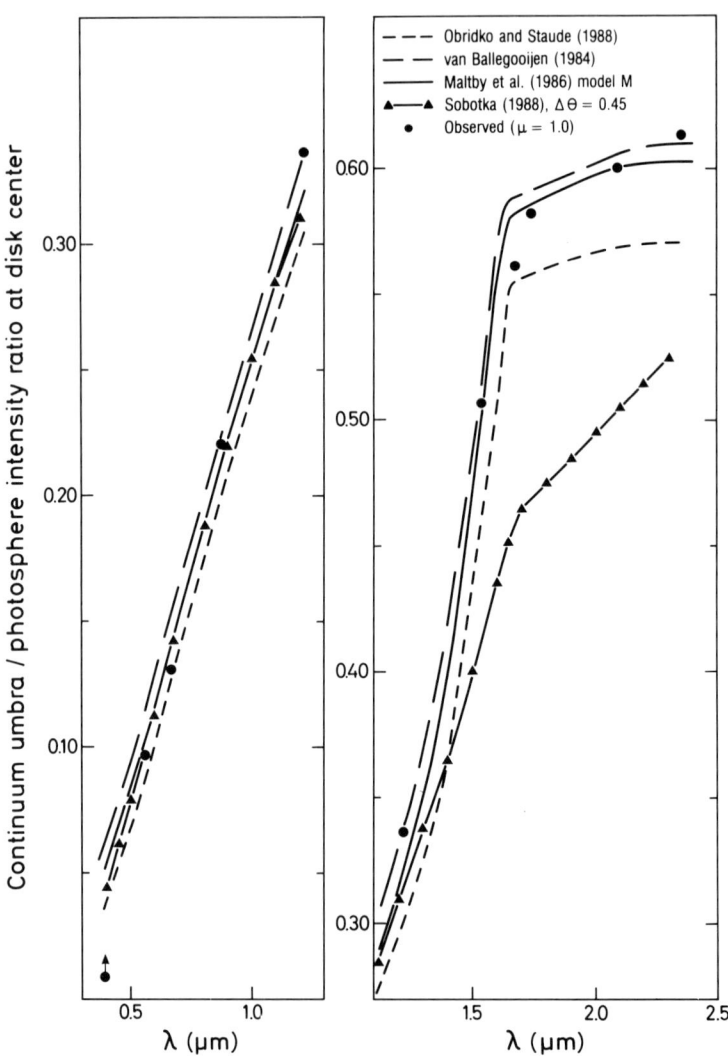

Fig. 1. Observed continuum umbra/photosphere intensity ratios at disk center compared with the predicted intensity ratios for four different umbral core models. The same computer code was used to obtain the four sets of intensity ratios.

the calculated and observed values are significant for the two other models, most apparently for the model by Sobotka (1988). Maltby (1992) compared the calculated radiative flux density, integrated over wavelength, as a function of optical depth at the reference wavelength (0.5 μm) for the umbral models. Whereas both the photospheric model and the umbral model by Maltby et al. (1986) vary relatively smoothly with depth, the models by van Ballegooijen (1984) and Obridko and Staude (1988) show more complicated variations. The transition from radiative to another form of energy transport occurs close to the solar surface.

The semi-empirical umbral and photospheric models give the stratification in the umbra and the photosphere, respectively. In order to make a comparison between semi-empirical and theoretical sunspot models meaningful, a value for the Wilson depression is needed. Although analytical magnetostatic models (Osherovich and Garcia, 1989) are of interest, the focus has shifted to numerical solutions (Jahn, 1989; Pizzo, 1990).

3.2. Two-Component Umbral Model

The two component model by Obridko and Staude (1988) combines one hot component with temperature close to that of the photosphere with an umbral core component. Several questions arise in connection with the two-component umbral models, in particular since the intensity of the bright component is uncertain. Furthermore, it is unlikely that the fraction covered with bright emission remains constant with height. We know that umbral dots in the central part of large sunspots show less contrast than peripheral umbral dots. It is possible that the distribution of umbral dot intensities is such that the bright component changes in temperature from one umbral dot to another and from one sunspot to another.

Since some umbral cores are free of umbral dots the energy flow in deep layers must be distributed laterally before reaching the visible layers. In order to get some insight into the time needed to distribute energy laterally by radiation we have calculated the cooling time for a hot filament embedded in a cold gas. For a hot filament with diameter 100 km, situated at an optical depth of 10 in the umbra, the cooling time is 50 s, increasing to 800 s for a diameter of 400 km. Hence, in order to account for the brightness of a featureless umbral core a fine structure is needed in subphotospheric layers.

3.3. Penumbral Models

The penumbra is much closer in temperature to the photosphere than is the umbra. The temperature of the penumbra may, in a first approximation, be described by a single parameter value of $\Delta\theta = $ constant, where now

$$\Delta\theta = 5,040 \, (1/T_\text{p} - 1/T_\text{ph})$$

and T_p and T_ph are respectively the temperature of the penumbra and the photosphere, as averaged over penumbral filaments and granules. The average value for penumbrae is $\Delta\theta = 0.055$. The one-component penumbral models of Yun et al. (1984) and Ding and Fang (1989) are primarily aimed at modeling the penumbral chromosphere; we note that these models are too cold in deeper layers to account for the continuum observations.

In the penumbra the filament structure is so evident that it is tempting to construct a two-component model. The model should agree with the observed intensities in dark (d) and bright (b) filaments. As a check on the model, the calculated spectral intensity as averaged over penumbral filaments may be compared with observations. A two-component model with $\Delta\theta_b = 0.010$, $\Delta\theta_d = 0.093$ and the ratio of the areas covered by bright and dark filaments equal to 0.75 was presented by Kjeldseth-Moe and Maltby (1974), based on observations of penumbral filament intensities by Muller (1973).

Evidently the intensity may change from one bright penumbral filament to another within the same penumbra so that the two-component model must be used with care. The construction of a two-component model also brings up the question of keeping the two atmospheres in balance in the horizontal direction over an extended height range. This implies that the sum of the gas pressure and the magnetic pressure must be the same in bright and dark filaments. The gas pressure stratification in the filament may be influenced by a variation in curvature of the magnetic field lines with depth. It is interesting that large fluctuations in inclination angle for the penumbral magnetic field have been reported (see section 4.4).

4. The Magnetic Field

The magnetic field observed on the Sun emerges from the solar interior and has already been broken up into separate flux tubes when reaching the visible surface. It is well known that the sunspots occur within activity belts. One third of all sunspots are members of "sunspot nests", *i.e.* sunspots that appear with practically the same coordinates in latitude and in longitude (Brouwer and Zwaan, 1990). Per unit area at the solar surface the probability is more than 25 times larger for emergence of new active regions within an already existing active region than it is for emergence elsewhere in the then available activity belt (Harvey, 1992). In a large bipolar active region at maximum development the typical distance between the centroids of opposite polarities is 150 Mm, a distance comparable to the depth of the convection zone. In fact, it has been suggested that the toroidal magnetic field system responsible for the sunspots is located at the interface between the convective zone and the radiative interior (*e.g.*, Schüssler, 1983; Zwaan, 1992). In the following we shall focus on the observable parts of the magnetic field in *single* sunspots, leaving aside magnetic elements that move radially in the sunspot moat. Studies of the network and intranetwork magnetic fields are also outside the scope of this review.

4.1. ZEEMAN AND PASCHEN–BACK EFFECTS

The magnetic field in sunspots are most commonly inferred from measurements of the polarization in spectral lines split by the Zeeman effect (see Zeeman, 1913). It may be of some interest to know that the splitting of the D_1 and D_2 lines of Na by a magnetic field was first found by Fievez (1885) a decade before the effect was discovered by Zeeman in 1896.

The fact that not all lines follow the Zeeman line splitting pattern was detected by Paschen and Back from investigations of the Li I 6708 Å resonance line. The reason is that for strong magnetic fields the magnetic splitting is comparable to the doublet separation. The splitting of the Li I resonance doublet by the sunspot magnetic field is discussed by Maltby (1971). A discussion of the Paschen–Back splitting of the Mg I line at 12.32 μm is given by Chang in this volume.

4.2. Line Formation Theory

The interpretation of the magnetic field measurements should be based on an understanding of the theory of spectral line formation in the presence of a magnetic field. The line formation theory in terms of Stokes parameters was formulated originally by Unno (1956) and extended by Rachkovsky (1962). Considerable effort has been devoted to the solution of the Unno-Rachkovsky equations; for a review see Semel et al. (1992). Although more general non-LTE formulations, including quantum interference and partial redistribution, exist (Landi degl'Innocenti, 1983), the practical applications are usually limited to electrical dipole transitions, LS-coupling, no quantum interference between the Zeeman states, complete redistribution on scattering, and steady state atmospheres.

Another limiting factor in the line formation theory is the lack of good correction methods for molecular and atomic blends as well as for depressions in the continuum caused by line haze. As an example, consider the Fe I 5250.22 Å line that has been extensively used for magnetic field measurements. One of the σ-components contains a blend from FeH at 5250.31 Å in the umbral spectrum (Wöhl et al., 1983). The blend in the π-component in the umbra was identified by Kjeldseth-Moe (1973) as the 5250.24 Å $P_1(71)$ line of TiO belonging to the (0,0) band of the α-system. His calculations showed that estimates of the inclination of the magnetic field vector will easily yield values 20° to 30° in error if the molecular blend is neglected. This suggests that the observed line haze in the spectra of umbrae may seriously influence the results of magnetic field measurements. Umbral spectra are so crowded with lines that care should be taken in all three cases of polarimetry measurements; i.e., broad-band, magnetograph and high spectral resolution observations. The broad-band polarization observed in sunspots (e.g. Leroy, 1962) has, so far, had limited diagnostic value. A better theoretical calculation of the spectrum, taking into account the blends present may improve the situation.

4.3. The Mesoscale Magnetic Structure

The magnitude of the magnetic field near the centre of the sunspot is between 2000 and 4000 G. The magnetic field strength is nearly independent of sunspot size (Brandt and Zwaan, 1982); a slow increase in B with increasing area has been found. Several authors (see review by Skumanich, 1992) have presented measurements of the magnetic induction $B(r)/B(0)$ as function of the distance r from the centre of the sunspot, measured in units of the penumbral radius. Recent studies suggest a steeper gradient and accordingly a smaller value of B at the rim of the penumbra than the relation $B(r) = B(0)/(1+r^2)$ derived by Beckers and Schröter (1969). The recent papers suggest an inclination of 60–70° close to the rim of the penumbra, whereas earlier papers found the field to be nearly horizontal in the outer parts of the penumbra. Improved accuracy in sunspot magnetic field determinations have been attained by using the magnetically sensitive infrared line at 1.5648 μm line (Solanki et al., 1992) and the Mg I line at 12.32 μm (Deming et al., 1988). Since sunspots in general do not show circular symmetry, there is an apparent need for infrared observations of the magnetic field distribution along different position angles within the sunspots, using detector arrays.

The field measurements at photospheric heights have also been studied with the intention to determine the vertical gradient of the magnetic field (e.g. Hofmann and Rendtel, 1989). A review of the extrapolation of photospheric magnetic fields into the corona is given by Semel et al. (1992).

4.4. HIGH SPATIAL RESOLUTION MEASUREMENTS

One problem we face comparing medium spatial resolution observations of the magnetic field with the direction of the gas flow in the penumbra is the following: the high electrical conductivity in the sunspot atmosphere makes it very likely that the flow is along the magnetic field lines. Whereas the flow is nearly horizontal at photospheric heights, the magnetic field measurements (see section 4.3) suggest an inclination of 60–70° close to the rim of the penumbra. High spatial-resolution observations of both the magnetic field and the flow field by Title et al. (1992) give insight into this problem and we would like to draw attention to two new findings: (1) There is a variation in inclination of the magnetic field in the penumbra of ±15–20° with a spatial scale of the penumbral fibrils. (2) The more horizontal magnetic fields occur in the regions of the Evershed flow.

Since there is a tendency for the Evershed flow to occur in darker penumbral structures, where the magnetic field is more horizontal, the new findings appear to be consistent with gas flow along the magnetic field lines, as expected from theoretical arguments. Observational arguments in favor of the rapid fluctuation in magnetic field inclination with position angle in the sunspot are given by Kalman (1991). We note that thin X-ray loops are observed to terminate in the penumbra (Golub et al., 1990). The orientation of the loops suggests an inclination close to 45° for the magnetic field. The high inclination found for some field lines in the sunspot region makes it easier to understand the observed velocity field in the chromosphere and transition region. The observed variation in inclination of the magnetic field in the penumbra with a spatial scale of the penumbral fibrils raises questions about the structure of magnetostatic atmospheres; for a general formulation see Low (1991).

5. Concluding Remarks

In this review attention has been given to the fact that the detection of the solar cycle variation of the sunspot intensity was based on observations in the infrared. By comparing different semi-empirical umbral models we have illustrated how continuum observations in the infrared have contributed considerably to our present understanding of the temperature structure of sunspots.

Using semi-empirical models one may determine the lateral pressure difference between the umbra and the photosphere. The largest contribution to the lateral gas pressure difference comes from the magnetic field, which acts both through the magnetic pressure, $B^2/2\mu$, and through the magnetic tension. Further comparisons with numerical models (Jahn, 1989; Pizzo, 1986) may be valuable.

The semi-empirical models indicate that the sunspot regions are in radiative equilibrium in the upper layers, but other energy transport processes take over close to the solar surface. In both the monolithic flux tube model and in the cluster model the umbral dots are regarded as evidence for the way energy is transported to the surface. If the umbral core is observed to be without umbral dots the energy must be distributed laterally by radiation in deeper layers. In order to account for the brightness of a featureless umbral core a fine structure is needed in subphotospheric layers.

The sunspot brightness is primarily determined by the energy transport processes, but the energy equation is mathematically coupled to the momentum equation. Along the magnetic field lines the gas is nearly in hydrostatic balance, but not quite since flows are observed. We have drawn attentional to recent observations of the magnetic field and the velocity field. These high-spatial-resolution observations show that the physical picture may change considerably when improved observations become available and suggest that infrared observations of the magnetic field with high spatial resolution would be valuable (see Rabin 1993, in this volume).

References

Adjabshirzadeh, A., and Koutchmy, S.: 1983, *Astron. Astrophys.* **122**, 1.
Albregtsen, F., Joràs, P.B., and Maltby, P.: 1984, *Solar Phys.* **90**, 17.
Albregtsen, F., and Maltby, P.: 1978, *Nature*, **274**, 41.
Albregtsen, F., and Maltby, P.: 1981, in L.E. Cram and J.H. Thomas (eds.), *The Physics of Sunspots*, Sacramento Peak Obs. Conf., Sunspot, NM, p. 127.
Beckers, J.M., and Schröter, E.H.: 1969, *Solar Phys.* **10**, 384.
Birkle, K., and Mattig, W.: 1965, *Zeitsch. Astrophys.* **60**, 204.
Brandt, J.J., and Zwaan, C. 1982, *Solar Phys.* **80**, 251.
Bray, R.J., and Loughhead, R.E.: 1964, *Sunspots*, Chapman and Hall, London.
Brouwer, M.P., and Zwaan, C.: 1990, *Solar Phys.* **129**, 221.
Bumba, V., Sobotka, M., and Simberová, S.: 1990, in L. Dezsö (ed.), *The Dynamic Sun*, Publ. Debrecen Heliophys. Obs. **7**, 84.
Chang, E.S.: 1993, these proceedings.
Collados, M., del Toro, J.C., and Vázquez, M.: 1987, in E.-H. Schröter, M. Vázquez, and A.A. Wyller (eds.), *Proc. Inaugural Workshop D-E-S Telescope Install. Canary Islands*, Cambridge Univ. Press, Cambridge, p. 214.
Cram, L.E. and Thomas, J.H.: 1981, *The Physics of Sunspots*, Sacramento Peak Obs. Conf., Sunspot, NM.
Cox, A.N., Livingston, W.C., and Matthews, M. (eds.): 1992, *The Solar Interior and Atmosphere*, Univ. Arizona Press, Tucson.
Deming, D., Boyle, R.J., Jennings, D.E., and Wiedemann, G.: 1988, *Astrophys. J.* **333**, 978.
Ding, M.D., and Fang, C.: 1989, *Astron. Astrophys.* **225**, 204.
Ekmann, G.: 1974, *Solar Phys.* **38**, 73.
Ewell, M.W.: 1992, *Solar Phys.* **137**, 215.
Fievez, Ch.: 1885, in F. Exner (ed.), *Repertorium der Physik*, München, Oldenbourg, **21**, 766.
Golub, L., Herant, M., Kalata, K., Lovas, I., Nystrom, G, Pardo, F., Spiller, E., and Wilczynski, J.: 1990, *Nature* **344**, 842.

Grossmann-Doerth, U., and Schmidt, W.: 1981, *Astron. Astrophys.* **95**, 366.
Grossmann-Doerth, U., Schmidt, W., and Schröter, E.H.: 1986, *Astron. Astrophys.* **156**, 347.
Harvey, K.: 1992, *Solar Phys.* (submitted).
Hofmann, A., and Rendtel, J.: 1989, *Astr. Nachr.* **310**, 61.
Iuell, P., and Staveland, L.: 1975, Inst. Theor. Ap. Oslo Rept. No. 43, pp.1-32.
Jahn, K.: 1989, *Astron. Astrophys.* **222**, 264.
Kalman, B.: 1991, *Solar Phys.* **135**, 299.
Kjeldseth-Moe, O.: 1973, *Solar Phys.* **33**, 393.
Kjeldseth-Moe, O., and Maltby, P.: 1974, *Solar Phys.* **36**, 101.
Krat, V.A., Karpinsky, V.N., and Pravdjuk, L.M.: 1972, *Solar Phys.* **26**, 305.
Kusoffsky, U., and Lundstedt, H.: 1986, *Astron. Astrophys.* **160**, 51.
Landi degl'Innocenti, E.: 1983, *Solar Phys.* **85**, 33.
Leroy, J.L.: 1962, *Ann. d'Astr.* **25**, 127.
Lites, B.W., Bida, T.A., Johannesson, A.,and Scharmer, G.B.: 1991, *Astrophys. J.* **373**, 683.
Livingston, W.: 1991, *Nature* **350**, 45.
Loughhead, R.E., Bray, R.J., and Tappere, E.J.: 1979, *Astron. Astrophys.* **79**, 128.
Low, C.B.: 1991, *Astrophys. J.* **370**, 427.
Maltby, P.: 1971, in R. Howard (ed.), 'Solar Magnetic Fields', *Proc. IAU Symp.* **43**, 141.
Maltby, P.: 1972, *Solar Phys.* **26**, 76.
Maltby, P.: 1992, in J.H. Thomas and N.O. Weiss (eds.), *Sunspots: Theory and Observations*, Kluwer Academic Publishers, Dordrecht, in press.
Maltby, P., Avrett, E.H. Carlsson, M., Kjeldseth-Moe, O., Kurucz, R.L., and Loeser, R.: 1986, *Astrophys. J.* **306**, 284.
Ming-de, D. and Cheng, F.: 1990, *Acta Astron. Sinica* **31**, 276.
Muller, R.: 1973, *Solar Phys.* **32**, 409.
Muller, R.: 1992, in J.H. Thomas and N.O. Weiss (eds.), *Sunspots: Theory and Observations*, Kluwer Academic Publishers, Dordrecht, in press.
Nordlund, A., and Stein, R.F.: 1990, in J.O. Stenflo (ed.) 'Solar Photosphere, Convection and Magnetic Fields', *Proc. IAU Symp. 138*, 191.
Obridko, V.N.: 1985, *Sunspots and Activity Complexes* (in Russian), Nauka, Moscow.
Obridko, V.N. and Staude, J.: 1988, *Astron. Astrophys.* **189**, 232.
Osherovich, V.A. and Garcia, H.A.: 1989, *Astrophys. J.* **336**, 468.
Parker, E.N.: 1979, *Astrophys. J.* **230**, 905.
Pizzo, V.J.: 1986, *Astrophys. J.* **302**, 785.
Rabin, D.M.: 1993, these proceedings.
Rachkovsky, D.N.: 1962, *Izv. Krymsk. Astr. Obs.* **28**, 259.
Semel, M, Maltby, P., Makita, M., Mouradian, Z., Rees, D.E., Sakurai, T., and Soru-Escaut, I.: 1992, in A.N Cox, W.C. Livingston and M. Matthews (eds.) *The Solar Interior and Atmosphere*, Univ. Arizona Press., Tucson, p. 844.
Rossbach, N., and Schröter, E.H.: 1970, *Solar Phys.* **12**, 95.
Schüssler, M.: 1980, *Nature* **288**, 150.
Schüssler, M.: 1983, in J.O. Stenflo (ed.), 'Solar and Stellar Magnetic Fields: Origins and Coronal Effects', *Proc. IAU Symp.* **102**, 213.
Skilling, J.: 1984, in J.H. Justice (ed.), *Maximum Entropy and Bayesian Methods in Applied Statistics*, Cambridge University Press, Cambridge p. 179.
Skumanich, A.: 1992, in J.H. Thomas and N.O. Weiss (eds.), *Sunspots: Theory and Observations*, Kluwer Academic Publishers, Dordrecht, in press.
Sobotka, M.: 1988, *Bull. Astr. Inst. Czech.* **39**, 236.
Solanki, S.K., Rüedi, I., and Livingston, W.: 1992, *Astron. Astrophys.* (submitted).
Title, A.M., Frank, Z.A., Shine, R.A., Tarbell, T.D., Topka, K.P., Scharmer, G., and Schmidt, W.: 1992, in J.H. Thomas and N.O. Weiss (eds.), *Sunspots: Theory and Observations*, Kluwer Academic Publishers, Dordrecht, in press.
Unno, W.: 1956, *Pub. Astron. Soc. Japan* **8**, 108.
Van Ballegooijen, A.A.: 1984, *Solar Phys.* **91**, 195.
Weiss, N.O., Brownjohn, D.P., Hurlburt, N.E., and Proctor, M.R.E.: 1990, *Mon. Not. Roy. Astron. Soc.* **245**, 434.
Wöhl, H., Engvold, O., and Brault, J.W.: 1983, Inst. Theor. Ap. Oslo Rept. No. 56, pp. 1-11.
Yoshimura, H.: 1983, *Solar Phys.* **87**, 251.

Yun, H.S., Beebe, H.A., and Baggett, W.E.: 1984, *Solar Phys.* **92**, 145.
Zeeman, P.: 1913, *Researches in Magneto-optics*, Macmillan, London.
Zirin, H., and Bopp, B.: 1989, *Astrophys. J.* **340**, 571.
Zwaan, C.: 1965,*Rech. Astr. Obs. Utrecht* **17**, Part 4, 1-182.
Zwaan, C.: 1992, in J.H. Thomas and N.O. Weiss (eds.), *Sunspots: Theory and Observations*, Kluwer Academic Publishers, Dordrecht, in press.

INFRARED MEASUREMENTS OF STELLAR MAGNETIC FIELDS

STEVEN H. SAAR

Center for Astrophysics, 60 Garden Street, Cambridge, MA 02138, U.S.A.

Abstract. I review the advantages, techniques, and results of measurement of magnetic fields on cool stars in the infrared (IR). These measurements have generated several important results, including the following: the first data on the magnetic parameters of dMe and RS CVn variables; evidence for field strength confinement by photospheric gas pressure; support for the correlation between magnetic flux and rotation, with possible saturation at high rotation rates; indications of horizontal and/or vertical magnetic field structure; and evidence of spatial variations in B over a stellar surface. I discuss these results in detail, and suggest future directions for IR magnetic field research.

Key words: infrared: stars – line: formation – stars: activity – stars: magnetic fields

1. Introduction

Welcome to probably one of the shorter reviews in this volume. The main reason for its brevity is that infrared measurements of stellar magnetic fields have only been made since 1982 and with only one instrument: the NOAO 4-m Fourier Transform Spectrometer (Hall *et al.* 1979). Until very recently, this was the only instrument in the world capable of the high resolution ($\lambda/\Delta\lambda \gtrsim 40,000$), low noise (S/N $\gtrsim$ 50) spectra needed to measure magnetic parameters from unpolarized absorption lines. The exposure times needed were long, however ($\approx$ 6 hours for $m_K = 5$), and consequently only a handful of stars have been studied to date. With the advent of cryogenic echelles and efficient IR arrays, there should be an explosion of new data in the near future. Thus, this review can be considered a summary of the "first generation" IR stellar field detections – few in number, but important in shaping our understanding of magnetic fields on solar-like stars.

Infrared observations to detect stellar magnetic fields on solar-like stars have several advantages. The most obvious of these is that the ratio of the Zeeman splitting to the Doppler width, $\Delta\lambda_B/\Delta\lambda_D$, is proportional to λ, making detection of typical stellar field strengths ($B \sim 1$ kG or more) considerably easier in the IR. Comparison with the half widths of typical moderate-strength lines ($\sim$3 km s^{-1}) makes this clear. The Zeeman width splitting (in km s^{-1}) is $v_B = 1.7 \times 10^{-4} g_{\mathrm{eff}} \lambda B$ (where λ is in Å, and B in kG), and so the improvement in detectability, in proportion to λ, changes a slight (and difficult to interpret) line broadening at optical wavelengths (*e.g.*, $v_B = 2.6$ km s^{-1} for $g_{\mathrm{eff}} = 2.5$ and $B = 1$ kG at 0.6 μm), to partial or full line *splitting* in the IR ($v_B = 9.4$ km s^{-1} for the same line at 2.2 μm). This increased sensitivity is particularly helpful in low-gravity RS CVn systems, where attempts at optical field measurements have mostly failed, probably because of low B values (Marcy and Bruning 1984; Saar 1990).

Other advantages of the IR include lower line density, so that blends are less likely to confuse a measurement, and that, for cooler stars at least, the near IR is close to the stellar flux maximum. Both of these are important for M dwarfs ($T_{\mathrm{eff}} \leq 3{,}750$ K), whose optical spectra are riddled with molecular bands. Since the difference between umbral and plage/network continuum levels is considerably lower in the IR, IR spectra also hold the potential for disentangling the various

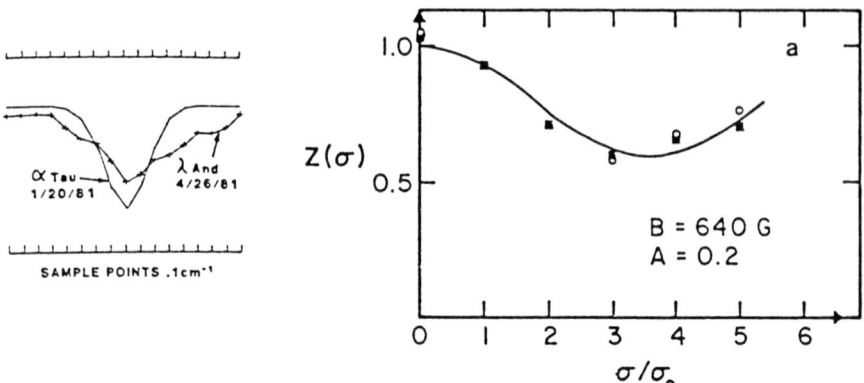

Fig. 1. Left: Comparison of the $g_{eff} = 3$, 1.56 µm line in λ And and α Tau; a Fourier ratio analysis yields $B = 1{,}290$ G and $f = 0.48$ (from Giampapa et al. 1983). Right: Fourier ratio $z(\sigma)$ vs. frequency, σ, of two Fe I lines (o), again from λ And, but at a different epoch (from Gondoin et al. 1985). A fit yielded $B \approx 600$ G and $f \approx 0.20$ (solid).

different *types* of magnetic regions expected on stars (optical spectra, due to the high contrast, can only detect the brighter magnetic regions). Finally, IR spectra offer new windows on the vertical structure of magnetic fields in stellar atmospheres, as the 1.6 µm opacity minimum permits B measurements deep into the photosphere, and the 12 µm Mg I lines sample the upper photosphere.

2. A Short History of IR Stellar Magnetic Field Research

Robinson, Worden, and Harvey (1980) made the first unambiguous detection of fields on a cool non-solar star (ξ Boo A), using the Fourier ratio method (Robinson 1980). This method assumes the lines are optically thin and treats a high-g_{eff} line as a convolution of three $g_{eff} = 0$ features (approximated by low-g_{eff} lines in the same spectrum). The ratio of the Fourier transforms of the lines then yields a cosine function whose amplitude is related to the fraction, f, of the stellar surface covered by fields, and whose period is proportional to the mean field strength, B, in these regions (see also Gray 1984). Three years later, Giampapa, Golub, and Worden (1983) made the first sortie into the IR, studying the popular 1.56 µm $g_{eff}=3.0$ Fe I line in the long period RS CVn star λ And (Fig. 1). Although they obtained several spectra (Giampapa and Worden 1982), only one showed clear Zeeman broadening in several lines (perhaps due to lower S/N in the other spectra). In a slight modification of the Robinson method, they took the same line from a different, low-activity star (α Tau) as their approximate $g_{eff} = 0$ line. From the one spectrum with a clear signal, they obtained $B = 1{,}290$ G and $f = 0.48$ – the first detection of magnetic fields on a giant star.

Two years later, Gondoin, Giampapa, and Bookbinder (1985) returned to λ And for further study (see Fig. 1). Perhaps because it becomes more strongly blended in cooler stars, and is thus difficult to use in a standard Fourier ratio analysis, the g_{eff} 3.0 line was not used in this work. Instead, Gondoin et al. studied pairs of high ($g_{\text{eff}} > 1.5$) and low g_{eff} lines with the basic Robinson (1980) method, and obtained a marginal detection for $B = 600$ G and $f = 0.20$. They attributed the difference between these values and those of Giampapa et al. (1983) to changing mixtures of spot (higher B) and plage (lower B) on the stellar surface at the two epochs. This is not unreasonable, as λ And is known to have large spotted regions on its surface, showing photometric amplitudes, ΔV, up to 0.28 (Strassmeier et al. 1988). If their interpretation is correct, it represents the first evidence for differing magnitudes of B in spot and plage regions on stars. Gondoin et al. (1985) also observed ξ Boo A and 61 UMa, detecting no clear evidence for fields on either. The null result for ξ Boo A is surprising, as most other measurements of the star have yielded positive detections. A list of further useful lines in the 1.6 μm region has been tabulated by Solanki, Biémont, and Mürset (1990).

Sun, Giampapa, and Worden (1987) explored the possible uses of IR lines in distinguishing between stellar spot and plage fields. Using 1.6 μm solar spectra (from the NSO McMath FTS), they combined varying flux-weighted fractions of umbral, plage, and quiet profiles, and analyzed the results as if they were stellar exposures. They found that spot umbrae dominate the derived f if $f_{\text{spot}} \geq 0.10$.

Meanwhile, others pushed further into the IR. Saar and Linsky (1985) noted a Ti I multiplet at 2.2 μm seen in umbral spectra (e.g., Hall 1973) with a helpfully large range of g_{eff} values (from $\sim$1.0 to $\sim$2.5, with the latter, at 4,480 cm^{-1}, having a Zeeman sensitivity $g_{\text{eff}}\lambda$ equal to that of the 1.56 μm line). They reasoned that this multiplet might be useful in studying fields on cool K and M stars, which are difficult to study in the optical due to blends. Their efforts were rewarded when only a six hour exposure of the active dM3.5e flare star AD Leo showed evidence for clear splitting of Zeeman components in several Ti I lines (Fig. 2). Using the simple Unno (1956) radiative transfer solution for Stokes I, and convolving the result with appropriate rotation and turbulent broadening functions, Saar and Linsky (1985) derived an average of $B \approx 3,800$ G and $f \approx 0.73$ from five Ti I lines. This detection was the first for an M dwarf, and the strongest field discovered up to that point on a cool star.

Continuing along these lines, Saar, Linsky, and Giampapa (1987) studied a number of late K and M dwarfs using the Ti I lines, plus two intermediate g_{eff} (1.0 and 1.33) Na I lines in the same bandpass. Only one of the six dM stars they studied showed a field (a marginal detection of the weak flare star GL 229), contrary to expectations of large f values on dM stars (Giampapa 1985; Fig. 3). On the other hand, all dMe stars showed strong ($B > 2$ kG), widespread ($f > 0.5$) fields, consistent with their high activity and flaring rates. The field strengths, though smaller than some predictions (Mullan 1985), followed the trend indicated by G and K dwarfs, namely larger B values with decreasing T_{eff} and increasing gas pressure, consistent with $B \propto P_{\text{gas}}^{0.5}$. The IR data also confirmed and extended the correlation between f or fB and rotation seen in optical magnetic data (Saar and Linsky 1986). No measurable field was detected on the old K5 dwarf 61 Cyg A, suggesting previous optical detections may have been misled by subtle blends.

Jennings et al. (1986) used a cryogenic postdisperser and a BIB detector with

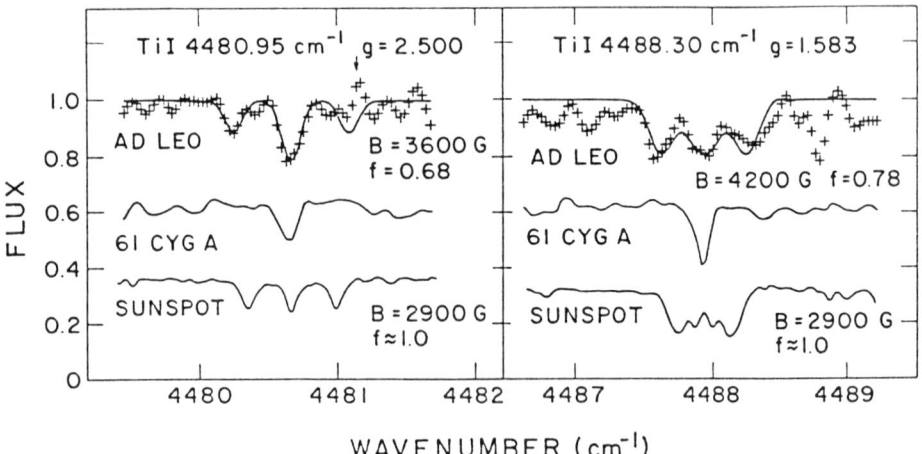

Fig. 2. Comparison of apodized 4-m FTS spectra of AD Leo (crosses), 61 Cyg A (an inactive K5V) and a umbra/photosphere ratio spectrum from Hall (1973). Fits to AD Leo for the indicated f and B values are also shown (solid). Areas of enhanced noise due to telluric line removal are marked by arrows (from Saar and Linsky 1985).

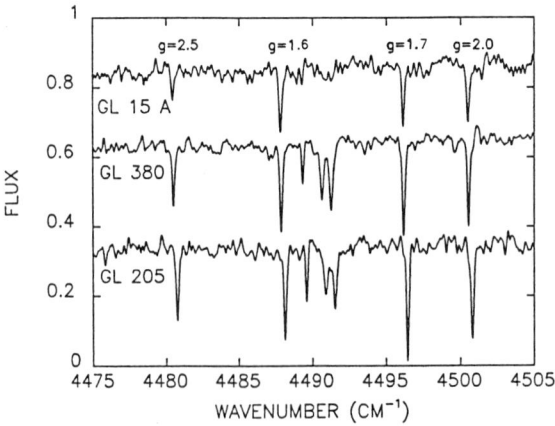

Fig. 3. Apodized 4-m FTS data for three inactive dM stars (from Saar et al. 1987). Note the lack of significant line broadening/splitting or differential shape changes in the Ti I multiplet indicated.

the 4-m FTS to push even further into the IR and detect the highly Zeeman-sensitive 12 μm Mg I lines in stars. As the instrument was only able to observe very bright, magnetically inactive stars (α Ori and α Tau), no field detections were made. Nevertheless, the lines are clearly worthy of further study once improved instruments are available, since they afford exceptional Zeeman sensitivity and a unique view of B in the upper photospheres of stars (*e.g.*, Deming *et al.* 1988; Solanki 1993).

IR lines were also used in passing by Saar (1988a) to demonstrate that ϵ Eri had relatively low magnetic flux (unlike occasional high values obtained earlier using non-RT methods). Basri and Marcy (1988), however, noted that the 2.2 μm Na I line modeled was quite optically thick and could actually be fit with a low f field using a more realistic model atmosphere. More sophisticated modeling methods generally agree on $B = 1$–2 kG and $f = 0.15$–0.30 for ϵ Eri (Marcy and Basri 1989; Saar 1990) with remaining differences probably explained by the difference in the heights of formation of various lines (Grossmann-Doerth and Solanki 1990).

More recently, an analysis of several co-added K band spectra of AD Leo showed anomalous broadening in the σ components of the Ti I lines (Fig. 4; Saar 1992a). Most non-magnetic broadening mechanisms could be eliminated: The lines are weak (ruling out opacity broadening), $v \sin i$ is low (~ 5 km s^{-1}), and > 10 km s^{-1} turbulence in magnetic regions seems implausible. Saar (1992a) suggested that the extra broadening as due to either a distribution of regions with differing B on the surface (*e.g.*, spot and plage) or a vertical gradient in B over the level of line formation. The first option suggests the possible separation of weak (network/plage) and strong (umbral) fields. The latter possibility is also interesting, as it permits the line to be fit with considerably smaller f values ($\sim 50\%$ lower) than traditional analyses. It is also consistent with the prediction of lowered canopy heights in the cooler, denser atmospheres of these stars (Solanki and Steiner 1990).

AD Leo was once more the focus of study in May 1991, when a large multiwavelength campaign was launched to monitor the star's activity, including IUE, HST, ROSAT, Ginga, radio, sub-mm, IR and optical measurements (Saar *et al.* 1991; Bookbinder, Walter, and Brown 1992). Five nights of IR FTS spectra taken during the campaign were combined to yield spectra at three rotational phases. Systematic shape changes visible in the Ti I line profiles suggested varying magnetic region geometry and/or coverage (Fig. 5). Approximate models of the Ti I lines with a vertical B gradient (assuming a homogeneous field distribution) supported this first impression, yielding tentative evidence for phase-modulated magnetic structures. We are continuing to analyze this large data set.

Finally, Saar (1992b) has made new measurements (and remeasurements using improved methods) of several KM dwarfs and RS CVn stars. These analyses have been made with an improved version of the Saar (1988a) code, which includes magneto-optical effects and a full disk-integration of the Stokes parameters (Saar *et al.* 1990). The improved models lower the derived filling factors somewhat, but f on most flare stars remains high ($\geq 40\%$). The analyses of RS CVn stars make use of the increase in equivalent width of optically thick lines in a magnetic field (see Leroy 1962; Basri, Marcy, and Valenti 1992), and are less precise; only the

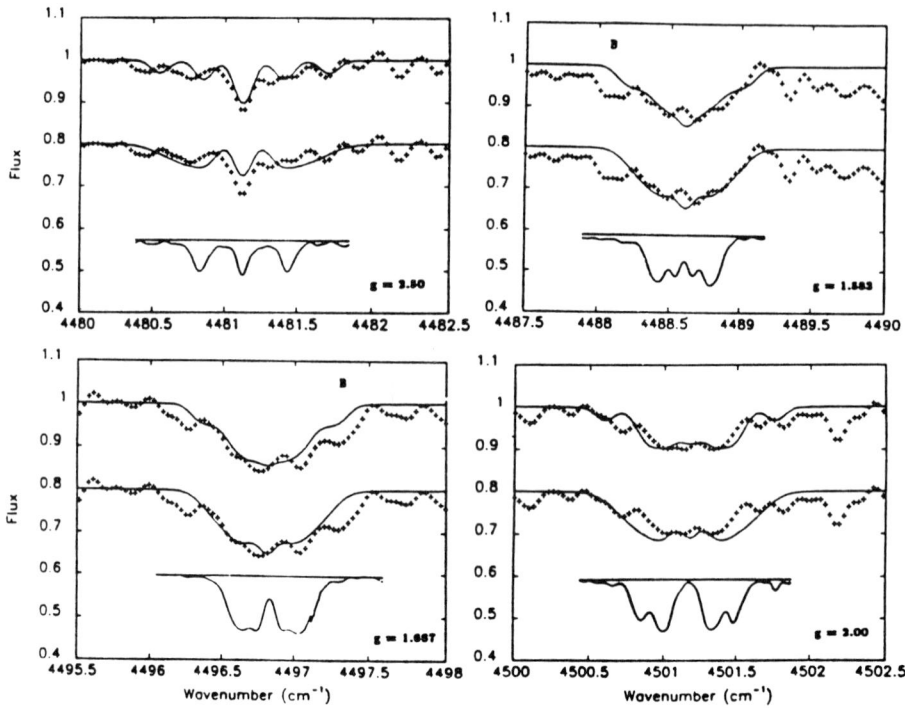

Fig. 4. Apodized, co-added Ti I line profiles from AD Leo (+) and a sunspot/quiet-Sun ratio spectrum (heavy solid; from Saar 1992a). The AD Leo lines have different shapes and appear more "blurred" than the umbral features (see text). Curves through the AD Leo spectra show best fits for two-component distributed field models (top; $f_1 = 0.45$, $B_1 = 2.4$ kG, $f_2 = 0.30$, $B_2 = 5.0$ kG) and an approximate vertical gradient model with $f = 0.45$, $\overline{B} = 3.0$ kG, and $\sigma_B = 1.6$ kG.

product fB is reasonably well determined. I refer the reader to Saar (1992b) for more details.

3. IR Measurements in Context: What Have We Learned?

The IR magnetic field measurements described above, while few in number, have played an important role in shaping the understanding of magnetic fields in cool stars. In this Section I combine these data with new IR $f \cdot B$ data (Saar 1993), some results from Saar et al. (1987) not in the Saar (1990) list, and the best optical determinations (compiled by Saar 1990, 1991); I then use these data to briefly explore the relationships between f, B, and other stellar parameters, and to show the role of IR data in defining them.

With detections of RS CVns at low B and M dwarfs at high B, IR data have been

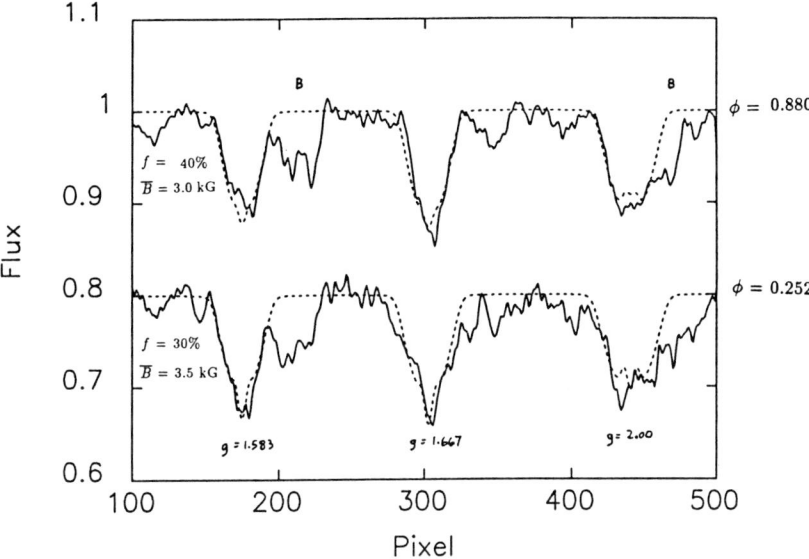

Fig. 5. Unapodized, smoothed profiles (4,488, 4,496, 4,501 cm^{-1}; 0.16 Å/pixel) from AD Leo at two rotational phases, ϕ. The lines have been "spliced together" for convenient display. The profiles appear to have systematically different shapes at the two phases, suggesting different magnetic field structures. Rough fits using an approximate ∇B model (Saar 1992a), yield the indicated $\overline{B}$ and f values.

instrumental in greatly extending the range of magnetic field strengths observed in cool stars. This, in turn, has strengthened the initial indications (Saar and Linsky 1986) that B is limited by photospheric gas pressure. Figure 6 shows the observed B for the combined data set plotted against $B_{eq} = (P_{gas,*}/P_{gas,\odot})^{0.5} \times B_\odot$, where P_{gas} is taken from an atmospheric model at $\tau_{5000} = 1$, and $B_\odot \equiv 1.5$ kG (e.g., Saar 1990). For the most part, $B \leq B_{eq}$, suggesting that the external P_{gas} plays an important role in confining B in stellar photospheres (outside of spots).

Most of the dMe flare stars are rapid rotators, and thus IR data are also important for defining the relation between f and fB and rotation (Saar and Linsky 1986; Marcy and Basri 1989: Saar 1990, 1991). Figure 7 shows the correlation between f and inverse Rossby number, $\tau_C\Omega$ (where τ_C is the convective turnover time, and I take $\Omega = P_{rot}^{-1}$). The IR data are especially helpful for $\tau_C\Omega \gtrsim 0.75$, where f appears to saturate. Below this value, $f \propto (\tau_C\Omega)^\alpha$, where $1.25 \lesssim \alpha \lesssim 2.0$, depending on the weight given the solar point. This relationship is somewhat steeper than determined before (e.g., Saar 1991), largely because here I have taken $f_\odot = 0.01$ (which is probably more realistic; e.g., Solanki 1993). The correlation is consistent with the rotational evolution models of MacGregor and Brenner (1991) and with several activity theories (e.g., Jordan 1991). One can also construct $\tau_C\Omega$ vs. fB, Ω vs. f and fB diagrams, with similar results (Saar 1990; 1991). Thus, the magnetic filling factor, f, is the primary quantity which varies as a star evolves and slowly

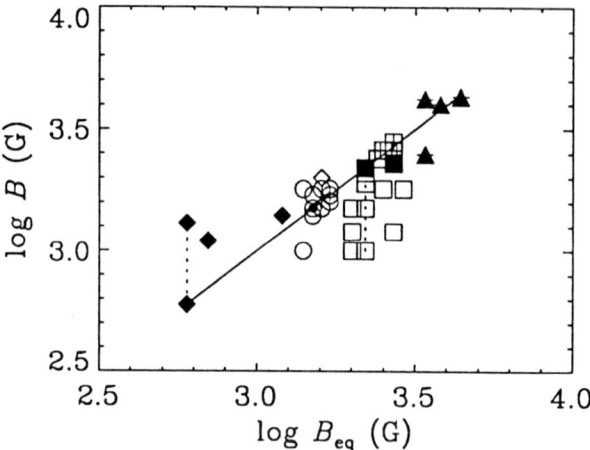

Fig. 6. Measured B vs. B_{eq}: circles, squares, triangles, and diamonds are G, K, M, and low-gravity stars (RS CVn, T Tau), respectively. IR measurements are filled, the Sun is designated by $\odot$ and multiple measurements are connected by dotted lines. Most points lie near or below the $B = B_{eq}$ line.

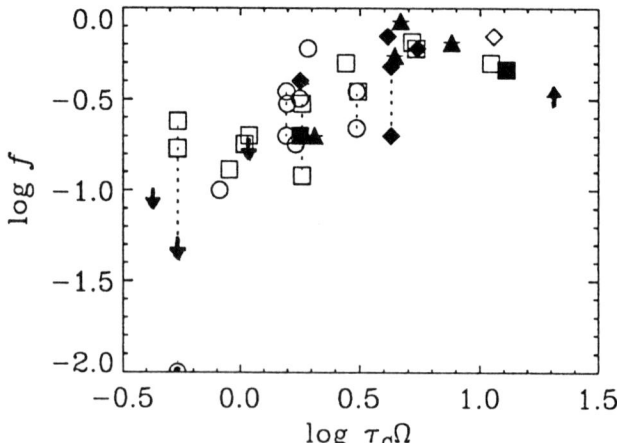

Fig. 7. Measured f vs. $\tau_C \Omega$; symbols are asigned as in Figure 6. A positive relation can be seen for $\tau_C \Omega \lesssim 0.75$, after which f appears to saturate.

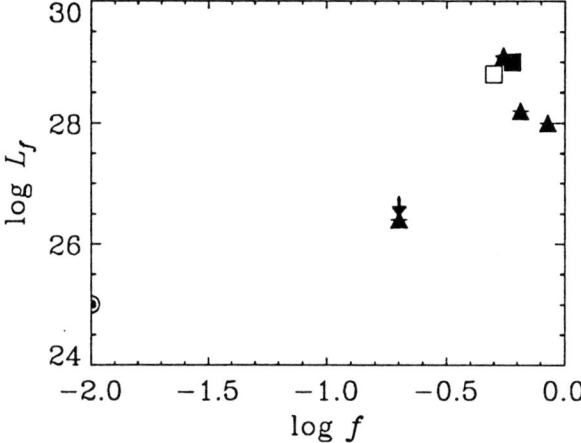

Fig. 8. Time averaged flare luminosity L_f, vs. f (symbols as in Fig. 6). No clear correlation is seen.

loses angular momentum.

One must be a bit careful in the interpretation of the above results, however: Optical fB measurements refer to lines formed at different depths in the stellar atmosphere than the IR features (e.g., Grossmann-Doerth and Solanki 1990). A range of formation heights (and hence B) may in fact account for much of the spread around the B–B_{eq} relation, for example. On the other hand, most of the optical measurements (made near 6,170 Å) are made at nearly the same continuum optical depth as the 2.2 μm Ti I data. More realistic atmospheric and B gradient models (Basri, Marcy, and Valenti 1990; Saar and Solanki 1992) are required to sort out precisely where the lines are formed. Only then can we say accurately which B_{eq} should be compared to a given B value, and what height f refers to (although the product fB will be constant throughout). There is also the question of the origin of the fields. Since the continuum contrast between spot and plage is so high in the optical, fB measurements made there surely derive from plage/network-like regions. In the IR the contrast is much smaller, so IR f and B values may have contributions from stellar umbrae (which may have quite strong fields; Mullan 1984). The low gravity stars (RS CVns – diamonds) tend to lie somewhat above the $B = B_{eq}$ line, perhaps precisely because larger numbers of (higher B) spots on their surfaces tend to increase the measured $\overline{B}$.

There are also good correlations between magnetic flux density fB and traditional signatures of magnetic "activity" in the chromosphere and corona (Saar and Schrijver 1987; Schrijver et al. 1989) and in the transition region (Saar 1988b; Schrijver 1990). The correlations, which also hold for individual active regions on the Sun, are approximately $F_{CHR} \propto (fB)^{0.5}$, $F_{TR} \propto (fB)^{0.75}$, and $F_{COR} \propto (fB)^{1.0}$, where F_{CHR}, F_{TR}, and F_{COR} are the flux densities in the chromosphere, transition region, and corona, respectively (see also Schrijver 1991).

Since all flare stars have large f values, and magnetic reconnection would sensibly increase with f, it seems reasonable to also look for correlations between, for example, integrated flaring energy and f. Figure 8 shows the correlation between L_f, the time averaged flare luminosity (from Doyle, Byrne, and Butler 1986; Shakhovskaya 1989; and Pettersen 1990) and f. There may be a weak trend with $L_f \propto f^2$, but there are, sadly, only a few points and surely no clear relationship.

Many of the ambiguities in the above results should be resolved in the near future by the application of improved models to large amounts of new data from IR echelles and array detectors. IR measurements of stellar magnetic parameters will then become as important as they already are in solar physics, providing crucial information on the true field strengths, filling factors, horizontal and vertical distribution and geometry, thermodynamics, and MHD properties of stellar magnetic regions.

Acknowledgements

This work is based on data obtained at NAOO. I thank G. Basri, M. Giampapa, J. Linsky, G. Marcy, and S. Solanki for enlightening discussions. This research is supported by the NASA grant NAGW-112, Interagency Transfer W-15130, and NSF grant INT-8900202.

References

Basri, G.S., Marcy, G.W., Valenti, J.: 1992, *Astrophys. J.* **390**, 622.
Basri, G.S., Marcy, G.W., Valenti, J.: 1990, *Astrophys. J.* **360**, 650.
Basri, G.S., Marcy, G.W.: 1988, *Astrophys. J.* **330**, 274.
Bookbinder, J.A., Walter, F.W., Brown, A.: 1992, in M. Giampapa and J. Bookbinder (eds.), *Cool Stars, Stellar Systems, and the Sun*, ASP Conf. Ser. 26, p. 27.
Deming, D., Boyle, R.J., Jennings, D.E., Wiedemann, G.: 1988, *Astrophys. J.* **333**, 978.
Doyle, J.G., Byrne, P.B., Butler, C.J.: 1986, *Astron. Astrophys.* **156**, 283.
Giampapa, M.S. 1985, *Astrophys. J.* **299**, 781.
Giampapa, M.S., Worden, S.P.: 1982, in J.O. Stenflo (ed.), *Solar and Stellar Magnetic Fields: Origins and Coronal Effects*, Reidel, Dordrecht, p. 29.
Giampapa, M.S., Golub, L., Worden, S.P.: 1983, *Astrophys. J. (Letters)* **268**, 121.
Gondoin, Ph., Giampapa, M.S., Bookbinder, J.A.: 1985, *Astrophys. J.* **297**, 710.
Gray, D.F.: 1984, *Astrophys. J.* **277**, 640.
Grossmann-Doerth, U., Solanki, S.K.: 1990, *Astron. Astrophys.* **238**, 279.
Hall, D.: 1973, *An Atlas of IR Spectra of the Solar Photosphere and Umbrae*, Kitt Peak National Observatory, Tucson, Arizona.
Hall, D.N.B., Ridgway, S., Bell, E.A., Yarborough, J.M. 1979, *S.P.I.E.*, **172**, 121.
Jordan, C.J.: 1991, in P. Ulmschneider, E.R. Priest and R. Rosner (eds.), *Mechanisms of Chromospheric and Coronal Heating*, Springer, Berlin, p. 300.
Jennings, D.E., Deming, D., Wiedemann, G., Keady, J.J.: 1986, *Astrophys. J. (Letters)* **310**, L39.
Leroy, J.L.: 1962, *Ann. d' Ap.*, **25**, 127.
MacGregor, K.B., Brenner, M.: 1991, *Astrophys. J.* **376**, 204.
Marcy, G.W., Basri, G.S.: 1989, *Astrophys. J.* **345**, 480.
Marcy, G.W., Bruning, D.H.: 1984, *Astrophys. J.* **281**, 286.
Mullan, D.J.: 1974, *Astrophys. J.* **279**, 746.
Mullan, D.J.: 1975, *Astron. Astrophys.* **40**, 41.
Pettersen, B.R.: 1990, in L.V. Miroyan, B. R. Pettersen, and M. K. Tsvetkov (eds.), *Flare Stars in Star Clusters, Associations, and the Solar Vicinity*, Kluwer, Dordrecht, p. 49.
Robinson, R.D.: 1980, *Astrophys. J.* **239**, 961.
Robinson, R.D., Worden, S.P., Harvey, J.W.: 1980, *Astrophys. J. (Letters)* **236**, L155.

Saar, S.H.: 1992a, in M. Giampapa and J. Bookbinder (eds.), *Cool Stars, Stellar Systems, and the Sun*, ASP Conf. Ser. 26, p. 252.
Saar, S.H.: 1993, these proceedings.
Saar, S.H.: 1991, in I. Tuominen, D. Moss, and G. Rüdiger (eds.), *The Sun and Cool Stars: Activity, Magnetism, Dynamos*, Springer, Berlin, p. 389.
Saar, S.H.: 1990, in J.O. Stenflo (ed.), *The Solar Photosphere: Structure, Convection, and Magnetic Fields*, Kluwer, Dordrecht, p. 427.
Saar, S.H.: 1988a, *Astrophys. J.* **324**, 441.
Saar, S.H.: 1988b, in R. Pallavicini (ed.), *Hot Thin Plasmas in Astrophysics*, Kluwer, Dordrecht, p. 139.
Saar, S.H., Bookbinder, J.A., Neff, J., Bromage, G., Bastian, T.: 1991, *Bull. Amer. Astron. Soc.* **23**, 1383.
Saar, S.H., Golub, L., Bopp, B.W., Herbst, W., Huovelin, J.: 1990, in E. Rolfe (ed.), *Evolution in Astrophysics*, ESA SP-130, p. 431.
Saar, S.H., Linsky, J.L., Giampapa, M.S.: 1987, in L. Delbouille and A. Monfils (eds.), *27th Liège Astrophys. Colloq.*, U. de Liège, Liège, p. 103.
Saar, S.H., Linsky, J.L.: 1985, *Astrophys. J. (Letters)* **299**, L47.
Saar, S.H., Linsky, J.L.: 1986, *Advances in Space Physics*, 6, No. 8, 235.
Saar, S.H., Schrijver, C.J.: 1987, in J. Linsky and R.E. Stencel (eds.), *Cool Stars, Stellar Systems, and the Sun*, Springer, New York, p. 38.
Saar, S.H., Solanki, S.K.: 1992, in M. Giampapa and J. Bookbinder (eds.), *Cool Stars, Stellar Systems, and the Sun*, ASP Conf. Ser. 26, p. 259.
Schrijver, C.J.: 1990, *Astron. Astrophys.* **234**, 315.
Schrijver, C.J.: 1991, in P. Ulmschneider, E. R. Priest and R. Rosner (eds.), *Mechanisms of Chromospheric and Coronal Heating*, Springer, Berlin, p. 257.
Schrijver, C.J., Coté, J., Zwaan, C., Saar, S.H.: 1989, *Astrophys. J.* **337**, 964.
Shakhovskaya, N.I.: 1989, *Solar Phys.* **121**, 375.
Solanki, S.K.: 1993, these proceedings.
Solanki, S.K., Biémont, E., Mürset, U.: 1990, *Astron. Astrophys. Suppl.* **83**, 307.
Solanki, S.K., Steiner, O.: 1990, *Astron. Astrophys.* **234**, 519.
Strassmeier, K.G., Hall, D. S., Zeilik, M., Nelson, E., Eker, Z., and Fekel, F. C.: 1988, *Astron. Astrophys. Suppl.* **72**, 291.
Sun, W.-H., Giampapa, M.S., Worden, S.P.: 1987 *Astrophys. J.* **312**, 930.
Unno, W.: 1956, *Pub. Astron. Soc. Japan* **8**, 108.

NEAR INFRARED IMAGING MAGNETOMETRY

DOUGLAS RABIN

National Solar Observatory, National Optical Astronomy Observatories,[*]
P. O. Box 26732, Tucson, Arizona 85726, U.S.A.

Abstract. Infrared array detectors are a new and promising tool for investigating the properties of magnetic field concentrations in the solar photosphere. Array measurements provide large statistical samples of polarized line profiles and display the spatial organization of the magnetic field. The wavelength region near 1.6 μm has important advantages for magnetometry: spectral lines with magnetic sensitivites ranging from low to very high; low continuum opacity; high continuum flux; and the possibility of sub-arcsecond angular resolution with existing telescopes. Initial results have extended earlier work on the distribution of field strength and flux in plages and revealed new properties specifically connected with spatial structure. The quality and flexibility of near infrared magnetographs can be expected to improve rapidly.

Key words: flux tubes – infrared: stars – instrumentation: polarimeters – Sun: faculae, plages – Sun: magnetic fields

1. Rationale

The 1–2.5 μm spectral region offers scientific and technical advantages for the measurement of photospheric magnetic fields. As discussed more fully elsewhere in this volume (Solanki, 1993), the ratio of Zeeman line splitting to non-magnetic line width varies approximately as $g\lambda$, where the Landé factor can attain $g = 3$ both in the visible and the near infrared. Also, lines in the continuum opacity window near 1.6 μm are formed deep in the photosphere – where the magnetic field is strong – and usually close to LTE. Contamination by stray light is smaller than at visible wavelengths, both because the continuum intensity is less sensitive to temperature and because there is typically less instrumental and atmospheric scattering; the relative freedom from stray light is particularly important in sunspots.

Here I shall address the specific rationale for *imaging* magnetometry in the near infrared. From the most general perspective, images with angular resolution $\sim 1''$ are in many ways the common language of observational solar physics. An image of this kind can be easily registered and compared to a wide variety of other images which measure various quantities and are often obtained on a routine basis. By contrast, a single-point measurement, particularly if it has crude angular resolution, raises questions of location and context. To which patch of plage, for example, does the measurement refer? How different would the measurement be if it were taken 5" away? How many measurements, distributed over what area, are needed to obtain a fair statistical sample? Imaging measurements may not eliminate these questions, but they make them easier to address. One may speculate that the difficulty of obtaining images with available technology was largely responsible for the hiatus in solar observations following the pioneering work of Harvey and Hall (Harvey and Hall, 1975; Harvey, 1977), although single-detector instruments could be used to begin the stellar measurements reviewed elsewhere in this volume (Saar, 1993). Stenflo, Solanki, and Harvey (1987) were able to make important

---

[*] Operated by the Association of Universities for Research in Astronomy, Inc., under cooperative agreement with the National Science Foundation.

progress in the solar case, despite the absence of imaging capability, by combining the extreme spectral resolution and accuracy of a Fourier transform spectrometer with full radiative transfer analysis.

If we grant the desirability of obtaining magnetic images, a strong case can be made for observing Fe I 6388.6 cm^{-1} (1.5649 μm). It has high magnetic sensitivity ($g = 3$) and is conveniently (for an array detector) near a line with a different Zeeman sensitivity (Fe I 6386.9 cm^{-1}, $g_{\text{eff}} = 1.53$). It is instructive to compare from a technical perspective the 1.56 μm lines with the emission line Mg I 12.32 μm ($g = 1$) and favorable visible-wavelength lines such as Fe I 5250.2 Å ($g = 3$) and 5247.1 Å; Solanki (1993, in these proceedings) makes a similar comparison from the viewpoint of line formation and profile fitting.

Angular Resolution. Magnetic flux tubes in plage and network areas are sub-arcsecond structures, with characteristic diameters of perhaps 200 km (Steiner, 1993). Resolving these structures will almost certainly require observations from space or adaptive optics on the ground. A lesser but still important goal is to study the properties of single flux tubes even when they are not resolved. For spaceborne observations, visible or ultraviolet wavelengths have an advantage because they can achieve resolutions well below 1″ with telescopes of aperture $\lesssim$ 1 m. On the ground, observations at 12 μm are diffraction-limited to 2″ resolution on the largest available (1.5-m) telescope, although, for the purpose of isolating flux tube properties, the great *magnetic* resolution of the 12-μm lines is a compensating advantage (Deming et al., 1988). Visible-light observations should be able to achieve $\sim$ 0″.2 resolution on a 75-cm telescope using adaptive optics, although the technical problems are formidable, especially for polarimetry. Observations at 1.6 μm offer an attractive middle ground: the diffraction limit is below 0″.3 on a 1.5-m telescope, and the adaptive optics needed to approach that limit are expected to be considerably less elaborate than at visible wavelengths (Roddier and Graves, 1993).

Photon Flux. Ground-based polarimetry is best done by rapidly modulating between complementary polarization states (*e.g.*, Stokes $I \pm V$) in order to minimize the signal-to-noise degradation due to seeing. Extracting information on the intrinsic magnetic field strength and direction – rather than just magnetic flux – calls for high spectral resolution, $\lambda/\Delta\lambda \gtrsim 10^5$. These two objectives, combined with the impetus toward high angular resolution, mean that photon flux is an important consideration for imaging magnetometry. The photon flux per linewidth (and per square arcsecond) is actually slightly higher for Fe I 1.565 μm than for Fe I 5250 Å; but the flux for Mg I 12.32 μm is lower by more than an order of magnitude.

Polarized Signal. The greatest barrier to vector magnetometry in the visible is the low levels of linear polarization (10^{-3}–10^{-4}) produced by incomplete Zeeman splitting. At 1.6 μm, and even more so at 12 μm, the prevalence of strong splitting means that Stokes Q and U are comparable in strength to Stokes V. On the other hand, infrared lines are generally weaker, reducing the amplitude of all Stokes components; and, *because* all the components are comparably strong, linear-to-circular crosstalk cannot be ignored as it often can be in the visible (although, at 12 μm, it may be possible to ignore instrumental polarization altogether).

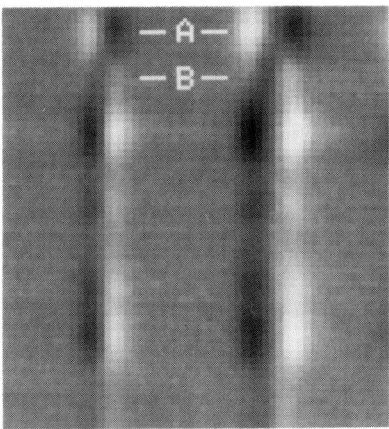

Fig. 1. Spatial–spectral Stokes V image in a plage region. The spectral lines are Fe I 6388.6 cm^{-1} (*right*) and Fe I 6386.9 cm^{-1} (*left*). The spatial extent (*top to bottom*) is 53″. The features marked A and B are discussed in the text.

Detectors. The standard detector for 0.3–1 μm is the CCD, which has high quantum efficiency and is readily available in 2048 × 2048 format. Several infrared array detectors that operate at 1.56 μm (HgCdTe, InSb, PtSi) are commercially available in formats ranging from 128 × 128 to 512 × 512. Recently, their formats have grown faster than CCD formats. A 512 × 512 format is already large enough to image an active region with good angular resolution. HgCdTe and InSb devices also have high quantum efficiency. At present, arrays for the 5–20 μm range are usually smaller in format and more expensive; but they too will enter a period of rapid growth. Many infrared arrays are designed to run at video or higher rates, which matches the need for rapid polarization modulation.

2. Initial Results

2.1. Spectra

Figure 1 is an example of a Stokes V spatial–spectral image at 1.565 μm, obtained in a plage region using the NSO Near Infrared Magnetograph – NIM (Rabin, 1992b). In the vicinity of the inversion line of the line-of-sight field (located between the features marked A and B), there are obvious changes in the splitting and central wavelength of the σ components. Elsewhere in these proceedings, Solanki (1993) discusses the interpretation of these V profiles in terms of a siphon flow in the flux-tube arches near the inversion line. However, simple Seares-profile fitting detects significant variations in field strength throughout this image and in other plage areas. Figure 2 illustrate fits to V profiles and the formal uncertainties in the

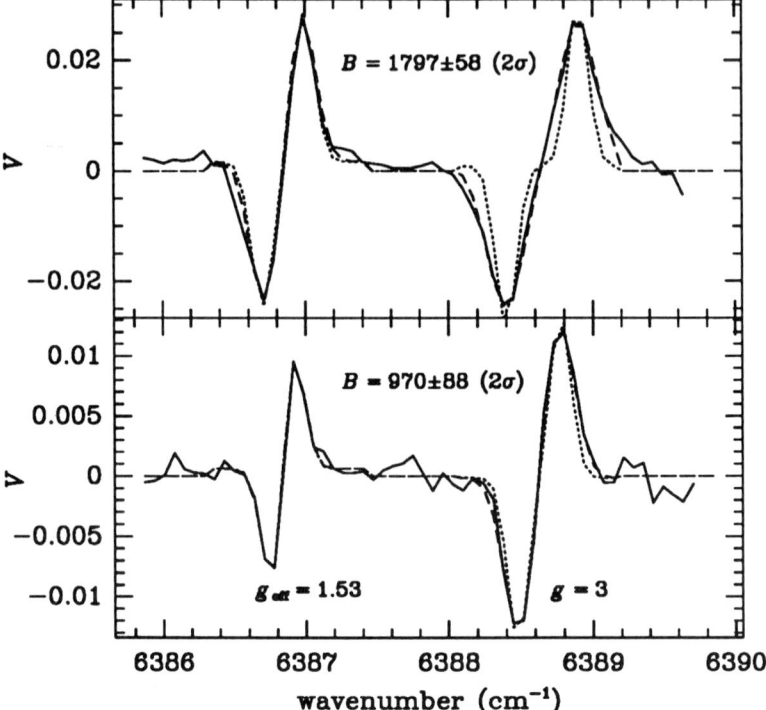

Fig. 2. Examples of observed and fitted V profiles. *Solid*: observed. *Dashed*: fitted with the width of the sigma components as a free parameter. *Dotted*: fitted with the width of the sigma components fixed at the quiet-Sun line width.

determination of field strength. Figure 2 also illustrates a pervasive feature of the plage profiles: the σ components are broader than what is obtained by simply splitting a quiet-Sun line profile. Zayer *et al.* (1989) demonstrated that the σ-broadening is magnetic in origin and that the degree of broadening is consistent with the vertical gradient of flux-tube field strength over the region of line formation.

2.2. MAGNETIC MAP

Figure 3 shows cospatial maps of magnetic flux and magnetic field strength in a plage region, derived from infrared spectra such as those shown in Figure 2. The bottom panel is a "true-field" magnetogram. The field-strength map shows that there are spatially coherent variations in $|\mathbf{B}|$ (at the level of line formation) throughout this plage region (Rabin, 1992a). Field strength and flux are related, but not one-to-one. For example, Region A includes the most intense flux in the field of view, but Regions B and C contain stronger magnetic fields, and patches of relatively strong field can occur in areas of weak flux (Region B). This relationship is explored further below.

Figure 3 can only indicate some of the potential of near-infrared magnetic map-

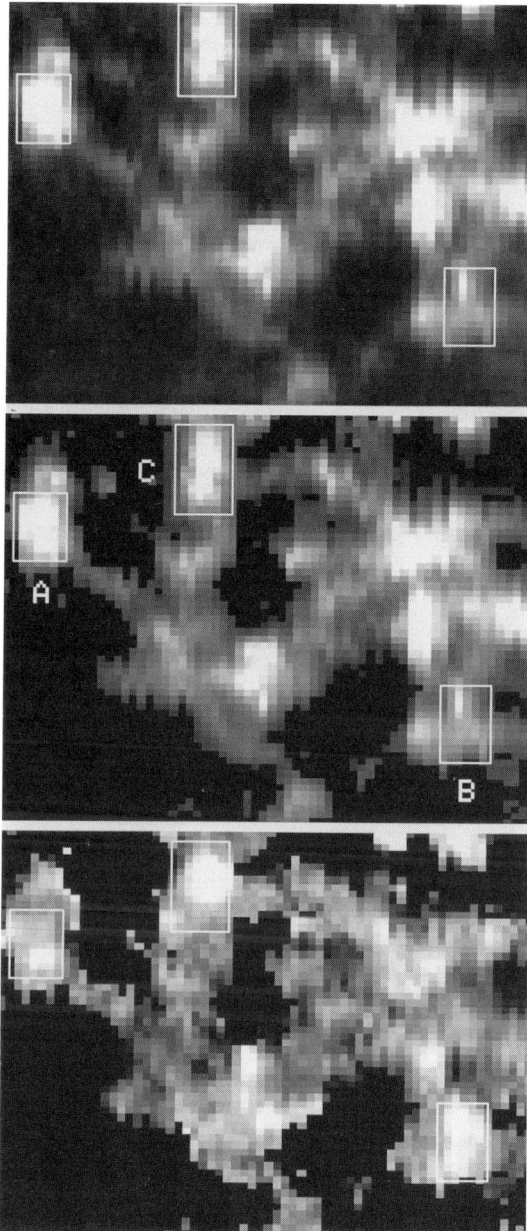

Fig. 3. Magnetic flux and magnetic field strength in a plage region. *Top*: model-independent magnetic flux, proportional to $\int |V| d\lambda / I_c$, where I_c is the unpolarized continuum intensity. *Middle*: magnetic flux as measured by the amplitude of the fitted profile. The grayscale (black-to-white) is linearly mapped to the range 0.00–0.03. *Bottom*: true magnetic field strength. The grayscale is linearly mapped to the range 1000–1600 Gauss. In the bottom two maps, pixels are blacked out where the profile did not satisfy the convergence and signal-to-noise criteria. The field of view is $74'' \times 55''$.

ping, which is in its infancy. Efforts are being directed toward creating vector-field maps from complete Stokes data. The Doppler information will be more fully utilized (McPherson, Lin, and Kuhn, 1992, give an example of a Doppler map around a sunspot derived from 1.565 μm data). It should be possible to create spatial images of σ-broadening and, in conjunction with visible-light diagnostics, of the thermal properties of flux tubes. In sunspots, the relationship between magnetic and thermal properties can be explored from infrared data alone (Kopp and Rabin, 1993, in these proceedings).

2.3. Statistical Properties

Even though the angular field of Figure 3 is small, it provides a much larger statistical sample than can be obtained practicably with point-by-point measurements. Figure 4 shows three histograms derived from this field of view: field strength B, amplitude factor A_f, and $\Delta\nu_E$, the Gaussian line width of the fitted σ components. A_f is the fitted amplitude of the V profile normalized by the maximum value it could achieve if the spectral line had its quiet-Sun central depth. To a first approximation, $A_f = fR\delta\cos\gamma$, where f (filling factor) is the fraction of the spatial resolution element occupied by the flux tube, γ is the inclination of the magnetic field to the line of sight, δ is the continuum contrast of the tube, and R is the ratio of the absorption-line depths inside and outside the tube after the profiles are separately normalized to unit continuum intensity. Because both the lines and the continuum are relatively insensitive to temperature, variations in δ and R are not expected to contribute much to the variation of A_f.

The width of the field-strength histogram is consistent with the range of plage field strengths measured by Harvey (1977). In future observations, it will be important to determine whether the distribution varies from region to region, and to explore the spatial environments of strong- and weak-field plages.

Figure 5 is a statistical representation of the relationship between field strength and flux (or V amplitude) illustrated by the labelled boxes in Figure 3. The most striking feature is the empty region in the upper left of the diagram. The origin of this property is not known. One may speculate that flux tubes in highly filled areas are relatively cool, because the efficiency of convective transport is locally reduced, and therefore relatively transparent, because the H$^-$ opacity is lower. The line is formed deep in the atmosphere, leading to high measured field strengths. In sparsely filled areas, the individual tubes may be collected into cool clumps (again with high measured field strengths), or instead spread out more uniformly. Quasi-isolated tubes would not suppress convective transport as effectively and would experience more radiative exchange with the surrounding atmosphere, resulting in higher temperature and pressure, increased opacity, and lower measured field strengths. Solanki suggests other possible interpretations based on a small sample of similar data (Solanki, 1993, Fig. 4).

3. Prospects

So far, near infrared imaging magnetometry has only scratched the surface of the scientific possibilities open to it. The following is a heterogeneous sample of expected

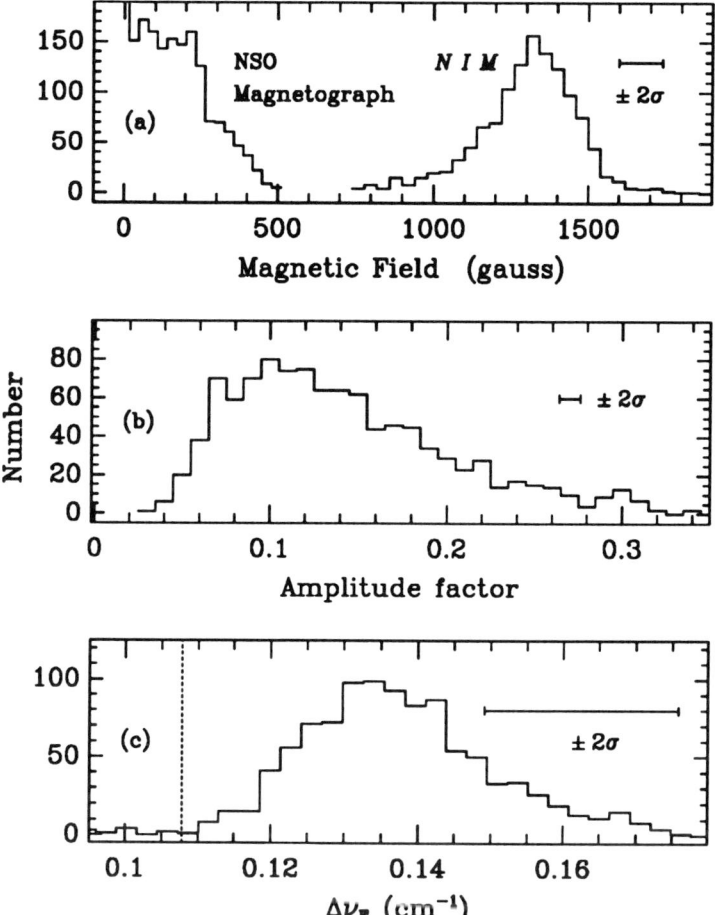

Fig. 4. Histograms of (a) true field strength, measured by NIM, compared to apparent field strength, measured by the NSO synoptic magnetograph; (b) A_f, the normalized amplitude of the V profile; (c) width of the sigma components. The *dotted line* in Panel c is the width of the line in the quiet Sun. All error bars indicate $\pm 2\sigma$ uncertainties in the *individual* measurements that contribute to the bins.

observational advances and what might be learned from them.

1. Spectral lines in the visible are often more temperature-sensitive than near-infrared atomic lines; the 1.56 μm lines offer a more sensitive determination of field strength at the level they are formed. Thus, simultaneous observations at 1.56 μm and in the visible should be a powerful probe of the flux-tube atmosphere.

2. Now that the formation of the 12 μm lines is well understood (Rutten and Carlsson, 1993; Avrett, Chang, and Loeser, 1993), simultaneous observations at

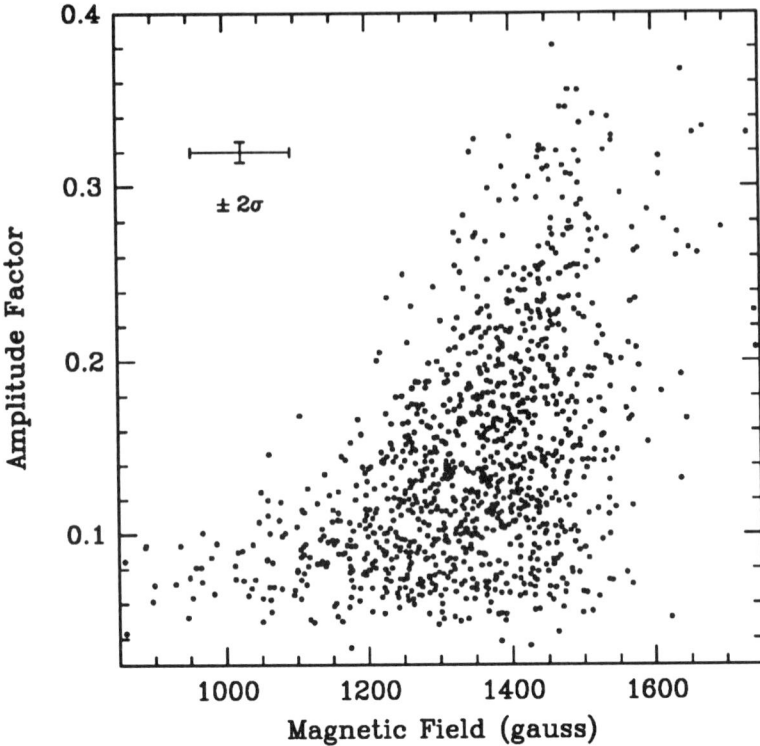

Fig. 5. Scatter plot of field strength versus A_f, the Stokes V amplitude factor. A typical $\pm 2\sigma$ uncertainty of fit along each axis is shown.

1.6 μm and 12 μm should allow an unambiguous measurement of field strength differences (flux-tube spreading) over the vertical span of the photosphere.

3. Complete Stokes polarimetry – *i.e.*, vector magnetometry – is, in the infrared, a relatively straightforward analogue of the circular polarimetry discussed above. Comparing 1.6 μm observations with visible-light data from, *e.g.*, the HAO-NSO Advanced Stokes Polarimeter (Lites *et al.*, 1991) will be particularly valuable in the periphery of sunspots, where the field geometry changes rapidly with height.

4. Buoyancy tends to drive an isolated flux tube to a vertical orientation in the atmosphere (Spruit, 1981). However, flux tubes are always, to some degree, part of a larger magnetic-field configuration. Near an inversion line, for example, we should find substantial inclinations. Because of the comparable magnitude of all the Stokes components, 1.56 μm observations in plage regions should provide relatively direct information about the inclination of flux tubes.

5. As mentioned earlier, Doppler information from infrared lines has barely been exploited. In sunspots, the infrared data will probe flow conditions deep in the photosphere. In plage flux tubes, the occurrence of siphon flows needs to be

confirmed and buttressed with time series, since such flows must be transient.

6. Bigger and better array detectors will help in many areas. It will no longer be necessary to compromise spectral resolution in order to include a usable wavelength range on the array. Lower dark current and readout noise will increase the polarization sensitivity, eventually to the point where we can study the strength of quiet network and internetwork fields.

7. NIM employs a nematic liquid-crystal optic as a voltage-controlled variable retarder for measuring the Stokes vectors. The present device requires as much as 20 ms to change state and stabilize, during which time a full data frame is contaminated and therefore discarded. A faster cadence is desirable to supress the noise introduced by rapid seeing variations. It is likely that both ferroelectric and nematic liquid-crystal retarders will soon be capable of good performance in the infrared at video rates.

8. NIM scans the slit of a large spectrograph across the solar image. Just as there are spectrograph- and filter-based magnetographs in the visible, each with distinct advantages, it makes sense to develop a filter-based infrared magnetograph. It will have a faster cadence, suitable for studies related to solar activity, and it will be adaptable to balloon- or spaceborne telescopes.

Acknowledgements

I am grateful to the NIM team – D. Jaksha, G. Kopp, C. Mahaffey, C. Plymate, and J. Wagner – for helping to make the concept a reality. This work was assisted by the NASA Supporting Research and Technology program in solar physics.

References

Avrett, E. H., Chang, E. S., and Loeser, R.: 1993, these proceedings.
Deming, D., Boyle, R.J., Jennings, D.E., and Wiedemann, G.: 1988, *Astrophys. J.* **333**, 978.
Harvey, J. W.: 1977, in E. A. Müller (ed.), *Highlights of Astronomy* **4**, 223.
Harvey, J. W., and Hall, D.: 1975, *Bull. Amer. Astron. Soc.* **7**, 459.
Kopp, G., and Rabin, D.: 1993, these proceedings.
Lites, B. W., Elmore, D., Murphy, G., Skumanich, A., Tomczyk, S., and Dunn, R. B.: 1991, in L. November (ed.), *Solar Polarimetry*, Proc. 11th Sacramento Peak Workshop, NSO, Sunspot, New Mexico, p. 3.
McPherson, M. R., Lin, H., and Kuhn, J. R.: 1992, *Solar Phys.* **139**, 255.
Rabin, D.: 1992a, *Astrophys. J. (Letters)* **390**, L103.
Rabin, D.: 1992b, *Astrophys. J.* **391**, 832.
Roddier, F., and Graves, J. E.: 1993, these proceedings.
Rüedi, I., Solanki, S. K., and Rabin, D. 1992, *Astron. Astrophys.* **261**, L21.
Rutten, R. J., and Carlsson, M.: 1993, these proceedings.
Saar, S.: 1993, these proceedings.
Solanki, S. K.: 1993, these proceedings.
Spruit, H.: 1981, in S. Jordan (ed.), *The Sun as a Star*, NASA SP-450, NASA, Washington, D.C., p. 385.
Steiner, O.: 1993, these proceedings.
Stenflo, J. O., Solanki, S. K., and Harvey, J. W.: 1987, *Astron. Astrophys.* **173**, 167.
Zayer, I., Solanki, S. K., and Stenflo, J. O.: 1989, *Astron. Astrophys.* **211**, 463.

FLUX TUBE SHREDDING AND ITS INFRARED SIGNATURE

M. BÜNTE

Institute of Astronomy, ETH Zentrum, CH-8092 Zürich, Switzerland

O. STEINER

Kiepenheuer-Institut für Sonnenphysik, Schöneckstrasse 6, D-7800 Freiburg, FRG

S.K. SOLANKI

Institute of Astronomy, ETH Zentrum, CH-8092 Zürich, Switzerland

and

V.J. PIZZO

San Juan Capistrano Research Institute, San Juan Capistrano, CA, U.S.A.

Abstract. The interchange instability of solar magnetic flux tubes and possible stabilization mechanisms are reviewed. Special attention is paid to the influence of magnetic tension forces and the internal atmosphere, both of which were neglected in earlier studies of this instability. It is found that whirl flows with velocities of only 2.2 km s^{-1} are strong enough to stabilize the flux tubes. However, their absence or the excitation of other instabilities might lead to a shredding of the tubes. The observability of such a scenario in the infrared is briefly discussed.

Key words: infrared: stars – MHD – Sun: magnetic fields

1. Introduction

Vertical magnetic flux tubes fan out with height due to the decreasing gas pressure of the stratified solar atmosphere. Concave magnetic configurations, however, are generally unstable to the interchange instability and should therefore easily break apart into smaller flux tubes, *i.e.*, become shredded. In contrast, the observed lifetimes of sunspots (weeks to months) and small magnetic elements (18 minutes or more) exceed the growth time of the instability by far ($\sim$ 1 h for sunspots, $\sim$ 20 s for small tubes), indicating that there exists a quasi-static phase in the life of magnetic flux tubes on the Sun. Meyer *et al.* (1977) found that tubes are stabilized by buoyancy forces if their magnetic fluxes exceed the *critical limit* of $10^{19} - 10^{20}$ Mx. Schüssler (1984) suggested that *smaller tubes* could be stabilized by *whirl flows*, which are expected to arise from the "bathtub effect" in granular downdrafts. However, for a certain range of fluxes ($5 \times 10^{17} - 10^{19}$ Mx) the whirl velocities found in Schüssler's analysis are unrealistically high (*i.e.*, well above 2 km s^{-1}) which led him to suggest that tubes with radii of 250–1,100 km (at $\tau_{5000} = 1, B \approx 1,600$ G) do not exist. We have extended the stability analysis to numerical models of solar magnetic flux tubes that include all tension forces and allow for a non-vanishing internal atmosphere.

2. The Stability Criterion

Consider a magnetic flux tube as depicted in Figure 1. In magnetohydrostatic equilibrium we have at every point of the surface S

$$p_i + \frac{B^2}{8\pi} = p_e ,$$

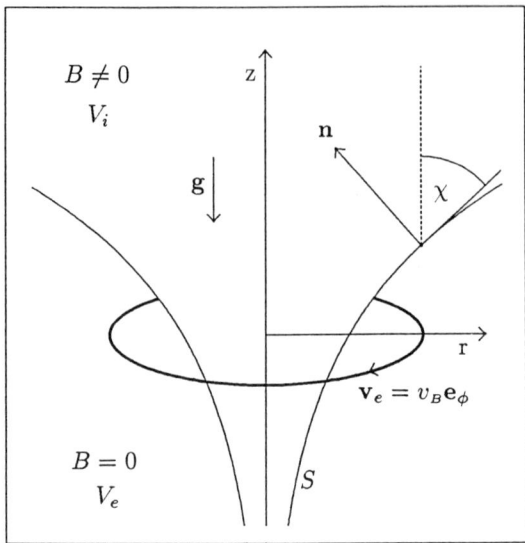

Fig. 1. Geometry of the models considered in this study. S is the surface of the tube, V_i is the ("internal") magnetic gas volume, V_e the ("external") non-magnetic volume in rotation about the tube axis (z-axis) at velocity v_B. Gravitational acceleration, g, the normal, n, and angle, χ, of the boundary to the vertical are indicated.

where p_e and p_i are external and internal gas pressures and B is the magnetic field strength at the tube surface. The system is stable to small surface displacements if at any point along S (Bernstein et al. 1958; Frieman and Rotenberg 1960)

$$\mathbf{n} \cdot \left[\nabla \left(p_i + \frac{B^2}{8\pi} \right) - \nabla p_e \right] > 0 \qquad (1)$$

where n is the unit normal pointing into the tube (see Fig. 1). We apply this criterion to a cylindrically symmetric, untwisted flux tube embedded in a vertically stratified atmosphere and surrounded by a purely azimuthal flow $\mathbf{v} = v_B \mathbf{e}_\phi$ (the "whirl flow"). Using the respective momentum equations we substitute the gradients in the stability criterion (1) and obtain as a condition for stability (Bünte et al. 1992)

$$-\frac{1}{4\pi} B_z \frac{dB_r}{dz}\bigg|_S + \rho_e v_B^2 \frac{1}{R} > 0 , \qquad (2)$$

where B_r and B_z are the magnetic field components at the tube boundary, ρ_e is the external density and R is the radius of the tube. The index S indicates that the derivative is evaluated along the surface S. From this criterion it follows that: (1) the system is susceptible to interchanges if the radial component of the magnetic field *increases* with height along the boundary, and (2) the surface can be stabilized by an *external* whirl flow surrounding the tube. In a given atmosphere (i.e., prescribed interior and exterior atmospheres) the only free parameter determining the structure of a flux tube is the total magnetic flux Φ. In the following we compute a magnetic flux tube and determine the v_B value which marginally stabilizes

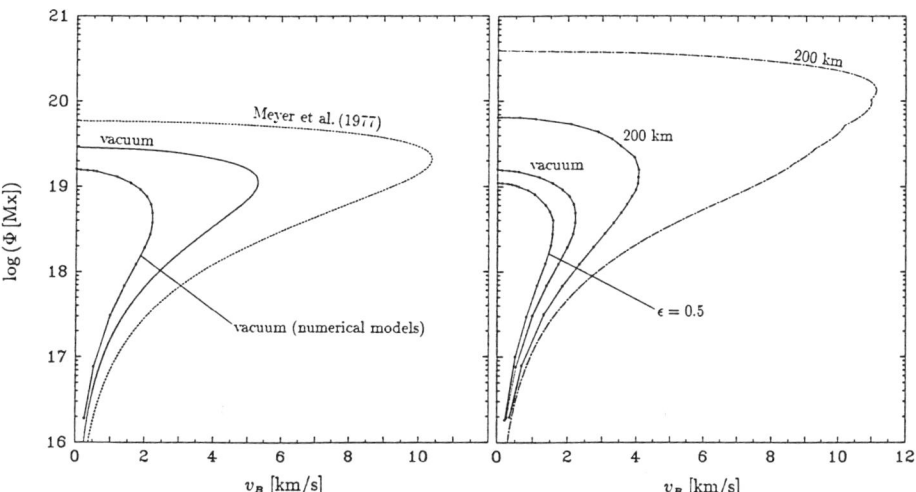

Fig. 2. Comparison of numerical models with thin tubes. Left: stability curves for evacuated tubes. –•–: numerical model tubes, embedded in a Spruit/VAL atmosphere; ——: corresponding thin tube result; - - -: thin tube result for the standard atmospheric model of Meyer et al. (1977) (see section 3.1). Right: stability curves for flux tubes with internal atmospheres. –•–: partially evacuated, "isothermal" tubes with $\epsilon = p_i/p_e = 0.5$, completely evacuated tubes, and tubes whose internal atmosphere corresponds to the external atmosphere shifted downwards by 200 km; – · –: result for thin tubes in the latter case (see section 3.2).

the tube. This corresponds to one point (v_B, Φ) in a flux vs. whirl velocity diagram. By repeated application to tubes of various fluxes we thus obtain a *"stability curve"* which separates stable and unstable regimes. The whirl velocity can serve as a "measure" of the instability in the sense that flows of moderate velocities, up to 2–3 km s^{-1} in the case of the Sun, appear likely to occur (Nordlund 1985), while much higher values indicate "severe" cases that are unlikely to be stabilized by this mechanism. A twist of the field would help to stabilize the tube. However, the twists needed for the stabilization of the interchange instability are in general so large that a quasistatic equilibrium is no longer possible (Bünte et al. 1992). Therefore, in the following we will concentrate on untwisted tubes.

3. Results of the Stability Analysis

We have calculated magnetic flux tubes embedded in a Spruit-VAL atmosphere (Spruit 1977) using the code of Pizzo (1990). First we have concentrated on evacuated tubes to isolate the influence of the magnetic tension forces in comparison with the thin tube results. In a second step we have considered two different cases: (1) the internal pressure is a constant fraction, ϵ, of the external one at equal geometric height, referred to as the "isothermal" case, and (2) the internal atmosphere is identical to the external one, but shifted downwards by 200 km.

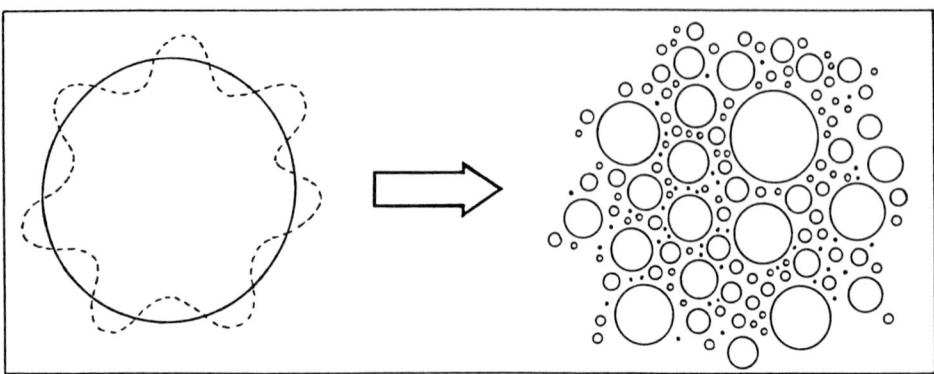

Fig. 3. Sketch of the shredding scenario: The undisturbed (——) cross section of a magnetic flux tube is deformed (- - -) due to fluting (left). If the system is interchange unstable the dashed contour is further distorted, leading to a fragmentation of the original tube (right). The smaller a fragment is the more efficiently it is heated by the radiation from the surrounding atmosphere.

3.1. The influence of magnetic tension forces

The results are summarized in the left diagram of Figure 2. The stabilization of thin tubes requires much higher whirl velocities than those of numerical model tubes including all tension forces. If the standard atmospheric model of Meyer et al. (1977) is used, the necessary velocities are even larger. Their atmosphere has a somewhat steeper temperature gradient in the critical layer at $\tau_{5000} = 1$. This demonstrates the great sensitivity of the stability criterion to changes in the atmospheric parameters.

3.2. The influence of an internal atmosphere

The results are summarized in the right diagram of Figure 2. An internal atmosphere can have either a stabilizing effect (in the isothermal case) or a destabilizing effect (in the shifted case) as compared to the vacuum case. For the latter case the thin tube result is also plotted (dot-dashed curve). Again, thin-tubes require much higher whirl velocities for the stabilization of the interchange instability than "real" tubes.

4. Discussion

We have shown that realistic (numerical) models of solar magnetic flux tubes are much easier to stabilize against fluting than thin-tube structures which do not take magnetic tension forces into account. A possible stabilization mechanism is whirl flows. Their velocities lie well within the range of values (2–3 km s^{-1}), reported from numerical simulations of solar granulation (Nordlund 1985), and are up to 60% lower when magnetic tension is included than for the thin-tube case. An internal atmosphere can either have a stabilizing or destabilizing effect depending upon the details of the assumed gas pressure stratification. Nevertheless, in the following cases the interchange instability might set in and lead to the shredding of flux tubes – (1) if strong enough whirl flows are absent, or (2) if other flute instabilities are excited, *e.g.*, the MHD Kelvin-Helmholtz instability driven by relative motion.

As shown in Figure 3, fluting-induced shredding of flux tubes produces ever smaller magnetic features. The smaller these features become the more effectively they are heated by radiation from the surrounding photosphere. The measurable field strength should then decrease because of two effects: (1) The increase in temperature shifts the $\tau_c = 1$ surface to higher atmospheric levels, *i.e.*, to lower field strengths (Steiner and Pizzo 1989), and (2) the heating increases the internal gas pressure and leads to an expansion of the tube. The latter effect could be denoted as the "inverse process" of the convective collapse. In fact, radiative effects have been found to reduce the efficiency of the convective collapse in the concentration of magnetic fields (Venkatakrishnan 1986).

Consequently, the onset of the instability leading to a rearrangement of magnetic flux may be tested by *measuring the field strength as a function of time*. Since very exact measurements of the field strength are required, they are best carried out using the $g = 3$ line at 1.5648 μm. Even better constraints can be set by combining time series of Stokes V observations of this line with Stokes V observations of a temperature diagnostic in the visible (*e.g.*, the well-studied set of lines around 5,250 Å). The expected signature is a drop in the field strength by a few hundred G over a time-scale of a few minutes coupled to a warming of the magnetic atmosphere. Also, if a compact flux tube with a diameter of a few hundred km is broken up into a number of fragments having a range of sizes and, consequently field strengths, then the σ-components of the V profile of the 1.5648 μm line should become anomalously broadened.

Acknowledgements

The work of M.B. was supported by grant No. 20-31'289.91 of the Swiss National Science Foundation, which is gratefully acknowledged.

References

Bernstein, I. B., Frieman, E. A., Kruskal, M. D., Kulsrud, R. M.: 1958, *Proc. R. Soc. London A* **244**, 17.
Bünte, M., Steiner, O., Pizzo, V.J.: 1992, *Astron. Astrophys.* (submitted).
Frieman, E. A., Rotenberg, M.: 1960, *Rev. Mod. Phys.* **32**, 898.
Meyer, F., Schmidt, H.U., Weiss, N.O.: 1977, *Mon. Not. R. Astron. Soc.* **179**, 741.
Nordlund, Å.: 1985, *Solar Phys.* **100**, 209.
Pizzo, V.J.: 1990, *Astrophys. J.* **365**, 764.
Schüssler, M.: 1984, *Astron. Astrophys.* **140**, 453.
Spruit, H.C.: 1977, *Ph.D. Thesis*, Univ. Utrecht.
Steiner, O., Pizzo, V.J.: 1989, *Astron. Astrophys.* **211**, 447.
Venkatakrishnan, P.: 1986, *Nature* **322**, 156.
Vernazza, J.E., Avrett, E.H., Loeser, R. (VAL): 1976, *Astrophys. J. Suppl.* **30**, 1.

THE STRUCTURE OF UMBRAL FLUXTUBES

D. DEGENHARDT and B. W. LITES

High Altitude Observatory, National Center for Atmospheric Research,
P.O. Box 3000, Boulder, CO 80307, U.S.A.

Abstract. Subsurface filamentation of sunspot magnetic fields has been postulated as a source of the visible small-scale structure of sunspot umbrae. We examine this possibility by investigating the magnetohydrodynamic structure of thin, vertical magnetized gas columns embedded in sunspot umbrae. These *umbral fluxtubes* are assumed to have weaker field strengths than the surrounding umbral atmosphere. The steady-state magnetohydrodynamic equations are solved numerically in the slender fluxtube approximation, thus allowing for stationary internal mass flows. We include the radiative exchange of heat between the umbral fluxtube and the ambient medium using a simple relaxation time approach.

The geometric shape of the steady flow solution is a gas column converging with height. We discuss the relationship of our results to observed properties of umbral brightenings (umbral dots). We show that, even if there is a large difference in magnetic field strength between the dot and the ambient medium in deeper layers, the field strengths are nearly equal in the observable layers, a result required by the observations. We also show that either high temperatures at the lower boundary of the dots or strong upflows are needed in order to produce bright continuum structures.

Key words: infrared: stars – MHD – Sun: magnetic fields – sunspots

1. Introduction

During the last decade the picture of an inhomogeneous subsurface structure of sunspots became popular. In this "cluster" model, the subsurface layers of the spot are fragmented into individual flux ropes which merge at a certain height to form a monolithic field structure at the surface (Parker 1979, Spruit 1981). This subsurface filamentation of the sunspot magnetic field would offer a natural explanation for the appearance of umbral dots, in the sense that umbral dots are caused by field-free plumes of gas pushing to the surface with high speed through a "magnetic valve" (Choudhuri 1986). However, contrary to prior theoretical treatment of this phenomenon, recent observations with very high spatial resolution indicate that the magnetic field strength is not greatly reduced within dots and they have only a very weak upflow, if any at all (Lites *et al.* 1991).

2. Theoretical Model

We change the Parker/Choudhuri configuration of umbral dots in a way that allows both for a (weaker) magnetic field and for a fluid flow in the regions they call "field-free" columns. Hence, we investigate the magnetohydrodynamic structure of thin, vertical magnetized gas columns ("umbral fluxtubes") embedded in a sunspot umbra, which is represented by model M of Maltby *et al.* (1986) permeated by a strong magnetic field.

We make the following assumptions: The basic geometry is such that the umbral fluxtubes are embedded in a cylinder of radius r_1 representing the ambient medium (see Fig. 1). At the lateral surface of this cylinder the external field is assumed to be vertical, that is the presence of the weak field tube is not felt beyond r_1. We

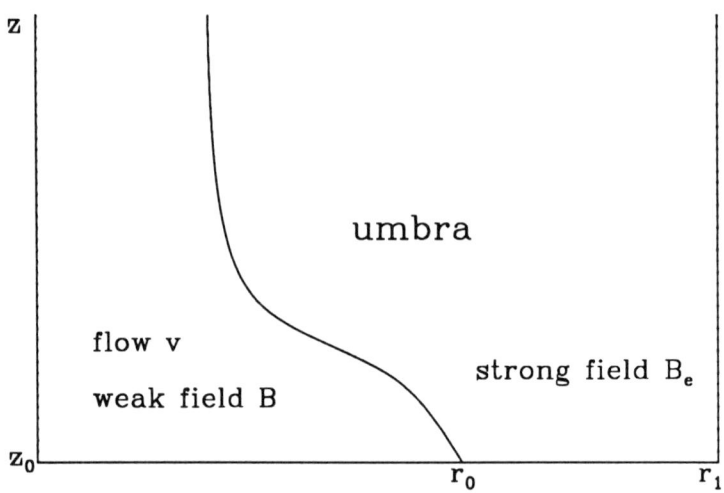

Fig. 1. Basic geometry of umbral fluxtubes. $r = 0$ is the axis of a cylinder of radius r_1, where the umbral field is assumed to be vertical.

assume that the fluxtube (carrying the fluid flow) and the surroundings remain two distinct entities.

2.1. Numerical Solution

We adopt the steady-state one-dimensional magnetohydrodynamic equations of motion and of energy,

$$\rho v \frac{dv}{dz} = -\frac{dp}{dz} - \rho g \tag{1}$$

$$v \frac{dT}{dz} = (1-\gamma)T(\frac{\chi_\rho}{\chi_T})\Delta_0 - \frac{1}{\rho c_v} \text{div } \boldsymbol{F}_R \tag{2}$$

which, together with the solenoidal condition for the magnetic field, the equation of continuity, and the ideal gas equation, are solved numerically using the slender fluxtube approximation (SFA) subject to the boundary condition

$$p + \frac{B^2}{8\pi} = p_e + \frac{B_e^2}{8\pi} \tag{3}$$

where the index "e" refers to the external atmosphere. In eqs. (1)–(3), Δ_0 is the zeroth order term of div v in the SFA, χ_ρ and χ_T describe the dependence of the mean molecular weight due to changes in ionization, $\boldsymbol{F}_R$ denotes the radiative flux and all other symbols have their usual meanings. We include the radiative exchange of heat between the umbral fluxtube and the ambient medium using a simple relaxation time approach (for references see Degenhardt 1991).

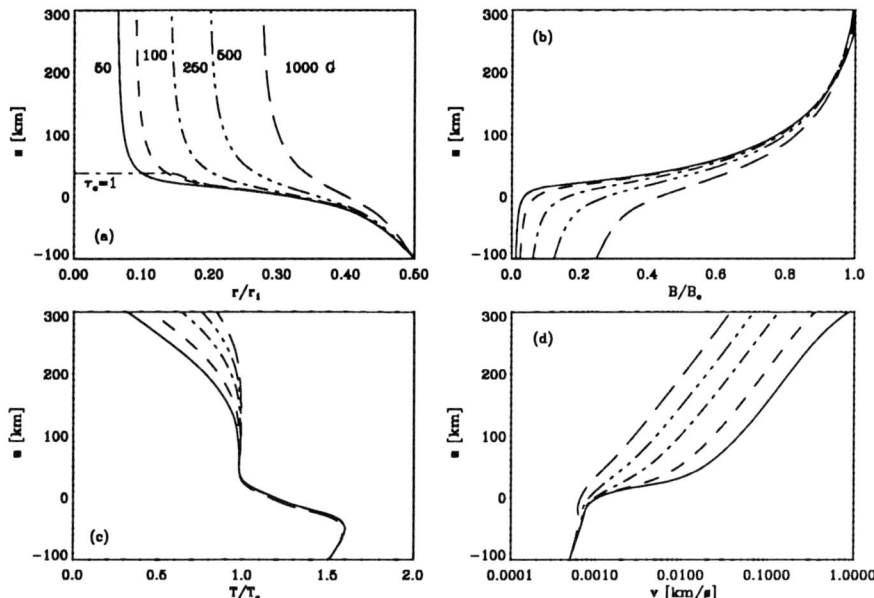

Fig. 2. (a) Magnetic field lines separating the umbral fluxtube and its surroundings, (b) field strength ratios, (c) temperature ratios and (d) flow speeds for varying ratios of magnetic field strengths at the lower boundary. The curves are labeled with the value of the internal magnetic field at the base.

The following model parameters can be prescribed at a certain height z_0 (see Fig. 1): the internal and external magnetic field strengths B_0 and B_{e0}, the radius of the fluxtube r_0, the radius of the total cylinder r_1 in which the tube is embedded, the ratio of internal to external temperature T_0/T_{e0}, and the velocity v_0.

Two different models are possible, corresponding to upflowing and downflowing gas, respectively. However, as there is no observational indication of a downflow in umbral dots, we only focus on models with upflowing material. The shape of the tube (and thus the external magnetic field) as well as the other physical quantities are calculated self-consistently during the computation.

3. Results

At first we investigate the influence of the ratio of magnetic field strengths at the lower boundary (see Fig. 2). We set the radius of the total cylinder to $r_1 = 1,000$ km and $r_0 = 500$ km. The umbral field strength at the lower boundary is chosen to be $B_{e0} = 4,000$ G and we assume that the internal material is hotter than the umbra by a factor $T_0/T_{e0} = 1.5$. This is based upon the assumption that the convective energy transport in deeper layers is suppressed in the strong-field region relative to the umbral fluxtube. We impose a very weak upflow of $v_0 = 0.5$ m s^{-1}. Internal field strengths at the base are $B_0 = 50, 100, 250, 500$ and $1,000$ G, respectively, as indicated.

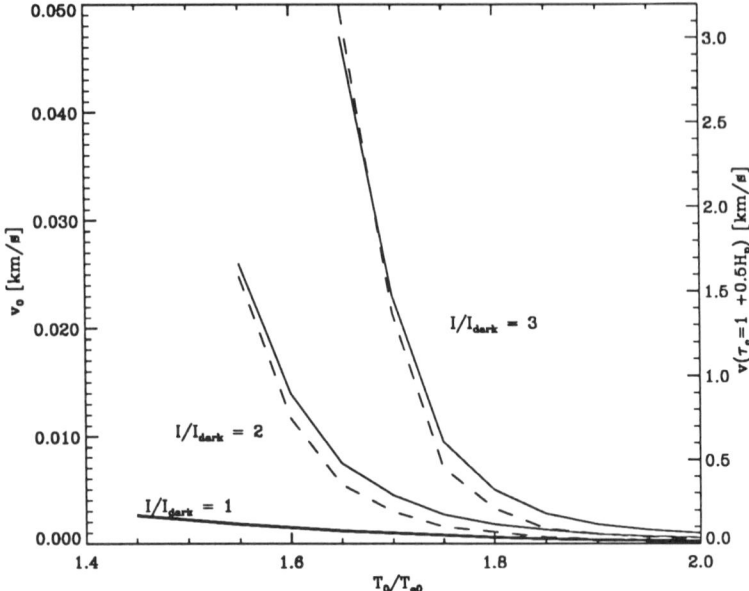

Fig. 3. Combinations of initial speed and temperature ratio which yield bright continuum structures, when observed from above, along the axis of the tube (see text).

The equilibrium structures are shown in the upper left panel of Figure 2. Due to the magnetic pressure of the umbral field the tubes with weaker field contract more strongly with height. The level $\tau_c = 1$ is indicated for the model with $B_0 = 250$ G. The upper right panel shows the variation of B/B_e with height. As can be noted, the field strengths become nearly equal in the observable regions, even if the internal field at the base is very weak.

The lower right panel shows the flow speed along the tube. There is a stronger flow in the narrower tubes due to the Bernoulli effect. This strong flow yields a significant adiabatic cooling due to stratification in the upper portions of the tubes, which may be seen in the lower left panel, where the run of the temperature ratio T/T_e is given.

Next, we examine what combinations of velocity v_0 and temperature ratio T_0/T_{e0} can produce bright structures in the continuum and may thus be identified with umbral dots. In principle, increasing the flow speed yields more heat advection into the upper portions of the tubes so that the hot, dense gas raises the $\tau_c = 1$ level. The same effect can be achieved by increasing the temperature at the bottom of the tube.

We adopt an umbral field $B_{e0} = 3,300$ G and an internal field $B_0 = 150$ G at the lower boundary. With $r_0 = 500$ km and $r_1 = 1,000$ km, these conditions result in a radius of $r \approx 125$ km at $z = 300$ km, a reasonable size for an umbral dot. At the same height the umbral field strength is $\sim 2,500$ G. We carry out a simple LTE radiation diagnostic in the continuum for a hypothetical observation along the axis of the tube. The results are shown in Figure 3.

The thick solid line indicates the combinations that result in structures as bright as the umbra itself. Choosing v_0 below this line leads to structures even darker than the umbra. The thinner solid lines give the combinations that result in structures twice or three times as bright as the umbra. We cannot simulate structures with $I/I_{\text{dark}} = 2$ for $T_0/T_{e0} < 1.55$, because the very strong flow needed results in a shock, which cannot be treated by our code.

The dashed lines (corresponding to the label on the right axis) show the velocities at a level 1/2 pressure scale height above the $\tau_c = 1$ level. We feel that this is a rough approximation for the speeds which would be observed. As can be seen, one does not need to impose large flows to produce bright structures if the temperature enhancement at the bottom is sufficiently large.

4. Conclusions

Assuming that the umbrae of sunspots are divided into "normal" umbral regions and "umbral fluxtubes" with weaker field strengths, we find that, in the observable layers, the field strengths are approximately the same within and without, no matter how weak the tube field is below $\tau_c = 1$ (in the umbra). Therefore, the best opportunity to observe reduced field strengths in umbral dots might be provided by the weak infrared Fe I lines at the opacity minimum around 1.56 μm. Bruls et al. (1991) show that these lines form deeper in the umbral photosphere than the Fe I lines in the visible spectrum commonly used as magnetic diagnostics.

In addition, we find that it is possible to construct umbral fluxtubes which appear as bright continuum structures (relative to the umbra) with upward flow speeds less than 0.5 km s^{-1}, provided the temperature enhancement at the lower boundary is large enough.

The results of this preliminary investigation should be confirmed by simulations including a more realistic energy equation, especially involving a non-local energy transfer by radiation and by a more detailed radiation diagnostic, especially involving the infrared lines around 1.56 μm.

Acknowledgements

We are grateful to K. Ferriere for her instructive comments on this manuscript.

References

Bruls, J. H. M. J., Lites, B. W., Murphy, G. A.: 1991, in L. J. November (ed.), *Solar Polarimetry*, Proc. 11th National Solar Observatory/Sacramento Peak Summer Workshop, p. 444.
Choudhuri, A. R.: 1986, *Astrophys. J.* **302**, 809.
Degenhardt, D.: 1991, *Astron. Astrophys.* **248**, 637.
Lites, B. W., Bida, T. A., Johannesson, A., Scharmer, G. B.: 1991, *Astrophys. J.* **373**, 683.
Maltby, P., Avrett, E. H., Carlsson, M., Kjeldseth-Moe, O., Kurucz, R. L., Loeser, R.: 1986, *Astrophys. J.* **306**, 284.
Parker, E. N.: 1979, *Astrophys. J.* **230**, 905.
Spruit, H. C.: 1981, in L. E. Cram and J. H. Thomas (eds.), *The Physics of Sunspots*, Sunspot, New Mexico, p. 98.

1.5 μm OBSERVATIONS AND THE DEPTH OF SUNSPOT PENUMBRAE

S. K. SOLANKI and I. RÜEDI

Institute of Astronomy, ETH-Zentrum, CH-8092 Zürich, Switzerland

W. LIVINGSTON

National Solar Observatory, NOAO,[*] P. O. Box 26732, Tucson, AZ 85726, U.S.A.

and

H. U. SCHMIDT

Max-Planck-Institute of Astrophysics, Garching, Germany

Abstract. The magnetic structure of a simple, relatively symmetric sunspot is determined using the extremely Zeeman sensitive Landé $g = 3$ line of Fe I at 1.5648 μm. From the measured strength and inclination of the magnetic field we estimate the fraction of the total magnetic flux of the sunspot passing through the solar surface in the penumbra. It is found that on average approximately 1/2–2/3 of the total magnetic flux of the spot emerges in the penumbra. Sunspot penumbrae are therefore deep, i.e., the $\tau = 1$ level does not correspond to the lower magnetic boundary of the spot in its penumbra.

Key words: infrared: stars – Sun: magnetic fields – sunspots

1. Introduction

Sunspots are the solar magnetic features most easily accessible to direct observations and have invited considerable attention. Nevertheless, much of their physics remains unresolved, including a number of global aspects of their magnetic fields. Here we mainly address the question: Are sunspot penumbrae deep or shallow?

In a *shallow penumbra* the current sheet bounding the sunspot roughly corresponds to the $\tau = 1$ surface in the penumbra, i.e., no (or very few) field lines cross the solar surface in the penumbra. In such a model the total magnetic flux of the sunspot emerging from the solar interior passes through the umbra. On the other hand, a significant number of field lines do cross the solar surface within a *deep penumbra*. We obtain a rough (indirect) measure of the "depth" of a penumbra by determining the total magnetic flux in the umbra, Φ_u, and comparing it with the magnetic flux in the penumbra, Φ_p. If $\Phi_p \ll \Phi_u$, then the umbra is shallow; if $\Phi_p \gtrsim \Phi_u$, it is deep. To determine the flux we need the magnetic field strength, B, and inclination angle to the vertical, γ', measured as a function of radial distance, r, from the center of the sunspot.

2. Observations and Analysis

The observed sunspot was close to solar disk center ($\theta = 10°$). Figure 1 shows a drawing of the umbral and penumbral boundaries of the spot with the approximate positions of the entrance apertures to the spectrograph overlaid. In all, 71 spectra of Fe I 1.5648 μm in Stokes I and V were obtained. Field strength, B, and magnetic

---

[*] Operated by the Association of Universities for Research in Astronomy, Inc., under cooperative agreement with the National Science Foundation.

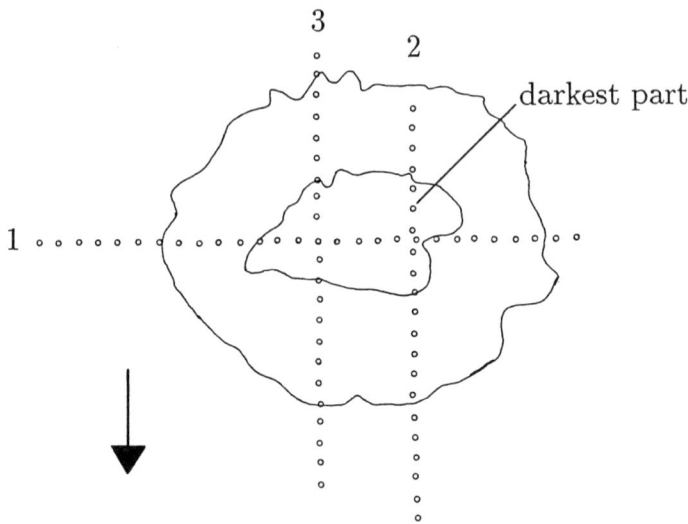

Fig. 1. Contours of the umbral and penumbral boundary of the sunspot. The small circles represent the positions at which spectra were obtained. The three scans through the sunspot are numbered near their starting positions. The arrow points toward the center of the solar disk.

inclination angle, γ, relative to the line of sight are determined by fitting the observed Stokes I and V profiles with numerically calculated synthetic profiles using an inversion code. Details of the code and the analysis are given by Solanki et al. (1992).

3. Field Strength and Inclination Angle

We combine the measurements of the three slices through the sunspot by plotting field strength and inclination angle vs. radial distance, r, from the geometrical center of the sunspot. The field strength, B, is plotted in Figure 2 as a function of r/r_p, where r_p is the radius of the sunspot. We wish to stress two points in Figure 2.

  a. Most of the scatter around the mean curve is intrinsic to the sunspot. To illustrate this we have represented the data points along each half of each slice by a different symbol. In some directions (e.g., dots) the field drops off more slowly with r/r_p than in others (e.g., open circles).
  b. The field strength at the outer penumbral boundary is accurate to approximately 50 G, so that the scatter there is also mainly solar in origin. Note also that the field strength values at $r/r_p > 1$ have been determined from the splitting of the V profiles (determined from the profile fits) and therefore represent the true field strengths in the superpenumbra.

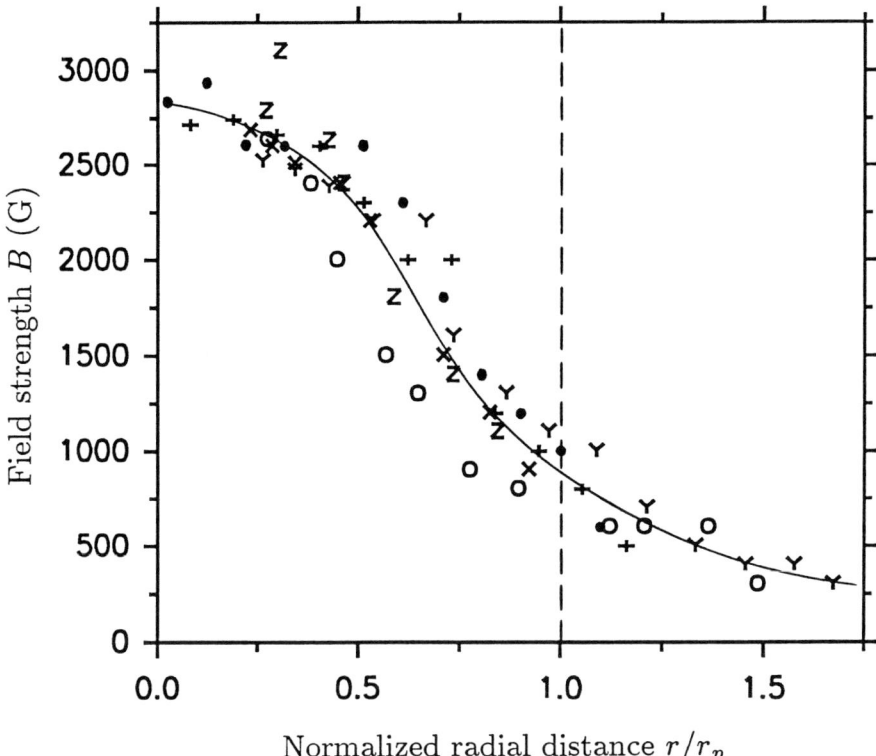

Fig. 2. Magnetic field strength, B, vs. r/r_p. Here r is the radial distance from the geometrical center of the sunspot and r_p is the outer penumbral radius r_p in the relevant direction (cf. Fig. 1). The symbols refer to different halves of the 3 sunspot slices shown in Figure 1.

In Figure 3 we plot the inclination angle, γ', of the field lines to the vertical. At the outer penumbral boundary we find $\gamma'(r/r_p = 1) = 82 \pm 4°$. The γ' values lying above the dashed line are more accurately determined than those lying below it.

4. Depth of the Penumbra

In addition to the B and γ' values plotted in Figure 3, we have used $B(r)$ and $\gamma'(r)$ values published by Beckers and Schröter (1969), Wittman (1974), Kawakami (1983), and Lites and Skumanich (1990) to determine $\Phi(r)$, the magnetic flux emerging within radius r. For simplicity we use azimuthally averaged curves of $B(r)$ and $\gamma'(r)$, such as the solid curves plotted in Figures 2 and 3. For details on how $\Phi(r)$ is determined see Solanki and Schmidt (1992).

Values of $\Phi(r)$ (normalized to $\Phi_{\max} = 1$), derived from the data in the literature and the 1.5 μm observations described above, are plotted vs. r/r_p in Figure 4.

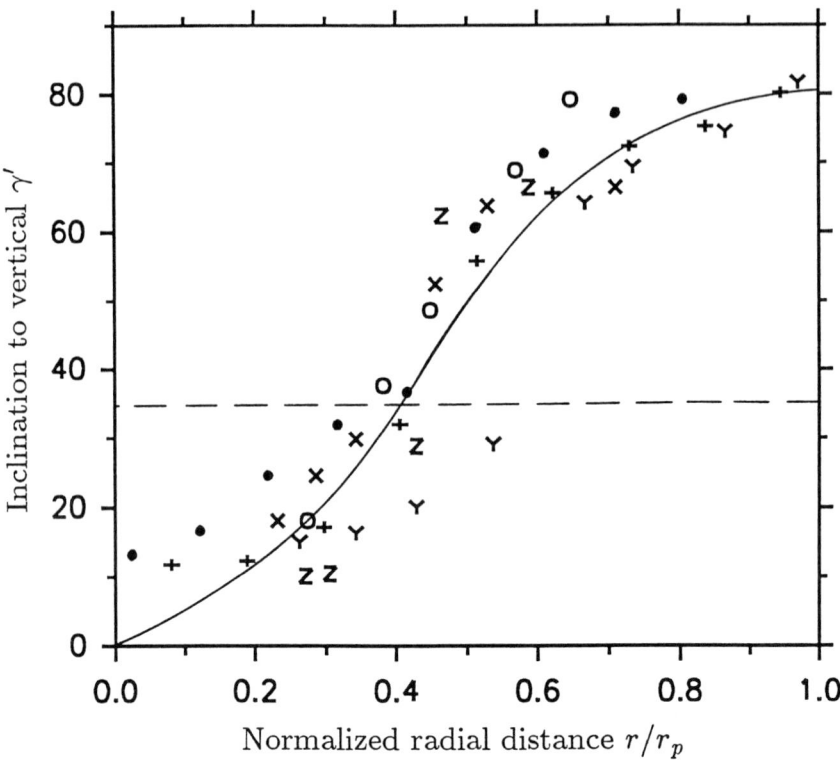

Fig. 3. Angle of inclination, γ', to the solar surface normal vs. r/r_p. Points below the horizontal dashed line are of lower accuracy than the rest.

For a shallow penumbra we expect the curves to be flat for $r/r_p \gtrsim 0.4$–0.5 (the umbral radius is approximately 0.4–0.45 of r_p). This is obviously not the case. Quantitatively, we find that, on average, 1/2–2/3 of the total magnetic flux of the spot emerges in the penumbra.

5. Conclusions

From an analysis of the radial dependence of the field strengths and inclination angles in nine sunspots (including one for which the field strength has been determined with great accuracy using the 1.5648 μm, $g = 3$ line) we conclude that a significant fraction (approximately 1/2–2/3) of the magnetic flux of a sunspot emerges in the penumbra – i.e., sunspot penumbrae are deep. This is in agreement with most magnetohydrostatic models of the sunspot magnetic field (e.g., Schlüter and Temesvary 1958, Deinzer 1965, Yun 1971, Pizzo 1986, and Jahn 1989). Models of a shallow penumbra (e.g., Schmidt et al. 1986, cf. Nordlund and Stein 1989) and observations suggesting a penumbral canopy (e.g., Giovanelli 1982) are not compatible with our analysis.

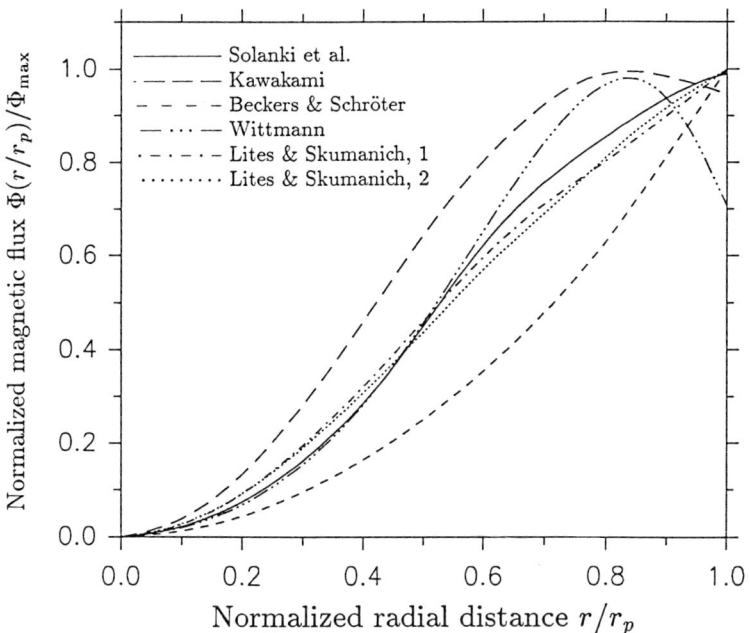

Fig. 4. Magnetic flux Φ emerging within radial distance r of the sunspot center, normalized to a maximum value of unity, Φ/Φ_{max}, vs. r/r_p.

References

Beckers, J. M., Schröter, E. H.: 1969, *Solar Phys.* **10**, 384.
Deinzer, W.: 1965, *Astrophys. J.* **141**, 548.
Giovanelli, R. G.: 1982, *Solar Phys.* **80**, 21.
Jahn, K.: 1989, *Astron. Astrophys.* **222**, 264.
Kawakami, H.: 1983, *Pub. Astron. Soc. Japan* **35**, 459.
Lites, B. W., Skumanich, A.: 1990, *Astrophys. J.* **348**, 747.
Nordlund, Å, Stein, R. F.: 1989, in R. J. Rutten and G. Severino (eds.), *Solar and Stellar Granulation*, Kluwer Academic Publishers, Dordrecht, p. 453.
Pizzo, V. J.: 1986, *Astrophys. J.* **302**, 785.
Schlüter, A., Temesvary, S.: 1958, in B. Lehnert (ed.), 'Electrodynamic Phenomena in Cosmical Physics', *IAU Symp.* **6**, 263.
Schmidt, H. U., Spruit, H. C., Weiss, N. O.: 1986, *Astron. Astrophys.* **158**, 351.
Solanki, S. K., Schmidt, H. U.: 1992, *Astron. Astrophys.*, submitted.
Solanki, Rüedi, I., Livingston, W. 1992, *Astron. Astrophys.* **263**, 312.
Wittmann, A. D.: 1974, *Solar Phys.* **36**, 29.
Yun, H. S.: 1971, *Solar Phys.* **16**, 398.

A MAGNETIC FIELD STRENGTH VS. TEMPERATURE RELATION IN SUNSPOTS

GREG KOPP and DOUGLAS RABIN

*National Solar Observatory, National Optical Astronomy Observatories,**
P.O. Box 26732, Tucson, AZ 85726, U.S.A.*

Abstract. The near infrared presents several new and powerful advantages in the diagnostics of sunspot atmospheres: (1) increased magnetic sensitivity in Zeeman-split lines, (2) increased sensitivity of umbral brightness to temperature, and (3) reduced scattered light and seeing disturbances due to atmospheric turbulence. This has revealed a strong and consistent relationship between sunspot brightness and magnetic field strength.

We have made spatial/spectral observations of sunspots in the highly sensitive ($g = 3$) Fe I line at $\lambda = 1.5649$ μm to compare field strengths with continuum intensities. We find a characteristic but nonlinear relationship between magnetic field strength, B, and brightness temperature, T_b, in sunspots. In umbrae there is an approximately linear relation between B^2 and T_b.

Key words: infrared: stars – Sun: magnetic fields – sunspots

1. Introduction

There are three primary reasons to observe in the infrared rather than the visible when making measurements of sunspot magnetic fields. The first is the extra magnetic sensitivity afforded by infrared lines, since Zeeman splitting as a fraction of linewidth increases linearly with wavelength. The second is the reduced stray light in the infrared, where instrumental scatter is lower (Pierce, 1991). Finally, the effects of stray light on intensity measurements are smaller in the infrared because of the greater umbral brightness. Whereas umbrae are typically 10% as bright as the mean quiet Sun in the visible, they are roughly 50% as bright in the near infrared, greatly reducing the effects of contamination by stray light from the surrounding photosphere.

We present sunspot observations using the magnetically-sensitive (Landé $g = 3$) Fe I line at 1.5649 μm. First used for magnetic work in 1975 (Harvey and Hall), this line has been useful in determining the magnetic fields of fluxtubes (Stenflo, Solanki and Harvey, 1987), plages (Rabin, 1992) and sunspots (McPherson, Lin, and Kuhn, 1992; Kopp and Rabin, 1992).

Zeeman splitting of this line is roughly three times that of high-g visible lines. Figure 1 shows an example of the splitting in this line for a 2,600 G sunspot. The magnetic field strength, not just magnetic flux, can be determined from the completely split π- and σ-components of the line; thus, these field strength determinations are insensitive to field inclination. The neighboring $g_{\text{eff}} = 1.53$ line at 1.5653 μm (Solanki, Biémont and Mürset, 1990) is obscured by umbral OH lines (Wallace and Livingston, 1992), making it useful for spectral calibration but not as an alternate measurement of field strength.

---

* Operated by the Association of Universities for Research in Astronomy, Inc., under cooperative agreement with the National Science Foundation.

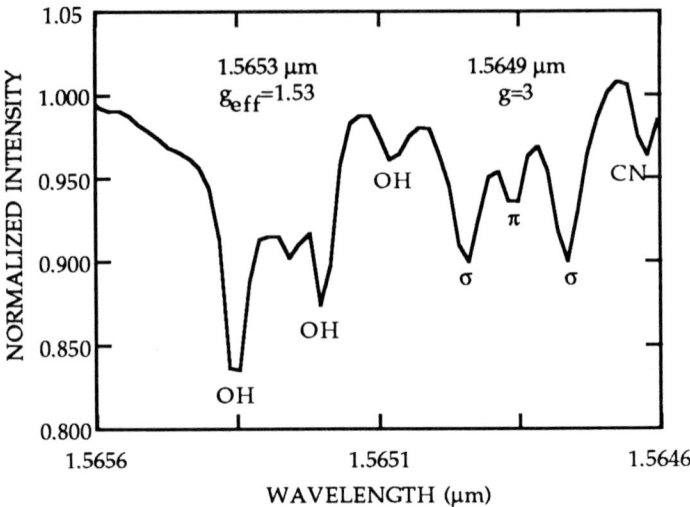

Fig. 1. Umbral spectrum of the Fe I lines at 1.5649 μm ($g = 3$) and at 1.5653 μm ($g_{\text{eff}} = 1.53$) for a field strength of 2,600 G. Umbral OH lines obscure the $g_{\text{eff}} = 1.53$ line.

2. Observations

Several sunspots, each close to disk center ($\mu = \cos\theta \geq 0.75$), were observed during two observing runs in May 1991 and January 1992. The observations were made using the McMath East Auxiliary Telescope and the 13.8-m vertical spectrograph. The KPNO 58 × 62 InSb infrared array (Fowler et al., 1987) was used to take spatial/spectral images of the 1.5649 μm line by scanning sunspots across the spectrograph. Large spots required multiple-swath scans. Swath registration was done using large-field K-line images taken before and after each scan. The spatial scale was 0.85″ per pixel, although the resolution was seeing-limited to roughly 2″. The spectral resolving power was limited by pixel sampling to ∼ 35,000.

Each spatial/spectral image, created by co-adding twenty 50-ms frames, corresponds to one chord in a sunspot admitted into the slit. The magnetic field strength is determined as a function of position across the spot from fits of blended Gaussians to the line profile. The error on the position of each completely split component of the line is typically ±20 G; the splitting is incomplete for fields weaker than 1,000 G. The continuum intensity is determined from a line-free region of the spectrum. A two-dimensional image of the continuum brightness of the spot and of its magnetic-field can be reconstructed from the spatial/spectral images in all of the swaths.

3. Results and Discussion

We find a non-linear relation between magnetic field strength, B, and continuum intensity, I_c, for the observed sunspots, although the relation is nearly linear within

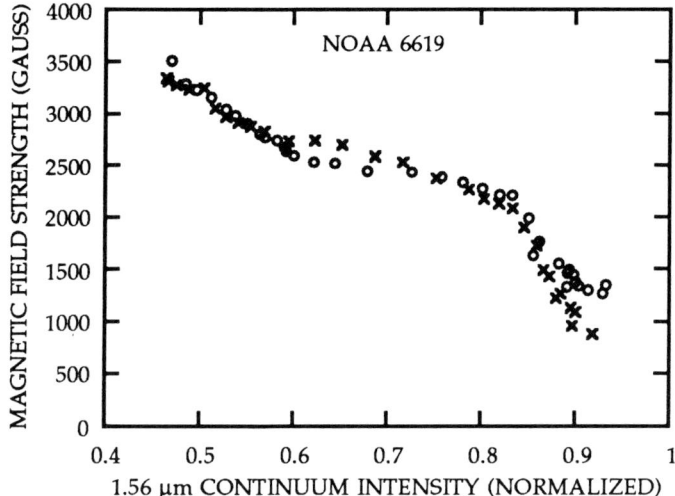

Fig. 2. Magnetic field strength *vs.* continuum intensity across a diameter of NOAA 6619, 13 May 1991. Opposite sides of spot center are identified by different symbols.

umbrae. Figure 2 shows this characteristic relation for a single large spot. The relationship appears linear in three distinct regions: for $B \gtrsim 2,500$ G (umbrae), field strength drops sharply with increasing intensity; for $2,000 < B < 2,500$ G (near the umbra/penumbra boundary), the field strength decreases little with increasing intensity; below 2,000 G the field strength again drops quickly with larger intensities. This region near the penumbral boundary has more scatter and the magnetic resolution of these observations begins to fail for $B \lesssim 1,000$ G.

Similar behavior is seen in other large spots. Small spots that do not reach field strengths larger than 2,500 G do not show the linear B vs. I_c dependence in this region (the intensity profile of a small spot may also be affected by blurring). A superposition of six spots observed in May 1991 is shown in Figure 3. Although no attempt was made to register the profiles of different spots, the same three-component behavior is evident in the superposition, although the scatter is larger than for a single spot. More recent data confirm this behavior.

Gurman and House (1981) proposed a linear B vs. I_c relation on the basis of visible-light observations, but there is apparently no magnetohydrodynamic basis for expecting linearity. Del Toro Iniesta *et al.*, 1991, suggest instead that a linear relationship holds between B^2 and the temperature T_b. If we plot B^2 vs. T_b, which is nearly proportional to I_c for the intensities observed near 1.6 μm, we again find a three-component shape, not a single linear relation.

A single linear relation between B or B^2 and I_c or T_b is inconsistent with most models of sunspot field profiles. The magnetic field in sunspots is typically modeled as a smooth function of distance from spot center, parameterized by only one length scale, usually the penumbral radius r_p (Beckers and Schröter 1969; Gurman and House 1981; Adam 1990). In contrast, the observed continuum intensity has

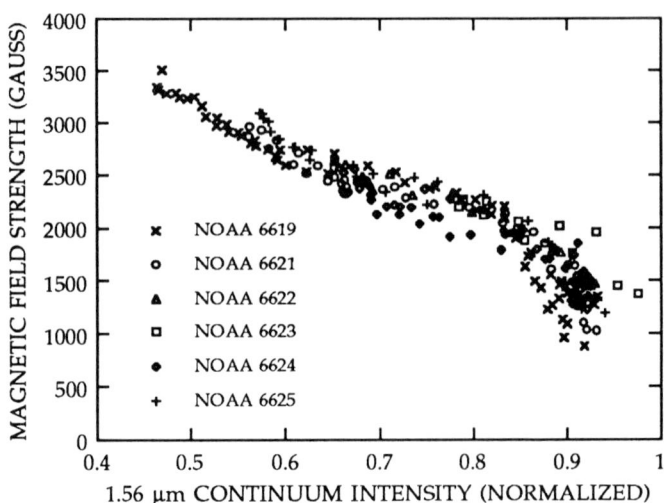

Fig. 3. Magnetic field strength *vs.* continuum intensity across 6 sunspots observed on 13 May 1991.

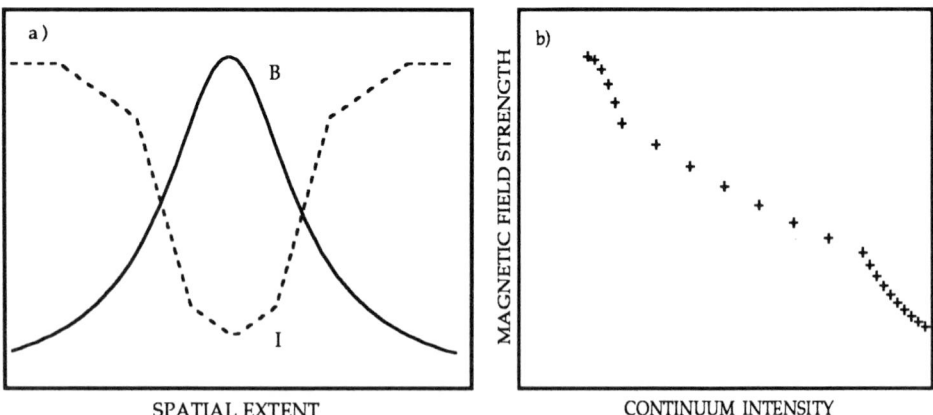

Fig. 4. a) Profiles of a hypothetical smooth magnetic field (solid curve) and a quickly changing intensity (dashed). b) Corresponding three-component nonlinear relation between field strength and intensity, similar to that observed.

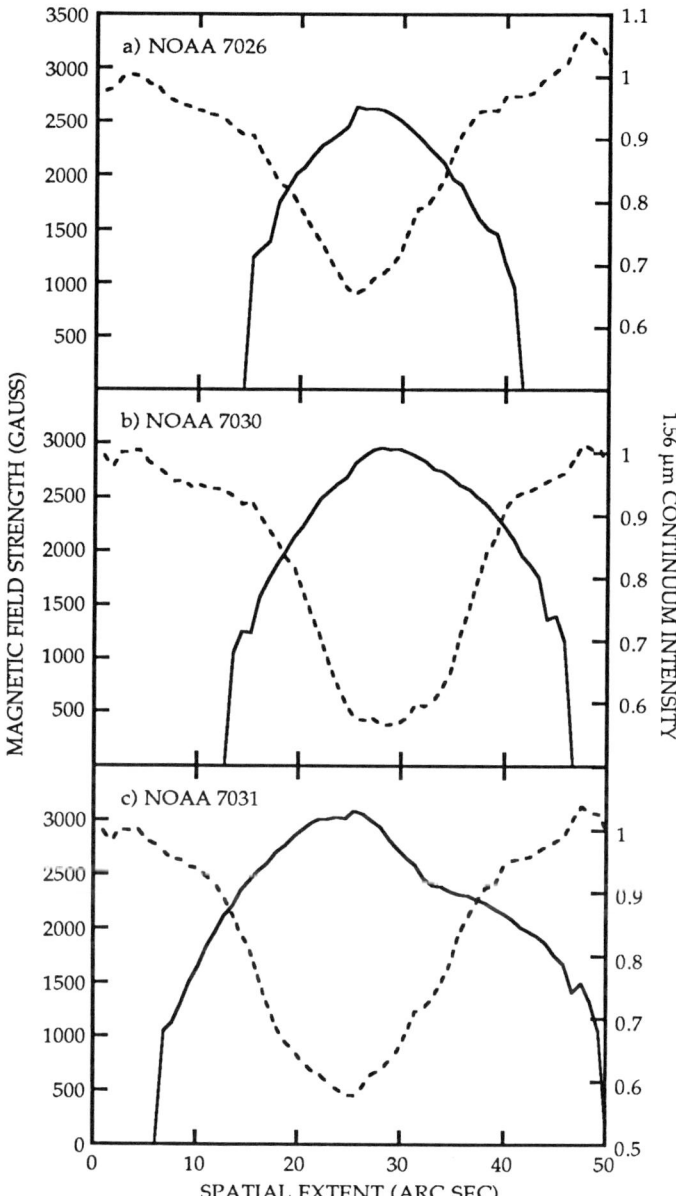

Fig. 5. Magnetic field (solid) and intensity (dashed) profiles for 3 spots observed on 26 Jan. 1992. Figures 5a and 5b show the field strength changing smoothly across the umbra/penumbra boundary, seen as a sudden change in slope of the intensity. Figure 5c shows that this is not always the case, as the boundary at 32″ on the abscissa is more evident in the magnetic field profile than in the intensity. Magnetic fields are only determined above 1,000 G.

rapid spatial variations, creating the effect of a sharp umbra/penumbra boundary and outer penumbral boundary. Smooth changes in B, in conjunction with sudden changes in I_c, naturally result in plots similar to those observed, because the magnetic field changes only slowly while the intensity changes rapidly near the umbra/penumbra boundary (see Fig. 4).

However, the observed field does not always vary smoothly with distance from spot center. Occasionally the magnetic field shows a sharper change at the umbra/penumbra transition than the intensity does, implying the magnetic field is affected by the transition. Figures 5a and 5b show typical cases where the magnetic field varies smoothly across the umbra/penumbra boundary, which is seen as a change in slope of the continuum intensity. Figure 5c, however, shows a spot where the magnetic field strength undergoes a marked change in slope near the transition, at 32″ on the abscissa, while the intensity shows only a small change. This result suggests that models of sunspot magnetic fields can be better parametrized by length scales including at least the umbral and penumbral radii r_u and r_p.

We conclude that the relation between magnetic field and continuum intensity or temperature is not linear, but behaves according to the three-component plots shown in Figures 2 and 3. The shape of this curve is largely a consequence of large variations in intensity with radial distance from spot center accompanied by small changes in magnetic field. Sunspot magnetic fields, however, are not always smooth across umbra/penumbra boundaries, and models of the fields further depend upon length scales r_u and r_p. Future work, combining better spectral resolution with full Stokes profiles of the 1.5649 μm line, will allow the determination of field orientation and filling factor as well as a sensitive measurement of field strength, allowing for vector magnetometry beyond the edge of sunspot penumbrae.

Acknowledgements

We thank D. Jaksha, J. Wagner, and C. Plymate for their dedication during the observations. This work was partially supported by NASA under Task 170-38-51-01-10.

References

Adam, M. G.: 1990, *Solar Phys.* **125**, 37.
Beckers, J. M., and Schröter, E. H.: 1969, *Solar Phys.* **10**, 384.
del Toro Iniesta, J.C., Martinez Pillet, V., and Vasquez, M.: 1991, in L. November (ed.), *Solar Polarimetry* (Proc. 11th Sacramento Peak Workshop), Sunspot, New Mexico, p. 224.
Fowler, A. M., Probst, R. G., Britt. J. P., Joyce, R. R., and Gillett, F. C.: 1987, *Opt. Eng.* **26**, 232.
Gurman, J., and House, L.: 1981, *Solar Phys.* **71**, 5.
Harvey, J. W., and Hall, D.: 1975, *Bull. Amer. Astron. Soc.* **7**, 459.
Kopp, G., and Rabin, D.: 1992, *Solar Phys.* **141**, 253.
McPherson, M. R., Lin, H., and Kuhn, J. R.: 1992, *Solar Phys.* **139**, 255.
Pierce, K.: 1991, *Solar Phys.* **133**, 215.
Rabin, D.: 1992, *Astrophys. J.* **391**, 832.
Solanki, S. K., Biémont, E., and Mürset, U.: 1990, *Astron. Astrophys. Suppl.* **83**, 307.
Stenflo, J. O., Solanki, S. K., and Harvey, J. W.: 1987, *Astron. Astrophys.* **173**, 167.
Wallace, L., and Livingston, W.: 1992, *An Atlas of a Dark Sunspot Umbral Spectrum from 1970 to 8640 cm^{-1} (1.16 to 5.1 μm)*, Kitt Peak National Observatory, Tucson.

THE IR CONTRAST OF MAGNETIC ELEMENTS OBTAINED FROM HIGH SPATIAL RESOLUTION OBSERVATIONS AT 1.6 μm

TRON A. DARVANN

Institute of Theoretical Astrophysics, University of Oslo,
P.O. Box 1029 Blindern, N-0315 Oslo 3, Norway
and
National Solar Observatory,* Sunspot, NM 88349, U.S.A.

and

SERGE KOUTCHMY

Institut d'Astrophysique, CNRS, 98 Bis, Blvd. Arago, F-75014 Paris, France
and
National Solar Observatory,* Sunspot, NM 88349, U.S.A.

Abstract. We report on new improved infrared (IR) imaging observations (1.6 μm) carried out with the National Solar Observatory's Vacuum Tower Telescope at Sacramento Peak, New Mexico (NSO/SP). Examples of high spatial resolution (up to 0.4'' images are shown, and results of comparisons between infrared and continuum at 526 nm and between infrared and the locations of enhanced magnetic field ("fluxtube regions" as defined by Mg I b1 line imaging) are given and discussed. Our results indicate that, close to disk center, magnetic elements have a *positive* contrast at the opacity minimum. This is contrary to the findings of several other authors (*e.g.*, Worden 1975, Foukal *et al.* 1990). We emphasize the necessity of multi-color, high spatial and temporal resolution observations. The potential of present "almost-on-a-routine-basis" IR observations utilizing a fast video acquisition system developed at NSO/SP is pointed out.

Key words: infrared: stars – Sun: faculae, plages – Sun: magnetic fields

1. Introduction

The first attempt to measure the contrast (brightness excess or deficit) of flux tube regions at the opacity minimum was carried out by Worden (1975) using a single element detector at 1.64 μm. From his analysis of the supergranulation network a 0.7% *negative* contrast (darker than average) in magnetic regions was inferred. Koutchmy (1989) reported indications of a *positive* contrast at 1.695 μm in a facular region close to a sunspot at $\cos\theta = 0.71$. A *positive* contrast was also (statistically) found in the quiet-Sun network (1.695 μm) by Koutchmy (1978). Finally, Foukal *et al.* (1990) (using a cooled PtSi CCD) reported examples of *negative* contrast, comparing 1.6 μm images of faculae ($\geq 2.5''$ resolution) with NSO/SP Ca K spectroheliograms. Measurements in the *visible* by del Toro Iniesta *et al.* (1990) may also be interpreted to indicate negative contrast in facular regions. A tendency for a reversed (red) color index in faculae was found by Keller and Koutchmy (1990), but at the same time their analysis shows a *positive* contrast in fluxtube regions at *subarcsecond* resolution. Their analysis suggests that the "observed" negative contrast could be a result of a spatial averaging, or, conversely, of an insufficient spatial resolution. Very high resolution (0.23'') observations by Auffret and Müller (1991) show a 23% *positive* contrast of network bright points at disk center. However, Topka *et al.* (1992) report a 3% *negative* contrast up to 20° heliocentric angle

---

* Operated by the Association of Universities for Research in Astronomy, Inc. (AURA) under cooperative agreement with the National Science Foundation.

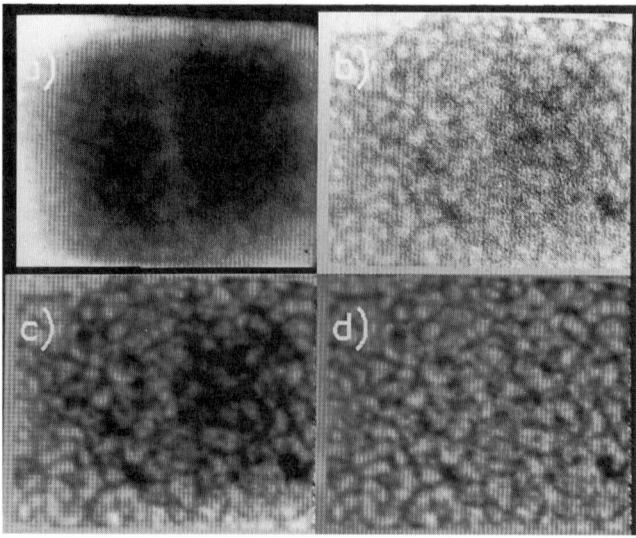

Fig. 1. *a)* Raw image after 8 bit digitization by CHIRP. *b)* Image after flat fielding by an average of 100 images taken with the telecope in motion. *c)* Image in *b* smoothed by simple 5 × 5 pixel bilinear interpolation (original pixel size 0.09"). *d)* Image in *b* after unsharp masking (Wiener filter with a low frequency cutoff at 0.067 arcsec^{-1}). The IR images were processed in this way before the contrast analysis (Section 3) was carried out.

(and thereafter *positive*) in fluxtube regions at 500 nm, using up to 0.3" spatial resolution. Recently, Zirin and Wang (1992) showed that their "micropores" are dark (in visible light, disk center) relative to the intergranular lanes.

2. Observations and Preprocessing

Our observations were carried out on Sept. 11, 1990 with the Vacuum Tower Telescope (VTT) of the National Solar Observatory at Sacramento Peak. Different wavelengths were observed sequentially, see Table I. Images (256 × 256 pixels) were taken with an IR Vidicon N2634 and video camera at the prime focus through a 1.64 μm interference filter. Real time digitization was carried out by use of a CHIRP image processing system, allowing a variable amount of processing before storage on an optical disk. Figure 1 illustrates different steps involved in subsequent preprocessing. The wavelengths other than IR were obtained with the Universal Birefringent Filter (UBF) and Technical Pan 2415 film. Digitization of the UBF images was performed with a video CCD camera.

3. Results: Contrast of Magnetic Elements

The region under study is shown at three different wavelengths (Table 1) in Figure 2, and is located in a faint plage area close to disk center. A porule (short lifetime) is seen in the field of view. We apply the Mg I b1 (518.41 nm) filtergram (lower

Fig. 2. Images (at 3 different wavelengths) used to study the contrast of magnetic elements. The images were obtained quasi-simultaneously (see the times given and Table I, below).

TABLE I
Sequence of Observations.

Time (UT) (h:m:s)	Spectral Region	Wavelength (nm)	FWHM (nm)	Exp. Time (s)	Line Blocking (%)
15:33:28	blue	445.124	0.0088	0.50	< 0.4
15:33:33	green	525.635	0.0131	0.10	< 0.2
15:33:38	red	606.950	0.0184	0.10	< 0.1
15:33:44	Mg I b1 + 0.4 pm	518.405	0.0127	0.30	58
15:33:51	Hα	653.281	0.0220	0.02	81
15:34:50	infrared	1640	100	0.033	< 5

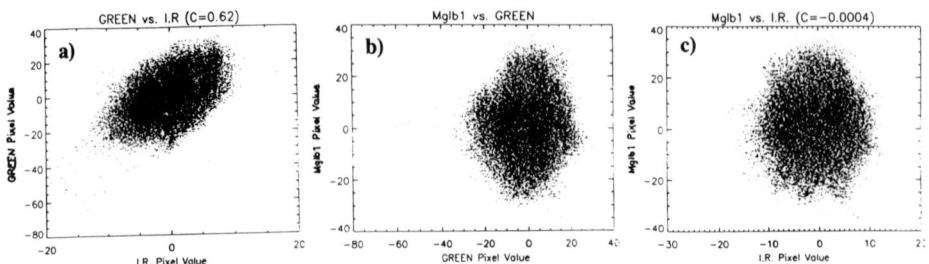

Fig. 3. Diagrams showing correlation between the 3 different wavelengths depicted in Figure 2.

right in Fig. 2) as a dependable proxy for the magnetic field strength (see Daras and Koutchmy 1983 and references therein). A correlation diagram between pixel values in the green continuum (525.63 nm) image and the IR image of Figure 2 is shown in Figure 3a, and the correlation coefficient is $c = +0.62$. Images were carefully aligned and scaled. We estimate the remaining differential image distortion due to seeing to be less than 0.5". Figures 3b and c show the (lack of) general correlation between Mg I b1 and green continuum ($c = +0.02$), and between Mg I b1 and the IR ($c = -0.0004$), respectively.

We investigate the properties of magnetic elements by selecting, and correlating, only those pixels in the 3 images that correspond to Mg I b1 brightness above a certain threshold. Figure 4a is a plot of the correlation computed between the visible (green continuum) image and the Mg I b1 image as the Mg I b1 brightness threshold is set successively higher. The correlation increases significantly as we leave out non-magnetic areas from the computation. This verifies the *positive* contrast (temperature excess) of magnetic elements at the photospheric level represented by the green continuum wavelength. Figure 4b shows a similar plot for 1.6 μm. We find that the contrast of the magnetic elements is *positive* also at this wavelength. In the Figure are plotted the results for 4 different IR images, all showing the same general behavior.

4. Conclusion

New high resolution analysis presented here indicates that the contrast of magnetic elements is *positive* at the opacity minimum (1.6 μm). Our result is contrary to most earlier measurements (Section 1). Reasons for the discrepancies may be differences due to resolution (smearing our Fig.-2 images to ∼ 4" produces an apparent negative contrast, by chance, in the region of the porule), different filter properties or properties of the regions being studied, or the type of magnetic field measurement. A larger statistical sample is needed, but can now more easily be provided as the near-IR window is being opened for high resolution imaging (Koutchmy 1993, in these proceedings; see also Koutchmy 1990).

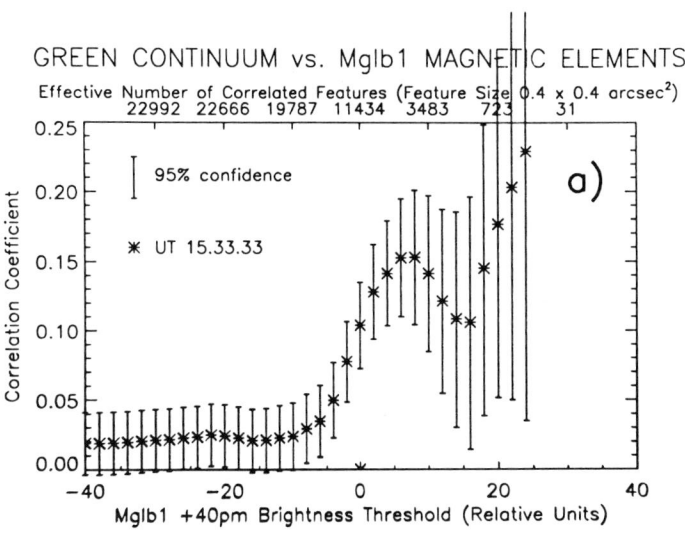

Fig. 4. *a)* Plots of correlation between IR and Mg I b1 as a successively higher threshold is selected for the Mg I b1 brightness. Points further to the right in the Figure are based on a subset of pixels corresponding to places where Mg I b1 brightness (and therefore magnetic field strength) is above the threshold (see Fig. 3). Error bars show a 95% confidence interval assuming the size of a resolution element to be $0.4'' \times 0.4''$. The error increases towards the right in the Figure due to a smaller number of features (see upper abscissa) being correlated.

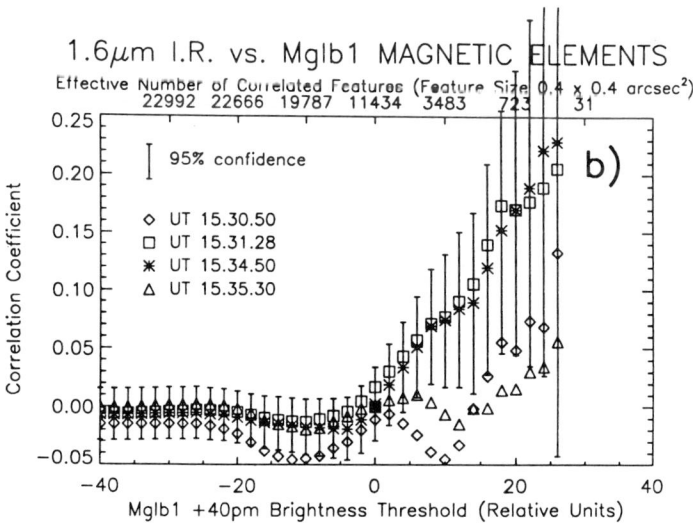

Fig. 4. *b)* Same as Figure 4a but now for correlation between green continuum and Mg I b1. The four different symbols in the Figure represent four different IR images.

Acknowledgements

The authors wish to acknowledge the important contribution to the NSO/SP IR program by the VTT observers under the leadership of Dick Mann. Fritz Stauffer carried out the programming of CHIRP, and Larry Wilkins provided necessary electronics for the IR camera.

References

Auffret, H., Müller, R.: 1991, *Astron. Astrophys.* **246**, 264.
Daras-Papamargaritis, H., Koutchmy, S.: 1983, *Astron. Astrophys.* **125**, 280.
del Toro Iniesta, J.C., Collados, M., Sánchez Almeida, J., Martinez Pillet, V., Ruiz Cobo, B.: 1990, *Astron. Astrophys.* **233**, 570.
Foukal, P., Little R., Graves, R., Rabin D., Lynch D.: 1990, *Astrophys. J.* **353**, 712.
Keller, C.U., Koutchmy, S.: 1991, *Astrophys. J.* **379**, 751.
Koutchmy S.: 1978, unpublished.
Koutchmy S.: 1989, in R. J. Rutten and G. Severino (eds.), *Solar and Stellar Granulation*, NATO ASI C-263, Kluwer Academic Publishers, Dordrecht, p. 253.
Koutchmy, S.: 1990, in J. O. Stenflo (ed.), 'The Solar Photosphere: Structure, Convection and Magnetic Fields', *Proc. IAU Symp.* **138**, 81.
Topka, K.P., Tarbell, T.D., Title, A.M.: 1992, "Properties of the Smallest Solar Magnetic Elements, I. Facular Contrast near Sun Center", preprint.
Worden, P.: 1975, *Solar. Phys.* **45**, 521.
Zirin, H., Wang, H.: 1992, *Astrophys. J.* **385**, L27.

DIAGNOSTIC TOOLS FOR SUNSPOTS: THE MOLECULES C_2, Mg H AND Ti O

K. SINHA

U.P. State Observatory, Manora Peak, Naini tal - 263 129, India.

Abstract. The aim of the present communication is to draw attention to the value of simultaneous observations of sunspot umbrae and the quiet Sun in selected molecular lines. It is felt that such observations may lead to an array of sunspot models which account for sunspot sizes, magnetic field strengths, and the solar activity cycle.

Key words: Sun: activity – Sun: magnetic fields – sunspots

1. Introduction

Because molecules are sensitive to their astrophysical environs, they serve as diagnostics for model atmospheres. The occurrence of the molecules C_2 and Ti O both in the quiet Sun and sunspots seems mutually exclusive. Also, the Mg H lines become stronger in sunspots than in the quiet Sun. Sunspot models that satisfy the C_2 and the Ti O lines should also be required to explain the Mg H line intensities. These molecules were selected for analysis for the additional reason that their lines fall into a narrow spectral region where the behavior of the opacity is well known. Careful, simultaneous observations of sunspots and quiet Sun will allow us to correct for scattered light—a common obstacle to sunspot investigations.

2. Results and Discussions

We shall only outline our procedures here, as the details are published elsewhere (Sinha and Tripathi, 1991 a, b). Under the influence of the magnetic field, the photospheric medium cools and changes in the spectrum become noticeable. However, in the literature, sunspot models with a range in magnetic field are sparse. Also, for obvious reasons, observations of large, stable umbrae are preferred. Stankiewicz (1967) constructed sunspot models as a function of magnetic field strength, and Sobotka (1985), using atomic lines, studied sunspots with different diameters and magnetic fields. His "hot", "intermediate", and "cool" sunspot models are referred to as models 12, 22, and 13 respectively. Maltby *et al.* (1986) proposed that sunspots formed in different phases of the solar activity cycle might be different from one another. If t is the time elapsed since the last cycle minimum, of duration t_0, then $t/t_0 = 0.1, 0.5$, and 0.9 refer to the early, middle, and late phases of the solar activity cycle, respectively. Accordingly, a set of three models, E, M, and L, have been presented by Maltby *et al.* (1986). To facilitate a comparison and to help us see how different these models can be, we present them in Figure 1.

Equivalent widths for molecular lines of C_2, Mg H, and Ti O are listed for these models in Table I. In actual observations one may wish to choose better quality lines. An inspection of Table I leads to the conclusion that in the Stankiewicz (1967) models the C_2 line is insensitive to the magnetic field strength, whereas the Mg H and Ti O lines are stronger in the cooler spots (*i.e.*, where the magnetic field is strong). Our calculations give larger equivalent widths than are observed

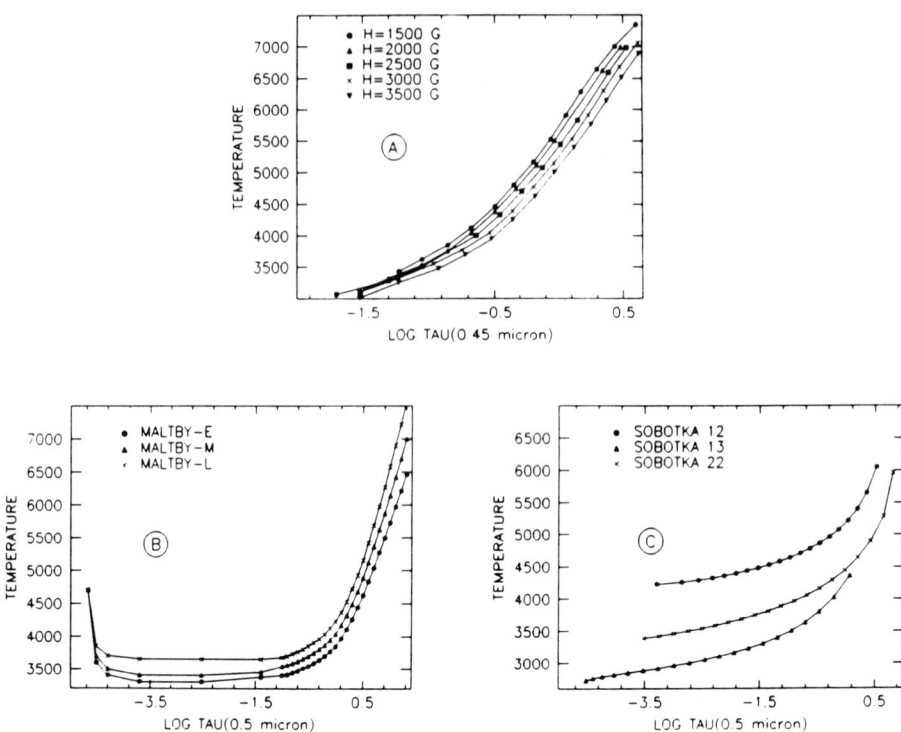

Fig. 1. Optical depth versus temperature for (A) the Stankiewicz (1967), (B) Maltby et al. (1986) models E, L, and M, and (C) Sobotka (1985) models 12, 13, and 22.

for C_2, which is not unexpected in view of the preliminary nature of these models. Moreover, these models are derived from a now-obsolete photospheric model due to Minnaert (1953).

Sobotka's (1985) models clearly demonstrate the mutual exclusiveness of the C_2 and TiO lines. As the sunspot cools, C_2 weakens and TiO strengthens. The MgH lines, as expected, intensify. It should be noted here that the lines of the (0-0) band of MgH are saturated and therefore may not be as sensitive for diagnostic purposes as the lines of the (0-1) band. Also, the equivalent widths for model 13 should be interpreted with caution, as this model does not go deeper than $\tau_{5000} = 0.06$.

The models of Maltby et al. (1986) provide ample proof that the umbrae of large, stable sunspots formed at different phases of the solar activity cycle are different in temperature structure. The mutually exclusive nature of the C_2 and TiO lines in cool atmospheres is again evident (cf. Table I). The behavior of the MgH lines is as generally expected.

Table I suggests that the C_2 lines are greatly reduced in strength in sunspots.

TABLE I
Equivalent Widths (mÅ) for Lines of C_2, MgH, and TiO.

Model	C_2 (0-0) Swan bands 5132.360 Å R_1 (18)	MgH (0-1) Green bands 5520.23 Å R_1 (13)	MgH (0-0) Green bands 5061.536 Å Q_2 (37)	TiO (0-0) α system 5189.80 Å P_1 (31)
Stankiewitz				
H = 1500 G	14.5	11.0	55.6	14.4
H = 2000 G	12.1	17.8	69.9	30.0
H = 2500 G	13.4	22.4	78.7	39.3
H = 3000 G	13.8	24.2	75.2	34.9
H = 3500 G	13.1	29.1	78.2	37.4
Sobotka				
Model 12	10.3	0.3	10.1	0.0
Model 22	3.3	8.1	60.4	3.2
Model 13	0.0	8.5	62.3	53.6
Maltby et al.				
Model E	0.7	29.3	86.5	42.2
Model M	1.6	24.7	91.6	44.6
Model L	3.1	16.9	82.8	15.2
Observations				
Quiet Sun[1]	8.0	–	2.9	–
Spot[2]	–	16	58	10

[1] Sinha (1984), Lambert et al. (1971).
[2] Sotirovski (1971).

However, they might still be observed if it is possible to detect lines as weak as about 3 mÅ in sunspots.

3. Conclusions

We believe that simultaneous observations of C_2, MgH, and TiO lines in the quiet Sun and in sunspot spectra should lead to a significant improvement in models of sunspots.

References

Lambert, D.L., Mallia, E.A., and Petford, A.D.: 1971, *Monthly Notices Roy. Astron. Soc.* **154**, 265.
Maltby, P., Avrett, E.H., Carlsson, M., Kjeldseth-Moe, O., Kurucz, R.L., and Loeser, R.: 1986, *Astrophys. J.* **306**, 306, 284.

Minnaert, M.: 1953, in G.P. Kuiper (ed.) *The Sun*, University of Chicago Press, Chicago, p. 126.
Sinha, K. : 1984, *Bull. Astron. Soc. India* **12**, 172.
Sinha, K., and Tripathi, B.M.: 1991a, *Bull. Astron. Soc. India* **19**, 13.
Sinha, K. and Tripathi, B.M.: 1991b, *Bull. Astron. Soc. India* **19**, 23.
Sobotka, M.: 1985, *Soviet Ast.* **29**, 576.
Sotirovski, P.: 1971, *Astron. Astrophys.* **14**, 319.
Stankiewicz, A.: 1967, *Acta Astr.* **17**, 341.

NEW INFRARED MEASUREMENTS OF MAGNETIC FIELDS ON COOL STARS

STEVEN H. SAAR

Center for Astrophysics, 60 Garden Street, Cambridge, MA 02138, U.S.A.

Abstract. I present a preliminary analysis of IR spectra of five K and M dwarfs and two RS CVn variables. Evidence for significant magnetic flux is found on several stars, a number of which are detected for the first time. Field strengths (B) on the RS CVn variables are lower than in the active dwarfs, consistent with the concept of pressure balance limiting B in stellar photospheres. I compare the results with previous measurements.

Key words: infrared: stars – stars: activity – stars: late-type – stars: magnetic fields

1. Introduction and Observations

Observations of stellar magnetic fields in the IR have three main advantages over the visible: (1) The ratio of the Zeeman splitting to the Doppler width, $\Delta\lambda_B/\Delta\lambda_D \propto \lambda$, increasing the ease of detection (*e.g.*, Giampapa et al. 1983); (2) the line density is lower, so that blends are less likely to confuse a measurement; and (3) for cooler stars, the near IR is close to the stellar flux maximum. Advantages (2) and (3) are important for M dwarfs, and (1) is particularly helpful for low-gravity RS CVn systems, where attempts at optical field measurements have mostly failed (Marcy and Bruning 1984; see, however, Bopp et al. 1989). I have been obtaining IR spectra with the NOAO 4-m Fourier Transform Spectrometer (4-m FTS; Hall et al. 1979) since 1985. I present some recent results of this program, including new measurements for several active K and M dwarfs and RS CVn variables.

Observations with the 4-m FTS were made in May 1991 (except for EV Lac, made in Decmber 1985) using the C+D detectors and an interference filter passing 4,400–4,600 cm^{-1} at an unapodized resolution of $\lambda/\Delta\lambda = 64,000$ (0.07 cm^{-1}). Of primary interest in this wavelength region is a Ti I multiplet with $g_{\rm eff}$ ranging from 1.125 to 2.5. These lines first appear in early K stars, grow strong by late K, and remain relatively strong at least to M5.

The spectra were reduced following Saar and Linsky (1985), with the exception that they were not apodized. While this leaves the spectra with a peculiar appearance due to the sinc function instrumental profile, it facilitates a more accurate extraction of telluric lines, and retains the full instrumental resolution. Spectra of inactive dwarfs and giants with similar spectral types (*e.g.*, GL 273, M4 V; μ Gem, M3 III) were used to help in distinguishing weak blends. For display purposes, all spectra and models have been smoothed with a five-point running mean to suppress sinc ringing.

2. Analysis and Discussion

Visual inspection of the Ti I lines in active M dwarfs (Fig. 1) reveals profiles that are much too broad for the (optically) measured $v \sin i$. Compare, for example, AU Mic (Fig. 1, left: 8 km s^{-1}; Vogt, Soderblom, and Penrod 1983) with HR 7275 (Fig. 2, right: $v \sin i = 15$ km s^{-1}; Strassmeier et al. 1988). The lines are also too

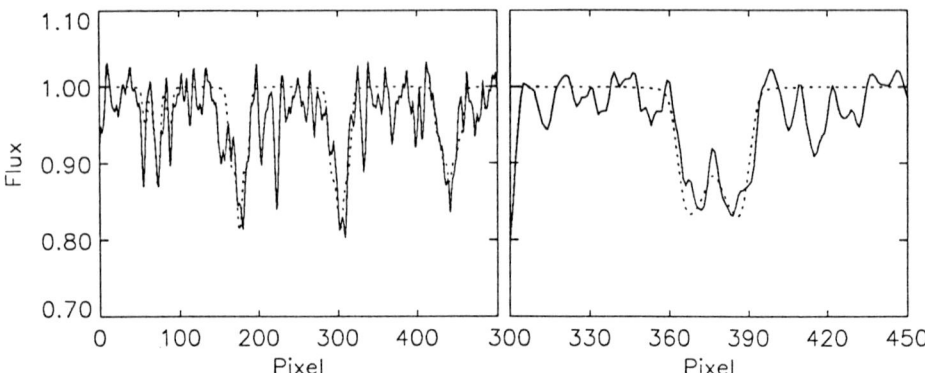

Fig. 1. Unapodized, smoothed line profiles from AU Mic (left; 4,480, 4,488, 4,496, 4,501 cm^{-1}; g_{eff} = 2.5, 1.58, 1.67, and 2.0, respectively; 0.16 Å/pixel) and EV Lac (right; 4,496 cm^{-1} only). The lines have been "spliced together" for convenient display. The fits (dashed) are for B = 4.2 kG with f = 0.55 (AU Mic) and B = 4.3 kG with f = 0.85 (EV Lac).

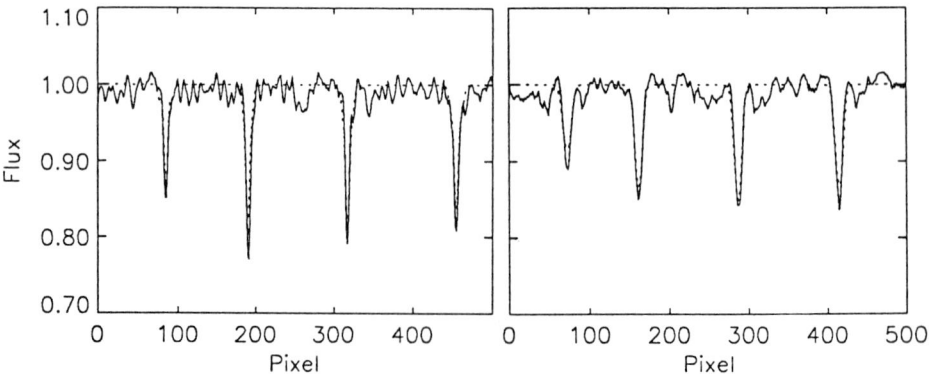

Fig. 2. Similar to Figure 1, but for ξ Boo B (left: fit with B = 2.3 kG and f = 0.20) and HR 7275 (right: fit with B = 1.4 kG and f = 0.40).

shallow to have significant opacity broadening, leaving magnetic fields as the only plausible explanation. In more rapidly rotating RS CVns, the effect of magnetic fields is subtle, appearing as slightly enhanced equivalent widths in lines with many and/or well-separated Zeeman components (e.g., Basri, Marcy, and Valenti 1992).

The models were calculated with a full-disk integration that partitions μ (the cosine of the emergence angle) into 5 intervals, and includes radial-tangential macroturbulence. Local line profiles were computed using the Unno-Rachkovsky radiative transfer solution (Unno 1956), including magneto-optical effects (see Saar et al. 1990). Line opacities were constrained to have their predicted (Kurucz and Peytremann 1975) ratios, and areas affected by blends (as determined from the comparison spectra) were given less weight. I adopted previous $v \sin i$ values where available. For ease of comparison with previous results, we used a simple two component model: The observed flux, F, is modeled as

$$F = fF_{\text{mag}}(B) + (1-f)F_{\text{nonmag}}(B=0),$$

where B is the average field strength of active regions and f is their fractional area coverage (filling factor). Fields were assumed to be homogeneously distributed and $F_{\text{mag}}(0) = F_{\text{nonmag}}(0)$.

No horizontal B distributions or vertical gradients were considered, although there is some evidence for both in dMe stars (Saar 1992a). The effect of a significant vertical gradient in B was tested for AD Leo, using a simple canopy, with B increasing exponentially with depth (Saar 1992a). In this model, the FWHM of the B distribution is $\sim$4 kG, about the same as B is in the model in which it is constant with height, although the average, $\overline{B}$, in this model is lower, $\sim$3.2 kG. The filling factor, f, on the other hand, is reduced by nearly a factor of two, from 65%, to 38%.

AU Mic, AD Leo, and EV Lac have had previous magnetic measurements and ξ Boo B had an upper limit (Saar and Linsky 1985; Saar, Linsky, and Giampapa 1987; Saar 1990). Using the new, better quality data now available and the more realistic line models described above, I have derived improved f and B values for these stars (Table 1 and Figs. 1 and 2). In general, f values are somewhat reduced from previous measurements. An upper limit can be set of f for HD 156026, a moderately active K dwarf rotating two times slower than ξ Boo B. Marginal detections of magnetic flux have been made in two RS CVn stars: σ Gem and HR 7275 (Fig. 2). No clear evidence for strong fields ($B \geq 4$ kG) are seen on these stars. Assuming a ratio $I_{C,\text{spot}}/I_{C,\text{quiet}} = 0.7$ (Saar and Linsky 1985), we can thus rule out high-B umbrae with $f_{\text{spot}} \gtrsim 0.20$ on either star. The equivalent width anomalies due to the magnetic intensification effect (see Leroy 1962; Basri et al. 1992) and slight line shape changes can be modeled well with fairly widespread fields in the 1–1.5 kG range. Due to their rapid rotation, f and B are poorly separated for the RS CVn stars; the product fB is the best determined quantity (see also Saar, Piskunov, and Tuominen 1992).

The stronger B seen in the M dwarfs relative to ξ Boo B is consistent with previous results (e.g., Saar and Linsky 1985; Saar et al. 1987). Taken together with the low B measured for the RS CVn stars, this result adds further evidence for a limit on B in the non-spot stellar active regions which is dependent on both T_{eff} and

TABLE I
New IR Measurements of Stellar Magnetic Parameters

Star	Sp. Type	R–I (B–V)	S/N[a]	B (kG)	f (%)	adopted B_{eq}(kG)	P_{rot} (days)	est. τ_c (days)
AU Mic	M1.6Ve	0.84	40	4.2	55	3.4	4.65	20
AD Leo	M3.5Ve	1.12	100	4.0	65	3.8	2.7	20
AD Leo[b]				3.2	38			
EV Lac	M4.5Ve	1.15	40	4.3	85	4.4	4.8	20
ξ Boo B	K4V	(1.17)	90	2.3	20	2.5	11	20
HD 156026	K5V	(1.16)	120	...	<20	2.5	18.5	20
σ Gem	K1IIIe	(1.12)	160	1.1:[c]	70:[c]	0.7	19.4	≈80
HR 7275	K1IV	(1.09)	190	1.4:[d]	40:[d]	1.2	≈28	≈50

[a] per resolution element
[b] ∇B model with $B_{min} = 2$ kG, $\overline{B} = 3.2$ kG, $B_{fwhm} = 4.3$ kG, $\sigma_B = 2$ kG.
[c] uncertain; $fB \approx 0.8$ kG is the best determined quantity
[d] uncertain; $fB \approx 0.55$ kG is the best determined quantity

luminosity class. The best explanation probably remains the following: B is limited by pressure balance with the ambient gas such that $B \leq B_{eq}$, where $B_{eq} \propto P_{gas}^{0.5}$ is the pressure equipartition field, (e.g., Galloway and Weiss 1981). I have estimated B_{eq} (at $\tau_{5000} = 1$) and find that B is *generally* near or below this value (see Table 1). The total magnetic flux ($\propto fB$) and f, on the other hand, both increase with angular velocity and inverse Rossby number, τ_c/P_{rot} (where τ_c is the convective turnover time, taken from Stepień 1989 and Basri 1987). This is again consistent with previous results (e.g., Saar 1990; 1991). The importance of the present results, however, is that they extend the relation to RS CVn's, more rapid rotators, and cool dMe stars. I refer the reader to Saar (1993) for a more complete analysis of these relationships.

Acknowledgements

This work is based on data obtained at NOAO. I thank K. Hinkle, W. Lenz, P. Russell, and D. Neff for help with the observations, and G. Basri, M. Giampapa, J. Linsky, G. Marcy, and S. Solanki for enlightening discussions. This research is supported by NASA grant NAGW-112, Interagency Transfer W-15130, and NSF grant INT-8900202.

References

Basri, G.S., Marcy, G.W., Valenti, J.: 1992, *Astrophys. J.* **390**, 622.
Basri, G.S., Marcy, G.W., Valenti, J.: 1990, *Astrophys. J.* **360**, 650.
Basri, G.S.: 1987, *Astrophys. J.* **316**, 377.

Bopp, B.W., Saar, S.H., Ambruster, C., Feldman, P., Dempsey, R., Allen, M., Barden, S.P.: 1989, *Astrophys. J.* **339**, 1059.
Galloway, D.J., Weiss, N.O. 1981, *Astrophys. J.* **243**, 945.
Giampapa, M.S. 1985, *Astrophys. J.* **299**, 781.
Giampapa, M.S., Golub, L., Worden, S.P.: 1983, *Astrophys. J. (Letters)* **268**, 121.
Hall, D.N.B., Ridgway, S., Bell, E.A., Yarborough, J.M. 1979, *S.P.I.E.*, **172**, 121.
Kurucz, R., Peytremann, E.: 1975, *A Table of Semiempirical gf Values*, SAO Rep. 362.
Leroy, J.L.: 1962, *Ann. d' Ap.*, **25**, 127.
Marcy, G.W., Bruning, D.H.: 1984, *Astrophys. J.* **281**, 286.
Saar, S.H.: 1992a, in *Cool Stars, Stellar Systems, and the Sun*, M. Giampapa and J. Bookbinder (eds.), ASP Conf. Ser. 26, p. 252.
Saar, S.H.: 1993, these proceedings.
Saar, S.H.: 1991, in *The Sun and Cool Stars: Activity, Magnetism, Dynamos*, I. Tuominen et al. (eds.), Springer, Berlin, p. 389.
Saar, S.H.: 1990, in *The Solar Photosphere: Structure, Convection, and Magnetic Fields*, J. O. Stenflo (ed.), Kluwer, Dordrecht, p. 427.
Saar, S.H., Golub, L., Bopp, B.W., Herbst, W., Huovelin, J.: 1990, in *Evolution in Astrophysics*, E. Rolfe (ed.), ESA SP-130, p. 431.
Saar, S.H., Linsky, J.L., Giampapa, M.S.: 1987, in *27th Liège Astrophys. Colloq.*, L. Delbouille and A. Monfils (eds.), U. de Liège, Liège, p. 103.
Saar, S.H., Linsky, J.L.: 1985, *Astrophys. J. (Letters)* **299**, L47.
Saar, S.H., Piskunov, N.E., Tuominen, I.: 1992, in *Cool Stars, Stellar Systems, and the Sun*, M. Giampapa and J. Bookbinder (eds.), ASP Conf. Ser. 26, p. 255.
Stepień, K.: 1989, *Astron. Astrophys.* **210**, 273.
Strassmeier, K.G., et al. : 1988, *Astron. Astrophys. Suppl.* **72**, 291.
Unno, W.: 1956, *Pub. Astron. Soc. Japan* **8**, 108.
Vogt, S.S., Soderblom, D.R., Penrod, G.D.: 1983, *Astrophys. J.* **269**, 250.

PART 6

THE INFRARED SPECTRUM

ATOMIC SPECTROSCOPY IN THE INFRARED*

E. BIÉMONT

Institut d'Astrophysique, Université de Liège, B-4000 Liège, Belgium

Abstract. Some recent developments of atomic spectroscopy in the infrared spectral range are briefly summarized. Different topics of interest for astrophysics, such as term analysis, far infrared measurements, isotope shifts, hyperfine structure observations and transition probability determination, are briefly considered.

Key words: atomic data – atomic processes – infrared: stars – line: identification

1. Introduction and Historical Comments

The interest of atomic spectroscopy for infrared solar physics can best be illustrated by a comparison of high resolution laboratory spectra obtained by Fourier Transform Spectroscopy (FTS) with recent recordings of the solar photospheric spectrum. As a specific example, Figure 1 shows that many infrared lines of the solar spectrum (Delbouille *et al.*, 1982) can easily be identified from a visual comparison with laboratory iron spectra (Biémont *et al.*, 1985a). In fact, the increasing resolution and signal-to-noise ratio which are now achieved for solar observations, and the possibility of disentangling the telluric and solar contributions (see *e.g.*, Farmer and Norton, 1989; Livingston and Wallace, 1991) require an increasing number of high quality atomic and molecular data.

Since the publication by Outred in 1978 of an extensive list of infrared lines (8885 lines belonging to 57 elements in the spectral range 1–4 μm), which was essentially a summary of the photographic and lead-sulfide observations available at that time, considerable progress has been made in infrared spectroscopy thanks to the use of FTS combined with improved detectors and more powerful computers. Further advances are also due to the use of more elaborate theoretical methods for investigating complex spectra and to the use of laser techniques in the far infrared region (Rydberg states spectroscopy and far-infrared laser magnetic resonance measurements).

The combination of FTS and hollow cathode (pulsed or not) has appeared to be very efficient not only for term analysis of neutral and singly ionized elements of solar interest but also for branching-ratio measurements, for isotope shifts, and hyperfine structure investigations.

2. Term Analysis

A recent summary has been published by Martin (1991). Due to space limitations, the references quoted in this section are restricted to neutral and singly ionized elements observed recently (in most cases after 1985) in the infrared region.

---

* Dedicated to the memory of Prof. M. Migeotte (Liège University), who died on February 25, 1992.

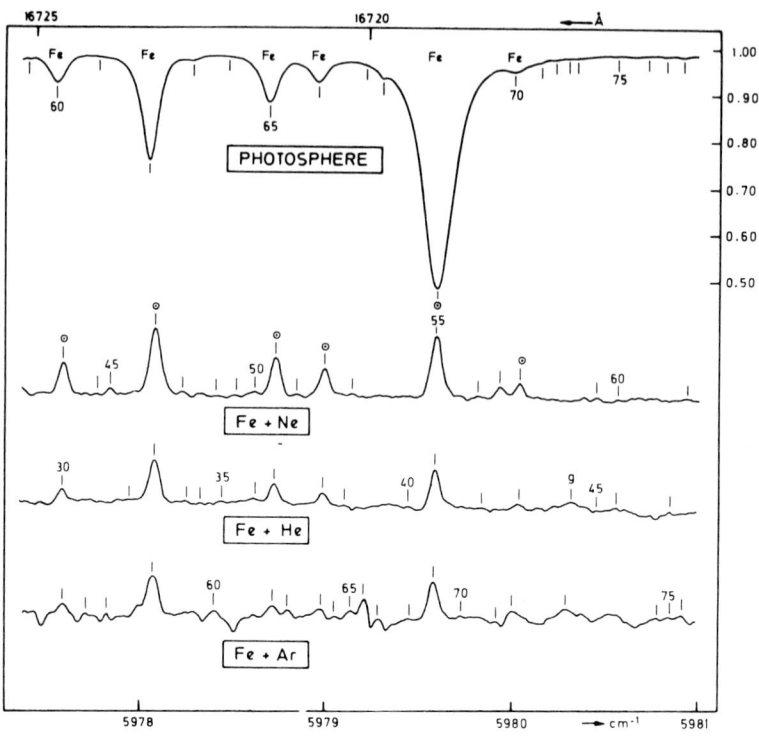

Fig. 1. A comparison of the photospheric absorption spectrum (top) (Delbouille *et al.*, 1982) and of the FTS emission hollow-cathode spectra of iron (bottom) (Biémont *et al.*, 1985a)

2.1. THE LIGHT ELEMENTS ($Z = 1$–19)

The compilations published by NIST during the past 7 years concern He I (Martin, 1987), Mg I–XII (Kaufman and Martin, 1991a), Al I–XIII (Kaufman and Martin, 1991b), P I–XV (Martin *et al.*, 1985), S I–XVI (Martin *et al.*, 1990) and K I–XIX (Sugar and Corliss, 1985). New observations of magnesium (Mg I, II) and aluminum spectra (Al I, II) have been performed in the infrared region (1.1–5.5 μm) by Biémont and Brault (1986, 1987) using a hollow-cathode source and the 1m FTS of the National Solar Observatory (Kitt Peak). An analysis of the O II spectrum in the photographic region has been carried out recently by Wenker (1990). The non-penetrating states of O I have been considered by Chang *et al.* (1988) and infrared laser absorption spectra of Rydberg transitions of the same ion have been reported by Brown *et al.* (1987).

2.2. The Iron Group Elements ($21 \leq Z \leq 28$)

Sugar and Corliss (1985) have summarized the energy level data for the elements K through Ni ($Z = 19$–28). In addition, updated line or level lists have been published for different ions [Fe I–XXVI : Wiese (1985); Sc I–XXI : Kaufman and Sugar (1988); Ti I–XXI, Cr I–XXIV, Ni I–XXVIII : Wiese and Musgrove, (1989)]. Recent review papers are due to Johansson and Cowley (1988) and Johansson (1991).

Infrared FTS observations of Sc I were carried out by Ben Ahmed and Vergès (1977) (up to 3.39 µm) and the theoretical analysis is due to Ben Ahmed (1977). The analysis of Sc II was extended by Johansson and Litzén (1980) (0.11–1 µm). The Ti I spectrum has been recently reconsidered by Forsberg (1991) in the region 0.19–5.5 µm using the Kitt Peak FT spectrometer. The analysis of Ti II was extended in the infrared (up to 1.1 µm) by Huldt et al. (1982). Infrared observations of V I (1.0–5.6 µm) were due to Davis et al. (1978). A monograph by Iglesias et al. (1988) includes all the data available for V II (visible and near infrared regions). New FTS observations of V I and V II in the infrared region (1.1–5.5 µm) have been obtained by J. Brault at the NSO. The spectra, obtained with a hollow-cathode and three different carrier gases (He, Ne and Ar), are presently being analyzed by E. Biémont (Liège) and P. Quinet (Mons).

The Cr I and II FTS infrared observations in the range 1.1–5.5 µm reported by Biémont et al. (1985b) extended the range of the available observations. Solar identifications only were considered by these authors but the analysis of the Cr I term system has been undertaken at Imperial College, London. For Cr II, the NIST compilation was based upon results obtained by Johansson (1983) from observations extending up to 5 µm. Since the publication of the NIST compilation (Sugar and Corliss, 1985), the spectrum of Mn I has been investigated by Taklif (1990a) in the region 0.82–1.97 µm using an electrodeless discharge tube.

The iron spectrum has been the subject of infrared observations (1–4 µm) by Litzén and Vergès (1976), and the term system has been extended by Litzén (1976). New FTS high resolution infrared observations have been reported by Biémont et al. (1985a) who prepared an extensive table of solar identifications. The term analysis of these observations is being carried out by Learner at ICL. A table of 360 Fe I infrared lines has been published by Johansson and Learner (1990). More recently, the $3d^6$ (^5D)$5g$ subconfiguration of Fe II has been established (0.9–1.1 µm) (Rosberg and Johansson, 1992). They complement earlier observations in the near infrared by Johansson (1978) using a pulsed hollow-cathode discharge.

The only infrared observations of Co I used in the NIST compilation are due to Russell et al. (1940) and do not extend further than 1.1895 µm. FTS spectra of Ni I and II registered at Kitt Peak have led to many new solar identifications (Biémont et al., 1986). The term analysis of Ni I is being carried out by Litzén (1992) and will be published soon. Approximately 625 lines have been identified in the region 1–5.5 µm. The most important new configurations involved in the observations are $3d^94f$, $5f$ and $5g$. Infrared observations of Ni II (1.8 µm) have been reported by Brault and Litzén (1983).

2.3. The Heavy Elements ($Z > 28$)

Several new compilations concerning all known spectra of Cu, Ge, Kr and Mo have been published (Sugar and Musgrove, 1990, 1992, 1991 and 1988). The analysis of Ga II has been extended by Isberg and Litzén (1985). Hollow-cathode spectra of yttrium have been registered in the wavelength range 0.1–4.8 µm and a new analysis of the Y II spectrum has been completed (Nilsson et al., 1991). Sb I has been the subject of two recent investigations in the lead-sulfide region by Hassini et al. (1988) (0.25–2.48 µm) and by George and Azhar (1990) (1.0–3.0 µm). The In I spectrum has been analyzed in the infrared by means of the FTS of the Laboratoire Aimé Cotton, Orsay (France) (George et al. 1990). Since the publication of the NBS tables for the lanthanides (Martin et al., 1978), results have been reported for Pr I (Ginibre, 1981), Pr II (Ginibre, 1989a,b, 1990), Nd II (Blaise et al., 1984), Eu I (Wyart, 1985) and Yb I (Aymar et al., 1984). New observations have been published for Pt I (Engleman, 1985; Reader et al., 1990) and Pt II (Reader et al., 1988, 1990). Results have also been reported for Bi I (George et al., 1985). A compilation of the actinides has been prepared by Blaise and Wyart (1992).

3. Wavelength Standards in the Infrared

Compilations of reference wavelengths in the spectral range 15 to 25000 Å, have been proposed by Kaufman and Edlén (1974) (the uncertainties lie between 0.0001 and 0.002 Å) and by Outred (1978). The new FTS measurements performed by Engleman (1985) for Pt I do not extend to wavelengths longer than 0.722 µm and the Pt II line list (Reader et al., 1988) contains only 3 infrared lines above 1 µm. A great deal of effort has been devoted to the production of standards for iron. An extensive, older, list (4000 lines) produced by Crosswhite (1975) covered the spectral range from 0.19 to 0.9 µm. The wavelength calibration of FTS Fe I emission spectra has been discussed in detail by Learner and Thorne (1988). They have shown that the accuracy of the calibration is limited by pressure shifts, by the inadequacy of available standards and by effects of illumination. The only line list available for the infrared region is due to Johansson and Learner (1990) (1.4–2.1 µm).

The classical reference for calibrating wavelengths in the infrared region remains the work of Norlén (1973) who reported accurate measurements for Ar I and Ar II using a Fabry-Perot interferometer in the vacuum; 76 Ar I reference lines ($4p$–$3d$ and $4p$–$5s$ transitions) cover the photoelectric region up to 2.4 µm. As underlined by Learner and Thorne (1988) from a comparison of their wavenumbers (Ar II) with those of Norlén (1973), pressure shifts have to be considered with care when calibrating the observed spectra. As the mean level shift in their work ($P = 3$ Torr) is found, when compared to Norlén's measurements ($P = 0.2$ Torr), to increase rapidly with the upper energy of the levels, the only lines suitable for calibration are those involving low excitation levels.

4. Far Infrared Measurements

The spectroscopy of atoms in the far infrared region (FIR) (i.e., roughly between 30 µm and 1 µm) is important because astrophysically abundant atoms (C, N, O,

Mg, Si, ...) or their isotopes can absorb or emit fine structure transitions in that region. The lines emitted can be used, for example, for the detection of atomic species in the cold interstellar medium or for the investigation of stratospheric composition. In fact atomic species such as C, O or Si can be studied very accurately only by far infrared (FIR) spectroscopy because their electronic transitions are not easily accessible to laser sources in the optical region.

FIR spectroscopy has seen a renewal during the past few years due to the development of new optically-pumped lasers [using CH_3OH, CH_2F_2 (etc.) molecules or their isotopes] and of liquid-helium-cooled bolometric detectors. A review paper on the subject has been published by Inguscio (1988). The FIR Laser Magnetic Resonance (LMR) technique is a sensitive, high resolution laboratory technique using an optically pumped laser as a source. The atoms to be investigated are situated inside the cavity of an optically pumped molecular laser. A change in the laser output is detected when a transition of the atom is tuned into coincidence with the laser transition by using an external magnetic field. The high sensitivity of the FIR-LMR technique has allowed spectroscopy of atomic fine structures, though it requires a close coincidence between an atomic transition and a known cw laser line and also the observation of M1 transitions (much weaker than the usual E1 transitions). The main fine-structure measurements obtained with that technique are summarized in Table 1. They appear considerably more accurate than the results determined from optical data. Though the FIR-LMR technique has been successful for the investigation of a number of atomic structures, the lack of close coincidences with laser lines has prevented the measurement of important fine structure intervals e.g., in the ground state of Si I, S I or N II.

5. Hyperfine Structures, Isotope Shifts and Zeeman Effect

The high resolution which is obtainable with the FTS in the infrared region makes possible the observation of hyperfine structures and isotope shifts. The consideration of such effects is important for investigating some solar profiles e.g., for deducing abundances, as underlined recently by Kurucz (1992). The importance of isotope shift for interpreting the infrared lines of nickel in the photospheric spectrum has been emphasized by Brault and Holweger (1981).

Many observed hypermultiplets are adequately represented by the well known Casimir formula. In Al II, it was shown by Biémont and Brault (1987) that the observed structures for $ns-n'p$ transitions were correctly reproduced by this formula while the observations for $np-n'd$, $nd-n'f$ and $nf-n'g$ transitions were not. Hyperfine structure profiles in the infrared have been the subject of recent experimental or theoretical investigations in Al II (Biémont and Brault, 1987), In I (George et al. 1990), Sb I (Hassini et al., 1988), Bi I (George et al. 1985), Pr II (Ginibre, 1989a,b) and Pt II lines (Engleman, 1989). Many hyperfine profiles are also observed in the vanadium spectra which are presently analyzed in Liège. An example is represented in Figure 2.

Zeeman patterns were considered for aiding in the identification work or for determining g factors in Zr I (Taklif, 1990b) and in Pr II (Ginibre, 1989a,b).

TABLE I
Far Infrared Measurements by the Laser Magnetic Resonance Technique

Ion	Transition	λ (μm)	σ (cm^{-1}) (FIR)		σ (cm^{-1}) (OPT)
^{12}C I	$2p^2\ ^3P_0 - {}^3P_1$	609.13536(7)	16.416712(2)	A	16.40a
		609.1348(14)	16.41673(4)	B	
		609.1333(9)	16.41677(2)	C	
	$^3P_1-{}^3P_2$	370.4144(4)	26.99679(3)	B	27.00a
		370.4139(13)	26.99683(9)	C	
^{13}C I	$2p^2\ ^3P_0-{}^3P_1$				
	(F, F'=1/2-3/2)	609.13088(9)	16.416833(2)	A	
	=1/2-1/2)	609.13599(21)	16.416695(6)	A	
	$^3P_0-{}^3P_1$	609.1327(25)	16.41678(7)	B	
		609.131(2)	16.41682(7)	C	
	$^3P_1-{}^3P_2$	370.4119(5)	26.99697(3)	B	
		370.413(2)	26.9969(1)	C	
^{12}C II	$2p\ ^2P^{\circ}_{1/2}-{}^2P^{\circ}_{3/2}$	157.74093(11)	63.39509(4)	D	63.42a
^{13}C II	$2p\ ^2P^{\circ}_{1/2}-{}^2P^{\circ}_{3/2}$	157.74019(17)	63.39538(7)	D	
	(F, F'=1-2)	157.74681(19)	63.39273(8)	D	
	=1-1)	157.7742(8)	63.3817(3)	D	
	=0-1)	157.70665(12)	63.40887(5)	D	
^{14}N I	$2p^3\ ^2D^{\circ}_{3/2}-{}^2D^{\circ}_{5/2}$				
	(F,F' =1/2-3/2)	1145.6209(4)	8.728891(3)	E	8.713b
	=3/2-5/2)	1146.4477(4)	8.722596(3)	E	
	=5/2-7/2)	1147.4900(2)	8.714673(2)	E	
^{14}N II	$2p^2\ ^3P_1-{}^3P_2$	121.89806(7)	82.03576(5)	F	82.1b
^{15}N II	$2p^2\ ^3P_1-{}^3P_2$	121.89750(9)	82.03614(6)	F	
^{16}O I	$2p^4\ ^3P_2-{}^3P_1$	63.183717(27)	158.26862(7)	G	158.265d
		63.183671(2)	158.268741(5)	H	
	$^3P_1-{}^3P_0$	145.52548(8)	68.71649(4)	J	68.712d
		145.525439(7)	68.716508(3)	H	
^{18}O I	$2p^4\ ^3P_1-{}^3P_0$	145.52474(8)	68.71684(4)	J	
^{24}Mg I	$3s3p\ ^3P^{\circ}_0-{}^3P^{\circ}_1$	498.592792(3)	20.0564472(1)	K	20.059e
	$^3P^{\circ}_1-{}^3P^{\circ}_2$	245.6157(6)	40.7140(1)	M	40.714e
^{25}Mg I	$3s3p\ ^3P^{\circ}_0-{}^3P^{\circ}_1$				
	(F, F' =3/2-5/2)	498.17611(1)	20.0732226(5)	N	
	=5/2-5/2)	498.46601(1)	20.0615484(5)	N	
	=7/2-5/2)	498.89415(1)	20.0443321(5)	N	
^{26}Mg I	$3s3p\ ^3P^{\circ}_0-{}^3P^{\circ}_1$	498.591377(3)	20.056504(1)	K	
^{28}Si I	$3p^2\ ^3P_0-{}^3P_1$	129.68173(4)	77.11187(2)	P	77.115c

References: a) Moore (1970); b) Moore (1975); c) Moore (1967); d) Moore(1976); e) Martin and Zalubas (1980) A) Yamamoto and Saito (1991); B) Cooksy et al. (1986c); C) Saykally and Evenson (1980); D) Cooksy et al. (1986a); E) Bogey et al. (1989); F) Cooksy et al. (1986b); G) Saykally and Evenson (1979); H) Zink et al. (1991); J) Davies et al. (1978); K) Bava et al. (1983); M) Inguscio et al. (1985); N) Godone et al. (1984); P) Inguscio et al. (1984).

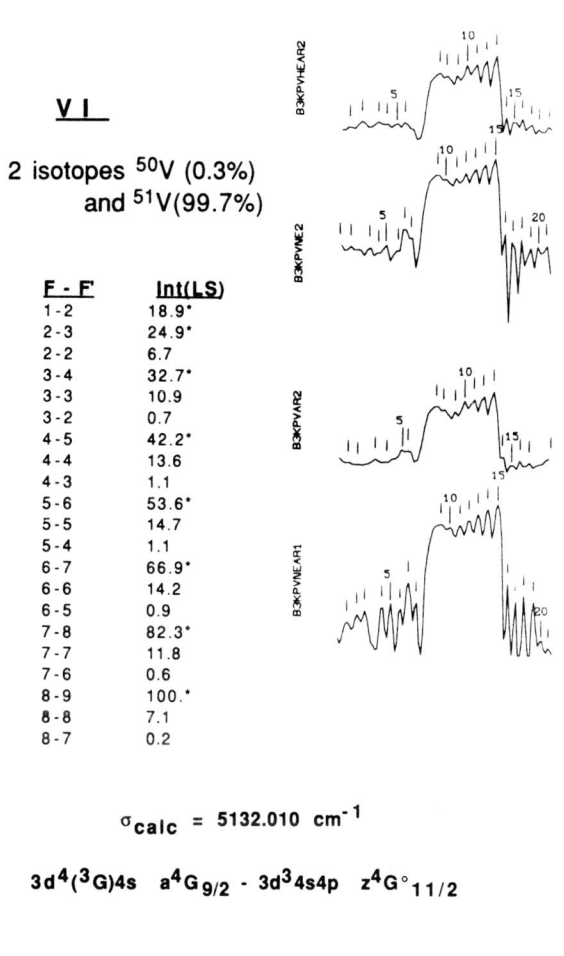

Fig. 2. Example of HFS in FTS vanadium spectra. The structure of the $3d^4(^3G)4s$ $a^4G_{9/2}$ − $3d^34s4p$ $z^4G°_{11/2}$ transition is illustrated. The most intense components are due to $\Delta F = 1$ transitions. The wavelength scale is in cm^{-1}.

6. Atomic Transition Probabilities

Transition probabilities for numerous infrared atomic lines have been calculated as part of two extensive projects developed during the past few years (Seaton, 1987, Kurucz, 1991). Through the combination of radiative lifetime measurements by laser techniques and of branching ratio determinations from FTS spectra, a large number of f-values have been obtained for Fe I (0.225–2.666 μm; O'Brian et al. 1991), Mo I (up to 1.0565 μm; Whaling and Brault, 1988) and Ce I (Bisson et al., 1991). In view of their astrophysical importance, N I and O I have been investigated in detail by Hibbert et al. (1991a,b) using the CIV3 method. Accurate oscillator strengths have been obtained for many infrared lines and the f values have been used for refining the solar abundances of these elements (Biémont et al., 1990, 1991) from a sample of near infrared lines. The results obtained ($A_N = 7.99 \pm 0.04$ and $A_O = 8.86 \pm 0.04$, in the usual logarithmic scale) are in agreement with the corresponding abundances deduced from molecular lines (see, e.g., Grevesse et al., 1992). Similar calculations for C I will be published soon.

Acknowledgements

The author thanks U. Litzén (Lund) for providing him with unpublished information concerning Ni.

References

Aymar, M., Champeau, R.J., Delsart, C., and Robaux, O.: 1984, *J. Phys.* **B17**, 3645.
Bava, E., DeMarchi, A., Godoni, A., Rovera, G.D., and Giusfredi, G.: 1983, *Opt. Commun.* **47**, 193.
Ben Ahmed, Z., and Vergès, J.: 1977, *Physica* **92C**, 113.
Ben Ahmed, Z.: 1977, *Physica* **92C**, 122.
Biémont, E., and Brault, J.W.: 1986, *Phys. Scr.* **34**, 751; 1987, *ibid.*, **35**, 286.
Biémont, E., Brault, J.W., Delbouille, L., and Roland, G.: 1985a, *Astron. Astrophys. Suppl.* **61**, 107; 1985b, *ibid.*, **61**, 185; 1986, *ibid.*, **65**, 21.
Biémont, E., Hibbert, A., Godefroid, M., Vaeck, N., and Fawcett, B.C.: 1991, *Astrophys. J.* **375**, 818.
Biémont, E., Froese Fischer, C., Godefroid, M., Vaeck, N., and Hibbert, A.: 1990, in J.E. Hansen (ed.), *Atomic Spectra and Oscillator Strengths for Astrophysics and Fusion Research*, North-Holland, p. 59.
Bisson, S.E., Worden, E.F., Conway, J.G., Comaskey, B., Stockdale, J.A.D., and Nehring, F.: 1991, *J. Opt. Soc. Am.* **B8**, 1545.
Blaise, J., Wyart, J.-F., Djerad, M.T., and Ahmed, Z.B.: 1984, *Phys. Scr.* **29**, 119.
Blaise, J., and Wyart, J.-F.: 1992, *Energy Levels and Spectral Lines of Actinide Atoms and Ions*, International Tables of Selected Constants Series, Vol. 20, Paris.
Bogey, M., Davies, P.B., Davis, I.H., and Destombes, J.L.: 1989, *Astrophys.J.* **339**, L49.
Brault, J.W., and Holweger, H.: 1981, *Astrophys. J.* **249**, L43.
Brault, J.W., and Litzén, U.: 1983, *Phys. Scr.* **28**, 475.
Brown, P.R., Davis, P.B., and Johnson, S.A.: 1987, *Chem. Phys. Lett.* **133**, 239.
Chang, E.S., Barowy, W.M., and Sakai, H.: 1988, *Phys. Scr.* **38**, 22.
Cooksy, A.L., Blake, G.A., and Saykally, R.J.: 1986a, *Astrophys. J.* **305**, L89.
Cooksy, A.L., Hovde, D.C., and Saykally, R.J.: 1986b, *J. Chem. Phys.* **84**, 6101.
Cooksy, A.L., Saykally, R.J., Brown, J.M., and Evenson, K.M.: 1986c, *Astrophys.J.* **309**, 828.
Crossswhite, H.M.: 1975, *J. Res. Natl. Bur. Stand.* **79A**, 17.
Davis, D., Andrew, K.L., and Vergès, J.: 1978, *J. Opt. Soc. Am.* **68**, 235.
Davies, P.B., Handy, B.J., Lloyd, E.K.M., and Smith, D.R.: 1978, *J. Chem. Phys.* **68**, 1135.

Delbouille, L., Roland,G., Brault, J., and Testerman, L.: 1982, *Photometric Atlas of the Solar Spectrum from 1,850 to 10,000 cm^{-1}*, Publ. Univ. Liège.
Engleman, R., Jr.: 1985, *J. Opt. Soc. Am.* **B2**, 1934; 1989, *Astrophys. J.* **340**, 1140.
Farmer,C.B., and Norton, R.H.: 1989, *A High-Resolution Atlas of the Infrared Spectrum of the Sun and the Earth Atmosphere from Space*, NASA Pub. 1224, Vol. I.
Forsberg, P.: 1991, *Phys. Scr.* **44**, 446.
George, S. and Azhar, A.: 1990, *J. Opt. Soc. Am.* **B7**, 697.
George, S., Munsee, J.H., and Vergès, J.: 1985, *J. Opt. Soc. Am.* **B2**, 1258.
George, S., Guppy, G., and Vergès, J.: 1990, *J. Opt. Soc. Am.* **B7**, 249.
Ginibre, A.: 1981, *Phys. Scr.* **23**, 260; 1989a, *ibid.* **39**, 694; 1989b, *ibid.* **39**, 710; 1990, *At. Data Nucl. Data Tables* **44**, 1.
Godone, A., Bava, E., and Giusfredi, G.: 1984, *Z. Phys.* **A318**, 131.
Grevesse, N., Sauval, J. and Blomme, R.: 1993, these proceedings.
Hassini, F., Ben Ahmed, Z., Robaux, O., Vergès, J., and Wyart, J.-F.: 1988, *J. Opt. Soc. Am.* **B5**, 2060.
Hibbert, A., Biémont, E., Godefroid, M., and Vaeck, N.: 1991a, *J. Phys.* **B24**, 3943; 1991b, *Astron. Astrophys. Suppl.* **88**, 505.
Huldt, S., Johansson, S., Litzén, U., and Wyart, J.-F.: 1982, *Phys. Scr.* **25**, 401.
Iglesias, L., Cabeza, M.I., and de Luis, B.: 1988, *Instituto de Optica "Daza de Valdès"*, Madrid, Publ. No. 47.
Inguscio, M., Evenson, K.M., Beltran-Lopez, V., and Ley-Koo, E.: 1984, *Astrophys.J.* **278**, L127.
Inguscio, M.: 1988, *Phys. Scr.* **370**, 699.
Inguscio, M., Leopold, K.R., Murray, J.S., and Evenson, K.M.: 1985, *J. Opt. Soc. Am.* **B2**, 1566.
Isberg, B., and Litzén, U.: 1985, *Phys. Scr.* **31**, 533.
Johansson, S.: 1978, *Phys. Scr.* **18**, 217; 1983, quoted by Sugar, J. and Corliss (1985); 1991, in C. Jaschek, Y. Andrillat (eds.), *The Infrared Spectral Region of Stars*, Cambridge Univ. Press, p. 87.
Johansson, S., and Cowley, C.R.: 1988, *J. Opt. Soc. Am.* **B5**, 2264.
Johansson, S., and Learner, R.C.M.: 1990, *Astrophys. J.* **354**, 755.
Johansson, S., and Litzén, U.: 1980, *Phys. Scr.* **22**, 49.
Kaufman, V., and Martin, W.C.: 1991a, *J. Phys. Chem. Ref. Data* **20**, 83; 1991b, *ibid.* **20**, 775.
Kaufman, V., and Sugar, J.: 1988, *J. Phys. Chem. Ref. Data* **17**, 1679.
Kaufman, V., and Edlén, B.: 1974, *J. Phys. Chem. Ref. Data* **3**, 825.
Kurucz, R.L.: 1991, in L.Crivellari, I. Hubeny, D.G. Hummer (eds.), *Stellar Atmospheres : Beyond Classical Models*, NATO ASI Series, Kluwer, Dordrecht.
Kurucz, R.L.: 1992, *Astrophys. J. Letters* (submitted).
Learner, R.C.M., and Thorne, A.P.: 1988, *J. Opt. Soc. Am.* **B5**, 2045.
Litzén, U.: 1976, *Phys. Scr.* **14**, 165; 1992, private communication.
Litzén, U., and Vergès, J.: 1976, *Phys. Scr.* **13**, 240.
Livingston, W. and Wallace, L.: 1991, *An Atlas of the Solar Spectrum in the Infrared from 1850 to 9000 cm^{-1} (1.1 to 5.4 μm)*, NSO Technical Report 91-001.
Martin, W.C.: 1987, *Phys. Rev.* **A36**, 3575; 1991, to be published in *Lecture Notes in Physics*, Proceedings of the IAU Meeting, Buenos Aires, August 1991.
Martin, W.C., Zalubas, R., and Hagan, L.: 1978, *Atomic Energy levels - The Rare-Earth Elements*, NSRDS-NBS 60.
Martin, W.C., and Zalubas, R.: 1980, *J. Phys. Chem. Ref. Data.* **9**, 1.
Martin, W.C, Zalubas, R., and Musgrove, A.: 1985, *J. Phys. Chem. Ref. Data.* **14**, 751; 1990, *ibid.*, **19**, 821.
Moore, C.E.: 1970, *Selected Tables of Atomic Spectra*, NSRDS-NBS 3, Section 3; 1975, *ibid.*, Section 5; 1967, *ibid.*, Section 2; 1976, *ibid.*, Section 7.
Nilsson, A.E., Johansson, S.,and Kurucz, R.L.: 1991, *Phys. Scr.* **44**, 226.
Norlén, G.: 1973, *Phys. Scr.* **8**, 249.
O'Brian, T.R., Wickliffe, M.E., Lawler, J.E., Whaling, W., and Brault, J.W.: 1991, *J. Opt. Soc. Am.* **B8**, 1185.
Outred, M.: 1978, *J. Phys. Chem. Ref. Data* **7**, 1.
Reader, J., Acquista, N., Sansonetti, C.J., and Sansonetti, J.E.: 1990, *Astrophys. J.* **72**, 831.
Reader, J., Acquista, N., and Sansonetti, C.J.: 1988, *J. Opt. Soc. Am.* **B5**, 2106.
Rosberg, M., and Johansson, S.: 1992, *Phys. Scr.* (in press).

Russel, H.N., King, R.B., and Moore, C.E.: 1940, *Phys. Rev.* **58**, 407.
Saykally, R.J. and Evenson, K.M.: 1979, *J. Chem. Phys.* **71**, 1564; 1980, *Astrophys. J.* **238**, L107.
Seaton, M.J.: 1987, *J. Phys.* **B20**, 6363.
Sugar, J., and Musgrove, A.: 1988, *J. Phys. Chem. Ref. Data* **17**, 155; 1990, *ibid.* **19**, 527; 1991, *ibid.*, **20**, 859; 1992, in preparation.
Sugar, J., and Corliss, C.: 1985, *Atomic Energy Levels of the Iron-group Period, Potassium through Nickel*, *J. Phys. Chem. Ref. Data* **14**, Suppl. 2.
Taklif, A.G.: 1990a, *Phys. Scr.* **42**, 69; 1990b, *ibid.*, **42**, 65.
Wenäker,I.: 1990, *Phys. Scr.* **42**, 667.
Whaling, W., and Brault, J.W.: 1988, *Phys. Scr.* **38**, 707.
Wiese, W.L.: 1985, *Spectroscopic Data for Iron, Atomic Data for Fusion*, ORNL-6089, US Dept. of Commerce.
Wiese, W.L., and Musgrove, A.: 1989, *Atomic Data for Titanium, Chromium and Nickel*, Vol. 1 Titanium, Vol. 2 Chromium, Vol. 3 Nickel, *Atomic Data for Fusion*, ORNL-6551, US Dept. of Commerce.
Wyart, J.-F.: 1985, *Phys. Scr.* **32**, 58.
Yamamoto, S., and Saito, S.: 1991, *Astrophys.J.* **370**, L103.
Zink, L.R., Evenson, K.M., Matsushima, F., Nelis, T., and Robinson, R.L.: 1991, *Astrophys.J.* **371**, L85.

THE *ATMOS* SOLAR ATLAS

CROFTON B. FARMER

Jet Propulsion Laboratory, California Institute of Technology, Pasadena, CA 91109, U.S.A.

Abstract. The *ATMOS* solar atlas covers the infrared spectrum of the Sun observed from space, over the frequency range from 650 to 4800 cm^{-1} at a resolution of 0.01 cm^{-1}. The spectrum reveals a large number of molecular and atomic features that are not visible in spectra taken from the ground. The acquisition of the spectra, details of their presentation in the atlas, and some highlights of the solar features are described.

Key words: artificial satellites, space probes – infrared: stars – instrumentation: interferometers – Sun: atmosphere

1. Introduction

The *ATMOS* solar infrared atlas (Farmer and Norton, 1989) was compiled from spectra obtained during the Spacelab-3 mission in April/May 1985. These observations, which were part of a continuing program of simultaneous measurements of the vertical profiles of the minor and trace gases in the earth's upper atmosphere, were made by recording the solar-telluric absorption spectrum at a spectral resolution of 0.01 cm^{-1} throughout the near and mid-infrared wavelength region, during sunrise and sunset occultation periods as seen from the Space Shuttle.

In order to be able to clearly identify the solar and any residual instrumental or spacecraft-associated spectral features, and remove them from the spectrum of the earth's atmosphere, the observations were designed to include – at the beginnings of the sunset occultations, and at the ends of the sunrises – a period during which spectra would be recorded with the tangent point of the Sun-spacecraft line well above the sensible atmosphere. After careful examination of the spectra, it was determined that no absorption features of atmospheric origin could be detected above about 150 km altitude, so that an altitude some two scale-heights above this (*i.e.*, 165 km) was defined as the top of the atmosphere for this purpose. Spectra obtained at tangent- point altitudes above 165 km were assumed to contain only features of solar or instrumental origin. Thus the program of detailed measurements of the infrared- active constituents of the earth's atmosphere provided the opportunity to record a large number of high resolution spectra of the Sun free from interfering telluric lines, covering the continuous wavelength range from 2 to 16 μm. These spectra, co-added to give a single average spectrum having a signal-to-noise ratio on the order of 2000 to 1, formed the basis of the solar reference spectra reproduced as Vol. 1 of the *ATMOS* atlas. A brief description of those aspects of the design of the instrument and the observations pertinent to the characteristics of the solar spectrum is given below.

2. The *ATMOS* Instrument

The requirement to make measurements of the vertical distributions of trace species throughout the stratosphere and mesosphere, with a vertical resolution of a few km, dictated the need to cover the 2–16 μm range at a spectral resolution of 10^{-2} cm^{-1}

in a scan time of about 1 second. (The rate of change of the tangent height of the Sun-spacecraft line of sight for a typical shuttle orbit is about 2 km s^{-1}.) The resulting instrument design specified a double-passed Michelson interferometer in which both retroreflectors ("cat's eyes") move in a parallel, reciprocal motion, symmetrical about the zero path difference (zpd) point. There are thus four single-sided interferograms in a total scan cycle of 4.2 s; data from the transformed single-sided scans sharing a common zpd were averaged, providing occultation sequences in which the effective altitudes of the individual spectra were spaced by about 4 km. While the minimum detectable equivalent width (MDEW) of features in the atmospheric spectra is not better than about 10^{-4} cm^{-1} (corresponding to a S/N ratio of 100:1), the relatively rapid scan time necessitated by the requirements of the atmospheric spectra works in favor of the solar spectra in the sense that many scans can be acquired in the time available at the ends of the occultations, resulting in an order of magnitude improvement in the MDEW of the final averaged solar spectra.

To optimize the measurements in terms of instrument performance and data rate under orbital conditions, the overall frequency range was divided into four narrower bands by means of optical filters. The bandpasses of the filters were chosen to be compatible with the alias limits of sampling at every second or every third fringe of the reference He:Ne laser, $i.e.$, 600–1200, 1100–2000, 1580–3400 and 3100–4700 cm^{-1}. The first three filters were used with a sampling interval of two laser fringes; for the fourth filter the interferogram was sampled at every third fringe. The instrument, which was built by Honeywell Electro-Optics Center, consists of four main parts: the suntracker, the foreoptics and frame camera, the interferometer and scan control system, and the electronics subunits, which include the closed-cycle detector cooler. Figure 1 is a schematic diagram of the optical system.

An important aspect of the instrument design, particularly in terms of its effect on the solar spectra, was the use of a thin aluminum cover which attached to the baseplate, enclosing all but the suntracker and camera. The pressure within the cover was maintained at ambient by means of a vent fitted with a dessicant filter designed to provide a clean, dry environment under all conditions (and, in particular, to protect the KBr beamsplitter and compensator during reentry into the atmosphere). This filter prevented complete evacuation of the air inside the cover, even after several days in orbit, so that residual lines of H_2O and CO_2 appear in all of the spectra. These lines fortuitously provide a valuable frequency reference for the rest frame against which the relative Doppler shifts of the solar lines could be verified. Although the numbers of molecules of water vapor and carbon dioxide were sufficient to cause the superimposed instrumental lines (about 2.5×10^{17} and 2.7×10^{15} cm^{-2}, respectively), the internal instrument pressure was too low to require an air-to-vacuum frequency correction to the data. The frequency stability of the instrument during the observations is estimated to be equivalent to an overall uncertainty of 3×10^{-4} cm^{-1} in the measurement of the frequency of a single line in a single spectrum.

In addition to the lines produced by the residual gas, several other absorption-like features of instrumental origin are superimposed on the solar and atmospheric spectra. These "artifacts", which are perhaps due to contamination of the surfaces

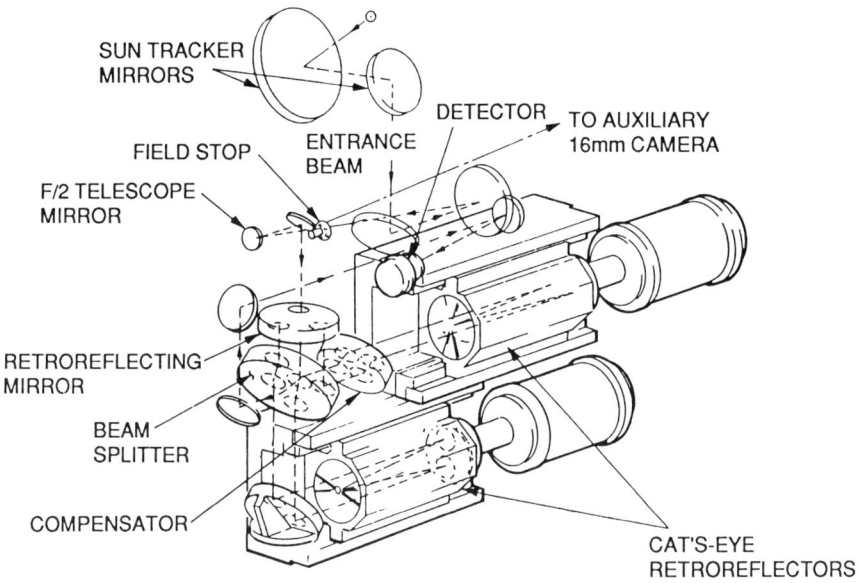

Fig. 1. Layout of the key optical components of the *ATMOS* interferometer.

of the optical components, can be distinguished from the residual gas lines by their width; whereas the latter appear at the instrument resolution (0.01 cm^{-1}), the artifacts are typically about 1 cm^{-1} wide.

The instrument views the Sun through a selectable field stop of 1, 2, or 4 mrad. which can be positioned with respect to the center of the disc in steps of 1 mrad. Thus coarse limb scans can be performed, or scans accumulated at chosen radial distances, although these options were not exercised during the first flight of the instrument. The suntracker is able to acquire and track the Sun over the entire hemisphere above the wing plane of the Shuttle; the tracking accuracy is 0.4 mrad, with a stability of 0.06 mrad.

3. Data Reduction

Details of the procedure used for transformation of the recorded interferograms to spectra are given in the Data Reduction section of Volume I of the Atlas. The point spacing of the transformed data is 0.00753 cm^{-1}, for filters 1, 2 and 3, and 0.00502 cm^{-1} for filter 4. The spectra were apodized (using function number 2 of Beer and Norton, 1976) and interpolated with four additional points between the primary points; the point spacings for the final spectra are thus 0.001506 cm^{-1} and 0.00100 cm^{-1}.

The variation of the Sun-spacecraft radial velocity during the acquisition of the solar scans was small, as these were taken at the highest tangent altitudes;

consequently, all of the solar spectra could be averaged together on a point-by-point basis to produce a solar reference spectrum for each occultation. The validity of this procedure was tested by cross-correlation of the individual spectra with the average. The high correlation coefficients found (see the descriptive text of the atlas for details of these results) provide a direct indication of the quality of the individual spectra making up the averages. A similar point-by-point averaging procedure was then used to generate "zonal average" spectra, or grand averages of all of the spectra obtained with each optical filter and at the same radial velocity (*i.e.*, a sunrise and a sunset average for each filter). Calculation of the cross-correlation coefficients between the occultation averages and the corresponding grand average again gave very high values (0.999944 in the worst case), because changes in the radial velocity resulting from changes in the orbital geometry over the two days during which data were acquired were too small to produce a significant spread in the frequency shift. Finally, in order to minimize any ambiguity in the identification of lines, the sunrise and sunset averages are displayed in the atlas with their frequencies corrected for the component of the relative velocity along the Sun-spacecraft line. In this way, lines whose positions appear shifted between the sunrise and sunset spectra are of instrumental origin and are easily distinguished.

4. Description of the *ATMOS* Solar Spectrum

In all, some 16,000 solar features appear in the *ATMOS* atlas, the majority of which are lines of the vibration-rotation bands of the diatomic molecular constituents of the photosphere: CO, CH, OH and NH. Of these, the $v = 1$ and $v = 2$ bands of $C^{12}O^{16}$ and of its isotopic variants dominate the spectrum as a whole, contributing more than a half of all of the lines of solar origin. In addition to the molecular lines, about 1700 atomic lines are present, due mainly to transitions in neutral Fe, Si, Mg, C, Ca, and Al. The atomic features represent about 11% of the total number of lines in this region of the infrared solar spectrum; almost 4000 lines (or 24% of the total) remain unidentified. The preparation of a key, listing the frequencies and identifications of the lines in the *ATMOS* spectrum, has been undertaken by M. Geller of JPL (Geller, 1992). This key has been submitted for publication as Volume III of the *ATMOS* atlas series. Table 1, taken from the key, summarizes the catalogued lines and their identifications. Note that $\Delta v = 1$ entries for CO include lines arising in $C^{12}O^{16}$, $C^{13}O^{16}$, $C^{12}O^{18}$; those for $\Delta v = 2$ are for $C^{12}O^{16}$ only.

4.1. MOLECULAR TRANSITIONS

Figure 2 shows an example of a portion of the spectrum in the region of the bandheads of the $v = 1$ bands of CO. The CO bands present a striking appearance in the solar spectrum seen without any interference from atmospheric absorptions, starting at 2328 cm^{-1} and progressing to lower frequencies, remaining visible to below 1400 cm^{-1}. With the instrumental CO$_2$ and H$_2$O lines clearly distinguishable, the high S/N ratio of the *ATMOS* spectra allows the sequences to be followed as far as $v'' = 20$ and J'' as high as 135. The increased range of rotational quantum values accessible in the spectrum has provided the basis for the revision of the

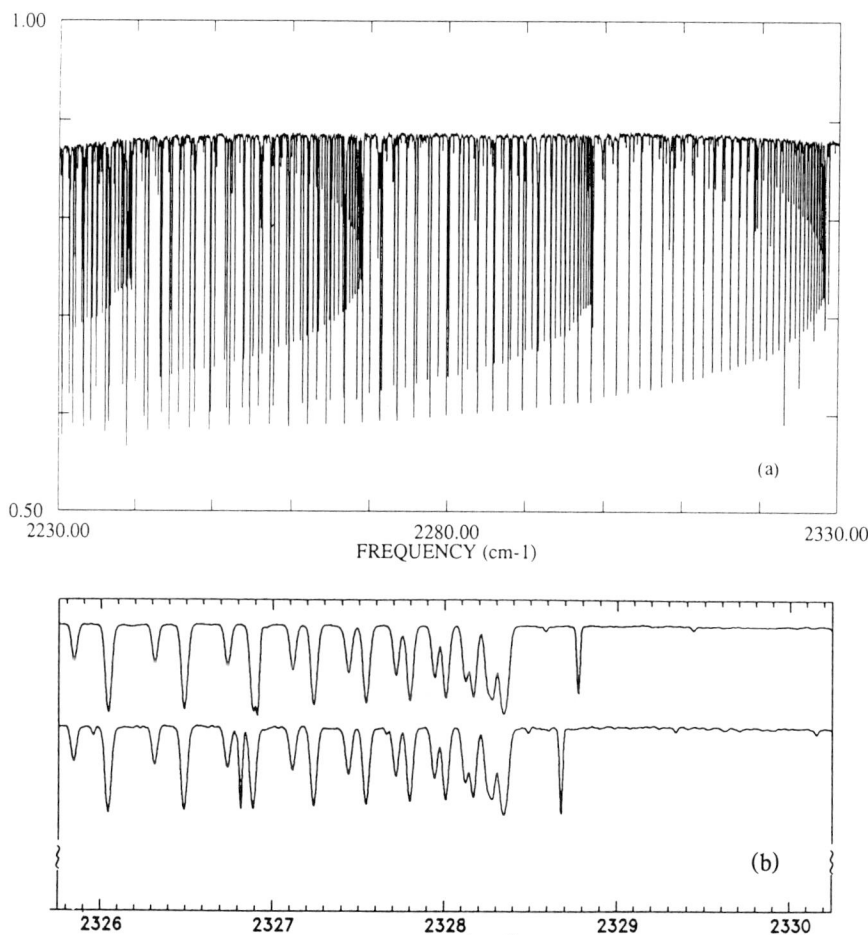

Fig. 2. (a) A low dispersion plot of a portion of the $\Delta v = 1$ sequence of CO bands, showing the first four bandheads; (b) A part of the 1-0 band reproduced at the dispersion of the atlas, illustrating the relative frequency shifts between the instrumental CO_2 lines in the sunset (above) and sunrise spectra.

TABLE I
ATMOS Solar Atlas
Summary of Catalogued Features and their Identifications (Geller, 1992)

Identification	Number of lines	Percent of total number
CO ($v = 1$)	6478	40.1
CO ($v = 2$)	2394	15.1
OH	743	4.7
CH	581	3.7
NH	181	1.1
Atomic	1709	10.8
Unidentified	3761	23.7

ground-state molecular constants for CO, by the use of the observed frequencies of the high-J lines seen in the spectrum together with available accurate laboratory measurements (Farrenq *et al.*, 1990). All of the C^{12} isotopes (O^{16}, O^{17}, and O^{18}), as well as $C^{13}O^{16}$, are observed in the $v = 1$ transitions. For the $v = 2$ bands, which cover the range from 4360 to 3330 cm^{-1} v'' as high as 16 (for $C^{12}O^{16}$) can be seen.

For OH, the *ATMOS* spectra reveal many more of the pure rotational lines (0-0 through 4-4 bands) in the 620 to 1100 cm^{-1} region than have been visible in ground- based solar spectra (Fig. 3). Some reservations regarding the quantitative fidelity in this region of the spectrum, perhaps resulting from uncorrected non-linearity in the detector, suggest that caution should be exercised in interpreting the intensities of the OH lines in terms of the solar oxygen abundance. This matter is being investigated at the present time. In addition to the 370 pure rotation lines, a similar number of vibration-rotation lines of OH have been identified between 1986 and 3508 cm^{-1} (Geller, 1992). As in the case of CO, the extended range of frequencies for OH transitions derived from the *ATMOS* spectrum is expected to be used to provide improved molecular constants for OH (Grevesse and Sauval, 1990).

Another diatomic molecule for which the fundamental vibration-rotation lines are excited to much higher rotational levels in the solar spectrum than is possible under laboratory conditions is CH. We have been able to measure a large number of new lines of the 1-0 through 4-3 bands; altogether some 558 lines have been used to derive new molecular constants for the $X^2\Pi$ ground state of CH (Mèlen *et al.*, 1989). From a selected set of these, Grevesse *et al.* (1991) have derived a solar carbon abundance (8.60 ± 0.05) which reduces considerably the overall uncertainty associated with values based on measurements in other spectroscopic regions. A portion of the spectrum containing several groups of lines of the fundamental CH stretching vibration, a region almost totally obscured from ground-based observations by the strong ν_3 band of CH$_4$, is shown in Figure 4. In the identification key to the *ATMOS* solar spectrum, Geller (1992) lists a total of 581 vibration-rotation lines of CH.

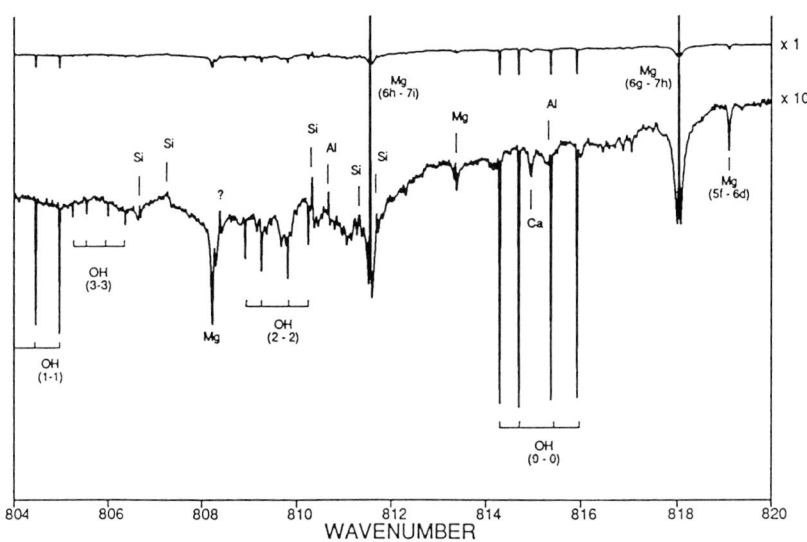

Fig. 3. Pure rotational lines of OH. The intensity scale of the lower trace has been expanded by a factor of 10.

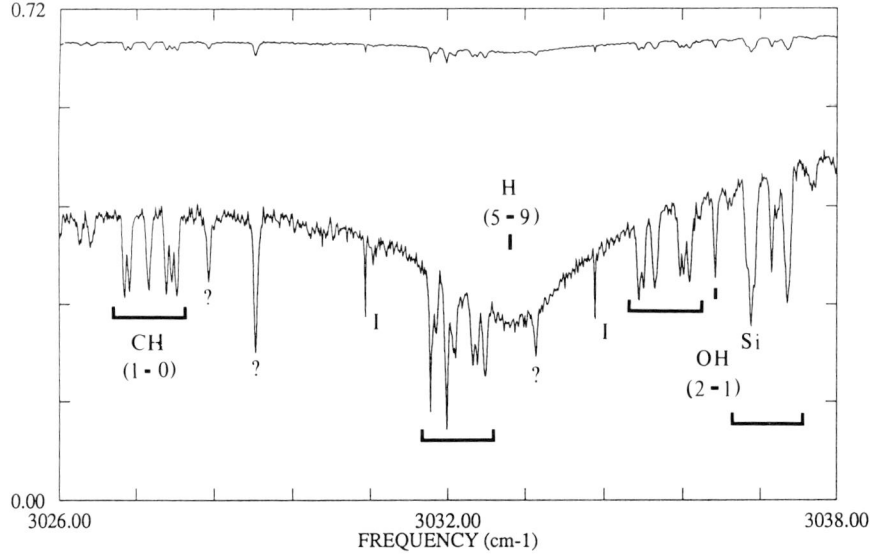

Fig. 4. CH manifolds in the 3 μm region. Lines due to residual gas in the instrument are marked with an I.

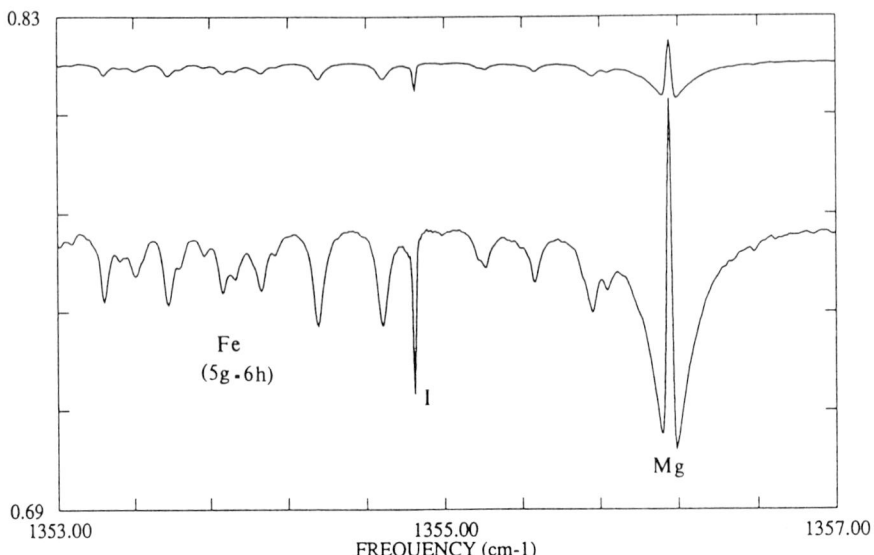

Fig. 5. A cluster of Fe lines in the vicinity of the Mg I $5g$–$6h$ emission feature.

Both vibration-rotation and pure rotational lines of NH have been observed for the first time in the *ATMOS* solar spectrum, as reported by Grevesse *et al.* (1990) and Geller *et al.* (1991), respectively. In total, 181 lines of NH (of which 36 are pure rotation lines) are identified in the key to the atlas.

4.2. ATOMIC TRANSITIONS

In addition to the molecular lines summarized above, more than 5000 other features, a large fraction of which remain unidentified, appear in the spectrum. The ability to examine the entire frequency range from 600 to 4700 cm^{-1} without the usual obscuration by broad absorptions in the earth's atmosphere, reveals that many of these features are grouped together in close proximity to the atomic hydrogen transitions. For example, groups of lines occur between 810 and 820 cm^{-1}, 1340 and 1360 cm^{-1} (shown in Fig. 5), and 2490 and 2530 cm^{-1}, which correspond to the hydrogen 6–7, 5–6, and 4–5 transitions at 808, 1340, and 2468 cm^{-1}, respectively. These include the prominent magnesium emission features observed in earlier ground-based spectra (Brault and Noyes, 1983) and suggested that many of the new lines could also be identified as atomic Rydberg transitions in the light elements. This was confirmed by Jefferies (1991) and Chang *et al.* (1991), who (along with the authors of several papers in these proceedings) have discussed both the relationship between the frequencies of these transitions and their hydrogenic analogues, and the formation of lines with characteristic emission cores.

In the initial attempts to identify the atomic features, Geller (1992) found that

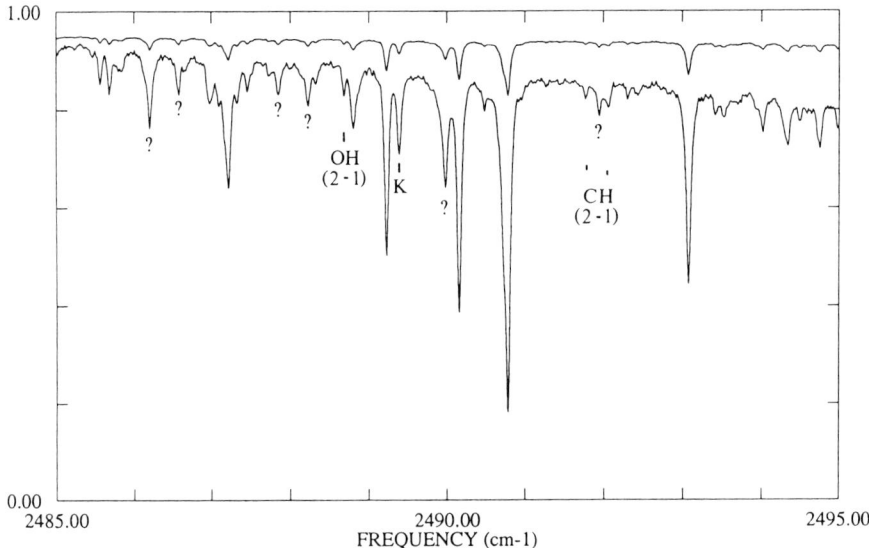

Fig. 6. The strong lines in this interval are Si transitions. Many of the smaller features are not yet identified.

the published compilations of data for the energy levels of the most probable elemental species were insufficient and that further calculations of atomic levels and transitions would have to be performed. As a result of these calculations, most notably for Fe and Si, by Geller and several collaborators, it has been possible to make unambiguous assignments for more than 1700 atomic transitions in the spectrum. (Details of the new energy levels and of work contributing to these new assignments by Chang, Grevesse, Jefferies, Johansson, Sauval and others can be found in Geller, 1992.)

Other examples of the spectra, in regions largely obscured from the ground, and in which several new atomic features occur, are shown in Figures 6 and 7. Figure 6 shows a group of strong Si lines together with a number of other unidentified features. Figure 7 shows a region dominated by $4f-5g$ transitions of Fe, with several smaller lines of the higher vibrational levels (4-3 and 3-2) of CH and OH. At the present time, of the total of almost 16,000 lines in the *ATMOS* solar spectrum, more than 3700 remain unidentified. Since many of the unidentified lines are relatively strong, it is probable that they are also due to transitions between undetermined

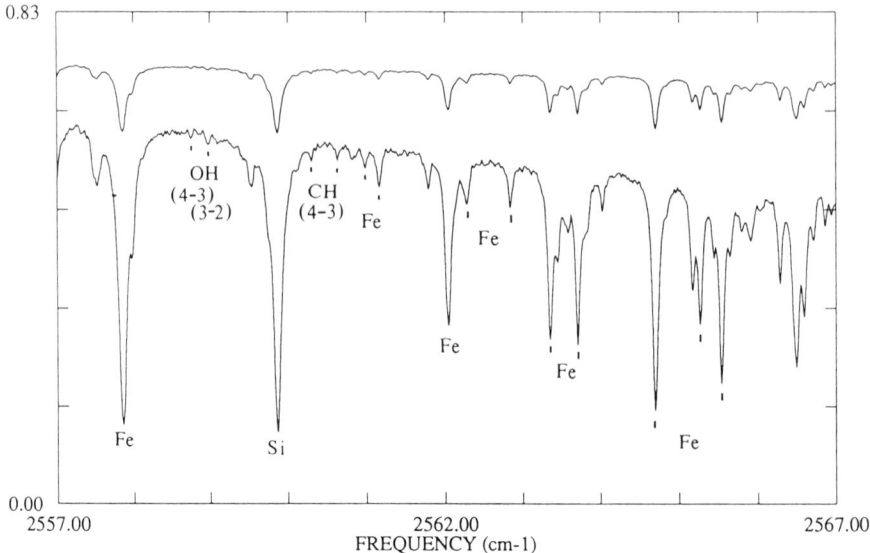

Fig. 7. In addition to the many Fe and Si lines in this region (all $4f$–$5g$ transitions), several CH and OH lines of the higher vibrational levels are found.

levels of the more abundant elements. The spectra will thus provide a valuable data base against which to validate further extensive calculations of energy levels in these species.

5. The Future

The solar infrared spectrum is rich in the number and variety of diagnostic features which occur in this wavelength region. This is particularly valuable because, even at high resolution, the spectrum is sufficiently uncluttered to allow detailed analysis of individual lines whose study bears on several important physical and chemical properties of the Sun. Amongst these are chemical abundances; thermal structure, radiative equilibrium and small-scale dynamics of the photosphere; and the distribution of electric and magnetic fields. However, a deeper understanding of the physics of the formation of these lines is necessary before their full potential as diagnostics of solar conditions can be realized. The *ATMOS* spectrum has provided the opportunity to view, for the first time, the continuous solar infrared spectrum, unbroken by the opaque regions of the Earth's atmosphere, and promises an exciting future for progress in these disciplines.

In the course of pursuing the long-term goals of the *ATMOS* experiment, NASA has planned to fly the instrument approximately once a year during the coming decade. *ATMOS* is currently included in the science payload of the first four Shuttle ATLAS missions, which began in March, 1992. From the point of view of solar infrared science these flights provide an opportunity to improve and extend the spectral data obtained from the Spacelab-3 flight, and to plan solar observations that make the best use of the capabilities of the instrument to serve the interests of solar physics. Further information regarding the availability of the *ATMOS* data and planning for future flight opportunities can be obtained from the Principal Investigator, Dr. M. R. Gunson, at JPL. It seems clear from the results obtained so far that future mission time devoted to acquiring more data to improve the S/N of the present spectra, and to making center-to-limb scans, would be of immediate value, while causing minimal perturbation to the primary atmospheric measurements. In addition, minor changes to the instrument, such as the use of smaller field apertures and a detector having a longer wavelength response (to the beamsplitter limit of 23 μm) would greatly enhance its potential value to solar science. The latter changes could be made without requiring any other modification to the interferometer or its electronic systems, and could perhaps be introduced for a single mission during which solar observations would be given priority. The cost effectiveness of this approach to obtaining data covering an unexplored wavelength region is compelling. It is hoped that for its future flights the full capabilities of the *ATMOS* instrument can be exploited to make observations designed expressly for these purposes.

References

Beer, R., and Norton, R. H.: 1976, *J. Opt. Soc. Am.* **66**, 3, 259.
Brault, J. W., and Noyes, R. W.: 1983, *Astrophys. J. (Letters)* **263**, 61.
Chang, E. S., Avrett, E. H., Mauas, P. J., Noyes, R. W., and Loeser, R.: 1991, *Astrophys. J. (Letters)* **379**, 79.
Farmer, C. B., and Norton, R. H.: 1989, *A High Resolution Atlas of the Infrared Spectrum of the Sun and the Earth Atmosphere from Space*, Vol. I: The Sun, NASA Ref. Pub. 1224.
Farrenq, R., Guelachvili, G., Sauval, A. J., Grevesse, N., and Farmer, C. B.: 1991, *J. Mol. Spectr.* **149**, 375.
Geller, M.: 1992, *A High Resolution Atlas of the Infrared Spectrum of the Sun and the Earth Atmosphere from Space*, Vol. III, in press.
Geller, M., Sauval, A. J., Grevesse, N., Farmer, C. B., and Norton, R. H.: 1991, *Astron. Astrophys.* **249**, 550.
Grevesse, N., Lambert, D. L., Sauval, A. J., van Dishoeck, E. F., Farmer, C. B., and Norton, R. H.: 1990, *Astron. Astrophys.* **232**, 225.
Grevesse, N., Lambert, D. L., Sauval, A. J., van Dishoeck, E. F., Farmer, C. B., and Norton, R. H.: 1991, *Astron. Astrophys.* **242**, 488.
Grevesse, N., and Sauval, A. J.: 1990, in C. Jaschek and Y. Andrillat (eds.), *The Infrared Spectral Region of Stars*, Proc. Montpellier Colloq. (October 1990), Cambridge University Press, Cambridge.
Jefferies, J. T.: 1991, *Astrophys. J.* **377**, 337.
Mèlen, F., Grevesse, N., Sauval, A. J., Farmer, C. B., Norton, R. H., Bredohl, H., and Dubois, I.: 1989, *J. Mol. Spect.* **134**, 305.

SYNTHETIC INFRARED SPECTRA

ROBERT L. KURUCZ

Harvard-Smithsonian Center for Astrophysics,
60 Garden St., Cambridge, MA 02138, U.S.A.

Abstract. The Sun is the star we can observe with the highest spectral resolution and signal-to-noise. From studying the infrared spectrum we can learn about the Sun, about stars in general, and about atomic and molecular spectroscopy. We discuss the computer programs for spectrum synthesis, the infrared flux and central intensity atlases of the solar spectrum, and the atomic and molecular line data. Considerable work is still required to improve the observations and to improve the line data.

Key words: atomic data – infrared: stars – line: formation – molecular data – Sun: atmosphere

1. The Importance of Studying the Solar Spectrum

The spectrum of a star consists of many thousands of lines blended together. Even at infinite resolution and signal-to-noise the blends are difficult to interpret. At low resolution and signal-to-noise a spectrum does not contain enough information for interpretation. Without a priori information from other sources, the analysis of such a spectrum is usually incorrect. I believe that we can learn more by studying the brightest stars with the highest possible resolution and signal-to-noise, than from any number of poor observations of fainter stars.

The Sun is the brightest star available to us. It is possible to observe the solar spectrum with a signal-to-noise of 10^4 and a resolving power of 10^6; but nobody has. In the Sun we can study contributions to blends at the 1 per mil level. Such lines can be better observed in the Sun than lines that are 1000 times stronger in a globular cluster star. There are many cases where lines can be seen in the Sun that have been difficult or impossible to see in the laboratory. The Sun is a unique spectroscopic source for studying atoms and molecules. Below I discuss the solar atlases that are available, but there is very little compared to what could be easily obtained, and there is very little compared to what is needed.

I have developed computer programs for producing model stellar atmospheres and for synthesizing spectra. I am collecting and computing data on all relevant atomic and molecular lines. I check the line gf values and damping constants by comparing the computed spectra to the observed spectra. Once I can compute realistic spectra for the Sun and the brightest stars these programs and data can be used to predict the spectra of stars that are too faint to observe well (or even stars from the early universe that no longer exist). Below I discuss these computer programs and the line data.

2. Spectrum Synthesis Programs

The spectrum synthesis computer programs have been under development since 1965 and have been described by Kurucz and Furenlid (1981) and by Kurucz and Avrett (1981). The algorithms for computing the total line opacity are extremely fast because maximum use is made of temperature and wavelength factorization and pretabulation. On a Cray computer a 500,000 point spectrum can be computed in one run. The same programs run on a VAX, only much more slowly. There is no limit to the number of spectrum lines that can be treated in LTE. I currently have 58,000,000. At present I can treat 50,000 lines including non-LTE effects. The line data are described below.

The spectrum calculations require a pre-existing model atmosphere that can be empirical, such as the Vernazza, Avrett, and Loeser (1981) solar models, or theoretical, such as the ones I describe below. The "model atmosphere" does not have to be stellar. It can be a disk, a planetary atmosphere, a laboratory source, etc. Quantities that need be computed only once for the model atmosphere are pretabulated. There can be a depth-dependent microturbulent velocity or a depth-dependent Doppler shift.

Line data are divided into two groups for treatment. In the first group, the lines must have a source function that is either the Planck function or some function that approximately accounts for non-LTE effects in the outer layers. The first group of lines is processed to produce a summed line absorption coefficient for the wavelength interval of interest, including radiative, Stark, and van der Waals broadening. The line center opacity is also saved for each line for subsequent computation of the central depth.

In the second group of lines, each line has its individual source function, which is taken to be the Planck function if the calculation is LTE, and which is determined from the departure coefficients in the model in a non-LTE calculation. This group of lines is processed by directly computing the line opacity and source function at every wavelength point.

The spectrum is computed with a version of the model atmosphere program ATLAS (Kurucz 1970) in which departure coefficients have been inserted in the partition functions, in the Saha and Boltzmann equations, and in the opacities. Departure coefficients for levels that are higher than have been computed are assumed to be the same as those for the ground state of the next higher stage of ionization. If the model atmosphere is in LTE the departure coefficients are all set to unity. The program computes the non-LTE opacity and source function, adds in the LTE opacity and source function, adds the continuum opacity and source functions, and then computes the intensity or flux at each wavelength point and for each line center. Photoionization continua are put in at their exact positions, each with its own cross-section and with the series of lines that merge into each continuum included so that there are no discontinuities in the spectrum.

Hydrogen line profiles are computed using a routine from Peterson (1979) that approximates the Vidal, Cooper, and Smith (1973) profiles, works to high n, and includes Doppler broadening, resonance broadening, van der Waals broadening, and fine-structure splitting. Autoionization lines have Shore-parameter Fano profiles. Other lines have Voigt profiles that are computed accurately for any value of the parameter a. A few strong lines can be treated with approximate partial redistribution effects but the computer cost increases dramatically.

To compute a rotationally broadened flux spectrum I first compute intensity spectra at 17 angles and then pass them through the rotation program. A grid of points is defined on the disk and, for the given $v \sin i$, the Doppler shift and angle are computed for each point. The intensity spectra are interpolated and summed over the disk to obtain the flux. In the rigid-body spherical approximation, symmetries are used to reduce the number of calculations, but the method works in the case of differential rotation as well.

To compute macroturbulent or instrumental broadening the broadening function is defined at integral values of the point spacing. Then the spectrum is read in, one wavelength at a time, redistributed among neighboring wavelengths, and added to a buffer for the new spectrum.

I also have a series of programs for computing the transmission of the spectrum through the Earth's atmosphere using the HITRAN database (Rothman *et al.* 1987) for the line data.

The most important step in the spectrum synthesis work is the final preparation of plots because I can display enough information to study the spectrum as a whole, to compare with one or more observed spectra, to study individual features in detail, and to identify lines and the relative composition of blends. Figures 1 and 2 show small sections of spectrum selected because they do not show dramatic discrepancies between the calculated and observed spectra. The figures show raw calculations, without adjustment. Usually the fits are much worse, with many missing lines.

3. Atlases

I have made a considerable effort to obtain observed spectra of the Sun and bright stars for testing my calculations. I have all the published atlases. Fortunately, Delbouille and Roland and the Griffins are committed to producing high quality atlases for the Sun and for bright stars, respectively. I have many solar FTS spectra from James Brault at Kitt Peak. In many cases I have had to take or reduce the spectra myself (Kohl, Parkinson, and Kurucz 1978; Kurucz and Avrett 1981; Kurucz and Furenlid 1981; Kurucz, Furenlid, Brault, and Testerman 1984), and projects are now underway for Sirius, Vega, and the Sun with a number of collaborators. Here I will describe a few of these atlases to give an impression of what is available in the infrared. In every case the wavelength coverage is incomplete and higher quality is possible and needed. I have a review paper (Kurucz 1992a) that shows each solar atlas plotted at very reduced scale.

The solar flux spectrum is important for its effects on atmospheric chemistry, on solar system objects, and on us, rather than for solar physics. In the flux spectrum much of the spatial and Doppler information about the solar atmosphere has been integrated away leaving a spectrum broadened and blended by the 2 km s^{-1} solar rotation. The flux spectrum is quite important for stellar physics, however, because the Sun serves as the "standard star". We can determine its properties much better than those of any other star. Solar flux spectra are required for planning and interpreting stellar and planetary observations because they have the resolution and signal-to-noise to show what is actually being observed.

As observations made from ground-based observatories include the atmospheric transmission spectrum, it is necessary to consider blending and blocking by terrestrial lines and to have resolution high enough to resolve their profiles. A solar flux spectrum observed from the ground is useful for indicating these problems. The spectrum should have a resolving power greater than 10^6 and a signal-to-noise

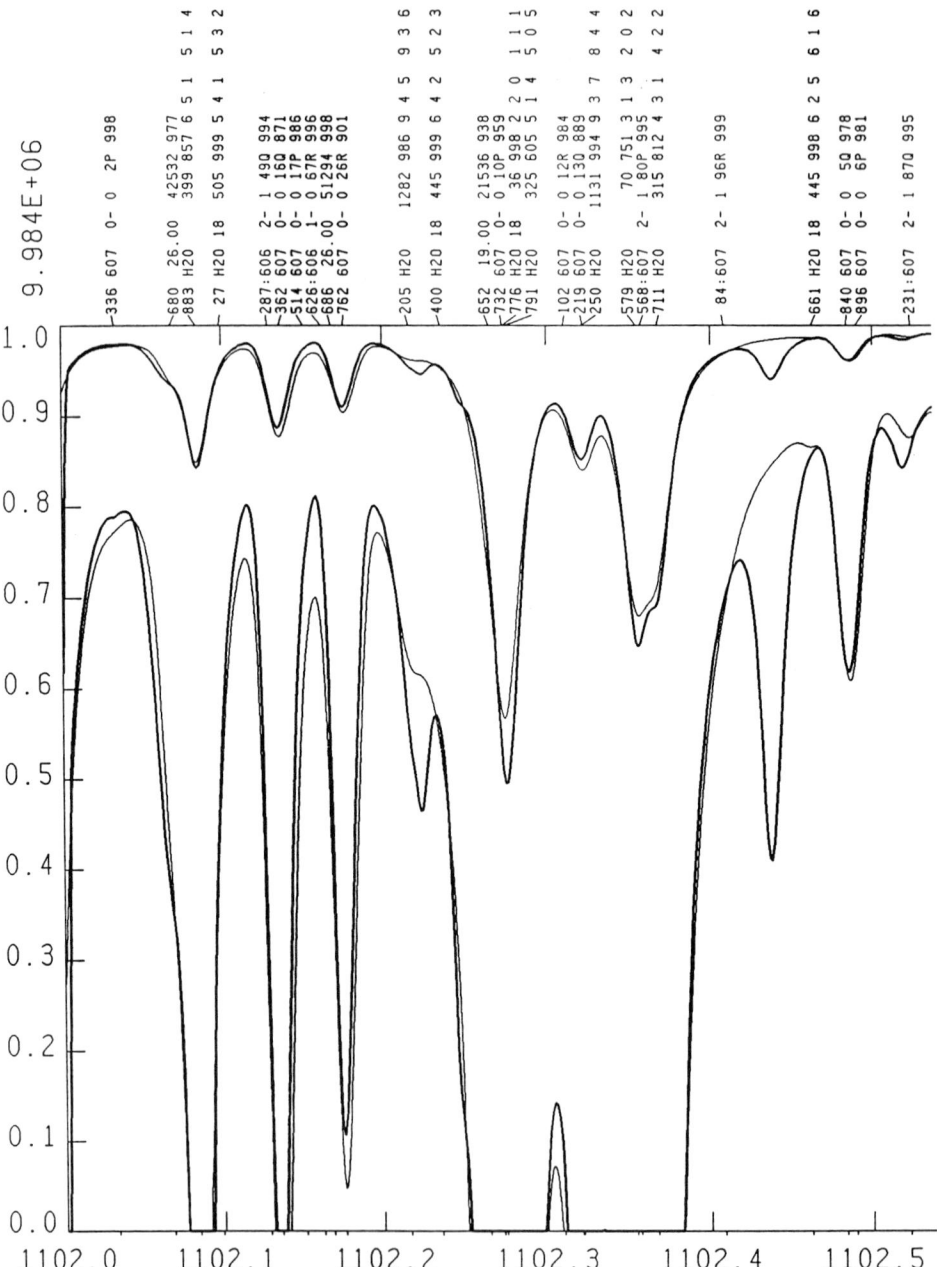

Fig. 1. A sample plot comparing a computed spectrum (thin line) to the 1.1 μm (1100 nm) central intensity spectrum observed by Brault (thick line). The spectra are shown twice, once at full scale, and once at 10 times scale. The long line labels are for terrestrial H_2O. The shorter labels are C_2, CN, Fe, and K lines. In each label the first 3 digits are the last 3 digits of the wavelength. The last 3 digits in the label are the per mil residual intensity of each line. Note the two obviously missing lines.

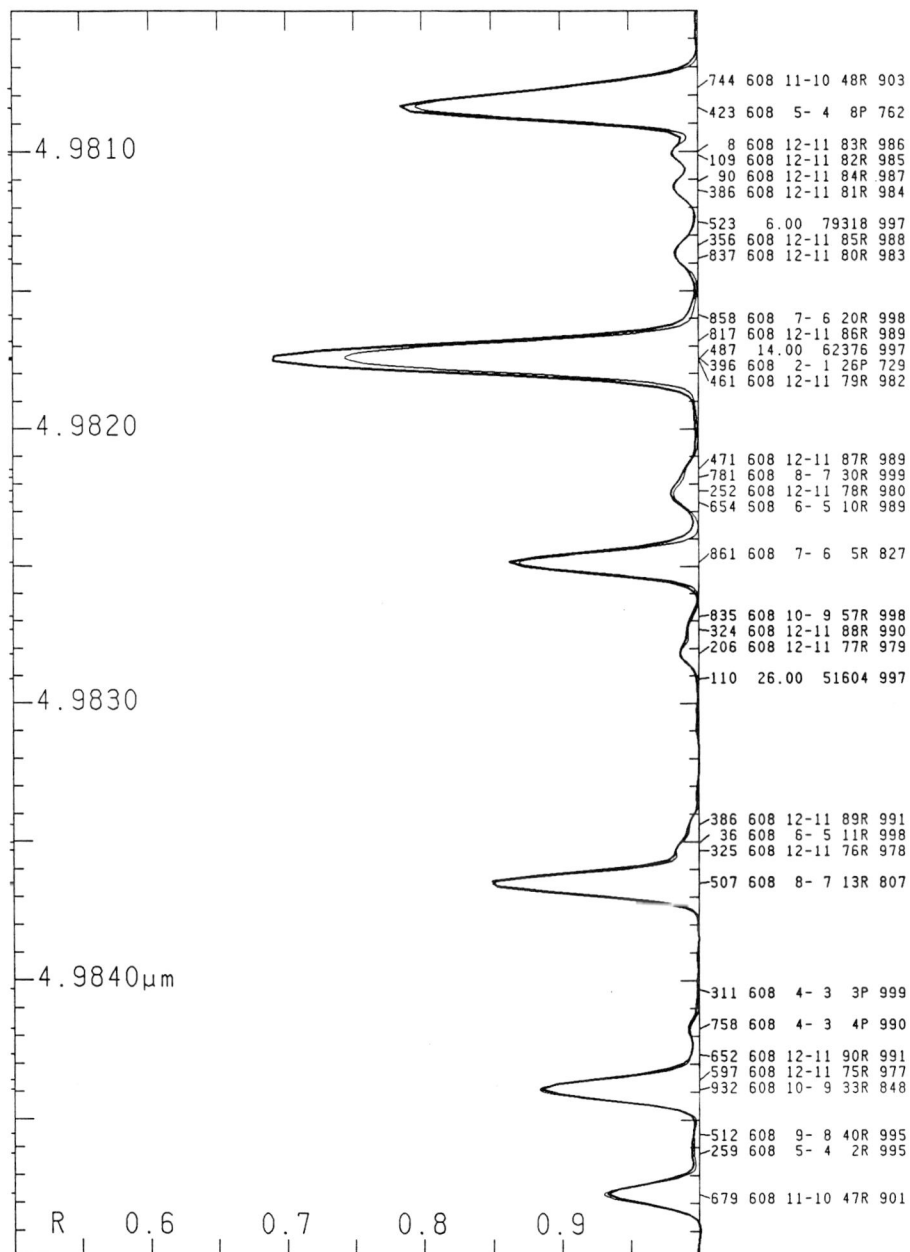

Fig. 2. A sample calculation near 5 μm (thin line) compared to the *ATMOS* central intensity spectrum described below (thick line). Except for single lines of C, Si, and Fe the lines are from highly excited bands of CO. In each label the first 3 digits are the last 3 digits of the wavelength. The last 3 digits in each label are the per mil residual intensity of each line.

greater than 10^4. For atmospheric chemistry and planetary and cometary atmospheres, and for space-based stellar observations, however, the true flux spectrum above the atmosphere is required.

The flux spectrum has been poorly observed. The existing atlas by Kurucz, Furenlid, Brault, and Testerman (1984) covers only the ground based spectrum up to 1.3 µm. It is very high quality by astronomical standards but still leaves considerable room for improvement. The atlas plots residual flux and also gives a table to convert to the absolute irradiance calibration by Neckel and Labs (1984). The spectrum was observed at Kitt Peak using the Fourier Transform Spectrograph on the McMath telescope at resolving power 522000 in the red and infrared. The resolution is not high enough to resolve the terrestrial lines so there is some ringing. The signal-to-noise varies from 2000 to 9000. The continuum level was estimated from high points and it is uncertain because of problems caused by broad structures in the atmospheric transmission produced by ozone and O_2 "dimer". In the infrared the O_2 "dimer" features are at 1.06 and 1.26 µm.

There are no high resolution flux atlases covering other wavelength regions in the infrared and there are none above the atmosphere. I do not expect there to be any improvement this century. In the meantime there are three approaches to approximating the flux spectrum. The first is to model the atmospheric transmission and then to divide the ground based spectrum by it. This should work quite well as long as the signal-to-noise is very high and the transmission is not near zero. The second method is semi-empirical: fitting a central intensity spectrum computed from a model to the observed central intensity spectrum and then using the derived line parameters to generate the flux spectrum. The problem is that a significant fraction of the lines in the spectrum have not been identified so they would have to be guessed. The third method is to compute a purely theoretical flux spectrum from the existing line data, but that is beyond the state of the art.

Intensity spectra are better for spectroscopy because there is no rotational broadening and so less blending. They are better for solar physics because they are determined by conditions in only a small region of the disk. Spectra-spectroheliograms show the spectrum at each resolution element, but they give almost too much information because they emphasize the instantaneous velocity field. The existing intensity atlases are space and time averages over a small area on the disk.

The Kitt Peak infrared central intensity atlas by Delbouille, Roland, Brault, and Testerman (1981) is the best available spectrum in the infrared. It is the combination of 9 FTS scans on the McMath telescope with resolving power about 400000 at 1 µm decreasing to about 130000 at 5 µm. The signal-to-noise varies from 3200 to 5200. Delbouille and Roland are redoing the atlas from Jungfraujoch to improve the resolution and signal-to-noise and especially to reduce the water vapor which is very bad on Kitt Peak.

Livingston and Wallace (1991) have just produced a new central intensity atlas for 1.1 to 5.4 µm with resolution 0.015 cm^{-1}. They observed at various airmasses and then reduced the spectra to 0 airmass where the atmospheric transmission was high enough to allow ratioing. They include many line identifications.

The JPL *ATMOS* experiment (Farmer and Norton 1989) was flown on the shuttle to obtain infrared FTS spectra of the atmosphere at sunset from which to measure trace molecules. Before sunset, solar intensity spectra were recorded. Wavelength coverage is 2 to 16 μm, resolution is 0.0147 cm^{-1}, and signal-to-noise varies from 1000 to 3000. These spectra show beautiful vibration-rotation bands of CO and hydrides. The two volume atlas of Farmer and Norton (1989) shows all the solar and terrestrial data. I have obtained the data tapes, reduced them, set the continuum level, and plotted the solar data on a very expanded scale. I can supply paper or magnetic tape copies of my version.

I plan to publish or republish atlases for the Sun and bright stars with the lines labeled, including terrestrial lines from the AFGL HITRAN line list (Rothman et al. 1987). I am synthesizing each spectrum and should be able eventually to deconvolve the blends and to deconvolve the atmospheric transmission where it is not near zero.

4. Atomic and Molecular Data Needs

All the calculations described above depend on having reliable gf values and damping constants for atomic and molecular lines, photoionization cross-sections, and, for non-LTE problems, collision cross-sections. My work on atomic and molecular line data and my line lists are described in Kurucz 1992b. Here I will concentrate on the present problems and on my future work.

Accurate energy levels, accurate wavelengths, accurate gf values, and accurate damping constants are required for computing spectra. Hyperfine and isotopic splitting are also required, and for magnetic structures, Landé g values are needed. We require completeness because we need to deconvolve blends. We need every level below the lowest ionization or dissociation energy. For molecules, every vibrational and rotational level is needed, not just the ones populated at low temperatures in the laboratory.

The procedure I use for generating line lists is straightforward and produces all the lines up to a specified cut off lower energy level. I start with all the known energy levels. I set up a model Hamiltonian that uses a Slater integral expansion for atoms or a rotational expansion for molecules. Then I do a least squares fit to determine the Slater integrals or the rotational constants. For atoms I use scaled Hartree-Fock starting guesses for the integrals. Once the fit has converged, I use the Hamiltonian to generate all possible eigenvalues and eigenvectors. The eigenvalues are replaced by the observed energies where they are known. For atoms I generate a scaled-Thomas-Fermi-Dirac wavefunction for each configuration and compute all the transition integrals. For molecules I compute the RKR potential and then all the vibrational wavefunctions. I then integrate over measured or computed transition moments taken from the literature to get the transition integrals. The transition integrals are divided into transition arrays in the adopted basis and are transformed to observed coupling using the eigenvectors. Given enough computer time, I can readily generate thousands of energy levels and millions of lines. Thus far I have produced atomic and molecular line lists with 58 million lines. Since the known energy levels are used when available, the line wavelengths are correct for lines between known energy levels. Lines to predicted levels are as accurate as the least squares fitting procedure. Radiative, Stark, and van der Waals damping constants and Landé g values are automatically produced for each line.

Problems arise when not all the "known" energy levels are really known. There are misassignments, typos, mistakes, etc. Also, the Hamiltonians are approximate, so any lines that occur only because of mixing may not be very reliable. If laboratory measurements exist for such lines, the laboratory measurements are always preferable to the calculation. I collect all published data on gf values and include them in the line list whenever they appear to be more reliable than the current data.

I hope to spend the next year improving the line data. I will extend the atomic calculation to elements lighter and heavier than the iron group which I have already computed. Several of the iron group calculations have already been revised. I will recompute the energy levels and line lists whenever new laboratory analyses become available and I will make the predictions available to laboratory spectroscopists. Because computers are now more powerful, I will increase the number of configurations treated. This should account for more of the missing infrared lines because they are usually transitions between highly excited levels.

The new complication I have recently discovered (Kurucz 1992c) is that isotopic splitting is important in the iron group. Hyperfine splitting of the odd iron group elements is well known. It turns out that the even elements have significant isotope splitting. In general, all lines are asymmetric because of substructure when resolution and signal-to-noise are high enough. Velocity measurements from bisectors and Fourier profile analysis to determine microturbulent and macroturbulent velocities cannot be reliable. Isotopic splitting introduces systematic errors in abundances and wavelengths because weak components will still be on the linear part of the curve of growth when the stronger components are becoming saturated. These effects can be very strong in the infrared because even a small energy shift can be significant compared to the transition energy. There are practically no published measurements of isotope splitting for the iron group. However, if you go into James Brault's office and look through the piles of infrared spectra, splittings are clearly visible. Brault and Holweger (1981) have published data for a few lines of Ni I in the infrared where the four isotopes are resolved. The infrared Ti lines that are used for Zeeman studies have isotopic components, so those studies must have systematic errors. Analysis of Brault's spectra and new laboratory measurements are urgently needed.

At the present time I am including hyperfine and isotopic splitting one level at a time from whatever laboratory data I can find. Given enough laboratory data, I think it should be possible to work semi-empiricially to generate the splittings for my whole computed transition arrays. The unfortunate result will be 10 times as many lines. For molecules, the isotopic splitting is so large that it is always treated, but there is also hyperfine splitting that is normally ignored in stars.

My molecular data are all for diatomic molecules and are electronic transitions except for the CO vibration-rotation bands. Most of the calculations were done more than 15 years ago. I need to include all the improvements in the laboratory analyses since that time and I need to add all the significant vibration-rotation bands. The newer analyses are based on FTS spectra and produce dramatic improvements in energy levels and line positions. However, they still do not go to high enough V

and J. Farrenq et al. (1990) have actually been able to use the solar spectrum itself in the *ATMOS* atlas to analyse CO to high J.

I plan to distribute tapes and CD-ROMs with files of energy levels, damping constants, Landé g values, lifetimes, branching ratios, and line gf values. These should be useful to both laboratory and astronomical spectroscopists.

Acknowledgements

This work is supported in part by NASA grant NSG-7054.

References

Delbouille, L., Roland, G., Brault, J., and Testerman, L.: 1981, *Photometric Atlas of the Solar Spectrum from 1850 to 10000 cm^{-1}*, Kitt Peak National Observatory, Tucson.
Farmer, C. B. and Norton, R. H.: 1989, *A High-Resolution Atlas of the Infrared Spectrum of the Sun and Earth Atmosphere from Space*, NASA Reference Pub. 1224.
Farrenq, R., Guelachvili, G., Sauval, A. J., Grevesse, N., Farmer, C. B.: 1991, *J. Molec. Spectrosc.* **149**, 375.
Kohl, J. L., Parkinson, W. H., and Kurucz, R. L.: 1978, *Center and Limb Solar Spectrum in High Spectral Resolution: 225.2 to 319.6 nm*, Harvard-Smithsonian Center for Astrophysics, Cambridge, MA.
Kurucz, R. L.: 1970, *Smithsonian Astrophys. Obs. Special Rep.* No. 309.
Kurucz, R. L.: 1992a, in A. N. Cox, W. C. Livingston, and M. Matthews (eds.), *The Solar Interior and Atmosphere*, University of Arizona Press, Tucson, p. 663.
Kurucz, R. L.: 1992b, *Rev. Mexicana Astron. Astrof.*, **23**, 45.
Kurucz, R. L.: 1992c, *Astrophys. J. (Letters)*, submitted.
Kurucz, R. L. and Avrett, E. H.: 1981, *Smithsonian Astrophys. Obs. Special Rep.* No. 391.
Kurucz, R. L. and Furenlid, I.: 1981, *Smithsonian Astrophys. Obs. Special Rep.* No. 387.
Kurucz, R. L., Furenlid, I., Brault, J., and Testerman, L.: 1984, *Solar Flux Atlas from 296 to 1300 nm*, National Solar Observatory, Sunspot, NM.
Neckel, H. and Labs, D.: 1984, *Solar Phys.* **90**, 205.
Livingston, W. and Wallace, L.: 1991, *National Solar Obs. Tech. Rep.* No. 91-001.
Peterson, D. M.: 1979, personal communication.
Pierce, A. K. and Breckinridge, J. B.: 1973, *Kitt Peak National Obs. Contribution* No. 559.
Rothman, L. S., Gamache, R. R., Goldman, A., Brown, L. R., Toth, R. A., Pickett, H. M., Poynter, R. L., Flaud, J.-M., Camry-Peyret, C., Barbe, A., Husson, N., Rinsland, C. P., and Smith, M. A. H.: 1987, *Appl. Optics* **26**, 4058.
Vernazza, J. E., Avrett, E. H., and Loeser, R.: 1981, *Astrophys. J. Suppl.* **45**, 635.
Vidal, C.R., Cooper, J., and Smith, E.W.: 1973, *Astrophys. J. Suppl.* **25**, 37.

LINE SHIFTS AND ASYMMETRIES IN THE IR SOLAR SPECTRUM

R. BLOMME and A. J. SAUVAL
Koninklijke Sterrenwacht van België/Observatoire Royal de Belgique,
B-1180 Brussels, Belgium

and

N. GREVESSE
Institut d'Astrophysique, Université de Liège
5, av. de Cointe, B-4000 Cointe-Liège, Belgium

Abstract. Line shifts and asymmetries of spectral lines have been found in the *ATMOS* spectra. These are due to the granulation at the solar surface. A two-component model for the solar photosphere is used to calculate theoretical profiles which are compared to the observations. The ATMOS lines provide additional constraints on models for the solar photosphere.

Key words: infrared: stars – Sun: granulation – Sun: photosphere

1. Introduction

Dravins *et al.* (1981, 1986) made a detailed study of line asymmetries observed in the Fe lines. The best way of presenting these asymmetries is through the use of the line bisector, *i.e.*, the line connecting the points midway between the blue and red side of the spectral line at equal intensity. These bisectors show a classical "C" shape instead of the vertical line that one would expect if the line were symmetric (see Fig. 1). When the deviations of the bisector are interpreted in terms of velocity, values of up to a few hundred m/s are observed.

Furthermore, when the laboratory wavelength of the line is known, one can compare it with the observed wavelength that has been corrected for the motion of the observer with respect to the sun, solar rotation and the gravitational redshift. The results show that the center of the line is shifted to the blue (see, *e.g.*, Dravins *et al.*, 1981; Nadeau, 1988).

Both phenomena can be explained as being due to the granulation present in the solar photosphere. Because the rising granules are hotter than the descending intergranular material, the Doppler shifts do not cancel exactly and an asymmetric line is formed; this is described in detail by de Jager (1959).

These asymmetries are not limited to lines in the visible region, but are also found in the infrared (see, *e.g.*, Nadeau, 1988). In this paper we try to fit the observed bisectors using a two component model for the granulation.

2. Observations

Preliminary experiments with our two-component model showed that when the calculated *stronger* lines showed a bisector similar to the observed ones then, for the same model, so did the *weaker* lines. We therefore decided to concentrate on the stronger lines. In the visible part of the spectrum, we selected 3 Fe I lines and 3 Fe II lines (see Table I).

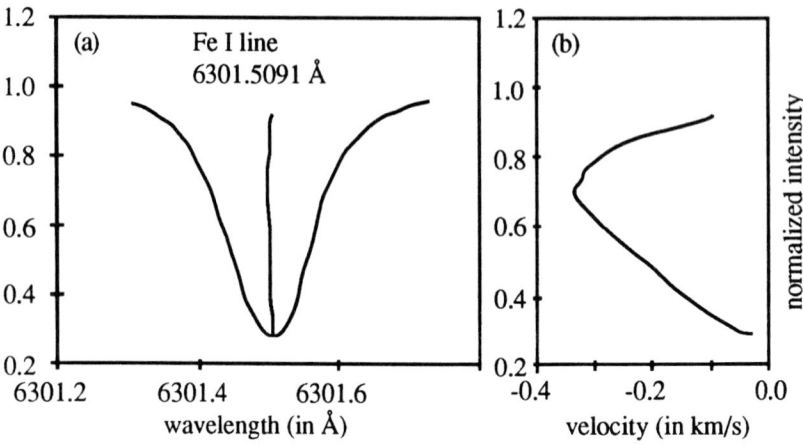

Fig. 1. (a) The observed line profile of an Fe I line (from Delbouille et al., 1973). (b) An enlargement of the bisector where the wavelength shifts have been interpreted as velocities.

TABLE I

The spectral lines used in this analysis. Solar wavelengths from Pierce and Breckinridge (1973) are given for Fe, and calculated wavenumbers from Farrenq et al. (1991) for CO.

Designation	Wavelength (Å)	Designation	Wavenumber (cm^{-1})
Fe I		CO	
$a^5P_1 \rightarrow y^5D_2^0$	6297.8013	4-3 R (18)	2128.2738
$z^5P_2^0 \rightarrow e^5D_2$	6301.5091	7-6 R (55)	2128.8350
$z^5P_1^0 \rightarrow e^5D_0$	6302.5017	5-4 R (28)	2129.3799
		7-6 R (65)	2140.9275
		4-3 R (28)	2156.6710
		1-0 R (8)	2176.2835
Fe II		3-2 R (73)	2262.7267
		3-2 R (75)	2264.1705
$a^6S_{5/2} \rightarrow z^6P_{3/2}^0$	4923.930	3-2 R (78)	2265.9903
$b^4F_{9/2} \rightarrow z^6P_{7/2}^0$	4993.3527	1-0 R (42)	2273.5527
$b^4F_{9/2} \rightarrow z^6F_{7/2}^0$	5100.6563	2-1 R (110)	2288.8991
		1-0 R (58)	2303.1663
		1-0 R (61)	2307.4949

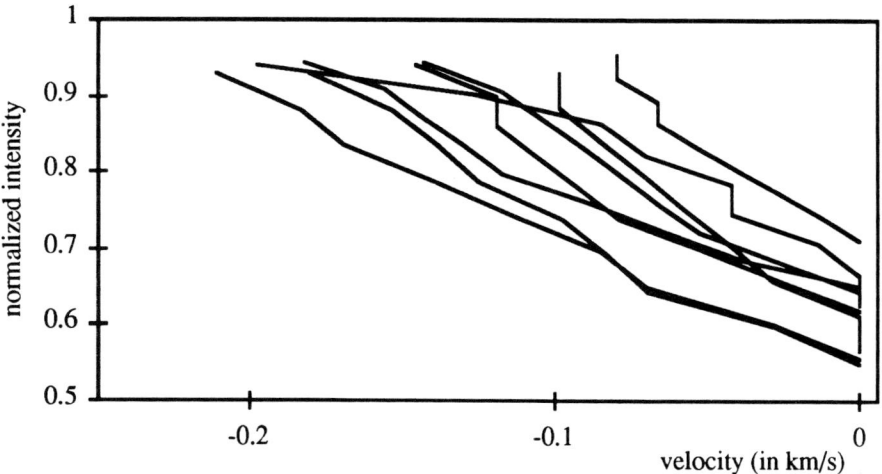

Fig. 2. Some representative examples of observed line bisectors for the CO lines. All bisectors have their footpoints shifted to zero.

In the infrared part of the spectrum we studied the spectra taken by the *ATMOS* experiment (Farmer and Norton, 1989). As not sufficient information is available to determine an absolute wavenumber scale, we cannot determine the absolute shifts of the line centers. The relative shifts show a very good correlation with the optical depth at which the line center is formed. Weaker lines (formed at larger depths) show a higher line shift than stronger lines (formed higher in the photosphere) (see also Grevesse and Sauval, 1991). For 13 CO lines that did not show any sign of blending (Table I), we determined the bisectors shown in Figure 2. They do not have the classical "C" shape as for the lines in the visible; instead the top part of the "C" seems to be missing, which is probably due to the different depths at which the continuum is formed for these wavelengths. Note that for these CO lines the observed asymmetries cannot be due to isotopic splitting – not even partially – contrary to what has been claimed for atomic lines (Kurucz, 1992).

3. Theoretical Model

We use a two-component model for the solar photosphere in which the hot component represents the rising granules and the cool component the descending intergranular material. We have to specify the relative surface areas of both components and their run of temperature as a function of depth. For the hot component we also specify the run of the velocity: the equation of mass conservation then gives the run of velocity for the cool component as well.

For each component the number densities of all molecules, atoms, and ions are calculated (in Local Thermodynamic Equilibrium) and the radiation transfer equation is solved for wavenumbers (or wavelengths) near the spectral lines of interest.

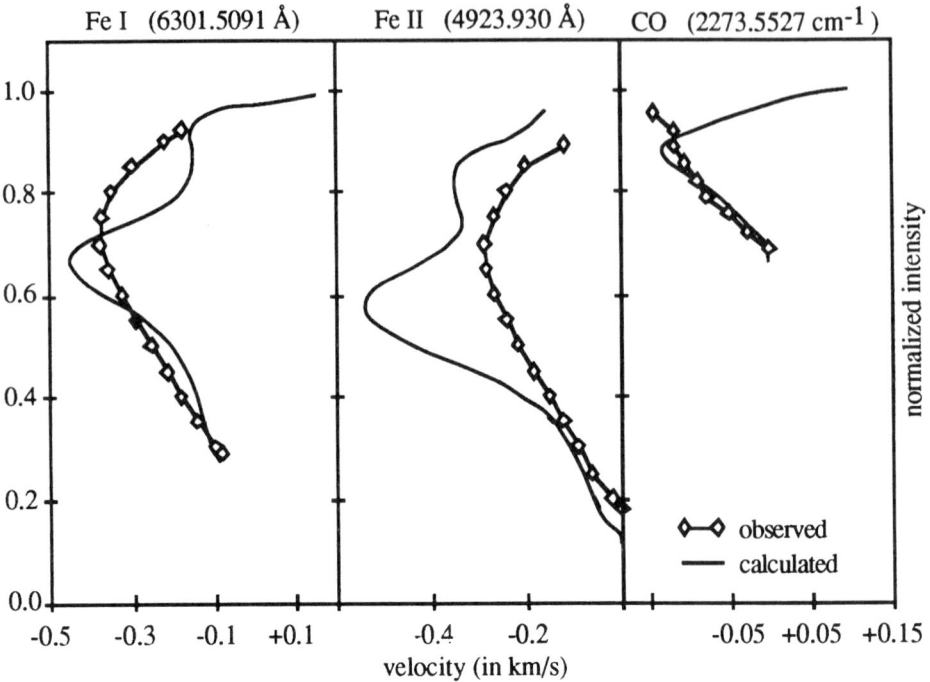

Fig. 3. The fit of our model calculation to the observed bisectors.

The sum of both line intensities (weighted by the relative surface areas) divided by the equivalent sum for the continua gives the resulting line profile. From this the bisector can be obtained and compared with the observed one.

4. Results

The best fit we have found so far uses the Holweger and Müller (1974) solar model photosphere for the temperature structure of the cool component. The hot component is based on this as well but the temperature was increased by a percentage that increases linearly from zero at $\log \tau_0 = -1.3$ up to 24% at $\log \tau_0 = +0.8$, where τ_0 is the optical depth at 5000 Å for the Holweger and Müller model. For deeper layers a constant value of 24% was taken. For the velocity law of the hot component we took an exponential function:

$$v = 1.472 \exp(-z/150)$$

where v is the velocity in km/s and z is the geometrical depth in km, measured from the point where $\tau_0 = 1$. The results for three typical lines are presented in Figure 3. While the agreement between theory and observations is reasonable, some improvement is needed, especially for the Fe II line.

5. Conclusions

It must be stressed that the present results are only preliminary. The parameter-space is very large and has not yet been exhaustively explored. Already some conclusions can be drawn, however. The data we have for the infrared CO lines give complementary information not found from the lines in the visible. This might have been expected *a priori* because the CO molecule is very sensitive to temperature and CO lines are formed over a very large range of formation depths. *A posteriori*, our model calculations confirmed this, as we sometimes found models that could explain the CO bisectors very well, but not the Fe ones, and *vice-versa*. This complementary information should allow us to derive a unique solution for the run of temperature and velocity (as discussed for example, by Kaisig and Durrant, 1982).

A solar photospheric model should explain not only the line bisectors but also the observed continuum intensities, line intensities and equivalent widths. In future work we shall also look at the weaker lines – which were left out here based on a qualitative argument. Furthermore, as has been pointed out above, complementary information can be obtained by using other atoms/molecules and by studying as large a spectral range as possible.

Finally one should also note that these techniques are not limited to the sun, but CO lines have also been used to study turbulence in cool stars (see, *e.g.*, Tsuji, 1991).

Acknowledgements

We thank Drs C.B. Farmer and M. Gunson (Jet Propulsion Laboratory) for permission to use the *ATMOS* infrared solar spectra before publication, for their hospitality and their generous help. We also thank W. Nijs for his help with implementing the computer code. All calculations were carried out on the Apollo computers at the Royal Observatory (Brussels).

References

De Jager, C.: 1959, in *Handbuch der Physik* **LII** , Springer, p. 80.
Delbouille, L., Neven, L., Roland, G.: 1973, *Photometric Atlas of the Solar Spectrum from* $\lambda 3000$ *to* $\lambda 10000$, Institut d'Astrophysique, Université de Liège.
Dravins, D., Larsson, B., Nordlund, Å: 1986, *Astron. Astrophys.* **158**, 83.
Dravins, D., Lindegren, L., Nordlund, Å: 1981, *Astron. Astrophys.* **96**, 345.
Farmer, C.B., Norton, R.H.: 1989, *A High-Resolution Atlas of the Infrared Spectrum of the Sun and the Earth Atmosphere from Space*, NASA Ref. Publ. 1224, Volume I.
Farrenq, R., Guelachvili, G., Sauval, A.J., Grevesse, N., Farmer, C.B.: 1991, *J. Molec. Spectrosc.* **149**, 375.
Grevesse, N., Sauval, A.J.: 1991, in C. Jaschek and Y. Andrillat (eds.), *The Infrared Spectral Region of Stars*, Cambridge University Press, Cambridge, p. 215.
Holweger, H., Müller, E.A.: 1974, *Solar Phys.* **39**, 19.
Kaisig, M., Durrant, C.J.: 1982, *Astron. Astrophys.* **116**, 332.
Kurucz, R.: 1992, *Astrophys. J. Letters*, preprint.
Nadeau, D.: 1988, *Astrophys. J.* **325**, 480.
Pierce, A.K., Breckinridge, J.B.: 1973, *The Kitt Peak Table of Photographic Solar Spectrum Wavelengths*, Contrib. No. 559, Kitt Peak National Observatory, Tucson, Arizona.
Tsuji, T.: 1991, *Astron. Astrophys.* **245**, 203

SOLAR ABUNDANCES OF C, N, AND O

N. GREVESSE

Institut d'Astrophysique, Université de Liège,
avenue de Cointe, 5 B-4000 Liège, Belgium

and

A. J. SAUVAL and R. BLOMME

Observatoire Royal de Belgique/Koninklijke Sterrenwacht van België,
B-1180 Brussels, Belgium

Abstract. We briefly review the many indicators, atoms as well as molecules, of the photospheric abundances of C, N and O and present preliminary updated values of these abundances.

Key words: C, N, O – infrared: stars – stars: abundances – Sun: abundances

1. Introduction

We briefly recall the main reasons why accurate values of the solar abundances of C, N and O are so important.

They contribute about 70 % to the metallicity Z; their detailed contribution to this metallicity is of crucial importance, as has been shown by the new opacity calculations for stellar interiors like OPAL (Iglesias and Rogers, 1991) and the OPACITY PROJECT (Seaton et al., 1992) and for stellar envelopes (Kurucz, 1991).

The solar photosphere is the only reliable source for the abundances of these elements. It is well known that they have partially escaped from meteorites. Furthermore, although their abundances can be derived from the coronal spectrum and from solar wind and solar energetic particles data, the fractionation process between the outer solar layers and the photosphere makes accurate comparisons difficult (see e.g. Anders and Grevesse, 1989).

It is also well known that the C/O ratio is a crucial parameter for the physicochemistry during the early phases of the evolution of the solar system.

2. Indicators of the Solar C, N and O Abundances

With the availability of the *ATMOS* infrared solar spectra obtained from space (Farmer and Norton, 1989) and covering the region from 2 to 16 μm, we have been able to use all the best indicators of the abundances of these elements, *i.e.*, atomic as well as molecular lines. The best indicators should have reliable atomic data (gf-values) and/or molecular data (transition probabilities, dissociation energies) as well as solar data (equivalent widths). We therefore disregarded some transitions in the visible spectrum which are too difficult to measure with accuracy in the solar spectrum because of the increasing blending problem as one goes to shorter wavelengths. The indicators we retained are the following: C I and [C I], N I, O I and [O I], CH (A-X), CH vibration-rotation, C_2 (Swan, Phillips and Ballik-Ramsay), NH vibration-rotation and pure rotation, OH vibration-rotation and pure rotation, CN red system, CO vibration-rotation ($\Delta v = 1$ and 2). Most of these indicators are in the infrared.

3. This Work

In several recent papers (Sauval et al., 1984; Grevesse et al., 1984; Grevesse et al., 1990, 1991; Geller et al., 1991; Grevesse and Sauval, 1991) we have analyzed these indicators and have shown how remarkably well they lead to the same results. It also became clear from these, and previous studies whose references are given in these papers, that permitted atomic lines are not the best indicators because of problems with the transition probabilities (although much progress has recently been made; see Biémont et al., 1991a; Hibbert et al., 1991) and of possible non-LTE effects. The best indicators are without any doubt the numerous molecular lines and particularly, the infrared lines.

We also showed (Grevesse and Sauval, 1991) how the numerous CO lines ($\Delta v = 1$ and 2), which can now be measured with high accuracy on the *ATMOS* infrared solar spectra, can be used to refine the solar photospheric model of Holweger and Müller (1974) that we have used throughout these analyses.

Very recently the solar photospheric abundance of iron – which was found to be higher ($A_{Fe} = 7.67$ (in the usual scale where log $N_H = 12.00$; Blackwell et al., 1984) than the meteoritic value (7.51; Anders and Grevesse, 1989) – was decreased down to the meteoritic value thanks to the use of higher excitation lines of Fe I, and to lines of Fe II for which accurate gf-values have recently been obtained (Holweger et al., 1990; Holweger et al., 1991; Biémont et al., 1991b; Hannaford et al., 1992; Johansson et al., 1993).

As Fe is a substantial electron donor, this decreased abundance has led to modifications in the electron and gas pressures. These modifications have non-negligible effects on the temperature structure as derived from the CO infrared lines, and therefore on the abundances derived from the many different atomic and molecular indicators of the C, N and O abundances described in Section 2.

Work is in progress to reanalyze all these effects in detail. Preliminary results indicate that our previous results should be decreased by about 0.05 dex. Thus, the preliminary recommended values of the solar abundances of C, N and O are:

$A_C = 8.55$
$A_N = 7.99$
$A_O = 8.87$.

The infrared CO vibration-rotation bands show numerous lines due to $^{13}C^{16}O$, $^{12}C^{18}O$, and even $^{12}C^{17}O$ which was identified for the first time in the solar photospheric spectrum. Isotopic ratios derived from these lines agree with the terrestrial ratios but, new more accurate, transition probabilities are urgently needed for these isotopic species.

Acknowledgements

We thank W. Nijs (Brussels) for his continuous help with the calculations carried out at the Apollo computers of the Royal Observatory. We also gratefully acknowledge the hospitality and help of the *ATMOS* Data Facility Team (C.B. Farmer and M. Gunson) at the Jet Propulsion Laboratory. We also thank the Belgian Fonds National de la Recherche Scientifique for financial support.

References

Anders, E., and Grevesse, N.: 1989, *Geochim. Cosmochim. Acta* **53**, 197.
Biémont, E., Hibbert, A., Godefroid, M., Vaeck, N., and Fawcett, B.C.: 1991a, *Astrophys. J.* **375**, 818.
Biémont, E., Baudoux, M., Kurucz, R.L., Ansbacher, W., and Prinnington, E.H.: 1991b, *Astron. Astrophys.* **249**, 539.
Blackwell, D.E., Booth, A.J., and Petford, A.D.: 1984, *Astron. Astrophys.* **132**, 236.
Farmer, C.B., and Norton, R.H.: 1989, *A High-Resolution Atlas of the Infrared Spectrum of the Sun and the Earth Atmosphere from Space*, Vol. 1, The Sun, NASA Ref. Pub. 1224, Washington, D.C.
Geller, M., Sauval, A.J., Grevesse, N., Farmer, C.B., and Norton, R.H.: 1991, *Astron. Astrophys.* **249**, 550.
Grevesse, N., Sauval, A.J., and van Dishoeck, E.F.: 1984, *Astron. Astrophys.* **141**, 10.
Grevesse, N., Lambert, D.L., Sauval, A.J., van Dishoeck, E.F., Farmer, C.B., and Norton, R.H.: 1990, *Astron. Astrophys.* **232**, 225.
Grevesse, N., Lambert, D.L., Sauval, A.J., van Dishoeck, E.F., Farmer, C.B., and Norton, R.H.: 1991, *Astron. Astrophys.* **242**, 488.
Grevesse, N., and Sauval, A.J.: 1991, in C. Jaschek and Y. Andrillat (eds.), *The Infrared Spectral Region of Stars*, Cambridge University Press, Cambridge, p. 215.
Hannaford, P., Lowe, R.M., Grevesse, N., and Noels, A., 1992, *Astron. Astrophys.* (in press).
Hibbert, A., Biémont, E., Godefroid, M., and Vaeck, N.: 1991, *Astron. Astrophys. Suppl.* **88**, 505.
Holweger, H., and Müller, E.A.: 1974, *Solar Physics* **39**, 19.
Holweger, H., Heise, C., and Kock, M.: 1990, *Astron. Astrophys.* **232**, 510.
Holweger, H., Bard, A., Kock, A., and Kock, M.: 1991, *Astron. Astrophys.* **249**, 545.
Iglesias, C., and Rogers, F.: 1991, *Astrophys. J.* **371**, 173 and 408.
Johansson, S., Nave, G., Geller, M., Sauval, A.J., and Grevesse, N.: 1993, these proceedings.
Kurucz, R.L.: 1991, in L. Crivellari, I. Hubeny and D.G. Hummer (eds.), *Stellar Atmospheres: Beyond Classical Models*, Kluwer, Dordrecht.
Sauval, A.J., Grevesse, N., Brault, J.W., Stokes, G.M., and Zander, R.: 1984, *Astrophys. J.* **282**, 330.
Seaton, M.J., Zeippen, C.J., Tully, J.A., Pradhan, A.K., Mendoza, C., Hibbert, A., and Berrington, K.A.: 1992, *Revista Mexicana de Astronomia y Astrofisica* **23**, 19.

ANALYSIS OF VERY HIGH EXCITATION Fe I LINES (4f–5g) IN THE SOLAR INFRARED SPECTRUM

S. JOHANSSON

Department of Physics, University of Lund, Sölvegatan, 14, S-223 62 Lund, Sweden

G. NAVE

Blackett Laboratory, Imperial College of Science, Technology and Medicine, London SW7 2BZ, England

M. GELLER

Jet Propulsion Laboratory, 4800 Oak Grove Drive, Pasadena, CA 91109, U.S.A.

A. J. SAUVAL

Observatoire Royal de Belgique, avenue circulaire, 3, B-1180 Brussels, Belgium

and

N. GREVESSE

Institut d'Astrophysique, Université de Liège, B-4000 Cointe-Liège, Belgium

Abstract. We present a detailed analysis of very high excitation lines (4f–5g) of Fe I which are present in the spectral region 2545–2585 cm^{-1} in high resolution spectra both in the laboratory and in the *ATMOS* solar spectra obtained from space. A value of the solar abundance of iron which agrees with the meteoritic value is derived.

Key words: infrared: stars – line: identification – Sun: abundances

1. Introduction

The identification of absorption lines in the near-IR solar spectrum as transitions between highly-excited Fe I levels (Litzén and Vergès, 1976; Johansson and Learner, 1990) has motivated further laboratory studies of even higher Fe I configurations for solar and stellar spectroscopy. The extraordinary quality of the solar IR spectrum obtained with the *ATMOS* space experiment (Farmer and Norton, 1989) offers the possibility to perform a fine analysis of hydrogenic lines of Fe I and to make a comparison between laboratory and solar spectra.

In the present paper we report on an analysis of the $3d^6 4s(^6D)4f - 3d^6 4s(^6D)5g$ supermultiplet based on laboratory FTS-spectra and on the *ATMOS* solar spectrum. The presence of 4f–5g lines in ground-based IR spectra of the sun and of α Tau (Ridgway et al., 1984) was first reported by Johansson et al. (1991). However, the *ATMOS* spectra allow a much more accurate analysis of the 4f–5g lines than can be obtained from ground-based spectra. The identifications have been confirmed by calculated line strengths in comparison with observed laboratory and solar intensities. A full presentation of all data, and the detailed analysis, will be published elsewhere (Johansson et al., 1992). The identifications we provide are confirmed in another paper in these proceedings by Schoenfeld et al. (1993).

2. Laboratory experiment and analysis

The laboratory spectrum used in our analysis of the 4f–5g supermultiplet has been recorded with the Fourier Transform Spectrometer (FTS) at Kitt Peak for

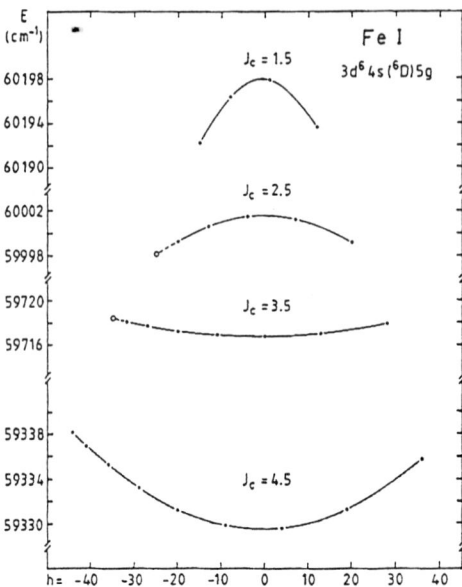

Fig. 1. The new $3d^6 4s(^6D)5g$ levels plotted as a function of $h = \mathbf{J_{itc}} \cdot \mathbf{l}$, where $\mathbf{J}_c$ is the total angular momentum of the ^6D parent level and $\mathbf{l}$ is the orbital angular momentum of the outer $5g$-electron. The curves are drawn through the experimental points (dots) and predicted c.g. values for unknown level pairs (circles).

an extensive analysis of the Fe I spectrum (see, e.g., Nave et al., 1992). A hollow-cathode lamp of pure iron was run with neon at 3.7 torr and at a DC-current of 1.4 A. The wavenumber resolution is 11.9 mK. Similar FTS-spectra from Kitt Peak have earlier been used in the IR for solar identifications (Biémont et al., 1985) and for laboratory and solar analyses (Johansson and Learner, 1990). However, it turns out that the spectrum used for the present work has a higher signal to noise ratio than earlier FTS-spectra, probably due to the water-cooling of the hollow-cathode. This has helped in the analysis of the $4f$–$5g$ transitions.

The analysis of the $3d^6 4s(^6D)5g$ subconfiguration has been performed in the same way as the $3d^6 4s(^6D)4f$ subconfiguration (Johansson and Learner, 1990), i.e., by means of the application of the quadrupole approximation. These configurations are very well described by the JK-coupling scheme, meaning that the fine structure splitting of the parent term $3d^6 4s\ ^6D$ in Fe II determines the gross structure. The electrostatic interaction between the outer $5g$ electron and the core separates level pairs built on the same parent level J_c. This separation is determined by the electrostatic parameter $F^2(3d, 5g)$. By plotting the energy of the level pairs as a function of the scalar product $h = \mathbf{J}_c \cdot \mathbf{l}$, all level pairs associated with a particular parent level fall on a parabola (see Fig. 1), as the diagonal coefficients of $F^2(d, g)$ are quadratic functions of h. The parabolas should be symmetric relative to $h = -1/2$. The change of the shape of the parabolas from "bowl-like" to "umbrella-like" was discussed by Johansson and Learner (1990). Probable deviations from a parabolic curve reveal either misidentifications or perturbations. In the case of $5g$, the deviations are smaller than 10 mK. Once the strongest transitions have been classified, the parabolas can be used to predict the rest of the levels. A few levels are still missing in the $5g$-subconfiguration due to missing levels in the $4f$-subconfiguration. We have indicated the position of the missing levels in $5g$ in Figure 1 with open circles.

We have also performed parametric calculations of the $3d^64s(^6D)5g$ subconfiguration by means of the Cowan code and calculated oscillator strengths. There is in general a very good agreement between laboratory intensities, calculated line strengths and oscillator strengths, derived from the solar spectrum by adopting an abundance of 7.51 for Fe in the usual logarithmic scale. Some small discrepancies still have to be investigated by a more thorough interpretation of possible level mixings or line blends.

A number of lines appearing in the 1350 cm^{-1} region in the *ATMOS* solar spectrum can certainly be identified by the next set of hydrogenic transitions in Fe I, viz. the $3d^64s(^6D)5g - 3d^64s(^6D)6h$ supermultiplet. There are no laboratory spectra of iron available in this wavelength region and the analysis has to be performed on the basis of the solar lines and the application of the quadrupole approximation.

3. Identifications in the Solar Spectrum

Between 2545 and 2585 cm^{-1}, more than 90% of the solar lines are due to $4f$-$5g$ Fe I transitions. All of these lines (about 100), which have excitation energies of 7.1 to 7.3 eV, have been identified without any doubt in the *ATMOS* solar infrared spectra. Figure 2 shows a comparison of laboratory and solar (observed and synthetic) spectra in the spectral region $\sigma = 2565-2570$ cm^{-1}.

4. Solar Analysis

These lines show typical shapes *i.e.*, they are broad with extended wings such as all other high excitation atomic lines. This is due to the expected increase of the damping constant with excitation energy. In Fe I, the profiles become nearly Lorentzian for excitation energies higher than about 6 eV.

Although relatively faint, these high excitation Fe I lines are very sensitive to the damping constants as shown in Figure 3 where we considered collisions with H atoms to be the main broadening mechanism. As it is well known that damping constants (γ_{coll}), calculated by means of Unsöld's approximation (1955), are too small (Blackwell *et al.*, 1984; Holweger *et al.*, 1991), the enhancement factor plays a crucial role.

The Stark broadening is expected to play a non-negligible role as the excitation energy increases (Chang and Schoenfeld, 1991; Carlsson *et al.*, 1992). In the absence of reliable data for our lines, we took it crudely into account using an approximate formula given by Cowley (1971) —see also Freudenstein and Cooper, (1978). Our results show that the Stark broadening is only about 1/3 of the Van der Waals broadening.

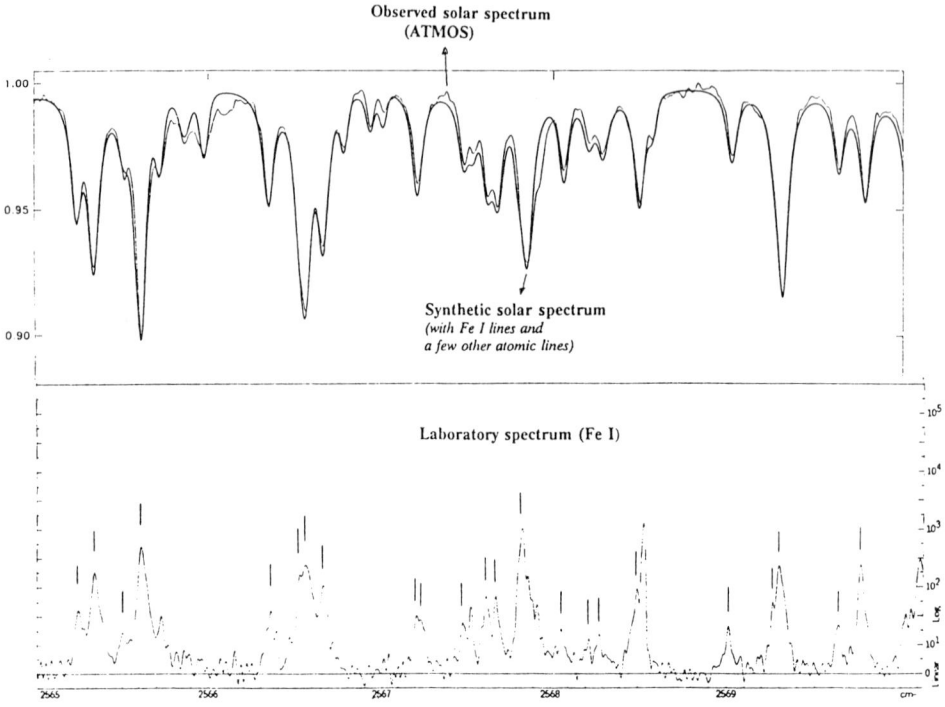

Fig. 2. Comparison of laboratory and (*ATMOS* and synthetic) solar spectra in the region 2565–2570 cm^{-1}.

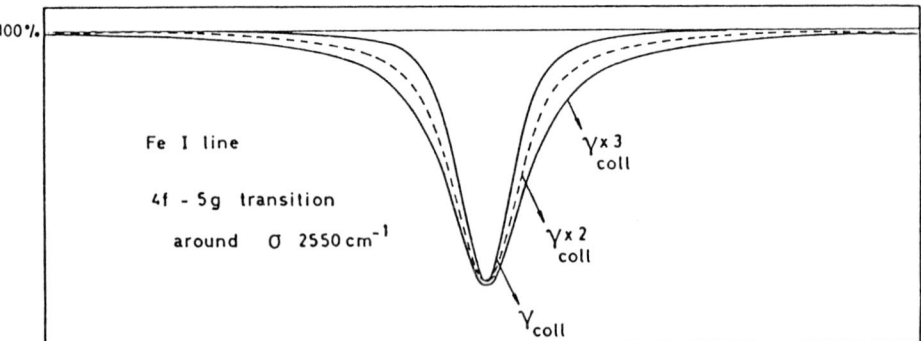

Fig. 3. Effect of an increase of the collisional damping constant (γ_{coll}) on the profiles of a typical $4f$–$5g$ iron line of about 5% central depth: there is a strong line broadening and the far wings become more and more important when increasing γ_{coll} by a factor 2 or 3.

We fitted synthetic line profiles to the observed solar spectrum for all our Fe I lines in the region $\sigma = 2545-2582$ cm^{-1}. Very good agreement is obtained, as can be seen in Figure 2 (which corresponds to $\sigma = 2565-2570$ cm^{-1}), for an enhancement factor of about 2 for the damping constant and for a solar abundance of iron, A_{Fe} = 7.51, a value that is in perfect agreement with the meteoritic value (Anders and Grevesse, 1989).

5. Conclusions

These very high excitation lines of Fe I are much less sensitive than lower excitation lines to temperature uncertainties and to departures from LTE. Based on theoretical transition probabilities described in Section 2, they lead to a photospheric abundance of Fe, that agrees with values recently derived by Holweger *et al.* (1990) using lower excitation Fe I lines and by Holweger *et al.* (1991), Biémont *et al.* (1991), Hannaford *et al.* (1992) using Fe II lines: all of these results being in agreement with the meteoritic abundance of Fe.

We note that Fe I lines of still higher excitation energies ($5g-6h$) have very recently been identified in the *ATMOS* solar spectrum around 1350 cm^{-1} (Schoenfeld *et al.*, 1993).

Acknowledgements

We thank Drs. C.B. Farmer and M. Gunson (Jet Propulsion Laboratory) for permission to use the *ATMOS* infrared solar spectra before publication and for their kind hospitality. We also want to thank W. Nijs (Brussels) for his help in all calculations using the Apollo computers at the Royal Observatory (Brussels).

References

Anders, E., and Grevesse, N.: 1989, *Geochim. Cosmochim. Acta* **53**, 197.
Biémont, E., Brault, J.W., Delbouille, L., Roland, G.: 1985, *Astron. Astrophys. Suppl.* **61**, 107.
Biémont, E,. Baudoux, M., Kurucz, R.L., Ansbacher, W., and Prinnington, E.H.: 1991, *Astron. Astrophys.* **249**, 539.
Blackwell, D.E., Booth, A.J., and Petford, A.D.: 1984, *Astron. Astrophys.* **132**, 236.
Carlsson, M., Rutten, R.J., and Shchukina, N.G.: 1992, *Astron. Astrophys.* **253**, 567.
Chang, E.S., and Schoenfeld, W.G.: 1991, *Astrophys. J.* **383**, 450.
Cowley, C.R.: 1971, *The Observatory* **91**, 139.
Farmer, C.B., and Norton, R.H.: 1989, *A High-Resolution Atlas of the Infrared Spectrum of the Sun and the Earth Atmosphere from Space*, Vol. I: The Sun, NASA Ref. Pub. 1224, Washington, D.C.
Freudenstein S.A., and Cooper, J.: 1978, *Astrophys. J.* **224**, 1079.
Hannaford, P., Lowe, R.M., Grevesse, N., and Noels, A.: 1992, *Astron. Astrophys.* (in press).
Holweger, H., Heise, C., and Kock, M.: 1990, *Astron. Astrophys.* **232**, 510.
Holweger, H., Bard, A., Kock, A., and Kock, M.: 1991, *Astron. Astrophys.* **249**, 545.
Johansson, S., and Learner, R.C.M.: 1990, *Astrophys. J.* **354**, 755.
Johansson, S., Nave, G., Learner, R.C.M., and Thorne, A.P.: 1991, in C. Jaschek and Y. Andrillat (eds.), *The Infrared Spectral Region of Stars*, Cambridge Univ. Press, Cambridge, p. 189.
Johansson, S., Nave, G., Geller, M., Sauval, A.J., Grevesse, N., Schoenfeld, W.G., Chang, E.S., and Farmer C.B.: 1992 (in preparation).
Litzén, U., and Vergès, J.: 1976, *Phys. Scripta* **13**, 240.
Nave, G., Learner, R.C.M., Murray, J.E., Thorne, A.P., and Brault, J.W.: 1992, *J. Physique II France* **2**, 913.

Ridgway S.T., Carbon, D.F., Hall, D.N.B., and Jewell, J.: 1984, *Astrophys. J. Suppl.* 54, 177.
Schoenfeld, W.G., Chang, E.S., and Geller, M.: 1993, these proceedings.
Unsöld, A.: 1955, *Physik der Sternatmosphären*, 2nd ed., Springer-Verlag, Berlin, p. 331.

THE SUN AS A LABORATORY SOURCE FOR IR MOLECULAR SPECTROSCOPY

A. J. SAUVAL

Observatoire Royal de Belgique, avenue circulaire, 3, B-1180 Brussels, Belgium

and

N. GREVESSE

*Institut d'Astrophysique, Université de Liège, avenue de Cointe, 5,
B-4000 Cointe-Liège, Belgium*

Abstract. The infrared solar spectrum is used to refine our knowledge of molecular constants of CH and CO and to test the accuracy of transition probabilities and dissociation energies of a few diatomic molecules.

Key words: : infrared: stars – molecular data – Sun: atmosphere

1. Introduction

In the past the Sun itself has proven to serve as a good laboratory source for atomic as well as molecular spectroscopy. Let us only quote two different investigations related to the CO molecule and based on the IR solar spectrum: the pioneering determination of the spectroscopic constants by Goldberg and Müller (1953) and the relative transition probabilities derived by Tsuji (1977). As accurate molecular data are still lacking for a few molecular species of astrophysical interest, we use the many molecular transitions present in the solar infrared spectrum to refine our knowledge. A summary of the advantages and disadvantages of using the solar spectrum is shown in Table I.

The spectroscopic data are deduced from infrared solar spectra recorded by the *ATMOS* experiment onboard the Space Shuttle in April/May 1985 (Spacelab 3 flight; Farmer and Norton, 1989) and from ground-based observations obtained at Kitt Peak (Delbouille *et al.*, 1981).

2. Spectroscopic Constants

2.1. CH $X^2\Pi$

The *ATMOS* solar spectra show the vibration-rotation lines of the 1-0, 2-1 and 3-2 bands to much higher N''-values (>30) than laboratory spectra ($N'' = 9$). The 4-3 band, which was not seen in the laboratory (in the days of our investigation), even shows up in the solar spectrum (Fig. 1).

We used solar wavenumbers together with laboratory wavenumbers to refine the molecular constants of the CH ground state (Mélen *et al.*, 1989).

2.2. CO $X^1\Sigma$

Over 2000 new very high J''-value solar lines ($J''_{\max} = 133$) for v''-values up to 19 of the fundamental and first-overtone bands, never seen in the laboratory ($J''_{\max} = 94$), clearly appear on the *ATMOS* solar IR spectra (σ 1350–4360 cm^{-1}) (Fig. 2).

TABLE I

The Sun – a very good source for spectroscopic studies

ADVANTAGES
- very stable source (permanent!)
- high temperature (high-excitation lines which are not seen in the laboratory)
- good representative model of the photospheric layers
- accurate determination of the physical conditions (T, P, N)
- equilibrium conditions (LTE): Boltzmann, Saha, Guldberg-Waage laws are valid

DISADVANTAGES
- broad solar lines as compared to IR laboratory lines
- large wavelength shifts
- relative motion Sun/observer
- rotation of the Sun
- Einstein gravitational redshift
- convective motions (intensity-dependent shift!)
- about 0.003 cm^{-1} between strong and weak CO lines
- gradual violet shift from strong to weak lines
- very accurate laboratory line positions needed to convert to rest wavelengths

Combining the new solar wavenumbers with the available laboratory data (14000 measurements) has permitted to derive a new set of 31 spectroscopic constants which allows to predict wavenumbers with high accuracy from low to high J''-values (Farrenq *et al.*, 1991).

2.3. OTHER MOLECULES

New molecular transitions, never observed in the laboratory, have been measured in the IR solar spectrum such as pure rotation lines of NH (X $^3\Sigma$, $v = 0, 1$) near 600–900 cm^{-1} (see Fig. 3, from Geller *et al.*, 1991).

3. Transition Probabilities and Dissociation Energies

The line absorption coefficient of a molecule AB depends on the line oscillator strength, f_{vJ}, on the dissociation energy, D_0^0, and on the number densities of the relevant atoms, N_A and N_B. From this well-known relationship in the solar photosphere, one can derive a very useful relation: $\Delta \log f_{vJ} \approx -\Delta D_0^0$ (eV)

3.1. C_2 PHILLIPS SYSTEM (A $^1\Pi_u$ - X $^1\Sigma_g^+$)

Both the dissociation energy of C_2 and the transition probability of the Phillips system are still rather uncertain. The combined use of C_2 (Swan and Phillips sys-

THE SUN AS A LABORATORY SOURCE FOR IR MOLECULAR SPECTROSCOPY 551

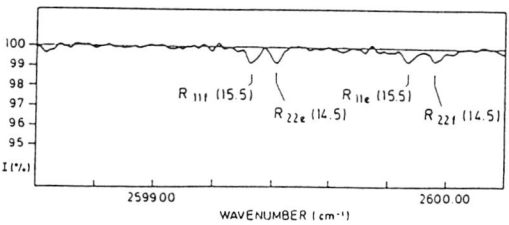

Fig. 1. Lines of the 4-3 band of CH in the *ATMOS* IR spectrum.

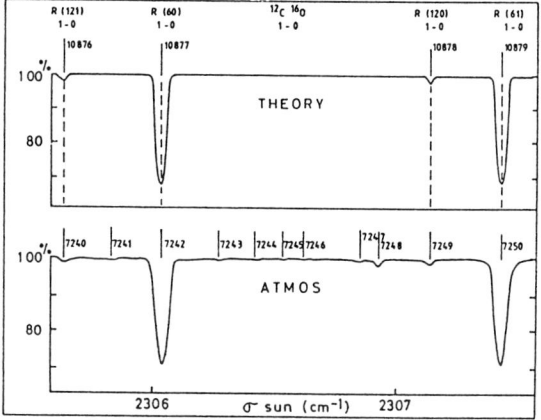

Fig. 2. $^{12}C^{16}O$ lines of the 1-0 band with $J'' = 60, 61, 120$ and 121 in the IR solar spectrum (*ATMOS*).

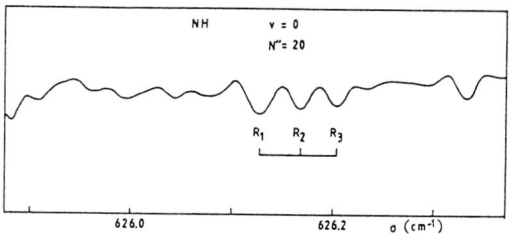

Fig. 3. First detection of pure rotation lines of NH ($X^3\Sigma$ $v = 0$) in the IR solar spectrum near 626 cm^{-1} (*ATMOS* spectra).

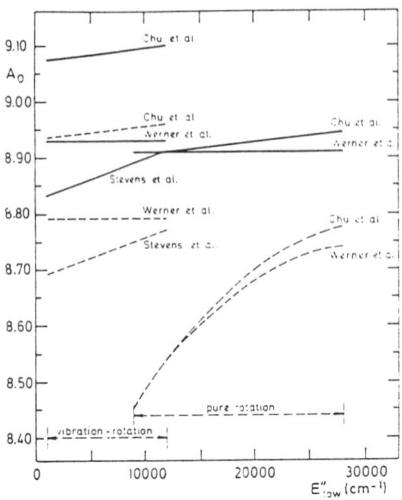

Fig. 4. The solar abundance of oxygen A_O must be independent of the excitation potential of OH lines (such as for Werner et al.'s EDMF). Results for two different photospheric models: Holweger-Müller 1974 (full lines) and Vernazza et al. 1976 (dashed lines) are shown. See Grevesse et al. (1984) for all references.

tems) together with other indicators of the solar abundance of carbon has enabled us (Grevesse et al., 1991) to derive $D_0^0 \simeq 6.23$ eV and f_{00} (Phillips system) = 2.28×10^{-3} in agreement with the most recent theoretical result of Langhoff et al. (1990), but 50% larger than the most recent experimental result (Bauer et al., 1986).

3.2. THE CN RED SYSTEM

The dissociation energy of CN is still a matter of debate. Combining the recent theoretical results for the transition probability that is supposed to be accurate, together with solar abundances of C and N derived independently, the solar intensities of our infrared CN lines are reproduced with $D_0^0 \simeq 7.75$ eV (Grevesse and Sauval, 1992), in agreement with two very recent measurements (Costes et al., 1990; Huang et al., 1992), and ruling out lower values of D_0^0 often used in the recent past.

3.3. TEST OF THE ELECTRIC DIPOLE MOMENT FUNCTION (EDMF) OF OH

We showed (Sauval et al., 1984; Grevesse et al., 1984) how sensitive the solar vibration-rotation and pure rotation lines are to the EDMF (Fig. 4).

Those lines have been remeasured in the *ATMOS* solar infrared spectra up to high rotational excitation.

The new EDMF of Nelson et al. (1990) leads to a very good agreement between observations and predictions from low to high N'' values.

3.4. CO $X^1\Sigma^+$

In the high resolution and low noise *ATMOS* solar spectra (Farmer and Norton, 1989) more than 7000 vibration-rotation lines of the fundamental $^1\Sigma^+$ state of four isotopic species of carbon monoxide ($^{12}C^{16}O$, $^{13}C^{16}O$, $^{12}C^{18}O$ which is about 500 times less abundant than $^{12}C^{16}O$; and eventually, for the first time in the solar photospheric spectrum, $^{12}C^{17}O$ which is about 2500 times less abundant than $^{12}C^{16}O$) are detected in the spectral range σ 1350–2328 cm^{-1} (fundamental bands, from 1-0 to 20-19) and σ 3410–4360 cm^{-1} (first overtone bands, from 2-0 to 14-12).

The dissociation energy of CO is known with high accuracy. But solar CO lines are extremely sensitive to the photospheric temperature structure. The available transition probabilities (Chackerian and Tipping, 1983) for medium and high J''-values observed in the *ATMOS* solar IR spectrum might not have the accuracy required for a precise comparison between observations and predictions. Abundances derived from the fundamental and first overtone bands slightly disagree. Very accurate transition probabilities are urgently needed up to high J''-values for the different isotopes present in the solar infrared spectrum.

3.5. OTHER MOLECULES

We have also tested the EDMF for vibration-rotation spectra of CH (X $^2\Pi$ 1-0, 2-1, 3-2, 4-3 bands) and of NH (X $^3\Sigma$ 1-0, 2-1 bands) and for pure rotation spectra of NH (X $^3\Sigma$, $v = 0, 1$) (see Grevesse *et al.*, 1990, 1991).

Acknowledgements

We thank Drs. C. B. Farmer and M. Gunson (Jet Propulsion Laboratory) for permission to use the *ATMOS* infrared solar spectra before publication and for their kind hospitality. We also thank W. Nijs and R. Blomme (Brussels) for their help with the calculations carried out at the Apollo computers of the Royal Observatory (Brussels). We also thank the Belgian Fonds National de la Recherche Scientifique for financial support.

References

Bauer, W., Becker, K.H., Bielefeld, M., Meuser, R.: 1986, *Chem. Phys. Letters* **123**, 33.
Chackerian, C., Jr., and Tipping, R.H. : 1983, *J. Mol. Spectrosc.* **99**, 431.
Costes, M., Naulin, C., and Dorthe, G.: 1990, *Astron. Astrophys.* **232**, 270.
Delbouille, L., Roland, G., Brault, J.W., and Testerman, L.: 1981, *Photometric Atlas of the Solar Spectrum from 1850 to 10000* cm^{-1}, Kitt Peak National Observatory, Tucson, Arizona.
Farmer, C.B., and Norton, R.H.: 1989, *A High-Resolution Atlas of the Infrared Spectrum of the Sun and the Earth Atmosphere from Space*, Vol. I: The Sun, NASA Ref. Pub. 1224, Washington, D.C.
Farrenq, R., Guelachvili, G., Sauval, A.J., Grevesse, N., and Farmer, C.B.: 1991, *J. Molec. Spectrosc.* **149**, 375.
Geller, M., Sauval, A.J., Grevesse, N., Farmer, C.B., and Norton, R.H.: 1991, *Astron. Astrophys.* **249**, 550.
Goldberg, L., and Müller, E.A.: 1953, *Astrophys. J.* **118**, 397.
Grevesse, N., Lambert, D.L., Sauval, A.J., van Dishoeck, E.F., Farmer, C.B., and Norton, R.H.: 1990, *Astron. Astrophys.* **232**, 225.

Grevesse, N., Lambert, D.L., Sauval, A.J., van Dishoeck, E.F., Farmer, C.B., and Norton, R.H.: 1991, *Astron. Astrophys.* **242**, 488.
Grevesse, N., Sauval, A.J., and van Dishoeck, E.F.: 1984, *Astron. Astrophys.* **141**, 10.
Grevesse, N., and Sauval, A.J.: 1992, *Revista Mexicana de Astronomia y Astrofisica* **23**, 71.
Holweger, H., and Müller, E.A.: 1974, *Solar Phys.* **39**, 19.
Huang, Y.H., Barts, S.A., and Halpern, J.B.: 1992, *J. Phys. Chem.* **96**, 425.
Langhoff, S.R., Bauschlicher, C.W.Jr., Rendell, A.P., and Komornicki, A.: 1990, *J. Chem. Phys.* **92**, 3000.
Mélen, F., Grevesse, N., Sauval, A.J., Farmer, C.B., Norton, R.H., Bredohl, H., and Dubois, I.: 1989, *J. Mol. Spectrosc.* **134**, 305.
Nelson, D.D., Schiffman, A., Nesbitt, D.J., Orlando, J.J., and Burkholder, J.B.: 1990, *J. Chem. Phys.* **93**, 7003.
Sauval, A.J., Grevesse, N., Brault, J.W., Stokes, G.M., and Zander, R.: 1984, *Astrophys. J.* **282**, 330.
Tsuji, T.: 1977, *J. Quant. Spectrosc. Rad. Transf.* **18**, 179.
Vernazza, J.E., Avrett, E.H., and Loeser, R.: 1976, *Astrophys. J. Suppl.* **30**, 1.

PART 7

INFRARED TECHNOLOGY AND THE FUTURE

PROSPECTS IN ADAPTIVE OPTICS FOR SOLAR APPLICATIONS

FRANÇOIS RODDIER AND J. ELON GRAVES

Institute for Astronomy, University of Hawaii at Manoa,
2680 Woodlawn Drive, Honolulu, HI 96822, U.S.A.

Abstract. We review the theoretical perspective for problems in adaptive optics, outline recent progress, and consider its application in the infrared. Techniques in adaptive optics are on the threshold of revolutionizing modern astronomy. These techniques are particularly applicable in the infrared, where refractive effects of turbulence are reduced, characteristic cell sizes are greater, and the isoplanatic patch diameter is increased. Adaptive techniques could be especially appropriate for modern large solar telescopes now under consideration that could operate in the infrared.

Key words: atmospheric effects – Sun: general – techniques: miscellaneous – telescopes

1. Introduction

The terms "active optics" and "adaptive optics" are commonly used. Active optics can be regarded as adjustments in telescope optics designed to correct for telescope induced aberrations, such as misalignment and mirror deformation. Adaptive optics refers to active optics to correct for atmospherically induced aberrations, *i.e.*, seeing. The basic operation of adaptive optics is illustrated conceptually in Figure 1. Light from the imaging system, which includes an adaptive optical corrector is imaged onto a wavefront sensor and a camera. The wavefront sensor assesses the aberrations and this information is converted by control electronics to electrical signals that drive the corrector. The camera simply records the corrected image.

2. Theoretical Review

In practice, it is only possible to correct the actual aberration approximately. The corrector is only capable of a finite number, N, of degrees of freedom. The wavefront perturbations the corrector can produce may be represented as a linear combination of modes that span a manifold of dimension N. These modes are most conveniently chosen to be functions of location in the system pupil, which can be chosen so as to be orthogonal under an appropriate inner product. Several systems of modes are used. For theoretical calculations one uses:

1. *The Karhunen-Loeve Modes*: the set of orthogonal functions which best matches the statistics of atmospheric turbulence – *i.e.*, the statistically-independent modes. For an adaptive optics system with N degrees of freedom the best possible compensation is achieved by compensating the first N Karhunen-Loeve modes.

2. *The Zernike Modes*: commonly used in optics because of their simple analytic expression.

The actual modes that characterize a certain physical system with are called the *system modes*.

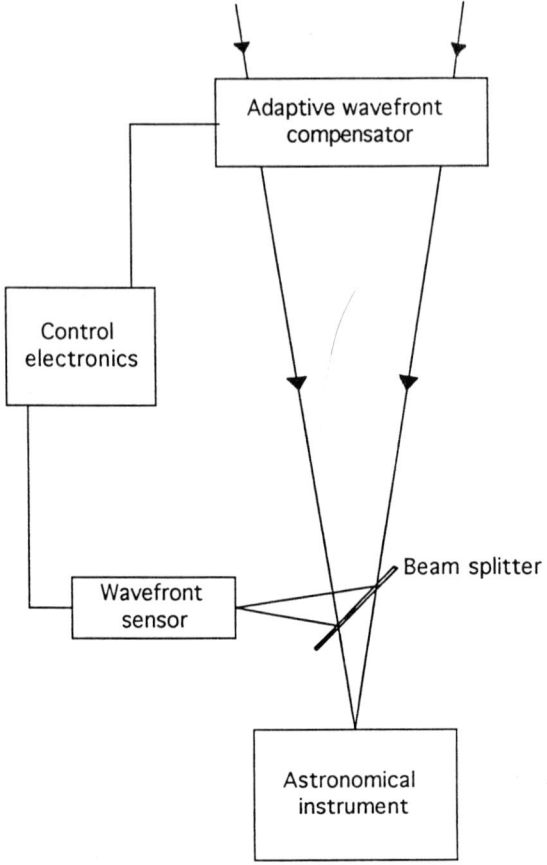

Fig. 1. Schematic of the general active- or adaptive-optical system.

Details regarding the Zernike modes are discussed by Born and Wolf (1975). Similar to cylindrical harmonics, they are represented by functions of the following form:

$$Z_{nm}(\rho, \theta) = P_{nm}(\rho)e^{im\theta}, \qquad (1)$$

where the radial functions, P_{nm}, are polynomials of finite degree, n. These "Zernike polynomials" share the property of Bessel functions that only terms of order $|m|+2l$ are included, where l must be an integer greater than or equal to zero. Thus, the limiting degree, n, must satisfy $n \geq |m|$, and $n - |m|$ is even. Table I lists how the lowest-order aberrations are characterized in terms of the Zernike modes.

Parameters customarily used for expressing the performance of an optical system are largely based on the "Strehl intensity", I_0, the central intensity in the degraded point-source image. The "Strehl ratio" is defined by $R \equiv I_0/I_{\max}$, the ratio of the Strehl intensity, I_0, to the central intensity in the diffraction-limited

TABLE I
The Zernike Aberrations

$n \backslash m$	0	1	2	3	4	5
0	piston					
1		tip-tilt				
2	defocus		astigmatism			
3		coma		(coma)		
4	spherical		(astigmatism)			

image. The "normalized Strehl ratio", R/R_{max}, is the Strehl ratio, R, of a particular adaptive-optics system being considered normalized by the Strehl ratio, R_{max}, of an uncompensated image observed through an infinitely large telescope. This normalized Strehl ratio, R/R_{max}, normally depends only on the ratio, D/r_0, of the telescope diameter, D, to Fried's seeing parameter, r_0. For the exact definition of r_0, we refer to Fried (1966); however r_0 can be roughly regarded as the separation in the entrance pupil over which the rms wavefront error approaches one wavelength, λ.

The theory implies that the statistics of atmospheric turbulence can be properly characterized as a function of scale size. A formalism for this was developed by Fried (1966), based on turbulence described by a "wave structure function" that expresses the correlation in wave perturbations over a separation r in the entrance pupil. Fried characterizes the wave structure function as proportional to $r^{5/3}$ based on the Kolmogoroff rule on the statistics of atmospheric turbulence. We note that r_0 is strongly dependent on wavelength, λ, since the scale over which a wavefront perturbation approaches a greater wavelength must be greater, according to the $r^{5/3}$ rule. It is then possible to express the Strehl ratio for a particular system for correcting wave-fronts as a function of the ratio D/r_0 alone.

Figure 2 shows computations of the dependence of the normalized Strehl ratio, R/R_{max}, on D/r_0, made by N. Roddier (1990) for Zernike corrections of various orders. Note the strong dependence of D/r_0 on λ, by comparing the upper and lower abscissa scales. The upper scale approximately applies to the McMath telescope under average seeing conditions. The normalized Strehl ratios for various Zernike orders of correction all rise initially in proportion to D/r_0, but eventually fall away from the diffraction limit, culminating at a critical D/r_0 and thence dropping rapidly toward the envelope for uncorrected optics. This behavior is summarized as follows:

1. A given adaptive-optics system provides a maximum image improvement at a critical ratio, $(D/r_0)_c$, at which the trade-off between diffraction-limited optics and seeing conditions is optimized.

2. At this optimum trade-off the Strehl ratio, R (un-normalized) is 0.3.

3. The number of degrees of freedom of the system can be much smaller than $(D/r_0)_c^2$ (typically 7 times smaller).

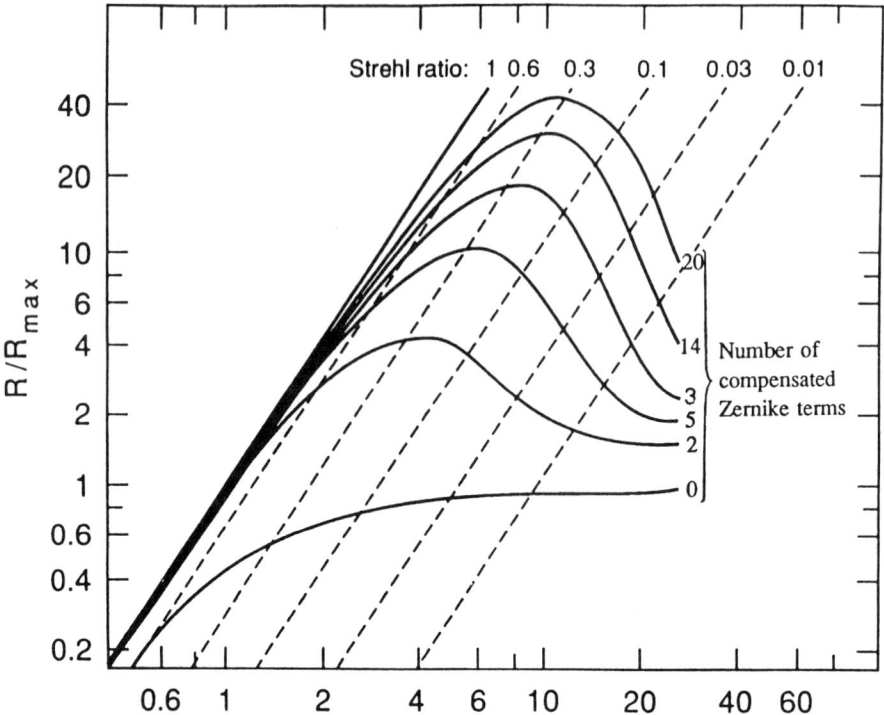

Fig. 2. Improvement in the Strehl Ratio (R/R_{max}) for various orders of Zernike correction.

4. A Strehl ratio higher than 0.3 can be obtained only at a high cost. As a consequence, the system bandwidth and the reference source brightness required also increase, while the isoplanatic patch size decreases.

5. In most cases, a Strehl ratio higher than 0.3 is not necessary to obtain diffraction-limited images. Images and data from imaging spectrographs can be easily deconvolved when a Strehl ratio of 0.3 is reached.

For moderate values of D/r_0, i.e., relatively small telescopes or good seeing conditions, corrections encompassing only a few degrees of freedom, perhaps only tip-tilt, can drastically improve image quality. Simple tip-tilt correction (2 degrees of freedom) is now a well-practiced art, particularly with relatively small telescopes, for which image motion alone may prevent attainment of the diffraction limit in extended exposures. For larger telescopes, requiring more degrees of correction, 5 to 10 degrees of freedom are now being tested with great success (Roddier 1991). While it is useful to further increase the flexibility of the correction as D/r_0 increases, and more sophisticated correctors can be expected in the future, the required degrees of freedom needed, eventually become oppressive for large D/r_0. The infrared, then offers a powerful handle for keeping D/r_0 within practical limits for larger apertures.

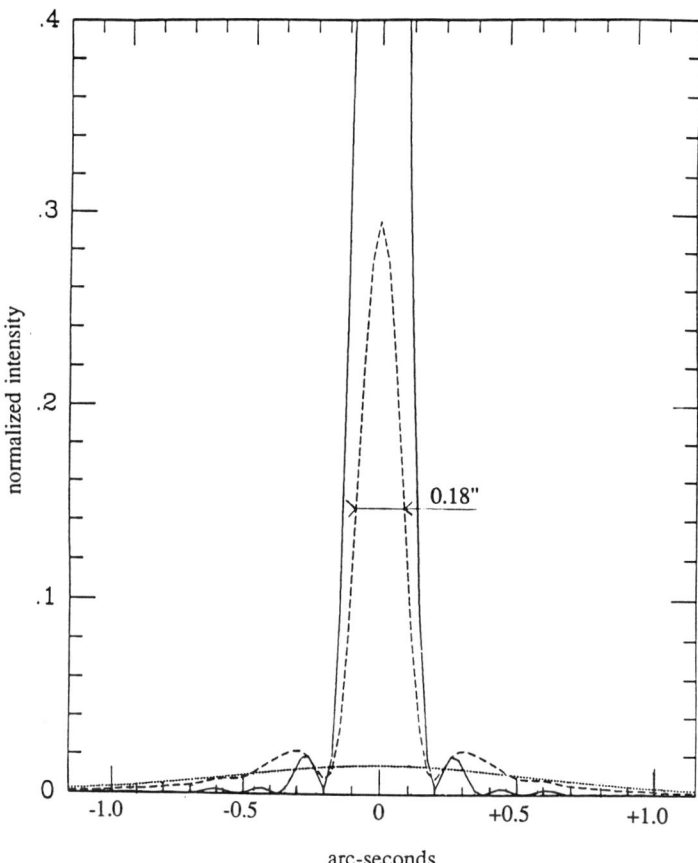

Fig. 3. Point source response functions for a 1.5 m telescope in 1.2 μm radiation for (dotted curve) uncompensated seeing that is 1.5″ in visible light and for (dashed curve) a 9-order Zernike correction. The solid curve shows the diffraction limited response of the telescope free from seeing aberrations.

3. An Example: Point Source Profiles

Figure 3 shows how the time-averaged point source profile looks for the 9-mode Zernike correction at its optimum D/r_0, for a Strehl ratio of 0.3. This simulation is made for a 1.5-m telescope at 1.2 μm in conditions under which the seeing in visible light is 1.5″. The dotted curve shows the point-source response of the telescope uncorrected for seeing aberrations. The dashed curve shows the profile with the 9-mode Zernike correction. The result is a substantial diffraction-limited core containing the expected 0.3-fraction of the core power of a perfect, diffraction-limited system (solid curve). Such a system would allow full diffraction-limited (0.18″) reconstruction of an image with only moderate losses in signal to noise. The foregoing

TABLE II
Projected Characteristics for an Adaptive Optics System for the McMath Telescope

λ	D/r_0	N	θ	B
0.5 μm	25	90	2.6"	250 Hz
0.8 μm	14	28	7"	125 Hz
1.2 μm	8.7	11	15"	75 Hz
1.6 μm	6.2	5	30"	50 Hz
2.2 μm	4.2	2	60"	30 Hz

theoretical considerations can be summarized by the following 5 points, which are discussed in greater detail by Roddier, Northcott, and Graves (1991). Further theoretical modeling in terms of residual wave-front errors is described by Chassat (1989).

The ratio D/r_0 serves as a useful guideline for anticipating system requirements for a particular telescope under typical seeing conditions as well as performance qualities that can be anticipated. System requirements include the number of degrees of freedom, N, and the electronic loop bandwidth, B, needed to process the correction. Performance qualities include the angular diameter, θ, of the compensated field of view (the isoplanatic patch). Table II summarizes these parameters for conditions typical at the McMath Telescope in terms of operation at the optimum Strehl ratio, $R = 0.3$, over a range of wavelength. System requirements are greatly relaxed in the infrared, while performance is greatly enhanced.

4. Practical Realization

The technical tasks confronting adaptive optics can be separated into two main categories: (1) wavefront sensing and (2) wavefront correction. Several schemes for wavefront sensing are under development, some having already attained considerable success, particularly in the case of stellar applications with objects that have sufficiently bright and sharp stellar references (see Fig. 4). Among these are the following:

The Shack-Hartmann Sensor: This device (Fig. 4a) analyzes the image displacement in individual subregions of the pupil, which is related to the wavefront slope in each subregion.

The Itek sensor: A Ronchi ruling moves across the image plane producing a modulation in the pupil image (Fig. 4b). The phase of the modulation at a given pupil point depends on where the corresponding rays are intercepted and gives a measure of the slope of the local wavefront.

The Curvature Sensor: This scheme (Fig. 4c) works by differencing pupil intensity maps measured in front of and behind the focus. The normalized difference,

$$S = \frac{I_1 - I_2}{I_1 + I_2},$$

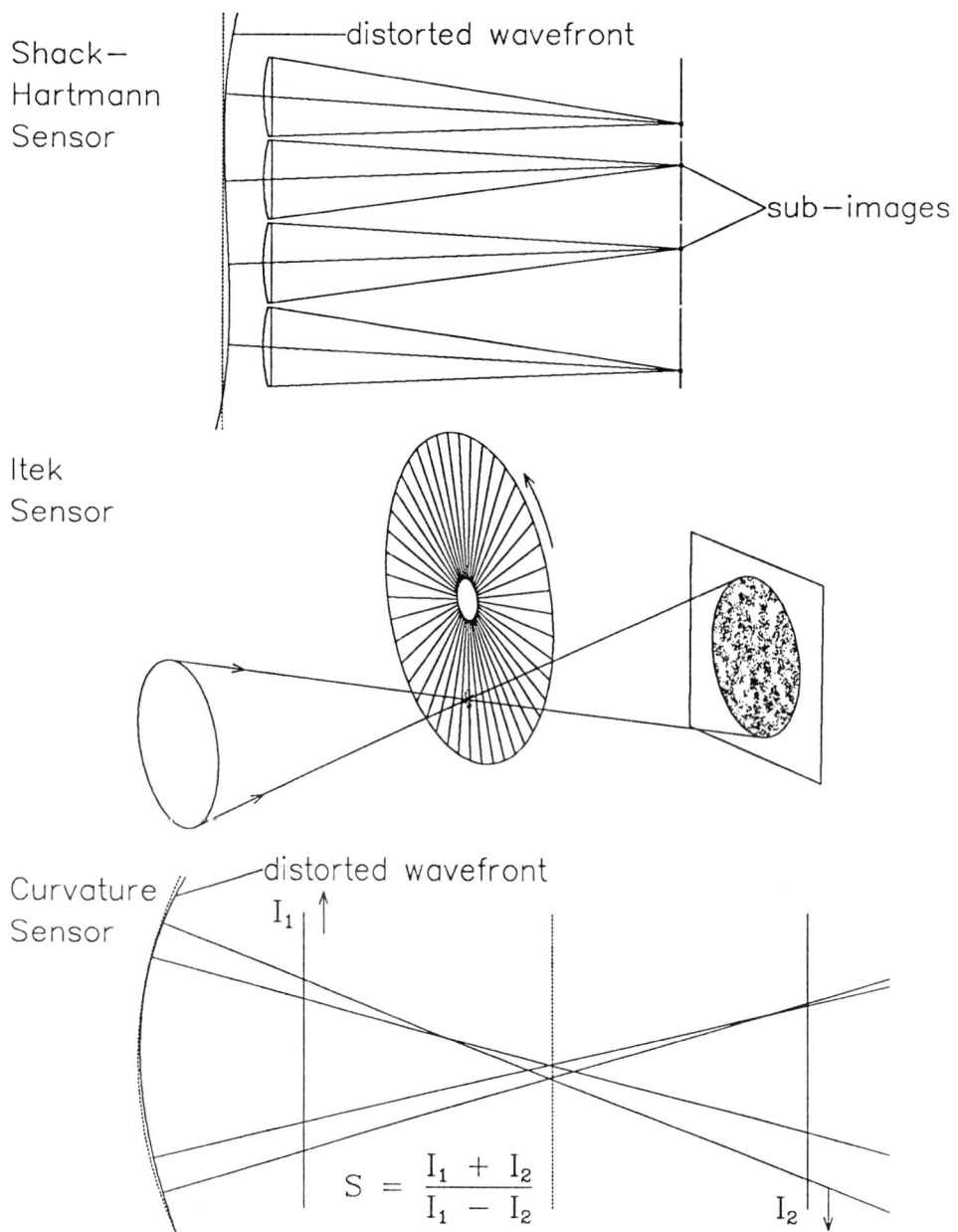

Fig. 4. Wavefront sensors for stellar applications.

gives a good approximation of curvature. This can be seen by considering that a positive perturbation in wavefront curvature draws the focus toward the near-side plane, resulting in a local enhancement of the intensity, I_1, in the pupil in this plane as compared with an intensity dilution of the intensity I_2, in the far-side plane. A negative perturbation in curvature acts *vice versa*.

Sensors built for stellar applications will also work for solar applications (although less efficiently) using a dark spot as a reference. This has already been demonstrated with both the ITEK and Shack-Hartmann sensors, and there is no doubt that similar performance can be obtained with a curvature sensor. Special wavefront sensors can be built for solar applications that work on extended sources, such as the solar granulation. Instead of measuring intensities, one has to measure the variance or covariance of intensity fluctuations. Current schemes under development include

- an array of correlation trackers (Sacramento Peak) (complex and expensive), and
- image sharpening with a dither mirror (LEST) (expensive and probably too slow).

The concept of curvature sensing could also be extended to variance measurements and provide a way to build a *simple* and *less expensive* high-speed solar wavefront sensor. This could be based simply on difference measurements in granular contrast on either side of the focal plane for the image formed by a particular region of the pupil analyzed to determine defocus due to wavefront curvature.

We have made considerable recent progress in wavefront correction. The scheme we have developed at the University of Hawaii is based on curvature sensing and compensation. The wavefront correction is based on a deformable piezo-bimorph mirror, made in collaboration with the Office National d'Etudes et de Recherches Aerospatiale in France. This is a thin double wafer of two materials of the same piezo-electric polarity, so that electric fields applied in opposite directions induce a differential warping of the two components, which results directly in an induced curvature. In our present application, the wafer, 6 cm in diameter, is partitioned into 13 electrically separate regions, as shown in Figure 5. The outer surfaces are uniformly covered with a thin conducting film that acts as a common ground electrode. The inner surface is also covered with a thin conducting layer, but partitioned into separate electrodes, each connected to its respective driver. In our application, the actual pupil occupies only the inner 7 regions. However, the outer regions of the membrane are needed for the function of setting appropriate outer boundary conditions. Figure 6 shows comparative time averaged-profiles obtained by this system in the laboratory.

5. Summary

Image enhancement based on wavefront sensing and correction by active optics is now making the transition from a very promising prospect to a broad range of immediate, practical scientific applications. This will probably revolutionize stellar astronomy. There is every reason to anticipate that the techniques now being

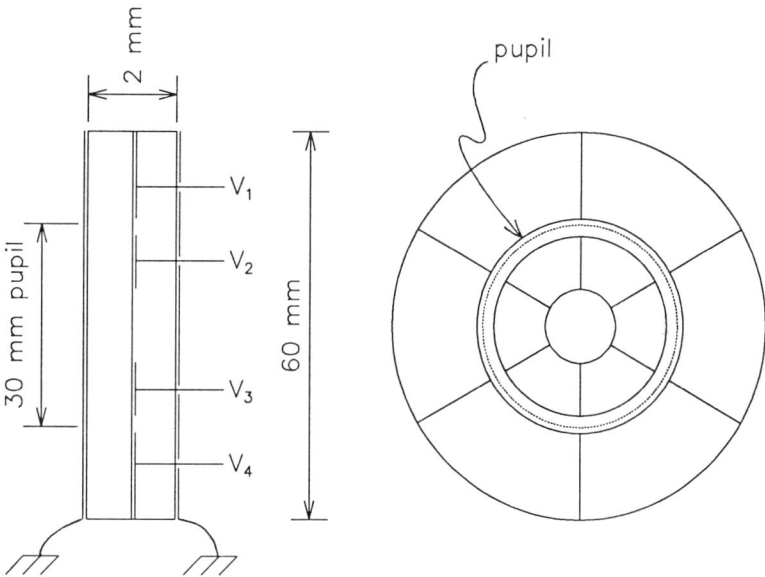

Fig. 5. The University of Hawaii deformable piezo-bimorph mirror.

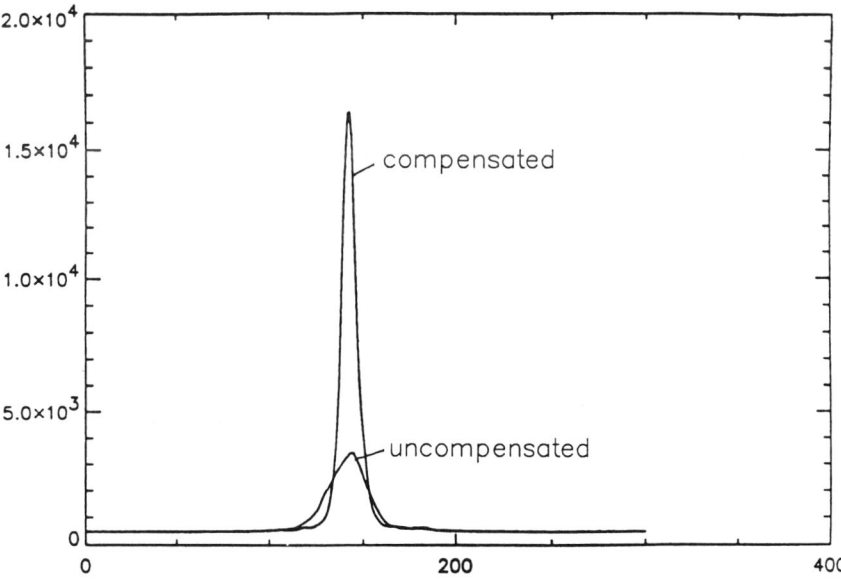

Fig. 6. Comparative time-averaged intensity profile of seeing-uncompensated and compensated point source images.

successfully applied in stellar observations can be extended to solar applications. Moreover, active optical techniques are *particularly* well suited to the infrared.

References

Born, M. and Wolf, E.: 1975, *Principles of Optics*, Pergamon Press, New York, p. 464.
Chassat, F.: 1989, *J. Opt.* (Paris) **20**, 13.
Fried, D. L.: 1966, *J. Opt. Soc. Am.* **56**, 1372.
Roddier, F., Northcott, M., and Graves, J. E.: 1991, *Pub. Astron. Soc. Pac.* **103**, 131.
Roddier, N.: 1990, *SPIE Conf. Proc.* **1237**, 668.

SOLAR OPTICAL INTERFEROMETRY

STEPHEN T. RIDGWAY

*Kitt Peak National Observatory, National Optical Astronomy Observatories,**
P. O. Box 26732, Tucson, AZ, U.S.A.

Abstract. Optical interferometry has numerous applications in stellar and extragalactic astronomy. It also offers the potential for unique solar observations. This review describes current and planned activity in these areas and some possibilities for the future.

Key words: infrared: stars – stars: imaging – Sun: general – techniques: interferometric – telescopes

1. Introduction

Astronomers recall fondly, in their community memory, Michelson's work at Mount Wilson, in which he achieved optical interference between two apertures to measure the angular diameters of several stars. His feat had long since attained the status of legend before it was revisited by radio astronomers and eventually elaborated into a powerful technique for radio imaging. Now it is the turn of the optical (and infrared) communities. The decreasing cost of computers and electronic control and improvements in detectors have made it feasible to build optical telescope arrays which achieve aperture synthesis and very high angular resolution, far beyond any plausible filled-aperture facility. The same techniques have potential applicability for ground- and space-based solar astronomy.

2. What is Interferometry?

Interferometrists work in Fourier space (the infamous u-v plane), use multiple telescopes or strange aperture masks, and are always worried about determining "phase". They are viewed with something between awe and suspicion, as it is not always clear whether they are specialists of an esoteric technique, or masters of obfuscation.

Perhaps it is worth recalling that image formation through a lens may be regarded as an interferometric process, with a series of Fourier transforms from the source to the telescope pupil and then to the image. Surely nobody would give up the simplicity of image formation with a lens or mirror without a good reason. But in fact there may be good reason, when the desired aperture can not be constructed, or when the atmosphere disturbs the image formation process. Then the only solution may be to disassemble the image formation process into its component parts, and exert greater control over them. Thus, to paraphrase von Clausewitz, interferometry is image formation by more vigorous means. A broad view of high resolution imaging, as seen by nighttime astronomers, will be found in *Diffraction-limited Imaging with Very Large Telescopes* (Alloin and Mariotti, 1989). The most recent major meeting on astronomical interferometry was held at ESO headquarters in München in 1991 (Beckers and Merkle, 1992).

* Operated by the Association of Universities for Research in Astronomy under cooperative agreement with the National Science Foundation.

To put solar optical interferometry in context, we will recall briefly the development of radio interferometry for lessons which may apply to optical interferometry and then describe developments in stellar interferometry and recommendations of the Astronomy and Astrophysics (Bahcall) Committee. We will mention some early experiments in solar interferometry, speculate on opportunities for ground-based and space solar interferometry, and describe a specific proposal for space optical interferometry, the SIMURIS mission.

3. Radio Interferometry

It is somewhat more than 4 decades since the first tentative steps toward radio interferometry (Ryle and Vonberg, 1946; Sullivan, 1991). The first efforts simply recorded fringes between two antennas. The first results were visibility curves, obtained from fringe modulation as a function of telescope separation. Visibility curves are difficult to interpret except for the simplest sources, and the first results were merely estimates of or limits on source angular sizes.

Once it was realized that radio interferometry was not particularly difficult to do, at least at longer wavelengths, progress was rather rapid. The addition of phase allowed proper inversion of visibility information to provide first one-dimensional and then two-dimensional maps. The information content of these maps was limited by the amount of information that could be obtained by a small number of telescopes in a reasonable length of time. It was understood, however, that the information content would increase with the square of the number of telescopes; this gave great motivation to construct many-telescope facilities, of which an outstanding example is the Very Large Array, with 27 telescopes in a reconfigurable layout (Thompson *et al.*, 1980). Another imaging radio telescope, familiar to the solar community, is the Culgoora Imaging Solar Radio Telescope in Australia (Wild, 1967).

Since the development of the VLA, much of the progress in observational techniques for radio astronomy has been in algorithms for image reconstruction from noisy data and incomplete coverage of the spatial frequency (u-v) plane. For example, the CLEAN algorithm (Clark, 1980) for deconvolving a "dirty" beam from an observed image has achieved great success in processing of radio images, and has been adopted for many other, non-radio-astronomy applications. Techniques of phase closure and other constraints on image formation have been exploited. The result is that the VLA is now capable of producing mega-pixel, high-dynamic-range maps which, by any definition, deserve to be called images, although they are synthesized in a computer one element at a time.

The "frontier" of radio astronomy now has moved on to ultra-high resolution (VLBI, VLBA) and to other wavelengths (millimeter and sub-millimeter arrays). Some radio interferometrists have even taken an interest in optical interferometry!

4. Optical-IR Stellar Interferometry

Optical interferometry in astronomy began with work by Michelson and Pease to measure the diameters of several bright stars (Pease, 1931). Michelson achieved an optical baseline of several meters, necessary to resolve stars like Aldebaran, by clamping a steel beam across the top of the Mount Wilson 150-cm Telescope and putting some mirrors on it. In a logical follow-up, Pease built a 15-m stand-alone system. It was abandoned after a number of years (and has remained, rusting, on Mount Wilson) apparently because flexures were too great for control techniques available at that time.

The most urgent scientific extension of Michelson's work was the determination of the diameters of some hot stars, and this was achieved by Hanbury Brown, who developed the intensity interferometer (Hanbury Brown et al., 1974). However, the high-order photon correlation observed in the intensity interferometer is too weak for any but the brightest and hottest stars, and is not generally applicable.

The next advance was accomplished by Antoine Labeyrie. Fresh from his triumph in the development of speckle interferometry, Labeyrie decided that interferometry between independent telescopes should be easy to achieve, and he proceeded to demonstrate that, if not trivial, it was at least neither difficult nor expensive.

In the subsequent decade, several groups entered the field with prototype optical and infrared interferometers. The typical prototype project began with a plan to deploy an array of 3 to 5 or more telescopes, but slipped back to 2 telescopes, "just to show feasibility and get the funding flowing." Unfortunately, funding has not flowed liberally for these programs, and more than one has closed its doors shortly after achieving successful operation.

Table 1 summarizes the major features of the optical interferometers which are currently in operation, or are in such an advanced stage of construction that completion is virtually certain. All entries down to SUSI have obtained interferometric operation, though some have since closed due to lack of funds or entered a phase of instrument improvements which precludes operations at this time.

I believe that none of these interferometers has operated on anything except stars (including circumstellar shells). The brightest extragalactic sources are within reach of a 1-m telescope, but may require some degree of wavefront correction – a technique which is still in its infancy.

Figure 1 is a photograph of the Sydney University Stellar Interferometer (SUSI). The white pipe is the enclosure for the optical beam propagation between telecopes. Conceived specifically for measurement of stellar angular diameters and binary orbits, SUSI has an optical baseline up to 640 m, giving an angular resolution of 0.2 *milliseconds* of arc. SUSI is currently operational with baselines to 50 m.

Now having experience with a variety of interferometric techniques at a considerable number of facilities, the interferometric community seems to have agreed with Labeyrie that it is relatively easy to obtain interference fringes between two telescopes. Thus, what might have seemed the greatest obstacle is not such. However, it is considerably more difficult to extract information in volume of adequate quality and with suitable calibration to address a range of scientific problems. Recent work in optical interferometry has emphasized systems issues related to efficient data collection, control of instrument configuration, and monitoring of atmospheric conditions. This work has met with considerable success, but with increased engineering costs.

The phase closure techniques, which are critical for interferometric imaging through the atmosphere at optical wavelengths, require three or more apertures.

TABLE I
Some contemporary optical interferometry projects

Facility Name	Location	Number of Telescopes*	Telescope Aperture	Baseline (Maximum)
I2T	Côte d'Azur	2	27 cm	140 m
GI2T	Côte d'Azur	2	150 cm	70 m
SOIRDETE	Côte d'Azur	2	100 cm	15 m
Mk III	Mt. Wilson	2	165 cm	34 m
ISI	Mt. Wilson	2	8 cm	32 m
IRMA	Wyoming	2	20 cm	19 m
COAST	Cambridge	4	40 cm	100 m
SUSI	Sydney	2	14 cm	640 m
IOTA	Mt. Hopkins	2	45 cm	45 m
BOA	Flagstaff	6	35 cm	470 m
NRL/USNO	Flagstaff	2	100 cm	40 m
VLT-VISA	Cerro Paranal	3	180/800 cm	200 m

*Number (to be) combined simultaneously

Fig. 1. The Sydney University Stellar Interferometer. The interferometer has many telescope stations, and combines the telescopes pair-wise to obtain one-dimensional visibility, but (initially) no image phase information.

Fig. 2. The Very Large Telescope and auxiliary array to be deployed on Cerro Paranal in northern Chile by the European Southern Observatory.

Such observations have been obtained optically, using masks over large telescope apertures. Somewhere between a simulation and a demonstration, these observations confirm that phase closure works at optical wavelengths and in fact radio astronomy reduction packages can be used for optical data. Unfortunately, no working interferometer yet provides for the combination of three independent telescope beams, owing primarily to limited funding. This demonstration is eagerly awaited, and expected within the next few years.

Of the numerous optical-interferometry projects in progress, the Very Large Telescope array, under construction by the European Southern Observatory, is an outstanding example. The array will consist of four 8-m telescopes, each equipped with adaptive optics to provide a coherent pupil. In addition, there will be (initially) three telescopes of approximately 1.8-m aperture, movable between stations to improve coverage of the u-v plane. Figure 2 shows a line drawing of a layout which is expected to correspond closely to the final design. Ground has been broken and funding for the entire array has been approved.

In spite of the many strong analogies between optical and radio interferometry, one dissimilarity remains, to optical interferometry's detriment. The existence of phase-coherent amplification for radio waves provides for great flexibility in a combination of telescopes. The non-existence of such amplifiers at visible wavelengths (perhaps in principle) means that optical interferometry of faint sources will "gain" only linearly with the number of telescopes.

The scientific programs carried out with optical-IR interferometers are too extensive to review here. A recent survey (Ridgway, 1992) describes work in several areas. The most immediate results are obtained in observations of binary stars, where the model is especially simple. Interferometric observations provide unprecedented precision for binary orbits (Armstrong, 1991). Stellar angular diameters are easily determined, and with careful work excellent results may be obtained (Mozurkewich et al., 1991). The choice of model, especially assumptions about limb darkening, is critical (Quirrenbach, 1992). Circumstellar envelopes may be studied with some ease, as the central "point" source gives a good phase reference. Recent results include observations of details of the dust shells in cool stars of several types (Danchi et al., 1990), and of the gaseous hydrogen disk in a B emission type star (Mourard et al., 1989). Interferometry has also advanced the state of the art in positional astrometry, using interference fringes with respect to a constant telescope baseline (Shao et al., 1990).

5. An Interferometry Program for the 1990's

In 1991 the National Academy of Sciences released the decade report of the Astronomy and Astrophysics Survey Committee. This committee took a close look at new optical technologies, and convened a panel specifically to study the opportunities offered by adaptive optics and interferometry. Both of these areas were prominent in the AASC recommendations (National Research Council, 1991).

In interferometry, two moderate scale programs were recommended. For the ground, the committee recommended support for several optical and infrared interferometers, particularly including one facility of 3-5 apertures of order 2 m, leading into a program for the next decade to build a very large optical array.

In space, the committee recommended an astrometric interferometry mission, with a capability of achieving positional accuracies of a few microarcsec. This was the only new mission specifically recommended by the AASC, which, however, strongly supported several missions continuing from the previous decade. In the longer term (and it seems further away every year) discussion of possible astrophysics observations from the lunar surface commonly mentions interferometry as a possibility, and it is indeed an excellent match, providing a large, firm platform with low temperatures, clear skies and long nights. The scientific case is very strong. Figure 3 shows a NASA artist's sketch of one concept for a short-baseline, ultra-sensitive interferometric optical array.

6. Motivation for Solar Optical Interferometry

Interferometric techniques will most likely be of interest for achieving high angular resolution. This could involve interferometric techniques used with a single, filled aperture, in order to achieve the full aperture-limited resolution. It could also be applied to an array of apertures.

Von der Lühe and Zirker (1988) have discussed the scientific goals for solar interferometry and some of the possible limitations. They note, for example, that

Fig. 3. A concept for a lunar optical interferometer. Not for this decade, but perhaps in the next century.

sub-arcsec resolution will be required to resolve the spatial scales of acoustic waves in the convection zone and that resolution of 0.1″ or better may be required to resolve magnetic flux tubes. They emphasize the requirements for high spectral resolution and polarimetric information. They also provide a very important discussion of the probable lifetime of spatial structure as a function of angular size and conclude that the observation time to construct a single image must be severely limited, perhaps to ∼10 s, to avoid change in the structure.

Von der Lühe and Zirker also compute the photon noise-limited angular resolution for several cases, and conclude that achievable resolutions may be only about an order of magnitude better than today's routine capability of about 1″. However, this conclusion is for the specific case of granulation and for $S/N = 1$ in the spatial power plane, which is probably a more severe condition than encountered in typical image structure.

7. Passive Interferometry of the Sun

Interferometric techniques have been used to study solar surface structure for several decades, with increasing instrumental sophistication. Harvey (1972) used short photographic exposures to record fringes formed between a pair of apertures defined with a pupil mask at the McMath solar telescope. Subsequent work with the multi-aperture Michelson technique emphasized understanding of the technique and calibration of the power spectra (Aime *et al.*, 1975, 1977; Aime, 1976).

There has also been extensive work with "speckle" analysis of short exposure images, including one-dimensional images obtained by rapid scanning and conventional electronic images. These observations have emphasized acquisition of properly calibrated spatial power spectra, especially for the study of solar granulation on a scale beyond that which is accessible in direct, conventional imaging. For a recent account with references to earlier work, see von der Lühe and Dunn (1987). Attempts to use speckle analysis to recover two-dimensional images of the solar surface do not appear to have been very productive. The absence of a point source within the isoplanatic angle is a serious drawback in comparison with many successful stellar speckle observations.

In the last few years, nighttime interferometrists have made considerable progress in the use of non-redundant aperture arrays, formed with masks over large telescopes, to obtain phase closure and reasonably effective image restoration. This concept has been explored in some detail for the solar case (Zirker and Brown, 1986; Zirker, 1987) with simulated data, but an effective practical demonstration is still needed. This would certainly appear to be possible, although the lack of a point reference may again make the case of the extended, low-contrast solar surface enormously more difficult than observations in the vicinity of a bright star.

8. Interferometric Ground-based Solar Telescopes

The largest solar telescope today is the McMath-Pierce 1.5-m telescope on Kitt Peak. At this conference we have heard about a proposal to upgrade this facility to a 4-m aperture (Livingston, 1993). A 4-m aperture would have a diffraction limited resolution at 2.2 μm of 0.1″, or about 100 km at the solar surface. Considering the advanced state of development of 4-m mirror technology at this time and the rapid progress of several strategies for monolithic or segmented 8–10 m apertures, it would seem that the filled aperture telescope would be the natural and preferred solution for high spatial resolution at visible and near infrared wavelengths.

In order to achieve 0.1″ resolution at 12 μm, a telescope diameter or baseline of 20 m would be required. Thus, very high resolution at long wavelengths is perhaps the most plausible motivation for very large telescope baselines. Livingston (1993) summarizes a number of interesting reasons for achieving high spatial resolution in the 10–12-μm spectral region.

A monolithic interferometric array, such as the MMT or binocular Columbus concept, would be possible at the 20-m scale, but perhaps better suited to a solar furnace than to a solar telescope, with the long focal lengths normally used for solar astronomy. A distributed array would be the natural solution to a requirement for optical baselines of order 10 m or larger. These baselines are very modest compared with the requirements for stellar astronomy (tens to hundreds of meters) and the optical tolerances at the longer wavelengths are somewhat more forgiving than in the visible. Therefore, it is reasonable to believe that a ground-based infrared solar interferometer would be a feasible project and perhaps easier to build than some of the interferometric systems already under construction for stellar astronomy.

8.1. LASER REFERENCE STARS

There is considerable interest in the nighttime astronomy community regarding the possible use of lasers to form artificial reference stars high in the atmosphere (Fugate et al., 1991; Humphreys et al., 1991). For example, a laser tuned to the wavelength of the sodium D line can be focused at the altitude of the layer of neutral sodium atoms in the atmosphere, about 100 km, to form a resonant scattering source with an apparent size from the ground of about 1″. Of course the Sun has a V magnitude of −26.7, which is considerably brighter than the laser spot. However, in intensity, the laser gains about 16 magnitudes from surface area, about 13 magnitudes from bandwidth, and 0.7 magnitudes in polarization, leaving the Sun at an equivalent magnitude of about +3 per laser spot size and bandwidth when the laser is on. This is brighter than current laser spots, but considerably fainter than the spot that would be produced if the sodium layer could be excited to saturation. Lower altitude, Rayleigh scattering spots could be brighter. So an application of lasers for solar adaptive optics and interferometry, while improbable, is not out of the question.

However, a subtlety of the laser guide star technique is that the laser spot gives no direct information about the tilt and piston errors, which probably must be determined from a source above the atmosphere. Also, the problem of isoplanatic angle remains to be addressed. Therefore use of lasers may potentially change some of the parameters, but it doesn't eliminate the fundamental requirement to register wavefronts in space and time using the solar image as a reference.

9. Solar Imaging from Space

In space, the atmospheric turbulence is eliminated, although flexure of spacecraft structure due to tidal and temperature effects may impose analogous disturbances. In space, apertures of even 2 m are currently considered to be large and expensive. An array of small apertures operated as an interferometer might offer an interesting alternative to a large, filled aperture. In fact, one such configuration has been studied in some detail.

9.1. SIMURIS

A concept for a four-aperture spatial interferometer for solar studies (Dame et al., 1987) has been developed in a pre-Phase-A study titled *Solar, Solar System and Stellar Interferometric Mission for Ultrahigh Resolution Imaging and Spectroscopy*, or SIMURIS (Coradini et al., 1991).

The concept incorporates four telescopes of 20-cm individual apertures, arranged in a non-redundant configuration with a maximum baseline of 200 cm. The proposed arrangement is a linear array. The combination of aperture and baseline dimensions is chosen to completely "fill" the u-v plane along one axis – that is, all possible baselines from 0 to 200 cm are available simultaneously in one dimension. Figure 4 shows the geometric cross-section of the aperture, clearly illustrating the extent to which the 2-m synthesized aperture is underfilled by the physical apertures. The instantaneous spatial frequency coverage of this pupil is shown in Figure 5.

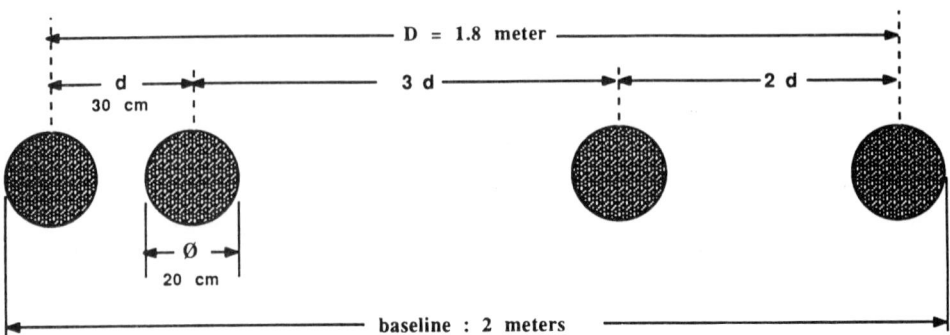

Fig. 4. The input pupil geometry of the SIMURIS high resolution imaging experiment.

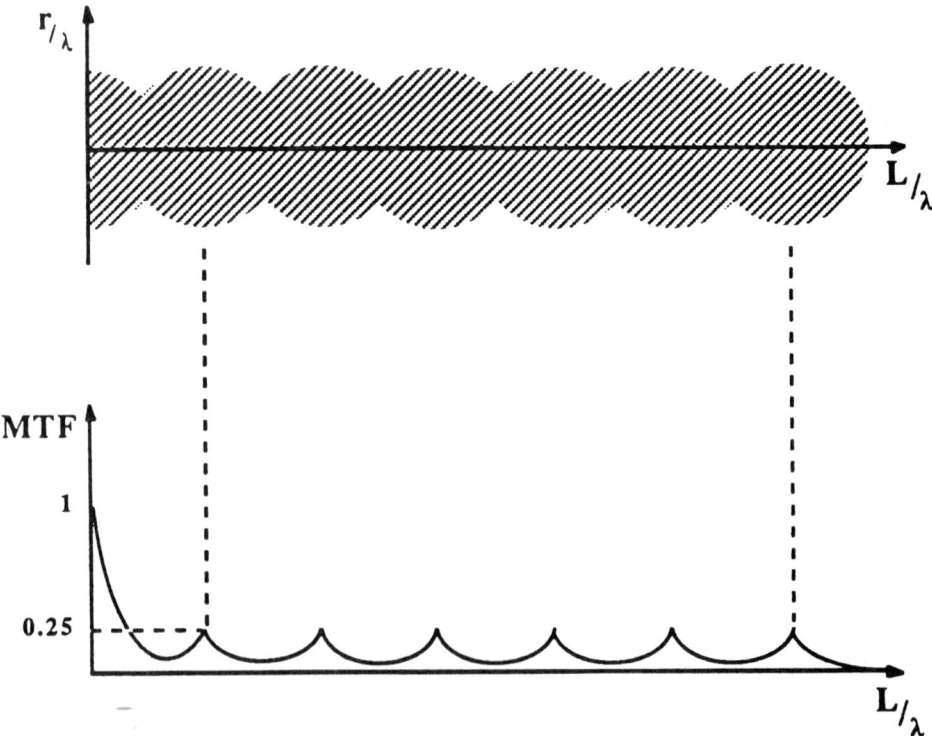

Fig. 5. The instantaneous spatial frequency coverage of the input pupil shown in Fig. 4.

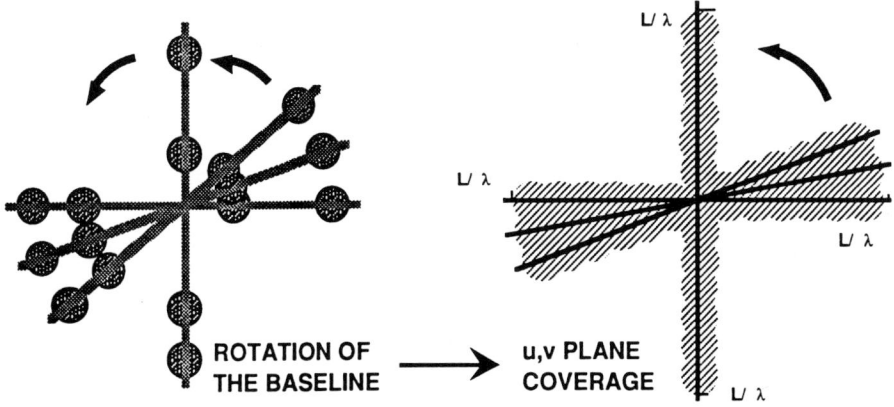

Fig. 6. Continuous baseline rotation fills the u-v plane enabling aperture synthesis.

This selection of geometry offers several advantages: Since complete spatial frequency coverage is available in one dimension in a single "snapshot", it is not necessary to change telescope separations, and there are no gaps in the spatial frequency map, which improves the concentration of the point spread function. Complete spatial frequency coverage to the array limit is achieved in two dimensions by rotating the array and recording data at a number of position angles (rotational aperture synthesis). The use of an array of small telescopes makes it possible to obtain the spatial resolution of a 2-m telescope in a package which is much smaller than would be required for a 2-m solar telescope with filled aperture – and hopefully also less expensive. Figure 6 illustrates the mapping of baseline rotation into u-v plane coverage. The spatial resolution expected is 30 km on the Sun in the visible, and 10 km in the UV. Observing time is 1 second for a snapshot (with the 20 cm × 200 cm aperture), and one minute for a rotational-synthesis, full-aperture observation.

(The SIMURIS concept also includes other systems and operating modes for imagery and spectroscopy, which are outside the subject of this discussion and will not be recounted here.)

The SIMURIS optical design incorporates beamsplitters to allow pair-wise recombination of the telescopes in 4 pairs. A phased condition is achieved by monitoring and controlling the optical path difference in each pair. In order to have high contrast interference fringes, the four component telescopes must point at the same location on the solar surface to within some fraction of the Airy disk diameter of the 20-cm apertures.

The SIMURIS technical study includes an interesting discussion of two alternate configurations for a space solar telescope: a filled aperture and an MMT concept. The difficulties of these two options are certainly imposing. The interferometric array doesn't have large or expensive mirrors, but it does have a large number of components. Interferometrists and solar astronomers will be interested to learn what the space agencies can do with this kind of system.

Acknowledgements

This article was written while the author was a visitor at the Meudon Observatory of the University of Paris, in a visit partially supported by NATO Collaborative Research Grant 86/0080.

References

Aime, C., Ricort, G., and Grec, G.: 1975, *Astron. Astrophys.* **43**, 313.
Aime, C.: 1976, *Astron. Astrophys.* **47**, 5.
Aime, C., Ricort, G., and Grec, G.: 1977, *Astron. Astrophys.* **54**, 505.
Alloin, D. M. and Mariotti, J.-M.: 1989, *Diffraction-limited Imaging with Very Large Telescope*, NATO ASI series **274**, Kluwer.
Armstrong, J. T.: 1991, in 'Complementary Approaches to Binary Star Research', *Proc. IAU Colloq.* **135** (in press).
Beckers, J. M., and Merkle, F. (eds.): 1992, *High-Resolution Imaging by Interferometry II*, ESO Conf. Proc. No. 39, ESO, Garching.
Clark, B. G.: 1980, *Astron. Astrophys.* **89**, 377.
Coradini, M., Damé, L., Foing, B., Haskell, G., Kassing, D., Olthof, H., Mersch, G., Rutten, R. J., Thorne, A. P., and Vial, J.-C.: 1991, *SIMURIS Scientific and Technical Study - Phase I*, European Space Agency SCI(91)7.
Damé, L., Aime, C., Faucherre, M., and Heyvaerts, J.: 1987, in N. Longdon and V. Davis (eds.), *Optical Interferometry in Space*, ESA SP **273**, 189.
Danchi, W. C., Bester, M., Degiacomi, C. G., McCullough, P. R., and Townes, C. H.: 1990. *Astrophys. J. (Letters)* **359**, L59.
Fugate, R. Q., Fried, D. L., Ameer, G. A., Boeke, B. R., Browne, S. L., Roberts, P. H., Ruane, R. E., Tyler, G. A. and Wopat, L. M.: 1991, *Nature* **353**, 144.
Hanbury Brown, R., Davis, J., Lake, R. J. W., and Thompson, R. T.: 1974, *Mon. Not. Roy. Astron. Soc.* **167**, 475.
Harvey, J. W.: 1972, *Nat. Phys. Sci.* **235**, 90.
Humphreys, R. A., Primmerman, C. A., Bradley, L. C., and Hermann, J.: 1991, *Opt. Let.* **16**, 1367.
Livingston, W. C.: 1993, these proceedings.
Mourard, D., Bosc, I., Labeyrie, A., Koechlin, L., and Saha, S.: 1989, *Nature* **342**, 520.
Mozurkewich, D.M., Johnston, K.J., Simon, R.S., Huffer, D.J., Colavita, M. M., Shao, M., and Pan, X. P.: 1991, *Astron. J.* **101**, 2207.
National Research Council: 1991, *The Decade of Discovery in Astronomy and Astrophysics*, National Academy Press, Washington, D.C.
Pease, F. G.: 1931, *Ergebn. Exact. Naturw.* **10**, 84.
Quirrenbach, A.: 1992, in J. M. Beckers and F. Merkle (eds.), *High-Resolution Imaging by Interferometry II*, ESO Conf. Proc. No. 39, p. 663.
Ridgway, S.: 1992, in J. M. Beckers and F. Merkle (eds.), *High-Resolution Imaging by Interferometry II*, ESO Conf. Proc. No. 39, p. 653.
Ryle, M., and Vonberg, D. D.: 1946, *Nature* **158**, 339.
Shao, M., Colavita, M. M., Hines, B. E., Hershey, J. L., Hughes, J. A., Hutter, D. J., Kaplan, G. H., Johnston, K. J., Mozurkewich, D., Simon, R. S. and Pan, X. P.: 1990, *Astron. J.* **100**, 1701.
Sullivan, W. T.: 1991, in T. J. Cornwell and R. A. Perley (eds.), *Radio Interferometry: Theory, Techniques and Applications*, Astron. Soc. Pac. Conf. Ser. **19**, 132.
Thomasson, P.: 1986, *Q. Journ. Roy. Astro. Soc.* **27**, 413.
Thompson, A. R., Clark, B. G., Wade, C. M., and Napier, P. J.: 1980, *Astrophys. J. Suppl.* **44**, 151.
von der Lühe, O., and Dunn, R. B.: 1987, *Astron. Astrophys.* **177**, 265.
von der Lühe, O., and Zirker, J. B.: 1988, in F. Merkle (ed.), *High Resolution Imaging by Interferometry*, ESO Conf. Proc. No. 29, p. 77.
Wild, J. P.: 1967, *Proc. Inst. Radio Electron. Eng. (Australia)* **28**, 277.

THE NEAR-INFRARED CAPABILITIES OF LEST

ODDBJØRN ENGVOLD

Institute of Theoretical Astrophysics, University of Oslo, N-0315 Oslo, Norway

Abstract. The *Large Earth-based Solar Telescope (LEST)* will be a powerful, next-generation telescope with unprecedented angular resolution, capable of highly accurate polarimetry of the Sun, covering the optical spectral range from about 300 nm into the near infrared to about 2.5 μm.

The telescope is a 2.4-m aperture, "polarization-free" concept based on a modified Gregorian optical system. A fast polarization modulator will be located close to the secondary focus of the system. An actively controlled NTT-type main mirror, a high precision pointing and tracking system, a helium-filled light path and a thin entrance window, together with an integrated adaptive optics system, will give the telescope near diffraction-limited performance in the visible. LEST will be sited on La Palma, in the Canary Islands, near the caldera rim on the Roque de los Muchachos Observatory, which often offers excellent seeing. A frequently occurring seeing parameter of $r_o = $ 15–20 cm in the visible will correspond to $r_o \geq 1$ m in the near IR.

The construction of LEST will begin in 1993, and the telescope is to be ready for "first light" in 1997. The telescope facility will accommodate a large number of focal plane instruments on a spacious instrument table. LEST will be made available for near-IR instrumentation from the start of its regular operation.

Key words: infrared: stars – instrumentation: polarimetric – telescopes – Sun: general

1. Introduction

1.1. INTRODUCTORY REMARKS

The Large Earth-based Solar Telescope (LEST) is not a telescope that has been designed for the infrared; yet, it will offer interesting opportunities for observations in the near-infrared out to wavelengths $\lambda \sim 2.5$ μm. LEST will be a unique facility for near-IR observations of the Sun because of its (*i*) large (2.4-m) light-collecting aperture (high photon flux), (*ii*) diffraction limited performance ($\theta \leq 0.25''$ in the near IR), and (*iii*) good polarimetric qualities.

1.2. SCIENCE WITH THE LEST

The unceasing scientific demand for yet sharper and more precise imaging of small-scale processes in the solar atmosphere is behind the joint international efforts to develop and build a next-generation, high-resolution solar observing facility. Observational and theoretical discoveries suggest that much of the dramatic variability of the Sun is tied to small-scale processes involving the magnetic fields in the solar atmosphere (MacQueen 1987). For instance, how does the presence of a magnetic field modify convection, and how does fluid motion alter the structure of the magnetic field? How is the fine-scale magnetic structure of the Sun's atmosphere related to the heating of the outer layers and to various forms of activity? Being the only star close enough for study of many of its phenomena and underlying processes, the Sun can be perceived as the key to understanding the physics of other stars.

We have been reminded by several speakers at this conference about the many unique and exciting possibilities in the infrared and near-infrared for diagnostics of the solar photosphere. Seeing is the basic limitation to spatial resolution in the visible, while at a good site resolution in the near IR may be telescope limited even for apertures exceeding 1 m. While in the visible one obtains magnetic field strengths by indirect methods (Stenflo 1989), in the near IR some lines are fully split (Solanki 1993) and thus provide direct measurements of magnetic field strengths. The richness of photospheric molecular and atomic lines in the IR, especially in spectra of sunspot umbrae, is demonstrated nicely by the new spectrum atlas of Livingston and Wallace (1991). In the near IR, one takes advantage of the minimum absorption around 1.6 μm to probe the deeper layer of the photosphere. The deep layers of the granulation and magnetic flux elements may be accessed here with high spatial resolution (Stenflo *et al.* 1987; Koutchmy 1993).

2. The LEST Project

2.1. An International Cooperation in Solar Physics

The ultimate realization of LEST, the world's largest solar telescope on a superb site in the Canary Islands, is expected to provide the global solar physics community with a next-generation scientific facility which will be uniquely able to tackle from the ground the difficult and pressing observational tasks in solar research for the 1990's and well into the 2000's.

The well-advanced development of the LEST project is being recognized in several national planning reports on future astronomical facilities. In the listing of future astronomical facilities by the Deutsche Forschungsgemeinschaft, the LEST project ranked high among projects recommended for national funding. In the report from the Joint Working Group on Ground-based Astronomy, which involved representatives from national research councils of Canada, France, Germany, Italy, Japan, United Kingdom, and the United States, the assessment was that "the LEST project is one of major importance for the solar physics community and one which can proceed only by international collaboration. The report *Astronomy in the U.S.A. for the 1990's* (Bahcall Report) and the European Physical Society (EPS) Report *Solar Physics in Europe* recommended a strengthening of solar ground-based research facilities. The priority for the project is very high within most of the countries of the LEST consortium."

2.2. Organization

The *LEST Foundation* is a non-profit organization created in 1983 to serve this major international activity and collaboration in solar physics and astronomy (Wyller 1991). It is domiciled, under Swedish law, in The Royal Swedish Academy of Sciences in Stockholm, and now comprises member organizations from nine nations; *Germany, Israel, Italy, Norway, Spain, Sweden, Switzerland,* and *USA*.

2.3. The Site of LEST in the Canary Islands

The LEST site is located on La Palma, in the Canary Islands, at altitude 2,350 m above sea level, in the neighborhood of the Swedish 50-cm optical solar telescope and a number of nighttime telescopes (Wyller 1991). The high optical quality of the La Palma site is well documented from extensive experience with the Swedish vacuum tower telescope, the site of which is practically identical to the LEST site. The image quality achieved with this telescope is widely recognized to be superior to that of any other solar telescope in the World, frequently providing diffraction limited performance with its 50-cm aperture (Wyller and Scharmer 1985; Scharmer 1989). This indicates the uniqueness of the site: The conditions for high-resolution observations are favorable, and the La Palma site satisfies the scientific requirements for a next-generation type 2.4-m telescope like LEST.

3. A Design for High Resolution Observation of the Sun with High Polarimetric Accuracy

3.1. DESIGN SPECIFICATIONS

LEST is conceived as a comprehensive facility for observations of the Sun in the visible and near-infrared, with angular resolution $\theta \leq 0.1''$ in the visible (which will require an aperture ≥ 2 m) and low instrumental polarization in order to measure polarization to an accuracy $\Delta p/p \approx 10^{-4}$. The large aperture will be needed also to ensure a high photon flux and thereby short integration times for measurements of small-scale solar structures. For more details, see Engvold and Andersen (1990a). Being a ground-based facility to serve a large international community, LEST will offer flexibility and ample space for a variety of both permanent and experimental post-focus instruments.

3.2. DEVELOPMENT OF THE LEST DESIGN

The design of LEST and its new-technology enclosure, is the result of extensive studies involving experts from participating institutes. The results from many such studies are described in the LEST Technical Report series, which presently count 54 reports. On the basis of a technical concept developed by Andersen et al. (1984), and the management plan by Engvold and Hillerud (1988), the LEST Foundation decided to contract a major part of the design work to the engineering team located at Risø, Denmark, that previously had designed and built the Nordic Optical Telescope in La Palma, in addition to doing consulting work for ESO's VLT. The LEST design, as described in some detail below, was adopted by the LEST Foundation in 1990 (Engvold and Andersen 1990a,b). Table I summarizes some of the basic design parameters.

3.3. THE TELESCOPE

The LEST telescope is a modified Gregorian type with an additional concave mirror behind the primary. More than 99% of the energy is absorbed by a heat-rejector at the primary focus of the parabolic f/2.3 main mirror. The third mirror forms an $f/75$ beam, and re-images the Gregorian focus at the large instrument table at the base of the telescope. The same mirror also forms an image of the pupil at the position of the flat M5, which will be a fast, agile guider mirror of this system. For the adaptive optics system of the LEST, the adaptive mirror will be inserted immediately above M5. Figure 1 shows the optical layout of the LEST.

TABLE I
Technical Data on LEST

Design Parameters	
Pupil diameter	240 cm
Focal length (effective)	176.5 m
f/D	73.5
Diffraction limited resolution	0.06" – 0.25" (visible - near IR)
Image scale	1.17 arcsec mm^{-1}
Entrance window (fused silica)	Ø250 cm × 30 mm
Wavelength range	0.3 - 2.5 μm
Field of view (non-rotating)	±1'
Number of exit ports	≥15
Tower height	~30 m

3.4. INSTRUMENT TABLE

A large, 19.5-m-diameter, rotating platform at the bottom of the tower provides adequate space for the LEST focal plane instruments (see Fig. 2). A 12-m deep, steel cylinder in the center of the instrument table provides space for large vertically-mounted instrumentation. Full compensation for field rotation is achieved through proper relative motion of the table and the mirror M6 (see Fig. 1).

3.5. TOWER AND DOME STRUCTURE

The compact telescope and integrated mounting of LEST is placed on top of a single conically-shaped, concrete tower. The tower is about 30 m tall, to minimize the influence of the ground turbulence. The compact, light-weight, spherical dome and tube protect the telescope from radiative heating and weather, and the aerodynamical shape serves to reduce wind buffeting. A cross section of tower and dome is seen in Figure 2.

3.6. POLARIMETRY

Because measurements of magnetic fields is a main scientific objective of LEST, the photon flux requirements of Stokes polarimetry have been a fundamental design consideration (*cf.* Lites 1987 a). The polarization modulator of the polarimetry system will be inserted into the light beam prior to the first inclined reflection. Its location is immediately before the secondary image plane of the modified Gregorian telescope. A high frequency (50–100 kHz) piezo-elastic modulator (PEM) will be used in conjunction with a demodulation scheme based on synchronous shifting of charges in a modified CCD in the instrument focus. The proposed system is being

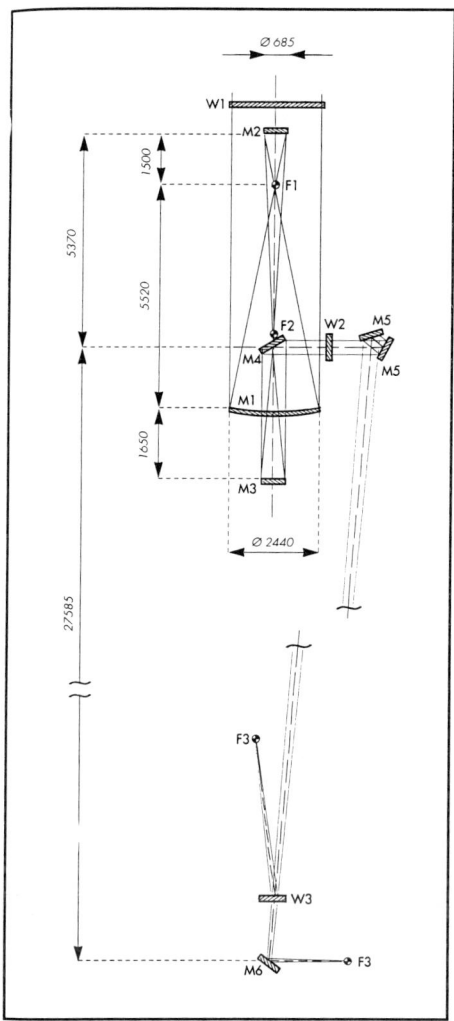

Fig. 1. Layout of the optical systems of LEST.

developed by Keller *et al.* (1989). An alternative modulation scheme is discussed by Lites (1987b).

The "straw-man" modulator package performs without problems over the whole near-IR wavelength range. Demodulation at the exit focus cannot, however, be done with the charge-shifting technique in the CCD chip, as foreseen for the observations in the visible, since IR-sensitive CCD's with this type of architecture are not yet available. For the near IR an optical demodulation scheme based on available technology is proposed, as described by Stenflo and Povel (1985) and Stenflo (1991).

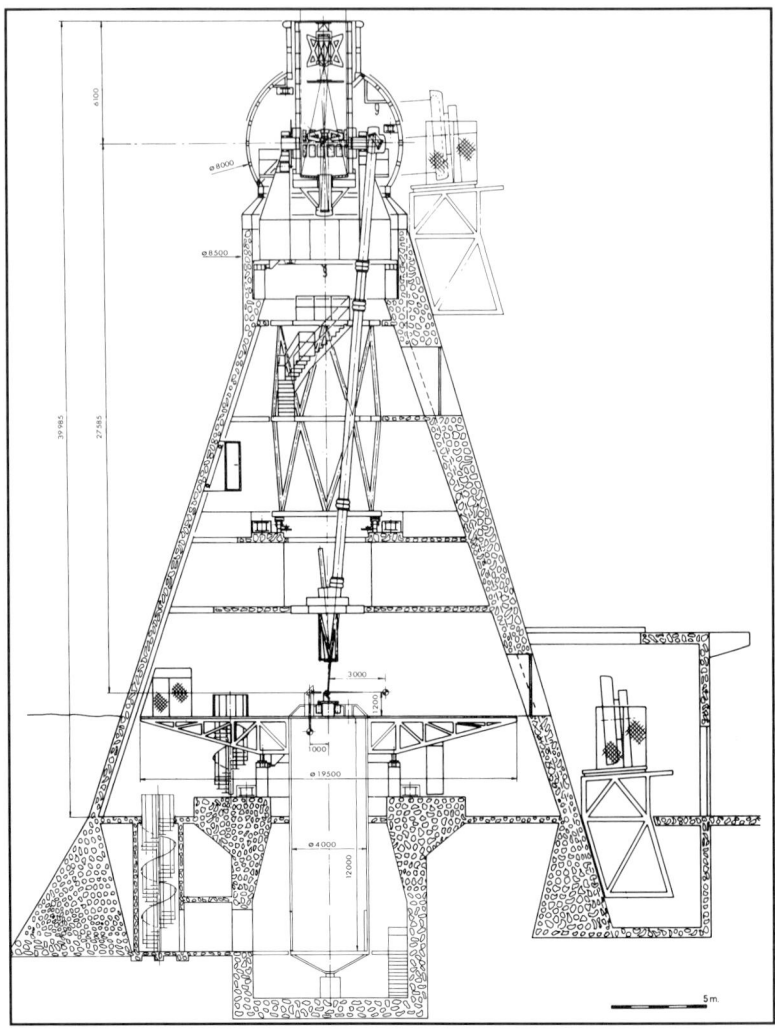

Fig. 2. Cross section of the telescope and tower structure of LEST.

3.7. ENTRANCE WINDOW

Most modern solar telescopes have an evacuated light path to eliminate internal seeing (Dunn 1985). However, the thermal and mechanical stresses in the entrance windows of vacuum telescopes are problematical and reasons for concern, particularly for polarimetric observations (Dunn 1984; Owner-Petersen 1991). Using a vacuum window will, in practice, limit the telescope aperture to less than 1 m, since the necessary thickness would otherwise be unacceptable with regard to optical quality. The idea of filling LEST with helium was explored through several practical studies (Engvold et al. 1983), and more recently by measurement in a

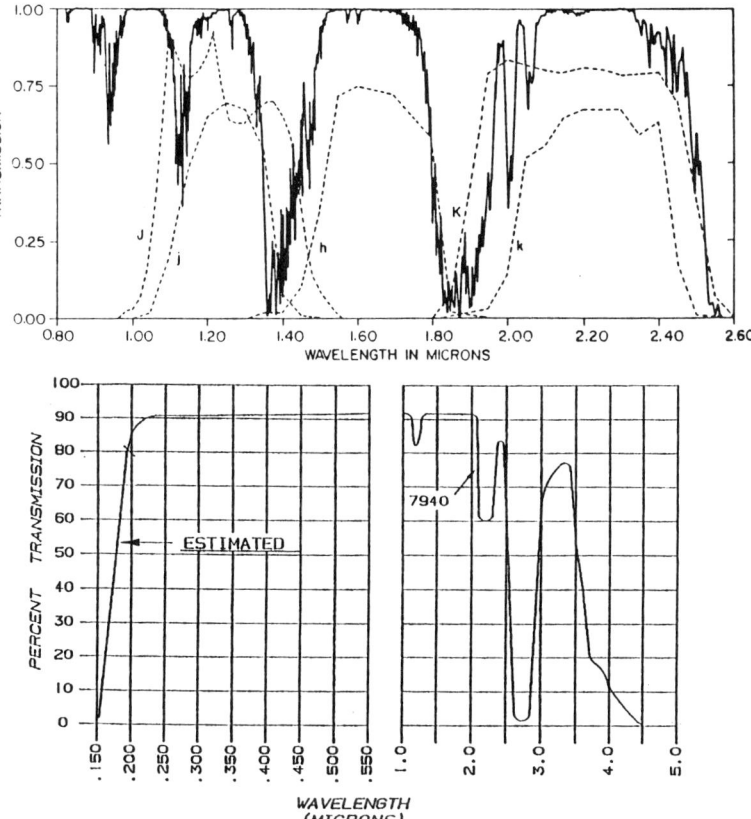

Fig. 3. Upper diagram: Optical transmission of the atmosphere in the near IR. Lower diagram: Optical transmission of a 1-cm thick window of fused silica, including surface losses.

full-scale model of the LEST telescope tank (Engvold and Andersen 1990a). These measurements have demonstrated that a circulating helium atmosphere in the tank, which allows for the use of a thin, high-quality optical window, eliminates the problem of telescope seeing.

The shortest wavelengths accessible to LEST will be limited by the atmosphere ($\lambda \sim 300$ nm) and the longest wavelengths by absorption in the entrance window (≤ 2.5 μm) (Fig. 3). The scientific gain from making LEST available for observations at wavelengths beyond ~ 2.5 μm is unquestionable. However, removing the entrance window would lead to a number of new problems. Exposure to weather and dust in the atmosphere would inevitably harm the many delicate components and high-quality optics inside the tube and, thus, severely impair the use of LEST in the visible. Moreover, mounting and dismounting of the fragile window would be extremely risky and difficult. For these reasons the LEST will not be used without its window.

3.8. HIGH-RESOLUTION IMAGING

The successful implementation of an active figure-controlled main mirror of ESO's NTT (Wilson 1990) has suggested use of a similar meniscus-type main mirror for LEST. The LEST main mirror will be supported axially by 81 individually regulated points and 3 actuators to control tilt, and in the transversal direction by 24 individually regulated points and again 3 actuators.

The tight positional and alignment tolerances of the primary and secondary mirrors call for a continuous monitoring and correcting system for LEST. For this purpose one will use a low spatial and temporal resolution Hartmann-Shack type wavefront sensor with correlation trackers. The wavefront sensor will use light reflected off the small exit window, W3, near focus F3 (see Fig. 1). It is foreseen that this system will be used to sense and update the mirror alignment, and to check the main mirror figure once every several minutes, without interfering with observations.

Adaptive optics (AO) systems are currently becoming available for astronomy. There is little doubt that an AO system will be available for LEST, the question is rather when. Recent tests made with the vacuum tower telescope at Sacramento Peak, using the AO system of the Lockheed group, have shown promising results (Acton 1990). An AO system will be an integral part of the LEST telescope (cf. Dunn 1987). The deformable mirror and the fast tilt-mirror will be located at the image of the pupil on the elevation axis outside the telescope tube.

LEST will be unsurpassed for high-resolution observations also in the near IR. A frequently-occurring seeing parameter $r_o = 15\text{--}20$ cm in the visible corresponds to $r_o \geq 1$ m in the near IR. With the good seeing quality of the La Palma site, LEST is expected to give near diffraction-limited imaging in the near IR, even without an adaptive optics system. Also, an AO system is going to work better in the near IR than in the visible, because of the more relaxed precision required at longer wavelengths.

4. Observing with LEST in the Near-Infrared

The high S/N made possible by the large aperture of LEST will allow the measurement of weak magnetic fields using ratios between different spectral lines in the near IR. The Stokes V signal is smaller in the near IR, since the lines are generally weaker and broader, and therefore a large aperture is needed. A high S/N ratio is also required to study granulation in the near IR, since the granulation contrast there is lower than in the visible.

Stokes Q, U, and V tend to be comparably strong in the IR (when there is complete Zeeman splitting of the lines). Therefore, any cross-talk that may occur between them (due to instrumental polarization) would be disastrous. The polarization-free property of the LEST is thus even more important in the IR than it is in the visible.

The use of the same modulator package at the LEST secondary focus for both the visible and near IR presents a fundamental advantage by allowing for simultaneous, or nearly simultaneous, observations in the visible and near IR. This is important, since the diagnostic information in these two wavelength regions are complementary and refer to different heights in the solar atmosphere. Such observations could allow the 3-D mapping of magnetic, thermal and dynamic structure of small-scale flux tubes.

In summary, LEST will offer the following features and possibilities for observations at near-IR wavelengths:

- Spatial resolution of 0.15–0.25".
- High S/N, made possible by large collecting aperture.
- The same polarization modulation package can be used in the visible and near IR, allowing nearly simultaneous observations in these two λ-regions.
- Fully split Zeeman lines for measurements of network elements (1200–1500 G).
- Low instrumental polarization – eliminating cross talk between Stokes Q, U, and V, which are comparably strong in the IR.
- Thermal control of outer and inner structures.
- Adequate space for a large number of focal-plane instruments.

5. Concluding Remarks

The Design Study has been reviewed and approved by the LEST Foundation, and the technical group responsible for this study is primed to embark upon the final Construction Phase of the LEST Telescope. With the requisite funding by various national agencies, the Construction Phase will begin in the fall of 1993 with a completion date in 1997.

The LEST Foundation will build the "core" telescope as described in Section 3. The telescope will be furnished with a polarimetry system, and later an AO system, in addition to such facilities as a high-resolution spectrograph and a universal narrow-band filter. The large instrument table and central tube provide space and focal-plane ports for a large number (≥ 15) of permanent and experimental set-ups. Standardized specification and interfacing will enable LEST members and other groups to bring and temporarily use a number of specialized instruments.

Since there is a large potential at LEST for near-IR research right from the start, due to less stringent requirements in optical precision and seeing conditions, the development of a near-IR spectrograph for LEST will be given high priority.

Acknowledgements

Helpful suggestions and input for this talk were given by Jan Olof Stenflo and Mette Owner-Petersen.

References

Acton, D.S.: 1990, *Real-Time Solar Imaging with a 19-segment Active Mirror System*, Ph.D. Thesis, Texas Tech University.
Andersen, T.E., Dunn, R.B., and Engvold, O.: 1985, *LEST Technical Report* No. 7.
Dunn, R.B.: 1984, *LEST Technical Report* No. 3.
Dunn, R.B.: 1985, *Solar Phys.* **100**, 1.
Dunn, R.B.: 1987, *LEST Technical Report* No. 28, p. 243.
Engvold, O., Dunn, R.B., Livingston, W.C., and Smartt, R.: 1983, *Appl. Optics* **22**, 10.

Engvold, O., and Hillerud, K.-I.: 1988, *LEST Technical Report* No. 29.
Engvold, O. and Andersen, T.: 1990a, *LEST Design*, Report of the LEST Foundation issued August 1990.
Engvold, O. and Andersen, T.: 1990b, *LEST Technical Report* No. 42.
Keller, C.U., Aebersold, F., Egger, U., Povel, H.P., Steiner, O., and Stenflo, J.O.: 1993, LEST Technical Report No. 53.
Koutchmy, S.: 1993, these proceedings.
Lites, B.W.: 1987a, LEST Technical Report No. 22.
Lites, B.W.: 1987b, LEST Technical Report No. 23.
Livingston, W. and Wallace, L.: 1991, *An Atlas of the Solar Spectrum in the Infrared from 1850 to 9000 cm^{-1} (1.1 to 5.4 μm)*, NSO Technical Report No. 91-001.
MacQueen, R.M.: 1987, LEST Technical Report No. 24.
Owner-Petersen, M.: 1991, LEST Technical Report No. 49.
Scharmer, G.B.: 1989, in R.J. Rutten and G. Severino (eds.), *Solar and Stellar Granulation*, NATO ASI Series, Kluwer Academic Publishers, p. 161.
Solanki, S.K.: 1993, these proceedings.
Stenflo, J.O.: 1989, *Astron. Astrophys. Rev.* **1**, 3.
Stenflo, J.O.: 1991, LEST Technical Report No. 44.
Stenflo, J.O. and Povel, H.P.: 1985, LEST Technical Report No. 12.
Stenflo, J.O., Solanki, S.K., and Harvey, J.W.: 1987, *Astron. Astrophys.* **173**, 167.
Wilson, R.N.: 1990, *The Messenger*, ESO, No. 59, p. 7.
Wyller, A.A., Scharmer, G.B.: 1985, *Vistas in Astronomy* **28**, 467.
Wyller, A.A.: 1991, *LEST - An International Solar Telescope*, Pub. of the LEST Foundation.

A 4-METER McMATH TELESCOPE FOR THE INFRARED

W. LIVINGSTON

National Solar Observatory, P. O. Box 26732, Tucson, AZ 85726, U.S.A.

Abstract. Having no window and a filled aperture, *i.e.*, no occlusion by secondary optics, the all-reflective McMath telescope is a proven IR facility. Beyond about 2 μm, it is diffraction limited, however. Engineering studies show that the McMath building could accommodate an increase to a 4-m aperture with a 6-m alt-azimuth feed, permitting sub-arcsec resolution to 12 μm. The use of cooled, solid aluminum mirrors would eliminate "mirror seeing", which plagues non-vacuum solar telescopes.

Key words: infrared: stars – Sun: general – telescopes

1. Introduction

1.1. WHY OBSERVE THE SUN IN THE INFRARED?

As well discussed in this symposium, the IR offers many advantages for observational study of the Sun. For example, at 1.6 μm the photospheric opacity reaches a minimum, and we see ~ 40 km deeper there, allowing the design and verification of convection models. Away from telluric absorption bands there is a true infrared continuum, as opposed to visible wavelengths, where line-blanketing is ubiquitous. Continuum opacity beyond 1.5 μm is almost entirely due to H^- free-free absorption, which is well understood.

Molecular bands of CO around 2.3 and 4.7 μm are excellent thermal probes of the temperature minimum and, at high spatial resolution, should permit the study of apparent thermal bifurcations in the outer layers that puzzle us.

Most exciting of all is the possibility of mapping the full vector form of surface magnetism. Zeeman splitting increases as λ^2, while Doppler broadening increases only as λ. In the visible, plage field strengths must be inferred by modeling. The $g=3$ line of Fe I 1.5648 μm is completely split in most plage regions allowing the direct measurement of field strength. Flux tubes of 1–2 kG are the rule, but there remains the question of the role of weak fields, how fields evolve, and the possible existence of a "turbulent" component. At 12.5 μm, telescope polarization becomes negligible and vector magnetic fields can be directly deduced, see Harvey (1985). Table I summarizes the various wavelengths and spectral lines now of interest in the infrared.

1.2. INFRARED FACILITY REQUIREMENTS

What are the desired features for a solar infrared telescope? Assuming the observational bandwidth to be 0.3 to 12.5 μm, we come up with the following needs, roughly in order of priority:

1. *All reflecting system*: For reasons of achromatic transmission, no windows are allowable. This rules out a vacuum-telescope configuration.
2. *Aperture $\geq$ 4 meters*: The diffraction limit, in arcsec, is given by $\Delta\theta \approx 0.25 \lambda/D$, where λ is in microns and D is the aperture, in meters. The present 1.5-m

TABLE I
Infrared Diagnostics of the Solar Atmosphere.

ID	λ (μm)	g	U	P	Application	References
FeH	1.006319	0	x		umbrae, pores	1
FeH	1.006270	~ 1	x		umbrae, pores, mag. field	
He	1.0830	1.2			high chromosphere	
H(Pβ)	1.2818	~ 1			chromosphere	2
Fe	1.564852	3	x	x	plage, umbrae mag. field	3,4,5,6
cont	1.63	-	x	x	opacity minimum	7
Na	2.208367	1.3	x	x	no π-component	8
Ti	2.227358	2.5	x		umbrae, pore mag field	8
CO	2.324	-	x	x	dT/dh in cool parts	9
H	4.0512	-	-	-	prominences, Stark effect	10
CO	4.665	-	x	x	oscillations, dT/dh	11
cont	10.0	-	x	x	limb dark, spots, seeing	12,13,14
He	10.879	-	-	-	prominences	15
OH	11.065	-		x	chromosphere oscillations	16
H	11.306	-	-	-	prominences	15
Mg	12.320	1.0		x	high photosphere mag. field	17,18

where g = Landé factor, U = umbra, P = photosphere, dT/dh implies thermal structure.

References:

1 Wallace 1991; 2 Livingston 1990; 3 Harvey and Hall 1975; 4 Solanki *et al.* 1990; 5 Stenflo *et al.* 1987; 6 Rabin *et al.* 1990; 7 Foukal *et al.* 1990; 8 Hall 1970; 9 Ayres and Testerman 1981; 10 Foukal and Hinata 1991; 11 Ayres and Brault 1990; 12 Boyd 1978a; 13 Kopp and Livingston 1991; 14 Turon and Lena 1970; 15 Zirker 1985; 16 Deming *et al.* 1984; 17 Deming*et al.* 1991; 18 Carlsson *et al.* 1992.

McMath, thus, has a diffraction limit of 2.1″ at 12.5 μm; a 4-m McMath would have 0.8″.

3. *Filled aperture*: In the thermal IR a cold aperture stop in the cryogenic detector should preferably see no telescope structure, ruling out Cassegrain optics.
4. *Long focal length*: A short-focus 4-m telescope becomes, in effect, a solar furnace. Even if the temperature of secondary mirrors could be adequately cooled, the radiation flux of an $f/1$ beam will cause heating of the air near focus with resulting turbulence and uncorrectable induced seeing. Filters will break, and other damage will be expected. The solution is a long-focal-length instrument which avoids any major concentration of heat.
5. *Adaptive optics*: Atmospheric seeing correction is much easier in the IR than visible. Refer to Table 2 in Roddier and Graves (1993), in these proceedings. At $\lambda = 0.5$ μm, to obtain the optimum Strehl ratio of 0.3 requires 90 compensating Zernike modes for a field of view (isoplanatic patch) of 2.6″. At $\lambda = 2.2$ μm, the same compensation is obtained with 2 modes (a simple tip-tilt correction) and the field of view realized is 60″. Roddier and Graves propose a wavefront sensor that works on out-of-focus granulation images. This is untested, however.

6. *Dry site*: Certain IR wavelengths are heavily absorbed by water vapor. By locating the facility on a high-elevation dry site, water vapor interference is minimized.

1.3. Strategy for Creating an IR Facility: a 4-m McMath

Why the McMath? The answer is, "because it is there." The existing McMath telescope on Kitt Peak already satisfies most of the demands for an IR solar facility: It is all reflective, has a filled aperture, a long focus throw, and plenty of room to install adaptive control. The site is relatively dry except in summer. The main limitation is its small, 1.5-m aperture. Fortunately, owing to conservative design, the building superstructure could easily accommodate an upgrade to a 4-m concave within the tunnel. A 6-m alt-azimuth flat as a light feed would replace the present 2-m heliostat. We believe this proposed upgrade to be especially cost-effective, providing a state-of-the-art telescope for a fraction of the cost of starting from scratch.

2. Upgrade of the McMath to 4 meters

Larry Barr has conducted an engineering study that indicates that it is feasible to modify the present structure to accommodate a 6-m feed, a 4-m concave and to provide mirror handling with existing cranes for periodic re-aluminizing. One unavoidable change would be an enlargement of the aluminizing room and the installation of a 6-m coating chamber.

2.1. The 6-m Light Feed

Especially challenging is this large assembly, Figure 1. Barr proposes that the 6-m flat be of solid aluminum, 0.15 m thick. Discussion with a potential vendor – Rozelot and Leblanc (1991) – indicates such a blank would be cast as subsections, which are then forged, electron-beam welded together, heat-treated by repeated cycling from cryogenic to +100 C temperatures, surfaced and ground, coated on both sides with a 150 μm nickel-sulfate amorphous layer capable of taking a polish, and then figured flat to ± 0.5 μm overall. Small-scale irregularity would be ± 0.1 μm. The aim here is about 0.5″ resolution in the visible, comparable to what the telescope achieves today.

The mirror would be supported by an array of hydraulic inter-connected assemblies. These push against its back surface in a cell under partial vacuum whose pressure depends on mirror attitude. Thus, the supporting force against the mirror is constant (except for any wind-buffeting component). Number and spacing of the arms has been optimized by computer modeling.

Liquid cooling would be applied to the back of the mirror so as to keep the front surface < 0.5 C from ambient. Even so, a front-to-rear temperature gradient will tend to induce a curvature. This will be countered by the hydraulic arms. The latter also correct for gravity.

We believe this use of a cooled metallic (aluminum) mirror, as opposed to an essentially insulating ceramic (glass) blank, may be of great benefit for non-vacuum

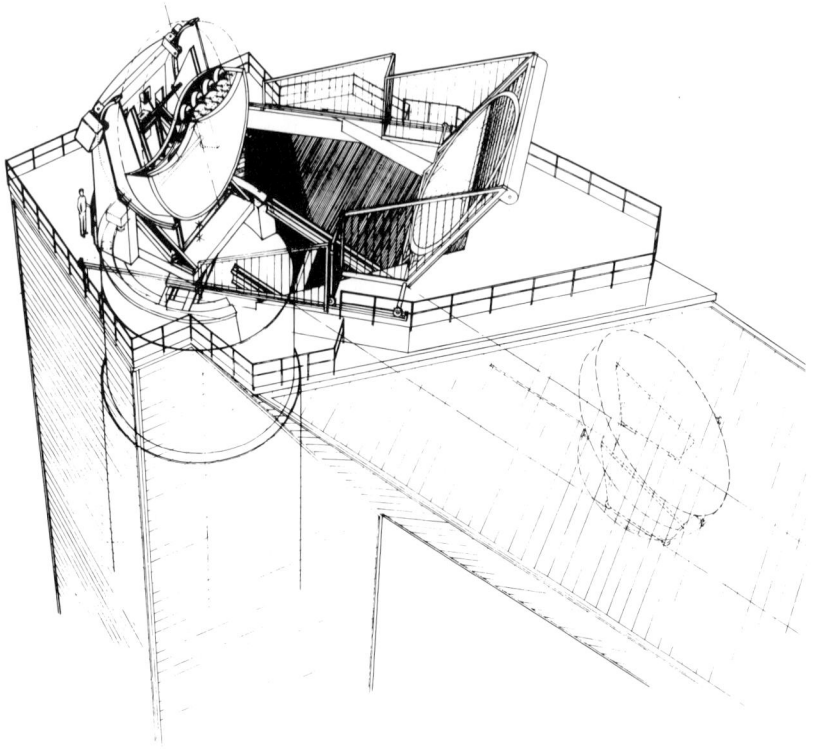

Fig. 1. Rendering of the top of the McMath and the 6-m alt-azimuth flat. Cutaway reveals some of the hydraulic support arms and the cooling channels. Wind fence is shown in its extended position away from the mirror.

solar telescopes. Within seconds after sunlight is incident on a 10% absorbing aluminum reflecting layer, that layer will rise in temperature if the blank is glass. A 1-C rise initiates "mirror seeing" (Barr et al. 1990; Iye et al. 1991). Experiments to document these thermal optical improvements are planned for this fall (1992).

The mirror and support cell, together with its bearing journals, will be about 30,000 kg. This moving component will be supported by hydrostatic pads in an alt-azimuth mount. For re-aluminizing, this moving component disengages from the main alt-azimuth column and is lowered down the shaft on self-contained wheels. All sky coverage to a declination limit of +50° (with 60% full aperture) is achieved. Thus, the McMath will become an important 4-m telescope for solar-stellar nighttime use (see Livingston and Barr 1992).

Wind and weather protection are important concerns. Our present wind-fence in front of the heliostat has proved remarkably effective in preventing image shake. Before the fence was installed, image motion was noticeable at about 6.5 m/s; at 18 m/s (not uncommon) the heliostat preload was inadequate to prevent tooth-to-tooth floating at the drive gear. With the wind-fence erected, image motion at 22 m/s is comparable to that at 6.5 m/s without it, at least for the most common condition of a south wind. For the 6-m, we plan an even more elaborate wind fence which can be moved as close to the mirror as the field of view permits. A bad-weather and closure mode would involve translating the fence, and rotating the mirror, to bring them into actual sealed contact (through a gasket).

2.2. THE 4-M CONCAVE

Here we envisage a 4-m meniscus similar to the 6-m, *i.e.*, of solid aluminum. It will be likewise cooled and actively supported, although its static orientation leads to certain simplifications. Like the present 1.5-m concave, it will have a focal length of approximately 86 m. An iris diaphragm will probably be used to stop down the mirror for certain visible-light applications.

Figuring a 4-m mirror of such a long focal length presents difficulties to most optical shops. We propose to do the final polishing in situ at the telescope. Testing would be against the present 2-m heliostat at the center-of-curvature.

2.3. THE 1.8-M FLAT

This mirror, which serves to direct the beam to one of several observing stations, is also of solid aluminum. It must be cooled and stressed to induce a slight cylindrical form in order to provide correction for astigmatism introduced by thermal curvature.

Of course, near the focus there will be further optical stops, all cooled. The unvignetted field on the solar image would be 6'.

3. Project Status

The feasibility study is now complete. A preliminary cost estimate has been prepared. Barr continues his work on the optimization of mirror-support schemes. As mentioned, we are now preparing for an experiment, using a surplus 127-cm aluminum "RCT" mirror, to measure the influence of temperature control on mirror seeing. Point source images will be recorded with and without incident solar radiation on the mirror and with and without temperature adjustment to ambient.

Acknowledgements

Larry Barr and the staff of NOAO Engineering and Technical Services are the source of most of the above concepts. Support and encouragement came from Sidney Wolff and John Leibacher.

References

Ayres, T. R. and Brault, J. W.: 1990, *Astrophys. J.* **363**, 705.
Ayres, T. R. and Testerman, L.: 1981, *Astrophys. J.* **245**, 1124.
Barr, L., Fox, J., Poczulp, G., and Roddier, C.: 1990, *SPIE Proc.* **1236**, p. 492.
Boyd, R. W.: 1978, *J. Opt. Soc. Am.* **68**, 877.
Carlsson, M., Rutten, R. J., and Shchukina, N. G.: 1992, *Astron. Astrophys.* **253**, 567.
Deming, D., Hewagama, T., Jennings, D. E., and Wiedemann, G.: 1991, in L. November (ed.), *Solar Polarimetry*, Proc. 11th Sacramento Peak Workshop, NSO, Sunspot, New Mexico, p. 341.
Deming, D., Hillman, J. J., Kostiuk, T., Mumma, M. J. and Zipoy, D. M.: 1984, *Solar Phys.* **94**, 57.
Foukal, P., Little, R., Graves, J., Rabin, D., and Lynch, D.: 1990, *Astrophys. J.* **353**, 712.
Hall, D. N. B.: 1970, *Doctoral Dissertation*, Harvard Univ., KPNO Contr. No. 556.
Harvey, J. and Hall, D. N. B.: 1975, *Bull. Amer. Astron. Soc.* **7**, 459.
Harvey, J.: 1985, in M. J. Hagyard (ed.) *Measurements of Solar Vector Magnetic Fields*, NASA Conf. Pub. 2374, NASA, Washington, D.C., 109.
Iye, M., Noguchi, T., Torii, Y., Mikami, Y., and Ando, H.: 1991, *Pub. Astron. Soc. Pacific* **103**, 712.
Kopp, G. and Livingston, W.: 1991, private communication.
Livingston, W. and Barr, L.: 1992 in M. S. Giampapa and J. A. Bookbinder (eds.), *Cool Stars, Stellar Systems, and the Sun, Seventh Cambridge Workshop*, Astron. Soc. Pacific Conf. Ser. Vol. 26, San Francisco, p. 604.
Rabin, D., Jaksha, D., Plymate, C., Wagner, J., and Iwata, K.: 1991 in L. November (ed.), *Solar Polarimetry*, Proc. 11th Sacramento Peak Workshop, NSO, Sunspot, New Mexico, p. 361.
Roddier, F. and Graves, J. E.: 1993, these proceedings.
Rozelot, J. P. and Leblanc, J.-M.: 1991, *SPIE Proc.*, **1994**, p. 481.
Solanki, S., Biémont, E., and Mürset, U.: 1990, *Astron. Astrophys. Suppl.* **83**, 307.
Stenflo, J. O., Solanki, S. K., and Harvey, J. W.: 1987, *Astrophys. J.* **173**, 167.
Turon, P. J. and Lena, P. J.: 1970, *Solar Phys.* **14**, 112.
Wallace, L.: 1991, private communication.
Zirker, J. B.: 1985, *Solar Phys.* **102**, 33-40.

THE APPLICABILITY OF A 5–18 μm ARRAY CAMERA TO SOLAR IMAGING

DANIEL Y. GEZARI

NASA/Goddard Space Flight Center, Infrared Astrophysics Branch, Code 685, Greenbelt, MD 20771, U.S.A.

Abstract. The anticipated requirements and operating conditions are considered for using a mid-infrared array camera in broad- and narrow-band solar imaging observations. The array camera system was designed for high-background 5–18 μm general astronomical imaging observations. The electronic and optical design of the camera, its photometric characteristics, examples of observational results, and the requirements for imaging in both high- and low-background solar applications are discussed.

Key words: infrared: general – instrumentation: detectors – Sun: general

1. Introduction

Solar imaging with an array camera at mid-infrared (5–20 μm) wavelengths introduces quite a different a set of concerns and requirements from those of either conventional infrared photometry or infrared array imaging observations of stellar sources. These include the problems of dealing with 1) high source signal levels in addition to the large 10 μm thermal background from the telescope optics and sky, 2) spatial chopping for background subtraction when observing a very extended source, 3) precise pointing, guiding and registration of images using an instrument capable of very high spatial resolution ($\sim 1''$) and relative astrometric precision ($\sim 0.1''$), and 4) assembling mosaic images from individual exposures of extended, low contrast infrared surface features which may not have visible counterparts.

In a $T \sim 270$ K observatory environment, the 10-μm thermal background photon flux in the Cassegrain focal plane of a large (~ 3 m) conventional telescope is about 10^9 photon s^{-1} m^{-2} μm^{-1} arcsec^{-2}, while the detector well capacity of infrared photoconductor arrays is typically 10^5–10^6 electrons. However, the small detector pixels, the optical efficiency of the camera instrument, the low photoconductive gain at which the detector can be operated, and the reasonably short integration time combine to reduce the number of electrons which actually accumulate in each detector well during an integration by about four orders of magnitude, making operation of the large-format array feasible.

2. The Array Camera Instrument

The infrared array camera system described here was developed for general high-background, diffraction-limited 5–18 μm astronomical imaging on large optical telescopes. The camera uses a 58 × 62 pixel Si:Ga (gallium doped silicon) photoconductor array detector manufactured by Hughes/Santa Barbara Research Center (SBRC). A full discussion of the array detector characteristics, camera optical and electronic design, and operating strategy has been presented by Gezari et al. (1992).

The detector array is a hybrid device, assembled from a wafer of Si:Ga detector material (nominally sensitive from 5–17 μm), bump-bonded to a Hughes CRC-228 direct readout (DRO) integrated circuit multiplexer chip (Hoffman 1987). The array pixels are read out serially, although the switched FET multiplexer design allows them to be sampled in any order, or polled non-destructively to determine the fullness of the well in low-signal, low-background applications (with the penalty only of added read noise). In the case of this Si:Ga photoconductor array, the photoconductive gain of the detector, $G_{\rm pc}$, is a function of net detector bias. $G_{\rm pc}$ can be reduced to about 0.1 by operating at a net detector bias of 4 volts. Since $G_{\rm pc}$ is a post-detection gain factor, reducing the photoconductive gain reduces the number of electrons generated per incident photon, but does not change the incident photon statistics or signal/noise of the observation. This characteristic of the device permits broad-band operation of the array at higher backgrounds, with better relative photon noise statistics.

The camera front-end electronics (Folz 1989) includes dual low-noise preamp modules, a regulated low-noise bias power supply, dual high-speed A/D converters, array clocking and pixel address timing generator, and a detector temperature monitor/regulator. Two 16-bit Analogics Corp. ADAM 826-1 analog-to-digital converter modules are used which can operate at 3 μs per conversion, permitting the camera frame rate to be limited by the intrinsic time constants of the array detector multiplexer. A Mercury ZIP-3216 array processor residing on the Q-bus of the DEC LSI-11/73 host computer system is used to acquire and co-add digital data from the A/D converters. The incoming image data from the two positions of the telescope chopping secondary mirror are sorted into two arrays by the ZIP, synchronized with the chopper drive signal. Data are ignored while the chopper mirror is moving between end positions. Two final images (source and sky) are down loaded to the LSI 11/73 host computer for image pair storage.

The camera optical design (Gezari 1989) uses a single off-axis parabolic mirror and a cold aperture stop (Fig. 1), providing achromatic performance, simple optical alignment, diffraction-limited images, and good suppression of out-of- field background radiation. The array plate scale is 0.26" pixel^{-1}, a field-of-view of 15.0" $\times$ 16.1" on the 3.0-m, $f/35$ NASA Infrared Telescope Facility (IRTF) on Mauna Kea. The camera optics are diffraction-limited at wavelengths $\lambda > 5$ μm and typically produce 1.0 ± 0.1" seeing- and tracking-limited stellar images in long integrations on the IRTF. Aberrations and distortion (pincushion, *etc.*) and field rotation are negligible, less than 1/2 pixel across the full field of the array (corresponding to $< 0.5°$ of field rotation). Two filter wheels contain a 4.8–14.0 μm, OCLI-Corp. circular variable filter (CVF), a set of OCLI-Corp. fixed interference filters (Silicate Filter Set) at 7.8, 8.7, 9.8, 10.3, 11.6 and 12.4 μm, and a set of neutral density filters for laboratory test work. The spectral bandwidth of the CVF is $\Delta\lambda/\lambda = 0.04$ at 10 μm (defined by the width of the cold aperture stop) with $\sim$50% transmission. The fixed filters have $\Delta\lambda/\lambda = 0.1$ bandwidths and $\sim$80% transmission. The combined instrumental optical efficiency is about 0.5 with the fixed filters, and is reduced to $\sim$0.3 using the CVF. An 18.1-μm fixed interference filter is also installed, and the camera operates with $\sim$50% efficiency beyond the nominal cut-off of the the Si:Ga detector, in a narrow spectral region between the cut-on of the 20 μm (Q band) atmospheric window at about 17 μm and the exponential fall-off of the Si:Ga detector quantum efficiency longward of 18 μm.

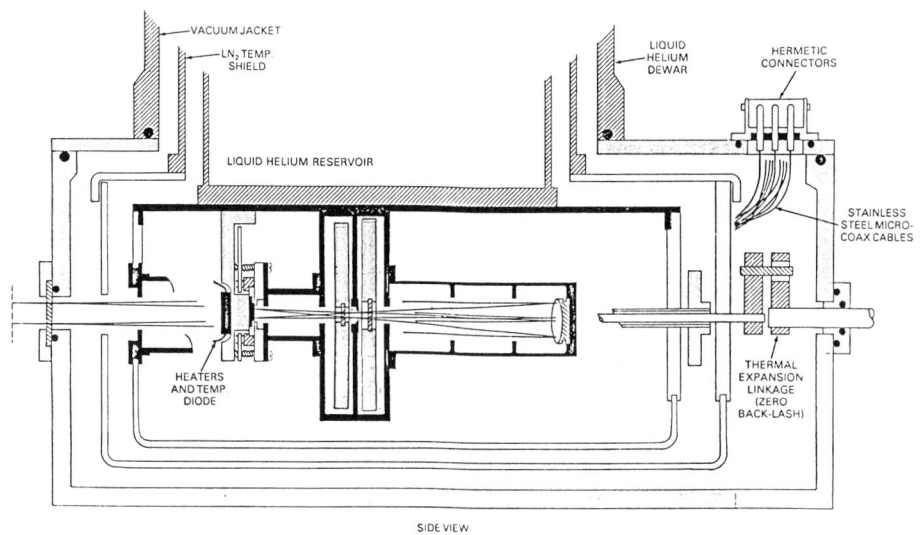
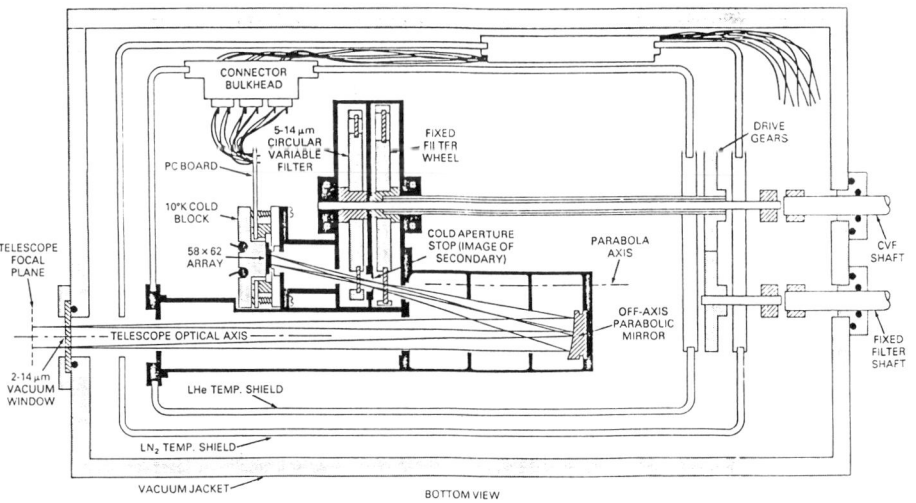

Fig. 1. Optical design and mechanical layout of the cryogenic dewar optical bench, showing the arrangement of the off-axis parabola, cold aperture stop at the image of the telescope secondary mirror, and the array at the de-magnified image of the telescope focal plane. The cold optics are surrounded by an optical baffle system at LHe temperature, within the outer LHe temperature radiation shield, to minimize light leaks.

The observational noise equivalent flux density (NEFD) on the IRTF using the relatively narrow-bandwidth CVF filter ($\Delta\lambda/\lambda = 0.04$, transmission ~ 0.5) at 10 µm is

$$\text{NEFD}(1\sigma) \approx 0.03 \text{ Jy pixel}^{-1}\text{min}^{-1/2},$$

(1 Jy $\equiv 1.0 \times 10^{-26}$ W m^{-2} Hz^{-1}). However, the more efficient fixed interference filters ($\Delta\lambda/\lambda = 0.1$, transmission ~ 0.8) are used for most observations, and the improved photon statistics due to the increased bandwidth and transmission result in

$$\text{NEFD}(1\sigma) = 0.010 \text{ Jy pixel}^{-1}\text{min}^{-1/2}$$

with the fixed filters, corresponding to a noise equivalent brightness

$$\text{NEB} = 0.15 \text{ Jy min}^{-1/2}\text{arcsec}^{-2}$$

(the surface brightness of an extended source yielding a 1σ detection). The image contrast we can detect in the background-limited image is about one part in 10^4.

3. Observational Considerations for Infrared Solar Imaging

In non-solar broadband ground-based applications at 10 µm, thermal background radiation completely dominates the incident photon flux. In most cases the raw image of an astronomical source (*i.e.*, the image before background subtraction) is indistinguishable from a raw image of blank sky (see Fig. 2). Blank sky images observed simultaneously using the telescope chopping-secondary mirror are used for precise background subtraction. The background flux can have both a temporal and spatial dependence. At present, slow (~ 1 Hz) chopping is standard procedure for high background observations at thermal infrared wavelengths. *In principle*, chopping should not be required for sky subtraction or flat-fielding of high-background array image data. Sky background subtraction could be done without chopping if some part of an image were known to contain blank sky or a well-defined signal level, and an accurate gain-matrix existed for that image. In practice, temporal variations in the sky flux level on a ~ 10 s time-scale make spatial chopping a basic part of the current operating procedure. Background subtraction, chopping and flat-fielding techniques are discussed in detail by Gezari *et al.* (1992).

A broadband mid-infrared observation of the Sun is not necessarily a "high background" measurement. The incident flux from the solar surface with $\Delta\lambda/\lambda \sim 1$ would saturate the array camera in its normal configuration by a factor of roughly 1,000. To cope with this large signal would require the use of either a cold neutral density filter inside the dewar, a faster readout rate, use of only a portion of the full array, or some combination of the three. Technically this is an inefficient observing mode since signal/noise is reduced, but is acceptable if narrowing the spectral bandpass is undesirable for scientific reasons (or simply impractical), especially since the Sun is such a strong infrared source. The observer pays no real penalty, because of the large relative continuum signal. The environmental background detected using a cold neutral density filter would be negligible compared to the solar flux in such a broadband observation, and could be thought of as a "low background" measurement. This observation might well be successful without spatial chopping for background subtraction (discussed below).

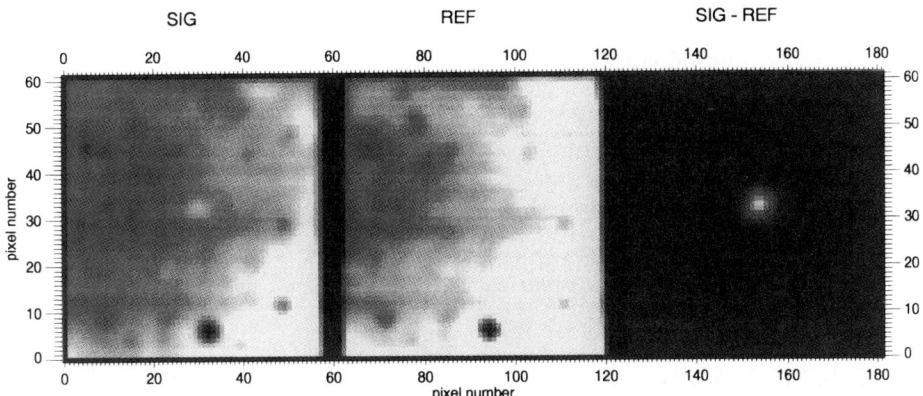

Fig. 2. The individual raw source (SIG) and reference (REF) images of the bright star α Boo, and the final sky-subtracted image (SIG - REF) obtained by taking the difference between the two. The star can barely be seen above the background in the SIG image. A gain gradient of about 30% can be seen across the field. Array gain defects of up to 25% of the mean can be seen resulting from defective bump-bonds. The gain variation over all of the pixels is 15% (1σ). Despite these gain non-uniformities, the sky-subtracted images can be flat-fielded to better than 0.015% (1σ) of the raw mean. Slight differences in the SIG and REF images shown here are due to the auto-ranging color display; the actual images appear essentially identical. If these image differences were real (due to sky background changes, *etc.*) the final sky-subtracted image would be useless.

Broadband observations of coronal structure would not require cold neutral density filters to attenuate the signal, and would thus fall into the "high background" category. However, in this case identifying a suitable featureless field-of-view to chop against (within a few arcmin of the observed position) for background subtraction might be a problem, as well might be scattered light from the nearby solar disk.

On the other hand, narrow bandwidth ($\Delta\lambda/\lambda < 10^{-3}$) imaging of solar surface features could be either a high background or low background observation, depending on whether the dispersing instrument is warm (an observatory facility spectrograph, for example) or cold (cooled spectrograph or Fabry-Perot cryogenically coupled directly to the array camera dewar). Using a cold dispersing instrument would dramatically improve the signal/noise ratio, and in this case the measurement would be read-noise- rather than background-limited. A narrow-band spectral line observation using a warm dispersing instrument would be a standard high-background measurement, with sensitivity limited by the noise level in the background from the warm instrument.

The environmental 10 μm background anticipated at the McMath Solar Telescope on Kitt Peak (altitude 2,100 m) can be compared to the 3-m Infrared Telescope Facility (IRTF) on Mauna Kea (altitude 4,200 m). Other things being equal, differences in ambient temperature (300 K *vs.* 270 K) and atmospheric precipitable water vapor (1 mm *vs.* ~3 mm in good weather) would combine to roughly double

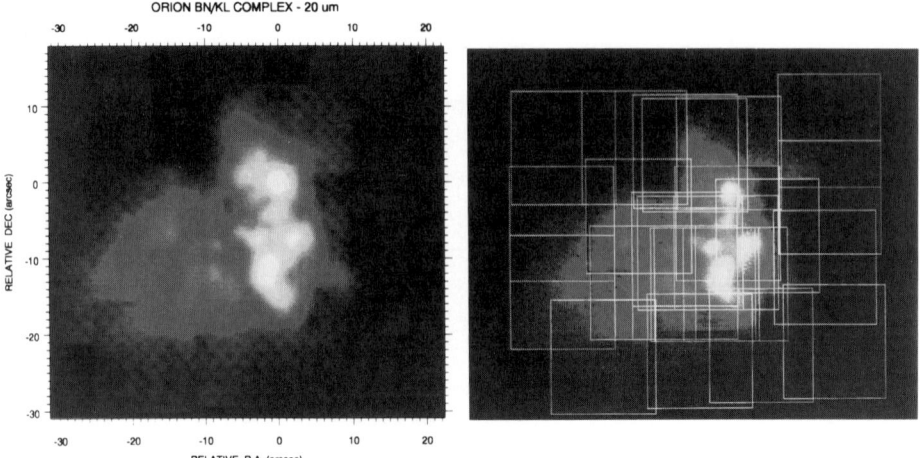

Fig. 3. 20-μm continuum mosaic of the Orion BN/KL complex obtained with the 58 × 62 array camera (Gezari et al. 1992) at the 3-m NASA IRTF Telescope on Mauna Kea. The mosaic was assembled from 23 overlapping 1-minute-integration frames (15″ × 16″ field of view, pixel size 0.26″) which were aligned, matched and co-added to make up the final mosaic, using our MOSAIC software package.

the 10 μm background at the McMath compared to the IRTF. But the slower focal ratio of the McMath (f/54 vs. f/35) would reduce the solid angle of warm telescope optics and sky seen by the array camera to 0.4 of that on the IRTF. The net result is that the 10 μm thermal background detected by the array camera at the two telescopes (with an appropriate change of cold aperture stop in the camera optics) should be roughly comparable. The array field of view on the McMath main beam would be 20″, corresponding to 0.33″ pixels, fully sampling the diffraction limited image at 10 μm ($\theta = 1.6″$ FWHM).

An example of the array camera imaging results on an extended source is given here to illustrate how large mosaic images of complex fields can be assembled successfully from many individual overlapping array images. Figure 3 shows the Orion BN/KL complex at 20 μm. The data were reduced and this mosaic was made using the MOSAIC image analysis software package we have developed in IDL (Varosi and Gezari 1992).

Several factors combine to make array camera observations more difficult than first impressions might suggest. NEFD *per pixel* numbers can seem deceptively good, but this is because the array pixels are small. One should look at the values of the NEFD *per square arcsec* for a more realistic picture. The NEFD is expressed as the 1σ noise level, but good images require $\sim 5\sigma$ results in most of the image, increasing integration time by 1–2 orders of magnitude over the NEFD time scale. Also, chopping more than doubles the elapsed observing time; overhead must be allowed for calibration observations; and flat fields (blank sky images for calibration) generally have to be observed to the same noise level as the image data. Despite these burdens, infrared array cameras can make simultaneous, well sampled, high-spatial-resolution images of low-contrast solar surface features in a reasonable period of time, which would be completely impractical to attempt by mapping with a single detector bolometer system.

Acknowledgements

We are grateful to Mary Hewitt of Hughes/Santa Barbara Research Center (SBRC) for her expert guidance and her extraordinary generosity during the development of the array camera system. This research is funded by NASA/OSSA (RTOP 188-44-23-08) and the NASA/Goddard Director's Discretionary Fund.

References

Folz, W.: 1989, NASA Internal Document (preprint).
Gezari, D. Y.: 1989, NASA Internal Document (preprint).
Gezari, D. Y., Backman, D. Werner, M. W., McKelvey, M., and McCreight, C.: 1992, *Pub. Astron. Soc. Pacific*, in preparation.
Gezari, D. Y., Folz, W. C., Woods, L. A., and Varosi, F.: 1992, *Pub. Astron. Soc. Pacific* **104**, 191.
Hoffman, A. W.: 1987, in C. G. Wynn-Williams and E. E. Becklin (eds.), *Infrared Astronomy with Arrays*, University of Hawaii Press, p. 29.
Varosi, F. and Gezari, D. Y.: 1992, in preparation.

NEAR-IR SOLAR CORONAL OBSERVATIONS WITH NEW-TECHNOLOGY REFLECTING CORONOGRAPHS

RAYMOND N. SMARTT

National Solar Observatory/Sacramento Peak,
National Optical Astronomy Observatories,* Sunspot, New Mexico 88349, U.S.A.

SERGE KOUTCHMY

Institut d'Astrophysique, CNRS, 98 bis, Boulevard Arago, F-75014 Paris, France

and

JACQUES-CLAIR NOËNS

Observatoires du Pic-du-midi et de Toulouse, 65200 Bagnères-de-Bigorre, France

Abstract. Emission-line and K-coronal observations in the IR have the significant advantage of reduced sky brightness compared with the visible, while the effects of seeing are also reduced. Moreover, strong lines are available in the near-IR. Examples of the current capabilities of IR coronal observations using conventional Lyot coronagraphs are discussed briefly. Photometric measurements using the two IR lines of Fe XIII (10,747 Å and 10,798 Å), together with the Fe XIII 3,388 Å line, have provided a valuable electron-density diagnostic, but with low-angular-resolution. The 10,747 Å line has high intrinsic polarization. It has been used for extensive coronal magnetic field measurements, but only the direction of the field, and that with modest angular resolution, has been achieved due basically to flux limitations. Such studies suffer from the lack of high angular resolution and high photon flux. Moreover, the chromatic properties of a singlet objective lens preclude simultaneous observations at widely-differing wavelengths of the important inner coronal region. A coronagraph based on a mirror objective avoids such problems. Further, comparatively high-resolution and high-sensitivity arrays are now available with quantum efficiencies up to 90%. Reflecting coronagraphs with advanced arrays then provide the possibility of obtaining high-resolution images in the infrared to carry out a wide variety of studies crucial to many of the outstanding problems in coronal physics. A program for the development of reflecting coronagraphs is described briefly, with an emphasis on applications to IR coronal studies.

Key words: infrared: stars – Sun: corona – telescopes

1. Introduction

Observations with emission-line and white-light ground-based coronagraphs have allowed extensive studies of the morphology of the inner coronal region. Such observations, usually with only modest angular resolution, have established the overall structural characteristics of the corona as well as typical changes that occur over time scales of minutes to hours to days, as well as over time scales of the order of a solar cycle (see, *e.g.*, Altrock 1988). Details have been investigated of localized coronal features such as loop oscillations (see, *e.g.*, Antonucci *et al.* 1984), the interaction of post-flare loops (Smartt and Zhang 1987) and unusual transient events that propagate through the coronal environment (Dunn 1971).

Space-based, white-light coronagraphs (MacQueen *et al.* 1974; Sheeley *et al.* 1980) have provided a large database of coronal transients, especially coronal mass ejections, but the important inner corona is not accessible, due to the vignetting

---

* Operated by the Association of Universities for Research in Astronomy, Inc. (AURA) under cooperative agreement with the National Science Foundation.

characteristics of externally-occulted coronagraphs and to the overall design constraints that must allow for the imprecision of the pointing system. Recent X-ray coronal observations from Yohkoh (Ogawara 1991) constitute a rich source of data that directly links the structure of surface-activity patterns with that of the more extended corona. Nevertheless, the angular resolution is necessarily limited, due to upper limits on the instrument dimensions. In general, space coronagraphs have the advantage of no sky background, of possible extended temporal coverage, and of access to wavelength regimes not observable from the ground.

Ground-based coronagraphs can be designed to observe the extreme inner corona down to a few arcsec above the limb, the limit determined by residual tracking errors and image displacements caused by seeing. Large-aperture systems are possible, together with high-spectral-resolution spectrographs and high-precision polarimeters. In principle, such large systems could be deployed in space; extremely high-precision tracking could realistically be achieved with a lunar-based coronagraph (see, e.g., Smartt 1992). However, large space-based systems would be extraordinarily expensive and probably would not be realized for at least several decades, since none are currently planned.

2. Infrared Coronal Observations

2.1. INSTRUMENTAL PERFORMANCE ADVANTAGES

The power of ground-based coronagraphs can be expanded significantly by carrying out observations in the near-infrared. In particular, the sky brightness decreases as a function of wavelength. Rayleigh scattering is approximately $\propto \lambda^{-4}$; hence, at a wavelength of 2.0 μm, the sky brightness due to Rayleigh scattering is ~ 0.004 of the brightness at 0.5 μm.

The angular and spectral scattering characteristics of aerosols depend on the size, shape, internal structure and complex refractive indices of the individual particles, and on their large-scale spatial distribution. Measurements of particles in the upper atmosphere (upper troposphere and stratosphere) indicate diameters typically < 0.2 μm (Bigg 1976). Such particles would likely have irregular shapes, but if hygroscopic, would tend to be spherical. In the limit of $p \equiv 2\pi a/\lambda \ll 1$, where a is the particle radius and λ the incident wavelength, the scattering of a spherical particle with a real refractive index has the same spectral dependence as that of Rayleigh scattering. For $p \gg 1$, the aerosol scattering coefficient can be assumed to be roughly constant with wavelength, and for p in the intermediate range the scattering coefficient can be assumed to be proportional to λ^{-n}, with n usually in the range 0.5 to 1.5. Even though particles of size larger than about 0.5 μm can potentially cause significant scattering for observations in the near infrared (~ 1 μm), in practice their number density is extremely small under "normal" conditions at mountain sites, with minimal degradation of clear-sky conditions (Halthore 1992).

Beyond the advantage of the reduced level of sky background, the scattering properties of a coronagraph objective improve as a function of wavelength. The relationship between the roughness of an optical surface and the resultant scattered radiation has been treated extensively in the literature (for a review, see Elson *et al.* 1979). In the case of an rms roughness, σ, where $\sigma \ll \lambda$, for radiation of wavelength, λ, and with a Gaussian distribution of roughness, the scattered component of reflectance at normal incidence, R_s, can be characterized by, $R_s \sim R_o(2k\sigma)^2$, where R_o is the reflectance of a completely smooth surface, and $k = 2\pi/\lambda$. Further, strong emission-lines are available in the near infrared, and with the availability of high-quantum-efficiency IR detector arrays, high-resolution imaging can be expected, especially since the seeing characteristics of the atmosphere also improve with wavelength – Fried's seeing parameter varies as $\lambda^{6/5}$. Hence, on this basis, an adaptive optics system operating in the infrared would require substantially fewer correcting elements, for a given telescope aperture, than one operating at much shorter wavelengths (the required number of elements is then approximately proportional to the ratio of the area of a seeing cell to that of the telescope aperture), and the bandwidth requirements would be less stringent.

2.2. OBSERVATIONAL ADVANTAGES

Observations of coronal emission lines in the near-IR also have some advantages over those at visible wavelengths. The two infrared lines, 10,747 Å and 10,798 Å, as well as the 3,388 Å line, all Fe XIII transitions, together provide a useful coronal density diagnostic. The line intensity ratios, $I(10,747)/I(10,798)$ and $I(3,388)/I(10,747)$, are functions of the electron density that are independent of the temperature in the equilibrium range of the Fe XIII line. The upper levels are excited by electron collisions (as well as by radiation); hence the populations for the two lines will be dependent on the electron density. Further, the radiative de-excitation coefficient is known. Significant uncertainties in the measurements are present when the density is relatively high. However, it has been established that this technique is useful as an electron density diagnostic in the corona provided that sufficient accuracy in the line intensity measurements can be achieved (Noëns *et al.* 1984).

Studies of the coronal magnetic field are, in general, more easily carried out in the infrared than the visible, with increased Zeeman sensitivity and increased linear-polarization sensitivity. For example, coronal-line emission is characterized by a linearly-polarized component, due predominantly to the mechanism of resonance fluorescence occurring in the presence of the anisotropic photospheric radiation field (House 1972). The strong 10,747 Å line of Fe XIII has high intrinsic polarization, since its transition ends in only one magnetic quantum sublevel. Neglecting collisions, the linear polarization value approaches unity towards the limit of radiation anisotropy, and is already 0.9 at a height of two solar radii (House 1977). By comparison, the corresponding value for the 5,303 Å line of Fe XIV is only 0.05. Therefore, the 10,747 Å line has an enormous advantage for magnetic field measurements as compared with the 5,303 Å line, given detectors of equal efficiency in these two spectral regions. Such observations have been carried out (Querfeld and Smartt 1984) using an infrared coronal emission-line polarimeter (Querfeld 1977). In this instrument, the analyzer consists of a chopper that modulates the intensity of the incoming light, followed by a linear-polarization sensor and a Wollaston prism. The result of the modulation is to produce 25-Hz intensity- and 100-Hz polarization-signals. The detectors (Querfeld 1982) are cooled (194 K) GaAsSb heterojunction photo-diodes that have a quantum efficiency of 0.9. The polarimeter has been used at the NSO/SP 40-cm aperture coronagraph. Measurements were made typically with a 5 s integration time, and a 1' aperture. Except under observing conditions characterized by extremely low sky-brightness and a relatively strong coronal signal, smaller apertures would result in an unacceptable noise level for the polarization measurements. This points to the need for larger-aperture coronagraphs that would allow high-angular-resolution, low-noise, measurements.

Other near-infrared coronal lines are also of interest. For example, the Si X line at 1.4305 μm was measured at the 12 November 1966 eclipse. The values obtained suggest that the intensity of this line might be approximately one order of magnitude greater than that of the 5,303 Å line (Münch et al. 1967).

3. Infrared Coronagraphs

A conventional, singlet-lens-objective Lyot coronagraph could be optimized for operation in the near IR, at least up to the spectral cut-off of the glass used for the objective, typically BKR7. But, this would preclude observations beyond about $\lambda = 2.5$ μm, and performance at visible wavelengths would be poor. Since the singlet-lens objective of a conventional coronagraph produces a chromatic image, the appropriate position of the occulting disk along the axis varies with wavelength. And for observations close to the limb, the size of the occulting disk must also be changed as a function of wavelength. The use of mirror-objective coronagraphs overcomes such problems.

3.1. Reflecting Coronagraphs

Mirror objectives have been applied successfully to small externally-occulted rocket- and balloon-borne coronagraphs (Kohl et al. 1978; Smartt 1979). Recent developments in the technology of extremely low-scatter mirrors have opened the possibility of using mirrors for internally-occulted coronagraphs. These mirrors rely on modern techniques for producing extremely smooth polished surfaces as measured by the level of residual micro-roughness over spatial scales ~0.01 mm, as well as extremely smooth evaporated reflecting films. A major advantage to using mirrors for coronagraph objectives is that the primary image is then achromatic. Hence a reflecting coronagraph is appropriate for measurements of both coronal-line emission and the K-corona. Although the K-corona brightness decreases with increasing wavelength beyond ~0.5 μm, extremely-low-scatter sky conditions could provide advantages for observing the K-corona in the near infrared. For an all-mirror system, infrared and ultraviolet observations are limited only by the spectral characteristics of the mirror surface, the detector's spectral response, and the spectral transmittance of the earth's atmosphere. The reflectance of an aluminized mirror is ~0.95 at $\lambda \sim 1.0$ μm, and increases uniformly with increasing wavelength, reaching ~0.98 at $\lambda = 10$ μm (Bennett et al. 1963). Further, simultaneous multi-wavelength observations can be carried out in the important inner-coronal region, while large apertures are possible. Also, for reasons pointed out above, the scattered radiation from a typical superpolished mirror should decrease with increasing wavelength, as $\sim \lambda^{-2}$.

A Mirror Advanced Coronagraph (MAC I) has been constructed at NSO/SP, the first of three prototype reflecting coronagraphs currently under development at SP (Smartt et al. 1990). The objective mirror in MAC I is a super-polished, spherical silicon mirror produced by Zeiss-FRG, with a focal length of 1 m, and a micro-roughness < 0.3 nm rms. Tests of this mirror at visible wavelengths indicate that its scattering is of the same order as that typical of good-quality coronagraph objective lenses. The optical system is simply off-axis reflection from the primary mirror to a secondary optical system that is a conventional Lyot coronagraph. This consists of an occulting disk at the primary image plane to block the image of the solar disk, followed by a field lens, Lyot stop, filter and detection system. A second, more advanced prototype instrument (MAC II), based on a 15-cm diameter super-polished, aluminized Zerodur mirror objective of 2.25-m focal length, has also been constructed. In a joint agreement with the Institut d'Astrophysique de Paris (IAP), part of the construction was carried out at IAP. A concave, annular field mirror, located near the primary focal plane, functions as an inverse occulting disk. The solar disk image, which passes through a central hole in the field mirror, is reflected via two plane mirrors externally from the coronagraph. A Lyot stop is located at the image of the objective, formed by the field mirror. This is followed by a collimating lens, a filter system, and a camera lens to form the final image.

A research-quality reflecting coronagraph (MAC III) with an aperture of 55 cm and focal length of 4.5 m is in the design phase. The small field mirror reflects only a limited part of the coronal field of $10' \times 10'$, and is rotated around the image of the solar disk to select different regions of the corona. This is an all-mirror design that achieves a diffraction-limited correction in the visible over the above field of view. With the incorporation of a correlation-tracker capability, this instrument should perform extremely well in the near IR. The design allows the incorporation of a spectrograph as an alternative to direct imaging at the final focal plane. Experience with the two smaller prototype instruments has provided key ideas that will be incorporated in this more advanced instrument. It is anticipated that this coronagraph will also be used for low-scattered-light observations of the solar disk. This would be accomplished simply by tilting the primary mirror such that the field mirror reflects the solar disk image along the secondary optical axis. MAC III will be mounted on the spar at the NSO/SP John W. Evans Solar Facility. A much larger reflecting coronagraph of the same basic design, with an aperture of two or more meters, is also planned. Such a major instrument would have both solar and nighttime applications. Nighttime IR studies would include observations of faint emission associated with planetary and stellar objects, as well as other galactic and extra-galactic objects.

4. Conclusion

There are several important instrumental performance advantages for IR observations of the solar corona as compared with visible wavelengths. The development of high-quantum-efficiency IR detector arrays together with large-aperture IR reflecting coronagraphs provides the possibility of obtaining high angular- and spectral-resolution IR observations of the solar corona. Moreover, there are considerable observational advantages in the IR as well. Strong spectral lines are available in the near-infrared that have high polarization sensitivity, suitable for magnetic field studies and as density diagnostics. In addition it should be possible to observe coronal images to much greater heights in the corona than with visible lines such as 5,303 Å. With the advent of new IR array detectors and the development of reflecting coronagraphs, it is clear that IR studies of the corona offer an extremely fruitful area of solar research. Finally, critical low-scattered-light IR observations of the solar disk as well as special nighttime studies are ideally suited to this new technology.

References

Altrock, R. C. (ed.): 1988, *Solar and Stellar Coronal Structure and Dynamics*, Proc. Ninth Sacramento Peak Summer Symposium, Sunspot, NM.
Antonucci, E., Gabriel, A. H., and Patchett, B. E.: 1984, *Solar Phys.* **93**, 85.
Bigg, E. K.: 1976, *J. Atmos. Sci.* **33**, 1080.
Bennett, H. E., Silver, M., and Ashley, E. J.: 1963, *J. Opt. Soc. Am.* **53**, 1089.
Dunn, R. B.: 1971, in Macris (ed.), *Physics of the Solar Corona*, D. Reidel Publishing Company, Dordrecht-Holland, p. 114.
Elson, J. M., Bennett, H. E., and Bennett, J. M.: 1979, in R. R. Shanon and J. C. Wyatt (eds.), *Applied Optics and Optical Engineering* **7**, Academic Press, New York, p. 191.
Halthore, R. N.: 1992 (private communication).
House, L. L.: 1972, *Solar Phys.* **23**, 103.
House, L. L.: 1977, *Astrophys. J.* **214**, 632.
Kohl, J. L., Reeves, E. M., and Kirkham, B.: 1978, in K. A. van der Hucht and G. Vaiana (eds.), *New Instrumentation for Space Astronomy*, Pergamon, New York, p. 91.
MacQueen, R. M., Gosling, J. T., Hildner, E., Munro, R. H., Poland, A. I., and Ross, C. L.: 1974, *S.P.I.E.* **44**, 207.
Münch, G., Neugebauer, G., and McCammon, D.: 1967, *Astrophys. J.* **149**, 681.
Noëns, J. C., Pageault, J., Ratier, G.: 1984, *Solar Phys.* **94**, 117.
Ogawara, Y., Takano, T., Kato, T., Kosugi, T., Tsuneta, S., Watanabe, T., Kondo, I., and Uchida, Y.: 1991, *Solar Phys.* **136**, 1 (see also complete issue of *Solar Phys.* **136/1**).
Querfeld, C. W.: 1977, *S.P.I.E.* **112**, 200.
Querfeld, C. W.: 1982, *AAS Photo. Bull.* **29**, 3.
Querfeld, C. W., and Smartt, R. N.: 1984, *Solar Phys.* **91**, 299.
Sheeley, N. R., Jr., Michels, D. J., Howard, R. A., and Koomen, M. J.: 1980, *Astrophys. J.* **237**, L99.
Smartt, R. N.: 1979, *S.P.I.E* **190**, 58.
Smartt, R. N. and Koutchmy, S.: 1992, in M. Giampapa and J. Bookbinder (eds.), *Seventh Cambridge Workshop on Cool Stars, Stellar Systems and the Sun*, ASP Conference Series Vol. 26, p. 660.
Smartt, R. N., Koutchmy, S. L., Colley, S. A., Caron, R., Schwenn, R., and Restaino, S. R.: 1990, *S.P.I.E.* **1236**, 206.
Smartt, R. N.: 1992, in *Space '92*, Am. Soc. Civil Eng., 31 May - 4 June, 1992, Denver, CO (in press).
Smartt, R. N., and Zhang, Z.: 1987, in G. Athay and D. S. Spicer (eds.), *Theoretical Problems in High Resolution Solar Physics II*, NASA Conf. Pub. 2483, p. 129.
Zuev, V. E.: 1970, *Atmospheric Transparency in the Visible and the Infrared*, Keter Press, Jerusalem.